Ticks

Widespread and increasing resistance to most available acaracides threatens both global livestock industries and public health. This necessitates better understanding of ticks and the diseases they transmit in the development of new control strategies. *Ticks: Biology, Disease and Control* is written by an international collection of experts and covers in-depth information on aspects of the biology of the ticks themselves, various veterinary and medical tick-borne pathogens, and aspects of traditional and potential new control methods. A valuable resource for graduate students, academic researchers and professionals, the book covers the whole gamut of ticks and tick-borne diseases from microsatellites to satellite imagery and from exploiting tick saliva for therapeutic drugs to developing drugs to control tick populations. It encompasses the variety of interconnected fields impinging on the economically important and biologically fascinating phenomenon of ticks, the diseases they transmit and methods of their control.

ALAN BOWMAN has worked at the Universities of Edinburgh, Oxford and Oklahoma State and is now at the University of Aberdeen. His research interests include tick physiology, bioactive factors in tick saliva, drug target development and ecological aspects of borreliosis. Funding for his tick research has come from national funding bodies and both large animal-health and small biotechnology companies for which he also acts as a consultant.

PAT NUTTALL is Director of the Centre for Ecology and Hydrology (CEH), the UK's centre of excellence for integrated research in land-based and freshwater environmental sciences, and part of the Natural Environment Research Council (NERC). She is Professor of Virology at the University of Oxford and a Fellow of Wolfson College, Oxford. She was awarded the Ivanovsky Medal for Virology in 1996 by the Russian Academy of Sciences, and the Order of the British Empire by the Queen in 2000 for services to environmental sciences.

Ticks

Biology, Disease and Control

Edited by
ALAN S. BOWMAN
University of Aberdeen

PATRICIA A. NUTTALL
Centre for Ecology and Hydrology, Wallingford

CAMBRIDGE UNIVERSITY PRESS

Cambridge, New York, Melbourne, Madrid, Cape Town, Singapore, São Paulo, Delhi

Cambridge University Press
The Edinburgh Building, Cambridge CB2 8RU, UK

Published in the United States of America by Cambridge University Press, New York

www.cambridge.org
Information on this title: www.cambridge.org/9780521867610

First published 2008

Printed in the United Kingdom at the University Press, Cambridge

A catalogue record for this publication is available from the British Library

Library of Congress Cataloguing in Publication data
Ticks : biology, disease, and control / edited by Alan S. Bowman, Patricia A. Nuttall.
   p.   cm.
Includes bibliographical references and index.
ISBN 978-0-521-86761-0 (hardback)
1. Ticks as carriers of disease.   2. Ticks.   I. Bowman, Alan S.   II. Nuttall, Patricia A.   III. Title.
[DNLM: 1. Tick-Borne Diseases.   2. Tick Control – methods.   3. Ticks – physiology. WC 600 T5565 2008]
RA641.T5T532   2008
614.4'33 – dc22   2008021955

ISBN 978-0-521-86761-0 hardback

Contents

Contributors

JENNIFER M. ANDERSON
Vector Molecular Biology Unit
Laboratory of Malaria and Vector Research NIAID
National Institute of Health
12735 Twinbrook Parkway
Room 2E-22
Rockville MD 20852 USA

ANDREW BALL
School of Biological Sciences
University of Aberdeen
Tillydrone Avenue
Aberdeen AB24 2TZ UK

STEPHEN C. BARKER
Parasitology Section
School of Molecular and Microbial
    Sciences
University of Queensland
Brisbane Qld 4072 Australia

RICHARD BISHOP
International Livestock Research Institute (ILRI)
P.O. Box 30709
Nairobi 00100 Kenya

EDMOUR F. BLOUIN
Department of Veterinary Pathobiology
250 McElroy Hall
Center for Veterinary Health Sciences
Oklahoma State University
Stillwater OK 74078 USA

RUSSELL E. BOCK
Tick Fever Centre
Biosecurity Queensland
Queensland Department of Primary Industries
    and Fisheries
280 Grindle Road
Wacol Qld 4076 Australia

ALAN S. BOWMAN
School of Biological Sciences
University of Aberdeen
Tillydrone Avenue
Aberdeen AB24 2TZ UK

MICHEL BROSSARD
Laboratory of Parasite Immunology
11 rue Emile Argand
CH-2007 Neuchâtel Switzerland

MILAN DANIEL
School of Public Health
Institute for Postgraduate Medical
    Education
100 05 Prague 10
Ruska 85 Czech Republic

RONALD B. DAVEY
Cattle Fever Tick Research Laboratory
USDA, ARS
Moore Air Base, Bldg. 6419
22675 N. Moorefield Road
Edinberg TX 78541 USA

JOSÉ DE LA FUENTE
Department of Veterinary Pathobiology
250 McElroy Hall
Center for Veterinary Health Sciences
Oklahoma State University
Stillwater OK 74078 USA
and
Instituto de Investigación en Recursos Cinegéticos IREC
    (CSIC-UCLM- JCCM)
Rhonda de Toledo s/n 13005
Ciudad Real
Spain

ALBERTUS J. DE VOS
Tick Fever Centre
Biosecurity Queensland
Queensland Department of Primary Industries
    and Fisheries
280 Grindle Road
Wacol Qld 4076 Australia

MALCOLM J. GARDNER
Seattle Biomedical Research Institute
307 Westlake Ave. N. Suite 500
Seattle WA 98109 USA

JOHN E. GEORGE
Knipling–Bushland US Livestock Insects
    Research Laboratory
USDA, ARS
2700 Fredericksburg Road
Kerrville TX 78028 USA

LISE GERN
Institut de Biologie
Emile-Argand 11
CH-2009 Neuchâtel Switzerland

HOWARD S. GINSBERG
Patuxent Wildlife Research Centre
US Geological Survey
University of Rhode Island
Woodward Hall–PLS
Kingston RI 02881 USA

ITAMAR GLAZER
Entomology and Nematology
ARO
The Volcani Centre
Bet Dagan
P.O. Box 6
Israel 50250

HEIDI K. GOETHERT
Division of Infectious Diseases
Cummings School of Veterinary Medicine
Tufts University
200 Westboro Road
North Grafton MA 01536 USA

R. GOTHE
Department of Biochemistry
University of Pretoria
Pretoria 0002 South Africa

LIBOR GRUBHOFFER
Biology Centre of the Academy of Sciences of the
    Czech Republic
Institute of Parasitology
Branisovska 31
370 05 Ceske Budejovice Czech Republic

ONDREJ HAJDUSEK
Faculty of Biological Sciences
University of South Bohemia
Branisovska 31
370 05 Ceske Budejovice Czech Republic

VACLAV HYPŠA
Faculty of Biological Sciences
University of South Bohemia
Branisovska 31
370 05 Ceske Budejovice Czech Republic

LOUISE A. JACKSON
Animal Research Institute
Biosecurity Queensland
Department of Primary Industries and Fisheries
Locked Mail Bag No. 4
Moorooka Qld 4105 Australia

WAYNE K. JORGENSEN
Animal Research Institute
Department of Primary Industries and Fisheries
Locked Mail Bag No. 4
Moorooka Qld 4105 Australia

W. REUBEN KAUFMAN
Z 606, Department of Biological Sciences
University of Alberta
Edmonton, Alberta T6G 2E9 Canada

KATHERINE M. KOCAN
Department of Veterinary Pathobiology
250 McElroy Hall
Center for Veterinary Health Sciences
Oklahoma State University
Stillwater OK 74078 USA

JAN KOLÁŘ
Department of Applied Geoinformatics
Faculty of Sciences
Charles University
128 43 Prague 2
Albertov 6 Czech Republic

VOJTECH KOVÁŘ
Biology Centre of the Academy of Sciences of the
    Czech Republic
Institute of Parasitology
Branisovska 31
370 05 Ceske Budejovice Czech Republic

MILAN LABUDA
Institute of Zoology
Slovak Academy of Sciences
Dubravska cesta 9
845 06 Bratislava Slovakia

BEN J. MANS
Old Main Building, Rm 35
Parasites, Vectors and Vector-Borne Diseases
Onderstepoort Veterinary Institute
Agricultural Research Council
Onderstepoort
0110
South Africa

SUBHASH P. MORZARIA
Food and Agriculture Organization (FAO)
39 Phra Atit Road
Bangkok 10200 Thailand

ANNA MURRELL
Parasitology Section
School of Molecular and Microbial Sciences
University of Queensland
Brisbane Qld 4072 Australia

ANTONY J. MUSOKE
Onderstepoort Veterinary Institute
Private Bag X5
Onderstepoort 0110 South Africa

VISHVANATH NENE
The Institute for Genomic Research (TIGR)
9712 Medical Center Drive
Rockville MD 20850 USA

ALBERT W. H. NEITZ
Department of Biochemistry
University of Pretoria
Pretoria 0002 South Africa

PATRICIA A. NUTTALL
Centre for Ecology and Hydrology
Maclean Building
Crowmarsh Gifford
Wallingford OX10 8BB UK

JAMES H. OLIVER JR
Georgia Southern University
Institute of Arthropodology and Parasitology
P.O. Box 8056
Statesboro GA 30460 USA

JOSEPH F. PIESMAN
CDC/DVBID
3150 Rampart Road
Fort Collins CO 80521 USA

MATHEWS POUND
Knipling–Bushland US Livestock Insects
  Research Laboratory
USDA, ARS
2700 Fredericksburg Road
Kerrville TX 78028 USA

SARAH E. RANDOLPH
Department of Zoology
University of Oxford
Tinbergen Building
South Parks Road
Oxford OX1 3PS UK

HUW H. REES
School of Biological Sciences
University of Liverpool
The Biosciences Building
Crown Street
Liverpool L69 7ZB UK

RYAN O. M. REGO
Laboratory of Zoonotic Pathogens
Rocky Mountain Laboratories
NIAID
NIH
903 South 4th Street
Hamilton MT 59840 USA

NATALIA RUDENKO
Biology Centre of the Academy of Sciences of the
  Czech Republic
Institute of Parasitology
Branisovska 31
370 05 Ceske Budejovice Czech
  Republic

MICHAEL SAMISH
Division of Parasitology
Kimron Veterinary Institute
Bet Dagan
P.O. Box 12
Israel 50250

JOHN SAUER
Department of Entomology and Plant
  Pathology
127 Noble Research Center

Oklahoma State University
Stillwater OK 74078 USA

ROBERT A. SKILTON
International Livestock Research Institute
  (ILRI)
P.O. Box 30709
Nairobi 00100 Kenya

DANIEL E. SONENSHINE
Department of Biological Sciences
45th Street and Elkhorn Avenue
Old Dominion University
Norfolk VA 2329 USA

SAM R. TELFORD
Division of Infectious Diseases
Cummings School of Veterinary Medicine
Tufts University
200 Westboro Road
North Grafton MA 01536 USA

JESUS G. VALENZUELA
Vector Molecular Biology Unit
Laboratory of Malaria and Vector Research
  NIAID
National Institutes of Health
12735 Twinbrook Parkway
Room 2E-22
Rockville MD 20852 USA

STEPHEN K. WIKEL
Department of Immunology
School of Medicine
University of Connecticut Health Center
263 Farmington Avenue, MC3710
Farmington CT 06030 USA

PETER WILLADSEN
CSIRO Livestock Industries
Queensland Bioscience Precinct
306 Carmody Road
St Lucia Qld 4067 Australia

PETR ZEMAN
State Veterinary Institute
165 03 Prague 6
Sidlistni 136/24 Czech Republic

Preface

Tick statistics are impressive. Some 907 tick species have been named. Their only food is blood, of which some ticks consume relatively vast quantities (several hundred times their unfed body weight). Some take 2 weeks or more to feed. Often they only feed three times during the whole of their life cycle (which may take 7 years to complete). They feed on mammals (including humans), birds and reptiles. Their geographical distribution ranges from sub-arctic through equatorial to antarctic regions, and habitats range from desert to rainforest. They even survive submersion in seawater as they feed on seabirds diving for fish. But the most important tick statistics concern their ability to transmit pathogens (disease-causing agents). And our greatest challenge is to devise efficient and effective means of controlling ticks and tick-borne pathogens.

Ticks transmit a great variety of disease-causing agents to humans (viral, bacterial and protozoal), including bacteria that cause Lyme disease, the reports of which increase in number year on year. About 80% of the world's cattle are infested with ticks. As a result, ticks are the most economically important ectoparasite of livestock. The impact of ticks on livestock producers in the developing world is a contributing factor to poverty.

In this book we have brought together experts from the tick world to express their views on the key advances in tick biology, diseases and control. Tick systematics and evolution highlight fundamental changes in our understanding, particularly for hard (ixodid) ticks, their life cycles and historical zoogeography (Barker & Murrell). While the ecology of ticks is a fundamental influence in pathogen transmission dynamics (Randolph), tick salivary glands perform a key function in survival (water balance) and pathogen transmission (Bowman, Ball & Sauer). For good reason, ticks have been called 'supreme pharmacologists', manipulating their hosts' attempts to get rid of them by secreting hundreds of antihaemostatic, anti-inflammatory, anaesthetic and immunomodulatory molecules in their saliva. Not surprisingly, the 'sialome' has become the frontier in understanding the role of tick saliva in blood-feeding and pathogen transmission (Anderson & Valenzuela). Saliva also contains toxins, a non-infectious cause of disease, though we know little of their functional significance (Mans, Gothe & Neitz). Besides saliva production, blood-feeding also enhances tick lectin activities, which play a role in defence reactions and pathogen transmission (Grubhoffer *et al.*), and triggers the endocrine system about which comparatively little is known (Rees). Similarly, the mechanisms used by male ticks to assure their paternity are largely virgin territory (Kaufman).

One of the reasons why ticks transmit so many pathogens is found in the dynamic interactions that occur at the tick–host–pathogen interface (Brossard & Wikel) where saliva assists pathogen transmission (Nuttall & Labuda). Because ticks transmit such a diversity of pathogens, we have had to be selective. For humans, the most common tick-borne infection is Lyme borreliosis (Piesman & Gern) though several tick-borne viruses cause human disease and even death (Labuda & Nuttall). More common are diseases of livestock, including babesiosis, the most economically important arthropod-borne disease of cattle (Bock *et al.*), theileriosis, a particular problem in developing countries (Bishop *et al.*), and anaplasmosis, caused by *Anaplasma marginale*, a bacterium (rickettsia) (Kocan, de la Fuente & Blouin). Few 'emerging' tick-borne infections are new to science (Telford & Goethert).

Controlling ticks and tick-borne pathogens requires new approaches, such as satellite-based remote sensing for landscape epidemiology to identify spatial and temporal distribution (Daniel, Kolář & Zeman). But the mainstay of tick and disease control remains acaricide use, despite the alarming problem of acaricide resistance (George, Pound & Davey). Although a commercial tick vaccine became available in 1994, progress in developing new and improved vaccines is slow (Willadsen). Development of biological agents to control ticks is still in its infancy (Samish, Ginsberg & Glazer), as is the use of pheromones and other semiochemicals (Sonenshine) although some show great promise.

This book follows on from the *Parasitology* Supplement, *Ticks: Biology, Disease and Control*, published in 2004. As a result of interest in the Supplement and requests from workers in the field, we went back to the authors and asked if they would update and revise their contributions. Where the book chapter has replaced valuable information in the Supplement, the appropriate Supplement reference has been cited. We hope this book inspires your interest in the remarkable world of ticks.

ALAN S. BOWMAN and PATRICIA A. NUTTALL
June 2008

1 • Systematics and evolution of ticks with a list of valid genus and species names

S. C. BARKER AND A. MURRELL

In recent years there has been much progress in our understanding of the phylogeny and evolution of ticks, in particular the hard ticks (Ixodidae). Indeed, a consensus about the phylogeny of the hard ticks has emerged which is quite different to the working hypothesis of 10 years ago. Several changes to the nomenclature of ticks have been made or are likely to be made in the near future. One subfamily, the Hyalomminae, should be sunk, while another, the Bothriocrotoninae, has been created (Klompen, Dobson & Barker, 2002). Bothriocrotoninae, and its sole genus *Bothriocroton*, have been created to house an early-diverging ('basal') lineage of endemic Australasian ticks that used to be in the genus *Aponomma*. The remaining species of the genus *Aponomma* have been moved to the genus *Amblyomma*. Thus, the name *Aponomma* is no longer a valid genus name. The genus *Rhipicephalus* is paraphyletic with respect to the genus *Boophilus*. Thus, the genus *Boophilus* has become a subgenus of the genus *Rhipicephalus* (Murrell & Barker, 2003). Knowledge of the phylogenetic relationships of ticks has also provided new insights into the evolution of ornateness and of their life cycles, and has allowed the historical zoogeography of ticks to be studied. Finally, we present a list of the valid genus and species names of ticks as at Febuary 2007.

INTRODUCTION

Hoogstraal & Aeschlimann (1982) were apparently the first people to publish a phylogenetic tree for the ticks (suborder Ixodida); however, hypotheses about the evolutionary relationships of ticks had been proposed well before this (e.g. Pomerantsev, 1948; Camicas & Morel, 1977). The Hoogstraal and Aeschlimann phylogeny was inferred from intuition about the relative 'primitiveness' of the morphology and life cycles of ticks, and their hosts. An alternative phylogeny was proposed by Filippova (1993, 1994), but the trees of Hoogstraal & Aeschlimann (1982) and Filippova

(1993, 1994) were not tested until the mid 1990s. The phylogeny of ticks was first studied with molecular characters in the 1990s; there have been over 30 papers on the molecular phylogeny and evolution of ticks, and, although they have not always agreed, a consensus on phylogenetic relationships of ticks has emerged: Wesson & Collins (1992); Wesson *et al.* (1993); Black & Piesman (1994); Caporale *et al.* (1995); McLain *et al.* (1995*a*, *b*); Rich *et al.* (1995); Crampton, McKay & Barker (1996); Klompen *et al.* (1996); Norris *et al.* (1996, 1997); Black, Klompen & Keirans (1997); Zahler *et al.* (1997); Barker (1998); Black & Roehrdanz (1998); Crosbie, Boyce & Rodwell (1998); Mangold, Bargues & Mas-Coma (1998*a*, *b*); Dobson & Barker (1999); Murrell, Campbell & Barker (1999); Norris, Klompen & Black (1999); Fukunaga *et al.* (2000); Klompen *et al.* (2000); Murrell, Campbell & Barker (2000, 2001*a*, *b*, 2003); Beati & Keirans (2001); Ushijima *et al.* (2003); Xu *et al.* (2003); Murrell *et al.* (2005); Shao *et al.* (2005); Szabo *et al.* (2005); and Miller *et al.* (2007). Cuticular hydrocarbon composition has also been used to infer phylogenies of populations of ticks (Estrada-Peña, Castellá & Morel, 1994; Estrada-Peña, Castellá & Moreno, 1994; Estrada-Peña *et al.*, 1997), and Hutcheson *et al.* (2000) reviewed progress in tick molecular systematics. At least nine papers have been published on the phylogeny and evolution of ticks inferred from morphology and other phenotypes: Klompen (1992); Klompen & Oliver (1993); Hutcheson *et al.* (1995); Klompen *et al.* (1997, 2000); Borges *et al.* (1998); Klompen (1999); Beati & Keirans (2001); and Murrell *et al.* (2001*b*).

The first part of this review draws together recent advances in our understanding of the phylogeny of ticks and shows how robust phylogenetic trees can help us to interpret the evolution of ticks and make informed changes to their taxonomy and nomenclature. Phylogenies of tick groups inferred from different sets of characters have not always been congruent; however, consensus has emerged

Ticks: Biology, Disease and Control, ed. Alan S. Bowman and Patricia A. Nuttall. Published by Cambridge University Press.
© Cambridge University Press 2008.

about many tick relationships. The second part of the review deals with the taxonomy and nomenclature of ticks. Table 1.1 (see below, pp. 3–25) is a list of the 907 valid genus and species names as at February 2007.

PHYLOGENETICS OF TICKS: RECENT ADVANCES

The sister-group of the ticks

Discovery of the sister-group, the nearest relatives, of the ticks will reveal much about the evolution of the ticks; then we will be able to root our phylogenetic trees with confidence. There are two main competing hypotheses: (1) that the sister-group of the Ixodida is the Order Holothyrida (Lehtinen, 1991) (((Ixodida, Holothyrida), Mesostigmata) Opilioacariformes); and (2) that the sister-group of the Ixodida is the Order Mesostigmata (Krantz, 1978) (((Ixodida, Mesostigmata), Holothyrida), Opilioacariformes). The presence of Haller's organ, the ability to retract the gnathosoma and a similar type of musculature at the base of the gnathosoma in ticks and holothyrid mites, which are putatively derived characters for the Acari, was the basis of Lehtinen's (1991) hypothesis that the sister-group of the ticks is the Holothyrida. All three of the tests of these hypotheses indicate that the sister-group of the Ixodida is the Holothyrida: Dobson & Barker (1999) and Murrell et al. (2005) (small subunit (SSU) rDNA); and Klompen et al. (2000) (total evidence analysis of morphology, SSU rDNA, large subunit (LSU) rDNA, 16S rDNA (mitochondrial)). However, it is still not certain that the sister-group of the ticks is the Holothyrida. More data are needed.

Phylogeny of the Ixodida

Here is our interpretation of the working hypothesis of the phylogeny of the subfamilies of ticks in use by many current tick systematists. This tree is based on information from the papers cited in the introduction to this chapter.

THE PHYLOGENETIC RELATIONSHIPS OF THE THREE TICK FAMILIES ARE UNRESOLVED

The phylogeny of the three families of ticks, Ixodidae (hard ticks), Argasidae (soft ticks) and Nutalliellidae, is still unresolved (Fig. 1.1). This is due to the fact that *Nuttalliella namaqua*, the only species in the Nutalliellidae, has not been collected for many years. Attempts to amplify DNA from museum specimens by the Black group and the Barker

group resulted only in the amplification of DNA from fungi that had infected the specimens either before or after their death (unpublished data).

THE RHIPICEPHALINAE IS PARAPHYLETIC

The subfamily Hyalomminae is embedded within the Rhipicephalinae (see below for a more detailed description of phylogenetic relationships in Rhipicephalinae).

MONOPHYLY OR PARAPHYLY OF *IXODES*?

There is evidence that the genus *Ixodes* has two main lineages, the Australasian *Ixodes* and the other *Ixodes* (Klompen, 1999; Klompen et al., 2000; Shao et al., 2005). Our working hypothesis has the genus *Ixodes* as a monophyletic lineage (Fig. 1.1). This is the traditional view; however, it is far from certain that this is correct. Indeed, morphological and molecular characters provide only weak evidence for monophyly of the genus *Ixodes* (Klompen et al., 1997, 2000; Dobson & Barker, 1999). The Australasian *Ixodes* may even be the sister-group to the rest of the Metastriata, but this idea is based at this stage on analysis of rDNA alone (Dobson & Barker, 1999; Klompen et al., 2000). The evolutionary relationships of most *Ixodes* species have not been studied so it is not known exactly how many of the extant *Ixodes* species belong to the Australasian *Ixodes* lineage. Analyses of morphology and nucleotides indicated that *I. tasmani*, *I. holocyclus* and *I. uriae* (= the *I. tasmani* group *sensu* Klompen et al., 2000) and *I. antechini* and *I. ornithorhynchi* belong to this lineage (Klompen et al., 2000). The presence of two control regions in their mitochondrial genomes indicates that *I. cordifer*, *I. cornuatus*, *I. hirsti*, *I. myrmecobii* and *I. trichosuri* also belong to this lineage (see below, and Shao et al., 2005). Until contrary evidence is found we presume that the remaining *Ixodes* that are endemic to and/or evolved in Australasia (at least Australia, New Guinea and New Zealand) also belong to this lineage, i.e. *I. amersoni*, *I. apteridis*, *I. australiensis*, *I. confusus*, *I. dendrolagi*, *I. eudyptidis*, *I. fecialis*, *I. hydromyidis*, *I. jacksoni*, *I. kohlsi*, *I. laysanensis*, *I. luxuriosus*, *I. priscollaris*, *I. steini*, *I. vestitus*, *I. victoriensis*, *I. zaglossi* and *I. zealandicus*. Thus, at present the Australasian *Ixodes* lineage has 28 extant species.

A NEW LINEAGE OF AUSTRALIAN TICKS

Dobson & Barker (1999) and Klompen, Dobson & Barker (2002) reported a new lineage of ticks that infest reptiles in Australia: five species of Bothriocrotoninae Klompen, Dobson & Barker, 2002. This group was first recognized by Kaufman (1972) as one of the three groups of *Aponomma*

Table 1.1 *A current list of valid genus and species names (in alphabetical order except for the five species that were previously in the genus* Boophilus *– these species are now at the top of the list of* Rhipicephalus *species)*

Family	Genus	Species
IXODIDA (907 valid species names)		
NUTTALLIELLIDAE (1 valid species name)	*Nuttalliella* (1 species)	*N. namaqua* Bedford, 1931[K,CHAM,HCK]
ARGASIDAE (186 valid species names)	*Argas* (58 species)	*A. abdussalami* Hoogstraal & McCarthy, 1965[K,CHAM,HCK]
		A. acinus (Whittick, 1938)[K,CHAM,HCK]
		A. africolumbae Hoogstraal, Kaiser, Walker, Ledger, Converse & Rice, 1975[K,CHAM,HCK]
		A. arboreus Kaiser, Hoogstraal & Kohls, 1964[K,CHAM,HCK]
		A. assimilis Teng & Song, 1983[K,CHAM,HCK]
		A. beijingensis Teng, 1983[K,CHAM,HCK]
		A. beklemischevi Pospelova-Shtrom, Vasil'eva & Semashko, 1963[K,CHAM,HCK]
		A. brevipes Banks, 1908[K,CHAM,HCK]
		A. brumpti Newmann, 1907[K,CHAM,HCK]
		A. bureschi Dryenski, 1957[K,CHAM,HCK]
		A. canestrinii Birula, 1895[K,HCK]
		A. cooleyi (Mclvor, 1941) *nec A. cooleyi* Kohls & Hoogstraal, 1960[K,CHAM,HCK]
		A. cooleyi Kohls & Hoogstraal, 1960 *nec A. cooleyi* (McIvor, 1941)[K,CHAM,HCK]
		A. cucumerinus Neumann, 1901[K,CHAM,HCK]
		A. dalei Clifford, Keirans, Hoogstraal & Corwin, 1976[K,CHAM,HCK]
		A. delanoei (Roubaud & Colas-Belcour, 1931)[K,CHAM,HCK]
		A. dulus Keirans, Clifford & Capriles, 1971[K,CHAM,HCK]
		A. eboris (Theiler, 1959)[K,CHAM]
		A. echinops Hoogstraal, Uilenberg & Blanc, 1967[K,CHAM,HCK]
		A. falco Kaiser & Hoogstraal, 1974[K,CHAM,HCK]
		A. foleyi (Parrot, 1928)[K,CHAM,HCK]
		A. giganteus Kohls & Clifford, 1968[K,CHAM,HCK]
		A. gilcolladoi Estrada-Peña, Lucientes & Sánchez, 1987[K,CHAM,HCK]
		A. hermanni Audouin, 1827[K,CHAM,HCK]
		A. himalayensis Hoogstraal & Kaiser, 1973[K,CHAM,HCK]
		A. hoogstraali Morel & Vassiliades, 1965[K,CHAM,HCK]
		A. japonicus Yamaguti, Clifford & Tipton, 1968[K,CHAM,HCK]
		A. keiransi Estrada-peña, Venzal, Gonzalez-Acuna & Gugliemone, 2003RD
		A. lagenoplastis Froggatt, 1906[K,CHAM,HCK]
		A. lahorensis (Neumann, 1908)[K,CHAM,HCK]
		A. latus Filippova, 1961[K,CHAM,HCK]

(*cont.*)

Table 1.1 (*cont.*)

Family	Genus	Species
		A. lowryae Kaiser & Hoogstraal, 1975[K,CHAM,HCK]
		A. macrostigmatus Filippova, 1961[K,CHAM,HCK]
		A. magnus Neumann, 1896[K,CHAM,HCK]
		A. miniatus Koch, 1844[K,CHAM,HCK]
		A. monachus Keirans, Radovsky & Clifford, 1973[K,CHAM,HCK]
		A. monolakensis Schwan, Corwin, & Brown, 1992[KR,CHAM,HCK]
		A. moreli Keirans, Hoogstraal & Clifford, 1979[K,CHAM,HCK]
		A. neghmei Kohls & Hoogstraal, 1961[K,CHAM,HCK]
		A. nullarborensis Hoogstraal & Kaiser, 1973[K,CHAM,HCK]
		A. peringueyi (Bedford & Hewitt, 1925)[K,CHAM,HCK]
		A. persicus (Oken, 1818)[K,CHAM,HCK]
		A. peusi (Schulze, 1943)[K,CHAM,HCK]
		A. polonicus Siuda, Hoogstraal, Clifford & Wassef, 1979[K,CHAM,HCK]
		A. radiatus Railliet, 1893[K,CHAM,HCK]
		A. reflexus (Fabricius, 1794)[K,CHAM,HCK]
		A. ricei Hoogstraal, Kaiser, Clifford & Keirans, 1975[CHAM,HCK]
		A. robertsi Hoogstraal, Kaiser & Kohls, 1968[K,CHAM,HCK]
		A. sanchezi Dugès, 1887[K,CHAM,HCK]
		A. streptopelia Kaiser, Hoogstraal & Homer, 1970[K,CHAM,HCK]
		A. striatus Bedford, 1932[K,CHAM,HCK]
		A. theilerae Hoogstraal & Kaiser, 1970[K,CHAM,HCK]
		A. transgariepinus White, 1846[K,CHAM,HCK]
		A. tridentatus Filippova, 1961[K,CHAM,HCK]
		A. vansomereni (Keirans, Hoogstraal & Clifford, 1977)[K,CHAM,HCK]
		A. vulgaris Filippova, 1961 (?=*A. delicatus* Neumann, 1910)[K,CHAM,HCK]
		A. walkerae Kaiser & Hoogstraal, 1969[K,CHAM,HCK]
		A. zumpti Hoogstraal, Kaiser & Kohls, 1968[K,CHAM,HCK]
	Antricola (3 species)	*A. delacruzi* Estrada-Pena, Venzal, Barros-Battesti, Onofrio, Trajano, Firmino 2004RD
		A. guglielmonei Estrada-Pena, Venzal, Barros-Battesti, Onofrio, Trajano & Firmino 2004RD
		A. inexpectata Estrada-Pena, Venzal, Barros-Battesti, Onofrio, Trajano & Firmino 2004RD
	Carios (88 species)	*C. amblus* (Chamberlain, 1920)[K,CHAM,HCK]
		C. aragaoi (Fonseca, 1960)[HCK,CHAM]
		C. armasi (de la Cruz & Estrada-Peña, 1995)[KR,CHAM,HCK]
		C. australiensis (Kohls & Hoogstraal, 1962)[K,CHAM,HCK]

Table 1.1 (*cont.*)

Family	Genus	Species
		C. azteci (Matheson, 1935)[K,CHAM,HCK]
		C. batuensis (Hirst, 1929)[K,CHAM,HCK]
		C. boueti (Roubaud & Colas-Belcour, 1933)[K,CHAM,HCK]
		C. brodyi (Matheson, 1935)[K,CHAM,HCK]
		C. camicasi (Sylla, Cornet & Marchand, 1997)[KR,HCK]
		C. capensis (Neumann, 1901)[K,CHAM,HCK]
		C. casebeeri (Jones & Clifford, 1972)[K,CHAM,HCK]
		C. centralis (de la Cruz & Estrada-Peña, 1995)[KR,CHAM,HCK]
		C. cernyi (de la Cruz, 1978)[K,CHAM,HCK]
		C. ceylonensis (Hoogstraal & Kaiser, 1968)[K,CHAM,HCK]
		C. cheikhi (Vermeil, Marjolet & Vermeil, 1997)[HCK]
		C. chironectes (Jones & Clifford, 1972)[K,CHAM,HCK]
		C. chiropterphila (Dhanda & Rajagopalan, 1971)[K,CHAM,HCK]
		C. clarki (Jones & Clifford, 1972)[K,CHAM,HCK]
		C. collocaliae (Hoogstraal, Kadarsan, Kaiser & Van Peenan, 1974)[K,CHAM,HCK]
		C. concanensis (Cooley & Kohls, 1941)[K,CHAM,HCK]
		C. confusus (Hoogstraal, 1955)[K,CHAM,HCK]
		C. coniceps (Canestrini, 1890)[K,CHAM,HCK]
		C. coprophilus (McIntosh, 1935)[K,CHAM,HCK]
		C. cordiformis (Hoogstraal & Kohls, 1967)[K,CHAM,HCK]
		C. cyclurae (de la Cruz, 1984)[K,CHAM,HCK]
		C. darwini (Kohls, Clifford & Hoogstraal, 1969)[K,CHAM,HCK]
		C. daviesi (Kaiser & Hoogstraal, 1973)[K,CHAM,HCK]
		C. denmarki (Kohls, Sonenshine & Clifford, 1965)[K,CHAM,HCK]
		C. dewae (Kaiser & Hoogstraal, 1974)[K,CHAM,HCK]
		C. dusbabeki (Cérny, 1967)[K,CHAM,HCK]
		C. dyeri (Cooley & Kohls, 1940)[K,CHAM,HCK]
		C. echimys (Kohls, Clifford & Jones, 1969)[K,CHAM,HCK]
		C. elongates (Kohls, Clifford & Sonenshine, 1965)[K,CHAM,HCK]
		C. eptesicus (Kohls, Clifford & Jones, 1969)[K,CHAM,HCK]
		C. faini (Hoogstraal, 1960)[K,CHAM,HCK]
		C. fischeri (Audouin, 1827)[CHAM]
		C. galapagensis (Kohls, Clifford & Hoogstraal, 1969)[K,CHAM,HCK]
		C. granasi (de la Cruz, 1973)[K,CHAM,HCK]
		C. habanensis (de la Cruz, 1976)[K,CHAM,HCK]
		C. hadiae Klompen, Keirans & Durden, 1995[KR,CHAM,HCK]
		C. hasei (Schulze, 1935)[K,CHAM,HCK]
		C. hummelincki (de la Cruz & Estrada-Peña, 1995)[KR,CHAM,HCK]

(*cont.*)

Table 1.1 (*cont.*)

Family	Genus	Species
		C. jerseyi Klompen & Grimaldi, 2001[HCK]
		C. jul (Schulze, 1940)[HCK,CHAM]
		C. kelleyi (Cooley & Kohls, 1941)[K,HCK]
		C. kohlsi (Guglielmone & Keirans, 2002)[HCK]
		C. macrodermae (Hoogstraal, Moorhouse, Wolf & Wassef, 1977)[K,CHAM,HCK]
		C. madagascariensis (Hoogstraal, 1962)[K,CHAM,HCK]
		C. marginatus (Banks, 1910)[K,CHAM,HCK]
		C. marinkellei (Kohls, Clifford & Jones, 1969)[K,CHAM,HCK]
		C. maritimus (Vermeil & Marguet, 1967)[K,CHAM,HCK]
		C. marmosae (Jones & Clifford, 1972)[K,CHAM,HCK]
		C. martelorum (de la Cruz, 1978)[K,CHAM,HCK]
		C. mexicanus (Hoffman, 1959)[K,CHAM,HCK]
		C. mimon (Kohls, Clifford & Jones, 1969)[K,CHAM,HCK]
		C. mormoops (Kohls, Clifford & Jones, 1969)[K,CHAM,HCK]
		C. muesebecki (Hoogstraal, 1969)[K,CHAM,HCK]
		C. multisetosus Klompen, Keirans & Durden, 1995[KR,CHAM,HCK]
		C. naomiae (de la Cruz, 1978)[K,CHAM,HCK]
		C. natalinus (Černý & Dusbábek, 1967)[HCK,CHAM]
		C. occidentalis (de la Cruz, 1978)[K,CHAM,HCK]
		C. papuensis Klompen, Keirans & Durden, 1995[KR,CHAM,HCK]
		C. peropteryx (Kohls, Clifford & Jones, 1969)[K,CHAM,HCK]
		C. peruvianus (Kohls, Clifford & Jones, 1969)[K,CHAM,HCK]
		C. piriformis (Warburton, 1918)[K,CHAM,HCK]
		C. puertoricensis (Fox, 1947)[K,CHAM,HCK]
		C. pusillus (Kohls, 1950)[K,CHAM,HCK]
		C. reddelli (Keirans & Clifford, 1975)[K,CHAM,HCK]
		C. rennellensis (Clifford & Sonenshine, 1962)[K,CHAM,HCK]
		C. rossi (Kohls, Sonenshine & Clifford, 1965)[K,CHAM,HCK]
		C. rudis (Karsch, 1880)[K,CHAM,HCK]
		C. salahi (Hoogstraal, 1953)[K,CHAM,HCK]
		C. sawaii (Kitaoka & Suzuki, 1973)[K,CHAM,HCK]
		C. setosus (Kohls, Clifford & Jones, 1969)[HCK,CHAM]
		C. siboneyi (de la Cruz & Estrada-Peña, 1995)[KR,CHAM,HCK]
		C. silvai (Černý, 1967)[K,CHAM,HCK]
		C. sinensis (Jeu & Zhu, 1982)[K,CHAM,HCK]
		C. solomonis (Dumbleton, 1959)[K,CHAM,HCK]
		C. spheniscus (Hoogstraal, Wassef, Hays & Keirans, 1985)[K,CHAM,HCK]
		C. stageri (Cooley & Kohls, 1941)[K,CHAM,HCK]
		C. tadaridae (Černý & Dusbábek, 1967)[K,CHAM,HCK]
		C. talaje (Guérin-Ménville, 1849)[K,CHAM,HCK]

Table 1.1 (*cont.*)

Family	Genus	Species
		C. tiptoni (Jones & Clifford, 1972)[K,CHAM,HCK]
		C. tuttlei (Jones & Clifford, 1972)[K,CHAM,HCK]
		C. vespertilionis Latreille, 1796[K,CHAM,HCK]
		C. viguerasi (Cooley & Kohls, 1941)[K,CHAM,HCK]
		C. yumatensis (Cooley & Kohls, 1941)[K,CHAM,HCK]
		C. yunkeri (Keirans, Clifford & Hoogstraal, 1984)[K,CHAM,HCK]
	Ornithodoros (37 species)	*O. alactagalis* Issaakjan, 1936[K,CHAM,HCK]
		O. antiquus Poinar, 1995[KR,CHAM,HCK] (known only as a fossil)
		O. apertus Walton, 1962[K,CHAM,HCK]
		O. arenicolous Hoogstraal, 1953[K,CHAM,HCK]
		O. asperus Warburton, 1918[K,CHAM,HCK]
		O. brasiliensis Aragão, 1923[K,CHAM,HCK]
		O. boliviensis (Kohls & Clifford, 1964)[K,CHAM]
		O. cholodkovskyi Pavlovsky, 1930[K,CHAM,HCK]
		O. compactus Walton, 1962[K,CHAM,HCK]
		O. coriaceus Koch, 1844[K,CHAM,HCK]
		O. eremicus Cooley & Kohls, 1941[K,CHAM,HCK]
		O. erraticus (Lucas, 1849)[K,CHAM,HCK]
		O. furcosus Neumann, 1908[K,CHAM,HCK]
		O. graingeri Heisch & Guggisberg, 1953[K,CHAM,HCK]
		O. grenieri Klein, 1965[K,CHAM,HCK]
		O. gurneyi Warburton, 1926[K,CHAM,HCK]
		O. hermsi Wheeler, Herms & Meyer, 1935[K,CHAM,HCK]
		O. indica Rau & Rao, 1971[K,CHAM,HCK]
		O. knoxjonesi Jones & Clifford, 1972[K,HCK]
		O. macmillani Hoogstraal & Kohls, 1966[K,CHAM,HCK]
		O. marocanus Velu, 1919[K,HCK]
		O. moubata (Murray, 1877)[K,CHAM,HCK]
		O. nattereri Warburton, 1927[K,HCK]
		O. nicollei Mooser, 1932[K,CHAM,HCK]
		O. normandi Larrousse, 1923[K,CHAM,HCK]
		O. parkeri Cooley, 1936[K,CHAM,HCK]
		O. porcinus Walton, 1962[K,CHAM,HCK]
		O. procaviae Theodor & Costa, 1960[K,CHAM,HCK]
		O. rostratus Aragão, 1911[K,CHAM,HCK]
		O. savignyi (Audouin, 1827)[K,CHAM,HCK]
		O. sonrai Sautet & Witkowski, 1943[K,CHAM,HCK]
		O. steini (Schulze, 1935)[K,HCK]
		O. tartakovskyi Olenev, 1931[K,CHAM,HCK]
		O. tholozani (Laboulbène & Mégnin, 1882)
		O. transversus (Banks, 1902)[K,CHAM,HCK]

(*cont.*)

Table 1.1 (*cont.*)

Family	Genus	Species
		O. turicata (Dugès, 1876)[K,CHAM,HCK]
		O. zumpti Heisch & Guggisberg, 1953
	Otobius (3 species)	O. lagophilus Cooley & Kohls, 1940[K,CHAM,HCK]
		O. mégnini (Dugès, 1883)[K,CHAM,HCK]
		O. sparnus (Kohls & Clifford, 1963)[K,CHAM,HCK]
IXODIDAE (720 species)	Amblyomma (143 species)	A. acutangulatum Neumann, 1899[CHAM]
		A. albolimbatum Neumann, 1907[K,CHAM,HCK]
		A. albopictum Neumann, 1899[K,CHAM,HCK]
		A. americanum (Linnaeus, 1758)[K,CHAM,HCK]
		A. antillorum Kohls, 1969[K,CHAM,HCK]
		A. arcanum Karsch, 1879[CHAM,HCK]
		A. argentinae Neumann, 1905[CHAM,HCK]
		A. arianae Keirans & Garris, 1986[KR]
		A. astrion Dönitz, 1909[K,CHAM,HCK]
		A. aureolatum (Pallas, 1772)[HCK]
		A. auricularium (Conil, 1878)[K,CHAM,HCK]
		A. australiense Neumann, 1905[K,CHAM,HCK]
		A. babirussae Schulze, 1933[K,CHAM,HCK]
		A. bibroni (Gervais, 1842)[CHAM]
		A. boulengeri Hirst & Hirst, 1910[K,CHAM,HCK]
		A. brasiliense Aragão, 1908[K,CHAM,HCK]
		A. breviscutatum Neumann, 1899[HCK]
		A. cajennense (Fabricius, 1787)[K,CHAM,HCK]
		A. calabyi Roberts, 1963[K,CHAM,HCK]
		A. calcaratum Neumann, 1899[K,CHAM,HCK]
		A. chabaudi Rageau, 1964[K,CHAM,HCK]
		A. clypeolatum Neumann, 1899[K,CHAM,HCK]
		A. coelebs Neumann, 1899[K,CHAM,HCK]
		A. cohaerens Dönitz, 1909[K,CHAM,HCK]
		A. colasbelcouri (Santos Dias, 1958)[K,HCK]
		A. compressum (Macalister, 1872)[K,CHAM,HCK]
		A. cooperi Nuttall & Warburton, 1908[K,HCK]
		A. cordiferum Neumann, 1899[K,CHAM,HCK]
		A. crassipes (Neumann, 1901)[K,HCK]
		A. crassum Robinson, 1926[K,CHAM,HCK]
		A. crenatum Neumann, 1899[K,CHAM,HCK]
		A. cruciferum Neumann, 1901[K,CHAM,HCK]
		A. curraca Schulze, 1936[CHAM]
		A. cyprium Neumann, 1899[K,HCK]
		A. darwini Hirst & Hirst, 1910[K,CHAM,HCK]
		A. decorosum (Koch, 1867)[K,CHAM]
		A. dissimile Koch, 1844[K,HCK]
		A. dubitatum Neumann, 1899[HCK]
		A. eburneum Gerstäcker, 1873[K,CHAM,HCK]

Table 1.1 (*cont.*)

Family	Genus	Species
		A. echidnae Roberts, 1953[K,HCK]
		A. elaphense (Price, 1959)[K,CHAM,HCK]
		A. exornatum Koch, 1844[K,CHAM,HCK]
		A. extraoculatum Neumann, 1899[K,CHAM,HCK]
		A. falsomarmoreum Tonelli-Rondelli, 1935[K,CHAM,HCK]
		A. fimbriatum Koch, 1844[K,CHAM,HCK]
		A. flavomaculatum (Lucas, 1846)[K,CHAM,HCK]
		A. fulvum Neumann, 1899[K,CHAM,HCK]
		A. fuscolineatum (Lucas, 1847)[K,CHAM,HCK]
		A. fuscum Neumann, 1907[CHAM]
		A. geayi Neumann, 1899[K,HCK]
		A. gemma Dönitz, 1909[K,CHAM,HCK]
		A. geochelone Durden, Keirans & Smith, 2002[HCK]
		A. geoemydae (Cantor, 1847)[K,CHAM,HCK]
		A. gervaisi (Lucas, 1847)[K,CHAM,HCK]
		A. glauerti Keirans, King & Sharrad, 1994[KR,CHAM,HCK]
		A. goeldii Neumann, 1899[K,CHAM,HCK]
		A. hainanense Teng, 1981[K,CHAM,HCK]
		A. hebraeum Koch, 1844[K,CHAM,HCK]
		A. helvolum Koch, 1844[K,CHAM,HCK]
		A. hirtum Neumann, 1906[HCK]
		A. humerale Koch, 1844[K,CHAM,HCK]
		A. imitator Kohls, 1958[K,CHAM,HCK]
		A. incisum Neumann, 1906[K,CHAM,HCK]
		A. inopinatum (Santos Dias, 1989)[HCK]
		A. inornatum (Banks, 1909)[K,CHAM,HCK]
		A. integrum Karsch, 1879[K,CHAM,HCK]
		A. javanense (Supino, 1897)[K,CHAM,HCK]
		A. komodoense (Oudemans, 1929)[K,CHAM,HCK]
		A. kraneveldi (Anastos, 1956)[K,CHAM,HCK]
		A. laticaudae Warburton, 1933[K,CHAM,HCK]
		A. latum Koch, 1844[K,CHAM,HCK]
		A. latepunctatum Tonelli-Rondelli, 1939[RD](refer to Labruna *et al.*, 2005)
		A. lepidum Dönitz, 1909[K,CHAM,HCK]
		A. limbatum Neumann, 1899[K,CHAM,HCK]
		A. loculosum Neumann, 1907[K,CHAM,HCK]
		A. longirostre (Koch, 1844)[K,CHAM,HCK]
		A. macfarlandi Keirans, Hoogstraal & Clifford, 1973[K,CHAM,HCK]
		A. macropi Roberts, 1953[K,CHAM,HCK]
		A. maculatum Koch, 1844[K,CHAM,HCK]
		A. marmoreum Koch, 1844[K,CHAM,HCK]

(*cont.*)

Table 1.1 (*cont.*)

Family	Genus	Species
		A. moreliae (Koch, 1867)[K,CHAM,HCK]
		A. moyi Roberts, 1953[K,CHAM,HCK]
		A. multipunctum Neumann, 1899[K,CHAM,HCK]
		A. naponense (Packard, 1869)[K,CHAM,HCK]
		A. neumanni Ribaga, 1902[K,CHAM,HCK]
		A. nitidum Hirst & Hirst, 1910[K,CHAM,HCK]
		A. nocens Robinson, 1912 sensu Theiler & Salisbury,1959[CHAM]
		A. nodosum Neumann, 1899[K,CHAM,HCK]
		A. nuttalli Dönitz, 1909[K,CHAM,HCK]
		A. oblongoguttatum Koch, 1844[K,CHAM,HCK]
		A. orlovi (Kolonin, 1992)[KR,CHAM,HCK]
		A. ovale Koch, 1844[K,CHAM,HCK]
		A. pacae Aragão, 1911[K,CHAM,HCK]
		A. papuanum Hirst, 1914[K,CHAM,HCK]
		A. parkeri Fonseca & Aragão, 1952[CHAM]
		A. parvitarsum Neumann, 1901[K,CHAM,HCK]
		A. parvum Aragão, 1908[K,CHAM,HCK]
		A. pattoni (Neumann, 1910)[K,CHAM,HCK]
		A. paulopunctatum Neumann, 1899[K,CHAM,HCK]
		A. pecarium Dunn, 1933[K,CHAM,HCK]
		A. perpunctatum (Packard, 1869)[CHAM]
		A. personatum Neumann, 1901[K,CHAM,HCK]
		A. pictum Neumann, 1906[K,CHAM,HCK]
		A. pilosum Neumann, 1899[K,CHAM,HCK]
		A. pomposum Dönitz, 1909[K,CHAM,HCK]
		A. postoculatum Neumann, 1899[K,CHAM,HCK]
		A. pseudoconcolor Aragão, 1908[K,HCK]
		A. pseudoparvum Guglielmone, Mangold & Keirans, 1990[K,CHAM,HCK]
		A. quadricavum (Schulze, 1941)[K,CHAM,HCK]
		A. rhinocerotis (de Geer, 1778)[K,CHAM,HCK]
		A. robinsoni Warburton, 1927[K,CHAM,HCK]
		A. rotundatum Koch, 1844[K,CHAM,HCK]
		A. sabanerae Stoll, 1890[K,CHAM,HCK]
		A. scalpturatum Neumann, 1906[K,CHAM,HCK]
		A. scutatum Neumann, 1899[K,CHAM,HCK]
		A. soembawensis (Anastos, 1956)[K,CHAM,HCK]
		A. sparsum Neumann, 1899[K,CHAM,HCK]
		A. sphenodonti (Dumbleton, 1943)[K,CHAM,HCK]
		A. splendidum Giebel, 1877[K,CHAM,HCK]
		A. squamosum Kohls, 1953[K,CHAM,HCK]
		A. striatum Koch, 1844[K,HCK]
		A. superbum Santos Dias, 1953[CHAM]

Table 1.1 (*cont.*)

Family	Genus	Species
		A. supinoi Neumann, 1905[K,CHAM,HCK]
		A. sylvaticum (de Geer, 1778)[K,CHAM,HCK]
		A. tachyglossi Roberts, 1953RD (refer to Andrews *et al.*, 2006)
		A. tapirellum Dunn, 1933[K,CHAM,HCK]
		A. testudinarium Koch 1844[K,CHAM,HCK]
		A. testudinis (Conil, 1877)[K,HCK]
		A. tholloni Neumann, 1899[K,CHAM,HCK]
		A. tigrinum Koch, 1844[K,CHAM,HCK]
		A. torrei Perez Viqueras, 1934[K,CHAM,HCK]
		A. transversale (Lucas, 1844)[K,CHAM,HCK]
		A. triguttatum Koch, 1844[K,CHAM,HCK]
		A. trimaculatum (Lucas, 1878)[K,CHAM,HCK]
		A. trinitatus Turk, 1948[CHAM]
		A. triste Koch, 1844[K,CHAM,HCK]
		A. tuberculatum Marx, 1894[K,CHAM,HCK]
		A. usingeri Keirans, Hoogstraal & Clifford, 1973[K,CHAM,HCK]
		A. varanense (Supino, 1897)[K,CHAM,HCK]
		A. variegatum (Fabricius, 1794)[K,CHAM,HCK]
		A. varium Koch, 1844[K,CHAM,HCK]
		A. vikirri Keirans, Bull, & Duffield, 1996[KR,HCK]
		A. williamsi Banks, 1924[K,CHAM,HCK]
	Anomalohimalaya (3 species)	*A. cricetuli* Teng & Huang, 1981[K,CHAM,HCK]
		A. lamai Hoogstraal, Kaiser & Mitchell, 1970[K,CHAM,HCK]
		A. lotozkyi Filippova & Panova, 1978[K,CHAM,HCK]
	Bothriocroton (6 species)	*B. auruginans* (Schulze, 1936)[CHAM,HCK]
		B. concolor (Neumann, 1899)[CHAM,HCK]
		B. glebopalma (Keirans, King, & Sharrad, 1994)[CHAM,HCK]
		B. hydrosauri (Denny, 1843)[CHAM,HCK]
		B. oudemansi (Neumann, 1910)[K,HCK,RD] (refer to Beati *et al.*, 2008)
		B. undatum (Fabricius, 1775)[HCK]
	Cosmiomma (1 species)	*C. hippopotamensis* (Denny, 1843)[K,CHAM,HCK]
	Cornupalpatum (1 species)	*C. burmanicum* Poinar & Brown, 2003RD
	Dermacentor (36 species)	*D. abaensis* Teng, 1963[CHAM,HCK]
		D. albipictus (Packard, 1869)[K,CHAM,HCK]
		D. andersoni Stiles, 1908[K,CHAM,HCK]
		D. asper Arthur, 1960[K,CHAM,HCK]
		D. atrosignatus Neumann, 1906[K,CHAM,HCK]
		D. auratus Supino, 1897[K,CHAM,HCK]
		D. circumguttatus Neumann, 1897[K,CHAM,HCK]
		D. compactus Neumann, 1901[K,CHAM,HCK]
		D. confractus (Schulze, 1933)[CHAM]

(*cont.*)

Table 1.1 (*cont.*)

Family	Genus	Species
		D. daghestanicus Olenev, 1928[CHAM]
		D. dispar Cooley, 1937[K,CHAM,HCK]
		D. dissimilis Cooley, 1947[K,CHAM,HCK]
		D. everestianus Hirst, 1926[K,CHAM,HCK]
		D. halli McIntosh, 1931[K,CHAM,HCK]
		D. hunteri Bishopp, 1912[K,CHAM,HCK]
		D. imitans Warburton, 1933[K,CHAM,HCK]
		D. latus Cooley, 1937[K,CHAM,HCK]
		D. marginatus (Sulzer, 1776)[K,CHAM,HCK]
		D. montanus Filippova & Panova, 1974[K,CHAM,HCK]
		D. nigrolineatus (Packard, 1869)[CHAM]
		D. nitens Neumann, 1897[K,CHAM,HCK]
		D. niveus Neumann, 1897[K,CHAM,HCK]
		D. nuttalli Olenev, 1928[K,CHAM,HCK]
		D. occidentalis Marx, 1892[K,CHAM,HCK]
		D. parumapertus Neumann, 1901[K,CHAM,HCK]
		D. pavlovskyi Olenev, 1927[K,CHAM,HCK]
		D. pomerantzevi Serdyukova, 1951[K,CHAM,HCK]
		D. raskemensis Pomerantsev, 1946[K,CHAM,HCK]
		D. reticulatus (Fabricius, 1794)[K,CHAM,HCK]
		D. rhinocerinus (Denny, 1843)[K,CHAM,HCK]
		D. silvarum Olenev, 1931[K,CHAM,HCK]
		D. sinicus Schulze, 1932[K,CHAM,HCK]
		D. steini Schulze, 1933[K,CHAM,HCK]
		D. taiwanensis Sugimoto, 1935[K,CHAM,HCK]
		D. ushakovae Filippova & Panova, 1987[K,KR,HCK]
		D. variabilis (Say, 1821)[K,CHAM,HCK]
	Haemaphysalis (166 species)	*H. aborensis* Warburton, 1913[K,CHAM,HCK]
		H. aciculifer Warburton, 1913[K,CHAM,HCK]
		H. aculeata Lavarra 1904[K,CHAM,HCK]
		H. adleri Feldman-Muhsam, 1951[K,CHAM,HCK]
		H. anomala Warburton, 1913[K,CHAM,HCK]
		H. anomaloceraea Teng & Cui, 1984[KR,HCK]
		H. anoplos Hoogstraal, Uilenberg & Klein, 1967[K,CHAM,HCK]
		H. aponommoides Warburton, 1913[K,CHAM,HCK]
		H. asiatica (Supino, 1897)[K,CHAM,HCK]
		H. atheruri Hoogstraal, Trapido & Kohls, 1965[K,CHAM,HCK]
		H. bancrofti Nuttall & Warburton, 1915[K,CHAM,HCK]
		H. bandicota Hoogstraal & Kohls, 1965[K,CHAM,HCK]
		H. bartelsi Schulze, 1938[K,CHAM,HCK]
		H. bequaerti Hoogstraal, 1956[K,CHAM,HCK]
		H. birmaniae Supino, 1897[K,CHAM,HCK]
		H. bispinosa Neumann, 1897[K,CHAM,HCK]

Table 1.1 (*cont.*)

Family	Genus	Species
		H. borneata Hoogstraal, 1971[K,CHAM,HCK]
		H. bremneri Roberts, 1963[K,CHAM,HCK]
		H. calcarata Neumann, 1902[K,CHAM,HCK]
		H. calva Nuttall & Warburton, 1915[K,CHAM,HCK]
		H. campanulata Warburton, 1908[K,CHAM,HCK]
		H. canestrinii (Supino, 1897)[K,CHAM,HCK]
		H. capricornis Hoogstraal, 1966[K,CHAM,HCK]
		H. caucasica Olenev, 1928[K,CHAM,HCK]
		H. celebensis Hoogstraal, Trapido & Kohls, 1965[K,CHAM,HCK]
		H. chordeilis (Packard, 1869)[K,CHAM,HCK]
		H. cinnabarina Koch, 1844[CHAM]
		H. colasbelcouri (Santos Dias, 1958)[CHAM,HCK]
		H. concinna Koch, 1844[K,CHAM,HCK]
		H. cooleyi Bedford, 1929[K,CHAM,HCK]
		H. cornigera Neumann, 1897[K,CHAM,HCK]
		H. cornupunctata Hoogstraal & Varma, 1962[K,CHAM,HCK]
		H. cuspidata Warburton, 1910[K,CHAM,HCK]
		H. dangi Phan Trong, 1977[CHAM,HCK]
		H. danieli Černý & Hoogstraal, 1977[K,CHAM,HCK]
		H. darjeeling Hoogstraal & Dhanda, 1970[K,CHAM,HCK]
		H. davisi Hoogstraal, Dhanda & Bhat, 1970[K,CHAM,HCK]
		H. demidovae Emel'yanova,[KR,CHAM,HCK]
		H. doenitzi Warburton, & Nuttall, 1909[K,CHAM,HCK]
		H. elliptica (Koch, 1844)[CHAM,HCK]
		H. elongata Neumann, 1897[K,CHAM,HCK]
		H. erinacei Pavesi, 1884[K,CHAM,HCK]
		H. eupleres Hoogstraal, Kohls & Trapido, 1965[K,CHAM,HCK]
		H. filippovae Bolotin, 1979[KR,HCK]
		H. flava Newmann, 1987[K,CHAM,HCK]
		H. formosensis Neumann, 1913[K,CHAM,HCK]
		H. fossae Hoogstraal, 1953[K,CHAM,HCK]
		H. fujisana Kitaoka, 1970[K,CHAM,HCK]
		H. garhwalensis Dhanda & Bhat, 1968[K,CHAM,HCK]
		H. goral Hoogstraal, 1970[K,CHAM,HCK]
		H. grochovskajae Kolonin, 1992[KR,CHAM,HCK]
		H. heinrichi Schulze, 1939[K,CHAM,HCK]
		H. himalaya Hoogstraal, 1966[K,HCK]
		H. hirsuta Hoogstraal, Trapido & Kohls, 1966[K,CHAM,HCK]
		H. hispanica Gil Collado, 1938[K,CHAM,HCK]
		H. hoodi Warburton & Nuttall, 1909[K,CHAM,HCK]
		H. hoogstraali Kohls, 1950[K,CHAM,HCK]
		H. houyi Nuttall & Warburton, 1915[K,CHAM,HCK]

(*cont.*)

Table 1.1 (*cont.*)

Family	Genus	Species
		H. howletti Warburton, 1913[K,CHAM,HCK]
		H. humerosa Warburton & Nuttall, 1909[K,CHAM,HCK]
		H. hylobatis Schulze, 1933[K,CHAM,HCK]
		H. hyracophila Hoogstraal, Walker & Neitz, 1971[K,CHAM,HCK]
		H. hystricis Supino, 1897[K,CHAM,HCK]
		H. ias Nakamura & Yajima, 1937[K,CHAM,HCK]
		H. indica Warburton, 1910[K,CHAM,HCK]
		H. indoflava Dhanda & Bhat, 1968[K,CHAM,HCK]
		H. inermis Birula, 1895[K,CHAM,HCK]
		H. intermedia Warburton & Nuttall, 1909[K,CHAM,HCK]
		H. japonica Warburton, 1908[K,CHAM,HCK]
		H. juxtakochi Cooley, 1946[K,CHAM,HCK]
		H. kadarsani Hoogstraal & Wassef, 1977[K,CHAM,HCK]
		H. kashmirensis Hoogstraal & Varma, 1962[K,CHAM,HCK]
		H. kinneari Warburton, 1913[K,CHAM,HCK]
		H. kitaokai Hoogstraal, 1969[K,CHAM,HCK]
		H. koningsbergeri Warburton & Nuttall, 1909[K,CHAM,HCK]
		H. kopetdaghica Kerbabaev, 1962[K,CHAM,HCK]
		H. kutchensis Hoogstraal & Trapido, 1963[K,CHAM,HCK]
		H. kyasanurensis Trapido, Hoogstraal & Rajagopalan, 1964[K,CHAM,HCK]
		H. lagostrophi Roberts, 1963[K,CHAM,HCK]
		H. lagrangei Larrousse, 1925 1978 [K,CHAM,HCK]
		H. laocayensis Phan Trong, 1977[HCK]
		H. leachi (Audouin, 1826)[K,CHAM,HCK]
		H. lemuris Hoogstraal, 1953[K,CHAM,HCK]
		H. leporispalustris (Packard, 1869)[K,CHAM,HCK]
		H. lobachovi Kolonin, 1995[KR,CHAM,HCK]
		H. longicornis Neumann, 1901[K,CHAM,HCK]
		H. luzonensis Hoogstraal & Parrish, 1968[K,CHAM,HCK]
		H. madagascariensis Colas-Belcour & Millot, 1948[K,CHAM,HCK]
		H. mageshimaensis Saito & Hoogstraal, 1973[K,CHAM,HCK]
		H. megalaimae Rajagopalan, 1963[K,CHAM,HCK]
		H. megaspinosa Saito, 1969[K,CHAM,HCK]
		H. menglaensis Pang, Chen & Xiang, 1982[KR,HCK]
		H. minuta Kohls, 1950[K,CHAM,HCK]
		H. mjoebergi Warburton, 1926[K,CHAM,HCK]
		H. montgomeryi Nuttall, 1912[K,CHAM,HCK]
		H. moreli Camicas, Hoogstraal & El Kammah, 1972[K,CHAM,HCK]
		H. moschisuga Teng, 1980[KR,CHAM,HCK]
		H. muhsamae Santos Dias, 1954[CHAM,HCK]

Table 1.1 (*cont.*)

Family	Genus	Species
		H. nadchatrami Hoogstraal, Trapido & Kohls, 1965[K,CHAM,HCK]
		H. nepalensis Hoogstraal, 1962[K,CHAM,HCK]
		H. nesomys Hoogstraal, Uilenberg & Klein, 1966[K,CHAM,HCK]
		H. norvali Hoogstraal & Wassef, 1983[K,CHAM,HCK]
		H. novaeguineae Hirst, 1914[K,CHAM,HCK]
		H. obesa Larrousse, 1925[K,CHAM,HCK]
		H. obtusa Dönitz, 1910[K,CHAM,HCK]
		H. orientalis Nuttall & Warburton, 1915[K,CHAM,HCK]
		H. ornithophila Hoogstraal & Kohls, 1959[K,CHAM,HCK]
		H. palawanensis Kohls, 1950[K,CHAM,HCK]
		H. papuana Thorell, 1883[K,CHAM,HCK]
		H. paraleachi Camicas, Hoogstraal & El Kammah, 1983[K,CHAM,HCK]
		H. paraturturis Hoogstraal, Trapido & Rebello, 1963[K,CHAM,HCK]
		H. parmata Neumann, 1905[K,CHAM,HCK]
		H. parva Neumann, 1897[K,CHAM,HCK]
		H. pavlovskyi Pospelova-Shtrom, 1935[K,HCK]
		H. pedetes Hoogstraal, 1972[K,CHAM,HCK]
		H. pentalagi Pospelova-Shtrom, 1935[K,CHAM,HCK]
		H. petrogalis Roberts, 1970[K,CHAM,HCK]
		H. phasiana Satto, Hoogstraal & Wassef, 1974[K,CHAM,HCK]
		H. pospelovashtromae Hoogstraal, 1966[K,CHAM,HCK]
		H. primitiva Teng, 1982[KR,CHAM,HCK]
		H. psalistos Hoogstraal, Kohls & Parrish, 1967[K,CHAM,HCK]
		H. punctaleachi Camicas, Hoogstraal & El Kammah, 1973[K,CHAM,HCK]
		H. punctata Canestrini & Fanzago, 1878[K,CHAM,HCK]
		H. quadriaculeata Kolonin, 1992[KR,CHAM,HCK]
		H. qinghaiensis Teng, 1980[KR,CHAM,HCK]
		H. ramachandrai Dhanda, Hoogstraal & Bhat, 1970[K,CHAM,HCK]
		H. ratti Kohls, 1948[K,CHAM,HCK]
		H. renschi Schulze, 1933[K,CHAM,HCK]
		H. roubaudi Toumanoff, 1940[K,CHAM,HCK]
		H. rugosa Santos Dias, 1956[K,CHAM,HCK]
		H. rusae Kohls, 1950[K,CHAM,HCK]
		H. sambar Hoogstraal, 1971[K,CHAM,HCK]
		H. sciuri Kohls, 1950[K,CHAM,HCK]
		H. semermis Neumann, 1901[K,CHAM,HCK]
		H. shimoga Trapido & Hoogstraal, 1964[K,CHAM,HCK]

(*cont.*)

Table 1.1 (*cont.*)

Family	Genus	Species
		H. silacea Robinson, 1912[K,CHAM,HCK]
		H. silvafelis Hoogstraal & Trapido, 1963[K,CHAM,HCK]
		H. simplex Neumann, 1897[K,CHAM,HCK]
		H. simplicima Hoogstraal & Wassef, 1979[K,CHAM,HCK]
		H. sinensis Zhang, 1981[KR,HCK]
		H. spinigera Neumann, 1897[K,CHAM,HCK]
		H. spinulosa Neumann, 1906[K,CHAM,HCK]
		H. subelongata Hoogstraal, 1953[K,CHAM,HCK]
		H. subterra Hoogstraal, El Kammah & Camicas, 1992[K,CHAM,HCK]
		H. sulcata Canestrini & Fanzago, 1878[K,CHAM,HCK]
		H. sumatraensis Hoogstraal, El Kammah, Kadarsan & Anastos, 1971[K,CHAM,HCK]
		H. sundrai Sharif, 1928[K,CHAM,HCK]
		H. suntzovi Kolonin, 1993[KR,CHAM,HCK]
		H. susphilippensis Hoogstraal, Kohls & Parrish, 1968[K,CHAM,HCK]
		H. taiwana Sugimoto, 1936[K,CHAM,HCK]
		H. tauffliebi Morel, 1965[K,CHAM,HCK]
		H. theilerae Hoogstraal, 1953[K,CHAM,HCK]
		H. tibetensis Hoogstraal, 1965[K,CHAM,HCK]
		H. tiptoni Hoogstraal, 1953[K,CHAM,HCK]
		H. toxopei Warburton, 1927[K,CHAM,HCK]
		H. traguli Oudemans, 1928[K,CHAM,HCK]
		H. traubi Kohls, 1955[K,CHAM,HCK]
		H. turturis Nuttall & Warburton, 1915[K,CHAM,HCK]
		H. verticalis Itagaki, Noda & Yamaguchi, 1944[K,CHAM,HCK]
		H. vidua Warburton & Nuttall, 1909[K,CHAM,HCK]
		H. vietnamensis Hoogstraal & Wilson, 1966[K,HCK]
		H. warburtoni Nuttall, 1912[K,CHAM,HCK]
		H. wellingtoni Nuttall & Warburton, 1908[K,CHAM,HCK]
		H. xinjiangensis Teng, 1980[KR,CHAM,HCK]
		H. yeni Toumanoff, 1944[K,CHAM,HCK]
		H. zumpti Hoogstraal & El Kammah, 1974[K,CHAM,HCK]
	Hyalomma (27 species)	*H. aegyptium* (Linnaeus, 1758)[K,CHAM,HCK]
		H. albiparmatum Schulze, 1919[K,CHAM,HCK]
		H. anatolicum Koch, 1844[K,CHAM,HCK]
		H. arabica Pegram, Hoogstraal & Wassef, 1982[K,CHAM,HCK]
		H. asiaticum Schulze & Schlottke, 1930[K,CHAM,HCK]
		H. brevipunctata Sharif, 1928[K,CHAM,HCK]
		H. detritum Schulze, 1919[K,CHAM,HCK]
		H. dromedarii Koch, 1844[K,CHAM,HCK]
		H. erythraeum Tonelli-Rondelli, 1932[K,CHAM,HCK]
		H. excavatum Koch, 1844[(refer to Apanaskevich & Horak, 2005)]

Table 1.1 (*cont.*)

Family	Genus	Species
		H. franchinii Tonelli-Rondelli, 1932[K,CHAM,HCK]
		H. glabrum Delpy, 1949 (refer to Apanaskevich & Horak, 2006)
		H. hussaini Sharif, 1928[K,CHAM,HCK]
		H. hystricis Dhanda & Raja, 1974[KR,CHAM,HCK]
		H. impeltatum Schulze & Schlottke, 1930[K,CHAM,HCK]
		H. impressum Koch, 1844[K,CHAM,HCK]
		H. kumari Sharif, 1928[K,CHAM,HCK]
		H. lusitanicum Koch, 1844[K,CHAM,HCK]
		H. marginatum Koch, 1844[K,CHAM,HCK]
		H. nitidum Schulze, 1919[K,CHAM,HCK]
		H. punt Hoogstraal, Kaiser & Pedersen, 1969[K,CHAM,HCK]
		H. rhipicephaloides Neumann, 1901[K,CHAM,HCK]
		H. rufipes Koch, 1844[K,HCK]
		H. schulzei Olenev, 1931[K,CHAM,HCK]
		H. sinaii Feldman-Muhsam, 1960[HCK]
		H. truncatum Koch, 1844[K,CHAM,HCK]
		H. turanicum Pomerantsev, 1946[HCK]
	Ixodes (249 species)	*I. abrocomae* Lahille, 1917[CHAM,HCK]
		I. acuminatus Neumann, 1901[K,CHAM,HCK]
		I. acutitarsus (Karsch, 1880)[K,CHAM,HCK]
		I. affinis Neumann, 1899[K,CHAM,HCK]
		I. albignaci Uilenberg & Hoogstraal, 1969[K,CHAM,HCK]
		I. alluaudi Neumann, 1913[K,CHAM,HCK]
		I. amarali Fonseca, 1935[K,CHAM,HCK]
		I. amersoni Kohls, 1966[K,HCK]
		I. anatis Chilton, 1904[K,HCK]
		I. andinus Kohls, 1956[K,CHAM,HCK]
		I. angustus Neumann, 1899[K,CHAM,HCK]
		I. antechini Roberts, 1960[K,CHAM,HCK]
		I. apteridis Maskell, 1897[CHAM]
		I. apronophorus Schulze, 1924[K,CHAM,HCK]
		I. arabukiensis Arthur, 1959[K,HCK]
		I. aragaoi Fonseca, 1935[CHAM,HCK]
		I. arboricola Schulze & Schlottke, 1930[K,CHAM,HCK]
		I. arebiensis Arthur, 1956[K,CHAM,HCK]
		I. asanumai Kitaoka, 1973[K,CHAM,HCK]
		I. aulacodi Arthur, 1956[K,CHAM,HCK]
		I. auriculaelongae Arthur, 1958[K,CHAM,HCK]
		I. auritulus Neumann, 1904[K,CHAM,HCK]
		I. australiensis Neumann, 1904[K,CHAM,HCK]
		I. baergi Cooley & Kohls, 1942[K,CHAM,HCK]
		I. bakeri Arthur & Clifford, 1961[K,CHAM,HCK]
		I. banksi Bishopp, 1911[K,CHAM,HCK]

(*cont.*)

Table 1.1 (*cont.*)

Family	Genus	Species
		I. bedfordi Arthur, 1959[K,CHAM,HCK]
		I. bequaerti Cooley & Kohls, 1945[K,CHAM,HCK]
		I. berlesei Birula, 1895[K,CHAM,HCK]
		I. bivari Santos Dias, 1990[KR,CHAM,HCK]
		I. boliviensis Neumann, 1904[K,HCK]
		I. brewsterae Keirans, Clifford & Walker, 1982[K,CHAM,HCK]
		I. browningi Arthur, 1956[K,CHAM,HCK]
		I. brumpti Morel, 1965[K,CHAM,HCK]
		I. brunneus Koch, 1844[K,CHAM,HCK]
		I. calcarhebes Arthur & Zulu, 1980[K,CHAM,HCK]
		I. caledonicus Nuttall, 1910[K,CHAM,HCK]
		I. canisuga Johnston, 1849[K,CHAM,HCK]
		I. capromydis Černý, 1966[K,CHAM,HCK]
		I. catherinei Keirans, Clifford & Walker, 1982[K,CHAM,HCK]
		I. cavipalpus Nuttall & Warburton, 1908[K,CHAM,HCK]
		I. ceylonensis Kohls, 1950[K,CHAM,HCK]
		I. chilensis Kohls, 1957[K,CHAM,HCK]
		I. colasbelcouri Arthur, 1957[K,CHAM,HCK]
		I. collocaliae Schulze, 1937[K,CHAM,HCK]
		I. columnae Takada & Fujita, 1992[KR,HCK]
		I. conepati Cooley & Kohls, 1943[K,CHAM,HCK]
		I. confusus Roberts, 1960[K,CHAM,HCK]
		I. cookei Packard, 1869[K,CHAM,HCK]
		I. cooleyi Aragão & Fonseca, 1951[K,CHAM,HCK]
		I. copei Wilson, 1980[K,CHAM,HCK]
		I. cordifer Neumann, 1908[K,CHAM,HCK]
		I. cornuae Arthur, 1960[K,CHAM,HCK]
		I. cornuatus Roberts, 1960[K,HCK]
		I. corwini Keirans, Clifford & Walker, 1982[K,CHAM,HCK]
		I. crenulatus Koch, 1844[K,CHAM,HCK]
		I. cuernavacensis Kohls & Clifford, 1966[K,CHAM,HCK]
		I. cumulatimpunctatus Schulze, 1943[K,CHAM,HCK]
		I. dampfi Cooley, 1943[K,CHAM,HCK]
		I. daveyi Nuttall, 1913[K,CHAM,HCK]
		I. dawesi Arthur, 1956[K,CHAM,HCK]
		I. dendrolagi Wilson, 1967[K,CHAM,HCK]
		I. dentatus Marx, 1899[K,CHAM,HCK]
		I. dicei Keirans & Ajohola, 2003
		I. diomedeae Arthur, 1958[K,CHAM,HCK]
		I. diversifossus Neumann, 1899[K,CHAM,HCK]
		I. djaronensis Neumann, 1907[K,CHAM,HCK]
		I. domerguei Uilenberg & Hoogstraal, 1965[K,CHAM,HCK]
		I. donarthuri Santos Dias, 1980[KR,CHAM]
		I. downsi Kohls, 1957[K,CHAM,HCK]

Table 1.1 (*cont.*)

Family	Genus	Species
		I. drakensbergensis Clifford, Theiler & Baker, 1975[K,CHAM,HCK]
		I. eadsi Kohls & Clifford, 1964[K,CHAM,HCK]
		I. eastoni Keirans & Clifford, 1983[K,CHAM,HCK]
		I. eichhorni Nuttall, 1916[K,CHAM,HCK]
		I. eldaricus Dzhaparidze, 1950[K,CHAM,HCK]
		I. elongatus Bedford, 1929[K,CHAM,HCK]
		I. eudyptidis Maskell, 1885[K,CHAM,HCK]
		I. euplecti Arthur, 1958[K,CHAM,HCK]
		I. evansi Arthur, 1956[K,CHAM,HCK]
		I. fecialis Warburton & Nuttall, 1909[K,CHAM,HCK]
		I. festai Tonelli-Rondelli, 1926[K,CHAM,HCK]
		I. filippovae Černý, 1961[CHAM,HCK]
		I. fossulatus Neumann, 1899[K,CHAM,HCK]
		I. frontalis (Panzer, 1798)[K,CHAM,HCK]
		I. fuscipes Koch, 1844[K,CHAM,HCK]
		I. galapagoensis Clifford & Hoogstraal, 1980[K,CHAM,HCK]
		I. ghilarovi Filippova & Panova, 1988[K,CHAM,HCK]
		I. gibbosus Nuttall, 1916[K,CHAM,HCK]
		I. granulatus Supino, 1897[K,CHAM,HCK]
		I. gregsoni Lindquist, Wu & Redner, 1999[HCK]
		I. guatemalensis Kohls, 1956[K,CHAM,HCK]
		I. hearlei Gregson, 1941[K,CHAM,HCK]
		I. heinrichi Arthur, 1962[K,CHAM,HCK]
		I. hexagonus Leach, 1815[K,CHAM,HCK]
		I. himalayensis Dhanda & Kulkarni, 1969[K,CHAM,HCK]
		I. hirsti Hassall, 1931[K,CHAM,HCK]
		I. holocyclus Neumann, 1899[K,CHAM,HCK]
		I. hoogstraali Arthur, 1955[K,CHAM,HCK]
		I. howelli Cooley & Kohls, 1938[K,CHAM,HCK]
		I. hyatti Clifford, Hoogstraal & Kohls, 1971[K,CHAM,HCK]
		I. hydromyidis Swan, 1931[K,CHAM,HCK]
		I. jacksoni Hoogstraal, 1967[K,CHAM,HCK]
		I. jellisoni Cooley & Kohls, 1938[K,CHAM,HCK]
		I. jonesae Kohls, Sonenshine & Clifford, 1969[K,CHAM,HCK]
		I. kaiseri Arthur, 1957[K,CHAM,HCK]
		I. kashimiricus Pomerantsev, 1948[K,CHAM,HCK]
		I. kazakstani Olenev & Sorokoumov, 1934[K,CHAM,HCK]
		I. kempi Nuttall, 1913[CHAM]
		I. kerguelenensis Andre & Colas-Belcour, 1942[K,CHAM,HCK]
		I. kingi Bishopp, 1911[K,CHAM,HCK]
		I. kohlsi Arthur, 1955[K,CHAM,HCK]
		I. kopsteini (Oudemans, 1926)[K,CHAM,HCK]

(*cont.*)

Table 1.1 (*cont.*)

Family	Genus	Species
		I. kuntzi Hoogstraal & Kohls, 1965[K,CHAM,HCK]
		I. laguri Olenev, 1929 sensu Olenev, 1931[K,CHAM,HCK]
		I. lasallei Mendez Arocha & Ortiz, 1958[K,CHAM,HCK]
		I. latus Arthur, 1958[K,CHAM,HCK]
		I. laysanensis Wilson, 1964[K,CHAM,HCK]
		I. lemuris Arthur, 1958[K,CHAM,HCK]
		I. lewisi Arthur, 1965[K,CHAM,HCK]
		I. lividus Koch, 1844[K,CHAM,HCK]
		I. longiscutatus Boero, 1944[K,CHAM,HCK]
		I. loricatus Neumann, 1899[K,CHAM,HCK]
		I. loveridgei Arthur, 1958[K,CHAM,HCK]
		I. luciae Sénevet, 1940[K,CHAM,HCK]
		I. lunatus Neumann, 1907[K,CHAM,HCK]
		I. luxuriosus Schulze, 1932[K,CHAM,HCK]
		I. macfarlanei Keirans, Clifford & Walker, 1982[K,CHAM,HCK]
		I. malayensis Kohls, 1962[K,CHAM,HCK]
		I. marmotae Cooley & Kohls, 1938[K,CHAM,HCK]
		I. marxi Banks, 1908[K,CHAM,HCK]
		I. maslovi[K,HCK]
		I. matopi Spickett, Keirans, Norval & Clifford, 1981[K,CHAM,HCK]
		I. mexicanus Cooley & Kohls, 1942[K,CHAM,HCK]
		I. minor Neumann, 1902[K,CHAM,HCK]
		I. minutae Arthur, 1959[K,CHAM,HCK]
		I. mitchelli Kohls, Clifford & Hoogstraal, Ensp; 1970[K,CHAM,HCK]
		I. monospinosus Saito, 1968[K,CHAM,HCK]
		I. montoyanus Cooley, 1944[K,CHAM,HCK]
		I. moreli Arthur, 1957[K,CHAM,HCK]
		I. moscharius Teng, 1982[KR,CHAM,HCK]
		I. moschiferi Nemenz, 1968[K,CHAM,HCK]
		I. muniensis Arthur & Burrow, 1957[K,CHAM,HCK]
		I. muris Bishopp & Smith, 1937[K,CHAM,HCK]
		I. murreleti Cooley & Kohls, 1945[K,CHAM,HCK]
		I. myospalacis Teng, 1986[K,CHAM,HCK]
		I. myotomys Clifford & Hoogstraal, 1970[K,CHAM,HCK]
		I. myrmecobii Roberts, 1962[K,CHAM,HCK]
		I. nairobiensis Nuttall, 1916[K,CHAM,HCK]
		I. nchisiensis Arthur, 1958[K,CHAM,HCK]
		I. nectomys Kohls, 1957[K,CHAM,HCK]
		I. neitzi Clifford, Walker & Keirans, 1977[K,CHAM,HCK]
		I. neotomae Cooley, 1944[K,CHAM,HCK,a]
		I. nesomys Uilenberg & Hoogstraal, 1969[K,CHAM,HCK]
		I. neuquenensis Ringuelet, 1947[K,CHAM,HCK]

Table 1.1 (*cont.*)

Family	Genus	Species
		I. nicolasi Santos Dias, 1982[KR,CHAM,HCK]
		I. nipponensis Kitaoka & Saito, 1967[K,CHAM,HCK]
		I. nitens Neumann, 1904[K,CHAM,HCK]
		I. nuttalli Lahille, 1913[K,CHAM,HCK]
		I. nuttallianus Schulze, 1930[K,CHAM,HCK]
		I. occultus Pomerantsev, 1946[K,CHAM,HCK]
		I. ochotonae Gregson, 1941[K,CHAM,HCK]
		I. okapiae Arthur, 1956[K,CHAM,HCK]
		I. oldi Nuttall, 1913[K,CHAM,HCK]
		I. ornithorhynchi Lucas, 1846[K,CHAM,HCK]
		I. ovatus Neumann, 1899[K,CHAM,HCK]
		I. pacificus Cooley & Kohls, 1943[K,CHAM,HCK]
		I. paranensis Barros-Battesti, Arzua, Pichorim & Keirans, 2003
		I. pararicinus Keirans & Clifford, 1985[K,CHAM,HCK]
		I. pavlovskyi Pomerantsev, 1946[K,CHAM,HCK]
		I. percavatus Neumann, 1906[K,CHAM,HCK]
		I. peromysci Augustson, 1940[K,CHAM,HCK]
		I. persulcatus Schulze, 1930[K,CHAM,HCK]
		I. petauristae Warburton, 1933[K,CHAM,HCK]
		I. philipi Keirans & Kohls, 1970[K,CHAM,HCK]
		I. pilosus Koch, 1844[K,CHAM,HCK]
		I. pomerantzi Kohls, 1956[K,CHAM,HCK]
		I. pomeranzevi Serdyukova, 1941[K,CHAM,HCK]
		I. priscicollaris Schulze, 1932[K,CHAM,HCK]
		I. procaviae Arthur & Burrow, 1957[K,CHAM,HCK]
		I. prokopjevi (Emel'yanova, 1979) [KR,CHAM,HCK]
		I. radfordi Kohls, 1948[K,CHAM,HCK]
		I. rageaui Arthur, 1958[K,CHAM,HCK]
		I. randriansoloi Uilenberg & Hoogstraal, 1969[K,CHAM,HCK]
		I. rangtangensis Teng, 1973[KR,HCK]
		I. rasus Neumann, 1899[K,CHAM,HCK]
		I. redikorzevi Olenev, 1927[K,CHAM,HCK]
		I. rhabdomysae Arthur, 1959[K,CHAM,HCK]
		I. ricinus (Linnaeus, 1758)[K,CHAM,HCK]
		I. rothschildi Nuttall & Warburton, 1911[K,CHAM,HCK]
		I. rotundatus Arthur, 1958[K,CHAM,HCK]
		I. rubicundus Neumann, 1904[K,CHAM,HCK]
		I. rubidus Neumann, 1901[K,CHAM,HCK]
		I. rugicollis Schulze & Schlottke, 1930[K,CHAM,HCK]
		I. rugosus Bishopp, 1911[K,CHAM,HCK]
		I. sachalinensis Filippova, 1971[K,HCK]
		I. scapularis Say, 1821[K,CHAM,HCK]

(*cont.*)

Table 1.1 (*cont.*)

Family	Genus	Species
		I. schillingsi Neumann, 1901[K,CHAM,HCK]
		I. schulzei Aragão & Fonseca, 1951[K,CHAM,HCK]
		I. sculptus Neumann, 1904[K,CHAM,HCK]
		I. semenovi Olenev, 1929[K,CHAM,HCK]
		I. shahi Clifford, Hoogstraal & Kohls, 1971[K,CHAM,HCK]
		I. siamensis Kitaoka & Suzuki, 1983[KR,HCK]
		I. sigelos Keirans, Clifford & Corwin, 1976[K,HCK]
		I. signatus Birula, 1895[K,CHAM,HCK]
		I. simplex Neumann, 1906[K,CHAM,HCK]
		I. sinaloa Kohls & Clifford, 1966[K,CHAM,HCK]
		I. sinensis Teng, 1977[K,CHAM,HCK]
		I. soricis Gregson, 1942[K,CHAM,HCK]
		I. spinae Arthur, 1958[K,CHAM,HCK]
		I. spinicoxalis Neumann, 1899[K,CHAM,HCK]
		I. spinipalpis Hadwen & Nuttall, 1916[K,CHAM,HCK]
		I. steini Schulze, 1932[K,CHAM,HCK]
		I. stilesi Neumann, 1911[K,CHAM,HCK]
		I. stromi Filippova, 1957[K,CHAM,HCK]
		I. subterranus Filippova, 1961[K,CHAM,HCK]
		I. taglei Kohls, 1969[K,CHAM,HCK]
		I. tamaulipas Kohls & Clifford, 1966[K,CHAM,HCK]
		I. tancitarius Cooley & Kohls, 1942[K,CHAM,HCK]
		I. tanuki Saito, 1964[K,CHAM,HCK]
		I. tapirus Kohls, 1956[K,CHAM,HCK]
		I. tasmani Neumann, 1899[K,CHAM,HCK]
		I. tecpanensis Kohls, 1956[K,CHAM,HCK]
		I. texanus Banks, 1909[K,CHAM,HCK]
		I. theilerae Arthur, 1953[K,CHAM,HCK]
		I. thomasae Arthur & Burrow, 1957[K,CHAM,HCK]
		I. tiptoni Kohls & Clifford, 1962[K,CHAM,HCK]
		I. tovari Cooley, 1945[K,CHAM,HCK]
		I. transvaalensis Clifford & Hoogstraal, 1966[K,CHAM,HCK]
		I. trianguliceps Birula, 1895[K,CHAM,HCK]
		I. trichosuri Roberts, 1960[K,CHAM,HCK]
		I. tropicalis Kohls, 1956[K,CHAM,HCK]
		I. turdus Nakatsuji, 1942[K,CHAM,HCK]
		I. ugandanus Neumann, 1906[K,CHAM,HCK]
		I. unicavatus Neumann, 1908[K,CHAM,HCK]
		I. uriae White, 1852[K,CHAM,HCK]
		I. uruguayensis Kohls & Clifford, 1967[K,CHAM]
		I. vanidicus Schulze, 1943[K,CHAM,HCK]
		I. venezuelensis Kohls, 1953[K,CHAM,HCK]
		I. ventalloi Gil Collado, 1936[K,CHAM,HCK]
		I. vespertilionis Koch, 1844[K,CHAM,HCK]

Table 1.1 (*cont.*)

Family	Genus	Species
		I. vestitus Neumann, 1908[K,CHAM,HCK]
		I. victoriensis Nuttall, 1916[K,CHAM,HCK]
		I. walkerae Clifford, Kohls & Hoogstraal, 1968[K,CHAM,HCK]
		I. werneri Kohls, 1950[K,CHAM,HCK]
		I. woodi Bishopp, 1911[K,CHAM,HCK]
		I. zaglossi Kohls, 1960[K,CHAM,HCK]
		I. zairensis Keirans, Clifford & Walker, 1982[K,CHAM,HCK]
		I. zealandicus (Dumbleton, 1953)[CHAM]
		I. zumpti Arthur, 1960[K,HCK]
	Margaropus (3 species)	*M. reidi* Hoogstraal, 1956[K,CHAM,HCK]
		M. wileyi Walker & Laurence, 1973[K,CHAM,HCK]
		M. winthemi Karsch, 1879[K,CHAM,HCK]
	Nosomma (1 species)	*N. monstrosum* (Nuttall & Warburton, 1908)[K,CHAM,HCK]
	Rhipicentor (2 species)	*R. bicornis* Nuttall & Warburton, 1908[K,CHAM,HCK]
		R. nuttalli Cooper & Robinson, 1908[K,CHAM,HCK]
	Rhipicephalus (79 species – includes the five species that were in the genus *Boophilus*. These are now in the subgenus *Rhipicephalus (Boophilus)*)	*R. (Boophilus) annulatus* (Say, 1821)[K,CHAM,HCK]
		R. (Boophilus) decoloratus (Koch, 1844)[K,CHAM,HCK]
		R. (Boophilus) geigyi (Aeschlimann & Morel, 1965)[K,CHAM,HCK]
		R. (Boophilus) kohlsi (Hoogstraal & Kaiser, 1960)[K,CHAM,HCK]
		R. (Boophilus) microplus (Canestrini, 1888)[K,CHAM,HCK]
		R. appendiculatus Neumann, 1901[WKH]
		R. aquatilis Walter, Keirans & Pegram, 1993[WKH]
		R. armatus Pocock, 1900[WKH]
		R. arnoldi Theiler & Zumpt, 1949[WKH]
		R. bequaerti Zumpt, 1949[WKH]
		R. bergeoni Morel & Balis, 1976[WKH]
		R. boueti Morel, 1957[WKH]
		R. bursa Canestrini & Fanzago, 1878[WKH]
		R. camicasi Morel, Mouchet & Rodhain, 1976[WKH]
		R. capensis Koch, 1844[WKH]
		R. carnivoralis Walker, 1966[WKH]
		R. complanatus Neumann, 1911[WKH]
		R. compositus Neumann, 1897[WKH]
		R. cuspidatus Neumann, 1906[WKH]
		R. deltoideus Neumann, 1910[WKH]
		R. distinctus Bedford, 1932[WKH]

(*cont.*)

Table 1.1 (*cont.*)

Family	Genus	Species
		R. duttoni Neumann, 1907[WKH]
		R. dux Dönitz, 1910[WKH]
		R. evertsi Neumann, 1897[WKH]
		R. exophthalmos Keirans & Walker, 1993[WKH]
		R. follis Dönitz, 1910[WKH]
		R. fulvus Neumann, 1913[WKH]
		R. gertrudae Feldman-Muhsam, 1960[WKH]
		R. glabroscutatum Du Toit, 1941[WKH]
		R. guilhoni Morel & Vassiliades, 1963[WKH]
		R. haemaphysaloides Supino, 1897[WKH]
		R. humeralis Tonelli-Rondelli, 1926[WKH]
		R. hurti Wilson, 1954[WKH]
		R. interventus Walker, Pegram & Keirans, 1995[WKH]
		R. jeanneli Neumann, 1913[WKH]
		R. kochi Dönitz, 1905[WKH]
		R. leporis Pomerantsev, 1946[WKH]
		R. longiceps Warburton, 1912[WKH]
		R. longicoxatus Neumann, 1905[WKH]
		R. longus Neumann, 1907[WKH]
		R. lounsburyi Walker, 1990[WKH]
		R. lunulatus Neumann, 1907[WKH]
		R. maculatus Neumann, 1901[WKH]
		R. masseyi Nuttell & Warburton, 1908[WKH]
		R. moucheti Morel, 1965[WKH]
		R. muehlensi Zumpt, 1943[WKH]
		R. muhsamae Morel & Vassiliades, 1965[WKH]
		R. neumanni Walker, 1990[WKH]
		R. nitens Neumann, 1904[WKH]
		R. oculatus Neumann, 1901[WKH]
		R. oreotragi Walker & Horak, 2000[WKH]
		R. pilans Schulze, 1935[WKH]
		R. planus Neumann, 1907[WKH]
		R. praetextatus Gerstäcker, 1873[WKH]
		R. pravus Dönitz, 1910[WKH]
		R. pseudolongus Santos Dias, 1953[WKH]
		R. pulchellus (Gerstäcker, 1873)[WKH]
		R. pumilio Schluze, 1935[WKH]
		R. punctatus Warburton, 1912[WKH]
		R. pusillus Gil Collado, 1936[WKH]
		R. ramachandrai Dhanda, 1966[WKH]
		R. rossicus Yakimov & Kol-Yakimova, 1911[WKH]
		R. sanguineus (Latreille, 1806)[WKH]
		R. scalpturatus Santos Dias, 1959[WKH]
		R. schulzei Olenev, 1929[WKH]

Table 1.1 (*cont.*)

Family	Genus	Species
		R. sculptus Warburton, 1912[WKH]
		R. senegalensis Koch, 1844[WKH]
		R. serranoi Santos Dias, 1950[WKH]
		R. simpsoni Nuttall, 1910[WKH]
		R. simus Koch, 1844[WKH]
		R. sulcatus Neumann, 1908[WKH]
		R. supertritus Neumann, 1907[WKH]
		R. theileri Bedford & Hewitt, 1925[WKH]
		R. tricuspis, Dönitz, 1906[WKH]
		R. turanicus Pomerantsev, 1936[WKH]
		R. warburtoni Walker & Horak, 2000[WKH]
		R. zambeziensis Walker, Norval & Corwin, 1981[WKH]
		R. ziemanni Neumann, 1904[WKH]
		R. zumpti Santos Diaz, 1950[WKH]

This list of valid names is a compilation of the genus and species names of Keirans (1992), Keirans & Robbins (1999), Walker *et al.* (2000) and Horak, Camicas & Keirans (2002). The major revisions of Klompen & Oliver (1993) and Klompen *et al.* (2002), and species-level revisions and name changes of Venzal *et al.* (2001), Guglielmone & Keirans (2002) and Murrell & Barker (2003) have been incorporated. Superscripts show the lists that contain each name: Keirans (1992)[K]; Camicas, Hervy, Adam & Morel (1998)[CHAM]; Keirans & Robbins (1999)[KR]; Walker, Keirans & Horak (2000)[WKH]; and Horak, Camicas & Keirans (2002)[HCK]. Subspecies and subgenera were ignored except for the new subgenus *Rhipicephalus* (*Boophilus*) – see text. The descriptions and redescriptions of species and one new genus since 2002 have the superscript [RD].

[a] *Ixodes neotomae* Cooley, 1944 is a valid species name. However, Norris *et al.* (1977) suggested that *I. neotomae* be reduced to a junior subjective synonym of *I. spinipalpis*. Although analyses of 16S and 12S rRNA failed to recover reciprocal monophyly for populations of *I. neotomae* and *I. spinipalpis* (Norris *et al.*, 1997) we conclude that insufficient evidence has been produced to reject the hypothesis of two separate species: i.e. *I. neotomae* and *I. spinipalpis*. In any case *I. neotomae* has not been formerly made a subjective synonym of *I. spinipalpis* so both names are in our list.

species the 'indigenous Australian *Aponomma* species', primarily on the basis of morphology. Analyses of SSU rDNA (Dobson & Barker, 1999) and evidence from morphological characters (Klompen *et al.*, 1997) indicate that this lineage is the sister-group to the rest of the Metastriata rather than being one of the three lineages of the genus *Aponomma*. Klompen *et al.* (2000) also had the endemic Australian *Aponomma* species in a separate lineage to the other *Aponomma* and *Amblyomma* species (Amblyomminae) but this lineage was the sister-group to three species of *Haemaphysalis* in their trees, rather than to the rest of the Metastriata.

PHYLOGENY OF THE ARGASIDAE

There has been little molecular study of the Argasidae, but Klompen (1992) and Klompen & Oliver (1993) studied the morphology and systematics of these ticks and confirmed that there were two lineages (= subfamilies Argasinae and Ornithodorinae). The species and genera in these subfamilies were revised by Klompen & Oliver (1993). Black & Piesman (1994) and Crampton *et al.* (1996) found evidence of paraphyly of the Argasidae from 16S rDNA, 18S (SSU) V4 rDNA and 28S (LSU) rDNA (the two studies differed in the type of paraphyly) but subsequent study with complete 18S (SSU) rDNA sequences indicates that the Argasidae is monophyletic (Black *et al.*, 1997).

PHYLOGENETIC RELATIONSHIPS WITHIN THE GENUS *IXODES*

Despite the fact that many *Ixodes* species are important parasites and vectors of pathogens to humankind, domestic

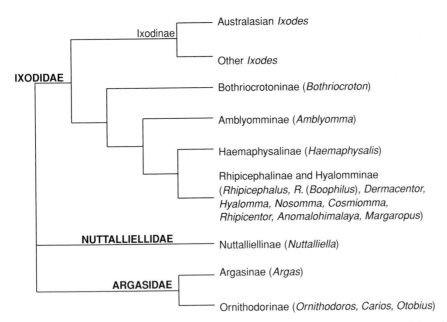

Fig. 1.1. Working hypothesis of the phylogeny of the subfamilies of ticks (Suborder Ixodida). This hypothesis was proposed from analyses of nucleotide sequences and phenotypes by Black & Piesman (1994); Crampton, McKay & Barker (1996); Black, Klompen & Keirans (1997); Dobson & Barker (1999); Klompen *et al.* (2000); and Murrell, Campbell & Barker (2001*b*).

animals and wildlife, there has been little study of the phylogeny of the group (16–18 subgenera, 249 valid species names). Klompen (1999) inferred the phylogeny of the subgenera of *Ixodes* from morphological characters whereas Fukunaga *et al.* (2000) inferred relationships among a few subgenera of *Ixodes* from internal transcribed spacer 2 (ITS2) rDNA sequences. It is not even certain that the genus is monophyletic (see above). However, robust phylogenies are known for some groups of closely related species of *Ixodes*, particularly some of the ticks that may transmit *Borrelia burgdorferi*: Caporale *et al.* (1995); McLain *et al.* (1995*a, b*); Rich *et al.* (1995, 1997); and Norris *et al.* (1996, 1997). Xu *et al.* (2003) found evidence that the 11 species in the *I. ricinus* complex were not monophyletic unless three other species, *I. muris*, *I. minor* and *I. granulatus*, were added to this complex.

PHYLOGENY AND EVOLUTION OF
RHIPICEPHALINE TICKS

There has been much recent progress in our understanding of the taxonomy and phylogeny of the Rhipicephalinae, so we review this work in detail. Murrell *et al.* (1999, 2000, 2001*a, b*) and Beati & Keirans (2001) used molecular and morpho-

logical characters to infer the phylogeny of the group. Here is our working hypothesis of the phylogeny of the Rhipicephalinae, after Murrell *et al.* (2001*b*) (Fig. 1.2).

We highlight five features of the tree: (1) *Anocentor nitens* is embedded in the genus *Dermacentor* Koch, 1844; (2) *Hyalomma aegyptium*, the only species in the subgenus *Hyalommasta* Schulze, 1930, is embedded in the lineage of the subgenus *Hy. (Hyalomma)* so the subgenus *Hyalommasta* should not be retained. *Hyalomma aegyptium* is the type species of the genus *Hyalomma* so the subgenus *Hyalommasta* is invalid (Robbins *et al.*, 1998); (3) molecular data are not yet available for *Cosmiomma* Schulze, 1919 (one species ex rhinoceros and antelopes), *Margaropus* Karsch, 1879 (three species ex horse, other livestock and giraffes) and *Anomalohimalaya* Hoogstraal *et al.*, 1970 (three species ex rodents, shrews and hares from Nepal, Tibet, China, Russia and environs); (4) there is strong molecular and morphological evidence that the five *Boophilus* species are monophyletic (Murrell *et al.*, 2001*a, b*); and (5) there is substantial molecular and morphological evidence that the sister-group to the subgenus *Rhipicephalus (Digineus)* (= *R. evertsi* group) plus the *R. pravus* group is the five *Boophilus* species, plus perhaps the genus *Margaropus* (Mangold *et al.*, 1998*b*; Murrell

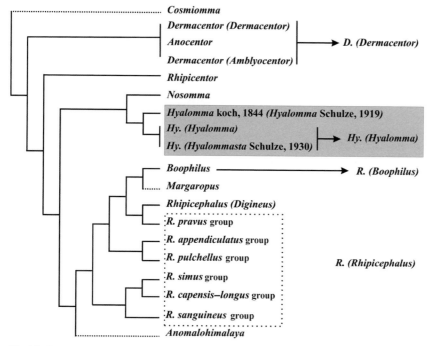

Fig. 1.2. Working hypothesis of the phylogeny of the Rhipicephalinae *sensu lato* from Murrell *et al.* (2001*b*). The taxa in plain text at the ends of broken lines have not been studied with molecular markers; their phylogenetic positions were inferred from the morphological analyses of Klompen *et al.* (1997). Shading indicates the subfamily Hyalomminae *sensu stricto*, which Murrell *et al.* (2001*b*) argued should be synonymized with the subfamily Rhipicephalinae. Arrows indicate taxonomic changes to the taxa that are apparently paraphyletic, which have been made (in the case of *Boophilus*) or which have been proposed (in the case of *Dermacentor* and *Hyalomma*). Taxa in brackets are subgenera. Six of the species-groups that constitute the subgenus *R. (Rhipicephalus)* are in a box (broken line). Reproduced with permission from Murrell *et al.* (2001*b*).

et al., 2000, 2001*b*; Beati & Keirans, 2001). This makes the genus *Rhipicephalus* paraphyletic with respect to the genus *Boophilus*. We note that the five *Boophilus* species are very different, morphologically, to the *Rhipicephalus* species. Furthermore, the *Boophilus* species are one-host ticks whereas the *Rhipicephalus* species are three-host ticks (rarely two–host ticks). But these differences are not necessarily relevant, phylogenetically. It seems that the *Boophilus* species are a highly derived (i.e. highly evolved) lineage within the *Rhipicephalus* lineage.

Phylogeny inferred from the mitochondrial genomes of ticks

Nucleotide sequences of three of the 37 genes of the mitochondrial genomes of ticks, 12S rRNA, 16S rRNA and cytochrome C oxidase I, have been a mainstay of tick molecular phylogenetics. The rest of the nucleotide sequence of the mitochondrial genome of tick species has been ignored.

Recent work shows, however, that mitochondrial genomes contain much unexploited potential for the inference of the phylogeny and evolution of ticks.

Mitochondrial genomes are circular and have 37 genes in most metazoa: 13 protein-coding genes, two ribosomal RNA genes, 22 transfer RNA genes and a non-coding (control) region. The mitochondrial genomes of most metazoa, including the ticks that have been studied so far, are 14–16 kb long (Shao & Barker, 2007). Two types of information from mitochondrial genomes have contributed, and are likely to continue to contribute, to our understanding of the phylogeny and evolution of ticks: (1) the nucleotide sequences of these genomes, or at least the nucleotide sequences of the 13 protein-coding genes and two rRNA genes, which together constitute 13–13.5 kb; and (2) idiosyncratic markers (*sensu* Murrell, Campbell & Barker, 2003). Translocations and inversions of genes are the best-known idiosyncratic markers. Duplications of genes and parts of genes, and changes to the secondary structures of tRNAs and rRNAs are other types of idiosyncratic markers. Nucleotide substitutions in tRNA motifs that are usually highly conserved are also markers that can be phylogenetically informative (Murrell *et al.*, 2003).

WHAT WE KNOW ABOUT THE MITOCHONDRIAL GENOMES OF TICKS

The mitochondrial genomes of 10 species of ticks have been sequenced entirely (Genbank numbers are given in parentheses): (1) *Rhipicephalus sanguineus* and *Ixodes hexagonus* (NC002010 and NC002074; Black & Roehrdanz, 1998); (2) *Haemaphysalis flava*, *Carios capensis* and *Ornithodoros moubata* (AB075954, AB075955 and AB073679; Shao *et al.*, 2004); (3) *I. persulcatus*, *I. holocyclus* and *I. uriae* (NC004470, AB075953 and AB087746; Shao *et al.*, 2005) and (4) *Amblyomma triguttatum* and *Ornithodoros porcinus* (AB113317 and AB105451). The mitochondrial genome of *Ixodes scapularis* has been assembled from the whole-genome shotgun sequence of this tick. The gene arrangement and nucleotide sequence will be available in 2007 (D. Hogenkamp, R. Shao, S.C. Barker, C. Hill & V. Nene, unpublished data). In addition, we know the arrangement of genes in the mitochondrial genome of *R. (B.) microplus* (Campbell & Barker, 1998, 1999) and part of the arrangement of these genes in the following 51 species of hard and soft ticks: *Amblyomma americanum*, *A. cajennense*, *A. hebraeum*, *A. fimbriatum*, *A. latum*, *A. maculatum*, *A. varanensis*, *A. variegatum*, *A. vikirri*, *Argas lagenoplastis*, *A. persicus*, *Bothriocroton concolor*, *B. glebopalma*, *B. undatum*, *Dermacentor andersoni*, *D. reticulatus*, *D. variabilis*, *Haemaphysalis humerosa*, *H. inermis*, *H. leporispalustris*, *H. longicornis*, *Hyalomma aegyptium*, *Hy. dromedarii*, *Hy. marginatum*, *Hy. truncatum*, *Ixodes affinis*, *I. auritulus*, *I. cookei*, *I. pilosus*, *I. ricinus*, *I. simplex*, *I. tasmani*, *Nosomma monstrosum*, *Ornithodoros turicata*, *Otobius megnini*, *Rhipicentor nuttalli*, *Rhipicephalus (B.) annulatus*, *R. appendiculatus*, *R. compositus*, *R. (B.) decoloratus*, *R. evertsi*, *R. (B.) geigyi*, *R. (B.) kohlsi*, *R. maculatus*, *R. pulchellus*, *R. punctatus*, *R. pravus*, *R. simus*, *R. turanicus*, *R. zambeziensis* and *R. zumpti* (Black & Roehrdanz, 1998; Campbell & Barker, 1998; Roehrdanz, Degrugillier & Black, 2002; Murrell *et al.*, 2003; Shao, Fukunaga & Barker, 2005).

TRANSLOCATION AND INVERSION OF GENES THAT ARE SYNAPOMORPHIES FOR GROUPS OF TICKS

Black & Roehrdanz (1998) and Campbell & Barker (1998) simultaneously discovered a major rearrangement of mitochondrial genes that was synapomorphic for *R. sanguineus*

and *B. microplus*: a section of the genome that contains the genes ND5, ND4, ND4L, ND6, Cyt b and five tRNAs has swapped positions with a section that contains ND1, 16S, 12S, CR and four tRNA genes. In addition, Black & Roehrdanz (1998), Campbell & Barker (1998, 1999) and Roehrdanz, Degrugillier & Black (2002) showed by PCR that this translocation was in fact synapomorphic for the known Metastriata since 28 other species from eight genera, which were from all of the subfamilies of metrastriate ticks, had this rearrangement too. The translocation of tRNA Leu (CUN) and the translocation and inversion of tRNA Cys are also synapomorphies for the Metastriata (27 species in eight genera: Black & Roehrdanz, 1998; Campbell & Barker, 1998, 1999; Murrell *et al.*, 2003).

OTHER IDIOSYNCRATIC MARKERS FROM THE MITOCHONDRIAL GENOMES OF TICKS

Murrell *et al.* (2003) assessed the value of idiosyncratic markers and changes to nucleotide sequences in tRNAs that are usually highly conserved for inference of the phylogeny of hard ticks. Many markers were informative. Moreover, parallel and convergent evolution of these markers was rare. Here are some examples of idiosyncratic markers that are synapomorphic for groups of ticks: (1) a region of tandemly repeated sequence (composed of tRNA Glu and 60 bp of the 3′ end of ND1) unites *R. (B.) microplus* and *R. (B.) annulatus* to the exclusion of the other three *Boophilus* species; (2) a region of 25 bp repeats between the 5′ end of ND1 and 16S that is synapomorphic for the Metastriata; and (3) a 15 bp insertion between tRNA Ala and tRNA Arg is synapomorphic for the genus *Hyalomma*.

Intriguingly, some species of hard ticks, but not others, have two control regions. On the one hand *I. hexagonus* (Black & Roehrdanz, 1998) and *I. acutitarsus*, *I. asanumai*, *I. bricatus*, *I. nipponensis*, *I. ovatus*, *I. pavlovskyi*, *I. persulcatus*, *I. pilosus*, *I. ricinus*, *I. scapularis*, *I. simplex* and *I. turdus* (Shao *et al.*, 2005) have one control region like the hypothetical ancestor of the arthropods. On the other hand, all eight of the 28 species examined from the putative Australasian *Ixodes* lineage (*I. antechini*, *I. cordifer*, *I. cornuatus*, *I. hirsti*, *I. holocyclus*, *I. myrmecobii*, *I. trichosuri* and *I. uriae*: Shao *et al.*, 2005), and all known metastriate ticks (Black & Roehrdanz, 1998; Campbell & Barker, 1998, 1999; Murrell *et al.*, 2003), have two control regions. Two control regions may be a synapomorphy for the Australasian *Ixodes*. Two control regions, together with certain gene rearrangements, may also be a synapomorphy for the metastriate ticks.

WHOLE MITOCHONDRIAL GENOME SEQUENCES

Whole-genome sequences, or at least the nucleotide or amino acid sequences of the 13 protein-coding genes (*c.* 12 kb), have been used to infer the phylogeny of a range of vertebrates (e.g. Broughton, Milam & Roe, 2001; Haring *et al.*, 2001; Inoue *et al.*, 2001; Maca-Meyer *et al.*, 2001; Miya, Kawaguchi & Nishida, 2001; Schmitz, Ohme & Zischler, 2002) and invertebrates (e.g. Black & Roehrdanz, 1998; Wilson *et al.*, 2000; Hwang *et al.*, 2001). Entire nucleotide sequences are available for 10 species of ticks (above). Once the nucleotide sequences of mitochondrial genomes from each of the major lineages of ticks are available the first whole mitochondrial genome phylogeny of the ticks will be attempted.

Using phylogeny to understand how ticks have evolved

Accurate phylogenies allow us to study the evolution of phenotypes (e.g. morphological and life history traits) and to infer historical zoogeography (e.g. where the first tick might have evolved). When traits are mapped onto a phylogeny we may discover, for example, how many times and in which environments a particular trait evolved. Below, we show how accurate phylogenies have contributed to our understanding of ornateness, life cycles, the archetypal host of ticks, where particular groups of ticks, and indeed the first tick, evolved, and the evolution of haematophagy in ticks.

EVOLUTION OF ORNATENESS IN RHIPICEPHALINE TICKS

A number of ticks in the subfamily Rhipicephalinae have ornate (patterned) scuta. By mapping ornateness onto the phylogeny of Murrell *et al.* (2001*b*) we can infer how ornateness might have evolved in the Rhipicephalinae *sensu lato* (Rhipicephalinae and Hyalomminae) (Fig. 1.3) based on the species in that study. We propose that an inornate scutum (not patterned) is the ancestral (plesiomorphic) character-state for the Rhipicephalinae since inornate scuta predominate in a putative sister-group of the Rhipicephalinae, the Haemaphysaline, and in the early-diverging ('basal') lineages of hard ticks. Even if the Amblyomminae (members of which are both ornate and inornate) is the sister-group of the Rhipicephalinae, as proposed by Klompen *et al.* (2000), the most parsimonious explanation is still that an inornate scutum is plesiomorphic for the Rhipicephalinae. If an inornate scutum is plesiomorphic for the Rhipicephalinae, then the simplest interpretation of Fig. 1.3 is that ornate scuta have evolved at least three times in the Rhipicephalinae s.l.: (1) in the ances-

tor of the *Dermacentor* and *Anocentor* species (all *Dermacentor* species are ornate but note that the most recent ancestor of *A. nitens* apparently reverted to the plesiomorphic state); (2) in *N. monstrosum*; and (3) in *R. maculatus* and *R. pulchellus*. Note that Klompen *et al.* (1997) found some support from morphological characters for *Cosmiomma* being the earliest-diverging lineage of Rhipicephalinae. If this is correct then it is just as parsimonious that ornateness is plesiomorphic for the Rhipicephalinae and that ornateness has been lost twice and secondarily gained twice in the Rhipicephalinae. In either scenario, ornateness has evolved more than once and been lost at least once in the rhipicephaline ticks.

The function, if any, of ornate scuta is unknown. It has been speculated, however, that these ticks are advertising something, perhaps that they are unpalatable. Indeed, it is possible that the ornate rhipicephaline species from Africa are mimicking other ornate ticks from the genus *Amblyomma*, which may be unpalatable to tickbirds (I. Horak, personal communication).

TRUNCATION OF LIFE CYCLES IN RHIPICEPHALINE TICKS

Most hard ticks have a three-host life cycle: each of the three mobile stages in the life cycle (larvae, nymphs and adults) leaves the host to moult to the next stage (Hoogstraal, 1978). In some ticks, moulting from larvae to nymphs takes place without leaving the host: the two-host ticks (all *Rhipicephalus* (*Digineus*) species and some *Hyalomma* species). In the most extreme case, all of the mobile stages remain on the one host: the one-host ticks (all *R.* (*Boophilus*) species, *A. nitens* and *D. albipictus*). There is much evidence that the plesiomorphic life cycle of the Rhipicephalinae s.l. is a three-host life cycle, since all known life cycles in the putative sister-groups of the Rhipicephalinae s.l., the Haemaphysalinae, and indeed in all other Ixodidae, are three-host life cycles. Murrell *et al.* (2001*b*) found that when the number of stages in the life cycle is mapped onto their phylogeny, a life cycle with a reduced number of hosts (two or one) has evolved as many as four times: in *R. evertsi*, in *A. nitens*, in some Hyalomma species (life cycles may vary within a species: Hoogstraal, 1978; Sonenshine, 1993) and in the *R.* (*Boophilus*) species (Fig. 1.3). It is unclear whether the evolution of a two-host life cycle in *Rhipicephalus* (*Digineus*) species (e.g. *R. evertsi*) and a one-host life cycle in *R.* (*Boophilus*) species are linked evolutionarily. Despite being phylogenetically close to *R. (Boophilus)* species, *R. evertsi* is apparently closer, phylogenetically, to species from the *R. pravus* group which are three-host ticks (refer to Fig. 1.2).

(a)   (b)

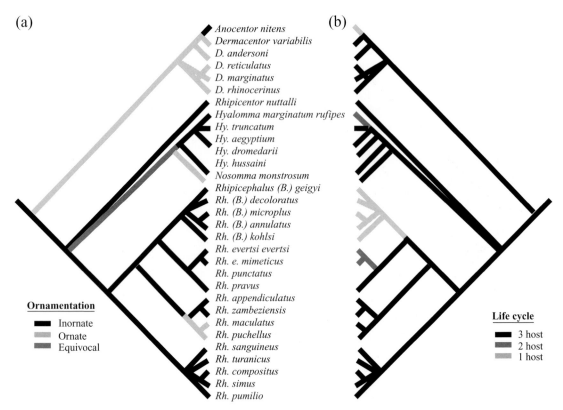

Anocentor nitens
Dermacentor variabilis
D. andersoni
D. reticulatus
D. marginatus
D. rhinocerinus
Rhipicentor nuttalli
Hyalomma marginatum rufipes
Hy. truncatum
Hy. aegyptium
Hy. dromedarii
Hy. hussaini
Nosomma monstrosum
Rhipicephalus (B.) geigyi
Rh. (B.) decoloratus
Rh. (B.) microplus
Rh. (B.) annulatus
Rh. (B.) kohlsi
Rh. evertsi evertsi
Rh. e. mimeticus
Rh. punctatus
Rh. pravus
Rh. appendiculatus
Rh. zambeziensis
Rh. maculatus
Rh. puchellus
Rh. sanguineus
Rh. turanicus
Rh. compositus
Rh. simus
Rh. pumilio

Ornamentation

■ Inornate
■ Ornate
■ Equivocal

Life cycle

■ 3 host
■ 2 host
■ 1 host

Fig. 1.3. A strict consensus of the eight shortest maximum parsimony trees from a total-evidence analysis of nucleotide sequences from four genes (12S, COI, ITS2, SSU rRNA) and morphology from Murrell *et al.* (2001*b*). Ornamentation of the scutum and the number of hosts required to complete life cycles have been mapped onto this tree. Note that *H. dromedarii* has one-, two- and three-host life cycles.

Hoogstraal (1978) discussed the evolution of reductions in the number of hosts in life cycles and concluded that the survival of larvae is more precarious than survival of other life stages, so the larvae are the 'weak link' in the life cycles of ticks. Hoogstraal (1978) concluded that selection pressure for two-host and one-host life cycles would be strong in ticks that are associated with hosts that wander, but weak in ticks with hosts that have a nest or localized breeding site. In ticks of hosts that wander, it would be more difficult for larvae and nymphs to find a host than for ticks with hosts that do not wander. As far as we know the only ixodid ticks with truncated life cycles, either with one or two hosts, are in the Rhipicephalinae.

HISTORICAL ZOOGEOGRAPHY OF
RHIPICEPHALINE TICKS

The regions inhabited by ticks may also be mapped onto the phylogeny of ticks. To demonstrate this approach we highlight the zoogeographical analysis of Murrell *et al.* (2001*b*).

This analysis shows that the *Dermacentor–Anocentor* clade, the *Rhipicephalus–Boophilus* species–*Hyalomma–Nosomma–Rhipicentor* clade, the *Rhipicephalus–Boophilus* species clade, the *Boophilus* species, and the hypothetical ancestors of the Rhipicephalinae s.l. probably evolved in the Afrotropical region (Fig. 1.4) but note that they did not study some lineages such as the oriental *Dermacentor* and many *Rhipicentor* species. Only a handful of rhipicephaline species seem to have evolved in a region other than the Afrotropical region: the clade with *Anocentor nitens* plus some *Dermacentor* species, *Hy. aegyptium* (ex tortoises), *Hy. dromedarii* (ex camels), *Hy. hussaini* (ex ungulates), *N. monstrosum* (ex buffalo and others), *R. (B). kohlsi* (ex goats and sheep) and *R. pumilio* (ex wide range of mammals). It is not clear where *R. sanguineus* (ex dogs) and *R. turanicus* (ex dogs and other large mammals) evolved. *Rhipicephalus turanicus* occurs in Africa and many European countries whilst *R. sanguineus* can now be found on dogs throughout the world.

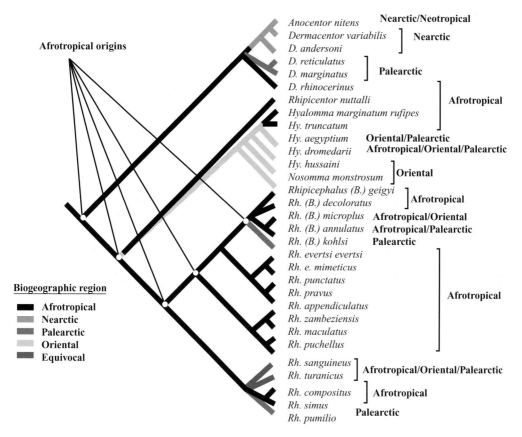

Anocentor nitens   **Nearctic/Neotropical**
Dermacentor variabilis] **Nearctic**
D. andersoni
D. reticulatus] **Palearctic**
D. marginatus
D. rhinocerinus]
Rhipicentor nuttalli   **Afrotropical**
Hyalomma marginatum rufipes
Hy. truncatum]
Hy. aegyptium   **Oriental/Palearctic**
Hy. dromedarii   **Afrotropical/Oriental/Palearctic**
Hy. hussaini   ] **Oriental**
Nosomma monstrosum
Rhipicephalus (B.) geigyi   ] **Afrotropical**
Rh. (B.) decoloratus
Rh. (B.) microplus   **Afrotropical/Oriental**
Rh. (B.) annulatus   **Afrotropical/Palearctic**
Rh. (B.) kohlsi   **Palearctic**
Rh. evertsi evertsi]
Rh. e. mimeticus
Rh. punctatus
Rh. pravus
Rh. appendiculatus   **Afrotropical**
Rh. zambeziensis
Rh. maculatus
Rh. puchellus
Rh. sanguineus] **Afrotropical/Oriental/Palearctic**
Rh. turanicus
Rh. compositus] **Afrotropical**
Rh. simus
Rh. pumilio   **Palearctic**

Afrotropical origins

Biogeographic region
- Afrotropical
- Nearctic
- Palearctic
- Oriental
- Equivocal

Fig. 1.4. Historical zoogeography of ticks of the subfamily Rhipicephalinae *sensu lato*. Information on the geographical distribution of ticks is from Camicas *et al.* (1998). Introductions of *R. (B.) microplus* and *R. (B.) annulatus* (cattle ticks) and *R. sanguineus* (dog ticks) to biogeographical regions by humans were excluded. Reproduced with permission from Murrell *et al.* (2001*b*).

Figure 1.4 also allows us to speculate about the history of the rhipicephaline ticks. Murrell *et al.* (2001*b*) proposed that: (1) the ancestor of the *Dermacentor–Anocentor* lineage evolved in the Afrotropical region. Then ticks of this lineage dispersed into Eurasia, probably in the Eocene (50 Mya) a period for which there is evidence for migration of mammals between Africa and Eurasia (Cox & Moore, 1993). After the Eocene, Africa became isolated and most dispersal and cladogenesis was in and between the Palearctic and Nearctic regions where most *Dermacentor* species are found today. Only two species from the *Dermacentor–Anocentor* lineage are found in Africa today, either because there has been little speciation in this lineage in Africa or because species from this lineage became extinct. During the Oligocene (*c.* 35 Mya) dispersal between Eurasia and the Nearctic was possible via the Bering land bridge and movement between Europe and the Nearctic via Greenland (Cox & Moore, 1993). Much later

(2.5 Mya) dispersal between the Nearctic and Neotropical regions was via the Isthmus of Panama.

Hypotheses of an African origin of the genus *Dermacentor* contrast with previous ideas about the origin of this genus which have centred around arguments over whether *Dermacentor* evolved in the Nearctic region and moved into the Palearctic and Oriental regions, or whether the *Dermacentor* species evolved in the Palearctic and then dispersed to Nearctic and Oriental regions (Balashov, 1994; Berdyev, 1989; Crosbie *et al.*, 1998); (2) The *Nosomma–Hyalomma* lineage appears to have evolved from an ancestor that lived in the Oriental region, perhaps in the early Miocene (ca. 19 Mya) when movement from Africa into Asia was possible for the ancestor of these species via a land bridge (Cox & Moore, 1993). Movement from Asia to Eurasia became possible and then there was probably dispersal back into Africa after Africa and Eurasia were joined by a land bridge at 14 Mya (Cox & Moore, 1993). This proposal is consistent with Balashov (1994); he proposed an Asian origin for the genus *Hyalomma*; (3) the *Boophilus* species–*Rhipicephalus* lineage probably evolved and radiated in Africa, i.e. when Africa was mostly isolated from the Palearctic and Oriental regions before the formation of the land bridge between Africa and Eurasia (14 Mya). Dispersal and radiation into Eurasia and Asia probably occurred after the land bridge formed between Africa and Eurasia in the Miocene. Balashov (1994) proposed that the genus *Rhipicephalus* evolved in Africa but thought it likely that the *Boophilus* species evolved in Europe. The phylogeny of Murrell *et al.* (2001*b*), however, indicates that the *Boophilus* species evolved in Africa (also see Murrell *et al.*, 2000); and (4) the *Rhipicentor* lineage (two species) appears to have evolved in, and then remained in, Africa, although the possibility that species from this lineage evolved in, or dispersed to, other regions but then became extinct in those regions cannot be ruled out. *Anomalohimalaya* species live in the Palearctic and Oriental regions but until the phylogenetic position of this genus is resolved its historical biogeography cannot be inferred.

WHERE DID THE FIRST HARD TICKS EVOLVE AND HOW DID THEIR DESCENDANTS SPREAD AROUND THE GLOBE?

Two main hypotheses have been proposed that address this question. On the one hand, Dobson & Barker (1999) proposed that hard ticks, and indeed all ticks, evolved in Australia, or more precisely in that part of Gondwana that became Australia, on early crocodile-like amphibians (labyrinthodonts, which are now extinct) in the Devonian ca. 390 Mya. On the other hand, Klompen *et al.* (1996, 2000) proposed that the first hard ticks evolved in Australia much later (120 Mya), when Australia and Antarctica had separated from the other land masses of Gondwana. The two proposals differ in the proposed time of origin of the hard ticks (*c.* 390 Mya versus *c.* 120 Mya). At 390 Mya Gondwana was intact, so land animals could have potentially moved between Australia, Antarctica, South America, Africa and India. However, by 120 Mya, Australia was connected to Antarctica but not to South America, Africa and India: these three land masses had become island continents. Klompen *et al.* (2000) argued against the origin of ticks at 390 Mya on labyrinthodont amphibians since modern-day Australian amphibians do not have ticks (we know of only one amphibian that is regularly infested with ticks: *Bufo marinus*, the cane toad infested by *Amblyomma rotundatum* in South America). However, the

labyrinthodont amphibians belonged to a 'distinct evolutionary radiation to that which produced our modern amphibian fauna' (Vickers-Rich & Rich, 1993). So we would not necessarily expect modern amphibians to be infested with the direct descendants of the ticks that infested labyrinthodont amphibians. Labyrinthodont amphibians lived alongside reptiles and mammal-like reptiles well into the Cretaceous, ca. 140–65 Mya. So there was ample time for ticks to infest lineages of reptiles and/or mammal-like reptiles before the labyrinthodont amphibians went extinct (see Vickers-Rich & Rich (1993) for an account of these terrestrial amphibians).

The evidence that the Ixodida evolved in that part of Gondwana that became Australia is that the putative sister-group of the ticks, the Holothyrida (Lehtinen, 1991), has a Gondwanan distribution (Australia, New Zealand, South and Central America, and islands in the Caribbean Sea and Indian Ocean) (Walter & Proctor, 1998, 1999) and that the most 'primitive' family of holothyrid mites is restricted to Australia and New Zealand (Allothyridae) (D. E. Walter, personal communication). The evidence that the Ixodidae evolved in Australia is that the earliest-diverging ('basal') lineage of the Metastriata, the Bothriocrotoninae, and one of the two putative lineages of the *Ixodes*, the Australasian *Ixodes*, live almost exclusively in Australia and New Guinea (but note that *I. uriae* of the Australasian *Ixodes* lineage has a worldwide distribution on seabirds). Questions about where the first ticks evolved, and indeed the first hard ticks, are still far from resolved.

WHAT WERE THE FIRST HOSTS OF TICKS: REPTILES, AMPHIBIANS OR BIRDS?

Alas, the oldest fossil ticks are only about 90–94 million years old (Cretaceous) (Klompen & Grimaldi, 2001; Poinar & Brown, 2003). Prior to Klompen & Grimaldi (2001) and Poinar & Brown (2003), records of ticks were restricted to the Miocene (20–15 Mya: Lane & Poinar, 1986), the Oligocene (*c.* 30 Mya) and the Eocene (40–35 Mya: reviewed in Klompen & Grimaldi, 2001 and in Fuente, 2003). So it seems that inference of the archetypal host will have to rely on estimates of the age of the Ixodida clade and knowledge of the potential hosts that were available at that time. Hoogstraal (1978) thought that the reptiles were the first hosts of ticks; Oliver (1989) proposed amphibians; whereas Stothard & Fuerst (1995) suggested birds. Dobson & Barker (1999) argued for labryrinthodont amphibians (which evolved in that part of Gondwana that became Australia) as the archetypal host (see also above).

HOW DID HAEMATOPHAGY (BLOOD-FEEDING BEHAVIOUR) EVOLVE IN TICKS?

To feed efficiently haematophagous arthropods adapt to the haemostatic system of their hosts (see Mans, Louw & Neitz, 2002). Mans, Louw & Neitz (2002) compared the sequences of inhibitors of blood coagulation and platelet aggregation in hard and soft ticks and insects and found that the inhibitors from hard and soft ticks did not share a common origin. This indicates independent adaptation to blood-feeding by hard and soft ticks, rather than adaptation to blood-feeding prior to the divergence of hard and soft ticks.

TAXONOMY AND NOMENCLATURE OF THE TICKS: THE INFLUENCE OF PHYLOGENY

Since Linnaeus described the first tick in 1746, a veritable army of biologists have contributed to the current taxonomic scheme of the ticks. Latreille was the first to classify the 'tiques' and in 1795 divided them into 11 genera, two of which were *Argas* and *Ixodes* (see Nuttall & Warburton, 1911). It is highly desirable that taxa are monophyletic and thus that classifications reflect accurately our knowledge of the evolutionary history (phylogeny) of organisms. A number of changes to the taxonomy of ticks have occurred or have been proposed recently to account for recent advances in our knowledge of the phylogeny of ticks.

Revision of the Argasidae

Klompen & Oliver (1993) revised the genera of the Argasidae. These authors reduced the number of genera from five to four: *Carios*, *Ornithodoros*, *Otobius* and *Argas*. Some species were moved from one genus to another and *Carios* was elevated from subgenus to genus. This revision is not universally accepted; however, a phylogeny inferred from SSU rDNA by Black *et al.* (1997) was consistent with the revision of Klompen & Oliver (1993). Ushijima *et al.* (2003) also concluded that their analyses of part of mitochondrial 16S rRNA of *Carios capensis* were consistent with Klompen & Oliver (1993). (However, we note that there was <50% bootstrap support for monophyly of *C. capensis*, *C. marginatus* and *C. mexicanus*.) We, like Horak, Camicas & Keirans (2002), incorporated the revision of Klompen & Oliver (1993) into our list of valid genus and species names (Table 1.1).

A new subfamily of hard ticks

A new subfamily, Bothriocrotoninae, and a new genus *Bothriocroton* were described for the lineage of five species of Australian endemic ticks that predominantly infest reptiles (Klompen, Dobson & Barker, 2002).

SYNONYMY OF *AMBLYOMMA* AND *APONOMMA*

The genus *Aponomma*, or remains thereof, was synonymized with the genus *Amblyomma* because *Amblyomma* is paraphyletic without the inclusion of *Aponomma* (Klompen *et al.*, 2002). Klompen *et al.* (2002) stated that 'the status of the "primitive *Aponomma*" remains unclear' and therefore *Aponomma sphenodonti* and *A. elaphense* were 'tentatively placed with the "typical *Aponomma*" in *Amblyomma*, until further evidence relating them to other ixodid lineages is generated'.

THE HYALOMMINAE AND RHIPICEPHALINAE SHOULD BE SYNONYMIZED

Many authors have suggested that the subfamily Hyalomminae Schulze, 1940 should be synonymized with the subfamily Rhipicephalinae Banks, 1908 because the Rhipicephalinae is paraphyletic without the inclusion of members of the Hyalomminae (Klompen *et al.*, 2000, 2002; Murrell *et al.*, 2001*b*). However, some authors still use this name (e.g. Horak, Camicas & Keirans, 2002) and thus, presumably, are not convinced by the evidence of Murrell *et al.* (2001*b*) and Klompen *et al.* (2000, 2002).

Rhipicephalus and *Boophilus* have been synonymized (Murrell & Barker, 2003). Murrell *et al.* (2001*b*) then Barker & Murrell (2002) proposed that the genus *Boophilus* Curtice, 1891 be relegated to a subgenus of *Rhipicephalus* Koch, 1844 because, as outlined in the previous section, the genus *Rhipicephalus* is paraphyletic without the inclusion of the species of *Boophilus*. The alternative was to elevate some species or species-groups of *Rhipicephalus* to the rank of genera. This was undesirable, first because it increases greatly the number of genus names and second because not enough is known about the phylogeny of the *Rhipicephalus* species to warrant splitting the genus into a number of new genera. By placing the *Boophilus* species in a subgenus of *Rhipicephalus*, *Rhipicephalus* (*Boophilus*), the name *Boophilus* can still be used for the five species that are presently in the genus *Boophilus*. It was desirable that the name *Boophilus* be available, since there is a lot of literature on the *Boophilus* species and hundreds, perhaps thousands, of people use these names regularly (Murrell *et al.*, 2001*b*).

A current list of the valid genus and species names of ticks

The most recent and pertinent taxonomic schemes and lists of names of ticks are: (1) Keirans (1992) for species names as at 1992; (2) Camicas *et al.* (1998) for species names as at the end of 1995; (3) Keirans & Robbins (1999), a list of species described between 1973 and 1997; (4) Horak, Camicas & Keirans (2002); and (5) Walker, Keirans & Horak (2000), which is a comprehensive taxonomy of 74 species of the genus *Rhipicephalus* (note that a subsequent revision by Murrell & Barker (2003) moved all five species from the genus *Boophilus* to the genus *Rhipicephalus* so there are now 79 species in the genus *Rhipicephalus*). Table 1.1 is a compilation of the genus and species names of Keirans (1992), Camicas *et al.* (1998), Keirans & Robbins (1999) and Horak, Camicas & Walker (2002) plus the seven species, including one new genus, *Cornupalpatum*, that have been described since 2002. For the *Rhipicephalus* species we list only the species in Walker, Keirans & Horak (2000) since this was a comprehensive taxonomic treatment of the group. We ignored the plethora of subgenera, species groups and subspecies with one exception: *Rhipicephalus* (*Boophilus*) – see discussion above. We also ignored the genus-level taxonomy of Camicas *et al.* (1998) because it differs markedly from the genus-level taxonomy of Keirans (1992), Keirans & Robbins (1999) and Horak, Camicas & Keirans (2002). For example, Camicas *et al.* (1998) has the *Ixodes* species of Keirans (1992), Keirans & Robbins (1999) and Horak, Camicas & Keirans (2002) in six different genera: *Ixodes, Ceratixodes, Eschatocephalus, Lepidixodes, Pholeoixodes* and *Scaphixodes*. Since Camicas is an author of the most recent of the above lists, Horak, Camicas & Keirans (2002), which ignores these genus names, this omission seems to be justified.

The superscripts in Table 1.1 show the checklist(s) in which each name appears: Keirans (1992)[K]; Camicas, Hervy, Adam & Morel (1998)[CHAM]; Keirans & Robbins (1999)[KR]; Walker, Keirans & Horak (2000)[WKH]; and Horak *et al.* (2002)[HCK]. You will notice that these lists differ. Most species names are in all possible lists for that particular species, e.g. *I. holocyclus* Neumann, 1899 which is in Keirans (1992), Camicas *et al.* (1998) and Horak *et al.* (2002). However, some names appear in two of three possible lists (e.g. *I. anatis* Chilton, 1904 which is in Keirans (1992) and Horak *et al.* (2002) but not in Camicas *et al.* (2002)) or one of three possible lists (e.g. *I. apteridis* Maskell, 1897 which is in Camicas *et al.* (1998) but not Keirans (1992), Keirans & Robbins (1999) or Horak *et al.* (2002)). We trust that our list will lay the foundation for a new list of valid names with evidence and arguments for the removal and/or addition of names from the literature. The species and genera described or reinstated since the last list of species names (Horak *et al.*, 2002) have the superscript [RD] (recent description). *Cornupalpatum burmanicum* Poinar & Brown, 2003 was described from tick larva found in amber from the Cretaceous Era (Poinar & Brown, 2003). There is only one species in the genus *Cornupalpatum*. This tick resembles some species of *Amblyomma* (formerly *Aponomma* species that infest reptiles). Note that *Argas cooleyi* (McIvor, 1941) is a homonym of *Argas cooleyi* Kohls & Hoogstraal, 1960, and vice versa. These homonyms were inadvertently created when Klompen & Oliver (1993) moved the subgenus *Ornithodoros* (*Alveonasus*) from the genus *Ornithodoros* to the genus *Argas*.

CONCLUSIONS

Taxonomic schemes provide the evolutionary framework that all tick biologists use to help interpret the biology and phenotypes of the species of ticks they study. Fortunately, there has been much progress in our understanding of the phylogeny of ticks since the mid 1990s when modern methods were first applied to the study of tick phylogeny. There is now consensus among many workers in the field about many clades of ticks, e.g. that the five species of endemic Australian ticks that were in the former genus *Aponomma*, which infest reptiles, are a distinct lineage of ticks; that the Hyalomminae is embedded in the Rhipicephalinae; and that some species of the genus *Rhipicephalus* are more closely related to the *Boophilus* species than they are to the other *Rhipicephalus* species. Yet there is disagreement, or more often a lack of information, about the relationships of other ticks (e.g. the phylogenetic position of the Haemaphysalinae and whether or not the genus *Ixodes* is monophyletic) and where and when the ticks evolved. Study of the mitochondrial genomes of ticks has also provided insight into the phylogeny and evolution of ticks. Entire mitochondrial genome sequences from each of the subfamilies of ticks and their potential sistergroups (holothyrid and mesostigmatid mites) are likely to reveal even more about the systematics of these fascinating arthropods. We have the tools to address many of the remaining outstanding questions but pivotal taxa may continue to be difficult to collect (e.g. *Nuttalliella namaqua* and *Anomalohimalaya* species). Nonetheless the tick systematics community will continue to seek the answers to longstanding questions about the evolution of these fascinating parasites.

ACKNOWLEDGEMENTS

We thank Claire Ellender and Maree Schabe for invaluable help with the word processing. Jim Keirans, Dave Kemp, Dave Walter and Renfu Shao criticized the manuscript for us and Tom Cribb provided advice on nomenclature. Two anonymous referees also provided valuable suggestions.

REFERENCES

Andrews, R. H., Beveridge, I., Bull, C. M., *et al.* (2006). Systematic status of *Aponomma tachyglossi* Roberts (Acari: Ixodidae) from echidnas, *Tachyglossus aculeatus*, from Queensland, Australia. *Systematic and Applied Acarology* 11, 23–39.

Apanaskevich, D. A. & Horak, I. G. (2005). The genus *Hyalomma* Koch, 1844. II. Taxonomic status of *H. (Euhyalomma) anatoloicum* Koch, 1844 and *H. (E.) excavatum* Kock, 1844 (Acari, Ixodidae) with redescriptions of all stages. *Acarina* 13, 181–197.

Apanaskevich, D. A. & Horak, I. G. (2006). The genus *Hyalomma* Koch, 1844. I. Reinstatement of *Hyalomma (Euhyalomma) glabrum* Delpy, 1949 (Acari, Ixodidae) as a valid species with a redescription of the adults, the first description of its immature stages and notes on its biology. *Onderstepoort Journal of Veterinary Research* 73, 1–12.

Balashov, Y. S. (1994). Importance of continental drift in the distribution and evolution of ixodid ticks. *Entomological Review* 73, 42–50.

Barker, S. C. (1998). Distinguishing species and populations of rhipicephaline ticks with ITS 2 ribosomal RNA. *Journal of Parasitology* 84, 887–892.

Barker, S. C. & Murrell, A. (2002). Phylogeny, evolution and historical zoogeography of ticks: a review of recent progress. *Experimental and Applied Acarology* 28, 55–68.

Beati, L. & Keirans, J. E. (2001). Analysis of the systematic relationships among ticks of the genera *Rhipicephalus* and *Boophilus* (Acari: Ixodidae) based on mitochondrial 12S ribosomal DNA gene sequences and morphological characters. *Journal of Parasitology* 87, 32–48.

Beati, L., Keirans, J. E., Durden, L. A. & Opiang, M. D. (2008). *Bothriocroton oudemansi* (Neumann, 1910) comb. nov. (Acari: Ixodida: Ixodidae), an ectoparasite of the long-beaked echidna in Papua New Guinea: redescription of the male and first description of the female and nymph. *Systematic Parasitology* 69, 185–200.

Berdyev, A. (1989). On the history of areas and ways of distribution of ticks of the genus *Dermacentor* Koch, 1844 (Parasitiformes, Ixodidae). *Parazitologiia* 23, 166–172.

Black, W. C. IV & Piesman, J. (1994). Phylogeny of hard- and soft-tick taxa (Acari: Ixodida) based on mitochondrial 16S rDNA sequences. *Proceedings of the National Academy of Sciences of the USA* 91, 10034–10038.

Black, W. C. IV & Roehrdanz, R. L. (1998). Mitochondrial gene order is not conserved in arthropods: prostrate and metastriate tick mitochondrial genomes. *Molecular Biology and Evolution* 15, 1772–1785.

Black, W. C. IV, Klompen, J. S. H. & Keirans, J. E. (1997). Phylogenetic relationships among tick subfamilies based on the 18S nuclear rDNA gene. *Molecular Phylogenetics and Evolution* 7, 129–144.

Borges, L. M. F., Labruna, M. B., Linardi, P. M. & Ribeiro, M. F. B. (1998). Recognition of the tick genus *Anocentor* Schulze, 1937 (Acari: Ixodidae) by numerical taxonomy. *Journal of Medical Entomology* 35, 891–894.

Broughton, R. E., Milam, J. E. & Roe, B. A. (2001). The complete sequence of the zebrafish (*Danio rerio*) mitochondrial genome and evolutionary patterns in vertebrate mitochondrial DNA. *Genome Research* 11, 1958–1967.

Camicas, J.-L. & Morel, P.-C. (1977). Position systématique et classification des tiques (Acarida: Ixodida). *Acarologia* 18, 410–420.

Camicas, J.-L., Hervy, J. P., Adam, F. & Morel, P. C. (1998). *Les Tiques du Monde: Nomenclature, Stades Décrits, Hôtes, Repartition [The Ticks of the World: Nomenclature, Described Stages, Hosts, Distribution]*. Pan's: Orstom.

Campbell, N. J. H. & Barker, S. C. (1998). An unprecedented major rearrangement in an arthropod mitochondrial genome. *Molecular Biology and Evolution* 15, 1786–1787.

Campbell, N. J. H. & Barker, S. C. (1999). The novel mitochondrial gene arrangement of the cattle tick, *Boophilus microplus*: five-fold tandem repetition of a coding region. *Molecular Biology and Evolution* 16, 732–740.

Caporale, D. A., Rich, S. M., Spielman, A., Telford, S. R. III & Kocher, T. D. (1995). Discriminating between *Ixodes* ticks by means of mitochondrial DNA sequences. *Molecular Phylogenetics and Evolution* 4, 361–365.

Cox, C. B. & Moore, P. D. (1993). *Biogeography: An Ecological and Evolutionary Approach*. Oxford, UK: Blackwell Science.

Crampton, A., McKay, I. & Barker, S. C. (1996). Phylogeny of ticks (Ixodida) inferred from nuclear ribosomal DNA. *International Journal for Parasitology* 26, 511–517.

Crosbie, P. R., Boyce, W. M. & Rodwell, T. C. (1998). DNA sequence variation in *Dermacentor hunteri* and estimated phylogenies of *Dermacentor* species (Acari: Ixodidae) in the New World. *Journal of Medical Entomology* 35, 277–288.

Dobson, S. J. & Barker, S. C. (1999). Phylogeny of the hard ticks (Ixodidae) inferred from 18S rRNA indicates that the genus *Aponomma* is paraphyletic. *Molecular Phylogenetics and Evolution* 11, 288–295.

Durden, L. A., Keirans, J. E. & Smith, L. L. (2002). *Amblyomma geochelone*, a new species of tick (Acari: Ixodidae) from the Madagascan ploughshare tortoise. *Journal of Medical Entomology* 39, 398–403.

Estrada-Peña, A., Venzal, J. M., Barros-Battesti, D. M., *et al.* (2004). Three new species of *Anticola* (Acari: Argasidae) from Brazil, with a key to the known species in the genus. *Journal of Parasitology* 90, 490–498.

Estrada-Peña, A., Castellá, J. & Morel, P. C. (1994). Cuticular hydrocarbon composition, phenotypic variability, and geographic relationships in allopatric populations of *Amblyomma variegatum* (Acari: Ixodidae) from Africa and the Caribbean. *Journal of Medical Entomology* 31, 534–544.

Estrada-Peña, A., Castellá, J. & Moreno, J. A. (1994). Using cuticular hydrocarbon composition to elucidate phylogenies in tick populations (Acari: Ixodidae). *Acta Tropica* 58, 51–71.

Estrada-Peña, A., Osácar, J. J., Calvete, C. & Estrada-Peña, R. (1997). Estimation of genetic affinities between sympatric populations of *Rhipicephalus pusillus* ticks (Acari: Ixodidae) by analysis of cuticular hydrocarbons. *Folia Parasitologica* 44, 147–154.

Estrada-Peña, A., Venzal, J. M., Gonzalez-Acuna, D. & Gugliemone, A. A. (2003). *Argas (Persicargas) keiransi* n. sp. (Acari: Argasidae), a parasite of the chimango, *Milvago c. chimango* (Aves: Falconiformes) in Chile. *Journal of Medical Entomology* 40, 766–769.

Filippova, N. A. (1993). Ventral skeleton of male [*sic*] of ixodid ticks of the subfamily Amblyomminae, its evolution and role for supergeneric taxonomy. *Parazitologiia* 27, 3–18.

Filippova, N. A. (1994). Classification of the subfamily Amblyomminae (Ixodidae) in connection with re-investigation of chaetotaxy of the anal valve. *Parazitologiia* 28, 3–12.

Fuente, J. (2003). The fossil record and the origin of ticks (Acari: Parasitiformes: Ixodida). *Experimental and Applied Acarology* 29, 331–344.

Fukunaga, M., Yabuki, M., Hamase, A., Oliver, J. H. Jr & Nakao, M. (2000). Molecular phylogenetic analysis of ixodid ticks based on the ribosomal DNA spacer, internal transcribed spacer 2, sequences. *Journal of Parasitology* 86, 38–43.

Guglielmone, A. A. & Keirans, J. E. (2002). *Ornithodoros kohlsi* Guglielmone and Keirans (Acari: Ixodida: Argasidae), a new name for *Ornithodoros boliviensis* Kohls and Clifford 1964. *Proceedings of the Entomological Society of Washington* 104, 822.

Haring, E., Kruckenhauser, L., Gamauf, A., Riesing, M. J. & Pinsker, W. (2001). The complete sequence of the mitochondrial genome of *Buteo buteo* (Aves, Accipitridae) indicates an early split in the phylogeny of raptors. *Molecular Biology and Evolution* 18, 1892–1904.

Horak, I. G., Camicas, J.-L. & Keirans, J. E. (2002). The Argasidae, Ixodidae and Nuttalliellidae (Acari: Ixodida): a world list of valid tick names. *Experimental and Applied Acarology* 28, 27–54.

Hoogstraal, H. (1978). Biology of ticks. In *Tick-Borne Diseases and their Vectors*, ed. Wilde, J. K. H., pp. 3–14. Edinburgh, UK: Centre for Tropical Veterinary Medicine.

Hoogstraal, H. & Aeschlimann, A. (1982). Tick–host specificity. *Bulletin de la Société Entomologique Suisse* 55, 5–32.

Hutcheson, H. J., Klompen, J. S. H., Keirans, J. E., *et al.* (2000). Current progress in tick molecular systematics. In *Proceedings, 3rd International Conference on Ticks and Tick-Borne Pathogens*, eds. Kazimirova, M. *et al.*, pp. 11–19. Bratislava, Slovak Republic: Institute of Zoology, Slovak Academy of Sciences.

Hutcheson, H. J., Oliver, J. H. Jr, Houck, M. A. & Strauss, R. E. (1995). Multivariate morphometric discrimination of nymphal and adult forms of the blacklegged tick (Acari: Ixodidae), a principal vector of the agent of Lyme disease in eastern North America. *Journal of Medical Entomology* 32, 827–842.

Hwang, U. W., Friedrich, M., Tautz, D., Park, C. J. & Kim, W. (2001). Mitochondrial protein phylogeny joins myriapods with chelicerates. *Nature* 413, 154–157.

Inoue, J. G., Miya, M., Tsukamoto, K. & Nishida, M. (2001). A mitogenomic perspective on the basal teleostean phylogeny: resolving higher-level relationships with longer DNA sequences. *Molecular Phylogenetics and Evolution* 20, 275–285.

Kaufman, T. S. (1972). A revision of the genus *Aponomma* Neumann, 1899 (Acarina: Ixodidae). Ph.D. dissertation, University of Maryland, College Park, MD.

Keirans, J. E. (1992). Systematics of the Ixodida (Argasidae, Ixodidae, Nuttalliellidae): an overview and some problems. In *Tick Vector Biology: Medical and Veterinary Aspects*, eds. Fivay, B. *et al.*, pp. 1–21. Berlin, Germany: Springer-Verlag.

Keirans, J. E. & Robbins, R. G. (1999). A world checklist of genera, subgenera, and species of ticks (Acari: Ixodida) published from 1973 to 1997. *Journal of Vector Ecology* **24**, 115–129.

Klompen, J. S. H. (1992). Comparative morphology of argasid larvae (Acari: Ixodida: Argasidae), with notes on phylogenetic relationships. *Annals of the Entomological Society of America* **85**, 541–560.

Klompen, J. S. H. (1999). Phylogenetic relationships in the family Ixodidae with emphasis on the genus *Ixodes* (Parasitiformes: Ixodidae). In *9th Acarology Symposium*, pp. 349–354. Colombus, OH: Ohio State University Press.

Klompen, H. & Grimaldi, D. (2001). First Mesozoic record of a parasitiform mite: a larval argasid tick in Cretaceous amber (Acari: Ixodida: Argasidae). *Annals of the Entomological Society of America* **94**, 10–15.

Klompen, J. S. H. & Oliver, J. H. Jr (1993). Systematic relationships in the soft ticks (Acari: Ixodida: Argasidae). *Systematic Entomology* **18**, 313–331.

Klompen, J. S. H., Black, W. C. IV, Keirans, J. E. & Norris, D. E. (2000). Systematics and biogeography of hard ticks: a total evidence approach. *Cladistics* **16**, 79–102.

Klompen, J. S. H., Black, W. C. IV, Keirans, J. E. & Oliver, J. H. Jr (1996). Evolution of ticks. *Annual Review of Entomology* **41**, 141–161.

Klompen, H., Dobson, S. J. & Barker, S. C. (2002). A new subfamily, Bothriocrotoninae n. subfam., for the genus *Bothriocroton* Keirans, King & Sharrad, 1994 status amend. (Ixodida: Ixodidae), and the synonymy of *Aponomma* Neumann, 1899 with *Amblyomma* Koch, 1844. *Systematic Parasitology* **53**, 101–107.

Klompen, J. S. H., Oliver, J. H., Jr, Keirans, J. E. & Homsher, P. J. (1997). A re-evaluation of relationships in the Metastriata (Acari: Parasitiformes: Ixodidae). *Systematic Parasitology* **38**, 1–24.

Krantz, G. W. (1978). *A Manual of Acarology*. Corvallis, OR: Oregon State University Press.

Labruna, M. B., Keirans, J. E., Camargo, L. M. A., *et al.* (2005). *Amblyomma latepunctatum*, a valid tick species (Acari: Ixodidae) long misidentified with both *Amblyomma incisum* and *Amblyomma scalpturatum*. *Journal of Parasitiology* **91**, 527–541.

Lane, R. S. & Poinar, G. O. (1986). First fossil tick (Acari: Ixodidae) in New World amber. *International Journal of Acarology* **12**, 75–78.

Lehtinen, P. T. (1991). Morphology and phylogeny. In *Modern Acarology* eds. Dusbabek, F. and Buvka, V., pp. 101–113. The Hague, the Netherlands: SPB Academic Publishing.

Lindquist, E. E., Wu, K. W. & Redner, J. H. (1999). A new species of the tick genus *Ixodes* (Acari: Ixodidae) parasitic on mustelids (Mammalia: Carnivora) in Canada. *Canadian Entomologist* **131**, 151–170.

Maca-Meyer, N., Gonzalez, A. M., Larruga, J. M., Flores, C. & Cabrera, V. M. (2001). Major genomic mitochondrial lineages delineate early human expansions. *BMC Genetics* **2**, 13.

Mangold, A. J., Bargues, M. D. & Mas-Coma, S. (1998*a*). 18S rRNA gene sequences and phylogenetic relationships of European hard-tick species (Acari: Ixodidae). *Parasitology Research* **60**, 31–37.

Mangold, J. J., Bargues, M. D. & Mas-Coma, S. (1998*b*). Mitochondrial 16S rDNA sequences and phylogenetic relationships of species of *Rhipicephalus* and other tick genera among Metastriata (Acari: Ixodidae). *Parasitology Research* **84**, 478–484.

Mans, B. J., Louw, A. I. & Neitz, A. W. H. (2002). Evolution of hematophagy in ticks: common origins for blood coagulation and platelet aggregation inhibitors from soft ticks of the genus *Ornithodoros*. *Molecular Biology and Evolution* **19**, 1695–1705.

McLain, D. K., Wesson, D. M., Collins, F. H. & Oliver, J. H. Jr (1995*a*). Evolution of the rDNA spacer, ITS 2, in the ticks *Ixodes scapularis* and *I. pacificus* (Acari: Ixodidae). *Heredity* **75**, 303–319.

McLain, D. K., Wesson, D. M., Oliver, J. H. Jr & Collins, F. H. (1995*b*). Variation in ribosomal DNA internal transcribed spacer 1 among eastern populations of *Ixodes scapularis* (Acari: Ixodidae). *Journal of Medical Entomology* **32**, 353–360.

Miller, H. C., Conrad, A. M., Barker, S. C. & Daugherty, C. H. (2007). Distribution and phylogenetic analyses of an endangered tick, *Amblyomma sphenodonti*. *New Zealand Journal of Zoology* **34**, 97–105.

Miya, M., Kawaguchi, A. & Nishida, M. (2001). Mitogenomic exploration of higher teleostean phylogenies: a case study for moderate-scale evolutionary genomics with 38 newly determined complete mitochondrial DNA sequences. *Molecular Biology and Evolution* **18**, 1993–2009.

Murrell, A. & Barker, S. C. (2003). Synonymy of *Boophilus* Curtice, 1891 with *Rhipicephalus* Koch, 1844 (Acari: Ixodidae). *Systematic Parasitology* **56**, 169–172.

Murrell, A., Campbell, N. J. & Barker, S. C. (1999). Mitochondrial 12S rDNA indicates that the Rhipicephalinae (Acari: Ixodida) is paraphyletic. *Molecular Phylogenetics and Evolution* **12**, 83–86.

Murrell, A., Campbell, N. J. & Barker, S. C. (2000). Phylogenetic analyses of the rhipicephaline ticks indicate that the genus *Rhipicephalus* is paraphyletic. *Molecular Phylogenetics and Evolution* **16**, 1–7.

Murrell, A., Campbell, N. J. H. & Barker, S. C. (2001*a*). Recurrent gains and losses of large (84–109 bp) repeats in the rDNA internal transcribed spacer 2 (ITS2) of rhipicephaline ticks. *Insect Molecular Biology* **10**, 587–596.

Murrell, A., Campbell, N. J. H. & Barker, S. C. (2001*b*). A total-evidence phylogeny of ticks provides insights into the evolution of life cycles and biogeography. *Molecular Phylogenetics and Evolution* **21**, 244–258.

Murrell, A., Campbell, N. J. H. & Barker, S. C. (2003). The value of idiosyncratic markers and changes to conserved tRNA sequences from the mitochondrial genome of hard ticks (Acari: Ixodida: Ixodidae) for phylogenetic inference. *Systematic Biology* **52**, 296–310.

Murrell, A., Dobson, S. J., Walter, D. E., *et al.* (2005). Relationships among the three major lineages of the Acari (Arthropoda: Arachnida) inferred from small subunit rRNA: paraphyly of the Parasitiformes with respect to the Opilioacariformes and relative rates of nucleotide substitution. *Invertebrate Systematics* **19**, 383–389.

Norris, D. E., Klompen, J. S. H. & Black, W. C. IV (1999). Comparison of the mitochondrial 12S and 16S ribosomal DNA genes in resolving phylogenetic relationships among hard ticks (Acari: Ixodidae). *Annals of the Entomological Society of America* **92**, 117–129.

Norris, D. E., Klompen, J. S. H., Keirans, J. E. & Black, W. C. IV (1996). Population genetics of *Ixodes scapularis* (Acari: Ixodidae) based on mitochondrial 16S and 12S genes. *Journal of Medical Entomology* **33**, 78–89.

Norris, D. E., Klompen, J. S., Keirans, J. E., *et al.* (1997). Taxonomic status of *Ixodes neotomae* and *I. spinipalpis* (Acari: Ixodidae) based on mitochondrial DNA evidence. *Journal of Medical Entomology* **34**, 696–703.

Nuttall, G. H. F. & Warburton, C. (1911). *Ticks: A Monograph of the Ixodoidea, Part II – Ixodidae*. Cambridge, UK: Cambridge University Press.

Oliver, J. H. Jr (1989). Biology and systematics of ticks (Acari: Ixodida). *Annual Review of Ecology and Systematics* **20**, 397–430.

Poinar, G. Jr & Brown, A. E. (2003). A new genus of hard ticks in Cretaceous Burmese amber (Acari: Ixodida: Ixodidae). *Systematic Parasitology* **54**, 199–205.

Pomerantsev, B. I. (1948). Basic directions of evolution in the Ixodoidea. *Parazitologicheskii Sbornik* **10**, 5–19.

Rich, S. M., Caporale, D. A., Telford, S. R. III, *et al.* (1995). Distribution of the *Ixodes ricinus*-like ticks of eastern North America. *Proceedings of the National Academy of Sciences of the USA* **92**, 6284–6288.

Rich, S. M., Rosenthal, B. M., Telford, S. R. III, *et al.* (1997). Heterogeneity of the internal transcribed spacer (ITS-2) region within individual deer ticks. *Insect Molecular Biology* **6**, 123–129.

Robbins, R. G., Karesh, W. B., Calle, P. P., *et al.* (1998). First records of *Hyalomma aegyptium* (Acari: Ixodida: Ixodidae) from the Russian spur-thighed tortoise, *Testudo graeca nikolskii*, with an analysis of tick population dynamics. *Journal of Parasitology* **84**, 1303–1305.

Roehrdanz, R. L., Degrugillier, M. E. & Black, W. C. IV (2002). Novel rearrangements of arthropod mitochondrial DNA detected with long-PCR: applications to arthropod phylogeny and evolution. *Molecular Biology and Evolution* **19**, 841–849.

Schmitz, J., Ohme, M. & Zischler, H. (2002). The complete mitochondrial sequence of *Tarsius bancanus*: evidence for an extensive nucleotide compositional plasticity of primate mitochondrial DNA. *Molecular Biology and Evolution* **19**, 544–553.

Shao, R. & Barker, S. C. (2007). Mitochondrial genomics of parasitic arthropods: implications for studies of population genetics and evolution. *Parasitology* **134**, 153–167.

Shao, R., Aoki, Y., Mitani, H., *et al.* (2004). The mitochondrial genomes of soft ticks have an arrangement of genes that has remained unchanged for over 400 million years. *Insect Molecular Biology* **13**, 219–224.

Shao, R., Barker, S. C., Mitani, H., Aoki, Y. & Fukunaga, M. (2005). Evolution of duplicate control regions in the mitochondrial genomes of Metazoa: a case study with Australasian *Ixodes* ticks. *Molecular Biology and Evolution* **22**, 620–629.

Shao, R., Fukunaga, M. & Barker, S. C. (2005). The mitochondrial genomes of ticks and their kin: a review plus the description of the mitochondrial genomes of *Amblyomma triguttatum* and *Ornithodoros porcinus*. In *Proceedings, 5th International Congress on Ticks and Tick-Borne Diseases*, Neuchatel, Switzerland, August 2005, pp. 210–214.

Sonenshine, D. E. (1993). *Biology of Ticks*, vol. 2. Oxford, UK: Oxford University Press.

Stothard, D. R. & Fuerst, P. A. (1995). Evolutionary analysis of the spotted fever and typhus group of *Rickettsia* using 16S rRNA gene sequences. *Systematic and Applied Microbiology* **18**, 52–61.

Szabo, M. P., Mangold, A. J., Joao, C. F., Bechara, G. H. and Guglielmone, A. A. (2005). Biological and DNA evidence of two dissimilar populations of the *Rhipicephalus sanguineus* tick group (Acari: Ixodidae) in South America. *Veterinary Parasitology* **130**, 131–140.

Ushijima, Y., Oliver, J. H. Jr, Keirans, J. E., *et al.* (2003). Mitochondrial sequence variation in *Carios capensis* (Neumann), a parasite of seabirds, collected on Torishima Island in Japan. *Journal of Parasitology* **89**, 196–198.

Venzal, J. M., Castro, O., Cabrera, P., *et al.* (2001). *Ixodes (Haemixodes) longiscutatum* Boero (new status) and *I. (H.) uruguayensis* Kohls & Clifford, a new synonym of *I. (H.) longiscutatum* (Acari: Ixodidae). *Memórias do Instituto Oswaldo Cruz* **96**, 1121–1122.

Vermeil, C., Marjolet, M. & Vermeil, F. (1997). *Ornithodoros (Alectorobius) cheikhi* n. sp. (Acarina, Ixodoidea, Argasidae, *Ornithodoros (Alectorobius) capensis* group) from tern nesting site in Mauritania. *Bulletin de la Société des Sciences Naturelles de l'ouest de la France* **19**, 66–76.

Vickers-Rich, P. & Rich, T. H. (1993). *Wildlife of Gondwana.* Chatswood, Australia: Reed.

Walker, J. B., Keirans, J. E. & Horak, I. G. (2000). *The Genus Rhipicephalus (Acari, Ixodidae): A Guide to the Brown Ticks of the World.* Cambridge, UK: Cambridge University Press.

Walter, D. E. & Proctor, H. C. (1998). Feeding behaviour and phylogeny: observations on early derivative Acari. *Experimental and Applied Acarology* **22**, 39–50.

Walter, D. E. & Proctor, H. C. (1999). *Mites: Ecology, Evolution and Behaviour.* Sydney, Australia: University of New South Wales Press.

Wesson, D. M. & Collins, F. H. (1992). Sequence and secondary structure of 5.8S rRNA in the tick *Ixodes scapularis. Nucleic Acids Research* **20**, 11.

Wesson, D. M., McLain, D. K., Oliver, J. H. Jr, Piesman, J. & Collins, F. H. (1993). Investigation of the validity of species status of *Ixodes dammini* (Acari: Ixodidae) using rDNA. *Proceedings of the National Academy of Sciences of the USA* **90**, 10221–10225.

Wilson, K., Cahill, V., Ballment, E. & Benzie, J. (2000). The complete sequence of the mitochondrial genome of the crustacean *Penaeus monodon*: are malacostracan crustaceans more closely related to insects than to branchiopods? *Molecular Biology and Evolution* **17**, 863–874.

Xu, G., Fang, Q. Q., Keirans, J. E. & Durden, L. A. (2003). Molecular phylogenetic analyses indicate that the *Ixodes ricinus* complex is a paraphyletic group. *Journal of Parasitology* **89**, 452–457.

Zahler, M., Filippova, N. A., Morel, P. C., Gothe, R. & Rinder, H. (1997). Relationships between species of the *Rhipicephalus sanguineus* group: a molecular approach. *Journal of Parasitology* **83**, 302–306.

2 • The impact of tick ecology on pathogen transmission dynamics

S. E. RANDOLPH

INTRODUCTION: THE IMPACT OF TICK ECOLOGY ON PATHOGEN TRANSMISSION DYNAMICS

The ecology of ticks, the outcome of their interactions with their natural environment, is fundamental to the spatial and temporal variation in the risk of infection by tick-borne pathogens. Due to the biology of ticks as blood-feeding parasites, their physical environment includes the host itself. This biotic environment reacts to the tick's presence in both the short and the long term in ways that the abiotic environment cannot do, imposing physiological, population and evolutionary pressures on ticks. Ticks, however, are only intermittent parasites, spending the greater part of their life cycle free within their habitat where they are at the mercy of abiotic factors such as habitat structure and climate. They take only one (ixodid ticks) or a few (argasid ticks) very large blood meals per life stage, as larvae, nymphs and adults, then develop to the next stage, which takes weeks, months or even years, depending on the ambient temperature. This inter-stadial period is usually passed off-host, although the relatively few two- and one-host ticks (e.g. *Hyalomma anatolicum excavatum* and *Rhipicephalus* (*Boophilus*) *microplus*, respectively) remain on the host for one or both of the interstadial periods. For simplicity, and because generally less is known about the ecology of argasid ticks, what follows will refer almost exclusively to ixodid ticks.

At the individual level, a tick's potential as a vector is enhanced by its habit of taking such large, slow blood meals, and by its adaptations to survive the host's defences (see Chapter 9). At the population level, vector potential is severely constrained by the limitations of taking so few meals. Vector–host contact rates can never exceed three per generation, and are reduced to only one per generation for one-host species that take all three meals on the same individual host. Furthermore, a tick of one stage that acquires pathogens in its blood meal from an infected host cannot transmit them to a new susceptible host until it feeds again as the next stage. Trans-stadial maintenance of infection (of which transovarial transmission, from female to larvae, is a special form), is the sine qua non of an ixodid tick's performance as a vector. Furthermore, the transmission cycle must involve at least two tick stages feeding on the same host individual (although not necessarily at the same time), one to transmit and the preceding stage to acquire. (Exceptions involve transmission by peripatetic male ticks of some species, either directly by their taking small meals on more than one host, or via the venereal route: Alekseev *et al.*, 1999.) This introduces a very long delay into the transmission process, accompanied by high tick mortality, both of which vary geographically and seasonally with climate. Climate-driven tick seasonal population dynamics therefore have an impact on the pace and force of transmission (the longer the delay in transmission, the slower the transmission cycle and the fewer ticks survive). Population dynamics also determine the seasonal abundance of active ticks questing for hosts and therefore the risk of infection.

In the case of some pathogens, the very existence of the specific transmission route also depends on a particular pattern of the vector's seasonal dynamics. Tick-borne encephalitis (TBE) virus (see Chapter 10), for example, is very short-lived in its principal rodent hosts, and commonly does not develop patent systemic infections (Labuda *et al.*, 1993*a*, *b*, 1996), so it must be transmitted directly from infected nymphal *Ixodes* ticks to uninfected larvae co-feeding on the same hosts. Only where large numbers of larvae feed together with nymphs, determined both by climate and host factors as described below, is there sufficient transmission to allow TBE virus to persist (Randolph *et al.*, 1999). For this, larvae and nymphs must have synchronous seasons of activity, which is characteristically seen within TBE foci, but not elsewhere (Randolph *et al.*, 2000). This synchrony may occur at any time of the year (see Figs. 2.1–2.4 in Randolph *et al.*, 2000), but will be most significant epidemiologically in the spring and summer when both tick stages are most abundant. This is also when most human TBE cases occur (Zabicka, 1994; Danielová *et al.*, 2002). Larval–nymphal synchrony and TBE foci are statistically significantly associated with a particular seasonal land surface temperature profile, characterized by a relatively rapid rate of cooling during the autumn (Randolph *et al.*, 2000). These empirical conclusions have been disputed on the grounds that TBE virus can also be transmitted from transovarially infected larvae to other co-feeding larvae (Danielová *et al.*, 2002). This transmission route cannot, however, account quantitatively for persistent cycles of TBE virus (see below), nor explain the absence of TBE in many places where *Ixodes ricinus* is abundant (Randolph *et al.*, 2000).

This system is just one example of a dominant theme of this chapter – the general influence of climate on tick–host–pathogen relationships and therefore on the epidemiology of tick-borne diseases. It also highlights another theme – that ecological (or epidemiological) systems can only be fully understood by quantifying the rates of the constituent processes. Differential rates and epidemiological patterns depend on specific biological features of pathogens, vectors and hosts. The longer the duration of infectivity of the pathogen in the vertebrate host, for example, the greater the window of opportunity for transmission and so the less specific is the pattern of tick seasonal dynamics necessary for persistent enzootic cycles. *Borrelia burgdorferi sensu lato*, the agent of Lyme borreliosis (see Chapter 11), survives in vertebrates for many weeks or months. Consequently, it is much more widespread, with higher infection prevalences, than TBE virus, despite being transmitted by the same vector species.

The ultimate aim of attempts to describe and explain any system must be to make testable predictions, ideally within the quantitative framework of a model. A tick population model can predict all the major risk factors for tick-borne diseases, the distribution (i.e. places where tick populations survive), abundance, seasonal dynamics and even host relationships of the vector tick. Ticks, however, pose a considerable challenge to population modellers. To be of any epidemiological use, the model must predict the abundance of three overlapping life stages simultaneously. This complexity must be captured in the simplest possible terms by focusing on the following four essential processes (Fig. 2.1). Firstly, there is entry into the population of unfed ticks of each stage by development from the previous stage. Secondly, there is a certain probability that an unfed tick will quest actively for

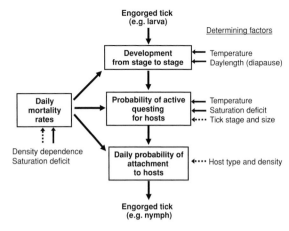

Fig. 2.1. The structure of a population model for the tick *Ixodes ricinus*. Factors which determine the rates of the processes described within each box may be abiotic (solid arrow) or biotic (dotted arrow).

a host on any day. These two terms describe the input into the questing tick population. Thirdly, if a tick is questing, it has a daily probability of attachment to a host. This will vary with host species and density, and so offers some insight into host relationships. Finally, there are daily rates of mortality, which are specific to each stage and state (questing, feeding and engorged). These latter two are output terms. Only analysis of field observational and experimental data can yield reliable measures of the quantitative relationships between these processes and the abiotic and biotic predictor variables.

Within this framework, the following review will examine how both the biotic and abiotic elements of the tick's environment give rise to the precise patterns of tick distribution, seasonal abundance and host relationships that determine pathogen transmission dynamics and therefore spatial and temporal variation in the risk of infection. This review does not aim to be exhaustive. Instead, I shall focus on studies that illustrate the hierarchy of processes and patterns inherent in the epidemiological risk posed by ticks as vectors. At the base is the physiological ecology of ticks, the adaptation of individuals to their natural habitats. As each tick is born, lives and dies, it contributes to the overall rates of the demographic processes that add up to the next level, population ecology. In this chapter I shall deal with tick–host relationships as properties of populations, rather than at the individual level as described in other chapters. Finally, I shall review recent attempts to describe and explain wide-scale patterns of epidemiological risk factors as a way towards biological, process-based predictive risk mapping. Most of the detailed accounts will use *I. ricinus* as a temperate exemplar and *Rhipicephalus appendiculatus* as a tropical exemplar, drawing on my first-hand experience of primary research on these species.

PHYSIOLOGICAL ECOLOGY: THE ADAPTATION OF TICKS TO THEIR NATURAL ENVIRONMENT

Natural outcomes of temperature-dependent development rates

Entry into a population is measured by the birth rate (and immigration rate where appropriate). Because a female ixodid tick lays a single large egg mass (typically several thousand eggs) over a short time period, a more useful demographic input parameter is the development rate from one stage to the next. In common with most physiological processes in poikilothermic animals, inter-stadial development rates of ticks increase non-linearly with increasing ambient temperature. To be of most use in predicting development periods under varying temperature conditions in different parts of a tick species' range, temperature-dependent development rates should be established under constant conditions in the laboratory. In equatorial regions with minimal seasonal temperature variation, however, field observations may be sufficiently precise.

The daily rates for both the tropical and temperate tick species may be adequately described by non-linear relationships with temperature (Ogden *et al.*, 2004) (Fig. 2.2), but there appear to be considerable differences in the precise rates for each species and each stage (the latter more marked for *R. appendiculatus* than *I. ricinus*) (Fig. 2.2). In Slovakia, a sympatric tick species, *Dermacentor reticulatus*, develops very much more rapidly than *I. ricinus* in the same habitat, for example passing from engorged larva to moulted nymph within one month during the summer rather than two or three. These differences suggest an element of adaptation evolved by each species, which may extend to each geographical strain of the same species. It is essential, therefore, to establish the correct rates for each species, laborious though this is. Quantitative inter-stadial linkages cannot be deduced reliably by assuming that development periods correspond to the interval between peak numbers of ticks seen questing in the field (see Figs. 2.4 and 2.5 below). In this and other ways, Randolph (1993) stands as an object lesson of what not to do!

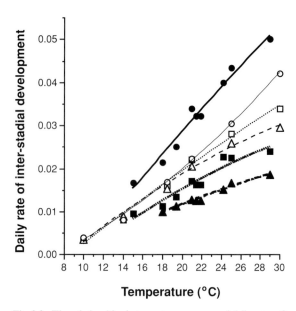

Fig. 2.2. The relationships between temperature and daily rates of inter-stadial development. Note the same pattern, but quantitative differences between stages (circles, larva–nymph; squares, nymph–adult; triangles, female–larva), and between species (filled symbols, *Rhipicephalus appendiculatus*; open symbols, *Ixodes ricinus*). Second-order polynomial relationships, R^2 values = 0.924–0.998. Redrawn from original data in Campbell, 1948; Branagan, 1973; Kaiser *et al.*, 1988.

Using a day–degree summation method, natural development periods in the field may be estimated by applying daily development rates to varying temperatures within the tick's habitat (Randolph, 1997). Development is complete once the proportional development per day sums to 1, after which there is a period of 'hardening off' before the emergent tick starts to quest. Because it is so important to describe this foundation layer of the tick's population performance accurately, temperature should be measured as closely as possible to the microhabitat in which the tick undergoes its development. Soil surface temperatures are buffered both diurnally and seasonally compared with air temperature, so uncorrected air temperatures predict development periods inaccurately. In the southern UK, for example, air temperature measured 50 cm above the ground predicts 10–40% faster development by *I. ricinus* larvae for most of the spring and summer months (Fig. 2.3), resulting in the emergence of nymphs 2–3 weeks earlier. Such inaccuracies could have a marked knock-on effect on the predicted phenology of tick populations, especially in continental Europe where air temperatures are even more variable. Only when we know

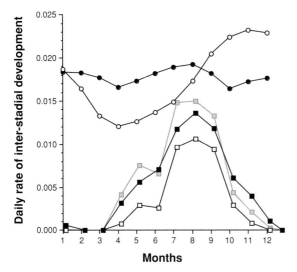

Fig. 2.3. Seasonal and geographical variation in daily rates of
larva–nymph inter-stadial development rates predicted by
day–degree summation models based on ambient temperatures.
Squares, *Ixodes ricinus* in Dorset, UK (■) and Wales, UK (□);
circles, *Rhipicephalus appendiculatus* in Kirundo, Burundi (●) and
Gulu, South Africa (○). The grey symbols (grey squares) show the
effect of basing the predictions for Dorset on air rather than soil
temperatures.

precisely which ticks, and therefore how many, of one stage
give rise to which and how many ticks of the next stage,
counted in the field, can we progress to further analysis of
the seasonal population dynamics of the tick and transmission dynamics of any pathogens.

The simple fact of geographical variation in absolute
ambient temperatures and their seasonal variability results
in contrasting patterns of recruitment to questing tick populations, potentially continuous in the tropics and even
in the subtropical regions of South Africa (Randolph,
1997), but intermittent in temperate regions. Where development rates drop very low and even to zero for large
parts of the year (Fig. 2.3), development periods are tele-
scoped and so emergence of unfed ticks becomes more
synchronized.

Diapause: significance in tropical and temperate tick-borne disease systems

`BEHAVIOURAL' DIAPAUSE IN TROPICAL TICKS
As an adaptation to ensure that emergence coincides with
optimal conditions for survival, many tick species have

incorporated some form of diapause into their life cycles.
In southern Africa, for example, recruitment of *R. appendiculatus* is not in fact continuous throughout the year (Rechav,
1981, 1982; Short & Norval, 1981; Pegram & Banda, 1990;
Randolph, 1997). This species illustrates nicely the plastic,
presumably adaptive nature of diapause. Diapause is confined to unfed adult ticks, which do not always quest and
feed as soon as they have hardened after emergence from the
engorged nymphal stage. In South Africa, those that emerge
after July each year do not start questing until some time after
November of the same year (Rechav, 1981; Randolph, 1997).
This diapause appears to be induced by short daylengths
(the southern winter solstice occurs in June) (Rechav, 1981;
Short & Norval, 1981; Pegram & Banda, 1990; Madder *et al.*,
1999). As a result, there are periods of near absence of each
tick stage sequentially during the year (Fig. 2.4a), and the
majority of the vulnerable egg-to-larval stage occurs during
the warm wet season (December–May), which would favour
rapid development and good survival, although excessively
wet conditions are correlated with increased mortality (see
below).

In contrast, in equatorial Africa, where long cold dry seasons do not occur, the same tick species does not show any
diapause (Branagan, 1973). All tick stages feed throughout
the year (Newson, 1978; Kaiser, Sutherst & Bourne, 1982,
1991; Kaiser *et al.*, 1988), resulting in continuous overlapping generations each of about 8–9 months duration (Fig.
2.4b). Marked declines in abundance of all stages occur
simultaneously during the dry season. Based on laboratory
experiments, Madder *et al.* (1999) concluded that there is a
latitudinal gradient in the propensity to diapause, with ticks
from Zimbabwe exhibiting diapause more consistently than
those from Zambia. While long daylength may trigger diapause termination in the south, further north in Zambia diapause may end gradually over periods of up to 5 months, perhaps in response to physiological ageing (Berkvens, Pegram
& Brandt, 1995).

This variable pattern of diapause is significant not only for
the survival of tick populations, but also for the epidemiology
of associated infections. Norval *et al.* (1991) concluded that
the occurrence of diapause in *R. appendiculatus* in southern
Africa may account for the absence there of the virulent strain
of *Theileria parva* that causes East Coast Fever in cattle in
eastern and central Africa (see Chapter 14). Only where there
is sufficient overlap between the more or less continuous
activity periods of the different tick stages can the virulent
strain of *T. parva* be acquired from infected cattle by larvae
and nymphs and transmitted onwards by nymphs and adults.

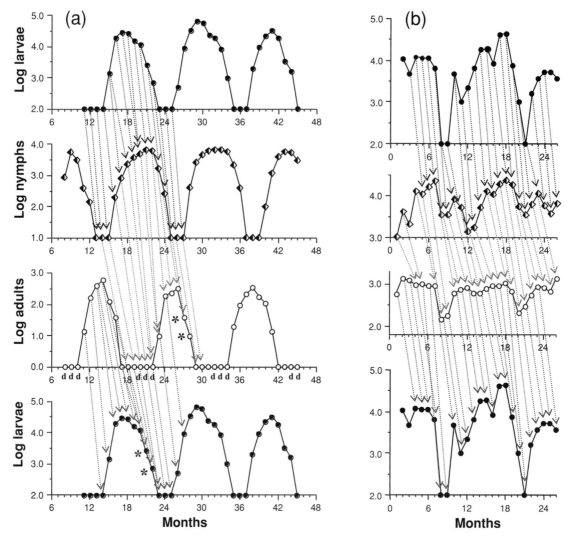

Fig. 2.4. Seasonal variation in the number of larval, nymphal and adult *Rhipicephalus appendiculatus* at (a) Gulu, South Africa (data from Rechav, 1982) and (b) Kirundo, Burundi (data from Kaiser *et al.*, 1988). The dotted arrows indicate which ticks of one life stage gave rise to which ticks of the next stage (according to Randolph, 1994*a*, 1997). In (a), **d** marks the months of diapausing adults, and asterisks mark those adults and larvae that were counted after recruitment had apparently ceased (see text).

If this is true, it suggests that inter-stadial transmission from larvae to nymphs, which do in fact overlap considerably more than nymphs and adults in South Africa (Fig. 2.4a), may not be sufficient to maintain cycles of virulent *T. parva*. The larval–nymphal route was indeed shown experimentally to be less efficient than the nymphal–adult route (Ochanda *et al.*, 1996). The less virulent strain of *T. parva* that causes January or Corridor disease survives for longer periods in carrier cattle, and so can be acquired by nymphs and transmitted

by adults where there is less inter-stadial overlap, hence its occurrence in southern Africa.

`MORPHOGENETIC´ AND `BEHAVIOURAL´ DIAPAUSE IN TEMPERATE TICKS

Temperate tick species have to deal with even more marked seasonality in environmental conditions. The inter-stadial periods between two or more stages of *Ixodes* species, including *I. rubicundus* from South Africa (Fourie, Belozerov &

Needham, 2001), may be prolonged (i.e. adjusted) by a variable delay in the onset of development (Campbell, 1948; Kemp, 1968), commonly called 'morphogenetic diapause' (Belozerov, 1982). As this occurs in *I. ricinus* that are collected from the field after July, fed in the laboratory and then held in either constant laboratory conditions (Campbell, 1948; Kemp, 1968) or quasi-natural field conditions (Chmela, 1969; Cerny, Daniel & Rosicky, 1974; Daniel *et al.*, 1976, 1977; Gray, 1982), diapause must be triggered by conditions experienced by the questing stages, or the difference between pre- and post-engorgement conditions. Belozerov identified photoperiod as the major cue for diapause, specifically the change from long (pre-feed) to short (post-feed) daylength for both Palaearctic (*I. ricinus* and *I. persulcatus*) and Nearctic (*I. scapularis*) members of this species complex (Belozerov, 1998*a*; Belozerov & Naumov, 2001).

The exact light sensitivity of the photoreceptor cells on the dorsolateral surface of all stages of *I. ricinus* (Perret *et al.*, 2003) is unknown, but ticks have now been shown to respond to simulated natural changes in daylength (C. Beattie & S. E. Randolph, unpublished data), which are much more gradual (about 20–30 minutes per week) than those used in Belozerov's experiments (typically an 8-hour difference within a few days, either side of feeding). Under natural conditions there is a latitudinal gradient in the date of diapause onset. It occurs in *I. ricinus* that feed from the end of July onwards in Scotland (*c.* 56° N), from mid-August in Ireland (*c.* 53° N) and from the end of August in the Czech Republic (*c.* 49° N). At that time of year, daylength is longer but declines faster with increasing latitude, suggesting that the rate of daylength change may be a critical cue. However, in light conditions that simulated rapid or slow daylength decrease equivalent to July–October at St Petersburg (60° N) and Valtice in the Czech Republic (49° N), respectively, and under 18 °C constant temperature conditions, ticks entered diapause as photoperiods reached the same absolute length, *c.* 13 hours light, irrespective of the rate of decrease (C. Beattie & S. E. Randolph, unpublished data). Therefore other factors, such as decreasing temperature, or even increasing age of the tick, both of which are known to enhance the diapause response (Belozerov, 1967, 1970; Fourie *et al.*, 2001), may be responsible for its earlier onset further north, although on their own they would offer less reliable signals to ticks.

There is clearer evidence that temperature is instrumental in breaking morphogenetic diapause in *I. ricinus*. Ticks exposed to even a brief period of 0 °C show non-delayed development, similar to those fed in the spring (i.e. pre-July) (Campbell, 1948). This ensures that all ticks, whenever they feed after July/August, start development once temperatures increase after the winter (Fig. 2.3). Those that feed before July develop rapidly without delay. Under a wide range of conditions across the UK, the combination of diapause and seasonal temperature-dependent development rates acts to synchronize the emergence of virtually all unfed ticks of each stage to within a couple of months every autumn (Randolph *et al.*, 2002) (which is likely to act as a brake on any impact of climate change). This conclusion was validated by examining the physiological state of questing ticks; emergence dates predicted from the day–degree summation development model (Fig. 2.5) coincided exactly with the first appearance in the field of questing nymphal and adult ticks with high fat content.

Fat is a source of energy derived from each blood meal, which can be used as a marker of physiological ageing in the field (Uspensky, 1995; Walker, 2001). After emergence, fat is used to fuel a tick's locomotory and physiological activity involved in locating and ascending vegetation stems from which to quest, and descending to moist conditions at the base of the vegetation to restore fluid content by passive and active uptake of atmospheric water (Knulle & Rudolph, 1982; Needham & Teel, 1991; Sonenshine, 1991; see Chapter 3). Its natural rate of usage therefore varies with seasonal activity and climatic conditions (Steele & Randolph, 1985; Randolph & Storey, 1999), allowing a distinction between the calendar age and physiological age of ticks. At three field sites in the UK, at the end of the summer all nymphs and adults of *I. ricinus* had very low fat contents until, in September or October, a large proportion of ticks had distinctly higher fat contents, typically about three times as much when corrected for tick size (Randolph *et al.*, 2002). At this point two overlapping cohorts of ticks were clearly present (Fig. 2.6). Thereafter modal fat contents decreased throughout the following spring. Increasing numbers of ticks became active in the spring, but they had lower fat contents than ticks seen in the previous autumn, indicating that they had endured a longer interval between emergence and questing than had the new recruits in the autumn. Through the summer, ticks still questing for hosts gradually exhausted their fat, so that virtually no low-fat ticks persisted beyond October.

So it appears that under a wide range of climatic conditions in the UK, a single cohort of each stage of *I. ricinus* is recruited each year in the autumn and then survives for one year until the ticks have either fed or died of fat exhaustion. The same appears to be true for *I. persulcatus*, despite a prolonged total life cycle (up to 6 years) in more extreme climates (Filippova, 1985; Uspensky, 1995; Korenberg, 2000).

Ticks counted per 100 m²

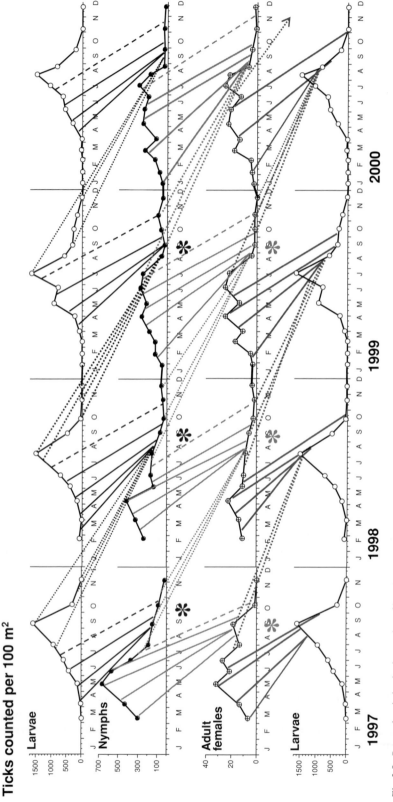

Fig. 2.5. Seasonal variation in the number of larval, nymphal and adult *Ixodes ricinus* in Dorset, UK. The lines indicate which ticks of one life stage gave rise to which ticks of the next stage (according to Randolph *et al.*, 2002): solid lines, ticks undergo direct development if fed any time before mid-July; dashed lines, ticks fed late July undergo direct development but may emerge either late in autumn of the same year or summer of next year; dotted lines, ticks fed after July undergo diapause over winter. Asterisks mark dates of first appearance of high-fat (newly recruited) nymphs and adults in the field.

Tick ecology and pathogen transmission dynamics

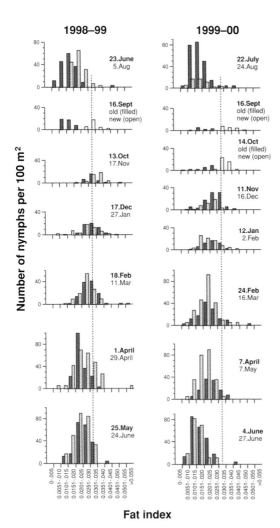

1998–99

1999–00

Number of nymphs per 100 m²

Fat index

Fig. 2.6. Seasonal changes in absolute frequency distributions of questing nymphal *Ixodes ricinus* with different fat content indices counted per 100 m² in Dorset, UK, starting with late summer at the top in each of two years. The fat index is calculated as (square root of fat content/fat-free weight) to correct for the effect of tick size. Data for two monthly samples are presented in each histogram, showing the earlier data of each pair as filled bars and the later date hatched. In September/October when old and new recruits occur within the same sample, new recruits are shown by open bars. The vertical dotted line is purely to guide the eye.

This timing of recruitment does not, however, coincide with the timing of peak numbers of ticks seen questing in the field. Most newly emergent ticks delay their questing activity until the following spring in many parts of Europe, *inter alia* (Lees & Milne, 1951; Daniel *et al.*, 1976; Gray, 1985; Randolph *et al.*, 2002). This may be true 'behavioural

diapause' (Belozerov, 1982), similar to that seen in adult *R. appendiculatus* and *I. pacificus* (Padgett & Lane, 2001), although the triggers will be species-specific. Belozerov (1982) concluded that unfed stages of *I. ricinus* entered diapause in response to short daylength. In the field in mild climates, however, ticks may be active from January onwards (Fig. 2.7), suggesting that decreasing daylength beyond a certain level, rather than absolute daylength, may be the cue for diapause. Ticks also vary their questing activity in response to their immediate climatic conditions (see below). All in all, the pattern of seasonal questing by ticks that diapause in the unfed stage is the product of behaviour (and mortality) superimposed on the phenology. It is time to move to the second box of the tick population model (Fig. 2.1) and consider determinants of the probability of questing.

Environmental constraints on host-questing activity and behaviour

TEMPERATURE AND DAYLENGTH

Abiotic factors impose constraints on when, where and how ticks quest for hosts. Interacting with the putative inhibitory effects of decreasing daylength, and the supposed converse permissive effects of increasing daylength, low winter temperatures in temperate regions inhibit tick activity. Mild winters in the southern UK allow us to disentangle daylength and temperature as determinants of tick questing probability. In contrast to the north of the UK, Switzerland and Wales, where questing activity by nymphs and adults of *I. ricinus* does not begin in the spring until the weekly mean maximum temperature has reached 7 °C (Macleod, 1936; Perret *et al.*, 2000; and Fig. 2.7c, respectively), in the south of the UK from 1998 to 2000 winter temperatures did not drop low enough to inhibit nymphal and adult questing activity (Fig. 2.7a, b). In the south, however, fewer questing nymphs (and adults – data not shown) were counted from October to December, when temperatures were well above threshold levels, than from January onwards. Thus we can propose that at the end of each year, decreasing daylength reduces the probability of questing and low temperatures may inhibit activity altogether (see Wales: Fig. 2.7c), while at the start of each year, increasing daylength is permissive, but only if temperatures are high enough. Gradual recruitment to the questing population will occur as temperatures reach this critical level in increasing parts of the ticks' overwintering microhabitats (Eisen, Eisen & Lane, 2002). Consistent with this, altitude influences the timing of the onset, more than the cessation, of tick activity (Jouda, Perret & Gern, 2004). Climate change may therefore affect tick activity in the spring more than

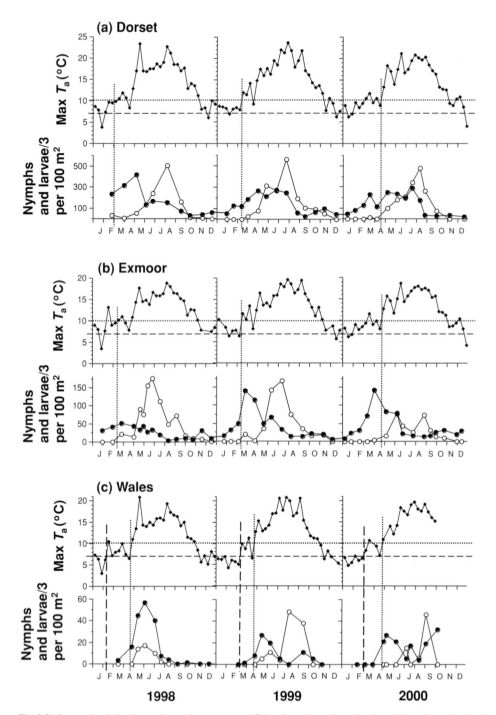

Fig. 2.7. Seasonal variation in maximum air temperature (T_a) and numbers of questing larval (○) and nymphal (●) *Ixodes ricinus* at three sites in the UK (note that larval density is presented at 33% of observed level). Dotted and dashed lines indicate the temperature thresholds for the onset of seasonal activity of larvae and nymphs, respectively.

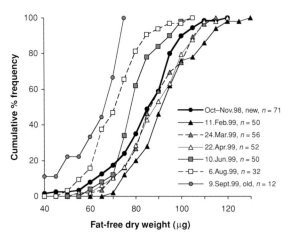

Fig. 2.8. Seasonal changes in the size (i.e. fat free weight) composition of populations of questing nymphal *Ixodes ricinus* during the lifespan of a single cohort on Exmoor, UK, 1998–9. Sample dates and sizes are shown in the key. Reproduced with permission from Randolph *et al.* (2002).

in the autumn, because diapause induced by daylength will continue unchanged. Experiments under controlled conditions are essential to test this proposed explanation, because the use of field sampling data to impute questing activity may be confounded by variable rates of disappearance from the questing population through variable feeding and/or mortality rates (see below).

TICK SIZE

Activity threshold temperatures vary interspecifically, inter-stadially and even intra-stadially, which last effect appears to be size dependent (Clark, 1995). In the UK, the onset of larval activity coincides with a threshold of 10 °C, rather than 7 °C, mean maximum temperature (Fig. 2.7). Furthermore, at our UK sampling sites, smaller nymphs (and adults – not shown) started questing later each year (Randolph *et al.*, 2002). Each new cohort of ticks that appeared on the vegetation in the autumn showed a normal size (fat-free weight) distribution, virtually symmetrical about a median of 75–85 μg, which then changed seasonally (Fig. 2.8). A subset of larger nymphs started questing in February, with only 15–30% <80 μg and almost none <60 μg. As the year progressed, the questing population was composed increasingly of smaller nymphs, as small ticks appeared from March onwards.

Figure 2.8 also shows that once recruitment has ceased beyond April/May and tick numbers fall, the largest ticks (>100 μg) disappear first and only the smallest ones persist to September. Larger ticks arrive sooner and may find hosts more rapidly if their relatively higher energy levels (Van Es,

Hillerton & Gettinby, 1998; Randolph *et al.*, 2002) permit them to quest for longer periods. Smaller ticks (including all larvae: Fig. 2.7) that arrive later may find hosts more slowly if they are forced into temporary inactivity during the summer by the other major constraint of questing probability, i.e. moisture stress.

MOISTURE STRESS

Despite generic adaptations permitting them to inhabit dry environments, water loss remains the Achilles heel of all terrestrial arthropods, including ticks. Ticks that live in xeric habitats show special adaptations. They tend to be larger than average, thereby reducing their relative surface area, but still they must shelter in cracks and under stones. They cannot sit in the open and ambush passing hosts, but must hunt them actively, travelling fast from shelter to host when the opportunity arises. Sonenshine (1991) describes how, 'camel ticks, *Hyalomma dromedarii*, emerge from the barren sand around desert caravansaries and run across the hot, exposed ground to attack people and livestock when these hosts were resting near their shelters'.

Other tick species live in far less challenging habitats, but have lower tolerance to water stress. They can sit on vegetation to await passing hosts, but must still limit their host-questing activity to maintain their water balance. Questing behaviour therefore varies diurnally and seasonally with climate (Lees & Milne, 1951; Belozerov, 1982). Prompted by the onset of darkness, ticks appear to make use of moist night-time conditions to walk, this activity being thought to be related to selecting suitable questing locations (Perret *et al.*, 2003). In experimental conditions, nymphs walked in the vertical plane further after periods of quiescence (median 43 cm, max 9.7 m) than after questing (median 17 cm, max 2.9 m), while those denied atmospheric moisture walked between 5 m and 31 m until they died. Walking in the horizontal plane was more likely by ticks with higher energy reserves, and was directed towards host odour only when humidity was high (Crooks & Randolph, 2006).

Under favourable conditions, ticks may remain in questing positions on the vegetation for periods of several days (Lees & Milne, 1951; Loye & Lane, 1988), but they descend much more frequently in response to increased saturation deficit (SD), a measure of the drying power of the atmosphere which depends on both temperature and relative humidity (RH) so that, commonly, fewest ticks quest during the middle of the day. Macleod (1935) observed that *I. ricinus* demonstrates positive geotropism at saturation deficits above 4.4 mm Hg (equal to 80% RH at 24 °C, or 71% RH at 18 °C). In the field, reductions in the questing *I. ricinus* population coincided with maximum SD above 4.4 mm Hg (Perret *et al.*, 2000), with adults less affected than nymphs. Likewise, experiments in quasi-natural arenas (Randolph & Storey, 1999) revealed that under such dry conditions questing activity was diminished more amongst larvae than nymphs, but for both stages it was reversible once moist conditions were restored, i.e. high SD, even up to 15–20 mm Hg, induced quiescence rather than direct mortality (Fig. 2.9). Nevertheless, high SD may shorten a tick's lifespan indirectly. In the dry arena, nymphs used up their fat twice as fast as those in the wet arena, which effect was presumably largely related to the increased metabolic costs of walking (Perret *et al.*, 2003) and active water absorption (Fielden & Lighten, 1996; Gaede & Knülle, 1997), thereby reducing the estimated maximum questing period from 4 to 2 months (Steele & Randolph, 1985; Randolph & Storey, 1999). If ticks run out of fat before finding a host they will die. Conversely, under favourable warm, moist conditions, prolonged questing will increase the probability of finding a host, thereby exhausting the questing tick population more rapidly. Standard field sampling data can rarely distinguish between tick quiescence/mortality and tick feeding as a cause for the end of the active questing season (Eisen *et al.*, 2002).

IMPACT ON HOST RELATIONSHIPS

Adult ticks adopt species-specific preferred questing heights, loosely correlated with the size of their principal host animal (Lees & Milne, 1951; Gigon, 1985; Loye & Lane, 1988). Laboratory experiments with artificial 'vegetation', such as glass rods or wooden dowels, suggest an inherent preference (Lees & Milne, 1951; Gigon, 1985; Loye & Lane, 1988), which may have been established under natural conditions through a response to scent-markings from glands on various parts of the host's body (see Chapter 21). Alternatively, the 'choice' of hosts may be determined purely mechanistically by the tick's questing height. A general feature for many tick species is vertical separation in questing positions between life stages, with subadults sitting nearest to the base of the vegetation, commonly with larvae lower than nymphs, and adults questing very much higher (Gigon, 1985). Inter-stadially, it seems as if ticks ascend as high as possible within the limitations of their size-related tolerance to moisture stress (Rechav, 1979), locomotory powers and costs, and energy reserves, but the benefits of height are not entirely obvious. Making contact with the larger circumference of the hosts' upper body parts will certainly facilitate attachment there (A. Lakos, personal communication), but large numbers of nymphal and larval *I. ricinus*, for example, attach and engorge successfully on the lower body and lower legs, respectively, of large ungulate hosts (Gilot *et al.*, 1994; Talleklint & Jaenson, 1994; Ogden, Hails & Nuttall, 1998). Any increased ability of later life stages to exploit higher, unoccupied feeding niches is offset by the lost opportunity to exploit smaller hosts, particularly rodents which are abundant and ubiquitous.

Stage-specific host relationships are affected by the impact of climate on questing heights. In the same arena experiments referred to above, dry conditions forced nymphs of *I. ricinus* to quest lower down the vegetation from where they contacted and attached to small rodents in greater numbers than in wet conditions (Fig. 2.9). Very few larvae in the same dry conditions fed on the rodents. Once the dry arena was watered, the same host relationships were established as in the wet arena, with the same low nymph : larva ratio on rodents as is typically seen in the wild. These climatic effects will clearly have an impact on the potential for pathogen transmission between nymphs and larvae feeding on rodents, and perhaps also on other trans-stadial routes via other host species, and involving other tick species.

Longevity at the expense of pathogen transmission

Ticks counterbalance their extraordinarily long feeding intervals by unusually low weight-specific metabolic rates, 12% less than those of ants, beetles or spiders (Lighton & Fielden, 1995). Diapause in both the engorged and unfed stages is accompanied by an even lower level of respiratory metabolism (Belozerov, 1998*b*). Unfed ticks of many species may survive for a year or more before they find a host, eking out the limited energy from their previous meal. Any period of diapause or short-term quiescence ensures optimally timed questing, as described above, and results in the longevity for which ticks are renowned, but reduces the pace (i.e. rate) of the tick's life cycle and therefore of the pathogen's transmission cycle. Transmission will occur with shorter or longer delays in different parts of the tick's seasonal cycles, depending on when the ticks feed.

POPULATION ECOLOGY: CONSEQUENCES OF DEMOGRAPHIC PROCESSES

Variable host attachment rates: impact on tick population size estimates

The risk of infection of humans by tick-borne pathogens depends chiefly on the numbers of infected questing ticks (Glass *et al.*, 1995; Nicholson & Mather, 1996; Dister *et al.*,

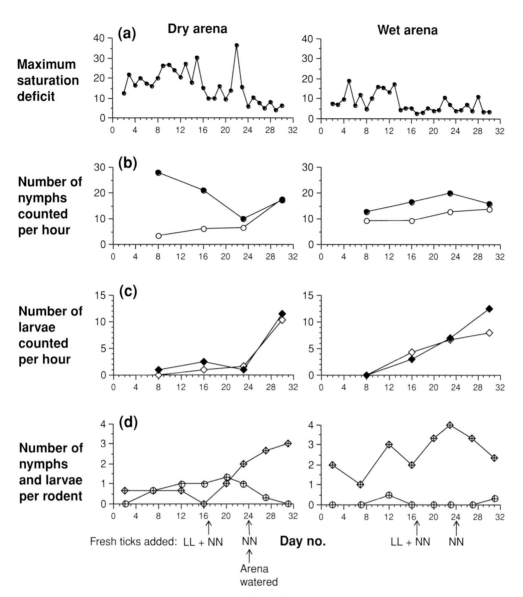

Fig. 2.9. Questing activity and attachment to rodents of *Ixodes ricinus* nymphs and larvae in relation to degrees of moisture stress in dry (left) and wet (right) experimental arenas, (a) maximum saturation deficit; (b) numbers of nymphs and (c) larvae counted per hour at 0900 and 2100 h (filled symbols) and 1200, 1500 and 1800 h (open symbols); (d) numbers of nymphs (circles) and larvae (diamonds) attached per rodent. Fresh ticks were added to the arenas on days 17 and 24 as indicated (LL, larvae; NN, nymphs), and the dry arena was watered on day 24, both of which showed that quiescence rather than death was the cause of low questing activity. Data taken from Randolph & Storey (1999).

1997; Kitron & Kazmierczak, 1997), which is principally a function of the population size of unfed ticks rather than their infection prevalence (Talleklint & Jaenson, 1996; Randolph, 2001). Indices of the standing crop of questing ticks provided by standard blanket-dragging or flagging methods (which count <10% of the ticks actually present: Daniels, Falco & Fish, 2000; Talleklint-Eisen & Lane, 2000) give us a measure of the epidemiological risk at any moment in time, but cannot yield comparative information on absolute tick population sizes from place to place unless relative

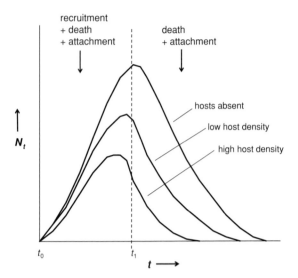

Fig. 2.10. Theoretical changes in the numbers of questing ticks (N_t) through time (t) in the absence of hosts or the presence of hosts at different densities. Seasonal recruitment (which may be directly through development, or by reactivation of ticks after a period of behavioural diapause or quiescence) starts at time t_0 and ceases at time t_1. Redrawn from Randolph & Steele, 1985.

feeding probabilities are known. This latter factor depends on host assemblages and densities. The standing crop of ticks depends on the balance of the rates of input through development and questing probability (above) and output through host attachment and death (Fig. 2.10), which we now go on to consider in turn. The higher the host density, the greater the probability of ticks attaching and progressing through their life cycle. Demographically, this is especially important for the reproductive stage, adult females, which is obviously the basis for the frequently observed positive relationship between densities of deer and ticks (e.g. *I. scapularis* and *I. ricinus*) (Wilson, Adler & Speilman, 1985; Gray *et al.*, 1992; Daniels & Fish, 1995; Falco, Daniels & Fish, 1995; Jensen, Hansen & Frandsen, 2000; Rand *et al.*, 2003). Yet seasonally, low host abundance may cause unfed ticks to accumulate on the vegetation, giving an impression of high tick abundance where hosts are very scarce (Allan, Keesing & Ostfeld, 2003). Conversely, the higher the host density, the smaller the proportion of the questing tick population available to be counted, and the earlier the onset of seasonal decline as ticks are picked up from the vegetation (Fig. 2.10). Decline in tick numbers may start before recruitment has completely ceased if the rates of death and attachment together exceed the rate of recruitment. An estimate of tick attachment rates is essen-

tial for a complete description of tick population dynamics, and any predictive model.

The best, perhaps the only, way to estimate this parameter is by experimental manipulation of natural systems, such as by the use of enclosures. On upland sheep grazings in the UK, rodents and other small hosts are very scarce so that sheep feed ca. 80% of larval and >99% of nymphal and adult *I. ricinus* (Ogden, Randolph & Nuttall, 1997). This convenient fact was exploited in Wales. When four, two or zero sheep were grazed in four identical contiguous 1-ha enclosures (Randolph & Steele, 1985), decreased host density resulted in more questing ticks being counted and a prolonged activity season of nymphs (but not larvae, which exhaust their fat reserves within about 2 months) (Fig. 2.11). A simple model describes the change in numbers of questing ticks after recruitment has finished:

$$\mathrm{d}N/\mathrm{d}t = -r\,NS - m\,N, \qquad (2.1)$$

where r is the rate of contact between ticks and sheep, N is the number of questing ticks, S is the number of sheep and m is the death rate of questing ticks. In the absence of sheep, the logarithmic decline in numbers of ticks counted after the cessation of recruitment gives a measure of the natural death rate. Once this is known, it is a matter of simple mathematics, based on the relationship between total numbers of ticks counted post-recruitment and host density, to calculate r (Randolph & Steele, 1985). This yielded an estimated contact rate of approximately 0.04 per host per day for nymphs (i.e. on average it takes 25 days for a questing nymph to find a sheep, so that 63% ($1 - \mathrm{e}^{-25r}$) of the population will have attached within 25 days), and only about one-tenth this value for larvae. As these estimates are based on counts of ticks that are, by definition, actually questing in the field, they refer to the daily probability of feeding per se. Although this is possibly a slight overestimate for nymphs (as it may include an element of mortality – see below and details in original paper), at least 80%, possibly well over 90%, of questing nymphs will have fed on sheep stocked at a density of 2 per hectare within 2 months, well before they have exhausted their energy reserves, compared with only 32% of larvae within their 2-month energy time-span. Furthermore, the total number of ticks fed per sheep per year was the same in each enclosure. This supports the idea of a characteristic contact rate for each tick stage, which allows each individual host to feed a certain proportion of the questing tick population. This must vary with host species and will also vary with climate (see above).

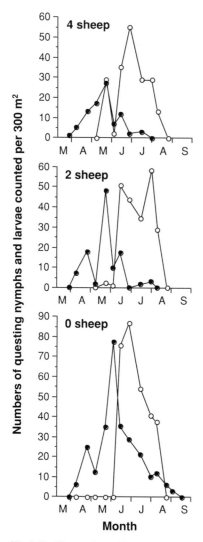

Fig. 2.11. The numbers of larvae (○) and nymphs (●) of *Ixodes ricinus* picked up by blanket-dragging in each of three enclosures on hill grazings in Wales, UK, with different densities of sheep, the only significant host. From Randolph & Steele, 1985.

Similar estimates of the period of persistence of questing ticks were derived from counts of *R. appendiculatus* feeding on cattle in South Africa. Because there was no element of experimental manipulation in this case, it was not possible to dissect mortality from host attachment as the cause of disappearance. Asterisks on Fig. 2.4a indicate that at the end of the feeding seasons, there are sometimes months when larvae and adults were counted despite an absence of adults and nymphs, respectively, that could have produced them *n*

months previously. These are interpreted as those ticks that persisted on the vegetation for up to 2 months beyond their recruitment until they eventually found a host. This must happen throughout the feeding season for each stage, but only becomes apparent at the end of each season. From the monthly rate of decline in the numbers of ticks counted after recruitment has ceased, the mean daily rate of disappearance was estimated as 0.043 for three farms, i.e. ticks persisted on the vegetation for an average of 23 days (range 15–34 days), with 7% remaining after 2 months. There was no discernible difference between tick stages; one interpretation is that, as mortality rates decrease with successive stages of this tick (see below), the questing success rate does in fact increase through the life cycle, as with *I. ricinus*, presumably because greater size, and thus resistance to environmental stress, permits prolonged exposure on the vegetation. Alternatively, but less probably, larvae may have a relatively high questing success rate because of a greater abundance of, and/or contact rate with, hosts for larval *R. appendiculatus* other than cattle (Macleod, 1970).

Thus we see that the rate of host attachment is perhaps the most difficult of the essential demographic processes (Fig. 2.1) to measure empirically, requiring a great deal of host-specific data necessarily collected under natural conditions. Fortunately, however, this parameter is one of the easiest to incorporate into a population model. It is a simple probability, which, even if not known accurately, can be assigned a realistic value and tweaked according to relative host availability.

Natural rates of tick mortality

NATURAL FIELD EXPERIMENTS

The final demographic process in any life history is death. Seasonally variable daily death rates are fixed by the specific local abiotic and biotic conditions, and then operate for the duration of each stage in the life cycle. While we commonly estimate daily rates empirically from observations on percent survival, in reality the latter is determined by the former together with the time-span, according to the simple equation:

$$\text{proportional survival} = e^{-mt}, \qquad (2.2)$$

where *m* is the daily mortality rate and *t* is the period in days over which that mortality acts. If, for example, hosts are scarce and host attachment rates are low, the period of questing will be prolonged; daily mortality will not increase (although possibly it will increase non-linearly with the age of the tick), but will operate for longer, and so the percentage survival will be reduced. To be of most versatile use, the daily probability of death should be quantified at as fine a resolution (temporal, spatial and phenomenological) as possible. Well-defined rates can always be combined subsequently for any particular use, but can rarely be dissected from data if the original measurements are too coarse. Ideally, seasonally variable daily mortality rates operating on each stage in each state (questing, feeding, developing) should be measured. This usually requires specific experimental, but natural, conditions, and has rarely been achieved.

By excluding sheep from one of the enclosures described above (Randolph & Steele, 1985), estimates could be made of the daily mortality rates for larval (0.015) and nymphal (0.031) *I. ricinus* questing on a Welsh hillside between June and August or September, respectively (Fig. 2.11). Normally, however, in the presence of hosts the nymphs would not suffer such high mortality, because this estimate includes some element of death by starvation for nymphs active from April onwards. For larvae not active until mid-June, this specific cause of mortality was probably minimal before mid-August, and the above estimate may be realistic – it would permit *c.* 40% of the unfed larval population to survive the 2-month questing season here. Although highly specific in place and time, these estimates highlight the potential importance of biotic agents of mortality, such as predators, parasites or microbial infections, in habitats where saturation deficit, and therefore abiotic stress, is typically low throughout the summer.

FIELD OBSERVATIONS WITHIN CONTAINERS

So that survivors may be recovered and counted, most estimates of tick mortality have involved the use of containers within natural habitats (Branagan, 1973; Daniel *et al.*, 1976; Gray, 1981; Daniels *et al.*, 1996; Bertrand & Wilson, 1997). These exclude most predators, but are designed so that ticks are fully exposed to ambient conditions through plastic or nylon mesh walls. Ticks, however, actively select their resting positions within their architecturally diverse microhabitats, which may well offer higher moisture and more buffered temperature conditions than are found within containers. An accident during observations on larval *R. appendiculatus* in Kenya (Branagan, 1973) illustrates this point convincingly. Larvae from one batch escaped from their container and survived about four times as long as those that did not. It appears that they positioned themselves relative to the stomata on the grass stems so as to benefit from the raised transpirational humidities both before and after temporary closure of stomata during brief periods of evaporative stress. These larvae eventually succumbed only during periods of more prolonged arrested transpiration. Branagan (1973) concluded: 'The association between the climatic recordings and survival within containers cannot therefore be related to the survival of emerged instars which have access to transpirational microclimates – i.e. those under natural conditions.'

Nevertheless, container-derived estimates are the most common currently available for comparing relative mortality rates for ticks at various stages of their development or activity, in different habitats and at different seasons. Examples from studies in the Czech Republic (Daniel *et al.*, 1976, 1977; and see Chapter 17) and Ireland (Gray, 1981) (Table 2.1) illustrate some general principles. For comparability across different seasons, durations and stages, the observations on percentage survival (or mortality) of each exposed group of ticks have been translated into daily mortality rates, using the equation described above. Where available, the estimates for each stage were very similar from the two sets of studies, apart from the very much higher over-winter mortality of developing larvae and developing eggs in the Czech Republic than in Ireland, presumably related to the harshness of the continental winter. Engorged nymphs appeared to suffer least from this, but this stage was consistently more robust (as are nymphs of *I. scapularis* when exposed to the cold: Vandyk *et al.*, 1996). Unfed larvae exposed in July suffered very high daily mortality (0.0067) until October, but much lower mortality (0.0018) over winter, as would be expected from both the greater tick activity and the moisture stress typical of late summer. Despite the very high percentage hatch of eggs laid by females in Ireland, only *c.* 40% of larvae became active and reached the top of the activity tubes a few weeks later (Gray, 1981, 1982). Also, as expected, the mortality rates of ticks exposed in open meadows, and even on forest edges, were consistently very much higher than those in forests (Daniel *et al.*, 1976, 1977).

Overall, the mortality rates estimated from these studies are much lower than the theoretical requirements for population regulation. A working model can be based on the expectation of decreasing natural mortality rates with successive, increasingly robust, life stages. This was indeed observed for *R. appendiculatus*, whose daily mortality rates were 0.019–0.070 for questing larvae, 0.017–0.029 for questing nymphs and 0.005–0.011 for questing adults confined in quasi-natural conditions in Zimbabwe and Zambia (calculated from survival data given in Short *et al.*, 1989;

Table 2.1 *Examples of mortality estimates for I. ricinus derived from observations on ticks within containers*

Life cycle stage	n	Period	Duration	Percentage survival	Daily mortality rate	Location	Reference
Larvae							
Unfed LL	118[a]	July–Oct	3 m (92 d)	54[b]	0.0067	Forest, Czech Rep.	Daniel et al. (1976)
		Oct–Apr	6 m (183 d)	72[b]	0.0018		
Engorged LL	7750	Nov–Apr	5 m (152 d)	74	0.0020	Forest, Czech Rep.	Daniel et al. (1976)
Engorged LL to emerged NN	687	Oct–Aug	10 m (305 d)	28	0.0042	Forest, Czech Rep.	Daniel et al. (1976)
(includes diapause)	1000	Aug–Aug	12.5 m (378 d)	50	0.0018	Farmland, Ireland	Gray (1981, 1982)
Engorged LL to emerged NN	583	May–Aug	3 m (92 d)	89	0.0013	Forest, Czech Rep.	Daniel et al. (1976)
(direct development)	2500	May–Sept	4 m (122 d)	88	0.0011	Farmland, Ireland	Gray (1981, 1982)
Nymphs							
Unfed NN	305	Nov–Apr	5 m (152 d)	73	0.0021	Forest, Czech Rep.	Daniel et al. (1976)
Engorged NN	777	Nov–Apr	5 m (152 d)	97	0.0002	Forest, Czech Rep.	Daniel et al. (1976)
Engorged NN to emerged AD	791	Aug–Aug	12 m (365 d)	76	0.0008	Forest, Czech Rep.	Daniel et al. (1976)
(includes diapause)	200	Oct–Sep	10.5 m (322 d)	82	0.0006	Farmland, Ireland	Gray (1981, 1982)
Engorged NN to emerged AD	135	Apr–Aug	4 m (122 d)	93	0.0006	Forest, Czech Rep.	Daniel et al. (1976)
(direct development)	300	May–Aug	3.5 m (105 d)	87	0.0013	Farmland, Ireland	Gray (1981, 1982)
Adults							
Unfed FF	98	Nov–Apr	5 m (152 d)	81	0.0014	Forest, Czech Rep.	Daniel et al. (1976)
Unfed MM	69	Nov–Apr	5 m (152 d)	93	0.0005	Forest, Czech Rep.	Daniel et al. (1976)
Engorged FF to hatched larvae (includes diapause)	102	Nov–July	8 m (244 d)	75[c]	0.0012	Forest, Czech Rep.	Daniel et al. (1976)
	40	Sep–Aug	10.6 m (324 d)	85	0.0005	Farmland, Ireland	Gray (1981, 1982)
Engorged FF to hatched larvae (direct development)	40	May–Sep	3.75 m (114 d)	99	0.00009	Farmland, Ireland	Gray (1981, 1982)

[a] Number of cages.

[b] Proportion of cages with live ticks present.

[c] Proportion of females that produced any hatched larvae.

LL, larvae; NN, nymphs; AD, adults; FF, females; MM, males.

Pegram & Banda, 1990), and is not contradicted by the above estimates for *I. ricinus*. We can therefore postulate that, on average, for a tick species with a mean fecundity of 2000 eggs, 5% survival from eggs to larvae, 10% survival from larvae to nymphs and 20% survival from nymphs to fully reproductive adults would theoretically allow the equilibrium state, i.e. one egg-laying female tick replacing herself per generation (Randolph, 1998). Although these are merely conveniently rounded numbers, they do allow some idea of the average inter-stadial mortality necessary to prevent exponential population increase in such fecund organisms: 20% survival requires a daily mortality rate of 0.0176 over a 3-month nymph–adult inter-stadial period, decreasing to 0.0044 if the inter-stadial period lasts 12 months. Likewise, 10% survival requires daily mortality rates of 0.0252–0.0063 for 3–12-month inter-stadial periods. These are approximately an order of magnitude greater than any observed for *I. ricinus* in container experiments (Table 2.1). Either mortality from abiotic causes within containers in temperate climates is much lower than occurs naturally, in contradiction to Branagan's (1973) observations in the tropics, or the vast majority of ticks die during processes prevented by containers, such as feeding or falling prey to predators.

It is clear that, with no realistic empirical estimates of absolute mortality rates, we lack a fundamental component for use in population models. Fortunately, the continuous production, more or less throughout the year, of new recruits into the tick populations of tropical and subtropical regions (Fig. 2.4) allows the calculation, not of absolute mortality rates, but of monthly relative indices of mortality. Standard census data on either questing or cattle-feeding *R. appendiculatus* of all stages in natural open-field situations at 11 sites in East and southern Africa yielded monthly inter-stadial mortality indices in the form of *k* values (Randolph, 1994a, 1997). These are estimated by subtracting the monthly $\log_{10}$(numbers + 1) of one stage of ticks (e.g. nymphs) from the $\log_{10}$(numbers + 1) of preceding stage (e.g. larvae) with the appropriate time interval equivalent to the development period (Varley, Gradwell & Hassell, 1973). By seeking correlations with these mortality indices, it was possible to identify the seasonally variable drivers of mortality, not only climate but also those biotic factors responsible for tick population regulation.

Natural population regulation in ticks

Ticks appear to be subjected to unusually intense mechanisms of natural population regulation. For organisms with such high birth rates, several thousand eggs per female, the potential for wide fluctuations in population size is large (Begon, Harper & Townsend, 2006). Yet regular sampling of various species over a number of years shows that ticks are typified by remarkably constant population sizes, varying annually by considerably less than one order of magnitude (see Figs. 2.4 and 2.5; see also references in Randolph, 1994a, 1997). Population regulation about a fairly constant equilibrium level can be brought about only if the rates of one or more of the demographic processes vary with population density. Death rates must increase, or birth rates must decrease, with increasing density. Only biotic agents can respond to population density in this way, although not all such agents necessarily do so.

Analyses of *R. appendiculatus* populations have shown that inter-stadial mortality does indeed increase with increasing density of ticks of the stage preceding that inter-stadial phase (Randolph, 1994a, 1997) (Fig. 2.12). Furthermore, population regulation occurred at a level rarely seen in free-living populations of arthropods. Both the amount of density dependence (the regression coefficient, *b*, when the mortality index is regressed on tick density) and its predictability (R^2 value for the same regression) are unusually high: of 22 observations of density-dependent mortality, 86% showed values of *b* between 0.51 and 2.0 (compared with 35% of observations on insects: Stubbs, 1977), while 73% gave R^2 values ≥ 0.5. Such intensity is more likely to arise from an intimate biological interaction involving host immunity rather than the more diffuse ecological interaction of predation or parasitism, although the latter may possibly also contribute.

Acquired resistance to feeding ticks can act in a density-dependent manner both in tick–cattle and tick–rodent interactions (Sutherst *et al.*, 1979; Randolph, 1979, 1994b) and has been observed in many natural tick–host interactions (reviewed in Rechav, 1992; see also Chapter 9). Its effects may operate both intra- and inter-stadially (Jongejan *et al.*, 1989; Walker, Fletcher & Todd, 1990), and vary with host type, such as breeds of cattle (Utech, Seifert & Wharton, 1978; Rechav, 1992; Moran, Nigarura & Pegram, 1996) and species of rodents. For example, woodmice (*Apodemus* spp.) do not but voles (*Clethrionomys glareolus*) do acquire effective resistance against feeding larvae of *Ixodes* spp., resulting in reduced engorged weight and reduced survival to the nymphal stage (Randolph, 1979, 1994b; Dizij & Kurtenbach, 1995).

As expected on theoretical grounds, the observed strength of density dependence operating in *R.*

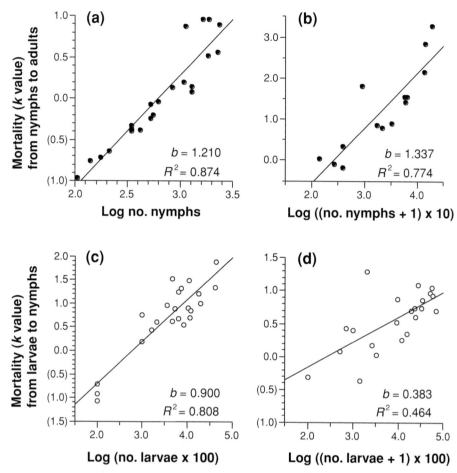

Fig. 2.12. Significant density-dependent relationships between the index of mortality from nymphs to adults and the log number of *Rhipicephalus appendiculatus* nymphs (●), and from larvae to nymphs and the log number of *R. appendiculatus* larvae (○), at Kirundo, Burundi (a) and (c) and Gulu, South Africa (b) and (d). Deviations from the regressions are significantly related to rainfall at Kirundo. Redrawn from Randolph, 1994a, 1997.

appendiculatus populations was weaker nearer to the edge of the tick's geographical range, where less favourable abiotic conditions cause greater deviations from the density-dependent relationship, and for earlier developmental phases because the smaller immature stages are more vulnerable to these adverse abiotic factors (Fig. 2.12). The absence of an apparent density-dependent relationship for the adult-to-larval stage (with one exception) is merely an extension of this pattern; it is not that acquired resistance, elicited either by adults or larvae, does not affect the numbers of eggs produced or the success of feeding by larval ticks, as they have been shown to do (Chiera, Newson & Cunningham, 1985; Fivaz & Norval, 1990; Walker *et al.*,

1990), but rather that even greater susceptibility of eggs and larvae to variable abiotic factors will swamp any biotic factors. These abiotic factors typically involved moisture availability to the eggs and larvae, measured in terms of rainfall or relative humidity, but probably more precisely dependent on soil moisture (Fig. 2.13) (see Guerra *et al.* (2002) for similar conclusions concerning *I. scapularis* in the Midwest of the USA). High mortality occurs where it is both too dry (typical of East Africa) and too wet (at the wettest site of South Africa), and also where development periods are prolonged in the coolest parts of South Africa (not shown). Sensitivity to moisture stress also introduces deviations from the density-dependent

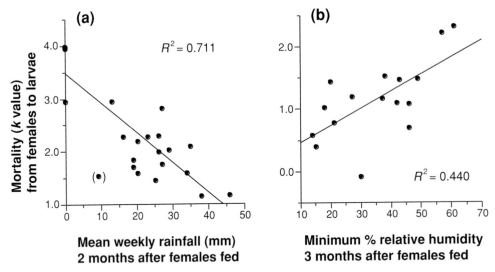

Fig. 2.13. Significant relationships between the index of mortality from females to larvae of *Rhipicephalus appendiculatus* and moisture conditions during egg development: (a) rainfall at Kirundo, Burundi and (b) minimum % relative humidity at Gulu, South Africa. The bracketed outlier in (a) refers to the mortality during egg development in a month of low rainfall in the middle of the wet season, when soil moisture was likely to be high. Redrawn from Randolph, 1994a, 1997.

regressions for the larva–nymph and nymph–adult stages, as dry seasons impose exceptionally high mortality on all stages simultaneously (Randolph, 1994a) (Fig. 2.12).

Thus, even in the absence of precise estimates of absolute mortality rates, we may combine the theoretical limits based on known tick fecundity with a quantitative understanding of the abiotic and biotic determinants, sufficient to generate a working population model (Randolph & Rogers, 1997) (and see below).

ECOLOGICAL AND EPIDEMIOLOGICAL CONSEQUENCES OF TICK–HOST INTERACTIONS

Interspecific patterns of tick–host relationships

Vertebrate hosts are absolutely necessary, of course, but not sufficient on their own for the presence of ticks. Animals upon which ticks can feed are more or less ubiquitous, and yet not all ticks are found on all apparently available hosts. Most tick species are less widely distributed than their principal hosts on both local and continental scales (Kitron *et al.*, 1992; Cumming, 1999). Although different stage-specific requirements may account for some of this, Cumming (1999, 2002) concluded that climate, rather than vegetation or host distributions, limits the species ranges of most African ticks.

Taking an explicitly evolutionary, global approach, Klompen *et al.* (1996) concluded that restrictions imposed by biogeography and ecological specificity, rather than the host specificity, lead to the observed host association patterns. This is not surprising, given the low specificity of many tick–host relationships. *Ixodes ricinus* in Europe and *I. scapularis* in the USA feed on virtually any vertebrate of any class (mammals, birds or reptiles, but not amphibia) with which they come into contact (Milne, 1949; Anderson, 1991; Apperson *et al.*, 1993). Other tick species show variable host specificity at each taxonomic level, some restricted merely to particular classes of vertebrates, others largely to particular orders within classes (e.g. the Carnivora, Artiodactyla, Rodentia among others), and some apparently species specific. Taking into account the many different sorts of biases introduced by collection practices and intensity (see also Klompen *et al.*, 1996), but based only on species records, Cumming (1998) argued that, although some may be classified as true specialists, the majority of the huge assemblage of African ticks are generalists to some degree. As contrasting examples of specialists, he cites *Dermacentor circumguttatus*, found on elephants but not on the many other sympatric mammals (i.e. host specificity), and *Ixodes matopi*, found only on granite kopjes where hyraxes and klipspringers are available as hosts (i.e. ecological specificity). Ivan Horak's uniquely detailed records of individual infestation levels (e.g. Horak *et al.*,

1991), however, reveal clear and marked differences in the probabilities of finding ticks of different species on sympatric hosts of different species.

In general, therefore, ticks may be seen as opportunists, limited by abiotic constraints on where they can quest. Superimposed on this are the physical, behavioural and immunological characteristics of vertebrates that determine the differential distribution of ticks amongst their host species. This has significant consequences for the ecology of the tick, and for the transmission potential of any dependent pathogens.

Because vertebrate species are differentially competent to support the transmission of various tick-borne pathogens, the distribution of ticks amongst different host species will have an impact on the basic reproduction number, R_0, of the pathogens. Non-competent hosts, for example, have been credited with having a zooprophylactic effect by decreasing the infection prevalence in the tick population (Spielman et al., 1985; Matuschka et al., 1992, 2000), but they may in fact have the opposite effect. Any host species that feeds enough ticks to reduce the overall infection prevalence in questing ticks of the next stage by diverting them away from transmission-competent host species, would be likely to increase the tick population density (and therefore the density of infected ticks – see above) by improving the chances of successful tick feeding. Deer seem to have played this role in relation to several tick-borne pathogens in northern temperate regions. In the USA and Denmark, the increase in incidence of Lyme borreliosis has been correlated with the increasing density and distribution of deer (Spielman et al., 1985; Jensen & Frandsen, 2000; Jensen et al., 2000). In parts of Europe, the risk of tick-borne encephalitis (TBE) has been positively related to the presence of deer (Zeman & Januska, 1999; Hudson et al., 2001), not because deer can transmit the virus, but because deer support I. ricinus populations. Similarly, new epidemics of Kyasanur fever disease in India appear to have been precipitated by a marked increase in human population during the 1950s: cattle newly introduced into forests, although not hosts for the virus, supported increased populations of ticks which infected humans as they gathered firewood (Hoogstraal, 1981).

The degree of zooprophylaxis depends on the relative proportion of the vector population fed by different host species, which is highly variable geographically, and may also change with human impacts on habitat structure (Allan, Keesing & Ostfeld, 2003). High mammalian diversity has been predicted to have a negative effect on the infection prevalence of Borrelia burgdorferi in nymphal I. scapularis (LoGiudice et al., 2003), but will only reduce the density of infected nymphs if the density of competent hosts species declines sufficiently with increasing host diversity. Where competent transmission hosts are more or less completely replaced, rather than augmented, by non-competent host species, enzootic cycles may be severely limited. This is thought to account for the much lower prevalence of B. burgdorferi sensu stricto in nymphal ticks in the southeast, compared with the northeast, of the USA (Piesman, 2002), where refractory lizards rather than rodents are the major hosts for immature I. scapularis (Apperson et al., 1993). Likewise, where pheasants are locally very abundant in southern England, they feed such a large proportion of the nymphal I. ricinus population that B. afzelii, which is not transmitted by birds, evidently cannot persist (Kurtenbach et al., 1998). In this case, the end point of $R_0 < 1$ is reached, thereby eliminating infection altogether. This particular case is largely artificial, created by unnaturally high densities of pheasants released from game farms in the interests of shooting. Usually, hosts large enough to carry very high individual infestation levels of ticks exist at low population densities and so do not necessarily feed any greater fraction of the total tick population than do smaller, abundant hosts such as rodents (Talleklint & Jaenson, 1994; Craine, Randolph & Nuttall, 1995). The western fence lizard in northern California, however, is both non-competent to transmit B. burgdorferi sensu lato and so abundant as to feed almost all of the nymphal I. pacificus in some areas, thereby significantly reducing the infection prevalence in adults compared with nymphs (Lane & Quistad, 1998).

Intraspecific patterns of tick–host relationships

CAUSES OF AGGREGATED DISTRIBUTIONS OF TICKS AMONGST THEIR HOSTS

Higher infestation levels on larger hosts are common to many sorts of parasites, and this applies not only between, but also within, host species (Milne, 1949; Matuschka et al., 1992; Craine et al., 1995; Moore & Wilson, 2002). Size is just one of a nexus of host characteristics that cause aggregated rather than random distributions of parasites amongst their hosts, seen habitually in macroparasites (reviewed by Shaw, Grenfell & Dobson, 1998). As mentioned above, these may be physical, behavioural or immunological, which frequently, but not invariably, operate synergistically. Thus within many wild populations, sexually mature male hosts tend to be larger, often range more widely in search of mates, and may be immunologically compromised by their hormones (Peters, 2000). All these features would promote higher tick infestation levels on this fraction of the host population, as has indeed been observed (Randolph, 1975; Matuschka *et al.*, 1992).

Underpinning these particular host features are variable levels of testosterone, and this hormone has been shown experimentally to have a direct impact on resistance to tick infestations (Hughes & Randolph, 2001*a*). Testosterone levels were manipulated in voles *C. glareolus* and woodmice *A. sylvaticus* within the normal physiological range seen in the wild. When exposed to repeated larval *I. ricinus* infestations, hosts with high hormone levels showed reduced innate and acquired resistance to tick feeding (Hughes & Randolph, 2001*a*). The innate resistance was manifested by the greater success with which larvae attached to high-testosterone hosts independent of previous tick exposure. Successive infestations on voles were accompanied by a decrease in tick feeding success and survival, and this was significantly more marked in ticks fed on control voles than in those fed on voles implanted with testosterone (Fig. 2.14). Mice, which do not acquire resistance to *I. ricinus*, showed no such effects. Moreover, when reduced feeding success had been induced, by vaccination with tick salivary gland extract or four successive infestations, implantation with testosterone partially reversed that acquired resistance.

These effects of testosterone will exacerbate the normal tendencies of acquired resistance to generate overdispersed distributions of parasites (Anderson & Gordon, 1982; Pacala & Dobson, 1988), especially as those hosts with low resistance are also prone to high contact with ticks. High testosterone levels are also associated with increased locomotory activity in rodents (Randolph, 1977; Rowsewitt, 1986). Therefore, the same individual hosts are likely both to pick up most ticks and to feed them successfully. The ubiquitous presence of high testosterone levels in one fraction of wild populations, at least during part of each year, could offer a common basis for the typical overdispersed distribution of diverse parasites amongst their hosts, and even amongst males themselves (Shaw *et al.*, 1998).

In the case of pheasants as hosts, a field experiment showed that high infestation levels of ticks may themselves exacerbate further exposure to ticks via a negative impact on territoriality in male birds (Hoodless *et al.*, 2002). Many sorts of parasites have an impact on male secondary sexual characteristics and female choice (reviewed in Clayton & Moore, 1997). When tick burdens on cock pheasants were

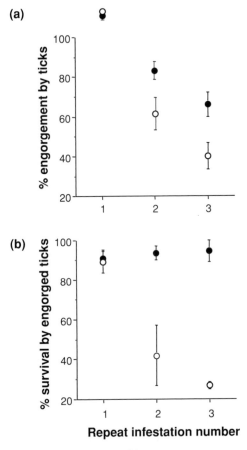

Fig. 2.14. Changes in mean (a) percentage engorgement and (b) percentage survival of *Ixodes ricinus* larvae with successive infestations on castrated voles (*Clethrionomys glareolus*) implanted with oil (○) or with testosterone (●). Data from Hughes & Randolph, 2001*a*.

reduced during March–August using a slow-release acaricide, the degree of wattle inflation was significantly improved (Fig. 2.15). This is an indicator of territorial status and a correlate of harem acquisition, and, correspondingly, harems were acquired by 44% of treated birds compared with only 22% of controls. Males that acquired females ranged over small areas on field edges, while those with no females ranged more widely in woods and the adjoining fields, thereby increasing their exposure to ticks. If non-territorial male pheasants do indeed feed a greater proportion of the tick population (which could not actually be observed in this experiment), *ipso facto* they support more tick-borne parasite transmission, which suggests therefore that ticks themselves

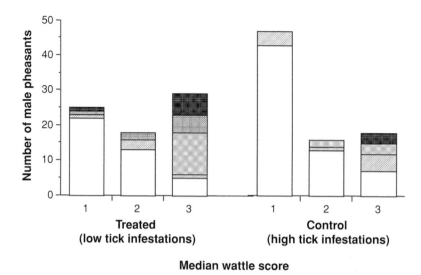

Fig. 2.15. Frequency distributions of the numbers of treated and control male pheasants that acquired different numbers of females, depending on their median wattle score (from three observations of each male in April). Score 1 = no inflation of wattle, pinnae (ear tufts) not visible; 2 = some inflation of wattles, pinnae discernible; 3 = wattles fully inflated, pinnae erect. Number of females: open bars, 0; diagonal hatch, 1; stippled, 2; cross-hatch, 3; black, 4–7. From Hoodless *et al.*, 2002.

could have a positive effect on the transmission dynamics of tick-borne parasites.

CONSEQUENCES OF AGGREGATED DISTRIBUTIONS
OF TICKS AMONGST THEIR HOSTS

The population consequences of aggregated distributions of parasites amongst their hosts have been well revisited since Crofton's seminal work (Crofton, 1971*a*, *b*). With host mortality rates that increase with parasite burdens, and density-dependent constraints (immunological and other) on parasite survival, overdispersion of parasites will stabilize the dynamical behaviour of host–parasite interactions (Anderson & May, 1978).

In addition, because ticks are also vectors, their overdispersion increases the transmission potential of the dependent pathogens (Hasibeder & Dye, 1988; Randolph, 1995; Randolph *et al.*, 1999) by focusing transmission efficiently within one highly infected fraction of the host population (Woolhouse *et al.*, 1997, 1998). Quantitatively, this is an essential element in the persistence of those tick-borne pathogens that have very low R_0 values. *Babesia microti*, for example, is maintained in rodent populations despite low mean infestation levels of the vector, *I. trianguliceps* (and possibly *I. ricinus*: Gray, von Stedingk & Gurtelschmid, 2002), and the very narrow window of transmission (Randolph, 1995). In this

system, testosterone has yet another effect, directly increasing the transmissibility of *B. microti* by causing prolonged and more intense infections (Hughes & Randolph, 2001*b*). This occurs in exactly the same fraction of the host population that carries most ticks through other, synergistic effects of testosterone.

Because tick-borne transmission of pathogens depends on at least two tick stages feeding on the same host, one to transmit and the other to acquire, we can predict that the transmission potential will be improved if both tick stages show coincident aggregated distributions. For pathogens such as TBE virus with short-lived, even non-systemic, infections, this coincidence must be more or less synchronous, whereas for longer-lived, systemic infections (e.g. *B. burgdorferi sensu lato*) it need not be so. Such a phenomenon would easily arise if the same causal factors of aggregated distributions apply to both tick stages, as would all intrinsic host factors. Analysis of tick infestations on over 2500 individual rodents trapped in western Slovakia over three years supported this prediction (Randolph *et al.*, 1999). The distributions of larval and nymphal *I. ricinus* were highly aggregated, and, rather than being independent, the distributions of each stage were coincident so that the same *c.* 20% of hosts fed about 75% of both larvae and nymphs (Fig. 2.16). As a result, twice the number

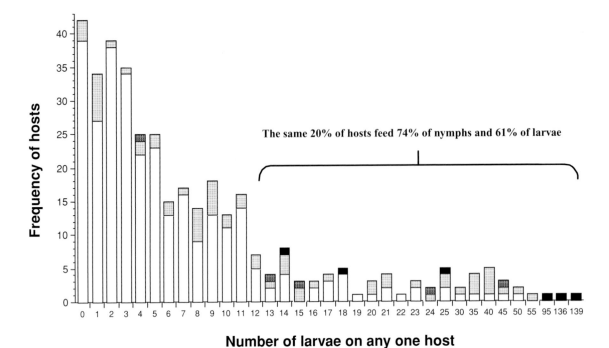

Fig. 2.16. The coincident aggregated frequency distributions of *Ixodes ricinus* larvae and nymphs on rodents (*Clethrionomys glareolus* and *Apodemus flavicollis*) from April to July in the Borská Nízina lowland, Slovakia. At each intensity of larval infestation (*x*-axis), the numbers of hosts coincidentally feeding zero (open bars), 1–2 (light stippled), 3–4 (dark stippled) or 5–23 (black) nymphs are shown. Mean number of larvae co-feeding with a nymph = 20. From Randolph *et al.*, 1999.

of infectible larvae fed alongside each potentially infected nymph compared with the null hypothesis of independent distributions, and estimated R_0 values were >1, as required for virus persistence.

The significance of this aspect of tick ecology in determining vector performance in nature is illustrated by the sympatric tick species in western Slovakia. Because of the different life cycle characteristics of *D. reticulatus*, the pattern of co-feeding larvae and nymphs does not apply to nearly the same extent as to *I. ricinus*. Development times of *D. reticulatus* are short, allowing the life cycle (apart from the egg stage) to be completed within a single year, with most larvae feeding 1 month before the nymphs. Thus, whereas only 3% of *I. ricinus* nymphs were recorded on hosts that were not carrying at least one larva, as many as 28% of *D. reticulatus* nymphs were not co-feeding with larvae of the same species (Randolph *et al.*, 1999). Alternatively, as both species are competent vectors of TBE virus in the laboratory (Korenberg & Kovalevskii, 1994), the virus could potentially be exchanged between ticks of each species where they coexist. These tick species, however, make differential use of

voles and mice as hosts: more *D. reticulatus* were recorded on *C. glareolus* than on *A. flavicollis*, while *I. ricinus* showed the reverse host association (Randolph *et al.*, 1999), so they do not co-feed to a significant extent. The principal host of *I. ricinus*, mice, are also the more efficient amplifiers of TBE virus (Labuda *et al.*, 1993*b*, 1997). Factors such as these may explain why, amongst 18 species competent to transmit TBE virus in the laboratory (Korenberg & Kovalevskii, 1994), *I. ricinus* in Europe and *I. persulcatus* in Eurasia appear to be the only vectors and reservoirs responsible for long-term survival of the TBE virus in nature (Nuttall & Labuda, 1994). This is ecological specificity, leading to quantitative vectorial non-competence.

Of course, intraspecific larvae are more likely to co-feed with other larvae than with nymphs, so could acquire the TBE virus from any trans-ovarially infected larva (Danielová *et al.*, 2002). This, however, can only be an accessory route, insufficient on its own for two reasons. Firstly, any resulting infected engorged larvae can only pass the infection on as they feed as infected nymphs on rodents, so we are back to nymphal–larval co-feeding transmission. Secondly, an essential quantitative element in transmission is amplification to offset the considerable natural mortality of ticks. If an infected nymph arising from a transovarially infected larva, as above, feeds instead on a non-competent host but retains the virus and passes it trans-stadially to the adult stage, there would only be enormous diminution of infection prevalence in the tick population. Each larva, whether infected or not, on average gives rise to 0.01 adult female ticks (see above). Unless this were completely offset by transovarial amplification, which is unlikely on the present evidence of 20% maternal transmission, 0.23–0.75% filial infection and the relatively low numbers of larvae that typically feed close together on any rodent (Danielová & Holubova, 1991; Danielová et al., 2002), TBE virus could not persist by this route. In fact, of 419 larval I. ricinus taken from 103 small mammals trapped in May and October within a TBE focus in the south of the Czech Republic, only three (one unfed and two fed) were infected with TBE virus (Danielová et al., 2002).

PREDICTIVE MAPPING OF THE RISK FACTORS OF TICKS AS VECTORS

Distribution, abundance, seasonal dynamics and host relationships

For certain tick-borne pathogens, the transmission pathways are so robust that more or less wherever populations of competent ticks exist the pathogen will circulate to a greater or lesser extent. This appears to be true, for example, of B. burgdorferi sensu lato in Eurasia. In this case mapping the risk of infection in humans effectively becomes an exercise in mapping the distribution and abundance of the vector ticks. The use of statistical, pattern-matching approaches for predicting tick distributions has become increasingly sophisticated over the past two decades (Daniel & Kolar, 1990; Hugh-Jones et al., 1992; Cumming, 1996, 2002; Daniel et al., 1998; Estrada-Peña, 1998, 1999; Guerra et al., 2002). For further information on the use of geographical information systems (GIS) for predicting tick and tick-borne diseases, see Chapter 17. Briefly, within GIS, statistical methods such as discriminant analysis, logistic regression or co-kriging can establish correlations between the presence of each tick species and selected environmental factors, including land cover features such as the presence and proximity of woodlands large enough to house the necessary host species. These factors are either measured directly on the ground or remotely sensed from satellites (reviewed in Randolph,

2000). Identified correlations may then be used to predict relative habitat suitability across wide areas, beyond the original observations of tick presence or absence, thereby generating maps of the relative probabilities of tick presence. A higher probability may reflect greater abundance, although this has not been tested. The remotely sensed normalized difference vegetation index (NDVI), a measure of plant photosynthetic activity, has proved to be a key predictor of tick presence in all these studies. This index reflects moisture availability on the ground, which, as we have seen above, is a major determinant of tick population performance and is indeed correlated with tick mortality rates (Randolph, 1994a).

For many tick-borne disease systems, however, the mere presence of the tick, even at high population densities, is not sufficient for a significant risk of infection. For the reasons outlined above, certain patterns of host relationships and tick seasonal dynamics may be required for robust enzootic cycles. In principle it should be possible to establish correlations between environmental conditions and tick–host relationships, because both vertebrate populations and tick questing habits are influenced by abiotic factors, but this has not yet been done on a wide scale. On the other hand, climatic correlates of particular patterns of tick seasonal dynamics have been identified and shown to be central to the accurate risk mapping of TBE (Randolph, 2000; Randolph et al., 2000). A relatively rapid spring warming is associated with both the continental distribution of TBE foci and the seasonal synchrony of larval and nymphal I. ricinus (Randolph & Šumilo, 2007). This replaces the original purely statistical rapid autumnal cooling correlate, bringing the statistics into line with new biological understanding – see above. This advance from a statistical prediction towards a process-based biological explanation for the precise way in which this seasonal temperature profile generates larval–nymphal synchrony is a step towards a much more versatile tool. This is obviously the role of a fully functional tick population model and the raison d'être of this chapter.

Such a model has not yet been fully achieved for the complex system of I. ricinus, largely because of unresolved questions about behavioural diapause, but a model driven by temperature and including morphogenetic diapause has given reasonable first results for I. scapularis in Canada (Odgen et al., 2005). Simulations driven by future climate scenarios predict range expansion further north in Canada by this species, and also a change in seasonal dynamics leading to greater overlap between larval and nymphal stages, with consequences for pathogen transmission (Ogden

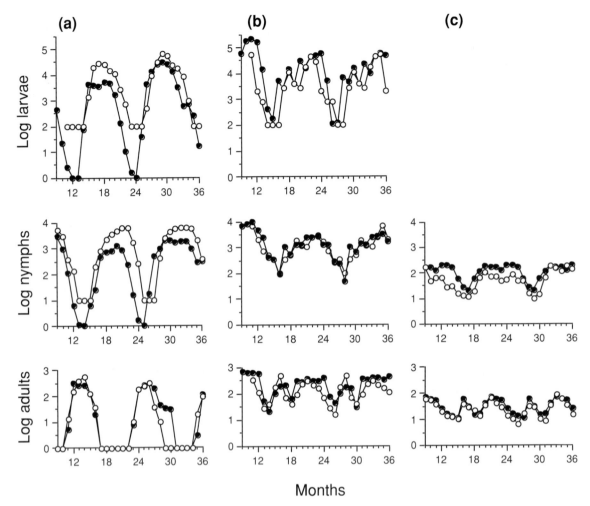

Fig. 2.17. Seasonal variation in the mean number of *Rhipicephalus appendiculatus* per host observed (○) and predicted by the population model (●); observed larvae × 100, observed nymphs × 10. (a) Gulu, South Africa: observations from Rechav (1981); (b) Baker's Fort, northern Uganda: observations from Kaiser, Sutherst and Bourne (1991); (c) Ukunda, coastal Kenya: observations from Newson (1978) did not include larvae. From Randolph & Rogers, 1997.

et al., 2006, 2007). Diapause induced by unchanging daylength could act as a brake on this effect of climate change.

Predictions of a model for the more tractable tropical system of *R. appendiculatus* in Africa (Randolph & Rogers, 1997) match the observed geographically variable patterns of seasonal dynamics closely (Fig. 2.17). This model is driven by just two climatic factors, temperature and rainfall. Therefore appropriate climatic data for any locality, accessible from an interpolated climate database (Hutchinson *et al.*, 1995) or from meteorological satellite imagery, can be inserted into the model to predict the distributional limits, abundance

and seasonal dynamics of the tick. Thus it should be possible to define the major risk factors for tick-borne diseases biologically in terms of population process rates, rather than relying on very much cruder statistical methods.

When this approach was applied to East Africa, the predicted spatial variation in relative numbers of adult ticks in Kenya, Uganda, Rwanda and Burundi showed vanishingly low *R. appendiculatus* abundance coinciding with the north-eastern edge of the recorded distribution (see Fig. 18a in Randolph, 2004). In Tanzania, however, although the model correctly predicted the observed declining abundance of ticks with increasing distance southwards and eastwards from Lake Victoria (Tatchell & Easton, 1986), in central Tanzania the model fit was less good. It did not capture the recorded tick distribution as accurately as did the earlier statistical attempts (Rogers & Randolph, 1993; J. F. Toomer, unpublished data) (see Fig. 18b in Randolph, 2004). Clearly there are still gaps in our knowledge that must be filled before this highly demanding biological approach can generate consistently reliable predictions over large geographical areas.

CONCLUSIONS

Counting ticks, of all stages, in the field is the essential starting point of understanding tick ecology and tick-borne disease epidemiology. All such data are grist to the mill of analysis. Models, sometimes unhelpfully dismissed as 'armchair ecology', are the equally essential framework for that analysis, defining and refining our questions as we plug some gaps in our knowledge, but uncover yet more. If model predictions sometimes appear to run ahead of the known facts, that is the very purpose of models. They direct the collection of good-quality data on vectors and vector-borne pathogens to match the increasingly sophisticated tools of analysis. The long, slow life cycle of ticks poses an extra challenge to tick ecologists working against the current insistence on short-term scientific productivity. The reward, a missing piece slotted neatly into the intricate jigsaw of the natural history of such important creatures, is worth the wait.

ACKNOWLEDGEMENTS

I am grateful to all my colleagues in many parts of the world who have contributed to my understanding of tick ecology. All past and present members of the Oxford Tick Research Group, particularly Chris Beattie, Noel Craine, Graeme Cumming, Rob Green, Andrew Hoodless, Vicki Hughes, Klaus Kurtenbach, Nick Ogden, Pat Nuttall and Mick Peacey, have helped to reveal the patterns and processes described here. David Rogers has been inspirational and operational in much of this work. I am grateful for financial support from the Royal Society, the Wellcome Trust and the Natural Environment Research Council of the UK.

REFERENCES

Alekseev, A. N., Dubinina, H. V., Rijpkema, S. G. T. & Schouls, L. M. (1999). Sexual transmission of *Borrelia garinii* by male *Ixodes persulcatus* ticks (Acari, Ixodidae). *Experimental and Applied Acarology* 23, 165–169.

Allan, B. F., Keesing, F. & Ostfeld, R. S. (2003). Effect of forest fragmentation on Lyme disease risk. *Conservation Biology* 17, 267–272.

Anderson, J. F. (1991). Epizootiology of Lyme borreliosis. *Scandinavian Journal of Infectious Diseases (Supplement)* 77, 23–34.

Anderson, R. M. & Gordon, D. M. (1982). Processes influencing the distribution of parasite numbers within host populations with special emphasis on parasite-induced host mortalities. *Parasitology* 85, 373–398.

Anderson, R. M. & May, R. M. (1978). Regulation and stability of host-parasite population interactions. I. Regulatory processes. *Journal of Animal Ecology* 47, 219–249.

Apperson, C. S., Levine, J. F., Evans, T. L., Braswell, A. & Heller, J. (1993). Relative utilization of reptiles and rodents as hosts by immature *Ixodes scapularis* (Acari: Ixodidae) in the coastal plain of North Carolina, USA. *Experimental and Applied Acarology* 17, 719–731.

Begon, M., Harper, J. L. & Townsend, C. R. (2006). *Ecology*, 4th edn. Oxford, UK: Blackwell Science.

Belozerov, V. N. (1967). [Larval diapause in the tick *Ixodes ricinus* L and its dependence on external conditions. IV. Interactions between exogenous and endogenous factors in the regulation of the larval diapause.] (In Russian) *Entomological Review* 46, 447–451.

Belozerov, V. N. (1970). [Nymphal diapause in the tick *Ixodes ricinus* L. (Ixodidae). III. Photoperiodic reaction in unfed nymphs.] (In Russian) *Parazitologiya* 4, 139–145.

Belozerov, V. N. (1982). Diapause and biological rhythms in ticks. In *Physiology of Ticks*, eds. Obenchain, F. D. & Galun, R., pp. 469–500. Oxford, UK: Pergamon Press.

Belozerov, V. N. (1998a). Role of two-step photoperiodic reaction in the control of development and diapause in the nymphs of *Ixodes persulcatus*. *Russian Journal of Zoology* 2, 414–418.

Belozerov, V. N. (1998b). Dynamics of gas exchange during development of Ixodid ticks. III. Dynamics of gas exchange in nymphs of the ixodid ticks *Hyalomma anatolicum* Koch (Acari, Ixodidae) during activie development and developmental diapause. *Entomological Review* 78, 197–205.

Belozerov, V. N. & Naumov, R. L. (2001). Photoperiodic control of nymphal diapause in the North American tick, *Ixodes (Ixodes) scapularis* Say (Acari: Ixodidae). In *4th European Workshop of Invertebrate Ecophysiology*, ed. Kipyatkov, V. E., p. 69. St Petersburg, Russia.

Berkvens, D. L., Pegram, R. G. & Brandt, J. R. A. (1995). A study of the diapausing behaviour of *Rhipicephalus appendiculatus* and *R. zambesiensis* under quasi-natural conditions in Zambia. *Medical and Veterinary Entomology* **9**, 307–315.

Bertrand, M. R. & Wilson, M. L. (1997). Microhabitat-independent regional differences in survival of unfed *Ixodes scapularis* nymphs (Acari: Ixodidae) in Connecticut. *Journal of Medical Entomology* **34**, 167–172.

Branagan, D. (1973). Observations on the development and survival of the ixodid tick *Rhipicephalus appendiculatus* Neumann, 1901 under quasi-natural conditions in Kenya. *Tropical Animal Health and Production* **5**, 153–165.

Campbell, J. A. (1948). The life history and development of the sheep tick *Ixodes ricinus* Linnaeus in Scotland, under natural and controlled conditions. Unpublished Ph.D. thesis, University of Edinburgh, UK.

Cerny, M., Daniel, M. & Rosicky, B. (1974). Some features of the developmental cycle of the tick *Ixodes ricinus* (L.) (Acarina: Ixodidae). *Folia Parasitologica* **21**, 85–87.

Chiera, J. W., Newson, R. M. & Cunningham, M. P. (1985). Cumulative effects of host resistance on *Rhipicephalus appendiculatus* Neumann (Acarina: Ixodidae) in the laboratory. *Parasitology* **90**, 401–408.

Chmela, J. (1969). On the developmental cycle of the common tick (*Ixodes ricinus* L.) in the north Moravian natural focus of tick-borne encephalitis. *Folia Parasitologica* **16**, 313–319.

Clark, D. D. (1995). Lower temperature limits for activity of several ixodid ticks (Acari: Ixodidae): effects of body size and rate of temperature change. *Journal of Medical Entomology* **32**, 449–452.

Clayton, D. H. & Moore, J. (eds.) (1997). *Host–Parasite Evolution*. Oxford, UK: Oxford University Press.

Craine, N. G., Randolph, S. E. & Nuttall, P. A. (1995). Seasonal variation in the role of grey squirrels as hosts of *Ixodes ricinus*, the tick vector of the Lyme disease spirochaete, in a British woodland. *Folia Parasitologica* **42**, 73–80.

Crofton, H. D. (1971*a*). A quantitative approach to parasitism. *Parasitology* **62**, 179–193.

Crofton, H. D. (1971*b*). A model of host–parasite relationships. *Parasitology* **63**, 343–364.

Crooks, E. & Randolph, S. E. (2006). Walking by *Ixodes ricinus* ticks: intrinsic and extrinsic factors determine the attraction of moisture or host odours. *Journal of Experimental Biology* **209**, 2138–2142.

Cumming, G. S. (1996). The evolutionary ecology of African ticks. Unpublished D.Phil. thesis, University of Oxford, UK.

Cumming, G. S. (1998). Host preference in African ticks (Acari: Ixodidae): a quantitative data set. *Bulletin of Entomological Research* **88**, 379–406.

Cumming, G. S. (1999). Host distributions do not limit the species ranges of most African ticks (Acari: Ixodida). *Bulletin of Entomological Research* **89**, 303–327.

Cumming, G. S. (2002). Comparing climate and vegetation as limiting factors for species ranges of African ticks. *Ecology* **83**, 255–268.

Daniel, M. & Kolar, J. (1990). Using satellite data to forecast the occurrence of the common tick *Ixodes ricinus* (L.). *Journal of Hygiene, Epidemiology, Microbiology and Immunology* **45**, 243–252.

Daniel, M., Cerny, V., Dusbabek, F., Honzakova, E. & Olejnicek, J. (1976). Influence of microclimate on the life cycle of the common tick *Ixodes ricinus* (L.) in thermophilic oak forest. *Folia Parasitologica* **23**, 327–342.

Daniel, M., Cerny, V., Dusbabek, F., Honzakova, E. & Olejnicek, J. (1977). Influence of microclimate on the life cycle of the common tick *Ixodes ricinus* (L.) in an open area in comparison with forest habitats. *Folia Parasitologica* **24**, 149–160.

Daniel, M., Kolar, J., Zeman, P., Pavelka, K. & Sadlo, J. (1998). Predictive map of *Ixodes ricinus* high-incidence habitats and a tick-borne encephalitis risk assessment using satellite data. *Experimental and Applied Acarology* **22**, 417–433.

Danielová, V. & Holubova, J. (1991). Transovarial transmission rate of tick-borne encephalitis virus in *Ixodes ricinus* ticks. In *Modern Acarology*, eds. Dusbabek, F. & Bukva, V., pp. 7–10. The Hague, Netherlands: SPB Academic.

Danielová, V., Holubova, J., Pejcoch, M. & Daniel, M. (2002). Potential significance of transovarial transmission in the circulation of tick-borne encephalitis viirus. *Folia Parasitologica* **49**, 323–325.

Daniels, T. J. & Fish, D. (1995). Effect of deer exclusion on the abundance of immature *Ixodes scapularis* (Acari: Ixodidae) parasitizing small and medium-sized mammals. *Journal of Medical Entomology* **32**, 5–11.

Daniels, T. J., Falco, R. C., Curran, K. L. & Fish, D. (1996). Timing of *Ixodes scapularis* (Acari: Ixodidae) oviposition and larval activity in southern New York. *Journal of Medical Entomology* **33**, 140–147.

Daniels, T. J., Falco, R. C. & Fish, D. (2000). Estimating population size and drag sampling efficiency for the Blacklegged Tick (Acari: Ixodidae). *Journal of Medical Entomology* **37**, 357–363.

Dister, S. W., Fish, D., Bros, S. M., Frank, D. H. & Wood, B. L. (1997). Landscape characterization of peridomestic risk for Lyme disease using satellite imagery. *American Journal of Tropical Medicine and Hygiene* **57**, 687–692.

Dizij, A. & Kurtenbach, K. (1995). *Clethrionomys glareolus*, but not *Apodemus flavicollis*, acquires resistance to *Ixodes ricinus* L., the main European vector of *Borrelia burgdorferi*. *Parasite Immunology* **17**, 177–183.

Eisen, L., Eisen, R. J. & Lane, R. S. (2002). Seasonal activity patterns of *Ixodes pacificus* nymphs in relation to climatic conditions. *Medical and Veterinary Entomology* **16**, 235–244.

Estrada-Peña, A. (1998). Geostatistics and remote sensing as predictive tools of tick distribution: a cokriging system to estimate *Ixodes scapularis* (Acari: Ixodidae) habitat suitability in the United States and Canada from Advanced Very High Resolution Radiometer satellite imagery. *Journal of Medical Entomology* **35**, 989–995.

Estrada-Peña, A. (1999). Geostatistics as predictive tools to estimate *Ixodes ricinus* (Acari: Ixodidae) habitat suitability in the western Palearctic from AVHRR satellite imagery. *Experimental and Applied Acarology* **23**, 337–349.

Falco, R. C., Daniels, T. J. & Fish, D. (1995). Increase in abundance of immature *Ixodes scapularis* (Acari: Ixodidae) in an emergent Lyme disease endemic area. *Journal of Medical Entomology* **32**, 522–526.

Fielden, L. J. & Lighton, J. R. B. (1996). Effects of water stress and relative humidity on ventilation in the tick *Dermacentor andersoni*: (Acari: Ixodidae). *Physiological Zoology* **69**, 599–617.

Filippova, N. A. (ed.) (1985). [Taiga tick *Ixodes persulcatus* Schulze (Acarina, Ixodidae).] (In Russian) Leningrad, USSR: Nauka.

Fivaz, B. H. & Norval, R. A. I. (1990). Immunity of the ox to the brown ear tick *Rhipicephalus appendiculatus*. *Experimental and Applied Acarology* **8**, 51–63.

Fourie, L. J., Belozerov, V. N. & Needham, G. R. (2001). *Ixodes rubicundus* nymphs are short-day diapause-induced ticks with thermolabile sensitivity and desiccation resistance. *Medical and Veterinary Entomology* **15**, 335–341.

Gaede, K. & Knulle, W. (1997). On the mechanism of water vapour sorption from unsaturated atmospheres by ticks. *Journal of Experimental Biology* **200**, 1491–1498.

Gigon, F. (1985). Biologie d' *Ixodes ricinus* L. sur le Plateau Suisse: une contribution à l'écologie de ce vecteur. Unpublished MS thesis, University of Neuchatel, Switzerland.

Gilot, B., Bonnefille, M., Degeilh, B., *et al.* (1994). La colonisation des massifs forestiers par *Ixodes ricinus* (Linné,

1758) en France: utilisation du chevreuil, *Capreolus capreolus* (L. 1758) comme marqueur biologique. *Parasite* **1**, 81–86.

Glass, G. E., Schwarz, B. S., Morgan, J. M. III, *et al.* (1995). Environmental risk factor for Lyme disease identified with geographic information systems. *American Journal of Public Health* **85**, 944–948.

Gray, J. S. (1981). The fecundity *of Ixodes ricinus* (L.) (Acarina: Ixodidae) and the mortality of its developmental stages under field conditions. *Bulletin of Entomological Research* **71**, 533–542.

Gray, J. S. (1982). The development and questing activity *of Ixodes ricinus* (L.) (Acari: Ixodidae) under field conditions in Ireland. *Bulletin of Entomological Research* **72**, 263–270.

Gray, J. S. (1985). Studies on the larval activity of the tick *Ixodes ricinus* L. in Co. Wicklow, Ireland. *Experimental and Applied Acarology* **1**, 307–316.

Gray, J. S., Kahl, O., Janetzki, C. & Stein, J. (1992). Studies on the ecology of Lyme disease in a deer forest in County Galway, Ireland. *Journal of Medical Entomology* **29**, 915–920.

Gray, J. S., Von Stedingk, L. V. & Gurtelschmid, M. E. A. (2002). Transmission studies of *Babesia microti* in *Ixodes ricinus* ticks and gerbils. *Journal of Clinical Microbiology* **40**, 1259–1263.

Guerra, M., Walker, E., Jones, C. J., *et al.* (2002). Predicting the risk of Lyme disease: habitat suitability for *Ixodes scapularis* in the north central United States. *Emerging Infectious Diseases* **8**, 289–297.

Hasibeder, G. & Dye, C. M. (1988). Population dynamics of mosquito-borne disease: persistence in a completely heterogeneous environment. *Theoretical Population Biology* **33**, 31–53.

Hoodless, A. N., Kurtenbach, K., Nuttall, P. A. & Randolph, S. E. (2002). The impact of ticks on pheasant territoriality. *Oikos* **96**, 245–250.

Hoogstraal, H. (1981). Changing patterns of tick-borne diseases in modern society. *Annual Review of Entomology* **26**, 75–99.

Horak, I. G., Fourie, L. J., Novell, P. A. & Williams, E. J. (1991). Parasites of domestic and wild animals in South Africa. XXVI. The mosaic of ixodid tick infestations on birds and mammals in the Mountain Zebra National Park. *Onderstepoort Journal of Veterinary Research* **58**, 125–136.

Hudson, P. J., Rizzoli, A., Rosa, R., *et al.* (2001). Tick-borne encephalitis virus in northern Italy: molecular analysis, relationships with density and seasonal dynamics of *Ixodes ricinus*. *Medical and Veterinary Entomology* **15**, 304–313.

Hugh-Jones, M., Barre, N., Nelson, G., *et al.* (1992). Landsat-TM identification of *Amblyomma variegatun*

(Acari: Ixodidae) habitats in Guadeloupe. *Remote Sensing of the Environment* **40**, 43–55.

Hughes, V. L. & Randolph, S. E. (2001a). Testosterone depresses innate and acquired resistance to ticks in natural rodent hosts: a force for aggregated distributions of parasites. *Journal of Parasitology* **87**, 49–54.

Hughes, V. L. & Randolph, S. E. (2001b). Testosterone increases the transmission potential of tick-borne parasites. *Parasitology* **123**, 365–371.

Hutchinson, M. F., Nix, H. A., McMahon, J. P. & Ord, K. D. (1995). *Africa: A Topographic and Climate Database.* Canberra, Australia: Centre for Resources and Environmental Studies, The Australian National University.

Jensen, P. M. & Frandsen, F. (2000). Temporal risk assessment for Lyme borreliosis in Denmark. *Scandinavian Journal of Infectious Diseases* **35**, 539–544.

Jensen, P. M., Hansen, H. & Frandsen, F. (2000). Spatial risk assessment for Lyme borreliosis in Denmark. *Scandinavian Journal of Infectious Diseases* **35**, 545–550.

Jongejan, F., Pegram, R. G., Zivkovic, D., *et al.* (1989). Monitoring of naturally acquired and artificially induced immunity to *Amblyomma variegatum* and *Rhipicephalus appendiculatus* ticks under field and laboratory conditions. *Experimental and Applied Acarology* **7**, 181–189.

Jouda, F., Perret, J.-L. & Gern, L. (2004). *Ixodes ricinus* density, and distribution and prevalence of *Borrelia burgdorferi* sensu lato infection along an altitudinal gradient. *Journal of Medical Entomology* **41**, 162–169.

Kaiser, M. N., Sutherst, R. W. & Bourne, A. S. (1982). Relationship between ticks and zebu cattle in southern Uganda. *Tropical Animal Health and Production* **14**, 63–74.

Kaiser, M. N., Sutherst, R. W. & Bourne, A. S. (1991). Tick (Acarina: Ixodidae) infestations on zebu cattle in northern Uganda. *Bulletin of Entomological Research* **81**, 257–262.

Kaiser, M. N., Sutherst, R. W., Bourne, A. S., Gorissen, L. & Floyd, R. B. (1988). Population dynamics of ticks on Ankole cattle in five ecological zones in Burundi and strategies for their control. *Preventive Veterinary Medicine* **6**, 199–222.

Kemp, D. H. (1968). Physiological studies on hard ticks Ixodidae. Unpublished Ph.D. thesis, University of Edinburgh, UK.

Kitron, U., Jones, C. J., Bouseman, J. K., Nelson, J. A. & Baumgartner, D. L. (1992). Spatial analysis of the distribution of *Ixodes dammini* (Acari: Ixodidae) on white-tailed deer in Ogle county, Illinois. *Journal of Medical Entomology* **29**, 259–266.

Kitron, U. & Kazmierczak, J. J. (1997). Spatial analysis of the distribution of Lyme disease in Wisconsin. *American Journal of Epidemiology* **145**, 558–566.

Klompen, J. S. H., Black, W. C., Keirans, J. E. & Oliver, J. H. (1996). Evolution of ticks. *Annual Review of Entomology* **41**, 141–161.

Knulle, W. & Rudolph, D. (1982). Humidity relationships and water balance of ticks. In *Physiology of Ticks*, eds. Obenchain, F. D. & Galun, R., pp. 43–70. Oxford, UK: Pergamon Press.

Korenberg, E. I. (2000). Seasonal population dynamics of *Ixodes* ticks and tick-borne encephalitis virus. *Experimental and Applied Acarology* **24**, 665–681.

Korenberg, E. I. & Kovalevskii, Y. V. (1994). A model for relationships among the tick-borne encephalitis virus, its main vectors, and hosts. *Advances in Disease Vector Research* **10**, 65–92.

Kurtenbach, K., Peacey, M. F., Rijpkema, S. G. T., *et al.* (1998). Differential transmission of the genospecies of *Borrelia burgdorferi* sensu lato by game birds and small rodents in England. *Applied and Environmental Microbiology* **64**, 1169–1174.

Labuda, M., Austyn, J. M., Zuffova, E., *et al.* (1996). Importance of localized skin infection in tick-borne encephalitis virus transmission. *Virology* **219**, 357–366.

Labuda, M., Jones, L. D., Williams, T., Danielova, V. & Nuttall, P. A. (1993a). Efficient transmission of tick-borne encephalitis virus between cofeeding ticks. *Journal of Medical Entomology* **30**, 295–299.

Labuda, M., Kozuch, O., Zuffova, E., *et al.* (1997). Tick-borne encephalitis virus transmission between ticks co-feeding on specific immune natural rodent hosts. *Virology* **235**, 138–143.

Labuda, M., Nuttall, P. A., Kozuch, O., *et al.* (1993b). Non-viraemic transmission of tick-borne encephalitis virus: a mechanism for arbovirus survival in nature. *Experientia* **49**, 802–805.

Lane, R. S. & Quistad, G. B. (1998). Borreliacidal factor in the blood of the Western Fence Lizard (*Sceloporus occidentalis*). *Journal of Parasitology* **84**, 29–34.

Lees, A. D. & Milne, A. (1951). The seasonal and diurnal activities of individual sheep ticks (*Ixodes ricinus*). *Parasitology* **41**, 180–209.

Lighton, J. R. B. & Fielden, L. J. (1995). Mass scaling of standard metabolism in ticks: a valid case of low metabolic rates in sit-and-wait strategists. *Physiological Zoology* **68**, 43–62.

LoGiudice, K., Ostfeld, R. S., Schmidt, K. A. & Keesing, F. (2003). The ecology of infectious disease: effects of host diversity and community composition on Lyme disease risk. *Proceedings of the National Academy of Sciences of the USA* **100**, 567–571.

Loye, J. E. & Lane, R. S. (1988). Questing behavior of *Ixodes pacificus* (Acari: Ixodidae) in relation to meteorological and seasonal factors. *Journal of Medical Entomology* **25**, 391–398.

Macleod, J. (1935). *Ixodes ricinus* in relation to its physical environment. II. The factors governing survival and activity. *Parasitology* **27**, 123–144.

Macleod, J. (1936). *Ixodes ricinus* in relation to its physical environment. IV. An analysis of the ecological complexes controlling distribution and activities. *Parasitology* **28**, 295–319.

Macleod, J. (1970). Tick infestation patterns in the Southern Province of Zambia. *Bulletin of Entomological Research* **60**, 253–274.

Madder, M., Speybroeck, N., Brandt, J. & Berkvens, D. (1999). Diapause induction in adults of three *Rhipicephalus appendiculatus* stocks. *Experimental and Applied Acarology* **23**, 961–968.

Matuschka, F.-R., Fischer, P., Heiler, M., Richter, D. & Spielman, A. (1992). Capacity of European animals as reservoir hosts for the Lyme disease spirochete. *Journal of Infectious Diseases* **165**, 479–483.

Matuschka, F.-R., Schinkel, T. W., Klug, B., Spielman, A. & Richter, D. (2000). Relative importance of European rabbits for Lyme disease spirochaetes. *Parasitology* **121**, 297–302.

Milne, A. (1949). The ecology of the sheep tick, *Ixodes ricinus* L.: host relationships of the tick. II. Observations on hill and moorland grazings in northern England. *Parasitology* **39**, 173–194.

Moore, S. L. & Wilson, K. (2002). Parasites as a viability cost of sexual selection in natural populations of mammals. *Science* **297**, 2015–2018.

Moran, M. C., Nigarura, G. & Pegram, R. G. (1996). An assessment of host resistance to ticks on cross-bred cattle in Burundi. *Medical and Veterinary Entomology* **10**, 12–18.

Needham, G. R. & Teel, P. D. (1991). Off-host physiological ecology of ixodid ticks. *Annual Review of Entomology* **36**, 659–681.

Newson, R. M. (1978). The life cycle of *Rhipicephalus appendiculatus* on the Kenyan coast. In *Tick-Borne Diseases and their Vectors*, ed. Wilde, J. K. H., pp. 46–50. Edinburgh, UK: Edinburgh University Press.

Nicholson, M. C. & Mather, T. N. (1996). Methods for evaluating Lyme disease risks using geographical information systems and geospatial analysis. *Journal of Medical Entomology* **33**, 711–720.

Norval, R. A. I., Lawrence, J. A., Young, A. S., *et al.* (1991). *Theileria parva:* influence of vector, parasite and host relationships on the epidemiology of theileriosis in southern Africa. *Parasitology* **102**, 347–356.

Nuttall, P. A. & Labuda, M. (1994). Tick-borne encephalitides. In *Ecological Dynamics of Tick-Borne Zoonoses*, eds. Sonenshine, D. E. & Mather, T. N., pp. 351–391. Oxford, UK: Oxford University Press.

Ochanda, H., Young, A. S., Wells, C., Medley, G. F. & Perry, B. D. (1996). Comparison of the transmission of *Theileria parva* between different instars of *Rhipicephalus appendiculatus*. *Parasitology* **113**, 243–253.

Odgen, N. H., Bigras-Poulin, M., O'Callaghan, C. J., *et al.* (2005). A dynamic population model to investigate the effect of climate on geographic range and seasonality of the tick *Ixodes scapularis*. *International Journal for Parasitology* **35**, 375–389.

Odgen, N. H., Bigras-Poulin, M., O'Callaghan, C. J., *et al.* (2007). Vector seasonality, host infection dynamics and fitness of pathogens transmitted by the *Ixodes scapularis*. *Parasitology* **134**, 209–227.

Ogden, N. H., Hails, R. S. & Nuttall, P. A. (1998). Interstadial variation in the attachment sites of *Ixodes ricinus* ticks on sheep. *Experimental and Applied Acarology* **22**, 227–232.

Ogden, N. H., Lindsay, L. R., Beauchamp, G., *et al.* (2004). Investigation of relationships between temperature and development rates of tick *Ixodes scapularis* (Acari: Ixodidae) in the laboratory and field. *Journal of Medical Entomology* **41**, 622–633.

Ogden, N. H., Maarouf, A., Barker, I. K., *et al.* (2006). Climate change and the potential for range expansion of the Lyme disease vector *Ixodes scapularis* in Canada. *International Journal for Parasitology* **36**, 63–70.

Ogden, N. H., Randolph, S. E. & Nuttall, P. A. (1997). Natural Lyme disease cycle maintained via sheep by co-feeding ticks. *Parasitology* **115**, 591–599.

Pacala, S. W. & Dobson, A. P. (1988). The relation between the number of parasites/host and host age: population dynamics causes and maximum likelihood estimation. *Parasitology* **96**, 197–210.

Padgett, K. A. & Lane, R. S. (2001). Life cycle *of Ixodes pacificus* (Acari: Ixodidae): timing of development processes under field and laboratory conditions. *Journal of Medical Entomology* **38**, 684–693.

Pegram, R. G. & Banda, D. S. (1990). Ecology and phenology of cattle ticks in Zambia: development and survival of free-living stages. *Experimental and Applied Acarology* 8, 291–293.

Perret, J.-L., Guérin, P. M., Diehl, P.-A., Vlimant, M. & Gern, L. (2003). Darkness favours mobiltiy and saturation deficit limits questing duration in *Ixodes ricinus*, the tick vector of Lyme disease in Europe. *Journal of Experimental Biology* 206, 1809–1815.

Perret, J.-L., Guigoz, E., Rais, O. & Gern, L. (2000). Influence of saturation deficit and temperature on *Ixodes ricinus* tick questing activity in a Lyme borrelosis-endemic area (Switzerland). *Parasitology Research* 86, 554–557.

Peters, A. (2000). Testosterone treatment is immunosuppressive in superb fairy-wrens, yet free-living males with high testosterone are more immunocompetent. *Proceedings of the Royal Society of London B* 267, 883–889.

Piesman, J. (2002). Ecology of *Borrelia burgdorferi* sensu lato in North America. In *Lyme Borreliosis Biology, Epidemiology and Control*, eds. Gray, J. S., Kahl, O., Lane, R. S. & Stanek, G., pp. 223–250. Wallingford, UK: CAB International.

Rand, P. W., Lubelczyk, C., Lavigne, G. R., *et al.* (2003). Deer density and the abundance of *Ixodes scapularis* (Acari: Ixodidae). *Journal of Medical Entomology* 40, 179–184.

Randolph, S. E. (1975). Patterns of distribution of the tick *Ixodes trianguliceps* Birula on its hosts. *Journal of Animal Ecology* 44, 451–474.

Randolph, S. E. (1977). Changing spatial relationships in a population of *Apodemus sylvaticus* with the onset of breeding. *Journal of Animal Ecology* 46, 653–676.

Randolph, S. E. (1979). Population regulation in ticks: the role of acquired resistance in natural and unnatural hosts. *Parasitology* 79, 141–156.

Randolph, S. E. (1993). Climate, satellite imagery and the seasonal abundance of the tick *Rhipicephalus appendiculatus* in southern Africa: a new perspective. *Medical and Veterinary Entomology* 7, 243–258.

Randolph, S. E. (1994a). Population dynamics and density-dependent seasonal mortality indices of the tick *Rhipicephalus appendiculatus* in east and southern Africa. *Medical and Veterinary Entomology* 8, 351–368.

Randolph, S. E. (1994b). Density-dependent acquired resistance to ticks in natural hosts, independent of concurrent infection with *Babesia microti*. *Parasitology* 108, 413–419.

Randolph, S. E. (1995). Quantifying parameters in the transmission of *Babesia microti* by the tick *Ixodes trianguliceps* amongst voles (*Clethrionomys glareolus*). *Parasitology* 110, 287–295.

Randolph, S. E. (1997). Abiotic and biotic determinants of the seasonal dynamics of the tick *Rhipicephalus appendiculatus* in South Africa. *Medical and Veterinary Entomology* 11, 25–37.

Randolph, S. E. (1998). Ticks are not insects: consequences of contrasting vector biology for transmission potential. *Parasitology Today* 14, 186–192.

Randolph, S. E. (2000). Ticks and tick-borne disease systems in space and from space. *Advances in Parasitology* 47, 217–243.

Randolph, S. E. (2001). The shifting landscape of tick-borne zoonoses: tick-borne encephalitis and Lyme borreliosis in Europe. *Philosophical Transactions of the Royal Society B* 356, 1045–1056.

Randolph, S. E. (2004). Tick ecology: processes and patterns behind the epidemiological risk posed by ixodid ticks as vectors. *Parasitology* 129, S37–S65.

Randolph, S. E. & Rogers, D. J. (1997). A generic population model for the African tick *Rhipicephalus appendiculatus*. *Parasitology* 115, 265–279.

Randolph, S. E. & Steele, G. M. (1985). An experimental evaluation of conventional control measures against the sheep tick *Ixodes ricinus* (L) (Acari: Ixodidae). II. The dynamics of the tick–host interaction. *Bulletin of Entomological Research* 75, 501–518.

Randolph, S. E. & Storey, K. (1999). Impact of microclimate on immature tick-rodent interactions (Acari: Ixodidae): implications for parasite transmission. *Journal of Medical Entomology* 36, 741–748.

Randolph, S. E. & Šumilo, D. (2007). Tick-borne encephalitis in Europe: dynamics of changing risk. In *Emerging Pests and Vector-Borne Disease in Europe*, eds. Takken, W. & Knols, B. G., pp. 187–206. Wageningen, The Netherlands: Wageningen University Publishers.

Randolph, S. E., Green, R. M., Hoodless, A. N. & Peacey, M. F. (2002). An empirical quantitative framework for the seasonal population dynamics of the tick *Ixodes ricinus*. *International Journal for Parasitology* 32, 979–989.

Randolph, S. E., Green, R. M., Peacey, M. F. & Rogers, D. J. (2000). Seasonal synchrony: the key to tick-borne encephalitis foci identified by satellite data. *Parasitology* 121, 15–23.

Randolph, S. E., Miklisova, D., Lysy, J., Rogers, D. J. & Labuda, M. (1999). Incidence from coincidence: patterns of tick infestations on rodents facilitate transmission of tick-borne encephalitis virus. *Parasitology* 118, 177–186.

Rechav, Y. (1979). Migration and dispersal patterns of three African ticks (Acari: Ixodidae) under field conditions. *Journal of Medical Entomology* 16, 150–163.

Rechav, Y. (1981). Ecological factors affecting the seasonal activity of the brown ear tick *Rhipicephalus appendiculatus*. In *Tick Biology and Control*, eds. Whitehead, G. B. & Gibson, J. D., pp. 187–191. Grahamstown, South Africa: Grahamstown University Press.

Rechav, Y. (1982). Dynamics of tick populations (Acari: Ixodidae) in the Eastern Cape Province of South Africa. *Journal of Medical Entomology* **19**, 679–700.

Rechav, Y. (1992). Naturally acquired resistance to ticks: a global view. *Insect Science and its Application* **13**, 495–504.

Rogers, D. J. & Randolph, S. E. (1993). Distribution of tsetse and ticks in Africa: past, present and future. *Parasitology Today* **9**, 266–271.

Rowsewitt, C. N. (1986). Seasonal variation in activity rhythms of male voles: mediation by gonadal hormones. *Physiology and Behaviour* **37**, 797–803.

Shaw, D. J., Grenfell, B. T. & Dobson, A. P. (1998). Patterns of macroparasite aggregation in wildlife host populations. *Parasitology* **117**, 597–610.

Short, N. J. & Norval, R. A. I. (1981). The seasonal activity of *Rhipicephalus appendiculatus* Neumann 1901 (Acarina: Ixodidae) in the high veld of Zimbabwe Rhodesia. *Journal of Parasitology* **67**, 77–84.

Short, N. J., Floyd, R. B., Norval, R. A. I. & Sutherst, R. W. (1989). Survival and behaviour of unfed stages of the ticks *Rhipicephalus appendiculatus*, *Boophilus decoloratus* and *B. microplus* under field conditions in Zimbabwe. *Experimental and Applied Acarology* **6**, 215–236.

Sonenshine, D. E. (1991). *Biology of Ticks*, vol. 1. Oxford, UK: Oxford University Press.

Spielman, A., Wilson, M. L., Levine, J. F. & Piesman, J. (1985). Ecology of *Ixodes dammini*-borne human babesiosis and Lyme disease. *Annual Review of Entomology* **30**, 439–460.

Steele, G. M. & Randolph, S. E. (1985). An experimental evaluation of conventional control measures against the sheep tick *Ixodes ricinus* (L) (Acari: Ixodidae). I. A unimodal seasonal activity pattern. *Bulletin of Entomological Research* **75**, 489–499.

Stubbs, M. (1977). Density dependence in the life-cycles of animals and its importance in *K*- and *r*-strategies. *Journal of Animal Ecology* **46**, 677–688.

Sutherst, R. W., Utech, K. B. W., Kerr, J. D. & Wharton, R. H. (1979). Density-dependent mortality of the tick *Boophilus microplus* on cattle: further observations. *Journal of Applied Ecology* **16**, 397–403.

Talleklint, L. & Jaenson, T. G. T. (1994). Transmission of *Borrelia burgdorferi s.l.* from mammal reservoirs to the primary vector of Lyme borreliosis, *Ixodes ricinus* (Acari: Ixodidae), in Sweden. *Journal of Medical Entomology* **31**, 880–886.

Talleklint, L. & Jaenson, T. G. T. (1996). Relationship between *Ixodes ricinus* density and prevalence of infection with *Borrelia*-like spirochetes and density of infected ticks. *Journal of Medical Entomology* **33**, 805–811.

Talleklint-Eisen, L. & Lane, R. S. (2000). Efficiency of drag sampling for estimating population sizes of *Ixodes pacificus* (Acari: Ixodidae) nymphs in leaf litter. *Journal of Medical Entomology* **37**, 484–487.

Tatchell, R. J. & Easton, E. (1986). Tick (Acari: Ixodidae) ecological studies in Tanzania. *Bulletin of Entomological Research* **76**, 229–246.

Uspensky, I. (1995). Physiological age of Ixodid ticks: aspects of its determination and application. *Journal of Medical Entomology* **32**, 751–764.

Utech, K. B. W., Seifert, G. W. & Wharton, R. H. (1978). Breeding Australian Illawara shorthorn cattle for resistance to *Boophilus microplus*. I. Factors affecting resistance. *Australian Journal of Agricultural Research* **29**, 411–422.

Vandyk, J. K., Bartholomew, D. M., Rowley, W. A. & Platt, K. B. (1996). Survival of *Ixodes scapularis* (Acari: Ixodidae) exposed to cold. *Journal of Medical Entomology* **33**, 6–10.

Van Es, R. P., Hillerton, J. E. & Gettinby, G. (1998). Lipid consumption in *Ixodes ricinus* (Acari: Ixodidae): temperature and potential longevity. *Bulletin of Entomological Research* **88**, 567–573.

Varley, G. C., Gradwell, G. R. & Hassell, M. P. (1973). *Insect Population Ecology: An Analytical Approach*. Oxford, UK: Blackwell Scientific Publications.

Walker, A. R. (2001). Age structure of a population of *Ixodes ricinus* (Acari: Ixodidae) in relation to its seasonal questing. *Bulletin of Entomological Research* **91**, 69–78.

Walker, A. R., Fletcher, J. D. & Todd, L. (1990). Resistance between stages of the tick *Rhipicephalus appendiculatus* (Acari: Ixodidae). *Journal of Medical Entomology* **27**, 955–961.

Wilson, M. L., Adler, G. H. & Speilman, A. (1985). Correlation between abundance of deer and that of the deer tick, *Ixodes dammini* (Acari: Ixodidae). *Annals of the Entomological Society of America* **78**, 172–176.

Woolhouse, M. E. J., Dye, C., Etard, J. F., *et al.* (1997). Heterogeneities in the transmission of infectious agents:

implications for the design of control programs. *Proceedings of the National Academy of Sciences of the USA* 94, 338–342.

Woolhouse, M. E. J., Etard, J.-F., Dietz, K., Ndhlovu, P. D. & Chandiwana, S. K. (1998). Heterogeneities in schistosome transmission dynamics and control. *Parasitology* 117, 475–482.

Zabicka, J. (1994). [Tick-borne encephalitis in Poland.] (In Polish) *Przegl Epidemiology* 48, 197–203.

Zeman, P. & Januska, J. (1999). Epizootiologic background of dissimilar distribution of human cases of Lyme borreliosis and tick-borne encephalitis in a joint endenic area. *Comparative Immunology, Microbiology and Infectious Diseases* 22, 247–260.

3 • Tick salivary glands: the physiology of tick water balance and their role in pathogen trafficking and transmission

A. S. BOWMAN, A. BALL AND J. R. SAUER

INTRODUCTION

Almost two millennia ago, the wonders of tick osmoregulation and excretion were commented upon by Pliny the Elder (AD 23–79) in his 37-volume *Historia Naturalis* when he wrote: 'a tick simply filled to bursting point with its victim's blood and then died because it had no anus' (from Hillyard, 1996). A millennium and a half later, the Reverend Dr Thomas Moufet (1553–1604) also noted in his *Insectorum sive Minimorum Animalium Theatrum* that '[*Ricinus*] is filled with food abundantly and yet there is no passage for any excrement'. Quite correctly, these ancient natural historians observed the tremendous increase in body size of feeding female ticks and then suddenly these engorged ticks would detach and fall to the ground, barely able to move with their enormous rounded bodies, and unwilling to reattach. However, the conclusions of Pliny and Moufet were incorrect that this apparent onset of tick ill-health was due to an inability to excrete caused by the lack of anus or that these inactive engorged females would die as a result. Indeed, if Pliny and Moufet had continued their observations of ticks for a few weeks, they would have seen that these engorged females (most likely *Ixodes ricinus*) would deposit several thousand eggs before dying, thus completing the life cycle. Ticks do possess an anus and excrete a small amount of nitrogenous waste (mainly guanine). However, whilst feeding on the host, ticks are actively excreting about 70% of their imbibed water and ions, but Pliny, Moufet or other observers would not see this hidden process since the tick returns the excess water and ions back into the host via its mouthparts. In ticks, it is the salivary glands that are the organs of osmoregulation not the Malpighian tubules (Fig. 3.1).

Studies of tick salivary glands are of major importance in tick biology because of their extraordinary physiology, and their role in pathogen transmission and secretion of bioactive products. The earliest work focused on structure, his-

tology and the striking growth and reorganization of the salivary glands in ixodid females during feeding (Till, 1961; Chinery, 1965; Binnington, 1978). From these studies it was shown that salivary glands occupy the anterolateral one-third to one-half of the tick's haemocoel with an abundance of oxygen supplying trachea. Three acini types, I, II and III, attached to a main and branching ducts in females and four (I–IV) in males, comprise the salivary glands of ixodid ticks. Two acini types (A and B or I and II) make up the salivary glands in argasid ticks (Coons & Roshdy, 1981). With the discovery of the salivary glands as organs of fluid excretion (Howell, 1966; Gregson, 1967; Tatchell, 1967) research was accelerated on understanding the control and mechanism of fluid secretion (Tatchell, 1969; Kaufman & Phillips, 1973). This research confirmed control by nerves and dopamine as the principal neurotransmitter at the neuroeffector junction controlling fluid secretion. Other studies by Rudolph & Knulle (1974) identified the salivary glands as the source of a hyperosmotic secretion that facilitates absorption of water from unsaturated air during protracted periods off the host by unfed ticks. With increasing awareness of the significance of the ixodid tick's capacity to remain attached and feeding on a host for relatively long periods, work has intensified on identifying factors in tick saliva that counter the haemostatic, immune and inflammatory responses of the host directed at the attached and feeding tick. This aspect of salivary gland biology aims to identify secretory proteins that may be targeted and used as vaccines for immunological control of ticks and tick-borne pathogens or possibly as molecules for designing novel pharmaceutical agents (see Chapters 4 and 10). Here we provide an overview of salivary gland structure and its physiology with an eye towards seeing how this information contributes to understanding the tick's unique success as an ectoparasite and its ability to be an efficient vector of pathogens. We discuss the application of new approaches, such as expressed sequence tag projects and RNA

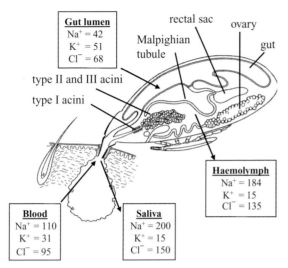

Fig. 3.1. Schematic of a feeding adult female tick (*Amblyomma americanum*) illustrating the relative position and size of the internal organs. The racemose-like structure of the salivary gland shows the small type I acini that are involved in off-host osmoregulation and the larger, more posterior, type II and III acini that are involved in on-host osmoregulation, are the site of synthesis and secretion of protein and lipid factors (anticoagulants, immunosuppressants etc.) and are the site of pathogen development and replication. A male tick is depicted on the underside of the female *in copula*, a process that takes about 48 hr as the spermatophore is inserted into the female genital pore by the male's mouthparts and involves copious salivation from the male-specific type IV acini (Feldman-Muhsam, Borut & Saliternik-Givant, 1970). The boxes indicate the composition of the main ions (units millimolar equivalents) in the various compartments as they are imbibed from the blood into the gut lumen, pass from the gut into the haemolymph and, finally, through the salivary gland back into the host in the saliva (data from Hsu & Sauer, 1975). (Artwork by Kerry Stricker, Oklahoma State University.)

interference (RNAi), to this important field of tick and tick-borne pathogen research.

SALIVARY GLAND MORPHOLOGY AND MORPHOGENESIS

The anatomy and morphogenesis of tick salivary glands during tick feeding has been previously reviewed (Sauer *et al.*, 1995; Coons & Alberti, 1999). An abbreviated description is included here to provide an overview of the tissue's complexity. The paired salivary glands are acinar, lying anterolaterally

on both sides of the tick's haemocoel (Fig. 3.1). Paired salivary ducts enter a shallow tube, the salivarium, lying above the pharynx. The salivarium empties into the buccal cavity where outward-flowing saliva alternates with inward-moving host tissue fluid.

Each acinus type in both ixodid and argasid ticks consists of multiple cell types (Fig. 3.2). The type I acinus is without a valve and directly attaches to the main duct via a short acinar duct and short distances along the two posterior branches of the main duct (Fig. 3.3) The agranular acinus type I in ixodid ticks consists of a single central cell, multiple pyramidal cells, a constrictor cell and peritubular cells surrounding the short acinar duct (Krolak, Ownby & Sauer, 1982). The cells do not change substantially in size or appearance during tick feeding and the number of cells remains constant (Barker *et al.*, 1984). The pyramidal cells are marked by tortuous plasma membrane infoldings with closely associated mitochondria that prompted some early investigators to propose that acinus type I performed a major role in fluid excretion during tick feeding (Kirkland, 1971). It is now generally believed that the principal function of the type I acinus is secretion of a hyperosmotic fluid for use in water vapour uptake in unfed ticks (see below).

Elegant ultrastructural and light microscopic descriptions of granular acini in salivary glands of both argasid and ixodid ticks have been published (Roshdy & Coons, 1975; Binnington, 1978; Megaw & Beadle, 1979; Coons & Roshdy, 1981; Walker, Fletcher & Gill, 1985; Fawcett, Binnington & Voight, 1986; Gill & Walker, 1987). The type II acinus in ixodid ticks is composed of granular 'a', 'b', 'c$_1$–c$_4$' cells, based on morphology and reactions to specific stains. The type II acinus also contains agranular abluminal interstitial cells and a single adluminal interstitial cell. The acinus has a cuticular-lined acinar duct with valve and surrounding neck cells. The acinus increases greatly in size during feeding even though by the end of feeding most of the granules in granular cells are depleted. The overall organization of the type III acinus is similar to type II but with only three granular cell types, 'd', 'e' and 'f'. The type III acinus also increases greatly in size during tick feeding and the 'f' cell undergoes extensive transformation with increasing plasma membranes with associated mitochondria. The transformed 'f' cell and proliferating abluminal interstitial cells are believed to be responsible for secreting the bulk of the fluid across the salivary glands to facilitate concentration of the blood meal. In support of this, the 'f' cells in males, which imbibe much smaller blood meals than females, do not increase in size comparably to females (Coons & Lamoreaux, 1986). The

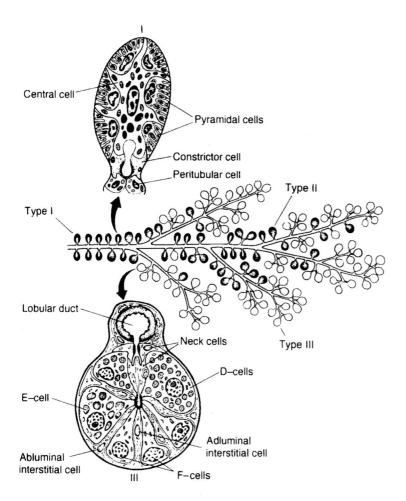

Fig. 3.2. Arrangement of the three acinus types of the ixodid tick female salivary gland (adapted from Binnington, 1978). The acinus types are distinguished by differential shading. Type I acini attach directly to the main salivary duct and to some principal branches of that duct. A diagrammatic section of a type I acinus with the cell types is shown (adapted from Barker *et al.*, 1984). Type II acini are more abundant in the proximal section and type III are more abundant in the distal section of the large branches. A schematic representation of the structure of the type III acinus and cell types from an unfed tick is depicted (from Fawcett, Binnington & Voight, 1986, with permission). The types II and IV acini have a similar organization to type III acini but they have different cell types. Type IV acini are found only in males. Type II and especially type III undergo remarkable and complex cytological transformation to meet the tick's changing physiological requirements (including fluid transport) during tick feeding (see Fawcett, Binnington & Voight, 1986). Nerve supply and tracheoles to the gland are not shown. (Reproduced from Sauer *et al.* (1996) with permission from CAB International.)

single adluminal interstitial cell in ixodid type II and III acini (Krolak *et al.*, 1982) is thought to function as a myoepithelial-like cell during tick feeding. During blood meal concentration, fluid accumulates in an expanding acinar lumen and is expelled by contraction of the adluminal cell winding in web-like fashion around the apical side of cells in acini II and III.

Salivary gland structure, development and degeneration in ixodid larvae and nymphs have been reviewed in Sauer *et al.* (1995). In general, acini types and cells are similar to those found in adults, except for the absence of type IV in larval ticks (Till, 1961; Chinery, 1965). A recent study of tick salivary gland development revealed that glands and ducts could be identified in 23-day-old embryo *I. ricinus* and

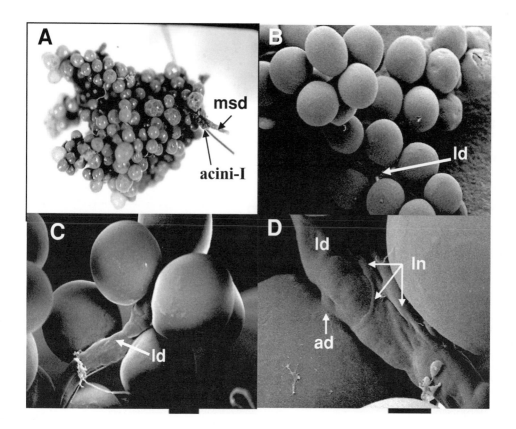

Fig. 3.3. Scanning electron micrographs of partially fed female *Ixodes ricinus* salivary glands. (A) Whole salivary gland showing the 'bunch of grapes' organization of the type II and III acini with the type I acini visible attached directly to the main salivary duct (**msd**). (B) and (C) Type II and III acini attached to the branching lobular salivary duct (**ld**). (D) Close-up of type II and III acini connected to a lobular salivary duct (**ld**) via the acinus duct (**ad**). Lobular nerves (**ln**) travel posteriorly down the lobular duct and innervate each acinus at the base of the acini where they control fluid, ion and protein secretion in the saliva by the release of dopamine.

that gland development was complete in 26-day-old embryos (Jasik & Buczek, 2004). Future studies on the development of embryonic salivary glands may give insight into the temporal and tissue movement of pathogens known to be efficiently transmitted transovarially (e.g. *Babesia* spp., *Rickettsia rickettsii*) and pathogens poorly transmitted transovarially (e.g. *Borrelia burgodorferi* sensu lato).

The salivary glands of argasid ticks consist of only two types of acini, A and B (Roshdy & Coons, 1975) or I and II (Coons & Roshdy, 1981; El Shoura, 1985). Type I (A) is similar in location and structure to the type I acinus of ixodid ticks. The type II (B) is granular, consisting of cell types 'a', 'b' and 'c'. A fourth type 'd' granular cell has been identified in the salivary glands of *Ornithodoros savignyi* (Mans *et al.*, 2004). The acinar lumen empties into a chitinous duct lacking the complex valvular structure of ixodid acini types II and III. Interstitial cells are present that develop complex, branched canaliculi with increased plasma membranes and mitochondria but not to the same extent as that seen in type III acini of ixodid ticks. The salivary glands of argasid ticks, in contrast to ixodid ticks, are surrounded by a myoepithelial sheath.

The salivary glands of ixodid females increase 25-fold in mass and protein content during tick feeding with acini types II and III exhibiting remarkable morphological change. Much of the granular material seen in salivary glands of unfed female ixodid ticks is absent from the salivary glands by the end of 7–14 days of feeding prior to repletion. Of note is the almost complete transformation of the type III acinus to a tissue with characteristics of a fluid-transporting epithelium.

The ixodid male salivary gland also increases in mass and protein content and exhibits expression of new genes during tick feeding but not to the same extent as that seen in females (Oaks *et al.*, 1991; Bior, Essenberg & Sauer, 2002). The presence or absence of females on a host had a marked effect of gene expression in the salivary glands of feeding males (Anyomi *et al.*, 2006). Perhaps these transcripts are present in the male-specific type IV acini and are involved in the process of transferring the spermatophore to the female which involves copious salivation by the male (Feldman-Muhsam, Borut & Saliternik-Givant, 1970).

SALIVARY GLAND DEGENERATION

Upon completion of the blood meal, the ixodid female tick drops to the ground, begins the slow regulated digestion of the blood meal, vitellogenesis and oocyte maturation; she oviposits and then dies. Within a few days of detachment, the glands degenerate with the appearance of notable autophagic whorls (Harris & Kaufman, 1981). This degeneration is neither necrotic nor pathological, but rather is a sequential and highly regulated physiological process of programmed cell death. Ecdysteroid treatment of glands in vitro or infusion into whole ticks causes gland degeneration (Harris & Kaufman, 1985). Haemolymph levels of ecdysteroids increase dramatically in the days immediately following detachment (Kaufman, 1991). An ecdysteroid-binding activity has been demonstrated in salivary glands of fed ticks (Mao, McBlain & Kaufman, 1995; Mao & Kaufman, 1998, 1999). Further, an ecdysteroid receptor (Guo *et al.*, 1997) and retinoid X receptor (Guo *et al.*, 1998) have been cloned and shown to be expressed in the salivary glands of *Amblyomma americanum*. Integumental tissue synthesizes ecdysteroid in response to a protein factor from the synganglion via an increase in cAMP levels (Lomas, Turner & Rees, 1997). A protein factor from the male gonad stimulates gland degeneration in virgin females (Lomas & Kaufman, 1992) and may act by targeting the synganglion to release the ecdysteroidogenic protein which subsequently acts at the integument. Recently, a combination of two recombinant proteins (termed 'voraxin') from an *A. hebraeum* male gonad cDNA library were shown to stimulate salivary gland degeneration in vivo (Weiss & Kaufman, 2004), though it awaits further investigation to establish if voraxin stimulates ecdysteroid synthesis by the integument. Taken together, these results suggest a model for the stimulus of tick salivary gland degeneration by which a competent ecdysteroid receptor complex in the gland responds to rising ecdysteroid levels at completion of the blood meal acquisition. This increased ecdysteroid sythesis is stimulated in some manner by a protein factor from the male gonad.

Though the endocrine stimulus for tick salivary gland degeneration has been delineated in some detail, the actual mechanism and machinery in this cell death is unknown. L'Amoreaux, Junaid & Trevidi (2003) reported significant DNA fragmentation in degenerating salivary glands of female *Dermacentor variabilis* using the *in situ* TUNEL technique. In *I. ricinus* salivary degeneration, we observed very limited TUNEL staining either in glands in vivo or cultured glands treated with ecdysteroid or several pro-apoptotic agents, though DNA laddering was obvious and degeneration was accompanied by increased caspase-3-like activity (B. Stewart & A. S. Bowman, unpublished data). Interestingly, the type I acini in *I. ricinus* did not exhibit any morphology of programmed cell death throughout the 12 days post-host observation period, during which period the cellular material of the type II and III acini was completely removed. This indicates that in *I. ricinus* the type I acini are likely to remain functional for water vapour uptake during the egg-laying period (>40 days). Similarly, DNA fragmentation and caspase-3-like activity occur in degenerating *Rhipicephalus (Boophilus) microplus* salivary glands (Freitas *et al.*, 2007) whilst the type I acini did not degenerate (Nunes, Mathias & Bechara, 2006). Recently, two caspase homologues were identified in *Haemaphysalis longicornis* most similar to caspase-2 and caspase-8 and localized in both the midgut and salivary glands (Tanaka *et al.*, 2007). The previous reports of caspase-3-like activity in *I. ricinus* and *B. microplus* in degenerating salivary glands were based on activity against relatively non-specific substrates and may well have been due to either of the two caspase homologues identified by Tanaka *et al.* (2007).

REGULATION OF FLUID SECRETION

The salivary glands are innervated with no evidence of control by hormones (Sauer, Essenberg & Bowman, 2000). Nerves from the synganglion impinge upon salivary gland cells with differing opinions on the number originating from the synganglion (Megaw, 1977; Binnington & Kemp, 1980; Kaufman & Harris, 1983; Fawcett, Binnington & Voight, 1986). The nerves travel posteriorly along the main salivary duct then divide into lobular nerves. They enter at the base of each acinus and terminate at synapses on glandular cells near the lumen (Fig. 3.3). Three types of synaptic vesicles have been identified in the axoplasm of nerve endings in the salivary glands: (i) large vesicles with granular content of moderate density that may contain peptides, (ii) small homogeneously dense vesicles that are thought to contain catecholamines, and (iii) non-granular vesicles (Fawcett, Binnington & Voight, 1986).

The neurotransmitter dopamine stimulates fluid secretion by isolated salivary glands maintained in culture medium and in vivo when injected into the haemocoel of partially fed female ticks (Kaufman, 1976; McSwain, Essenberg & Sauer, 1992). A dopamine, D1-like receptor has been found in the salivary glands of female *A. americanum* linked to activation of adenylate cyclase (Schmidt, Essenberg & Sauer, 1981, 1982). GTP and non-hydrolysable Gpp(NH)p stimulate adenylate cyclase in a washed particulate fraction of the salivary gland indicating a G-protein-coupled dopamine receptor (Sauer *et al.*, 1986). Adenylate cyclase, Na^+/K^+-ATPase activity and fluid secretion rates increase with increased tick feeding and are highest in salivary glands of mating, rapidly feeding females (reviewed by Sauer & Essenberg, 1984). Cyclic AMP also stimulates fluid secretion by isolated salivary glands maintained in culture medium (Needham & Sauer, 1975, 1979). Dopamine has been identified in the salivary glands and salivary gland nerves of *B. microplus* and *A. hebraeum* (Binnington & Stone, 1977; Kaufman & Harris, 1983). Finally, phenylethylamines stimulate adenylate cyclase and fluid secretion in the rank order of dopamine > noradrenaline > adrenaline > isoproterenol (Schmidt, Essenberg & Sauer, 1982; Kaufman & Wong, 1983). The increase in cAMP stimulates a salivary gland protein kinase which has been partially purified from salivary glands of *A. americanum* (Mane *et al.*, 1985; Mane, Sauer & Essenberg, 1988). Genes for catalytic subunit isoforms of cAMP-dependent protein kinase have been sequenced and localized in the salivary glands for *A. americanum* (Palmer *et al.*, 1999) and *A. hebraeum* (Tabish *et al.*, 2006). Incubation of salivary glands with dopamine or cAMP and theophylline (inhibitor of phosphodiesterase) increased the phosphorylation of at least 12 salivary gland proteins. The function and cellular location of the phosphoproteins are unknown (McSwain, Essenberg & Sauer, 1985; McSwain *et al.*, 1987). Overall, fluid secretion is controlled via a cAMP-dependent protein phosphorylation cascade following salivary gland stimulation by dopamine released from nerve endings.

Other receptors, however, and/or other pathways appear to be involved in controlling secretion. Dopamine antagonists spiperone, pimozide and haloperidol potentiate the stimulatory effect of dopamine without having a stimulatory effect on their own (Wong & Kaufman, 1981). The potentiating effect of spiperone is abolished by sulpiride. The authors postulated two possibilities to explain the results: (a) spiperone allosterically affects the dopamine receptor to increase secretion or (b) spiperone inhibits an inhibitory pathway that responds to dopamine. Gamma-aminobutyric acid (GABA) also potentiates the effect of dopamine on stimulating fluid secretion (Lindsay & Kaufman, 1986). GABA antagonists picrotoxin and biculline block GABA- and spiperone-induced potentiation of fluid secretion. GABA-induced potentiation of secretion by dopamine was also blocked by sulpiride. It was speculated that sulpiride and spiperone may interact with the GABA receptor. Salivary gland fluid secretion is also stimulated by ergot alkaloids (Kaufman & Wong, 1983). Ergot alkaloid stimulation is inhibited by sulpiride which is ineffective in reducing stimulation by dopamine. This was an unexpected result because ergot alkaloids interact with 5-hydroxytryptamine (5-HT) receptors but 5-HT has no effect on stimulating fluid secretion by isolated salivary glands (Needham & Sauer, 1975; Kaufman, 1977). Clearly, more work is required to understand fully all pathways involved in controlling fluid secretion. Overall, fluid secretion is controlled via a cAMP-dependent protein phosphorylation cascade following salivary gland stimulation by dopamine released from nerve endings. A major task remaining is to verify the cellular location and mechanism of fluid secretion in the salivary glands.

Salivation can be studied in vivo by injection of a dopamine solution into the haemocoel of partially fed ticks recently removed from the host. Alternatively, a solution of pilocarpine, a cholinomimetic agent, can be applied to the surface of the tick through which it is rapidly absorbed and causes the release of dopamine from the salivary nerves resulting in salivation. In vitro studies of isolated tick salivary gland secretion allow researchers to control conditions more than the in vivo approach and have led to most advances in our understanding of tick salivary gland secretion. The original in vitro method was a modification of the Ramsay technique utilized for studying fluid secretion of insect Malpighian tubules. Tick salivary glands are placed in a small volume of medium under liquid paraffin in a Petri dish and the main salivary duct drawn from the medium into the liquid paraffin. Agonists (e.g. dopamine) or antagonists can be added to the media causing changes in the rate of fluid secretion by the gland expressed as a droplet at the end of the salivary duct which can be measured with a microscope and ocular micrometer and converted to the volume secreted (e.g. Needham & Sauer, 1975). The alternative in vitro method is the ligated duct gravimetric method in which the main

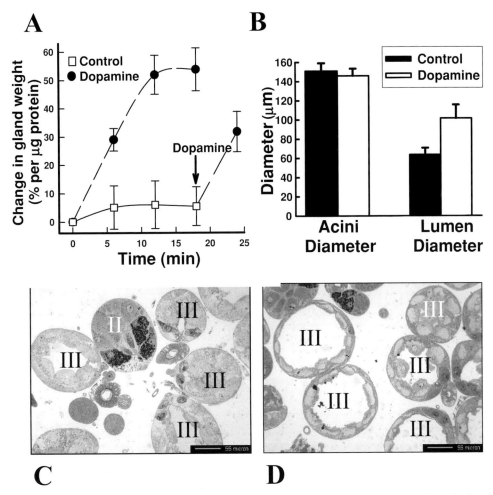

A

B

C

D

Fig. 3.4. Demonstration of dopamine-stimulated fluid uptake by partially fed female *Ixodes ricinus* salivary glands and involvement of type III acini. (A) The main salivary duct of isolated glands was ligated and the glands incubated in media (control) or media supplemented with 10 μM dopamine. The change in weight was recorded on an analytical balance at 6-min intervals (mean ± S.E.M., $n = 4$). Control glands were subjected to dopamine stimulation at 18 min (arrow) to check viability of these glands. (B) These ligated glands were then fixed (1% osmium tetroxide), processed, sectioned (semi-thin, 0.5 μm), and stained with toluidine blue. The diameter of the acini and the lumen were determined using Image-Pro Plus (Media Cybernetics). No difference was observed between treatments for the total acini size ($P > 0.05$), but the acini lumens were greater ($P < 0.05$) following dopamine treatment (Mean ± S.E.M., $n = 30$). (C) Type III acini in the ligated control glands show no or small lumens. A type II acinus with its more prominent secretory granules is also visible in this section. (D) The type III acini of dopamine-stimulated glands show large fluid filled lumens with the acini cells pushed back towards the basal lamima. In both C and D, scale bar = 55 μm.

salivary ducts of isolated tick glands are occluded by ligation with a very fine strand of silk or hair, placed in media containing the agents of interest and changes in weight recorded on an analytical balance (e.g. Harris & Kaufman, 1985). The ligation method measures fluid uptake by the gland rather than true secretion but has the advantage of allowing several glands to be studied simultaneously.

Recently, we investigated the stimulation of fluid uptake in isolated *I. ricinus* salivary glands by dopamine using the ligated duct technique (S. Whitelaw & A. S. Bowman, unpublished data). As expected, dopamine caused an increase in gland weight (≡ fluid uptake) (Fig. 3.4A). When the histology of these glands was examined, we observed that there was no increase in the acini diameter, but that the lumens of the

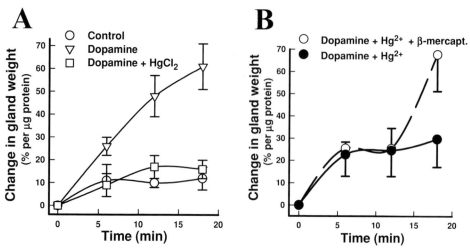

Fig. 3.5. Evidence for the role of a mercurial-sensitive aquaporin in fluid uptake by *I. ricinus* salivary glands. (A) Ligated salivary glands were incubated in either media (control), or media supplemented with 10 μM dopamine, or media supplemented with 10 μM dopamine and 10 μM mercuric chloride. Uptake of fluid was assessed gravimetrically, as described for Fig. 3.4. Mean ± S.E.M., $n = 9$. (B) Ligated glands were incubated in media containing 10 μM mercuric chloride for 10 min. Half the glands were transferred to media containing 10 μM dopamine with 10 μM mercuric chloride. The other half were transferred to media containing 10 μM dopamine with 10 μM mercuric chloride and 50 μM β-mercaptoethanol. Uptake of fluid was assessed gravimetrically, as described for Fig. 3.4. Mean ± S.E.M., $n = 4$.

type III acini of the dopamine-treated glands were greatly increased (Fig. 3.4B) with the cytoplasm of the cells forced back toward the basal lamina (Fig. 3.4C–D). It is believed that the adluminal interstitial cell in the type III acinus functions as a myoepithelial cell and contracts to eject the fluid from the expanded acinus into connecting ducts (Coons *et al.*, 1994; Lamoreaux, Needham & Coons, 1994). We observed no histological change in type I to implicate these acini in the fluid secretion, as has previously been suggested by some workers (e.g. Kirkland, 1971). These studies clearly demonstrate that the type III acini are indeed the acini involved with bulk fluid movement in tick salivary glands as has been hypothesized for 30 years.

Using the in vitro assay techniques, dopamine-stimulated secretion was shown to be inhibited by oubain (70%) and bafilomycin (25%) (McSwain *et al.*, 1997), indicating a role for a Na^+/K^+-ATPase and V-ATPase in supplying the electromotive force in salivary gland secretion or modification. But what is known about the fluid (i.e. water) transport in tick salivary glands? Historically, transport across biological lipid membranes was thought to be by simple diffusion, primarily driven by osmotic forces generated by ion fluxes. The landmark discovery of a water channel in human red blood cell membranes (Preston *et al.*, 1992) changed our understanding of water transport physiology. The water channels

were renamed aquaporins (AQPs) (for reviews see Borgnia *et al.*, 1999; Verkman, 2002). There are 13 (AQP0–12) well-characterized mammalian AQPs with additional AQPs found in fish, plants, insects, protists, yeast and bacteria. Many AQPs are inhibited by mercurial reagents.

Mammalian salivary glands contain several AQP types (for review see Ishikawa & Ishida, 2000). AQP1, 3 and 5 have been immunolocalized in human salivary glands: AQP5 (apical membrane) and AQP3 (basolateral membrane) in the salivary gland acinar cells and AQP1 in the endothelial cells and red blood cells associated with the glands (Gresz *et al.*, 2001). AQP5 is sensitive to Hg^{2+} mercury. AQP5 is shuttled from intracellular vesicles to the apical membrane within seconds in response to pilocarpine and adrenaline (Matsuzaki *et al.*, 1999; Ishikawa & Ishida, 2000) in a process that is dependent on $[Ca^{2+}]_i$ (intracellular) and $[Ca^{2+}]_e$ (extracellular). It should be noted that there are many similarities between the mouse and tick salivary gland: pilocarpine and adrenaline cause salivation in tick glands and this is dependent on Ca^{2+}.

We investigated evidence of a role for an AQP in *I. ricinus* salivary glands (S. Whitelaw & A. S. Bowman, unpublished data). Using the ligated salivary duct technique, the dopamine-stimulated secretion in *I. ricinus* salivary glands was completely inhibited by co-incubation with 10 μM $HgCl_2$ (Fig. 3.5A), but there was no difference in the

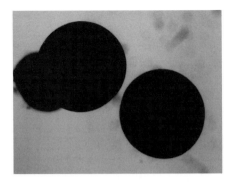

Fig. 3.6. Water permeability of an aquaporin cloned from *R. sanguineus* salivary glands and then expressed in *Xenopus laevis* oocytes. Control oocytes exhibit negligible water permeability in hypo-osmotic buffer (right) whereas oocytes expressing the tick AQP rapidly swell and explode (left).

adenylate cyclase activity in these glands (data not shown) indicating the inhibition of secretion by Hg^{2+} was not due to non-specific toxicity. Further, incubation of the Hg^{2+}-treated glands with 50 μM β-mercaptoethanol rescued their ability to secrete (Fig. 3.5B). Our results suggested that an Hg^{2+}-sensitive AQP might be involved in tick salivary gland water transport.

Using sequence data from an expressed sequence tag (EST) project on the salivary glands of *Rhipicephalus appendiculatus*, we recently sequenced two AQPs from the salivary glands of *R. sanguineus* (A. Ball & A. S. Bowman, unpublished data). *Xenopus* oocytes injected with cRNA coding for either of the *R. sanguineus* AQPs swelled and eventually burst when placed in hypo-osmotic media, whereas control oocytes exhibited negligible swelling (Fig. 3.6) demonstrating these two tick AQPs readily transport water and, thus, render the oocytes permeable. The permeability of oocytes expressing one of the *R. sanguineus* AQPs was greatly reduced by Hg^{2+} and may explain our findings of the Hg^{2+}-sensitivity in our secretion studies in *I. ricinus* (Fig. 3.5). Further investigation of AQPs in tick salivary glands should prove fruitful in our attempts to better understand the movement of water and other non-polar solutes into the saliva.

PROSTAGLANDINS IN TICK SALIVARY GLANDS AND SALIVA

Prostaglandins (PGs) are an important group of biologically active lipid molecules in mammals and invertebrates. Tick saliva and salivary glands contain unusually high amounts of PGE_2 and $PGF_{2\alpha}$. Prostaglandins of the 2-series (i.e. PGE_2, PGD_2, $PGF_{2\alpha}$, PGI_2) are derived from the polyunsaturated

fatty acid, arachidonate (AA; 20:4, *n*-6), which constitutes approximately 8% of the fatty acids in salivary glands of partially fed female *A. americanum* increasing from about 2% in unfed ticks (Shipley *et al.*, 1993*a*) and increasing more than any other fatty acid (Shipley *et al.*, 1993*b*). The increase in AA content in the salivary glands is not only a reflection of the AA in the blood meal because the amount in salivary glands is far higher than other internal tick tissues (Bowman *et al.*, 1993). Most animals are capable of synthesizing AA from essential fatty acid linoleate by a series of desaturation and elongation reactions. However, female *A. americanum* have an extremely degenerate fatty acid synthesis capability lacking the ability to desaturate or elongate fatty acids with more than one double bond (Bowman *et al.*, 1995*a*). All fatty acids in the salivary glands with more than one double bond were shown to be sequestered from the host blood meal. Radiolabelled AA when fed to *A. americanum* via a capillary tube was incorporated only into phosphatidylcholine and phosphatidylethanolamine in a ratio that was different from other fatty acids (Bowman *et al.*, 1995*b*).

AA is released by an intracellular phospholipase A_2 (PLA_2). Treatment of isolated salivary glands with Ca^{2+} ionophore A23187 increased the level of free AA (Bowman *et al.*, 1995*c*). A23187 increases free levels of intracellular Ca^{2+} and, typically, cytosolic PLA_2s are stimulated by Ca^{2+}. The increase in free AA was reduced in a dose-dependent manner by the PLA_2 inhibitor oleyloxyethylphosphorylcholine (OPC) indicating that the increased free AA was dependent upon the activity of PLA_2 (Bowman *et al.*, 1995*c*). Dopamine was also shown to increase free AA and its ability to do so was inhibited by verapamil suggesting that dopamine stimulates PLA_2 through the opening of a voltage-dependent Ca^{2+} channel. One of the difficulties associated with studying PGs in tick salivary glands has been the problem of showing biosynthesis of PGs by isolated salivary glands (Bowman *et al.*, 1995*d*; Pedibhotla, Sauer & Stanley-Samuelson, 1997) despite the inordinately high levels in saliva and the high levels of AA in the salivary glands. However, PG synthesis was shown in whole ticks when dopamine-induced saliva collected from ticks fed [³H]-AA was shown to contain radiolabelled PGs (Bowman *et al.*, 1995*d*). The attempts to demonstrate synthesis in vitro used a labelled substrate and standard incubation conditions employed for mammalian tissues. Robust synthesis by isolated salivary glands was demonstrated at high concentrations of exogenous AA (Aljamali *et al.*, 2002) that previously had not been employed (Bowman *et al.*, 1995*d*; Pedibhotla, Sauer & Stanley-Samuelson, 1997). The reasons why a high AA concentration is required are unknown but one possibility to explain the results could be that salivary glands have a very high capacity for the uptake and incorporation of AA into phospholipids by isolated salivary glands (Bowman et al., 1995d; Madden et al., 1996).

FUNCTIONS OF PROSTAGLANDINS IN SALIVARY GLAND PHYSIOLOGY

As noted, the saliva of several tick species contains extremely high levels of PGs and much higher levels than those found in mammalian inflammatory exudates (Bowman, Dillwith & Sauer, 1996). Prostaglandins of the 2-series exhibit antihaemostatic, vasodilatory, immunosuppressive and anti-inflammatory activities but the exact functions of PGs in tick saliva are presumed and await further study. In contrast, much is known about the physiology of PGE_2 in tick salivary gland secretion (Sauer et al., 2000).

Incubation of isolated salivary glands of partially-fed female A. americanum with the PLA_2 inhibitor OPC reduced dopamine-induced fluid secretion and cAMP levels in the glands (Qian et al., 1997). Inhibition of secretion by OPC was reversed by PGE_2 and its analogue, 17-phenyl-trinor PGE_2. The results suggested that PGE_2 has an autocrine or paracrine effect in modulating tick salivary gland secretion. This hypothesis was confirmed by identification of a PGE_2-specific receptor in the plasma membrane fraction which exhibits a single, high affinity PGE_2-binding site that is saturable, reversible, specific for PGE_2 and coupled to a cholera-toxin-sensitive guanine nucleotide regulatory protein (Qian et al., 1997). PGE_2 does not stimulate adenylate cyclase activity in isolated salivary gland membranes (Qian et al., 1998) suggesting that the noted PGE_2 effects on cAMP levels and fluid secretion are indirect, probably through its ability to mobilize intracellular Ca^{2+}. PGE_2 stimulates an increase in intracellular inositol tris-phosphate (IP_3) and stimulates an efflux of $^{45}Ca^{2+}$ from pre-loaded salivary gland acini (Qian et al., 1998). Protein secretion from dispersed salivary gland acini was shown to be specific for PGE_2, as compared to $PGF_{2\alpha}$ or the thromboxane analogue U-46619, in accordance with their binding affinities for the PGE_2 receptor (Yuan et al., 2000). In mammals, PGE_2 receptors are classified into four subtypes: EP1, EP2, EPS, EP4 (Coleman et al., 1990; Negishi, Sugimoto & Ichikawa, 1993). EP1 receptors mobilize Ca^{2+} and the other receptors affect adenylate cyclase (Coleman et al., 1990). It appears that the PGE_2 receptor in tick salivary glands is EP1-like. The mammalian PGE_2 EP1 receptor agonist 17-phenyl-trinor PGE_2 was as effective as PGE_2 in stimulating secretion of anticoag-

ulant protein by dispersed salivary glands (Yuan et al., 2000). This finding is consistent with the previous finding that 17-phenyl-trinor PGE_2 partially reverses the small inhibition of dopamine-stimulated salivary gland fluid secretion by PLA_2 inhibitor OPC. PGE_2 and the non-hydrolysable analogue of GTP, GTPγS, were shown to activate phospholipase C (PLC) directly in a membrane-enriched fraction of the salivary glands (Yuan et al., 2000). TMB-8, an inhibitor of IP3 receptors on the endoplasmic reticulum, inhibited PGE_2-stimulated secretion of anticoagulant protein. Overall, the results support the hypothesis that PGE_2 stimulates secretion of tick salivary gland protein via a phosphoinositide signalling pathway and mobilization of intracellular Ca^{2+} (Yuan et al., 2000).

A model was proposed whereby dopamine released at the neuroeffector junction binds to a dopamine, D1-like receptor on the salivary glands to effect fluid secretion (Sauer et al., 2000) (Fig. 3.7). The dopamine receptor is coupled to a G-protein to activate adenylate cyclase and increase cAMP. The increase in cAMP controls fluid secretion (Sauer et al., 1986). Dopamine also opens a voltage-gated Ca^{2+} channel allowing an influx of extracellular Ca^{2+} to stimulate a cytosolic PLA_2 to liberate free AA. The AA is then converted by the cyclooxygenase pathway to PGE_2 and $PGF_{2\alpha}$ and small amounts of PGI_2 (Aljamali et al., 2002). High levels of PGE_2 and $PGF_{2\alpha}$ are secreted into the host via saliva (Ribeiro et al., 1992; Aljamali et al., 2002). In addition, PGE_2 interacts with an EPl-like receptor coupled to a G-protein to activate PLC and increase intracellular IP_3 (Qian et al., 1997, 1998; Yuan et al., 2000). The increased Ca^{2+} mobilized by IP_3 regulates exocytosis of anticoagulant protein from secretory vesicles into the saliva.

The mechanisms for intracellular trafficking of secretory vesicles and exocytosis are complex; however, highly conserved SNARE complex proteins have been identified in all neuronal and non-neuronal secretory cells studied. SNARE terminology is based upon the requirement of N-ethylmaleimide (NEM-sensitive fusion protein – NSF) and soluble NSF attachment proteins (SNAPs) involvement in the process. SNAREs are defined as the receptors for SNAPs. SNARE proteins associated with vesicles are referred to as v-SNAREs and those associated with the plasma membrane as t-SNAREs. NSF uses energy from ATP hydrolysis to dissociate SNARE complexes after membrane fusion, enabling SNARE proteins to be recycled for subsequent rounds of fusion (May, Whiteheart & Weis, 2001). GTP-binding proteins (e.g. Rab 3A) are localized to the cytosolic face of specific intracellular secretory vesicle membranes where they

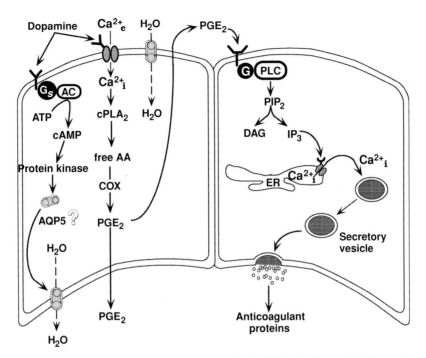

Fig. 3.7. Known and hypothesized factors and events controlling secretion in ixodid female salivary glands. Dopamine released at the neuroeffector junction interacts with a dopamine (D1 subtype) receptor coupled to a stimulatory guanine nucleotide protein (G_s) that activates adenylate cyclase leading to increased levels of cAMP. Numerous gland proteins are phosphorylated by a cAMP-dependent protein kinase. We speculate a water channel protein (AQP5-like?) is phosphorylated, moves to the cell membrane and is responsible for fluid transport observed in dopamine- and cAMP-stimulated glands. Dopamine also opens a voltage-gated Ca^{2+} channel allowing an influx of extracellular Ca^{2+} (Ca^{2+}_e), thus increasing intracellular Ca^{2+} (Ca^{2+}_i) levels and stimulating a cytosolic phospholipase A_2 (cPLA$_2$). The free arachidonic acid (AA) liberated by the cPLA$_2$ is converted by the cyclooxygenase pathway (COX) into a variety of prostaglandins, including PGE$_2$. High levels of PGE$_2$ are secreted into the host via saliva. Additionally, on the same or neighbouring cells, PGE$_2$ interacts with an EP1-subtype receptor coupled to a guanine nucleotide protein (G) which activates phospholipase C (PLC) and generates diacylglycerol (DAG) and IP$_3$. The IP$_3$ interacts with and opens the IP$_3$-receptor channel on the endoplasmic reticulum (ER) releasing Ca^{2+}_i into the cytosol. The increased Ca^{2+}_i appears to regulate the exocytosis of anticoagulant protein from secretory vesicles into the saliva. Reprinted from Sauer, Essenberg & Bowman (2000) *Journal of Insect Physiology*, with permission from Elsevier.

are thought to function in vesicle trafficking. Additional components exist that inhibit fusion of vesicle and plasma membranes until the cell receives an appropriate signal, typically a rise in intracellular Ca^{2+}. A major Ca^{2+}-signalling protein is synaptotagmin (Geppert *et al.*, 1994). Proteins in the salivary glands of partially fed female *A. americanum* cross-react individually with antibodies to SNARE complex proteins synaptobrevin-2, syntaxin-1A, syntaxin-2 and SNAP-25, synaptotagmin, Rab 3A and nSec1 (Karim *et al.*, 2002). Significantly, antibodies to these SNARE complex proteins inhibit PGE$_2$-stimulated secretion of anticoagulant protein in permeabilized tick salivary glands (Karim *et al.*, 2002). Further, employing the technique of double-stranded

RNA interference (dsRNAi) to target and 'knock down' the v-SNARE protein synaptobrevin, PGE$_2$-stimulated protein secretion was inhibited in isolated *A. americanum* salivary glands (Karim *et al.*, 2004a).

When salivary gland nSec1 mRNA transcript and protein expression levels were reduced by dsRNAi, protein exocytosis was also inhibited and, importantly, repletion weight was modestly reduced by 12%, but the time to attain repletion was extended by 33% (Karim *et al.*, 2004b). Further, dsRNAi knockdown of synaptobrevin in female *A. americanum* caused death or detachment of ticks so that only 20% attained repletion and those that did reach repletion had body weights of only 33% those of control ticks (Karim *et al.*, 2005). The results strongly suggest that SNARE and cell trafficking regulatory proteins are present and functioning in the process of PGE_2-stimulated, Ca^{2+}-regulated protein secretion in tick salivary glands. The control and mechanism of vesicle transport are likely to be important for pathogen transmission. Lyme disease spirochaetes are believed to move across cells by a transcytotic mechanism, entering by endocytosis and, exiting by exocytosis (Hechemy et al., 1992; Kurtti et al., 1994). It seems likely that pathogen delivery to the host may be facilitated by the tick's mechanisms of transport.

The control mechanisms of fluid and protein secretion in argasid salivary glands are much less well studied than those of ixodid ticks. However, recent studies on the argasid Ornithodoros savignyi salivary glands followed the release of apyrase as the marker protein and, as seen in ixodids, exocytosis was stimulated by dopamine and involved verapamil-sensitive Ca^{2+} influx, but, in contrast to ixodids, PGE_2 had no effect on the protein exocytosis in O. savignyi glands (Maritz-Olivier, Louw & Neitz, 2005)

FUNCTION OF THE SALIVARY GLANDS IN UNFED TICKS

The ability of unfed ticks to survive prolonged periods without feeding is a hallmark of tick biology and an attribute that is assisted by the tick's ability to absorb water from unsaturated air (Needham & Teel, 1991). The lowest relative humidity (RH) at which water vapour uptake is possible in ticks and a few other arthropods is defined by the critical equilibrium humidity (CEH). In most unfed ticks the CEH is in the range of 85% to 90% RH. In a simple but elegant study, Rudolph & Knulle (1974) identified the mouth as the site of water vapour uptake and the salivary glands as a likely source of hygroscopic material secreted into the oral cavity. Fluid in the salivarium of rehydrating A. americanum females was shown to have a melting point of −10 to −12 °C; this indicates a solution whose water molecules would be in equilibrium at an RH of ≈90% (Sigal, Needham & Machin, 1991). These authors postulated that as ticks become dehydrated, the type I acini secrete salts from increasingly concentrated haemolymph onto the mouthparts until rehydration and subsequent reingestion is possible. The type I acinus is probably the source of the hygroscopic material. Megaw & Beadle (1979) compared the ultrastructural features of the 'pyramidal' cells to those of cells found in nasal salt glands of birds and reptiles. Type I acini are found in argasids and ixodids and in larvae, nymphs and adults all of which possess the ability to take water from unsaturated air. Rudolph

& Knulle (1979) noted that heavy infestations of the protozoan Theileria annulata damaged the acini types II and III of Hyalomma anatolicum and left type I intact. Infected ticks were still capable of taking up water at 93% RH. More recently, the wholly solute-driven mechanism for explaining water vapour uptake was questioned by Gaede & Knulle (1997). The authors demonstrated that water from unsaturated air condenses on the hydrophilic cuticle on the hypostome of A. variegatum. The authors hypothesized that only a slightly hyperosmotic secretion from type I acini would be sufficient to change the water affinity at the adsorbing cuticle surface and release adsorbed water. The solution would then be drawn into the mouth by suction of the pharynx. The authors further suggested that another mechanism occurs in unfed ticks with a large water deficit. Here type I acini secrete ions which are removed from the haemolymph and produce hyperosmotic secretion (consistent with storage excretion as proposed by Needham & Teel, 1986). A crystalline substance remains and when the humidity increases above the CEH, the substance dissolves and is ingested, providing another means for recovering lost water (Gaede & Knulle, 1997).

ROLE OF SALIVARY GLANDS IN PATHOGEN TRANSMISSION

The salivary glands play a critical role in pathogen transmission at several levels. Factors in the saliva or salivary gland homogenate are known to greatly enhance pathogen transmission and establishment and are discussed elsewhere in this book (see chapter 10). For many pathogens the salivary gland is the site of development and replication. The following accompanying chapters give details on specific pathogens: Anaplasma marginale (Chapter 15), Theileria spp. (Chapter 14); Babesia spp. (Chapter 13) and Borrelia burgdorferi (Chapter 11). For some pathogens (e.g. viruses), there appears to be little cell specificity either within the salivary glands or the tick body per se. Conversely, some protozoan pathogens have high specificity. For example, Theileria annulata only infects cell type 'e' of type III acini: the mechanism or function of this high specificity is unknown, but elucidating the pathogen–gland ligands may well uncover effective transmission-blocking targets. In this respect, research on tick salivary glands would greatly benefit from the development of methods for specific cell type isolation or cell lines.

The application of genetic manipulation of tick-borne pathogens for the characterization of ligand–receptor interactions at the arthropod–pathogen interface was reported by

Pal *et al.* (2004) and powerfully demonstrated what is now possible in this field. The outer surface protein C (OspC) of *Borrelia burgdorferi sensu stricto* bound strongly to *I. scapularis* salivary gland both in an ELISA-like assay using gland homogenate and to whole glands. An OspC-deficient mutant *B. burgdorferi* did not bind to the salivary glands, but the OspC-deficient strain reconstituted to contain a plasmid encoding wild-type OspC did bind to salivary glands. It was shown in vivo that OspC was required for invasion of the salivary gland and subsequent transmission to mice. Further, IgG F(ab$_2$) fragments to OspC were effective in blocking *B. burgdorferi* colonization of the salivary glands (Pal *et al.*, 2004). A similar approach was employed to show that OspC is required for *B. afzelii*'s migration from the *I. ricinus* midgut and colonization of the salivary gland (Fingerle *et al.*, 2007). Though the experiment was not performed, one would anticipate that antibodies directed to the salivary gland 'receptor' that binds OspC would also effectively block entry of *Borrelia* into the glands. These studies demonstrate both the intricacy and specificity of the interaction between the pathogen and the salivary gland and also the potential for interrupting these receptor–ligand interactions to block transmission once either receptor or ligand is identified.

Recently, it was reported that *B. burgdorferi* caused a 13-fold increase in the expression of an *I. scapluaris* salivary gland protein (Salp15), a T-cell immunosuppressant (Ramamoorthi *et al.*, 2005). The effect of *B. burgdorferi* on the salivary gland appeared not to be a global increase in gene expression, but specific to Salp15. Within the salivary gland, *B. burgdorferi* coats itself with Salp15 via binding to OspC which greatly increased its establishment in the host following transmission (Ramamoorthi *et al.*, 2005). A similar scenario occurs for the causative agent of human anaplasmosis in which the rickettsial pathogen *Anaplasma phagocytophilum* induces the specific expression of Salp16 in the salivary gland of *I. scapularis* during feeding (Sukumaran *et al.*, 2006). There is now clear evidence that at least two tick-borne pathogens cause specific expression of proteins by the tick salivary glands, but as yet it is not known which acinus type or cell type is being orchestrated by the pathogens nor by what mechanism.

FUTURE DIRECTIONS

Tick salivary glands have been studied by various groups worldwide for over 40 years reflecting the huge importance of this organ in the transmission of various pathogens to humans and domestic animals. With the information garnered over this period as a platform, what are the likely fruitful areas and future directions in tick salivary gland research? Undoubtedly, the great advances in molecular biology technologies will have a tremendous effect on our understanding of tick salivary gland physiology. There have been several EST projects on salivary glands of various tick species (reviewed in Chapter 4). The primary reason for these salivary gland projects was driven by the interest in the secreted components of the salivary gland transcriptome that allow ticks to overcome host defence mechanisms and to obtain the prolonged blood meal (see Chapter 4). However, these EST projects also yield important sequence information on genes involved in salivary gland cell biology. Now we have so much salivary gland sequence information available that we can already look forward to the 'post-genomic era' and the era of 'functional genomics'. It is here where recently acquired sequence data and the information on salivary gland physiology and cell biology can be brought together to great synergistic benefit. The new sequence information can fill in gaps where anticipated components were implicated by previous functional assays. An example would be the finding of an aquaporin partial sequence in the *R. appendiculatus* EST project. Sauer *et al.* (2000) had previously predicted a tick salivary gland aquaporin (Fig. 3.7) and functional data had been acquired for a role of an aquaporin in fluid secretion by isolated *I. ricinus* salivary glands (Fig. 3.5). This has led to the initial characterization of aquaporins from *R. sanguineus* salivary glands (Fig. 3.6).

It is possible to suppress or 'knock down' the expression of selected genes in tick salivary glands both in vivo and in isolated glands by the technique of dsRNAi. Whilst tick researchers do not have the luxury of gene knockout strains of the *Drosophila* and *C. elegans* research communities, the dsRNAi technique does allow us to assess the importance of the gene in question in these 'phenotypes'. The first studies using dsRNAi in tick salivary glands were on the secreted salivary components of a histamine-binding protein (Aljamali *et al.*, 2003) and an anticoagulant (Narasimhan *et al.*, 2004), but the mechanism of PGE$_2$-stimulated protein exocytosis has also been studied by dsRNAi knockdown of a synaptobrevin homologue (Karim *et al.*, 2004a, b, 2005). It is very likely that the dsRNAi knockdown approach will be extensively employed in tick salivary gland research in the future. As we look to the salivary gland as a potential 'Achilles' heel' for acaracide development, the importance, or otherwise, of postulated targets or the existence of unforeseen salvage pathways could be assessed by dsRNAi experiments. Likewise, development of potential vaccines able to block the binding and uptake of pathogens to the salivary gland can be tested by dsRNAi knockdown of the ligand–receptor.

It must be noted that despite the many tick salivary gland ESTs sequenced, there will still be a large requirement for standard gene hunting and identification studies for specific genes of interest. For example, throughout the available tick salivary gland EST databases and the ongoing *I. scapularis* genome project to date, there are extremely few G-protein-coupled receptors identified or transcripts that contain any of the characteristic motifs. Specifically, the dopamine receptor and PGE$_2$ receptor still remain elusive. Potentially rare transcripts such as these receptors might not be expected to be sequenced in EST projects, but it is surprising that no prostaglandin synthase (or cyclooxygenase) gene has been identified in the EST projects when one considers the exceedingly high quantities of prostaglandin detected in the saliva or salivary gland homogenate. We have tried several approaches to clone and sequence the dopamine receptor and prostaglandin synthase without success. It is feasible that these gene sequences are already in the EST databases, but are so different from the expected sequences that the current bioinformatic programs do not recognize them as such. This would explain the difficulty of homology cloning we have encountered. Whilst this is frustrating on the one hand, it would indicate that such divergent genes might prove attractive acaracide targets, as well proving academically interesting.

Overall, the explosion of tick salivary gland transcript sequence information that has become available over the past 10 years can be combined with fundamental information about tick salivary gland physiology that has been gained over the past 45 years in a synergistic manner. The glands are vital to tick survival both on and off the host and are critical to pathogen development and transmission and, thus, present a target for acaracide development and pathogen transmission-blocking vaccines. The future for tick salivary gland research will now move into functional genomics utilizing the available sequence data and techniques such as dsRNAi to elucidate critical pathways and promising targets. It would appear that we are entering perhaps the most exciting time in tick salivary gland physiology in which tremendous advances are within our grasp.

ACKNOWLEDGEMENTS

This work was supported in part by the Leverhulme Trust (grant F/00152/C to ASB), the BBSRC (grant BB/C517833/1 to ASB) and the National Science Foundation (grant IBN 9974299 to JRS). We also thank the Wellcome Trust for a Vacation Scholarship (grant to ASB) to fund Sandra Whitelaw's aquaporin work.

REFERENCES

Aljamali, M. N., Bior, A. D., Sauer, J. R. & Essenberg, R. C. (2003). RNA interference in ticks: a study using histamine binding protein dsRNA in the female tick *Amblyomma americanum*. *Insect Molecular Biology* **12**, 299–305.

Aljamali, M. N., Bowman, A. S., Dillwith, J. W., *et al.* (2002). Identity and synthesis of prostaglandins in the lone star tick, *Amblyomma americanum* (L.), as assessed by radio-immunoassay and gas chromatography/mass spectrometry. *Insect Biochemistry and Molecular Biology* **32**, 331–341.

Anyomi, F. M., Bior, A. D., Essenberg, R. C. & Sauer, J. R. (2006). Gene expression in male tick salivary glands is affected by feeding in the presence of females. *Archives of Insect Biochemistry and Physiology* **63**, 159–168.

Barker, D. M., Ownby, C. L., Krolak, J. M., Claypool, P. L. & Sauer, J. R. (1984). The effects of attachment, feeding and mating on the morphology of the type I alveolus of the salivary glands of the lone star tick, *Amblyomma americanum* (L.). *Journal of Parasitology* **70**, 99–113.

Binnington, K. C. (1978). Sequential changes in salivary gland structure during attachment and feeding of the cattle tick, *Boophilus microplus*. *International Journal for Parasitology* **8**, 97–115.

Binnington, K. C. & Kemp, D. H. (1980). Role of tick salivary glands in feeding and disease transmission. *Advances in Parasitology* **18**, 315–319.

Binnington, K. C. & Stone, B. F. (1977). Distribution of catecholamines in the cattle tick *Boophilus microplus*. *Comparative Biochemistry and Physiology* **58C**, 21–28.

Bior, A. D., Essenberg, R. C. & Sauer, J. R. (2002). Comparison of differentially expressed genes in the salivary glands of male ticks, *Amblyomma americanum* and *Dermacentor andersoni*. *Insect Biochemistry and Molecular Biology* **32**, 645–655.

Borgnia, M., Nielsen, S., Engel, A. & Acre, P. (1999). Cellular and molecular biology of the aquaporin water channels. *Annual Review of Biochemistry* **68**, 425–458.

Bowman, A. S., Dillwith, J. W., Madden, R. D. & Sauer, J. R. (1995*b*). Uptake, incorporation and redistribution of arachidonic acid in isolated salivary glands of the lone star tick. *Insect Biochemistry and Molecular Biology* **25**, 441–447.

Bowman, A. S., Dillwith, J. W., Madden, R. D. & Sauer, J. R. (1995*c*). Regulation of free arachidonic acid levels in isolated salivary glands from the lone star tick: a role for dopamine. *Archives of Insect Biochemistry and Physiology* **29**, 309–327.

Bowman, A. S., Dillwith, J. W. & Sauer, J. R. (1996). Tick salivary prostaglandins: presence, origin and significance. *Parasitology Today* **12**, 388–396.

Bowman, A. S., Sauer, J. R., Neese, P. A. & Dillwith, J. W. (1995*a*). Origin of arachidonic acid in the salivary glands of the lone star tick, *Amblyomma americanum*. *Insect Biochemistry and Molecular Biology* **25**, 225–233.

Bowman, A. S., Sauer, J. R., Shipley, M. M., *et al.* (1993). Tick salivary prostaglandins: their precursors and biosynthesis. In *Host-Regulated Developmental Mechanisms in Vector Arthropods,* vol. 3, eds. Borovsky, D. & Spielman, A., pp. 169–177. Vero Beach, FL: University of Florida Institute of Food and Agriculture.

Bowman, A. S., Sauer, J. R., Zhu, K. & Dillwith, J. W. (1995*d*). Biosynthesis of salivary prostaglandins in the lone star tick, *Amblyomma americanum*. *Insect Biochemistry and Molecular Biology* **25**, 735–741.

Chinery, W. A. (1965). Studies on the various glands of the tick, *Haemophysalis spingera* Neumann 1987. III. The salivary glands. *Acta Tropica* **22**, 321–349.

Coleman, R. A., Kennedy, I., Humphrey, P. P. A., Bunce, K. & Lumley, P. (1990). Prostanoids and their receptors. In *Comprehensive Medicinal Chemistry*, vol. 3, *Membranes and Receptors*, ed. Emmett, J. C., pp. 643–714. Oxford, UK: Pergamon Press.

Coons, L. B. & Alberti, G. (1999). The Acari – Ticks. In *Microscopic Anatomy of Invertebrates*, vol. 8B, *Chelicerate Arthropoda*, eds. Harrison, F. W. & Foelix, R., pp. 267–514. New York: Wiley-Liss.

Coons, L. B. & Lamoreaux, W. J. (1986). Developmental changes in the salivary glands of male and female *Dermacentor variabilis* (Say) during feeding. In *Host-Regulated Developmental Mechanisms in Vector Arthropods*, vol. 1, eds. Borovsky, D. & Spielman, A., pp. 86–92. Vero Beach, FL: University of Florida, Institute of Food and Agriculture.

Coons, L. B., Lessman, C. A., Ward, M. W., Berg, R. H. & Lamoreaux, W. J. (1994). Evidence of a myoepithelial cell in tick salivary glands. *International Journal for Parasitology* **24**, 551–562.

Coons, L. B. & Roshdy, M. A. (1981). Ultrastructure of granule secretion in salivary glands of *Argas (Persicargas) arboreus* during feeding. *Zeitschrift für Parasitenkunde – Parasitology Research* **65**, 225–234.

El Shoura, S. (1985). Ultrastructure of salivary glands of *Ornithodorus (Ornithodorus) moubata* (Ixodoidae: Argasidae). *Journal of Morphology* **186**, 45–52.

Fawcett, D. W., Binnington, K. C. & Voight, W. R. (1986). The cell biology of the ixodid tick salivary gland. In *Morphology, Physiology and Behavioral Biology of Ticks*, eds. Sauer, J. R. & Hair, J. A., pp. 22–45. Chichester, UK: Ellis Horwood.

Feldman-Muhsam, B., Borut, S. & Saliternik-Givant, S. (1970). Salivary secretion of the male tick during copulation. *Journal of Insect Physiology* **16**, 1945–1949.

Fingerle, V., Goettner, G., Gern, L., Wilske, B. & Schulte-Spechtel, U. (2007). Complementation of *Borrelia afzelii* OspC mutant highlights the crucial role of OspC for dissemination of *Borrelia afzelii* in *Ixodes ricinus*. *International Journal of Medical Microbiology* **287**, 97–107.

Freitas, D. R. J., Rosa, R. M., Moura, D. J., *et al.* (2007). Cell death during preoviposition period in *Boophilus microplus* tick. *Veterinary Parasitology* **144**, 321–327.

Gaede, H. & Knulle, W. (1997). On the mechanism of water vapour sorption from unsaturated atmospheres by ticks. *Journal of Experimental Biology* **200**, 1491–1498.

Geppert, M., Goda, Y., Hammer, R. E., *et al.* (1994). Synaptotagmin I: a major calcium sensor for transmitter release at a central synapse. *Cell* **79**, 717–727.

Gill, H. S. & Walker, A. R. (1987). The salivary glands of *Hyalomma anatolicum anatolicum*: structural changes during attachment and feeding. *International Journal for Parasitology* **17**, 1381–1392.

Gregson, J. D. (1967). Observations on the movement of fluids in the vicinity of the mouthparts of naturally feeding *Dermacentor andersoni* Stiles. *Parasitology* **57**, 1–8.

Gresz, V., Kwon, T. H., Hurley, P. T., *et al.* (2001). Identification and localization of aquaporin water channels in human salivary glands. *American Journal of Physiology* **281**, G247–G254.

Guo, X., Harmon, M. A., Laudet, V., Mangelsdorf, D. J. & Palmer, M. J. (1997). Isolation of a functional ecdysteroid receptor homologue from the ixodid tick, *Amblyomma americanum* (L.). *Insect Biochemistry and Molecular Biology* **27**, 945–962.

Guo, X., Xu, Q., Harmon, M., *et al.* (1998). Isolation of two functional retinoid X receptor subtypes from the ixodid tick, *Amblyomma americanum* (L.). *Molecular and Cellular Endocrinology* **139**, 45–60.

Harris, R. A. & Kaufman, W. R. (1981). Hormonal control of salivary gland degeneration in the ixodid tick, *Amblyomma hebraeum*. *Journal of Insect Physiology* **27**, 241–248.

Harris, R. A. & Kaufman, W. R. (1985). Ecdysteroids: possible candidates for the hormone which triggers salivary gland degeneration in the tick, *Amblyomma hebraeum. Experientia* **41**, 740–741.

Hechemy, K. E., Samsonoff, W. A., Harris, H. L. & McKee, M. (1992). Adherence and entry of *Borrelia burgdorferi* in Vero cells. *Journal of Medical Microbiology* **36**, 229–238.

Hillyard, P. D. (1996). *Ticks of North-West Europe*. London: Linnean Society.

Howell, C. J. (1966). Collection of salivary gland secretion from the argasid, *Ornithodoros savignyi* Audouin (1827) by use of a pharmacological stimulant. *Journal of the South African Veterinary Medical Association* **37**, 36–39.

Hsu, M.-H. & Sauer, J. R. (1975). Ion and water balance in the feeding lone star tick. *Comparative Biochemistry and Physiology* **52A**, 269–276.

Ishikawa, Y. & Ishida, H. (2000). Aquaporin water channel in salivary glands. *Japanese Journal of Pharmacology* **83**, 95–101.

Jasik, K. & Buczek, A. (2004). Development of the salivary glands in embryos of *Ixodes ricinus* (Acari: Ixodidae). *Experimental and Applied Acarology* **32**, 219–229.

Karim, S., Essenberg, R. C., Dillwith, J. W., et al. (2002). Identification of SNARE and cell trafficking regulatory proteins in salivary glands of the lone star tick. *Insect Biochemistry and Molecular Biology* **32**, 1711–1721.

Karim, S., Miller, N. J., Valenzuela, J., Sauer, J. R. & Mather, T. N. (2005). RNAi-mediated gene silencing to assess the role of synaptobrevin and cystatin in tick blood feeding. *Biochemical and Biophysical Research Communications* **334**, 1336–1342.

Karim, S., Ramakrishnan, V. J., Tucker, J. S., Essenberg, R. C. & Sauer, J. R. (2004a). *Amblyomma americanum* salivary glands: double stranded RNA-mediated gene silencing of synaptobrevin homologue and inhibition of PGE_2 stimulated protein secretion. *Insect Biochemistry and Molecular Biology* **34**, 407–413.

Karim, S., Ramakrishnan, V. J., Tucker, J. S., Essenberg, R. C. & Sauer, J. R. (2004b). *Amblyomma americanum* salivary gland homolog of nSec1 is essential for saliva protein secretion. *Biochemical and Biophysical Research Communications* **324**, 1256–1263.

Kaufman, W. R. (1976). The influence of various factors on fluid secretion by *in vitro* salivary glands of ixodid ticks. *Journal of Experimental Biology* **64**, 727–742.

Kaufman, W. R. (1977). The influence of adrenergic agonists and their antagonists on isolated salivary glands of ixodid ticks. *European Journal of Pharmacology* **45**, 61–68.

Kaufman, W. R. (1991). Correlation between haemolymph ecdysteroid titre, salivary gland degeneration and ovarian development in the ixodid tick, *Amblyomma hebraeum* Koch. *Journal of Insect Physiology* **37**, 95–99.

Kaufman, W. R. & Harris, R. A. (1983). Neural pathways mediating salivary fluid secretion in the ixodid tick, *Amblyomma hebraeum. Canadian Journal of Zoology* **61**, 1976–1980.

Kaufman, W. R. & Phillips, J. E. (1973). Ion and water balance in the ixodid tick *Dermacentor andersoni*. I. Routes of ion and water excretion. *Journal of Experimental Biology* **58**, 523–536.

Kaufman, W. R. & Wong, D. L. P. (1983). Evidence for multiple receptors mediating fluid secretion in salivary glands of ticks. *European Journal of Pharmacology* **87**, 43–52.

Kirkland, W. L. (1971). Ultrastructural changes in the nymphal salivary glands of the rabbit tick, *Haemaphysalis leporispalustris* during feeding. *Journal of Insect Physiology* **17**, 1933–1946.

Krolak, J. M., Ownby, C. L. & Sauer, J. R. (1982). Alveolar structure of salivary glands of the lone star tick, *Amblyomma americanum* (L.): unfed females. *Journal of Parasitology* **68**, 61–82.

Kurtti, T. J., Munderloh, U. G., Hayes, S. F., Krueger, D. E. & Ahlstrand, G. G. (1994). Ultrastructural analysis of the invasion of tick cells by Lyme disease spirochetes (*Borrelia burgdorferi*) *in vitro. Canadian Journal of Zoology* **72**, 977–994.

Lamoreaux, W. J. L., Junaid, L. & Trevidi, S. (2003). Morphological evidence that salivary gland degeneration in the American dog tick, *Dermacentor variabilis* (Say), involves programmed cell death. *Tissue and Cell* **35**, 95–99.

Lamoreaux, W. J., Needham, G. R. & Coons, L. B. (1994). Fluid secretion by isolated tick salivary glands depends on an intact cytoskeleton. *International Journal for Parasitology* **24**, 563–567.

Lindsay, P. J. & Kaufman, W. R. (1986). Potentiation of salivary fluid secretion in ixodid ticks: a new receptor system for γ-aminobutyric acid. *Canadian Journal of Physiology and Pharmacology* **64**, 1119–1126.

Lomas, L. O. & Kaufman, W. R. (1992). An indirect mechanism by which a protein from the male gonad hastens salivary gland degeneration in the female tick, *Amblyomma hebraeum. Archives of Insect Biochemistry and Physiology* **21**, 169–178.

Lomas, L. O., Turner, P. C. & Rees, H. H. (1997). A novel neuropeptide–endocrine interaction controlling ecdysteroid production in ixodid ticks. *Proceedings of the Royal Society of London, B* **264**, 589–596.

Madden, R. D., Sauer, J. R., Dillwith, J. W. & Bowman, A. S. (1996). Dietary modification of host blood lipids affects reproduction in the lone star tick, *Amblyomma americanum* (L.). *Journal of Parasitology* **82**, 203–209.

Mane, S. D., Darville, R. G., Sauer, J. R. & Essenberg, R. C. (1985). Cyclic AMP-dependent protein kinase from the salivary glands of the tick, *Amblyomma americanum*: partial purification and properties. *Insect Biochemistry* **15**, 777–787.

Mane, S. D., Sauer, J. R. & Essenberg, R. C. (1988). Molecular forms and free cAMP receptors of the cAMP-dependent protein kinase catalytic subunit isoforms from the lone star tick, *Amblyomma americanum* (L.). *Insect Biochemistry* **29**, 43–51.

Mans, B. J., Venter, J. D., Coons, L. B., Louw, A. I. & Neitz, W. H. (2004). A reassessment of argasid tick salivary gland ultrastructure from an immuno-cytochemical perspective. *Experimental and Applied Acarology* **33**, 119–129.

Mao, H. & Kaufman, W. R. (1998). DNA binding properties of the ecdysteroid receptor in the salivary gland of the female ixodid tick, *Amblyomma hebraeum*. *Insect Biochemistry and Molecular Biology* **28**, 947–957.

Mao, H. & Kaufman, W. R. (1999). Profile of the ecdysteroid hormone and its receptor in the salivary gland of the adult female tick, *Amblyomma hebraeum*. *Insect Biochemistry and Molecular Biology* **29**, 33–52.

Mao, H., McBlain, W. A. & Kaufman, W. R. (1995). Some properties of the ecdysteroid receptor in the salivary gland of the ixodid tick, *Amblyomma hebraeum*. *General and Comparative Endocrinology* **99**, 340–348.

Maritz-Olivier, C., Louw, A. I. & Neitz, A. W. H. (2005). Similar mechanisms regulate protein exocytosis from the salivary glands of ixodid and argasid ticks. *Journal of Insect Physiology* **51**, 1390–1296.

Matsuzaki, T., Suzuki, T., Koyama, H. & Takata, K. (1999). Aquaporin-5 (AQP-5), a water channel protein, in the rat salivary and lacrimal glands: immunolocalization and effect of secretory stimulation. *Cell and Tissue Research* **295**, 513–521.

May, A. P., Whiteheart, S. W. & Weis, W. I. (2001). Unraveling the mechanism of the vesicle transport ATPase NSF, the *n*-ethylmaleimide sensitive factor. *Journal of Biological Chemistry* **276**, 21991–21994.

McSwain, J. L., Essenberg, R. C. & Sauer, J. R. (1985). Cyclic AMP mediated phosphorylation of endogenous proteins in salivary glands of the lone star tick, *Amblyomma americanum* (L.). *Insect Biochemistry* **15**, 789–802.

McSwain, J. L., Essenberg, R. C. & Sauer, J. R. (1992). Oral secretion elicited by effectors of signal transduction pathways in the salivary glands of *Amblyomma americanum* (Acari: Ixodidae). *Journal of Medical Entomology* **29**, 41–48.

McSwain, J. L., Luo, C., Desilva, G. A., *et al.* (1997). Cloning and sequence of a gene for a homologue of the C subunit of the V-ATPase from the salivary gland of the tick *Amblyomma americanum* (L.). *Insect Molecular Biology* **8**, 67–76.

McSwain, J. L., Schmidt, S. P., Claypool, D. M., Essenberg, R. C. & Sauer, J. R. (1987). Subcellular location of phosphoproteins in salivary glands of the lone star tick, *Amblyomma americanum* (L.). *Archives of Insect Biochemistry and Physiology* **5**, 29–43.

Megaw, M. W. J. (1977). The innervation of the salivary gland of the tick, *Boophilus microplus*. *Cell and Tissue Research* **184**, 551–558.

Megaw, M. W. J. & Beadle, M. W. J. (1979). Structure and function of the salivary glands of the tick, *Boophilus microplus* Canestrini (Acarina: Ixodidae). *International Journal of Insect Morphology and Embryology* **8**, 67–83.

Narasimhan, S., Montgomery, R. R., Deponte, K., *et al.* (2004). Disruption of *Ixodes scapularis* anticoagulation by using RNA interference. *Proceedings of the National Academy of Sciences of the USA* **101**, 1141–1146.

Needham, G. R. & Sauer, J. R. (1975). Control of fluid secretion by isolated salivary glands of the lone star tick. *Journal of Insect Physiology* **21**, 1893–1898.

Needham, G. R. & Sauer, J. R. (1979). Involvement of calcium and cyclic AMP in controlling ixodid tick salivary fluid secretion. *Journal of Parasitology* **65**, 531–542.

Needham, G. R. & Teel, P. D. (1986). Water balance by ticks between bloodmeals. In *Morphology, Physiology and Behavioral Biology of Ticks*, eds. Sauer, J. R. & Hair, J. A., pp. 100–151. Chichester, UK: Ellis Horwood.

Needham, G. R. & Teel, P. D. (1991). Off-host physiological ecology of ixodid ticks. *Annual Review of Entomology* **36**, 659–681.

Negishi, M., Sugimoto, Y. & Ichikawa, A. (1993). Prostanoid receptors and their biological action. *Progress in Lipid Research* **32**, 417–434.

Nunes, E. T., Mathias, M. L. C. & Bechara, G. H. (2006). Structural and cytochemical changes in the salivary glands of the *Rhipicephalus (Boophilus) microplus* (Cannestrini, 1887) (Acari: Ixodidae) tick female during feeding. *Veterinary Parasitology* **140**, 114–123.

Oaks, J. F., McSwain, J. L., Bantle, J. A., Essenberg, R. C. & Sauer, J. R. (1991). Putative new expression of genes in ixodid tick salivary gland development during feeding. *Journal of Parasitology* 77, 378–383.

Pal, U., Yang, X., Chen, M., *et al.* (2004). OspC facilitates *Borrelia burgdorferi* invasion of *Ixodes scapularis* salivary glands. *Journal of Clinical Investigation* 113, 220–230.

Palmer, M. J., McSwain, J. L., Spatz, M. D., *et al.* (1999). Molecular cloning of cAMP-dependent protein kinase catalytic subunit isoforms from the lone star tick, *Amblyomma americanum* (L.). *Insect Biochemistry and Molecular Biology* 29, 43–51.

Pedibhotla, V. K., Sauer, J. R. & Stanley-Samuelson, D. W. (1997). Prostaglandin biosynthesis by salivary glands isolated from the lone star tick *Amblyomma americanum*. *Insect Biochemistry and Molecular Biology* 27, 255–261.

Preston, G. M., Carroll, T. P., Guggino, W. B. & Agre, P. (1992). Appearance of water channels in *Xenopus oocytes* expressing red cell CHIP28 protein. *Science* 256, 385–387.

Qian, Y., Essenberg, R. C., Dillwith, J. W., Bowman, A. S. & Sauer, J. R. (1997). A specific prostaglandin E_2 receptor and its role in modulating salivary secretion in the female tick, *Amblyomma americanum* (L.). *Insect Biochemistry and Molecular Biology* 27, 387–395.

Qian, Y., Yuan, J., Essenberg, R. C., *et al.* (1998). Prostaglandin E_2 in the salivary glands of the female tick *Amblyomma americanum* (L.): calcium mobilization and exocytosis. *Insect Biochemistry and Molecular Biology* 28, 221–228.

Ramamoorthi, N., Narasimhan, S., Pal, U., *et al.* (2005). The Lyme disease agent exploits a tick protein to infect the mammalian host. *Nature* 436, 573–577.

Ribeiro, J. M. C., Evans, P. M., McSwain, J. L. & Sauer, J. R. (1992). *Amblyomma americanum:* characterization of salivary prostaglandins E_2 and $F_{2\alpha}$ by RP-HPLC/ bioassay and gas chromatography–mass spectrometry. *Experimental Parasitology* 74, 112–116.

Roshdy, M. A. & Coons, L. B. (1975). The subgenus *Persicargas* (Ixodoidae: Argasidae: *Argas*). XXIII. Fine structure of the salivary glands of unfed *A. (P.) arboreus* Kaiser, Hoogstraal and Kohls. *Journal of Parasitology* 61, 743–752.

Rudolph, D. & Knulle, W. (1974). Site and mechanism of water vapour uptake from the atmosphere of ixodid ticks. *Nature* 249, 84–85.

Rudolph, D. & Knulle, W. (1979). Mechanisms contributing to water balance in non-feeding ticks and their ecological implications. In *Recent Advances in Acarology*, vol. 1, ed. Rodriguez, J. G., pp. 375–383. New York: Academic Press.

Sauer, J. R. & Essenberg, R. C. (1984). Role of cyclic nucleotides and calcium in controlling tick salivary gland function. *American Zoologist* 24, 217–227.

Sauer, J. R., Bowman, A. S., McSwain, J. L. & Essenberg, R. C. (1996). Salivary gland physiology of blood-feeding arthropods. In *The Immunology of Host–Ectoparasitic Arthropod Relationships*, ed. Wikel, S. K., pp. 62–84. Wallingford, UK: CAB International.

Sauer, J. R., Essenberg, R. C. & Bowman, A. S. (2000). Salivary glands in ixodid ticks: control and mechanism of secretion. *Journal of Insect Physiology* 46, 1069–1078.

Sauer, J. R., Mane, S. D., Schmidt, S. P. & Essenberg, R. C. (1986). Molecular basis for salivary secretion in ixodid ticks. In *Morphology, Physiology and Behavioral Biology of Ticks*, eds. Sauer, J. R. & Hair, J. A., pp. 55–74. Chichester, UK: Ellis Horwood.

Sauer, J. R., McSwain, J. L., Bowman, A. S. & Essenberg, R. C. (1995). Tick salivary gland physiology. *Annual Review of Entomology* 40, 245–267.

Schmidt, S. P., Essenberg, R. C. & Sauer, J. R. (1981). Evidence for a D1 dopamine receptor in the salivary glands of *Amblyomma americanum* (L.). *Journal of Cyclic Nucleotide Research* 7, 375–384.

Schmidt, S. P., Essenberg, R. C. & Sauer, J. R. (1982). Dopamine sensitive adenylate cyclase in the salivary glands of the lone star tick. *Comparative Biochemistry and Physiology* 72, 9–14.

Shipley, M. M., Dillwith, J. W., Bowman, A. S., Essenberg, R. C. & Sauer, J. R. (1993*b*). Changes in lipids of salivary glands of the lone star tick, *Amblyomma americanum*, during feeding. *Journal of Parasitology* 79, 834–842.

Shipley, M. M., Dillwith, J. W., Essenberg, R. C., Howard, R. W. & Sauer, J. R. (1993*a*). Analysis of lipids in the salivary glands *of Amblyomma americanum* (L.): detection of a high level of arachidonic acid. *Archives of Insect Biochemistry and Physiology* 23, 37–52.

Sigal, M. D., Needham, G. R. & Machin, J. (1991). Hyperosmotic oral fluid secretion during active water vapour absorption and during desiccation-induced storage excretion by the unfed tick *Amblyomma americanum*. *Journal of Experimental Biology* 157, 585–591.

Sukumaran, B., Narasimhan, S., Anderson, J. F., *et al.* (2006). An *Ixodes scapularis* protein required for survival of *Anaplasma phagocytophilum* in tick salivary glands. *Journal of Experimental Medicine* 203, 1507–1517.

Tabish, M., Clegg, R. A., Turner, P. C., *et al.* (2006). Molecular characterization of cAMP-dependent protein kinase (PK-A) catalytic subunit isoforms in the male tick,

Amblyomma hebraeum. Molecular and Biochemical Parasitology **150**, 330–339.

Tanaka, M., Liao, M., Zhou, J. L., *et al.* (2007). Molecular cloning of two caspase-like genes from the hard tick *Haemaphysalis longicornis. Journal of Veterinary Medical Science* **69**, 85–90.

Tatchell, R. J. (1967). Salivary secretions in the cattle tick as a means of water elimination. *Nature* **213**, 940–941.

Tatchell, R. J. (1969). The ionic regulatory role of salivary secretions of the cattle tick, *Boophilus microplus. Journal of Insect Physiology* **15**, 1421–1430.

Till, W. M. (1961). A contribution to the anatomy and histology of the brown ear tick *Rhipicephalus appendiculatus* Neumann. *Memoirs of the Entomological Society of South Africa* **6**, 1–124.

Verkman, A. S. (2002). Aquaporin water channels and endothelial cell function. *Journal of Anatomy* **200**, 617–627.

Walker, A. R., Fletcher, J. D. & Gill, H. S. (1985). Structural and histochemical changes in the salivary glands of *Rhipicephalus appendiculatus* during feeding. *International Journal for Parasitology* **15**, 81–100.

Weiss, B. L. & Kaufman, W. R. (2004). Two feeding-induced proteins from the male gonad trigger engorgement of the female tick *Amblyomma hebraeum. Proceedings of the National Academy of Sciences of the USA* **101**, 5874–5879.

Wong, D. L. P. & Kaufman, W. (1981). Potentiation by spiperone and other butyrophenones of fluid secretion by isolated salivary glands of ixodid ticks. *European Journal of Pharmacology* **73**, 163–173.

Yuan, J., Bowman, A. S., Aljamali, M., *et al.* (2000). PGE$_2$ stimulated secretion of protein in the salivary glands of the lone star tick via a phosphoinositide signaling pathway. *Insect Biochemistry and MolecularBiology* **30**, 1099–1106.

4 • Tick saliva: from pharmacology and biochemistry to transcriptome analysis and functional genomics

J. M. ANDERSON AND J. G. VALENZUELA

INTRODUCTION

When a tick attaches to a mammalian host to obtain a blood meal it must counteract the well developed haemostatic, inflammatory and immune systems which function to avoid blood loss and to reject unwanted guests. Ticks have been in the blood-feeding business for millions of years and have acquired potent pharmacologically active molecules found in their saliva that can disarm and counteract the haemostatic system of the mammalian host (Ribeiro, 1987b, 1995) and alter the host inflammatory and immune responses (Gillespie *et al.*, 2000; Wikel, 1999). The types of molecules present in tick saliva range from lipids to small peptides and large proteins; each is capable of altering the physiology of the feeding site, consequently affecting pathogen transmission (Ribeiro, 1995; Valenzuela, 2002b; see also Chapter 10). Adaptation of ticks to their natural hosts resulted in the ability of ticks to modulate host immune and haemostatic responses with their saliva. However, tick feeding on non-natural hosts often results in an immune and allergic response, presumably to the injected salivary proteins, resulting in tick rejection (Ribeiro, 1989). Furthermore, in some cases, an immune response to tick feeding confers protection against the pathogens ticks transmit.

Because of the importance of tick saliva, there is increasing interest in the identification and isolation of the molecules in saliva responsible for these effects. The small amount of protein and other biological material present in tick salivary glands has made this a difficult task for many years. Recently, modern molecular approaches utilizing a small amount of starting material (RNA for example), the relatively low cost of DNA sequencing, and customized computational biology tools have made the identification and isolation of a vast repertoire of transcripts and proteins from the salivary glands of different ticks possible. The isolation of these molecules contributes to the understanding of the role of saliva in blood-feeding and pathogen transmission, and may help in the identification of potential vaccine candidates to control tick-borne diseases.

ANTIHAEMOSTATIC ACTIVITIES OF TICK SALIVA

The three branches of the haemostatic system (vasoconstriction, platelet aggregation, and the blood coagulation cascade) pose a real threat to ticks during blood meal ingestion. These three branches are highly interconnected making haemostasis a redundant system, thereby decreasing the chances for blood-feeders to acquire a blood meal.

Ticks have been successful blood-feeders for at least 120 million years. This success is almost certainly due to adaptations that counteract the host haemostatic, inflammatory and immune systems by creating a repertoire of potent bioactive salivary molecules containing vasodilatory, antiplatelet and anticoagulant activities with which the tick is able to battle the specificity and redundancy of the haemostatic system. A summary of these molecules is presented in Table 4.1.

Vasodilatory activities

Vasodilators are molecules that increase blood flow by antagonizing vasoconstrictors produced by the haemostatic system following tissue injury. Vasodilators act directly or indirectly on smooth muscle cells. To date, all known tick salivary vasodilators are lipid derivatives. The hard tick *Ixodes scapularis* vasodilatory molecules are the arachidonic acid lipid derivative, prostacyclin (PGI_2) (Ribeiro *et al.*, 1988) and prostaglandin E_2 (PGE_2) (Ribeiro *et al.*, 1985). Prostaglandin E_2 is a short-acting vasodilator as well as an inhibitor of platelet aggregation and exerts its effect by increasing cAMP in smooth muscle cells resulting in vasorelaxation. Saliva of the lone star tick

Ticks: Biology, Disease and Control, ed. Alan S. Bowman and Patricia A. Nuttall. Published by Cambridge University Press.
© Cambridge University Press 2008.

Table 4.1 *Antihaemostatic components identified in tick saliva*

Tick species	Vasodilatory factor	Anticlotting factor	Antiplatelet factor
Amblyomma americanum	PGE_2, $PGF_{2\alpha}$[a] (Aljamali et al., 2002a; Ribeiro et al., 1992)	Factor Xa inhibitor (Zhu et al., 1997c); americanin (antithrombin) (Zhu et al., 1997a)	Americanin (antithrombin) (Zhu et al., 1997a)
Dermacentor andersoni		Inhibitor of FV and FVII (Gordon & Allen, 1991)	
Dermacentor variabilis			GPIIa–IIIb antagonist[a] (Wang et al., 1996)
Haemaphysalis longicornis		Madanin 1 and 2 (Iwanaga et al., 2003)	Madanin 1 and 2 (Iwanaga et al., 2003)
Hyalomma truncatum		FXa inhibitor (Joubert et al., 1995)	
Ixodes scapularis	PGI_2(prostacyclin)[a] (Ribeiro et al., 1988)	Tissue factor pathway inhibitor (Ixolaris)[a] (Francischetti et al., 2002b); pentalaris (Francischetti et al., 2004); factor Xa inhibitor[a] (Narasimhan et al., 2002)	Apyrase (Ribeiro et al., 1985)
Ornithodorus moubata		Tick anticoagulant peptide (TAP)[a] (Hawkins & Hellmann, 1966); ornithodorin (antithrombin)[a] (van de Locht et al., 1996)	Apyrase (Ribeiro et al., 1991); disagregin[a] (Karczewski et al., 1994); moubatin[a] (Keller et al., 1993); tick adhesion inhibitor (TAI) (Karczewski et al., 1995)
Ornithodorus savignyi		Savignin (antithrombin)[a] (Nienaber et al., 1999); factor Xa inhibitor[a] (Joubert et al., 1998); BSAP1, BSAP2 (Ehebauer et al., 2002)	Savignygrin (disintegrin)[a] (Mans et al., 2002b); apyrase (Mans et al., 2000); savignin (antithrombin)[a] (Nienaber et al., 1999)
Rhipicephalus appendiculatus		Factor Xa inhibitor (Limo et al., 1991)	
Rhipicephalus (Boophilus) microplus	PGE_2[a] (Dickinson et al., 1976)	Antithrombin (Horn et al., 2000)	Antithrombin (Horn et al., 2000)

[a] Isolated and characterized at the molecular level.

Amblyomma americanum also contains the vasodilator PGE_2 and $PGF_{2\alpha}$ (Aljamali *et al.*, 2002*a*; Ribeiro *et al.*, 1992). The presence of PGE_2 has also been reported in *Rhipicephalus (Boophilus) microplus*, *Dermacentor variabilis* (Dickinson *et al.*, 1976), *Haemaphysalis longicornis* and *I. holocyclus* (Inokuma *et al.*, 1994).

Antiplatelet factors

Platelets are activated by diverse agonists including thrombin, collagen, and adenosine diphosphate (ADP). Following activation, they aggregate, promote clotting and release vasoconstrictor substances to form the platelet plug. Ticks are able to prevent platelet aggregation by secreting salivary components that either hydrolyse or bind platelet agonists such as ADP, or by inhibiting the actions of other agonists such as thrombin or collagen.

The strategy employed by most blood-feeders to block platelet aggregation is to destroy or hydrolyse the platelet agonist, ADP. This is achieved through the presence of a salivary enzyme named apyrase. This enzyme (EC 3.6.1.5) hydrolyses the phosphodiester bonds of nucleoside tri- and diphosphates but not monophosphates. Apyrase activity has been reported in the saliva of many ticks, including *I. scapularis*, *Ornithodoros moubata* (Ribeiro *et al.*, 1985, 1991) and *O. savignyi* (Mans *et al.*, 2000). Two classes of apyrases have been isolated and characterized at the molecular level from haematophagic arthropods, the 5'-nucleotidase family of apyrases, present in mosquitoes such as *Anopheles gambiae* (Arca *et al.*, 1999) and *Aedes aegypti* (Champagne *et al.*, 1995), and the Cimex family of apyrases that is present in bed bugs (Valenzuela *et al.*, 1998) and sandflies (Valenzuela *et al.*, 2001). In ticks, apyrases have not been characterized at the molecular level; however, there is a report of a transcript encoding a secretory 5'-nucleotidase in the salivary glands of *I. scapularis* (Valenzuela, 2002a) that may encode the proteins with the apyrase activity detected in the saliva of this tick.

Ticks also use other strategies to prevent platelet aggregation, for example *O. moubata* produces a salivary antiplatelet factor named moubatin (Keller *et al.*, 1993). Structurally, moubatin belongs to a family of lipocalins, which are beta-barrel structures that, in general, bind small hydrophobic molecules (Keller *et al.*, 1993; Mans *et al.*, 2003). Salivary proteins with lipocalin structure have been described in the kissing bug *Rhodnius prolixus* (Francischetti *et al.*, 2002a) and in the tick *Rhipicephalus appendiculatus* (Paesen *et al.*, 1999). Moubatin is a protein of 17 kDa which inhibits platelet aggregation induced by collagen; however, the mechanism of action is not clear. Because of structural similarities between moubatin and the platelet inhibitor RPAI-1 from *R. prolixus*, it is possible that moubatin binds ADP with high affinity, thereby preventing platelet aggregation caused by ADP released from collagen-activated platelets.

Thrombin is a potent agonist of platelet activation; thus inhibition of thrombin activity is of central importance to blood-feeders. Salivary antithrombins from the soft tick *O. savignyi* have been characterized as anticoagulants as well as inhibitors of platelet aggregation (Nienaber *et al.*, 1999). Ornithodorin, described from the soft tick *O. moubata* (van de Locht *et al.*, 1996), is highly similar to savignin, and therefore probably shares similar antihaemostatic characteristics.

Ticks have developed more complex strategies to prevent platelet aggregation that involve preventing or blocking platelet–platelet interactions. These interactions are mediated by the integrin αIIbβ3 (glycoprotein IIb–IIIa, GP IIb–IIIa) which is an inactive receptor on resting platelets. Upon platelet activation this glycoprotein mediates platelet–fibrinogen–platelet interactions (Ferguson & Zaqqa, 1999). The soft ticks *O. moubata* and *O. savignyi* contain small molecules of about 7 kDa named disagregin and savignygrin, respectively, which bind to GPIIb–IIIa on platelets and prevent platelet–fibrinogen–platelet interactions, therefore preventing platelet aggregation (Karczewski *et al.*, 1994; Mans *et al.*, 2002b). Savignyrin uses the classical Arg-Gly-Asp (RGD) motif to bind to GPIIb–IIIa while disagregin uses a different motif. *Dermacentor variablilis* saliva also contains a molecule of 4.9 kDa named variabilin that contains an RGD motif and blocks this receptor. However, this peptide has little sequence homology to other GPIIb–IIIa antagonists (Wang *et al.*, 1996). It is interesting to note that even when different ticks use the same strategy to counteract a biological activity they may use different salivary proteins for the same target.

Anticoagulant activities

The blood coagulation cascade involves the sequential activation of pro-enzymes ultimately resulting in thrombin activation, which in turn cleaves fibrinogen into fibrin. Polymerization of fibrin results in blood clot formation. Inhibitors of the blood coagulation cascade are the most characterized activities from the saliva of ticks. The tick anticoagulant peptide (TAP) isolated from the saliva of the tick *O. moubata* (Hawkins & Hellmann, 1966; Waxman *et al.*, 1990) is a specific inhibitor of factor Xa (FXa) with a molecular mass of 6977 Da. FXa is involved in the activation of thrombin, highlighting the importance of blocking the activity of this protease. The structure of this protein has been determined by means of nuclear magnetic resonance (NMR) (Antuch *et al.*, 1994) as well as X-ray crystallography (St Charles *et al.*, 2000). The saliva of *O. savignyi* additionally contains a FXa inhibitor with 46% identity to TAP (Joubert *et al.*, 1998). The salivary anticoagulant activity of *R. appendiculatus* also inhibits the activation of FXa in human plasma (Limo *et al.*, 1991) but it does not inhibit the activity of FXa towards a chromogenic substrate. This inhibitor is probably directed to the exosite of FXa. FXa inhibitors were also reported from the saliva of *A. americanum* (Zhu *et al.*, 1997*c*) and from the saliva of *Hyalomma truncatum* (Joubert *et al.*, 1995).

Thrombin is the last enzyme of the blood coagulation cascade and is a strong agonist for platelet aggregation. Thrombin inhibitors have been described in the saliva of the ticks *O. savignyi* (Nienaber *et al.*, 1999), *O. moubata* (van de Locht *et al.*, 1996), *A. americanum* (Zhu *et al.*, 1997*a*), *R. (Boophilus) microplus* (Horn *et al.*, 2000) and *Haemaphysalis longicornis* (Iwanaga *et al.*, 2003; Nakajima *et al.*, 2006). Americanin, the salivary antithrombin from *A. americanum*, is a specific, reversible and slow tight-binding inhibitor of thrombin (Zhu *et al.*, 1997*a*). Madanin 1 and 2, 7-kDa proteins identified from partially fed salivary glands of *H. longicornis*, prevent thrombin from cleaving fibrinogen without inhibiting thrombin amidolytic activity (Iwanaga *et al.*, 2003) whereas chimadanin, a 93-amino-acid protein also from *H. longicornis* salivary glands, appears to inhibit thrombin amidolytic activity suggesting it functions as an anticoagulant (Nakajima *et al.*, 2006). The cattle tick *R. (Boophilus) microplus* contains two antithrombins, BmAP (Horn *et al.*, 2000) and microphilin (Ciprandi *et al.*, 2006), 60 kDa and 1770 Da, respectively. Microphilin inhibits fibrinocoagulation and thrombin-induced platelet aggregation by blocking thrombin at exosite I.

The salivary antithrombin from *O. savignyi* is a 12.4-kDa protein named savignin (Nienaber *et al.*, 1999). It is a slow, tight-binding inhibitor of thrombin and interacts with the active site as well as with the binding exosite of this protease (Mans *et al.*, 2002*b*). Savignin is 63% identical to ornithodorin, the salivary antithrombin from *O. moubata* (van de Locht *et al.*, 1996). The mechanism of action of these antithrombins differs from the classical mechanism found in the Kunitz family of protease inhibitors. Soft tick antithrombins insert their N-terminal residues into the thrombin active site inhibiting the activity of this protease, whereas traditional Kunitz-type inhibitors use a central, reactive loop (Mans *et al.*, 2002*a*). Two other anticoagulants have been identified from the salivary glands of the tick *O. savignyi*. The anticoagulants BSAP1 and BSAP2 have molecular masses of 9.3 and 9.2 kDa, respectively, and are inhibitors of the extrinsic pathway of the blood coagulation cascade (Ehebauer *et al.*, 2002).

Ixolaris is a novel salivary anticoagulant from *I. scapularis* which contains extensive sequence homology to tissue factor pathway inhibitors (TFPI). It is a small protein of 140 amino acids containing 10 cysteines and two Kunitz-type domains (Francischetti *et al.*, 2002*b*). It specifically inhibits the activation of FX by tissue factor/FVIIa. This salivary protein

interacts with FX and FXa but not with inactive FVII. One of the Kunitz domains of ixolaris appears to bind to the exosite of FXa and the second domain binds to FVIIa but only when FVIIa is complexed with tissue factor (TF/VIIa) which functions as scaffolding for the inhibition of FX activation by ixolaris (Francischetti *et al.*, 2002*b*). Recently, ixolaris was shown to reduce, in a dose-dependent manner, the production of thrombus formation in vivo (Nazareth *et al.*, 2006). Another novel tissue factor pathway inhibitor named pentalaris was also characterized from the salivary glands of *I. scapularis*. As its name suggests, this 35-kDa protein contains five tandem Kunitz domains and 12 cysteine bridges. In the same way as ixolaris, it inhibits the activation of FX by FVIIa/TF complex (Francischetti *et al.*, 2004).

An additional type of anticoagulant that inhibits the activity of FXa was reported from the salivary glands of *I. scapulari* (Narasimhan *et al.*, 2002). The isolated protein matched a cDNA that codes for a protein homologous to Salp14, a previously characterized antigenic protein with a molecular weight of 14 kDa present in the saliva of this tick (Das *et al.*, 2001). Recombinant Salp14 was shown to inhibit FXa but not other proteases. RNA silencing of Salp14 and its paralogs produced a 60–80% reduction of anti-FXa as well as a 50–70% reduction in tick feeding based on engorgement weight (Narasimhan *et al.*, 2004).

A novel anticoagulant was recently identified in the salivary glands of the tick *Haemaphysalis longicornis*. This anticoagulant is a Kunitz-type protein termed haemaphysalin which inhibits the activation of the plasma kallikrein–kinin system by interfering with the reciprocal activation of FXII and prekallikrein. Furthermore, this anticoagulant binds directly to the cell-binding domains of FXII and kininogen (HK); these interactions are Zn^{2+} dependent (Kato *et al.*, 2005).

Other antihaemostatic activities

Other biological activities that may be related to haemostasis have been described in the saliva of ticks. Phospholipase A_2 activity was detected in the saliva of *A. americanum* (Bowman *et al.*, 1997). This salivary activity hydrolyses arachidonyl phospatidylcholine, is activated by submicromolar calcium and has a predicted molecular mass of 55 kDa. It has been suggested that this phospholipase may be involved in producing PGE_2 from host substrates and may also be responsible for the haemolytic activity reported in *A. americanum* saliva (Zhu *et al.*, 1997*b*). Gelatinase and fibrinolytic activities were identified in the saliva of *I. scapularis* (Francischetti *et al.*, 2003). These proteolytic activities are metal-dependent and target gelatin, fibrin, fibrinogen and fibronectin, but not collagen or laminin. Gelatinase or fibrinolytic activities may confer additional anticoagulant activity by preventing the formation of the fibrin clot or dissolving the already formed blood clot.

ANTI-INFLAMMATORY ACTIVITIES OF TICK SALIVA

Ticks remain attached to the host for a long period of time as compared to other blood-feeding arthropods (Binnington & Kemp, 1980). Ticks damage the host by inserting their chelicerae into the host skin and injecting salivary components through a central rod-like structure called the hypostome. The natural consequence of this action is an acute inflammatory response by the host which can impair blood-feeding and lead to rejection of the tick (Ribeiro, 1989). Inflammation is a response to localized injury involving neutrophils, macrophages, mast cells, basophils, eosinophils and lymphocytes as well as chemokines, plasma enzymes, lipid inflammatory mediators and cytokines. The main function of neutrophils is to engulf and kill invading microorganisms by secreting inflammatory mediators and oxygen radicals. *Ixodes scapularis* saliva has been reported to inhibit key pro-inflammatory activities of neutrophils such as aggregation following activation by anaphylotoxins, the release of enzymes, production of oxygen radicals and the engulfing of bacteria (Ribeiro *et al.*, 1990). Cell recruitment during the inflammatory response is triggered by chemokines, which are primarily derived from macrophages. The chemokine interleukin-8 (IL-8) is a potent chemoattractant for neutrophils. Interleukin-8 triggers a G-protein-mediated signalling activation that results in neutrophil adhesion and subsequent transendothelial migration to the injury site. Anti-IL-8 activity was recently reported from saliva of various ticks including *D. reticulatus*, *A. variegatum*, *R. appendiculatus*, *H. inermis* and *I. ricinus* (Hajnicka *et al.*, 2001). Comparison of salivary gland extracts from three ixodid tick species (*D. reticulatis*, *A. variegatum* and *I. ricinus*) found anti-cytokine activity to seven cytokines – IL-8, MCP-1, MIP-1 α, RANTES, eotaxin, IL-2 and IL-4 – suggesting that ticks can manipulate the cytokine network and thus facilitate blood-feeding and avoid rejection by the host (Hajnicka *et al.*, 2005). Recently, anti-tumour necrosis factor-alpha (TNFα) activity was identified from salivary gland extracts from *I. ricinus* (Konik *et al.*, 2006) as well as an inhibitor of IL-10 (Hannier *et al.*, 2003).

Antihistamine and antiserotonin activities

Histamine is a very potent inflammatory mediator and a vasoactive factor that binds to histamine receptors, causing oedema and erythema by dilating and increasing the permeability of small blood vessels. Histamine is released from mast cells and basophils, primarily, but not exclusively, via an IgE-dependent mechanism and is also released by platelets of many mammals. Additionally, histamine is a regulator of T-cell responses, yet binding to the H2 receptor results in inhibition of Th1 and Th2 responses (Jutel *et al.*, 2001). *Rhipicephalus appendiculatus* has a set of novel salivary histamine-binding proteins (Paesen *et al.*, 1999). The structure of this high-affinity histamine-binding protein is a lipocalin. Most lipocalins are beta-barrel structures with only one binding site for hydrophobic molecules. Interestingly, histamine binding proteins from *R. appendiculatus* have two binding sites, one displaying higher affinity for histamine than the other. Tick histamine-binding proteins were shown to outcompete histamine receptors for the ligand and decrease the effects of histamine. A protein homologous to the *R. appendiculatus* histamine-binding protein was identified in the salivary glands of the ticks *I. scapularis* (Valenzuela *et al.*, 2002) and *A. americanum* (Bior *et al.*, 2002) suggesting these ticks may have similar antihistaminic activities.

Serotonin, another mediator of the inflammatory response, is secreted by tissue mast cells (in rodents) and has similar activities to those of histamine. A serotonin-binding protein was isolated from the salivary glands of *D. reticulatus* (Sangamnatdej *et al.*, 2002). This protein is similar in structure to the *R. appendiculatus* histamine-binding protein. It has a predicted molecular weight of 22 kDa and contains two binding sites, one of which binds histamine, while the other is slightly larger and is able to accommodate and bind serotonin.

Anticomplement activity

Another important part of the inflammatory response is the alternative pathway of the complement cascade which is important for the evasion of pathogens (Joiner, 1988). This cascade produces two inflammatory mediators, C5a and C3a (Hugli & Muller-Eberhard, 1978). These two anaphylotoxins are chemotactic for neutrophils and can cause histamine release from mast cells and basophils. The final step of the complement cascade is the formation of the membrane attack complex, which causes lysis of invading organisms.

Ixodes scapularis anticomplement, isac, is a novel 18-kDa protein which inhibits the formation of C3 convertase, thereby acting as a regulator of the complement cascade (Ribeiro, 1987a; Valenzuela *et al.*, 2000). Isac inhibits the interaction of factor Bb to C3b, a mechanism resembling those used by regulators of the complement cascade. Recently, Soares *et al.* (2005) reported the silencing of isac by oral delivery of double-stranded RNA (dsRNA) to *I. scapularis* nymphs and observed a concurrent reduction of tick engorgement weight as compared to control groups, suggesting that isac is relevant for blood-feeding in these ticks. Recently, a novel complement inhibitory protein was characterized from the soft tick *O. moubata*, named OmC1, which inhibits C5 activation (Nunn *et al.*, 2005). In *I. ricinus*, salivary gland homogenate inhibited the complement cascade by inhibiting C3a generation and factor B cleavage (Lawrie *et al.*, 2005). Anticomplement activity has also been found in other tick species including *I. ricinus*, *I. hexagonus* and *I. uriae* (Lawrie *et al.*, 1999).

Kininase activity

Bradykinin is an important mediator of the inflammatory response. When coagulation FXII is activated following tissue injury, it converts prekallikrein to kallikrein. Kallikrein activates high-molecular-weight kininogen to release bradykinin. This nonapeptide is a mediator of pain and causes oedema formation by increasing capillary permeability. *Ixodes scapularis* saliva contains an activity that proteolytically cleaves bradykinin (Ribeiro *et al.*, 1985) and is reportedly a metallo dipetidyl carboxypeptidase (Ribeiro & Mather, 1998).

Cystatins

Sialostatin L, the salivary cystatin from *I. scapularis*, has anti-inflammatory and immunosuppressive activities (Kotsyfakis *et al.*, 2006). This protein inhibits the activity of specific proteases including cathepsin L (cysteine protease). Sialostatin L also inhibits the proliferation of CTTL-2 cells, causing a decreased inflammatory reaction produced by carrageenan injections in the foot pad of animals as well as inhibition of myeloperoxidase activity (Kotsyfakis *et al.*, 2006).

Cystatins have also been identified in other tick species, including *A. americanum* (Karim *et al.*, 2005), *O. moubata* (Grunclova *et al.*, 2006), and *R. (Boophilus) microplus* (Lima *et al.*, 2006). Although derived from a midgut specific cDNA library, a cystatin from the soft tick *O. moubata*,

Om-cystatin 2, appears to be expressed in the salivary glands of unfed ticks, based on reverse transcriptase-polymerase chain reaction (RT-PCR) and immunofluorescent detection. This cystatin appears to inhibit cathepsin B and H (Grunclova *et al.*, 2006).

MOLECULAR APPROACHES TO THE STUDY OF TICK SALIVARY TRANSCRIPTS AND NOVEL COMPONENTS

Classical biochemical and molecular biology protocols have enabled the identification, isolation and characterization of a number of tick salivary proteins. Gene discovery has relied on the identification of a biological activity, isolation and partial sequencing of the corresponding protein, and application of this information to design DNA probes or primers that can be used to screen a salivary gland cDNA library to identify and isolate a single transcript or gene. At the turn of the twenty-first century, this approach changed dramatically, switching from a one-protein–one-gene strategy to the use of high-throughput approaches to simultaneously identify a large number of molecules present in the saliva of ticks. Some of the strategies and subsequent novel molecules characterized by these approaches are described below.

Tick salivary gland cDNA library screening

One of the first approaches to isolate tick salivary gland genes or transcripts was reported by Das *et al.* (2001). This strategy was based on the construction of an expression cDNA library from the salivary glands of the black-legged tick, *I. scapularis*. The cDNA library was screened using serum from rabbits infested repeatedly with adult *I. scapularis*. The screening resulted in the isolation of 47 clones; of these, 14 different tick salivary cDNAs were identified. From this initial list of salivary components, functional genomic approaches were performed and demonstrated that the salivary protein Salp15 blocks CD4+ T-cell activation (Anguita *et al.*, 2002). This protein was also shown to bind *Borrelia burgdorferi* and protect the bacteria against antibody-mediated killing (Ramamoorthi *et al.*, 2005). Recently, it was shown that Salp15 binds specifically to the CD4+ T-cell co-receptor on mammalian cells and inhibits T-cell antigen receptor (TCR) ligation thus disrupting the activation of T-cells (Garg *et al.*, 2006). Another molecule identified in this early study, namely Salp16, appears to be involved in the transmission of *Anaplasma phagocytophilum*. RNA interference silencing of Salp16 gene expression disrupted the survival of *A. phagocytophilum* by, presumably, preventing the migration of the pathogen to the salivary glands (Sukumaran *et al.*, 2006).

Polymerase chain reaction subtraction

Another strategy used to isolate tick genes was reported by Leboulle *et al.* (2002*b*) and is based on PCR subtraction and the construction of a full-length cDNA library from salivary glands of engorged ticks. Using this strategy, 27 different cDNA (the majority of which were novel) were identified from 96 randomly selected clones. Only three cDNAs had significant homologies to other proteins: human tissue factor pathway inhibitor (TFPI), snake metalloproteases and a human thrombin inhibitor (which also had significant similarities to a monocyte/neutrophil elastase inhibitor). Using this approach, the first salivary immunosuppressive protein from the salivary glands of *I. ricinus* (iris) was identified (Leboulle *et al.*, 2002*a*).

Salivary gland transcriptome analysis

Another approach to study the repertoire of tick salivary molecules is based on the construction of a high-quality, full-length cDNA library, massive sequencing of a large set of transcripts and customized analysis to create a databank of tick salivary genes (Valenzuela *et al.*, 2002). This tissue specific method relies on high-quality starting material (mRNA) and can be accomplished with as few as two pairs of adult glands (Francischetti *et al.*, 2005). In general, cDNA libraries are constructed using reverse transcriptase with an oligo dT primer which anneals to the poly-A tail of the eukaryotic mRNA. Second-strand synthesis incorporates restriction enzyme sites for ease of cloning into either a bacterial plasmid or phage. Size fractionation of the transcripts before cloning into the appropriate vector appears to be critical for complete coverage of small to large transcripts and to reduce the bias toward sequencing of small cDNA fragments. The construction of cDNA libraries has been simplified as a result of several commercially available kits. Unfortunately, computational analysis of the resulting expressed sequence tag (EST) data has not been developed commercially and currently each laboratory has developed its own method (Holt *et al.*, 2002; Ribeiro *et al.*, 2006) for sequence evaluation using published bioinformatics algorithms such as Phred/Phrap (Lee & Vega, 2004), ClustalX (Thompson *et al.*, 1997), CAP3 assembler (Huang & Madan, 1999) and

Blast (Altschul & Gish, 1996). Transcriptome analysis has proved to be a very powerful tool, generating much more data than other techniques together with the acquisition of knowledge from tissue-specific molecules. Using these combined approaches, salivary gland transcriptomes have been published from four genera of ticks to date: *Ixodes*, *Dermacentor*, *Amblyomma* and *Rhipicephalus*. A brief description of each published library follows.

Ixodes scapularis

Following incrimination of the black-legged tick *I. scapularis* as the vector of the Lyme disease spirochaete, salivary components of this North American tick have been studied extensively. One of the first tissue-specific cDNA libraries was constructed from 25 *I. scapularis* partially fed (3–4 days) female salivary glands (Valenzuela *et al.*, 2002). From this library, 735 clones were randomly sequenced and grouped into 383 clusters or families of related transcripts. From these clusters, 87 transcripts contained a putative secretory signal peptide. The full-length transcripts were sequenced and grouped according to their abundance in the cDNA library. Seven distinct groups of secreted proteins were identified, including a number of novel sequences that predicted activities never previously reported in ticks. Expression of many of the most abundant transcripts was confirmed in the saliva as well as salivary gland homogenates from fed ticks by SDS-PAGE electrophoresis and *N*-terminal protein sequencing followed by comparison with the cDNA database.

An updated analysis of the *I. scapularis* transcriptome analysed 8000 transcripts using customized bioinformatic and computational biology tools (Valenzuela *et al.*, 2002; Ribeiro *et al.*, 2006). The transcripts were from cDNA libraries of salivary glands of partially fed (3 days) *B. burdorferi* infected nymphs, partially fed (3 days) uninfected nymphs, unfed adults, and adults fed for 6–12 h, 18–24 h and 3–4 days. Salivary glands were dissected from 20–30 females or from 100 nymphs per library. This extensive analysis resulted in the identification of 500 new predicted protein sequences that were divided into 27 groups (Table 4.2). The transcripts were compared to transcripts of *I. ricinus* and *I. pacificus*, as well other acarine sequences published on GenBank. Of the 27 groups of proteins, three appeared to be unique to *I. scapularis* (IS6 family, 26-kDa family and the SRAEL family).

The analysis was the first to directly compare complete cDNA library transcripts from different life stages (adult and

Table 4.2 *Secreted salivary protein families based on the* Ixodes scapularis *transcriptome*

Group[a]	Group name	Molecular weight (kDa)	Putative function[a]
1	Basic tail polypeptides	13–14, 18–19	Anticoagulant
2	Basic tailless polypeptides	10–11	Unknown
3	Kunitz domain containing proteins	12	Anticoagulant
4	6–8-kDa proline- and glycine-rich peptides	6–8	Platelet aggregation inhibitor
5	Similar to *I. scapularis* 18.7-kDa protein	18.7	Unknown
6	Similar to *I. scapularis* 5.3-kDa peptide	5.3	Antimicrobial defence mechanism
7	9- and 7-kDa family of peptides	9–7	Unknown
8	Metalloproteases	36	Antihaemostatic
9	GPIIb–IIIa antagonists	4.1	Antiplatelet
10	Ixostatin family, short cysteine-rich peptides,	9–11	Antiplatelet
11	Lipocalins	20	Antihaemostatic
12	Neuropeptide-like with GGY repeats and antimicrobial peptides	4.7–13	Antimicrobial defence mechanism
13	Oxidant metabolism	15	Anti-inflammatory
14	Anticomplement (Isac) proteins	20	Immunomodulatory
15	WC-10 peptide	9–11	Unknown
16	LPTS peptide	12–16	Unknown
17	Antiprotease (serpins, cystatins, trypsin inhibitors)	–	Anti-inflammatory
18	Mucins	–	Tick mouthpart maintenance
19	IS6 family	9–12	Vasodilators, neurotoxins or antimicrobial
20	12-kDa family	12–13	Unknown
21	26-kDa family	23–26	Weakly related to cytotoxins
22	30-kDa family	30	Unknown
23	Toxin like	8–9	Weakly related to toxins
24	SRAEL family	16–22	Unknown
25	Other enzymes	–	Variable
26	Peptides of unknown function	–	Unknown
27	H proteins	–	Housekeeping functions

[a] Grouping and putative functional assignment based on Ribeiro *et al.* (2006).

nymphs) and feeding stages (unfed to 3-day fed ticks), and between *B. burgdorferi* infected and uninfected ticks. This comparative analysis provided several interesting observations about salivary gland component complexity, temporal expression, redundancy, allelic variation and response to infection, which together highlight the need for further tissue-specific and genome-wide analysis to fully understand tick salivary gland function. Through comparison of infected and uninfected nymphal ticks, a 5.3-kDa protein of unknown function was found to be significantly over-expressed in infected salivary glands suggesting a role in antimicrobial defence mechanisms. Without the benefit of a direct comparative analysis the potential role of this protein may have been overlooked. Finally, the significant differential expression of proteins found among the adult *Ixodes* ticks as a function of time post-attachment emphasizes the need for other molecular studies, such as microarray analysis, and also the need of confirmatory assays such as RT-PCR and saliva bioactivity assays to verify the lack (or presence) of protein families in other tick species.

Amblyomma variegatum

Nene *et al.* (2002) reported the isolation of 2109 non-redundant transcripts from the salivary glands of the tropical African bont tick *Amblyomma variegatum*, the vector for pathogenic rickettsia that cause heartwater in ruminants. Their approach was based on the construction of a cDNA library from 400 salivary gland pairs and the sequencing of randomly obtained plasmids. The resulting sequences were analysed using the TIGR gene index system (Quackenbush *et al.*, 2001). In addition to an abundance of housekeeping gene transcripts, a large set of proteinases was identified including cathepsin L-like protease, a probable zinc protease, serine protease inhibitor 4 and serine protease inhibitor 2. Other transcripts coding for glutathione S-transferase, calreticulin and proteins similar to some of the immunodominant antigens from the saliva of *I. scapularis* (Das *et al.*, 2001) were also identified. A large group of proteins rich in glycine were identified and separated into 11 distinct families. These proteins are probably part of the cement secreted during attachment of the tick to the host. A cDNA homologous to the immunosuppressant protein from *D. andersoni* (Bergman *et al.*, 2000) was also identified.

Rhipicephalus appendiculatus

The brown ear tick *Rhipicephalus appendiculatus*, a hard tick found on cattle and other livestock, is a major pest where it is endemic in Africa (southern Sudan and eastern Zaire to South Africa and Kenya). *Rhipicephalus appendiculatus* transmits several diseases, including East Coast fever (*Theileria parva*). Much attention has been given to the saliva from this tick for anti-tick vaccine production based on secreted cement proteins (Trimnell *et al.*, 2002, 2005). Using the strategy described above for the *A. variegatum* transcriptome, a total of 19 046 EST sequences were derived and analysed from a directional cDNA library (plasmid) constructed from 500 *T. parva* infected or uninfected partially fed (4 days) female ticks (Nene *et al.*, 2004). The sequences were clustered into 2543 contigs and 4797 singletons creating a database of 7359 non-redundant sequences. Unlike the *Ixodes* transciptomes, the most abundant contigs (highest level of EST redundancy) appear to be proteins with housekeeping functions. From the initial analysis, salivary-gland-derived histamine-binding proteins, a serpin family of serine proteases, metalloproteases, carboxypeptidases and Kunitz-domain-containing proteins, as well as other immunodominant proteins, have been identified. While secreted salivary proteins were identified based on homology, the authors do not clearly delineate secreted proteins from non-secreted contigs, so that direct comparison with other tick salivary transcriptomes is not possible. However, to date, this is the largest collection of ESTs from a tick species made publicly available, and can be further analysed and mined by the scientific community.

Ixodes pacificus

Analysis of the *I. pacificus* salivary gland transcriptome was recently reported (Francischetti *et al.*, 2005). *Ixodes pacificus*, the western black-legged tick, is the vector of the aetiologic agent of Lyme disease in western United States. The salivary gland transcriptome was generated from two partially fed (3 days) salivary gland pairs. A total of 1068 transcripts were sequenced and clustered into 557 clones of which 138 appeared to be secreted based on the presence of a putative signal peptide, and 87 appeared to be novel sequences. From this analysis, 16 different families of proteins were identified including basic tail proteins, Kunitz-like proteins, proline-rich peptides, metalloproteases, disintegrins, ixostatins, anti-complements, neuropeptide-like proteins and three novel families of proteins with unknown function. The *I. pacificus* disintegrin, termed ixodegrin, which has homologues in the *I. scapularis* salivary gland transcriptome, is related to other tick GPIIB–IIIa antagonists such as variabilin and disagregin from *A. americanum* and *O. moubata*, respectively. Ixodegrin is related to a family of short neurotoxin proteins which includes cobra toxin, snake toxins and mamba cardiotoxins (Francischetti *et al.*, 2005). Comparatively, only 58% sequence identity was found among the 66 contigs that matched peptide sequences from *I. scapularis*, even though these are closely related tick species (both phylogenetically and geographically) within the same genus.

Dermacentor andersoni

The Rocky Mountain wood tick *Dermacentor andersoni* salivary gland transcriptome is the first report of the repertoire of salivary molecules from the genus *Dermacentor* (Alarcon-Chaidez *et al.*, 2007). *Dermacentor andersoni* is an important vector of pathogens that cause Rocky Mountain spotted fever, Colorado tick fever and tularaemia in the United States. A total of 1440 transcripts were sequenced and bioinformatics analysis was conducted on 1299 quality sequences. A total of 762 contigs (of which 544 were singletons) were obtained from 24 partially fed (18–24 h) female salivary gland pairs. Approximately 56% of the contigs were similar to genes of known function. From this work a number of novel sequences were identified such as transcripts coding for molecules associated with extracellular matrix, immune suppression, tumour suppression and wound healing. Other isolated transcripts had similarities to transcripts reported in other tick species. One notable difference found in the *D. andersoni* cDNA library as compared with the *I. scapularis* cDNA library was the abundance of silk-related and glycine-rich proteins which are associated with cement. *Dermacentor* species do not insert their mouthparts as deeply into the skin as other ticks, so that this putative excess of cement could aid in stable and prolonged attachment to the host.

Haemaphysalis longicornis

The bush tick *H. longicornis* is an economically important ectoparasite and a major concern for the cattle industry in Australia and Southern Asia. Because of this, a considerable amount of knowledge concerning the genus *Haemaphysalis* has been produced using this species. Nakajima *et al.* (2005) constructed a cDNA library from 30 salivary gland pairs from partially fed adult female *H. longicornis*. A total of 826 clones were analysed from which 633 quality sequences were derived and grouped into 213 clusters of which 157 clusters contained only one clone (singleton). From the original 633 sequences, 36% had homology to known genes of which only 83 clones coded for functions other than housekeeping. The majority of proteins (60/83) predicted to be biologically active were related to protease inhibitors, namely mandin 1 and 2 thrombin inhibitor, Kunitz-type protease inhibitors, von Willebrand factor and ixodidin. Clones with homologies to trypsin-inhibitor-like domain, metalloproteases and immunosuppressant proteins were also found. One of the most interesting discoveries of this annotated catalogue, ixodidin, was originally described from *B. microplus* haemocytes (Fogaca *et al.*, 2004) and has recently been identified as a novel antibacterial peptide (Fogaca *et al.*, 2006). The presence of this molecule in the salivary glands could have intriguing consequences; therefore, confirmatory analysis should be conducted to verify salivary expression of this molecule.

FUNCTIONAL GENOMICS OF TICK SALIVARY TRANSCRIPTS

As recently as the year 2000, only 11 tick salivary proteins had been deposited in GenBank whereas by the beginning of 2007 over 4500 proteins had been submitted. The amazing pace of this field is even more astounding considering that approximately 143 ESTs were submitted to GenBank prior to 2000 and today there are >92 500 sequences. This dramatic increase in tick genomic and proteomic research can be attributed to high-throughput and other molecular approaches that have led to the identification and isolation of a large set of genes or transcripts from the salivary gland of different ticks.

Functional genomics is a term coined for the approaches that can be utilized to obtain biological information from the massive number of sequencing data generated by high-throughput approaches. These approaches include biochemical assays, genetic manipulations, RNA interference and the expression of recombinant proteins in mammalian, insect, yeast and bacterial systems. Other approaches include DNA microarray analysis as well as proteomic analysis including mass spectroscopy and Edman degradation analysis. In practice, the functional genomics approach of choice depends very much on the area of research and the hypothesis to be tested. In most genomic approaches (including those for ticks) about 50% of the transcripts are similar to proteins of known function, including, of course, housekeeping genes. The other 50% are transcripts of unknown function. Although the information generated by these approaches can be judged as purely descriptive, it generates a large number of hypotheses that can be immediately tested. The 'low-hanging-fruits' or the obvious transcripts that are predicted to possess a particular activity are historically the first ones to be tested. For tick transcripts, the expression of recombinant proteins in heterologous systems and the testing of their potential biological activities have been very successful. For example, the identification of ixolaris as a tissue factor pathway inhibitor was accomplished by the expression of the recombinant protein in insect cells (Francischetti *et al.*, 2002*b*). The immunosuppressor from the salivary glands of *D. andersoni* was also expressed using insect cells (Alarcon-Chaidez *et al.*, 2003). However, when the same protein was expressed in bacteria it lost its biological activity. These results suggest that a eukaryotic expression system may be a better option for expressing certain proteins when the functional activity is dependent on correct folding and post-translational modifications. An example of the utility of eukaryotic expression systems comes from a novel transcript found in the salivary glands of *I. ricinus* named iris (*Ixodes ricinus* immunosuppressor) (Leboulle *et al.*, 2002*a*). When expressed in a mammalian cell line (CHO-K1 cells) the recombinant protein was able to modulate T-cell and macrophage responsiveness and inhibited secretion of pro-inflammatory cytokines (Leboulle *et al.*, 2002*a*). In another example, an antigenic protein, Salp15, from the salivary glands of *I. scapularis*, codes for a protein with weak homology to inhibin A, a member of transforming growth factor beta (TGFβ) superfamily, suggesting Salp15 may have some immunomodulatory activities. Expression of Salp15 in *Drosophila* cells resulted in an active protein capable of blocking CD4+ T-cell activation by binding to the receptor of CD4+ T-cells (Anguita *et al.*, 2002; Garg *et al.*, 2006). This protein was also shown to bind to the bacterium *Borrelia burgdorferi*, and to inhibit antibody-mediated killing (Ramamoorthi *et al.*, 2005). Expression of functional recombinant tick salivary proteins has also been achieved using bacterial expression systems. For example, sialostatin L, the salivary cystatin from *I. scapularis*, was expressed in *E. coli* and shown to inhibit cysteine proteases as well as inflammatory responses (Kotsyfakis *et al.*, 2006).

Currently there are a number of expression vectors with high-throughput cloning capabilities. The problem is the lack of a robust expression system which can accommodate the large repertoire of expression constructs containing a myriad of folding or post-translational requirements. Insect cell and mammalian cell expression systems appear to be better at producing certain functional proteins. However, the amount of protein produced is still very limited, but recent advances in eukaryotic cell expression systems can now produce fairly substantial quantities of recombinant protein. Protein expression in bacteria or cell-free expression systems can produce very large amounts of material but with less likelihood of obtaining a soluble, functional protein. Baculovirus expression systems produce large amounts of conformationally active material and with recent technological advancements and the ease of the cloning process this is becoming an attractive system to pursue. Technological developments in recombinant protein expression and purification are significantly impacting functional genomics in tick research.

Another functional genomic tool, RNA interference (RNAi), relies on the ability to silence the expression of a gene by degrading a targeted mRNA in vivo (Brooks & Isaac, 2002). RNA interference has been successfully demonstrated in several tick species (Aljamali *et al.*, 2002*b*; Karim *et al.*, 2005; Ramakrishnan *et al.*, 2005). Expression of a transcript coding for a histamine-binding protein from the salivary glands of *A. americanum* was reduced by injecting ticks with dsRNA from this clone. Furthermore, ticks treated with dsRNA resulted in lower histamine-binding ability and increased oedema around the feeding sites (Aljamali *et al.*, 2003). Silencing of isac, the tick salivary anticomplement molecule, was also achieved (Soares *et al.*, 2005) as well as the silencing of cystatin (Karim *et al.*, 2005). The RNAi approach may help identify the function and significance of tick salivary gene transcripts with no clear predicted function. A clear downside to this approach is the apparent redundancy found in tick saliva (Ribeiro *et al.*, 2006). Most RNAi studies, to date, target one tick salivary molecule and assume homologous transcripts will be silenced. However, the lack of complete functional gene knockout found in most studies indicates that this does not occur. An understanding of the complexity of the saliva of many tick species, achieved through transcriptome analysis, will improve the design of RNAi studies.

Because tick salivary proteins have become an attractive target for vaccines to control vector-borne diseases, the resulting gene transcripts generated using high-throughput approaches are ready to be evaluated for their potential use as vaccines. The primary issue is how to select the right candidates and then how to test them (see Chapter 19). Regardless of the anti-tick vaccine strategy used, the findings of the *I. scapularis* comparative analysis (Ribeiro *et al.*, 2006) should not be overlooked. The complexity, redundancy and variability of tick salivary gland proteins must be a major consideration during vaccine development and evaluation.

CONCLUSIONS

Tick saliva research has taken a great leap forward in recent years. Molecular biology and high-throughput approaches (PCR subtraction, massive cDNA sequencing, proteomics and computational biology) continue to increase our knowledge of the repertoire of proteins present in the salivary glands of ticks. This new information, together with the vast knowledge acquired over the last three decades regarding the pharmacology of tick saliva and immune responses to tick salivary proteins, will open new avenues to understanding the involvement of tick saliva in blood-feeding, host immunity and pathogen transmission. Ultimately, this information should generate enhanced strategies for the use of tick salivary antigens as vaccines to control vector-borne diseases.

REFERENCES

Alarcon-Chaidez, F. J., Muller-Doblies, U. U. & Wikel, S. (2003). Characterization of a recombinant immunomodulatory protein from the salivary glands of *Dermacentor andersoni. Parasite Immunology* **25**, 69–77.

Alarcon-Chaidez, F. J., Sun, J. & Wikel, S. K. (2007). Transcriptome analysis of the salivary glands of *Dermacentor andersoni* Stiles (Acari: Ixodidae). *Insect Biochemistry and Molecular Biology* **37**, 48–71.

Aljamali, M. N., Bior, A. D., Sauer, J. R. & Essenberg, R. C. (2003). RNA interference in ticks: a study using histamine binding protein dsRNA in the female tick *Amblyomma americanum. Insect Molecular Biology* **12**, 299–305.

Aljamali, M. N., Bowman, A. S., Dillwith, J. W., *et al.* (2002*a*). Identity and synthesis of prostaglandins in the lone star tick, *Amblyomma americanum* (L.), as assessed by radio-immunoassay and gas chromatography/mass spectrometry. *Insect Biochemistry and Molecular Biology* **32**, 331–341.

Aljamali, M. N., Sauer, J. R. & Essenberg, R. C. (2002*b*). RNA interference: applicability in tick research. *Experimental and Applied Acarology* **28**, 89–96.

Altschul, S. F. & Gish, W. (1996). Local alignment statistics. *Methods in Enzymology* **266**, 460–480.

Anguita, J., Ramamoorthi, N., Hovius, J. W., *et al.* (2002). Salp15, an *Ixodes scapularis* salivary protein, inhibits CD4(+) T cell activation. *Immunity* **16**, 849–859.

Antuch, W., Guntert, P., Billeter, M., *et al.* (1994). NMR solution structure of the recombinant tick anticoagulant protein (rTAP), a factor Xa inhibitor from the tick *Ornithodoros moubata. FEBS Letters* **352**, 251–257.

Arca, B., Lombardo, F., De Lara Capurro, M., *et al.* (1999). Trapping cDNAs encoding secreted proteins from the salivary glands of the malaria vector *Anopheles gambiae. Proceedings of the National Academy of Sciences of the USA* **96**, 1516–1521.

Bergman, D. K., Palmer, M. J., Caimano, M. J., Radolf, J. D. & Wikel, S. K. (2000). Isolation and molecular cloning of a secreted immunosuppressant protein from *Dermacentor andersoni* salivary gland. *Journal of Parasitology* **86**, 516–525.

Binnington, K. C. & Kemp, D. H. (1980). Role of tick salivary glands in feeding and disease transmission. *Advances in Parasitology* **18**, 315–339.

Bior, A. D., Essenberg, R. C. & Sauer, J. R. (2002). Comparison of differentially expressed genes in the salivary glands of male ticks, *Amblyomma americanum* and *Dermacentor andersoni. Insect Biochemistry and Molecular Biology* **32**, 645–655.

Bowman, A. S., Gengler, C. L., Surdick, M. R., *et al.* (1997). A novel phospholipase A$_2$ activity in saliva of the lone star

tick, *Amblyomma americanum* (L.). *Experimental Parasitology* **87**, 121–132.

Brooks, D. R. & Isaac, R. E. (2002). Functional genomics of parasitic worms: the dawn of a new era. *Parasitology International* **51**, 319–325.

Champagne, D. E., Smartt, C. T., Ribeiro, J. M. & James, A. A. (1995). The salivary gland-specific apyrase of the mosquito *Aedes aegypti* is a member of the 5′-nucleotidase family. *Proceedings of the National Academy of Sciences of the USA* **92**, 694–698.

Ciprandi, A., De Oliveira, S. K., Masuda, A., Horn, F. & Termignoni, C. (2006). *Boophilus microplus*: its saliva contains microphilin, a small thrombin inhibitor. *Experimental Parasitology* **114**, 40–46.

Das, S., Banerjee, G., Deponte, K., *et al.* (2001). Salp25D, an *Ixodes scapularis* antioxidant, is 1 of 14 immunodominant antigens in engorged tick salivary glands. *Journal of Infectious Diseases* **184**, 1056–1064.

Dickinson, R. G., O'Hagan, J. E., Schotz, M., Binnington, K. C. & Hegarty, M. P. (1976). Prostaglandin in the saliva of the cattle tick *Boophilus microplus. Australian Journal of Experimental Biology and Medical Science* **54**, 475–486.

Ehebauer, M. T., Mans, B. J., Gaspar, A. R. & Neitz, A. W. (2002). Identification of extrinsic blood coagulation pathway inhibitors from the tick *Ornithodoros savignyi* (Acari: Argasidae). *Experimental Parasitology* **101**, 138–148.

Ferguson, J. J. & Zaqqa, M. (1999). Platelet glycoprotein IIb/IIIa receptor antagonists: current concepts and future directions. *Drugs* **58**, 965–982.

Fogaca, A. C., Almeida, I. C., Eberlin, M. N., *et al.* (2006). Ixodidin, a novel antimicrobial peptide from the hemocytes of the cattle tick *Boophilus microplus* with inhibitory activity against serine proteinases. *Peptides* **27**, 667–674.

Fogaca, A. C., Lorenzini, D. M., Kaku, L. M., *et al.* (2004). Cysteine-rich antimicrobial peptides of the cattle tick *Boophilus microplus*: isolation, structural characterization and tissue expression profile. *Developmental and Comparative Immunology* **28**, 191–200.

Francischetti, I. M., Andersen, J. F. & Ribeiro, J. M. (2002*a*). Biochemical and functional characterization of recombinant *Rhodnius prolixus* platelet aggregation inhibitor 1 as a novel lipocalin with high affinity for adenosine diphosphate and other adenine nucleotides. *Biochemistry* **41**, 3810–3818.

Francischetti, I. M., Mather, T. N. & Ribeiro, J. M. (2003). Cloning of a salivary gland metalloprotease and characterization of gelatinase and fibrin(ogen)lytic activities in the saliva of the Lyme disease tick vector *Ixodes scapularis*.

Biochemical and Biophysical Research Communications **305**, 869–875.

Francischetti, I. M., Mather, T. N. & Ribeiro, J. M. (2004). Penthalaris, a novel recombinant five-Kunitz tissue factor pathway inhibitor (TFPI) from the salivary gland of the tick vector of Lyme disease, *Ixodes scapularis*. *Thrombosis and Haemostasis* **91**, 886–898.

Francischetti, I. M., My Pham, V., Mans, B. J., *et al.* (2005). The transcriptome of the salivary glands of the female western black-legged tick *Ixodes pacificus* (Acari: Ixodidae). *Insect Biochemistry and Molecular Biology* **35**, 1142–1161.

Francischetti, I. M., Valenzuela, J. G., Andersen, J. F., Mather, T. N. & Ribeiro, J. M. (2002*b*). Ixolaris, a novel recombinant tissue factor pathway inhibitor (TFPI) from the salivary gland of the tick, *Ixodes scapularis*: identification of factor X and factor Xa as scaffolds for the inhibition of factor VIIa/tissue factor complex. *Blood* **99**, 3602–3612.

Garg, R., Juncadella, I. J., Ramamoorthi, N., *et al.* (2006). Cutting edge: CD4 is the receptor for the tick saliva immunosuppressor, Salp15. *Journal of Immunology* **177**, 6579–6583.

Gillespie, R. D., Mbow, M. L. & Titus, R. G. (2000). The immunomodulatory factors of blood feeding arthropod saliva. *Parasite Immunology* **22**, 319–331.

Gordon, J. R. & Allen, J. R. (1991). Factors V and VII anticoagulant activities in the salivary glands of feeding *Dermacentor andersoni* ticks. *Journal of Parasitology* **77**, 167–170.

Grunclova, L., Horn, M., Vancova, M., *et al.* (2006). Two secreted cystatins of the soft tick *Ornithodoros moubata*: differential expression pattern and inhibitory specificity. *Biological Chemistry* **387**, 1635–1644.

Hajnicka, V., Kocakova, P., Slavikova, M., *et al.* (2001). Anti-interleukin-8 activity of tick salivary gland extracts. *Parasite Immunology* **23**, 483–489.

Hajnicka, V., Vancova, I., Kocakova, P., *et al.* (2005). Manipulation of host cytokine network by ticks: a potential gateway for pathogen transmission. *Parasitology* **130**, 333–342.

Hannier, S., Liversidge, J., Sternberg, J. M. & Bowman, A. S. (2003). *Ixodes ricinus* tick salivary gland extract inhibits IL-10 secretion and CD69 expression by mitogen-stimulated murine splenocytes and induces hyporesponsiveness in B lymphocytes. *Parasite Immunology* **25**, 27–37.

Hawkins, R. I. & Hellmann, K. (1966). Factors affecting blood clotting from the tick *Ornithodoros moubata*. *Journal of Physiology* **185**, 70.

Holt, R. A., Subramanian, G. M., Halpern, A., Sutton, G. G., Charlab, R., Nusskern, D. R., Wincker, P., Clark, A. G., Ribeiro, J. M., Wides, R., Salzberg, S. L., Loftus, B., Yandell, M., Majoros, W. H., Rusch, D. B., Lai, Z., Kraft, C. L., Abril, J. F., Anthouard, V., Arensburger, P., Atkinson, P. W., Baden, H., De Berardinis, V., Baldwin, D., Benes, V., Biedler, J., Blass, C., Bolanos, R., Boscus, D., Barnstead, M., Cai, S., Center, A., Chaturverdi, K., Christophides, G. K., Chrystal, M. A., Clamp, M., Cravchik, A., Curwen, V., Dana, A., Delcher, A., Dew, I., Evans, C. A., Flanigan, M., Grundschober-Freimoser, A., Friedli, L., Gu, Z., Guan, P., Guigo, R., Hillenmeyer, M. E., Hladun, S. L., Hogan, J. R., Hong, Y. S., Hoover, J., Jaillon, O., Ke, Z., Kodira, C., Kokoza, E., Koutsos, A., Letunic, I., Levitsky, A., Liang, Y., Lin, J. J., Lobo, N. F., Lopez, J. R., Malek, J. A., Mcintosh, T. C., Meister, S., Miller, J., Mobarry, C., Mongin, E., Murphy, S. D., O'Brochta, D. A., Pfannkoch, C., Qi, R., Regier, M. A., Remington, K., Shao, H., Sharakhova, M. V., Sitter, C. D., Shetty, J., Smith, T. J., Strong, R., Sun, J., Thomasova, D., Ton, L. Q., Topalis, P., Tu, Z., Unger, M. F., Walenz, B., Wang, A., Wang, J., Wang, M., Wang, X., Woodford, K. J., Wortman, J. R., Wu, M., Yao, A., Zdobnov, E. M., Zhang, H., Zhao, Q., Zhao, S., Zhu, S. C., Zhimulev, I., Coluzzi, M., Della Torre, A., Roth, C. W., Louis, C., Kalush, F., Mural, R. J., Myers, E. W., Adams, M. D., Smith, H. O., Broder, S., Gardner, M. J., Fraser, C. M., Birney, E., Bork, P., Brey, P. T., Venter, J. C., Weissenbach, J., Kafatos, F. C., Collins, F. H. & Hoffman, S. L. (2002). The genome sequence of the malaria mosquito *Anopheles gambiae*. *Science* **298**, 129–149.

Horn, F., Dos Santos, P. C. & Termignoni, C. (2000). *Boophilus microplus* anticoagulant protein: an antithrombin inhibitor isolated from the cattle tick saliva. *Archives of Biochemistry and Biophysics* **384**, 68–73.

Huang, X. & Madan, A. (1999). CAP3: a DNA sequence assembly program. *Genome Research* **9**, 868–877.

Hugli, T. E. & Muller-Eberhard, H. J. (1978). Anaphylatoxins: C3a and C5a. *Advances in Immunology* **26**, 1–53.

Inokuma, H., Kemp, D. H. & Willadsen, P. (1994). Prostaglandin E_2 production by the cattle tick (*Boophilus microplus*) into feeding sites and its effect on the response of bovine mononuclear cells to mitogen. *Veterinary Parasitology* **53**, 293–299.

Iwanaga, S., Okada, M., Isawa, H., *et al.* (2003). Identification and characterization of novel salivary thrombin inhibitors from the ixodidae tick, *Haemaphysalis longicornis*. *European Journal of Biochemistry* **270**, 1926–1934.

Joiner, K. A. (1988). Complement evasion by bacteria and parasites. *Annual Review of Microbiology* **42**, 201–230.

Joubert, A. M., Crause, J. C., Gaspar, A. R., *et al.* (1995). Isolation and characterization of an anticoagulant present in the salivary glands of the bont-legged tick, *Hyalomma truncatum. Experimental and Applied Acarology* **19**, 79–92.

Joubert, A. M., Louw, A. I., Joubert, F. & Neitz, A. W. (1998). Cloning, nucleotide sequence and expression of the gene encoding factor Xa inhibitor from the salivary glands of the tick, *Ornithodoros savignyi. Experimental and Applied Acarology* **22**, 603–619.

Jutel, M., Watanabe, T., Klunker, S., *et al.* (2001). Histamine regulates T-cell and antibody responses by differential expression of H1 and H2 receptors. *Nature* **413**, 420–425.

Karczewski, J., Endris, R. & Connolly, T. M. (1994). Disagregin is a fibrinogen receptor antagonist lacking the Arg-Gly-Asp sequence from the tick, *Ornithodoros moubata. Journal of Biological Chemistry* **269**, 6702–6708.

Karczewski, J., Waxman, L., Endris, R. G. & Connolly, T. M. (1995). An inhibitor from the argasid tick *Ornithodoros moubata* of cell adhesion to collagen. *Biochemical and Biophysical Research Communications* **208**, 532–541.

Karim, S., Miller, N. J., Valenzuela, J., Sauer, J. R. & Mather, T. N. (2005). RNAi-mediated gene silencing to assess the role of synaptobrevin and cystatin in tick blood feeding. *Biochemical and Biophysical Research Communications* **334**, 1336–1342.

Kato, N., Iwanaga, S., Okayama, T., *et al.* (2005). Identification and characterization of the plasma kallikrein–kinin system inhibitor, haemaphysalin, from hard tick, *Haemaphysalis longicornis. Thrombosis and Haemostasis* **93**, 359–367.

Keller, P. M., Waxman, L., Arnold, B. A., *et al.* (1993). Cloning of the cDNA and expression of moubatin, an inhibitor of platelet aggregation. *Journal of Biological Chemistry* **268**, 5450–5456.

Konik, P., Slavikova, V., Salat, J., *et al.* (2006). Anti-tumour necrosis factor-alpha activity in *Ixodes ricinus* saliva. *Parasite Immunology* **28**, 649–656.

Kotsyfakis, M., Sa-Nuñes, A., Francischetti, I. M., *et al.* (2006). Anti-inflammatory and immunosuppressive activity of sialostatin L, a salivary cystatin from the tick *Ixodes scapularis. Journal of Biological Chemistry* **281**, 26298–26307.

Lawrie, C. H., Randolph, S. E. & Nuttall, P. A. (1999). *Ixodes* ticks: serum species sensitivity of anticomplement activity. *Experimental Parasitology* **93**, 207–214.

Lawrie, C. H., Sim, R. B. & Nuttall, P. A. (2005). Investigation of the mechanisms of anti-complement activity in *Ixodes ricinus* ticks. *Molecular Immunology* **42**, 31–38.

Leboulle, G., Crippa, M., Decrem, Y., *et al.* (2002a). Characterization of a novel salivary immunosuppressive protein from *Ixodes ricinus* ticks. *Journal of Biological Chemistry* **277**, 10083–10089.

Leboulle, G., Rochez, C., Louahed, J., *et al.* (2002b). Isolation of *Ixodes ricinus* salivary gland mRNA encoding factors induced during blood feeding. *American Journal of Tropical Medicine and Hygiene* **66**, 225–233.

Lee, W. H. & Vega, V. B. (2004). Heterogeneity detector: finding heterogeneous positions in Phred/Phrap assemblies. *Bioinformatics* **20**, 2863–2864.

Lima, C. A., Sasaki, S. D. & Tanaka, A. S. (2006). Bmcystatin, a cysteine proteinase inhibitor characterized from the tick *Boophilus microplus. Biochemical and Biophysical Research Communications* **347**, 44–50.

Limo, M. K., Voigt, W. P., Tumbo-Oeri, A. G., Njogu, R. M. & Ole-Moiyoi, O. K. (1991). Purification and characterization of an anticoagulant from the salivary glands of the ixodid tick *Rhipicephalus appendiculatus. Experimental Parasitology* **72**, 418–429.

Mans, B. J., Coetzee, J., Louw, A. I., Gaspar, A. R. & Neitz, A. W. (2000). Disaggregation of aggregated platelets by apyrase from the tick, *Ornithodoros savignyi* (Acari: Argasidae). *Experimental and Applied Acarology* **24**, 271–282.

Mans, B. J., Louw, A. I. & Neitz, A. W. (2002a). Evolution of hematophagy in ticks: common origins for blood coagulation and platelet aggregation inhibitors from soft ticks of the genus *Ornithodoros. Molecular Biology and Evolution* **19**, 1695–1705.

Mans, B. J., Louw, A. I. & Neitz, A. W. (2002b). Savignygrin, a platelet aggregation inhibitor from the soft tick *Ornithodoros savignyi*, presents the RGD integrin recognition motif on the Kunitz–BPTI fold. *Journal of Biological Chemistry* **277**, 21371–21378.

Mans, B. J., Louw, A. I. & Neitz, A. W. (2003). The major tick salivary gland proteins and toxins from the soft tick, *Ornithodoros savignyi*, are part of the tick lipocalin family: implications for the origins of tick toxicoses. *Molecular Biology and Evolution* **20**, 1158–1167.

Nakajima, C., Da Silva Vaz, I. Jr, Imamura, S., *et al.* (2005). Random sequencing of cDNA library derived from partially-fed adult female *Haemaphysalis longicornis* salivary gland. *Journal of Veterinary Medical Science* **67**, 1127–1131.

Nakajima, C., Imamura, S., Konnai, S., *et al.* (2006). A novel gene encoding a thrombin inhibitory protein in a cDNA library from *Haemaphysalis longicornis* salivary gland. *Journal of Veterinary Medical Science* **68**, 447–452.

Narasimhan, S., Koski, R. A., Beaulieu, B., *et al.* (2002). A novel family of anticoagulants from the saliva of *Ixodes scapularis*. *Insect Molecular Biology* **11**, 641–650.

Narasimhan, S., Montgomery, R. R., Deponte, K., *et al.* (2004). Disruption of *Ixodes scapularis* anticoagulation by using RNA interference. *Proceedings of the National Academy of Sciences of the USA* **101**, 1141–1146.

Nazareth, R. A., Tomaz, L. S., Ortiz-Costa, S., *et al.* (2006). Antithrombotic properties of ixolaris, a potent inhibitor of the extrinsic pathway of the coagulation cascade. *Thrombosis and Haemostasis* **96**, 7–13.

Nene, V., Lee, D., Kang'A, S., *et al.* (2004). Genes transcribed in the salivary glands of female *Rhipicephalus appendiculatus* ticks infected with *Theileria parva*. *Insect Biochemistry and Molecular Biology* **34**, 1117–1128.

Nene, V., Lee, D., Quackenbush, J., *et al.* (2002). AvGI, an index of genes transcribed in the salivary glands of the ixodid tick *Amblyomma variegatum*. *International Journal of Parasitology* **32**, 1447–1456.

Nienaber, J., Gaspar, A. R. & Neitz, A. W. (1999). Savignin, a potent thrombin inhibitor isolated from the salivary glands of the tick *Ornithodoros savignyi* (Acari: Argasidae). *Experimental Parasitology* **93**, 82–91.

Nunn, M. A., Sharma, A., Paesen, G. C., *et al.* (2005). Complement inhibitor of C5 activation from the soft tick *Ornithodoros moubata*. *Journal of Immunology* **174**, 2084–2091.

Paesen, G. C., Adams, P. L., Harlos, K., Nuttall, P. A. & Stuart, D. I. (1999). Tick histamine-binding proteins: isolation, cloning, and three-dimensional structure. *Molecular Cell* **3**, 661–671.

Quackenbush, J., Cho, J., Lee, D., *et al.* (2001). The TIGR Gene Indices: analysis of gene transcript sequences in highly sampled eukaryotic species. *Nucleic Acids Research* **29**, 159–164.

Ramakrishnan, V. G., Aljamali, M. N., Sauer, J. R. & Essenberg, R. C. (2005). Application of RNA interference in tick salivary gland research. *Journal of Biomolecular Techniques* **16**, 297–305.

Ramamoorthi, N., Narasimhan, S., Pal, U., *et al.* (2005). The Lyme disease agent exploits a tick protein to infect the mammalian host. *Nature* **436**, 573–577.

Ribeiro, J. M. (1987*a*). *Ixodes dammini*: salivary anti-complement activity. *Experimental Parasitology* **64**, 347–353.

Ribeiro, J. M. (1987*b*). Role of saliva in blood-feeding by arthropods. *Annual Review of Entomology* **32**, 463–478.

Ribeiro, J. M. (1989). Role of saliva in tick/host interactions. *Experimental and Applied Acarology* **7**, 15–20.

Ribeiro, J. M. (1995). How ticks make a living. *Parasitology Today* **11**, 91–93.

Ribeiro, J. M. & Mather, T. N. (1998). *Ixodes scapularis*: salivary kininase activity is a metallo dipeptidyl carboxypeptidase. *Experimental Parasitology* **89**, 213–221.

Ribeiro, J. M., Alarcon-Chaidez, F., Francischetti, I. M., *et al.* (2006). An annotated catalog of salivary gland transcripts from *Ixodes scapularis* ticks. *Insect Biochemistry and Molecular Biology* **36**, 111–129.

Ribeiro, J. M., Endris, T. M. & Endris, R. (1991). Saliva of the soft tick, *Ornithodoros moubata*, contains anti-platelet and apyrase activities. *Comparative Biochemistry and Physiology A* **100**, 109–112.

Ribeiro, J. M., Evans, P. M., McSwain, J. L. & Sauer, J. (1992). *Amblyomma americanum*: characterization of salivary prostaglandins E_2 and $F_{2\alpha}$ by RP-HPLC/bioassay and gas chromatography-mass spectrometry. *Experimental Parasitology* **74**, 112–116.

Ribeiro, J. M., Makoul, G. T., Levine, J., Robinson, D. R. & Spielman, A. (1985). Antihemostatic, antiinflammatory, and immunosuppressive properties of the saliva of a tick, *Ixodes dammini*. *Journal of Experimental Medicine* **161**, 332–344.

Ribeiro, J. M., Makoul, G. T. & Robinson, D. R. (1988). *Ixodes dammini*: evidence for salivary prostacyclin secretion. *Journal of Parasitology* **74**, 1068–1069.

Ribeiro, J. M., Weis, J. J. & Telford, S. R. III (1990). Saliva of the tick *Ixodes dammini* inhibits neutrophil function. *Experimental Parasitology* **70**, 382–388.

Sangamnatdej, S., Paesen, G. C., Slovak, M. & Nuttall, P. A. (2002). A high affinity serotonin- and histamine-binding lipocalin from tick saliva. *Insect Molecular Biology* **11**, 79–86.

Soares, C. A., Lima, C. M., Dolan, M. C., *et al.* (2005). Capillary feeding of specific dsRNA induces silencing of the *isac* gene in nymphal *Ixodes scapularis* ticks. *Insect Molecular Biology* **14**, 443–452.

St Charles, R., Padmanabhan, K., Arni, R. V., Padmanabhan, K. P. & Tulinsky, A. (2000). Structure of tick anticoagulant peptide at 1.6 Å resolution complexed with bovine pancreatic trypsin inhibitor. *Protein Science* **9**, 265–272.

Sukumaran, B., Narasimhan, S., Anderson, J. F., *et al.* (2006). An *Ixodes scapularis* protein required for survival of *Anaplasma phagocytophilum* in tick salivary glands. *Journal of Experimental Medicine* **203**, 1507–1517.

Thompson, J. D., Gibson, T. J., Plewniak, F., Jeanmougin, F. & Higgins, D. G. (1997). The CLUSTAL X windows interface: flexible strategies for multiple sequence alignment aided by quality analysis tools. *Nucleic Acids Research* **25**, 4876–4882.

Trimnell, A. R., Davies, G. M., Lissina, O., Hails, R. S. & Nuttall, P. A. (2005). A cross-reactive tick cement antigen is a candidate broad-spectrum tick vaccine. *Vaccine* **23**, 4329–4341.

Trimnell, A. R., Hails, R. S. & Nuttall, P. A. (2002). Dual action ectoparasite vaccine targeting 'exposed' and 'concealed' antigens. *Vaccine* **20**, 3560–3568.

Valenzuela, J. G. (2002*a*). Exploring the messages of the salivary glands of *Ixodes ricinus*. *American Journal of Tropical Medicine and Hygiene* **66**, 223–224.

Valenzuela, J. G. (2002*b*). High-throughput approaches to study salivary proteins and genes from vectors of disease. *Insect Biochemistry and Molecular Biology* **32**, 1199–1209.

Valenzuela, J. G., Belkaid, Y., Rowton, E. & Ribeiro, J. M. (2001). The salivary apyrase of the blood-sucking sand fly *Phlebotomus papatasi* belongs to the novel Cimex family of apyrases. *Journal of Experimental Biology* **204**, 229–237.

Valenzuela, J. G., Charlab, R., Galperin, M. Y. & Ribeiro, J. M. (1998). Purification, cloning, and expression of an apyrase from the bed bug *Cimex lectularius*: a new type of nucleotide-binding enzyme. *Journal of Biological Chemistry* **273**, 30583–30590.

Valenzuela, J. G., Charlab, R., Mather, T. N. & Ribeiro, J. M. (2000). Purification, cloning, and expression of a novel salivary anticomplement protein from the tick, *Ixodes scapularis*. *Journal of Biological Chemistry* **275**, 18717–18723.

Valenzuela, J. G., Francischetti, I. M., Pham, V. M., *et al.* (2002). Exploring the sialome of the tick *Ixodes scapularis*. *Journal of Experimental Biology* **205**, 2843–2864.

Van de Locht, A., Stubbs, M. T., Bode, W., *et al.* (1996). The ornithodorin–thrombin crystal structure, a key to the TAP enigma? *EMBO Journal* **15**, 6011–6017.

Wang, X., Coons, L. B., Taylor, D. B., Stevens, S. E. Jr & Gartner, T. K. (1996). Variabilin, a novel RGD-containing antagonist of glycoprotein IIb–IIIa and platelet aggregation inhibitor from the hard tick *Dermacentor variabilis*. *Journal of Biological Chemistry* **271**, 17785–17790.

Waxman, L., Smith, D. E., Arcuri, K. E. & Vlasuk, G. P. (1990). Tick anticoagulant peptide (TAP) is a novel inhibitor of blood coagulation factor Xa. *Science* **248**, 593–596.

Wikel, S. K. (1999). Tick modulation of host immunity: an important factor in pathogen transmission. *International Journal for Parasitology* **29**, 851–859.

Zhu, K., Bowman, A. S., Brigham, D. L., *et al.* (1997*a*). Isolation and characterization of americanin, a specific inhibitor of thrombin, from the salivary glands of the lone star tick *Amblyomma americanum* (L.). *Experimental Parasitology* **87**, 30–38.

Zhu, K., Dillwith, J. W., Bowman, A. S. & Sauer, J. R. (1997*b*). Identification of hemolytic activity in saliva of the lone star tick (Acari: Ixodidae). *Journal of Medical Entomology* **34**, 160–166.

Zhu, K., Sauer, J. R., Bowman, A. S. & Dillwith, J. W. (1997*c*). Identification and characterization of anticoagulant activities in the saliva of the lone star tick, *Amblyomma americanum* (L.). *Journal of Parasitology* **83**, 38–43.

5 • Tick toxins: perspectives on paralysis and other forms of toxicoses caused by ticks

B. J. MANS, R. GOTHE AND A. W. H. NEITZ

INTRODUCTION

Tick toxicosis has been a research focus for almost 80 years and during this time, several excellent reviews on this subject have been written that cover the history of toxicosis research as well as its aetiology and pathology (Gregson, 1943, 1973; Stampa, 1959; Neitz, 1962; Murnaghan & O'Rouke, 1978; Gothe, Kunze & Hoogstraal, 1979; Gothe, 1984, 1999; Wikel, 1984; Gothe & Neitz, 1991; Sonenshine, 1993; Masina & Broady, 1999). A comprehensive monograph on tick toxicoses of all forms has also been published (Gothe, 1999).

FUNCTIONAL SIGNIFICANCE OF TICK TOXINS

While arthropods such as spiders and scorpions are notoriously venomous organisms that utilize their toxins for protection as well as predation, the advantages for ticks being toxic is unclear. It has been suggested that tick paralysis may be a vestigial function conserved in ticks, when ticks evolved a parasitic lifestyle (Stone et al., 1989). Paralysis toxins have been attributed to functional significance during feeding of the tick, in that host mobility and grooming is impaired. This might be relevant, as tick paralysis sets in at the later stages of tick engorgement, when the tick is most liable to be killed by grooming practices. Paralysis would also affect the respiratory system leading to elevated breathing rates and an increase in carbon dioxide expiration. This together with pheromone secretion could attract ticks to the paralysed animal, which accelerates the finding and feeding of ticks. It has been argued that this might be true for most ticks, even though no clinical symptoms can be observed in the majority of feeding events (Gothe, 1984). Toxins might also exert local anaesthesia, prevent blood coagulation or act as a general feeding stimulant (Stone et al., 1989). A role as a regulator of protein synthesis has also been suggested for

the paralysis toxin from *Rhipicephalus evertsi evertsi* based on localization to chromatin in the nuclei (Crause et al., 1993). Problems with attributing functional significance to tick toxins lie in the fact that tick paralysis is not a widespread phenomenon found in all tick species.

TOXICOSES FROM AN EVOLUTIONARY PERSPECTIVE

Another consideration is whether tick toxins have a common ancestor shared with toxins from other toxic arthropods as this would assign biological significance or function to tick toxins. Toxins might also have specific functions related to their toxicity that were specifically acquired during adaptation to a blood-feeding environment. Toxicity could also be a by-product of proteins occurring in a novel environment and recognition of host targets, a chance event. To investigate these possibilities, it is important to delineate clearly various forms of toxicosis and find their shared properties or differences (e.g. mechanism of pathogenesis, homology) as this will give valuable information as to their origins. Such comparative studies are one way in which a holistic view of tick toxicosis will be attained. There is no consensus yet whether the paralysis toxins from different tick species are homologous. Based on sequence similarity of holocyclotoxin with scorpion toxins (data not yet published) it was speculated that other paralysis toxins from ticks might also be related (Masina & Broady, 1999). On the other hand, neurotoxins from predatory mites, *Pyemotes tritici* (Superorder: Acariformes) are unique with molecular masses of ~30 kDa (Tomalski & Miller, 1991; Tomalski et al., 1993). Toxins from different arachnid subclasses (spiders and scorpions) also do not all fall into the same protein families (Menez, 1998; Escoubas, Diochot & Corzo, 2000; Rash & Hodgson, 2002). Ticks are also more closely related to the non-toxic

Ticks: Biology, Disease and Control, ed. Alan S. Bowman and Patricia A. Nuttall. Published by Cambridge University Press.
© Cambridge University Press 2008.

ricinulei (tick-like spiders) than to spiders and scorpions (Lindquist, 1984; Shultz, 1990). Phylogenetic analyses have also shown that the sister-group to ticks are the Holothyrida (Dobson & Barker, 1999). Holothyrida is a group of free-living scavenging mites which mainly live on body fluids of dead arthropods. It has been suggested that ticks shared this same trait before adaptation to a blood-feeding environment (Walter & Proctor, 1998). An evolutionary origin for tick toxins shared with other toxic arthropods would thus seem to be a remote possibility. The conclusion derived from this is that evolutionary relationships between toxins from the different arthropod classes cannot yet be inferred with certainty. Another consideration is that ticks originated either ∼390 or ∼120 million years ago (Klompen *et al.*, 1996; Dobson & Barker, 1999). This was at a time when most modern hosts that are affected by paralysis toxins did not exist. If tick toxins do confer survival advantage, this character probably only emerged within the tick lineage during the adaptation of ticks to a blood-feeding environment or even later when modern hosts were encountered.

SOURCES OF TICK TOXINS

The origins of tick toxicoses should be distinguished from the source of the toxin. The former relates to the evolutionary history of toxins and the latter to its more immediate history and the organisms or tissues it derives from. A number of different possibilities exist as to the source of tick toxins (Fig. 5.1). They can be a natural tick product that is either in itself toxic or is transformed into a toxic component in the host, or toxins are derived from breakdown of host tissues or a product of a symbiotic organism in the tick (Gregson, 1973). Various other possibilities have also been reviewed (Neitz *et al.*, 1983). An updated version of possible toxin sources is given in Fig. 5.1: (A) Toxins might derive directly from a pathogen or symbiotic organism living within the tick. In the case of ticks, pathogenic or symbiotic organisms can be ruled out as possible sources, due to the fact that toxicosis or paralysis could not be transferred when healthy animals were inoculated with blood from affected animals. Toxicosis in this case would only be transferred if toxicoses were associated with an infectious agent. If tick toxins are synthesized by a symbiotic organism within the tick and transported to the salivary glands where they accumulate and are concentrated to sufficient concentration, then inoculation with affected host-derived material would not produce toxicosis.

This same rationale follows for non-toxic components from symbiotic organisms being transformed within the tick

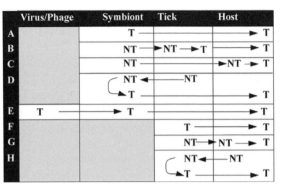

Fig. 5.1. Possible sources of tick toxins. (A) Toxins (T) might derive directly from a pathogen or symbiotic organism living within the tick. (B) A non-toxic (NT) component from a symbiotic organism is transformed by the tick to a toxic form. (C) A non-toxic component from a symbiotic organism is transformed by the host to a toxic form. (D) A non-toxic component from a tick is transformed by a symbiotic organism to a toxic form. (E) A toxin could derive from a bacteriophage or virus that infects a tick symbiont or pathogen. (F) A toxin might derive directly from the tick. (G) The tick secretes a non-toxic component which is transformed within the host to a toxin. (H) A non-toxic component from the host is ingested by the tick and transformed to a toxic form before resecretion into the host.

(B) or the host (C) to toxins, or for non-toxic components from ticks being transformed to toxins by a symbiotic organism (D), or toxins derived from a virus or bacteriophage that infects the tick symbiont or pathogen (E).

In some cases ample evidence indicates that some tick toxins are derived from the tick itself (F). These include the sand tampan toxins that are part of a number of highly abundant proteins found within the tick salivary glands and group by phylogenetic analysis within the tick lipocalin family (Mans *et al.*, 2001, 2002, 2003). A further example is tick paralysis which occurs after a slow feeding period that lasts for 5–7 days; it is associated with rapid engorgement indicating that toxins are secreted during this phase. This suggests that paralysis toxins are synthesized by ticks per se. The toxin from *R. evertsi evertsi* has been purified from salivary glands and has been localized to secretory granules and the nucleus suggesting that this protein is from a tick origin. However, no paralysis could be induced with tick salivary gland extracts obtained from fed ticks, although feeding ticks caused paralysis in sheep. The same phenomenon was observed for *Dermacentor variabilis* where induction of paralysis by salivary gland extracts failed. This suggests that, in these cases, the toxins secreted by these ticks are either inactivated by components not secreted by the tick or that toxins are activated during the secretory process, or else that the toxin is not derived from the tick itself.

Non-toxic components from the tick can also be converted into toxins in the host (G). Non-toxic material from the host could also be ingested by the tick during feeding, transformed into a toxic component and be resecreted into the host (H). This possibility might not be an impossible scenario, as resecretion of host proteins during the rapid engorgement phase has been indicated for hard ticks (Wang & Nuttall, 1994). It could explain why removal of ticks leads to rapid recovery of affected hosts. If we hold to the principle of Occam's razor, the most simple route (F) should also be the most likely route. The other possibilities should also be borne in mind when investigating tick toxicoses as the enigma of toxin origins for most tick toxins still exists.

VENOMOUS AND TOXIC ARTHROPODS

A distinction should be made between venomous organisms (those that actively secrete toxic components) and those that contain toxins (toxic components present in biological tissues). In ticks both forms of toxins are recognized. Certain tick species can cause pathological changes in their host by inoculation of non-infectious components during feeding. On the basis of this, tick toxins have been identified in whole tick extracts, salivary gland secretions (SGS), salivary gland extracts (SGE) and tick eggs. Tick toxins can, however, only be assigned a venomous status if they are secreted during feeding. The major form of tick toxicosis associated with feeding is tick paralysis, while various other forms of toxicoses induced by individual tick species include sand tampan toxicoses from *Ornithodoros savignyi*, *Hyalomma truncatum* toxicoses that include sweating sickness, Mhlosinga and Magudu, necrotic stomatitis nephrosis syndrome and toxicoses from *Rhipicephalus (Boophilus) microplus*, *Dermacentor marginatus*, *Rhipicephalus appendiculatus*, *Ixodes redikorzevi* and *O. gurneyi* (Gothe, 1999). Hypersensitivity and other immunological reactions can also be included within the scope of general tick toxicoses (Wikel, 1984). It is also clear that many tick saliva components, e.g. proteases, protease inhibitors, hyaluronidases, anticoagulants, platelet aggregation inhibitors and haemolytic agents, could have toxic effects at the attachment site of the tick on the host (Vermeulen & Neitz, 1987). For practical reasons a distinction is made in this review between tick paralysis and tick toxicosis. The latter is regarded to include all toxicoses, excepting paralysis.

TICK PARALYSIS

Tick paralysis was the earliest form of toxicosis described for ticks and records describing death caused by ticks date back to 1824 (cited in Standbury & Huyck, 1945). For an interesting history on the subject, the reader is referred to the thorough account by Gregson (1973). Paralysis is the most important tick toxicosis for veterinary and human medicine (Gothe & Neitz, 1991). Of the approximately 869 tick species, paralysis has been described for 55 species of hard ticks and 14 soft tick species (Gothe, 1999). However, for many of these species only a few, often insufficient or uncertain records regarding the actual toxicity exist in literature (Tables 5.1 and 5.2). It would thus seem as if species with toxins are rather the exception than the norm. The most important paralysis-inducing ticks include the hard ticks *I. holocyclus* (Australia), *D. andersoni*, *D. variabilis* (North America), *I. rubicundus* and *R. evertsi evertsi* (South Africa). Soft ticks for which paralysis has been described, demonstrated or suspected include *Argas walkerae*, *A. arboreus*, *A. lahorensis* (Eurasia), *O. capensis*, *O. savignyi* (see section on sand tampan toxicoses) and *Otobius megnini* (South Africa). No logical pattern can be observed for paralysis ticks within the Acari (Fig. 5.2) and it can only be concluded that more ixodid tick species cause paralysis than argasid species. In fact, paralysis only occurs after prolonged periods of feeding and in soft ticks this type of feeding behaviour is only observed for larvae or nymphae, which feed for several days in contrast to other stages, which complete feeding within minutes.

Paralysis is associated with definite feeding phases

In experimentally induced paralysis, secretion of neurotoxin coincides with a definite repletion phase and in hard ticks is limited to females only. For *R. evertsi evertsi* paralysis, toxicity is associated with a short period between day 4 and 5 of feeding and a tick body mass of 15–21 mg (Gothe & Lämmler, 1982; Neitz & Gothe, 1986). Paralysis induced by the tick *I. holocyclus* sets in after 4–5 days of feeding, while paralysis with *D. variabilis* is only detected approximately 6–8 days after attachment (Gregson, 1973; Masina & Broady, 1999). In the soft tick *A. walkerae*, it is only larvae that cause paralysis, which occurs after 5–6 days of feeding (Gothe, 1984; Gothe & Neitz, 1991). In all instances, paralysis coincides with the rapid engorgement phase which is marked by the production and secretion of numerous protein products by the salivary glands. Paralysis is exhibited as an ascending flaccid tetraplegia due to an impaired functioning of the

Table 5.1 *Ticks that have been implicated in paralysis. Definite confirmation of paralysis includes paralysis induced under experimental conditions, or where numerous reports from various sources implicated the specific tick species, or where the host recovered from paralysis symptoms after identification and removal of the tick*

Paralysis tick	Country	Main hosts implicated to be paralysed
Argas		
A. africolumbae	Africa	Chick
A. arboreus	Africa	Avian
A. miniatus	South America	Chick
A. persicus	South Africa	Avian
A. radiatus	USA	Avian
A. sanchezi	USA	Avian
A. walkerae	Southern Africa	Avian
A. lahorensis	Eurasia	Sheep, bovine
Ixodes		
I. brunneus	USA	Avian
I. gibbosus	Europe, Israel	Sheep, calves, horse, goats
I. holocyclus	Eastern Australia	Dog, sheep, humans etc.
I. rubicundus	South Africa	Sheep
Amblyomma		
A. cajannense	South America	Bovine, sheep, goats
A. maculatum	USA, South America	Humans, dog
A. ovale	South America	Humans
A. testudinis	Argentinia, Peru	Reptiles
Dermacentor		
D. andersoni	USA	Humans, sheep, horse, bovine
D. occidentalis	Pacific Coast USA	Horse
D. rhinocerinus	Africa	Rabbit
D. variabilis	USA	Humans, dog
Haemaphysalis		
H. kutchensis	India, Pakistan	Rabbits
H. punctata	Britain, Europe, Japan	Goats, sheep, chickens
Hyalomma		
H. truncatum	Africa	Humans, sheep
Rhipicephalus		
R. evertsi evertsi	Southern Africa	Sheep
R. evertsi imeticus	Africa	Sheep
R. exophthalmos	Southern Africa	Rabbit
R. warburtoni	South Africa	Goats, sheep

Source: Data were compiled from Gothe (1999).

Table 5.2 *Ticks that have been implicated in paralysis by case studies but for which no conclusive experimental data exist*

Tick	Country	Main hosts implicated to be paralysed
Argas		
A. monolakensis	Mono Lake, USA	Gull
A. reflexus	Palearctic	Avian and mammal
A. robertsi	Asia/Australia	Avian
Ornithodoros		
O. capensis	Oceans worldwide	Seabird
O. savignyi	Africa, India	Bovine
Otobius		
O. megnini	Worldwide	Mammal
Ixodes		
I. arboricola	Europe	Avian
I. cookei	USA	Humans
I. cornuatus	Tasmania, Southern Australia	Dog, cat, child
I. crenulatus	USSR	Sheep
I. frontalis	Europe	Dove
I. hexagonus	Europe	Humans
I. hirsti	Tasmania, Australia	Cat
I. muris	USA	Dog, cat
I. pacificus	Western USA	Dog
I. redikorzevi	Israel	Humans
I. ricinus	Europe	Humans, sheep
I. scapularis	USA	Humans, dog
I. tasmani	Tasmania, Australia	Koala
I. tancitarius	Mexico	Humans
Amblyomma		
A. americanum	USA	Humans, dog, wolf
A. hebraeum	Southern Africa	Humans, sheep, goats
A. variegatum	Africa	Sheep
Dermacentor		
D. albipictus	USA, Canada, Mexico	Horse, elk
D. auratus	South East Asia	Humans
D. marginatus	Europe, Asia, Africa	
D. nuttalli	Asia	
D. reticulates	Eurasia	Sheep
D. silvarum	Mongolia	Sheep

(*cont.*)

Table 5.2 (*cont.*)

Tick	Country	Main hosts implicated to be paralysed
Haemaphysalis		
H. chordeilis	Canada	Humans
H. cinnabarina	Brazil	
H. concinna	Yugoslawia	Ruminants
H. inermis	Europe	Goats, sheep, calves
H. parva	Middle-East	Sheep
H. sulcata	Europe, India	Ruminants
Hyalomma		
H. scupense	Europe, Kazackstan, Yogod	Sheep
H. aegyptium	Mediterranean	Sheep, tortoise
Boophilus (*Rhipicephalus*)		
B. annulatus	Europe, Africa, Mexico	Humans
Ripicentor		
R. nuttalli	Africa	Dog
Rhipicephalus		
R. bursa	Mediterranean	Sheep
R. praetextatus	North East Africa	Humans
R. sanguineus	Global	Humans, dog
R. simus	Africa	Humans
R. tricuspis	Africa	Sheep, bovine

Source: Data were compiled from Gothe (1999).

nervous system (Gothe & Neitz, 1991). While these are generally observed symptoms, most neurotoxins have specific characteristics not necessarily shared with toxins from other tick species.

PARALYSES AND TOXICOSES OF THE ARGASID TICK FAMILY

Gregson (1973) postulated that paralysis induced by soft ticks is distinct from paralysis of hard ticks in that paralysis is only caused by the immature stages. The only argasids for which definite paralysis is observed are ticks in the genus *Argas*.

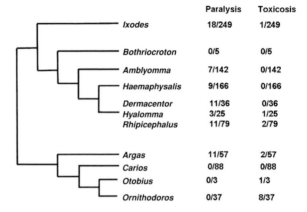

Fig. 5.2. The phylogenetic distribution of the major paralysis-inducing tick genera. Indicated are the number of species that are implicated in causing paralysis or toxicoses and the total number of species within the genus. The phylogenetic tree and species data were updated and compiled from data of Barker and Murrell (2004) and data on toxins from Gothe (1999).

Paralysis and toxicoses of the genus *Argas*

Argas larvae that cause paralysis of fowls under laboratory conditions include *A. africolumbae*, *A. arboreus*, *A. lahorensis*, *A. persicus*, *A. radiatus*, *A. sanchezi* and *A. walkerae* (Gothe, 1984). In all cases, paralysis symptoms coincide with the rapid engorgement phase (5–6 days) and persist until all larvae attain a maximal state of engorgement or until termination of the parasitic phase. Symptoms abate as the number of larvae diminishes, with total recovery after all larvae have fallen off the host. *Argas. miniatus* has also been implicated in causing paralysis (Hoogstraal, 1985). Although no paralysis is caused in humans by *Argas* species, severe irritation has been associated with the bites of *A. reflexus* (Hoogstraal, 1985). This is most probably related to the recent upsurge in described allergic reactions of humans against bites by *A. reflexus* in Europe (Sirianni *et al.*, 2000; Rolla *et al.*, 2004; Hilger *et al.*, 2005; Quercia *et al.*, 2005; Kleine-Tebbe *et al.*, 2006; Spiewak *et al.*, 2006). Of interest, is that one of the major allergens has been identified as a lipocalin (Arg r1), which shows homology to soft tick lipocalins from *O. savignyi*, some of which were implicated in toxicoses (Hilger *et al.*, 2005; Mans, 2005). In terms of pathology of *Argas* paralysis, *A. walkerae* has been studied in most detail.

PARALYSIS BY *A. WALKERAE*
Electromyographical studies indicated that the fast-conducting nerve fibres of the peripheral nervous system

are affected and the paralysis can be classified as a motor polyneuropathy that does not affect the afferent paths (Gothe & Kunze, 1971; Gothe *et al.*, 1971; Gothe & Neitz, 1991). The toxin seems to affect the liberation of acetylcholine as well as its receptor's sensitivity at the myoneural synapse (Gothe & Neitz, 1991). A monoclonal antibody directed against the toxin from *R. evertsi evertsi* also recognizes protein complexes of molecular mass 60–70 kDa from crude *A. walkerae* extracts and prevents paresis (partial paralysis) of day-old chicks (Crause *et al.*, 1994). The toxic fraction was purified using a bioassay to detect toxic activity, based on injection of day-old chicks (Viljoen *et al.*, 1990). The purified fraction showed two bands with molecular masses of 32 and 60 kDa using reducing SDS-PAGE while one band (pI ∼4.5) was obtained by isoelectric focusing. Macromolecular complexes (80–100 kDa) were observed using size exclusion chromatography (Viljoen *et al.*, 1990). Recently, the monoclonal antibody directed against the toxin from *R. evertsi evertsi* was used in an attempt to purify the neurotoxin from *A. walkerae* extracts and whilst a 68-kDa toxin was detected using Western blot analysis, an 11-kDa protein was purified that showed cross-reactivity with the mAb using enzyme-linked immunosorbent assay (ELISA), although not being detected during Western blot analysis (Maritz *et al.*, 2000). Crude *A. walkerae* larval extracts inhibited potassium-stimulated and veratridine-evoked release of [³H] glycine from rat brain synaptosomes, suggesting that the toxin might be targeting ion channels involved in depolarization (Maritz *et al.*, 2001).

PARALYSIS BY *A. LAHORENSIS*

Paralysis of sheep and cattle by *A. lahorensis* has been reported in Yugoslavia, Macedonia, the Caucasus, Kazakhstan, Central Asia, Turkmenistan and possibly Turkey (Gothe, 1999). Paralysis is caused by high numbers (100–200) of nymphal ticks of the slow-feeding third stage. Paralysis progresses rapidly from the rear to the front of the body accompanied by occasional convulsions. Death occurs on the third or fourth day after the initial onset of paralysis (Gregson, 1973).

Paralysis and toxicoses of the genus *Ornithodoros*

In the genus *Ornithodoros*, no clear-cut evidence for paralysis exists. Paralysis of marine birds by *O. capensis* has been doubted (Hoogstraal, 1985). Other members of this genus can, however, cause severe reactions in the host and symptoms ranging from pain, blisters, local irritation, oedema, fever, pruritus, inflammation and systemic disturbances have

been indicated for *O. amblus*, *O. capensis*, *O. coniceps*, *O. coriaceus*, *O. gurneyi*, *O. muesbecki*, *O. savignyi* and *O. rostratus* (Hoogstraal, 1985). From these the only form of toxicosis investigated in depth is sand tampan toxicosis caused by *Ornithodoros savignyi*.

SAND TAMPAN TOXICOSIS AND PARALYSIS

One of the first indications of toxicosis due to sand tampans (*O. savignyi*) was a report by Kone (1948) that described the death of 10 bovines from a herd of 98 cattle within 6 hours of exposure to the tampans. Their ages varied from 18 months to 3 years (Kone, 1948). These cattle were en route to Nigeria and were being vaccinated at a vaccination post in N'Guigmi. Before vaccination, they were tethered within a pen that housed numerous sand tampans. Within 2 hours of confinement the first animal succumbed and others soon followed. The symptoms displayed were indicative of agonizing pain and their rapid development before death suggested anaphylactic shock. Rousselot (1956), who was part of the French delegation at the joint FAO/OIE meeting on the control of tick-borne diseases of lifestock, reported the involvement of several hundred cattle on the borders of Lake Chad in which a high mortality rate from asphyxiation was indicated. Unfortunately, this report was included under the heading of tick paralysis in the meeting report and might have led to subsequent confusion regarding the involvement of *O. savignyi* as an agent of tick paralysis (Neitz, 1962; Hoogstraal, 1985; Gothe & Neitz, 1991; Gothe, 1999). Exsanguination was another cause of death initially attributed to mortality of cattle, sheep and camels induced by *O. savignyi* (Hoogstraal, 1956; Du Toit & Theiler, 1964). Howell (1966) stated that feeding of only three ticks kills a guinea pig, suggesting that a toxin is secreted by this tick species.

Purification of sand tampan toxins

It was subsequently shown that salivary gland secretion (SGS) obtained from *O. savignyi* by pilocarpine stimulation has an LD_{50} of ∼200 μl/kg when injected subcutaneously into 10-g albino mice (Howell, Neitz & Potgieter, 1975). Undiluted SGS (50 μl ∼5 salivary gland equivalents) kills a mouse within 8 minutes. The toxic activity is temperature stable for at least 15 minutes at 80 °C. An acidic toxin (pI ∼5) was purified from SGS and shown to be a highly abundant protein (∼9% of the total salivary gland secretion protein) (Neitz, Howell & Potgieter, 1969). It had a molecular mass of ∼15 400 Da and in this purified form could kill a mouse within 90 minutes if injected at a concentration of 400 μg/10 g mouse. A non-toxic component (Mr ∼16 000 Da) that showed N-terminal amino acid sequence similarity to the toxin was also described (Neitz, 1976). Recently, the acidic toxin (tick salivary gland protein 2 – TSGP2), a non-toxic homologue (TSGP3) and a basic toxin (TSGP4, pI ∼8) were purified from SGE (Mans et al., 2001, 2002). The toxins, non-toxic homologue and another non-toxic protein (TSGP1) are the most highly abundant proteins in the SGE (TSGPs) and comprise individually ∼4–5% of the total SGE protein (Mans et al., 2001). Based on this high abundance it was suggested that the TSGPs may be involved in tick salivary gland granule biogenesis. The N-terminal sequences and molecular masses obtained for the acidic toxin and its homologue corresponded very well with each other and the sequences obtained before. It was also shown that 24 μg of the acidic toxin or 34 μg of the basic toxin in purified form are sufficient to kill a 20-g mouse within 30 minutes (Mans et al., 2002). The difference observed between the toxicities of the original and more recent toxin preparations could be ascribed to the fact that the toxins in purified form seem to be labile and that the current success in obtaining such active toxic fractions can be ascribed to modern high-performance liquid chromatography technology that allows purification of the toxins within a few hours, compared to several days for the previous isolation attempts.

Clinical pathology of sand tampan toxicosis

In controlled experiments where animals (sheep, rats and mice) were injected subcutaneously with SGS, visible symptoms were minimal (Mans et al., 2002). This was also observed in an experiment where a 300-kg bull was confined to a camp that contained approximately 2000 tampans for 2 hours on four consecutive days. Most animals show symptoms of shock just prior to death. In large animals the histopathology indicates congestion and oedema in the myocardium, integument, spleen, kidneys, lungs and lymph glands. Haemorrhage within the lungs also occurs in smaller animals such as rats and mice. No other pathological changes can normally be observed and Howell suggested that animals die of heart failure (Mans et al., 2002). Addition of SGE to a rat heart perfusion system causes arrhythmia and bradycardia, followed by cardiac arrest. Monitoring of mouse electrocardiograms after subcutaneous injection of purified toxins showed that ventricular tachycardia (TSGP2) and a Mobitz-type ventricular block (TSGP4) are induced (Mans et al., 2002). This suggests that the pathogenicity of sand tampan toxicosis is due to a targeting of the host's cardiac system and as such is distinct from tick paralysis.

The origins of sand tampan toxicosis

It was shown that the TSGPs belong to the lipocalin protein family (Mans, Louw & Neitz, 2003). In haematophagous organisms lipocalins perform a variety of functions that include the regulation of inflammation during feeding of the hard tick R. appendiculatus by sequestration of histamine (Paesen et al., 1999, 2000). Lipocalins from the blood-sucking bugs Rhodnius prolixus and Triatoma pallidipenis inhibit various haemostatic processes within the host. The toxic TSGP2 and non-toxic TSGP3 show high sequence identity (46%) with moubatin, a platelet aggregation inhibitor from the closely related non-toxic tick, O. moubata (Keller et al., 1993; Waxman & Connolly, 1993). No anti-platelet activity or anti-blood coagulation capabilities are associated with the TSGPs. Phylogenetic and Western blot analysis suggests that the toxins and the non-toxic TSGP3 only evolved after the divergence of O. savignyi and O. moubata by gene duplication from existing lipocalins. This gives convincing evidence of a recent origin for specific tick toxins within a single genus and species. This has important implications as to the origins of all types of tick toxicoses and suggests that various forms of tick toxicoses originated at different times independently. Of interest is the fact that SGE from O. moubata also shows a high abundance of proteins with molecular masses that corresponds with the lipocalins from O. savignyi. Their distribution in terms of mass and isoelectric point are, however, slightly different (Mans et al., 2004). This indicates that O. moubata probably has its own set of lipocalin proteins that are non-toxic and which might be more closely related to the original toxin ancestral lipocalins. It should be pointed out that numerous new soft tick lipocalins have been deposited in the sequence databases, for O. porcinus, O. moubata and A. monolakensis. Phylogenetic analysis of these and other lipocalins shows that TSGP2 and TSGP4 have closely related homologues (Mans, 2005) (Fig. 5.3). OmCI, a complement inhibitor described for O. moubata (Nunn et al., 2005), groups particularly close to TSGP2/TSGP3 and raises the question whether one of these proteins might also have this function. It also raises the question why TSGP2 and TSGP4 are toxic while other closely related lipocalins in other ticks are not. Phylogenetic analysis also indicates that TSGP2 and TSGP4 belong to two different groups of lipocalins, which can be distinguished by their unique disulphide bond patterns and their seemingly deep separation in phylogenetic space (Fig. 5.3). As such this would suggest that these two toxins did not have a toxic ancestor, but evolved toxic properties separately. It would be of interest to elucidate their respective molecular

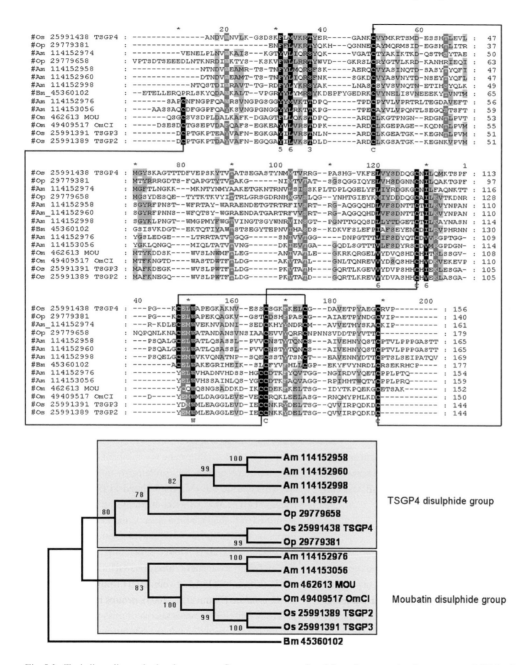

Fig. 5.3. Toxic lipocalins and related sequences. Sequences were retrieved from the non-redundant sequence (NR) database by doing a single round of BLASTP analysis for TSGP2 and TSGP4, respectively or a TBLASTN analysis against the expressed sequence tag (EST) database limiting the search against tick sequences. Shown is the alignment of the retrieved sequences and their predicted disulphide bond patterns. Sequences are annotated by species abbreviation (Bm – *B. microplus*, Om – *O. moubata*, Op – *O. porcinus*, Os – *O. savignyi*, Am – *A. monolakensis*) followed by the GenBank accession number and in some cases a general abbreviation (MOU – moubatin, OmCI – complement inhibitor). Also indicated is a unrooted tree from a neighbour-joining analysis based on sequence identity. Values are for 10 000 bootstraps and gapped positions were not used in the analysis.

mechanisms of action, to determine whether any similarities exist.

Paralysis and the genus *Otobius*

Paralysis-like symptoms of a 16-month old boy caused by *Otobius megnini* have been reported (Peacock, 1958). Symptoms included neck retraction and respiratory difficulty, although no ankle or knee jerks or sensory loss were observed. Ear infestations of numerous humans by *O. megnini* did not result in paralysis although irritation was common (Chellappa, 1973; Eads & Campos, 1984). In horses, death following infestation by *O. megnini* has been observed, although symptoms observed are not those commonly associated with paralysis (Rich, 1957). While a neurological pathology that includes muscle tremors and muscle contractions is observed, electromyographic measurements suggest these may be due to increased motor unit activity (Madigan *et al.*, 1995). No conclusive evidence supports the classification of *O. megnini* as a paralysis tick. The fact that this tick feeds within the ears of its hosts where inflammatory reactions could affect the balance of the host and lead to symptoms that could be interpreted as being neurological in origin should be considered.

PARALYSES OF THE IXODID FAMILY

Paralysis has been described for almost all the genera of the ixodid family (Fig. 5.2, Table 5.1). It is generally agreed that paralysis caused by *I. holocyclus* differs from paralysis caused by *Dermacentor* and *Rhipicephalus* species.

Paralyses and the genus *Ixodes*

By far the most implicated tick species in paralysis are from the genus *Ixodes* (Tables 5.1 and 5.2; Fig. 5.2). Paralysis ticks from this genus are distributed worldwide. Definite confirmation of paralysis exists for most *Ixodes* species. In the case of *I. redikorzevi* paralysis probably has been misinterpreted as a toxicosis for which symptoms of fever and torticollis have been reported (Kassis *et al.*, 1997).

Ixodes holocyclus

Ixodes holocyclus has been reported to paralyse dogs, cats, cattle, horses and humans. Adult ticks and nymphs have been associated with paralysis, while larvae cause local irritation only (Masina & Broady, 1999). Salivary gland extract affects dogs and mice, although only suckling mice (4–5 g) are generally affected, while adult mice (20–25 g) do not show paralysis symptoms (Stone & Binnington, 1986). An increase in blood pressure occurs during paralysis, in contrast to paralysis caused by *D. andersoni* where blood pressure remains normal. Original purification attempts of the neurotoxin from *I. holocyclus* showed that the paralysis toxin was associated with high-molecular-mass complexes (40–80 kDa) (Masina & Broady, 1999). Recently, the paralysis toxin was purified and was shown to have a molecular mass of ∼5 kDa; it binds to rat synaptosomes in a temperature-dependent manner (Thurn, Gooley & Broady, 1992). This temperature dependence coincides with earlier observations that there is a temperature-dependent inhibition of evoked acetylcholine release during paralysis (Cooper & Spence, 1976). It was also shown that three different isoforms (HT-1, HT-2 and HT-3) of the toxin exist (Thurn, Gooley & Broady, 1992). The cloning of the holocyclotoxin gene for HT-1 has been reported (Masina & Broady, 1999). The sequence includes a signal peptide (18 residues), start and stop codons, the polyadenylation signal and a poly-A tail. The calculated molecular mass (5–9 kDa) corresponds to the mass obtained for the native toxin and a basic isoelectric point (pI ∼8.86) is predicted. The arrangement of cysteines suggests that the disulphide bond pattern of holocyclotoxin might be similar to that of scorpion toxins and could indicate structural similarity (Masina & Broady, 1999). Antibodies raised against the recombinant HT-1 protected neonatal mice against the native toxin thereby confirming the sequence identity (Masina & Broady, 1999). A toxic lethal fraction (Mr <20 kDa) without paralysing activity was also identified in *I. holocyclus* and it was suggested that this might be the agent of cardiovascular failure previously attributed to holocyclotoxin (Stone, Doube & Binnington, 1979). It was recently shown that ticks cause long QT syndrome in dogs and bats (Campbell and Atwell, 2002; Campbell *et al.*, 2003). In rats, positive inotropic responses were induced in heart papillary muscles that caused arrhythmias in the right atria and prolonged the action potential duration of repolarization in left ventricular papillary muscles and delayed ventricular papillary muscle relaxation (Campbell *et al.*, 2004). It was concluded that toxins from saliva of *I. holocyclus* affect cardiac and vascular K^+ channels. It would be of interest to know whether the cardiotoxic effects are due to a paralysis-like toxin or something more similar to the toxins described for sand tampan toxicoses. The reader is referred to the recent review for a more thorough account of the biology of *I. holocyclus* (Masina & Broady, 1999) and to Gothe (1999) for a historical overview of the paralysis induced by this tick species.

Paralysis and the genus *Haemaphysalis*

Eight out of 150 species of *Haemaphysalis* have been implicated in paralysis. However, paralysis has only been confirmed for *H. kutchensis* and *H. punctata* (Table 5.1). In most other cases, additional tick species known to cause paralysis were also present on the host. Such mixed infestations are one of the major problems encountered when evaluating clinical reports on tick paralysis.

Paralysis and the genus *Amblyomma*

Paralysis caused by *Amblyomma* species has been confirmed by more than one report (Table 5.1). However, as with most other species responsible for paralysis, not all individuals within a given *Amblyomma* species will necessarily cause paralysis. All the reports concerning paralysis caused by *Amblyomma* ticks are of a clinical nature and no experimental investigations have yet been conducted to confirm paralysis-inducing capabilities.

Paralysis and the genus *Dermacentor*

While 10 species from the genus *Dermacentor* have been implicated in paralysis, extensive records exist for only *D. andersoni*, *D. variabilis* and *D. occidentalis* (Table 5.1). The pathological mechanisms of only *D. andersoni* have been extensively studied.

PARALYSIS CAUSED BY *DERMACENTOR ANDERSONI*
Infestation by *D. andersoni* affects motor neurons of the efferent pathway and not the afferent. The neuromuscular junction of peripheral nerves is targeted through inhibition of acetylcholine release from the synapse, suggesting a pre-synaptic target (Gothe & Neitz, 1991). Feeding ticks affect dogs, sheep, cattle, guinea pigs, hamsters and humans, but not cats, rabbits, rats and mice. In those animals affected the symptoms could be fairly rapidly reversed on removal of ticks, except for marmots which frequently do not recover (Emmons & McLennan, 1980). Gross symptoms of paralysis in marmots include the loss of the animal's normal piercing cry (paralysis of the vocal cords) followed by an ataxia and weakness of the hind limbs. The condition progresses until the fore limbs are paralysed and the animals are unable to move and lie on their sides; there is retention of urine and

faeces (Emmons & McLennan, 1980). No paralysis could be observed when SGE from fast-feeding ticks were injected subcutaneously into mice and lambs and intraspinally into puppies (Gregson, 1943). Fractionated extracts also failed to produce any symptoms. However, paralysis was observed when saliva from fed females was continuously injected subcutaneously into marmots and hamsters (Gregson, 1973). Variation in the ability of individual ticks to cause paralysis was also observed. This highlights the problems associated with paralysis by *D. andersoni*. It also points out why progress on this specific toxicosis has been so slow in recent years. No development of immunity has yet been reported for *D. andersoni*, while immunity to the holocyclotoxin from *I. holocyclus* is well established. Hyperimmune serum against holocyclotoxin also fails to relieve *D. andersoni* paralysis (Gregson, 1973). This indicates that no cross-reactivity exists between these toxins and suggests that *D. andersoni*- and *I. holocyclus*-derived toxins are evolutionarily distant or not at all related.

Paralysis and toxicoses of the genus *Hyalomma*

Hyalomma truncatum is the only species from this genus definitely implicated in paralysis (Table 5.1). In addition, it is also implicated as a vector of sweating sickness, a non-paralytic toxicosis.

SWEATING SICKNESS, MHLOSINGA AND MAGUDU
Sweating sickness occurs in central, eastern and southern Africa and has been recorded in Sri Lanka and India. In nature it affects only cattle, especially young calves, but can also be transmitted to other artiodactyls such as sheep, goats or pigs (Neitz, 1959, 1962). The last species is particularly useful due to the ease of diagnosis of sweating sickness symptoms in pigs such as hyperaemia of the skin and the development of pharyngeal and laryngeal lesions that manifest as changes in the tone and pitch of squeals during handling (Neitz, 1956). *Hyalomma truncatum* has been implicated as being responsible for sweating sickness in the Kwa-Zulu Natal region of Southern Africa (Neitz, 1954, 1956). Sweating sickness positive (SS+) and sweating sickness negative (SS−) tick strains are found. Exposure and recovery from the disease leads to immunity. Two similar, but milder forms of toxicosis named Mhlosinga and Magudu are also induced by *H. truncatum*. It appears that Magudu is more closely related to typical sweating sickness as animals recovered from sweating sickness were resistant to Magudu but not to Mhlosinga. In Kenya, sweating sickness and *Hyalomma*

ticks go by the same name of 'Ol masheri'. Other names used to describe sweating sickness include wet calf disease, vuursiekte (Afrikaans), Schwitzkrankheit (German), la hydrose tropicale (French) and Foma (Swazi) (Neitz, 1959).

Sweating sickness symptoms and clinical pathology

The name sweating sickness derives from the profuse moist eczema that is the most noticeable symptom. Symptoms are normally observed after 5–7 days of tick feeding. Other symptoms include fever, pyrexia, anorexia, hyperaemia and hyperaesthesia of the skin and mucous membranes, salivation, lachrymation, serous or crupous rhinitis, epistaxis, diarrhoea and diphtheroid stomatitis, pharyngitis, laryngitis, oesophagitis, vaginitis or posthitis. Subepicardial and subendocardial petechiae, hyperaemia and oedema of the lungs, liver and kidney congestion, atrophy of the spleen, fatty degeneration of the liver and general inflammation and necrosis of the mucous membranes of the buccal cavity, larynx, pharynx and oesophagus have been observed (Neitz, 1959). During the symptomatic period the fibrin content in the plasma increases but decreases again with recovery (Neitz, 1959). Addition of a few drops of citrated plasma of sick animals to Hayem's fluid (mercuric chloride, sodium chloride, sodium sulphate) causes haemagglutination. Magudu and Mhlosinga are characterized by pyrexia, anorexia and listlessness. The fibrin content of the plasma also increases over time and haemagglutination of citrated plasma is also observed as seen for sweating sickness (Neitz, 1962). The only difference between Magudu and Mhlosinga is that no immunological cross-reactivity can be observed (Neitz, 1962).

Sweating sickness as a tick-derived toxicosis

Tick-derived toxins were suspected as the causal agents since the disease could not be transferred from sick to healthy animals. Animals were, however, protected by hyperimmune sera. The disease could be transmitted over 15 generations of ticks (Neitz, 1962), after which toxicity was suddenly lost (W. O. Neitz, personal communication). Injecting animals with salivary gland secretions from partially or fully engorged SS+ ticks before challenge with SS+ ticks failed to protect against sweating sickness. Animals injected with partially fed or fully engorged SS+ tick suspension before challenge with SS+ ticks were protected against sweating sickness, although the tick suspension itself did not induce any sweating sickness symptoms (Bezuidenhout & Malherbe, 1981). This suggests that the salivary glands are not the organs from which the toxins originate and that tick sus-

pensions in themselves probably do not contain the active toxin or sufficient quantities of toxin to induce sweating sickness, although enough of the toxin must be present to allow development of a toxin neutralizing response. Toxicosis could be associated with several novel proteins found in the positive strain that are absent in the negative strain (Burger et al., 1991; Spickett et al., 1991). These include three non-immunogenic proteins with molecular masses of approximately 24, 26 and 42 kDa and three immunogenic proteins with molecular masses of 20, 30 and 32 kDa (Fig. 5.4). Immune responses against these unique immunogenic proteins were observed in cattle that were treated with hyperimmune sera at an advanced stage of sweating sickness (day 7) and rechallenged on day 27. The immunogenic bands became prominent at day 35 (Spickett et al., 1991). Treatment with hyperimmune sera protected against sweating sickness and the recovered animals showed no sweating sickness symptoms upon rechallenge.

The presence of rickettsiae in sweating-sickness-inducing strains (SS+) and their absence in negative strains cast doubt on a tick-derived origin for the toxins and suggested that toxicosis might be associated with this pathogen or symbiont (Bezuidenhout & Malherbe, 1981). As yet, the involvement of the novel proteins identified for SS+ strains, or their association with rickettsial organisms identified in SS+ ticks, have not been ascertained.

Paralysis and toxicoses of the genus *Rhipicephalus*

Most cases of paralysis caused by *Rhipicephalus* have been confirmed (Table 5.1). It is of interest that *B. annulatus* is the only tick from the genus *Boophilus* (now part of *Rhipicephalus*: Barker & Murrell, 2004) implicated in paralysis, although this has not yet been confirmed. Within *Rhipicephalus*, both *R. appendiculatus* and *B. microplus* have been implicated in other forms of toxicoses. In terms of the molecular nature of paralysis induced by *Rhipicephalus*, most studies have been conducted on *R. evertsi evertsi*.

BROWN TICK TOXICOSIS

Rhipicephalus appendiculatus causes a leukocytotropic disease in cattle. Symptoms are prolonged fever, oedema of subcutaneous tissues of the ears, eyes, jowls and dewlap, a swelling of the palpable lymphatic glands, anorexia, lachrymation, serous nasal discharge, listlessness and general weakness (Neitz, 1962). This is normally followed by relapses of other tick-borne diseases such as babesiosis, spirochaetosis, anaplasmosis and heartwater. Recovered animals are immune

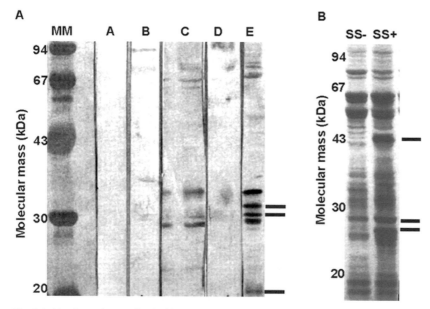

Fig. 5.4. Novel proteins associated with sweating sickness positive strains of *H. truncatum*. (A) Western blot analysis of salivary gland extracts (SGE) obtained from positive and negative strains of *H. truncatum*. Lane A is naïve serum on SS+ SGE. Lane B is naïve serum on SS− SGE. Lane C is SS+ antisera on SS− SGE. Lane D is SS− antisera on SS− SGE. Lane E is SS+ antisera on SS+ SGE. (B) SDS-PAGE analysis of SS− and SS+ SGE. Indicated are unique protein bands observed in the SS+ SGE.

against this toxicosis. Animals resistant to tick-borne diseases from *R. appendiculatus*-free areas that are introduced into areas where *R. appendiculatus* is prevalent succumb to this disease, known also as 'Tzaneen disease' (De Kock *et al.*, 1937). Treatment with antibiotics known to counter heartwater and anaplasmosis failed to protect animals against this relapse (Thomas & Neitz, 1958). It was suggested that this form of toxicosis weakens the immune system of the affected host to such a degree that no protection against parasitic relapse is present.

TOXICOSIS INDUCED BY *B. MICROPLUS*

Cattle exposed to *B. microplus* show a weight loss related to anaemia and loss of appetite (Gothe, 1999). It is considered that this may be due to a disruption in the metabolic processes of the hosts. The possible presence of egg-derived toxins secreted by larvae during feeding via the salivary glands may also have an influence on this phenomenon (see section on tick egg toxins).

PARALYSIS INDUCED BY *R. EVERTSI EVERTSI*

Rhipicephalus evertsi evertsi affects the peripheral nervous system by inducing a motor polyneuropathy in sheep that is appropriately known as spring lamb paralysis (Gothe &

Kunze, 1982). Ticks fed on laboratory animals affect mice, rats, hamsters, guinea pigs and rabbits only slightly or not at all (Gothe & Lämmler, 1982). Injection of SGE into sheep, mice and chickens failed to elicit a paralysis response (Viljoen *et al.*, 1986). A very sensitive in vitro assay using a sciatic nerve–gastrocnemius muscle preparation was developed to characterize this toxin (Viljoen *et al.*, 1986). In this assay the dissected nerve was bathed with SGE or purified neurotoxin in a specially constructed nerve bath with a volume of 60 µl. In contrast to the in vivo tests, both SGE and neurotoxin preparations effectively paralysed muscle contraction (Viljoen *et al.*, 1986; Crause *et al.*, 1994). It could thus be argued that the neurotoxin affects the nerve and not the neuromuscular junction. Large quantities of SGE (400–900 µg protein) were used to elicit a response and it could be disputed whether this was a truly specific response. However, total inhibition of nerve impulse propagation was observed with purified preparations (74 µg/ml), which is probably much closer to physiological conditions (Viljoen *et al.*, 1986). The toxin from *R. evertsi evertsi* that purified as a ∼68-kDa protein (Viljoen *et al.*, 1986) was later shown to be the trimeric form of a ∼20-kDa protein (Crause *et al.*, 1994). A monoclonal antibody directed against this toxin showed cross-reactivity with both non-paralysis-inducing

(*R. appendiculatus*, *Hyalomma marginatum rufipes*, *Boophilus decoloratus* and a non-paralysing strain of *R. evertsi evertsi*) as well as paralysis-inducing ticks (*I. rubicundus*, *A. walkerae* and a paralysis-inducing strain of *R. evertsi evertsi*). Significant was the fact that only paralysis-inducing ticks seem to possess a ~68-kDa reactive antigen (Crause *et al.*, 1994).

OTHER FORMS OF TOXICOSIS CONSIDERED AS NON-PARALYTIC

The fact that some forms of tick toxicosis are considered to be distinct from paralysis indicates that a study into the toxic mechanisms of ticks should take this into account. If toxicoses can be shown to differ clearly in their mechanisms of action, it would provide a specific reference point to catalogue the different toxicoses forms.

Tick egg toxins (ixovotoxins)

Regendanz & Reichenow (1931) postulated that the source of tick paralysis toxin resides in the ovaries of the tick and enters the salivary glands at a late stage of engorgement, only when egg development starts. To corroborate this they produced paralysis-like symptoms in a dog injected with egg extracts from *R. sanguineus*. This laid the foundation for the investigations into tick egg toxins and their relationship with tick paralysis. The name 'ixovotoxins' was proposed for egg-derived toxins and their relationship to paralysis toxins questioned based on the fact that ticks that possess egg toxins do not necessarily cause paralysis during feeding (Oswald, 1938). The name ixovotoxin is particularly apt since egg toxins seem to be limited to hard ticks: tick egg extracts from 17 ixodid species tested were toxic, while extracts from five argasid species (*A. persicus*, *O. coriaceus*, *O. lahorensis*, *O. moubata* and *O. savignyi*) were not (Riek, 1957). Cross-reactivity was also observed between hard tick extracts, with antisera against hard tick egg extracts that did not cross-react with soft tick extracts (Riek, 1958). Egg toxins have been identified in most hard ticks investigated including *A. hebraeum*, *A. moreliae*, *A. triguttatum*, *A. variegatum*, *Aponomma hydrosauri*, *B. calcaratus*, *B. decloratus*, *B. microplus*, *D. albipictus*, *D. reticulatus*, *D. sinicus*, *D. variabilis*, *H. bispinosa*, *H. leachi*, *H. dromedarii*, *H. scupense*, *H. truncatum*, *I. hexagonus*, *I. holocyclus*, *I. pilosus*, *I. ricinus*, *I. rubicundus*, *R. bursa*, *R. eversti evertsi* and *R. sanguineus* (Regendanz & Reichenow, 1931; Hoeppli & Feng, 1933; Oswald, 1938; De Meillon, 1942; Gregson, 1941; Steinhaus, 1942; Riek, 1957, 1959; Neitz *et al.*, 1981; Viljoen *et al.*, 1985).

Investigations into the relationship between ixovotoxins and paralysis toxins showed that egg toxins differ from paralysis toxins in terms of pathology and molecular properties (Hoeppli & Feng, 1933; Gregson, 1941; Steinhaus, 1942; De Meillon, 1942; Riek, 1957, 1959; Neitz *et al.*, 1981; Viljoen *et al.*, 1985). Research into the ixovotoxins serves the purpose of making the point that deleterious effects caused by tick-derived components within the vertebrate host cannot necessarily be ascribed functional significance.

CLINICAL SYMPTOMS AND HISTOPATHOLOGY OF IXOVOTOXINS

The pathological effects of ixovotoxins on guinea pigs have been studied in various species (Riek, 1957; Neitz *et al.*, 1981; Vermeulen *et al.*, 1984; Vermeulen & Neitz, 1987; Viljoen *et al.*, 1985). Mortality of mice and rabbits injected with ixovotoxins was also demonstrated (Riek, 1957). All toxins cause similar clinical symptoms within guinea pigs and include hyperaesthesia and anorexia, serous nasal and eye discharge accompanied by conjunctivitis and rhinitis, apparent paresis (differing from ascending paralysis) and a loss of voice over a 15–36-hour period. Histopathology includes necrosis of the liver and kidneys and oedema of the urinary bladder, lungs and skin at the site of injection. Necrosis is accompanied by elevated calcium levels within the cytoplasm which lead to mineralization within the cytoplasm and mitochondria. This suggests that the ion permeability of these cells is compromised and that lesions are probably all of vascular origin. Lipofuscin also accumulates within necrotic cells within the 36-hour period of induced toxicosis. This is of interest as lipofuscin is normally associated with ageing and an impaired lysosomal function, which is intimately associated with damaged mitochondria (Brunk & Terman, 2002). In the case of tick egg toxins, it probably indicates that disruption of selective membrane permeability to calcium leads to mitochondrial damage, which in turn places an oxidative stress on the system and impairs lysosomal function, thereby allowing rapid accumulation of lipofuscin.

IXOVOTOXINS ARE PROTEASE INHIBITORS

All egg toxins investigated so far function as serine protease inhibitors although no causal link has yet been established between protease inhibition and toxicosis (Vermeulen & Neitz, 1987; Vermeulen *et al.*, 1988). Vertebrate proteases are involved in various important physiological functions such as digestion, maturation of hormones, immune responses, inflammation, blood coagulation, fibrinolysis and morphogenic response (Holzer & Heinrich, 1980). It should therefore come as no surprise that inhibition of proteases within the vertebrate host will lead to pathological states. The concentrations of the tick egg toxins and protease inhibitors within tick eggs are more than 1000 times higher than their respective K_i values with trypsin and chymotrypsin (Vermeulen & Neitz, 1987; Vermeulen et al., 1988). This indicates that they probably have a physiological function within the eggs. Several functions have been proposed which include involvement in egg development via the regulation of egg proteases during hatching, as antihaemostatic components and as antimicrobial agents that protect the egg against microbial invasion (Willadsen & Riding, 1979, 1980; Neitz et al., 1981; Viljoen et al., 1985; Vermeulen & Neitz, 1987; Vermeulen et al., 1988). Toxins and anti-proteases within tick eggs could even protect them against microbial, insect and arachnid scavengers and predators.

ANTI-PROTEASE KINETICS OF IXOVOTOXINS

The ixovotoxin (Mr $\sim$ 10 kDa) from A. hebraeum shows specific non-competitive fast-binding inhibition of trypsin (K_i $\sim$255 nM), but not chymotrypsin (Neitz et al., 1981; Vermeulen & Neitz, 1987; Vermeulen et al., 1988). A non-toxic inhibitor (Mr $\sim$8400 Da) from A. hebraeum shows specific fast-binding competitive inhibition for trypsin (K_i $\sim$25 nM) (Vermeulen et al., 1984, 1988; Vermeulen & Neitz, 1987). Ixovotoxin from R. evertsi evertsi (Mr $\sim$5–6 kDa) is a specific competitive fast tight-binding inhibitor of trypsin (Viljoen et al., 1985; Vermeulen & Neitz, 1987). The purified ixovotoxins from B. microplus (Mr $\sim$30–35 kDa), B. decoloratus (Mr $\sim$40 kDa) and H. truncatum (Mr $\sim$27 kDa) are all competitive slow-binding inhibitors of trypsin. The toxins from B. decoloratus and H. truncatum also inhibit chymotrypsin competitively via a fast tight-binding mechanism, while B. microplus toxin does not inhibit chymotrypsin (Viljoen et al., 1985; Vermeulen & Neitz, 1987; Vermeulen et al., 1988). The toxins from B. microplus and B. decoloratus showed cross-reactivity during Ouchterlony double diffusion, while toxins from R. evertsi evertsi and H. truncatum showed no cross-reactivity with any of these toxins (Vermeulen et al., 1988). This is particularly interesting, considering the molecular mass and kinetic differences between B. microplus and B. decoloratus toxins. Immunization against egg extracts from H. bispinosa induced sensitivity that could also be observed with egg extracts from H. dromedarii and A. variegatum, while B. microplus, I. ricinus, I. holocyclus and I. hexagonus did not show as large a sensitivity reaction (Riek, 1958). Immunization did give cross-protection against egg extracts from B. decoloratus, B. microplus, I. holocyclus, I.

ricinus and H. dromedarii, but not against paralysis induced by I. holocyclus (Riek, 1957).

IXOVOTOXINS AND PROTEASE INHIBITORS
FROM B. MICROPLUS

The toxic component from the eggs of B. microplus was shown to occur in larvae, but not in nymphs and adults (Riek, 1957). Presence of egg toxins in larvae was also confirmed when immunization with larval extracts protected against toxicoses by egg extracts (Riek, 1958). Multiple toxic fractions that cross-reacted with larval proteins were subsequently identified in tick extracts (Riek, 1959). Toxic as well as non-toxic fractions from larvae induced hypersensitivity reactions (Riek, 1958). A double-headed protease inhibitor, which inhibits trypsin (K_i <0.002 μM) and chymotrypsin (K_i $\sim$0.2 μM), was purified from larvae of B. microplus and caused an immediate hypersensitivity reaction when injected intradermally into bovines exposed to this tick species (Willadsen & Riding, 1979). It was subsequently shown that this inhibitor occurs in large amounts in tick eggs and in the initial stages of the larvae before disappearing in the later stages (Willadsen & Riding, 1980). The proteins from tick eggs and larvae, while closely related, differ significantly in terms of their kinetic properties (Willadsen & McKenna, 1983). Two different trypsin–chymotrypsin inhibitors (20 800 and 15 800 Da) were isolated from tick eggs (Willadsen & McKenna, 1983). A toxic component (Mr $\sim$30–35 kDa) was also purified from B. microplus eggs that specifically inhibit trypsin (K_i $\sim$4.6 nM) but not chymotrypsin (Viljoen et al., 1985; Vermeulen & Neitz, 1987). A larval inhibitor (22 500 Da) was given functional significance during feeding as it was shown that this inhibitor can inhibit bovine plasmin (K_i $\sim$0.1 μM) and pig pancreatic kallikrein (K_i $\sim$0.33 μM), that it inhibited both extrinsic and intrinsic pathways of the coagulation cascade as well as complement-dependent cell lysis but did not stimulate lymphocyte proliferation (Willadsen & Riding, 1980). The fact that this inhibitor disappears after the initial larval stage suggests that its biological roles are important within the egg and for the larvae during their initial feeding stages. Recently, a double-headed member (BmTI-A) of the basic pancreatic trypsin inhibitor (BPTI) family was described from B. microplus larvae (Tanaka et al., 1999). It inhibited trypsin (K_i $\sim$3 nM), chymotrypsin (K_i $\sim$33 nM), elastase (K_i $\sim$1.4 nM), plasmin (K_i $\sim$590 nM) and human plasma kallikrein (K_i $\sim$120 nM) as well as the intrinsic blood coagulation cascade. It did not, however, inhibit porcine pancreatic kallikrein. These studies indicate that tick proteins may have more than one function within the tick and the host. Immunization of cattle with BmTI-A led to a reduction in total tick number, egg weight and female engorged weight although the number of eggs was not affected (Andreotti *et al.*, 2002). Another trypsin inhibitor (BmTI-B – 10 kDa) was also purified from larvae (Tanaka *et al.*, 1999).

While specific egg toxins have been isolated from different tick species, it appears that egg extracts are complex mixtures in which more than one toxic principle might reside. Cross-reactivity assays and similar biochemical data suggest that more than one protein family is involved in toxicosis and antiproteolytic activity. The identification of these components could assist in the study of the origins of all forms of toxicosis, if it is considered that tick toxins might be evolutionary related. The study of protein structure, function and evolution has shown that proteins can exist as single domain or multiple domain structures. As such, a protein fold involved in tick toxicoses can occur within a small protein or even as part of a larger protein. It can be concluded that tick egg toxins seem to be distinct from other forms of toxicosis and due to their source cannot be described as being venomous. They thus serve as evidence of toxic molecules derived from ticks that can influence potential hosts in deleterious ways, although toxicity may not be their main property.

CONCLUSIONS

Tick toxicosis and tick paralysis in particular are still one of the most enigmatic of pathogenesis caused by ticks. The difficult nature of the research into tick toxicoses will continue to hamper us in years to come. However, a comprehensive view of tick–host interactions cannot be compiled without a thorough understanding of the mechanisms of tick toxicoses and their origins. Elucidation of the molecular mechanisms by which tick toxins perform their action should expand our understanding of tick evolution.

ACKNOWLEDGEMENTS

Part of this work was funded by the National Research Foundation of South Africa and the University of Pretoria.

REFERENCES

Andreotti, R., Gomes, A., Malavazi-Piza, K. C., *et al.* (2002). BniTI antigens induce a bovine protective immune response against *Boophilus microplus* tick. *International Immunopharmacology* 2, 557–563.

Barker, S. C. & Murrell, A. (2004). Systematics and evolution of ticks with a list of valid genus and species names. *Parasitology* 129, S15–S36.

Bezuidenhout, J. D. & Malherbe, A. (1981). Sweating sickness: a comparative study of virulent and avirulent strains of *Hyalomma truncatum*. In *Tick Biology and Control*, eds. Whitehead, G. B. & Gibson, J. D., pp. 7–12. Grahamstown, South Africa: Tick Research Unit, Rhodes University.

Brunk, U. T. & Terman, A. (2002). The mitochondrial–lysosomal axis theory of aging: accumulation of damaged mitochondria as a result of imperfect autophagocytosis. *European Journal of Biochemistry* 269, 1996–2002.

Burger, D. B., Crause, J. C., Spickett, A. M. & Neitz, A. W. H. (1991). A comparative study of proteins present in sweating-sickness-inducing and non-inducing strains of *Hyalomma truncatum* ticks. *Experimental and Applied Acarology* 13, 59–63.

Campbell, F. E. & Atwell, R. B. (2002). Long QT syndrome in dogs with tick toxicity (*Ixodes holocyclus*). *Australian Veterinary Journal* 80, 611–616.

Campbell, F., Atwell, R., Fenning, A., Hoey, A. & Brown, L. (2004). Cardiovascular effects of the toxin(s) of the Australian paralysis tick, *Ixodes holocyclus*, in the rat. *Toxicon* 43, 743–750.

Campbell, F. E., Atwell, R. B. & Smart, L. (2003). Effects of the paralysis tick, *Ixodes holocyclus*, on the electrocardiogram of the Spectacled Flying Fox, *Pteropus conspicillatus*. *Australian Veterinary Journal* 81, 328–331.

Chellappa, D. J. (1973). Note on spinose ear tick infestation in man and domestic animals in India and its control. *Madras Agricultural Journal* 60, 656–658.

Cooper, B. J. & Spence, I. (1976). Temperature-dependent inhibition of evoked acetylcholine release in tick paralysis. *Nature* 263, 693–695.

Crause, J. C., Van Wyngaardt, S., Gothe, R. & Neitz, A. W. H. (1994). A shared epitope found in the major paralysis inducing tick species of Africa. *Experimental and Applied Acarology* 18, 51–59.

Crause, J. C., Verschoor, J. A., Coetzee, J., *et al.* (1993). The localization of a paralysis toxin in granules and nuclei of prefed female *Rhipicephalus evertsi evertsi* tick salivary gland cells. *Experimental and Applied Acarology* 17, 357–363.

De Kock, G., Van Heerden, C. J., Du Toit, R. & Neitz, W. O. (1937). Bovine theileriosis in South Africa with special reference to *Theileria mutans. Onderstepoort Journal of Veterinary Science and Animal Industry* 8, 9–125.

De Meillon, B. (1942). A toxin from the eggs of South African ticks. *South African Journal of Medical Science* 7, 226–235.

Dobson, S. J. & Barker, S. C. (1999). Phylogeny of the hard ticks (Ixodidae) inferred from 18S rRNA indicates that the genus *Aponomma* is paraphyletic. *Molecular Phylogenetics and Evolution* 11, 288–295.

Du Toit, R. & Theiler, G. (1964). Ticks and tick-borne diseases in South Africa. *Scientific Bulletin of the Department of Agriculture and Technical Services of the Republic of South Africa* 364.

Eads, R. B. & Campos, E. G. (1984). Human parasitism by *Otobius megnini* (Acari: Argasidae) in New Mexico, USA. *Journal of Medical Entomology* 21, 244.

Emmons, P. & Mclennan, H. (1980). Some observations on tick paralysis in marmots. *Journal of Experimental Biology* 37, 355–362.

Escoubas, P., Diochot, S. & Corzo, G. (2000). Structure and pharmacology of spider venom neurotoxins. *Biochimie* 82, 893–907.

Gothe, R. (1984). Tick paralyses: reasons for appearing during ixodid and argasid feeding. In *Current Topics in Vector Research*, ed. Harris, K. F., vol. 2, pp. 199–223. New York: Praeger.

Gothe, R. (1999). *Zecken Toxikosen*. Munich, Germany: Hieronymus.

Gothe, R. & Kunze, K. (1971). Stimulus conduction of efferent and afferent peripheral nerve fibers in fowl tick paralysis caused by *Argas (Persicargas) persicus* larvae. *Zeitschrift für Tropenmedizin und Parasitologie* 22, 292–296.

Gothe, R. & Kunze, K. (1982). Action potentials and conduction velocities of the tibial nerve in sheep paralysis caused by *Rhipicephalus evertsi evertsi. Zentralblatt für Veterinaermedizin* B29, 186–192.

Gothe, R. & Lämmler, M. (1982). Sensitivity of laboratory animals to *Rhipicephalus evertsi evertsi* paralysis. *Zentralblatt für Veterinaermedizin* B29, 249–252.

Gothe, R. & Neitz, A. W. H. (1991). Tick paralysis: pathogenesis and etiology. *Advances in Disease and Vector Research* 8, 177–204.

Gothe, R., Hager, H., Jehn, E., Kunze, K. & Thoenes, W. (1971). Pathological–anatomical studies of peripheral nerves in fowl tick paralysis caused by *Argas (Persicargas) persicus* larvae. *Zeitschrift für Tropenmedizin und Parasitologie* 22, 285–291.

Gothe, R., Kunze, K. & Hoogstraal, H. (1979). The mechanisms of pathogenicity in tick paralyses. *Journal of Medical Entomology* 16, 357–369.

Gregson, J. D. (1941). The discovery of an ixovotoxin in *Dermacentor andersoni* eggs. *Proceedings of the Entomological Society of British Columbia* 37, 9–10.

Gregson, J. D. (1943). The enigma of tick paralysis. *Proceedings of the Entomological Society of British Columbia* 40, 19–23.

Gregson, J. D. (1973). *Tick Paralysis: An Appraisal of Natural and Experimental Data*, Monograph No. 9. Canadian Department of Agriculture.

Hilger, C., Bessot, J. C., Hutt, N., *et al.* (2005). IgE-mediated anaphylaxis caused by bites of the pigeon tick *Argas reflexus*: cloning and expression of the major allergen Arg r1. *Journal of Allergy and Clinical Immunology* 115, 617–622.

Hoeppli, R. & Feng, L. C. (1933). Experimental studies on ticks. *Chinese Medical Journal* 47, 29–43.

Holzer, H. & Heinrich, P. C. (1980). Control of proteolysis. *Annual Review of Biochemistry* 49, 63–91.

Hoogstraal, H. (1956). African Ixodoidea. I. Ticks of the Sudan (with special reference to Equatoria Province and with preliminary reviews of the genera *Boophilus, Margaropus* and *Hyalomma*), Research Report NM 005050.29.07. Washington, DC: US Government Printing Office.

Hoogstraal, H. (1985). Argasid and nuttalliellid ticks as parasites and vectors. *Advances in Parasitology* 24, 135–238.

Howell, C. J. (1966). Collection of salivary gland secretion from the argasid *Ornithodoros savignyi* Audouin (1827) by the use of a pharmacological stimulant. *Journal of the South African Veterinary Medical Association* 37, 236–239.

Howell, C. J., Neitz, A. W. H. & Potgieter, D. J. J. (1975). Some toxic and chemical properties of the oral secretion of the sand tampan, *Ornithodoros savignyi* Audouin (1825). *Onderstepoort Journal of Veterinary Research* 43, 99–102.

Kassis, I., Ioffe-Uspensky, I., Uspensky, I. & Mumcuoglu, K. Y. (1997). Human toxicosis caused by the tick *Ixodes redikorzevi* in Israel. *Israel Journal of Medical Science* 33, 760–761.

Keller, P. M., Waxman, L., Arnold, B. A., *et al.* (1993). Cloning of the cDNA and expression of moubatin, an inhibitor of platelet aggregation. *Journal of Biological Chemistry* 268, 5450–5456.

Kleine-Tebbe, J., Heinatz, A., Graser, I., *et al.* (2006). Bites of the European pigeon tick (*Argas reflexus*): risk of IgE-mediated sensitizations and anaphylactic reactions. *Journal of Allergy and Clinical Immunology* 117, 190–195.

Klompen, J. S. H., Black, W. C. IV, Keirans, J. E. & Oliver, J. H. (1996). Evolution of ticks. *Annual Reviews in Entomology* **41**, 141–161.

Kone, K. (1948). Accidents mortels chez les zébus causés par des piqûres d'*Ornithodoros*. *Bulletin des Services de l'Elevage et des Industries de A.O.F.* **2**, 25–26.

Lindquist, E. E. (1984). Current theories on the evolution of major groups of Acari and on their relationships with other groups of Arachnida, with consequent implications for their classification. In *Acarology VI*, vol. 1, eds. Griffiths, D. A. & Bowman, C. E., pp. 28–62. New York: John Wiley.

Madigan, J. E., Valberg, S. J., Ragle, C. & Moody, J. L. (1995). Muscle spasms associated with ear tick (*Otobius megnini*) infestations in five horses. *Journal of the American Veterinary Medical Association* **207**, 74–76.

Mans, B. J. (2005). Tick histamine-binding proteins and related lipocalins: potential as therapeutic agents. *Current Opininion in Investigative Drugs* **6**, 1131–1135.

Mans, B. J., Gothe, R. & Neitz, A. W. (2004). Biochemical perspectives on paralysis and other forms of toxicoses caused by ticks. *Parasitology* **129**, S95–S111.

Mans, B. J., Louw, A. I. & Neitz, A. W. H. (2003). The major tick salivary gland proteins and toxins from the soft tick, *Ornithodoros savignyi*, are part of the lipocalin family: implications for the origins of tick toxicoses. *Molecular Biology and Evolution* **20**, 1158–1167.

Mans, B. J., Steinmann, C. M., Venter, J. D., Louw, A. I. & Neitz, A. W. H. (2002). Pathogenic mechanisms of sand tampan toxicoses induced by the tick, *Ornithodoros savignyi*. *Toxicon* **40**, 1007–1016.

Mans, B. J., Venter, J. D., Vrey, P. J., Louw, A. I. & Neitz, A. W. H. (2001). Identification of putative proteins involved in granule biogenesis of tick salivary glands. *Electrophoresis* **22**, 1739–1746.

Maritz, C., Louw, A. I., Gothe, R. & Neitz, A. W. (2000). Detection and micro-scale isolation of a low molecular mass paralysis toxin from the tick, *Argas (Persicargas) walkerae*. *Experimental and Applied Acarology* **24**, 615–630.

Maritz, C., Louw, A. I., Gothe, R. & Neitz, A. W. (2001). Neuropathogenic properties of *Argas (Persicargas) walkerae* larval homogenates. *Comparative Biochemistry and Physiology A* **128**, 233–239.

Masina, S. & Broady, K. W. (1999). Tick paralysis: development of a vaccine. *International Journal for Parasitology* **29**, 535–541.

Menez, A. (1998). Functional architectures of animal toxins: a clue to drug design? *Toxicon* **36**, 1557–7152.

Murnaghan, M. F. & O'Rouke, F. J. (1978). Tick paralysis. In *Handbook of Experimental Pharmacology*, vol. 48, *Arthropod Venoms*, eds. Bettini, S., pp. 419–464. Berlin: Springer-Verlag.

Neitz, W. O. (1954). *Hyalomma transiens* Schulze: a vector of sweating sickness. *Journal of the South African Veterinary Medical Association* **25**, 19–20.

Neitz, W. O. (1956). Studies on the aetiology of sweating sickness. *Onderstepoort Journal of Veterinary Research* **27**, 197–203.

Neitz, W. O. (1959). Sweating sickness: the present state of our knowledge. *Onderstepoort Journal of Veterinary Research* **28**, 3–38.

Neitz, W. O. (1962). The different forms of tick toxicoses: a review. In 2nd *Meeting of the FAO/OIE Expert Panel on Tick-Borne Diseases of Livestock*, Cairo, U.A.R., 3–10 December.

Neitz, A. W. H. (1976). Biochemical investigation into the toxic salivary secretion of the tick *Ornithodoros savignyi*. Unpublished D.S.c (Agric) thesis, University of Pretoria, South Africa.

Neitz, A. W. & Gothe, R. (1986). Changes in the protein pattern in the salivary glands of paralysis inducing female *Rhipicephalus evertsi evertsi* during infestation. *Journal of Veterinary Medicine B* **33**, 213–220.

Neitz, A. W. H., Bezuidenhout, J. D., Vermeulen, N. M. J., Potgieter, D. J. J. & Howell, C. J. (1983). In search of the causal agents of tick toxicoses. *Toxicon* **S3**, 317–320.

Neitz, A. W. H., Howell, C. J. & Potgieter, D. J. J. (1969). Purification of the toxic component in the oral secretion of the sand tampan *Ornithodoros savignyi* Audouin (1827). *Journal of the South African Chemical Industry* **22**, 142–149.

Neitz, A. W. H., Prozesky, L., Bezuidenhout, J. D., Putterill, J. F. & Potgieter, D. J. (1981). An investigation into the toxic principle in eggs of the tick *Amblyomma hebraeum*. *Onderstepoort Journal of Veterinary Research* **48**, 109–117.

Nunn, M. A., Sharma, A., Paesen, G. C., *et al.* (2005). Complement inhibitor of C5 activation from the soft tick *Ornithodoros moubata*. *Journal of Immunology* **174**, 2084–2091.

Oswald, B. (1938). A review of work published in Yugoslavia on the tick problem and research on toxins in the eggs of ticks. *Annales de Parasitologie Humaine et Comparée* **16**, 548–559.

Paesen, G. C., Adams, P. L., Harlos, K., Nuttall, P. A. & Stuart, D. I. (1999). Tick histamine-binding proteins: isolation, cloning, and three-dimensional structure. *Molecular Cell* **3**, 661–671.

Paesen, G. C., Adams, P. L., Nuttall, P. A. & Stuart, D. L. (2000). Tick histamine-binding proteins: lipocalins with a second binding cavity. *Biochimica et Biophysica Acta* **1482**, 92–101.

Peacock, P. B. (1958). Tick paralysis or poliomyelitis. *South African Medical Journal* **32**, 201–202.

Quercia, O., Emiliani, F., Foschi, F. G. & Stefanini, G. F. (2005). Anaphylactic shock to *Argas reflexus* bite. *Allergy and Immunology (Paris)* **37**, 66–8.

Rash, L. D. & Hodgson, W. C. (2002). Pharmacology and biochemistry of spider venoms. *Toxicon* **40**, 225–254.

Regendanz, P. & Reichenow, E. (1931). Über Zeckengift und Zeckenparalyse. *Archiv für Schiffs- und Tropen-Hygiene* **35**, 255–273.

Rich, G. B. (1957). The ear tick, *Otobius megnini* (Dugès) (Acarina: Argasidae), and its record in British Columbia. *Candian Journal of Comparative Medicine* **21**, 415–418.

Riek, R. F. (1957). Studies on the reactions of animals to infestation with ticks. *Australian Journal of Agricultural Research* **8**, 215–223.

Riek, R. F. (1958). Studies on the reactions of animals to infestation with ticks. *Australian Journal of Agricultural Research* **9**, 830–841.

Riek, R. F. (1959). Studies on the reactions of animals to infestation with ticks. *Australian Journal of Agricultural Research* **10**, 604–613.

Rolla, G., Nebiolo, F., Marsico, P., *et al.* (2004). Allergy to pigeon tick (*Argas reflexus*): demonstration of specific IgE-binding components. *International Archives for Allergy and Immunology* **135**, 293–295.

Rousselot, R. (1956). Meeting report No. 1956/18, *Report of the Joint FAO/OIE Meeting on the Control of Tick-Borne Diseases of livestock*. Rome: Food and Agriculture Organization.

Shultz, J. W. (1990). Evolutionary morphology and phylogeny of Arachnida. *Cladistics* **6**, 1–38.

Sirianni, M. C., Mattiacci, G., Barbone, B., *et al.* (2000). Anaphylaxis after *Argas reflexus* bite. *Allergy* **55**, 303.

Sonenshine, D. E. (1993). *Biology of Ticks*, vol. 2. Oxford, UK: Oxford University Press.

Spickett, A. M., Burger, D. B., Crause, J. C., Roux, E. M. & Neitz, A. W. H. (1991). Sweating sickness: relative curative effect of hyperimmune serum and precipitated immunoglobin suspension and immunoblot identification of proposed immunodominant tick salivary gland proteins. *Onderstepoort Journal of Veterinary Research* **58**, 223–226.

Spiewak, R., Lundberg, M., Johansson, G. & Buczek, A. (2006). Allergy to pigeon tick (*Argas reflexus*) in Upper Silesia, Poland. *Annals of Agricultural and Environmental Medicine* **13**, 107–112.

Stampa, S. (1959). Tick paralysis in the Karoo areas of South Africa. *Onderstepoort Journal of Veterinary Research* **28**, 169–227.

Standbury, J. B. & Huyck, J. H. (1945). Tick paralysis: a critical review. *Medicine (Baltimore)* **24**, 219–242.

Steinhaus, E. A. (1942). Note on a toxic principle in eggs of the tick, *Dermacentor andersoni* Stiles. *United States Public Health Report* **57**, 1310–1312.

Stone, B. F. & Binnington, K. C. (1986). The paralyzing toxin and other immunogens of the tick *I. holocyclus* and the role of the salivary glands in their biosyntheses. In *Morphology, Physiology and Behavioral Biology of Ticks*, eds. Sauer, J. R. & Hair, J. A., pp. 75–99. Chichester, UK: Ellis Horwood.

Stone, B. F., Binnington, K. C., Gauci, M. & Aylward, J. H. (1989). Tick/host interactions for *Ixodes holocyclus*: role, effects, biosynthesis and nature of its toxic and allergenic oral secretions. *Experimental and Applied Acarology* **7**, 59–69.

Stone, B. F., Doube, B. F. & Binnington, K. C. (1979). Toxins of the Australian paralysis tick *Ixodes holocyclus*. *Recent Advances in Acarology* **1**, 347–356.

Tanaka, A. S., Andreotti, R., Gomes, A., *et al.* (1999). A double-headed serine proteinase inhibitor–human plasma kallikrein and elastase inhibitor from *Boophilus microplus* larvae. *Immunopharmacology* **45**, 171–177.

Thomas, A. D. & Neitz, W. O. (1958). Rhipicephaline tick toxicosis in cattle: its possible aggravating effects on certain diseases. *Journal of the South African Veterinary Medical Association* **29**, 29–50.

Thurn, M. J., Gooley, A. & Broady, K. W. (1992). Identification of the neurotoxin from the paralysis tick, *Ixodes holocyclus*. In *Recent Advances in Toxicology Research*, vol. 2, eds. Gopalakrishnakone, P. & Tan, C. K., pp. 243–256. Singapore: Venom and Toxin Research Group, National University of Singapore.

Tomalski, M. D. & Miller, L. K. (1991). Insect paralysis by baculovirus-mediated expression of a mite neurotoxin gene. *Nature* **352**, 72–75.

Tomalski, M. D., Hutchinson, K., Todd, J. & Miller, L. K. (1993). Identification and characterization of tox21A: a mite cDNA encoding a paralytic neurotoxin related to TxP-I. *Toxicon* **31**, 319–326.

Vermeulen, N. M. J. & Neitz, A. W. H. (1987). Biochemical studies on the eggs of *Amblyomma hebrauem*. *Onderstepoort Journal of Veterinary Research* **54**, 451–459.

Vermeulen, N. M. J., Neitz, A. W. H., Potgieter, D. J. J. & Bezuidenhout, J. D. (1984). Anti-protease from *Amblyomma hebraeum*. *Insect Biochemistry* **14**, 705–711.

Vermeulen, N. M. J., Viljoen, G. J., Bezuidenhout, J. D., Visser, L. & Neitz, A. W. H. (1988). Kinetic properties of toxic protease inhibitors isolated from tick eggs. *International Journal of Biochemistry* **20**, 621–631.

Viljoen, G. J., Bezuidenhout, J. D., Oberem, P. T., *et al.* (1986). Isolation of a neurotoxin from the salivary glands of female *Rhipicephalus evertsi evertsi*. *Journal of Parasitology* **72**, 865–874.

Viljoen, G. J., Neitz, A. W. H., Prozesky, L., Bezuidenhout, J. D. & Vermeulen, N. M. J. (1985). Purification and properties of tick egg toxic proteins. *Insect Biochemistry* **15**, 475–482.

Viljoen, G. J., Van Wyngaardt, S., Gothe, R., *et al.* (1990). The detection and isolation of a paralysis toxin present in *Argas (Persicargas) walkerae*. *Onderstepoort Journal of Veterinary Research* **57**, 163–168.

Walter, D. E. & Proctor, H. C. (1998). Feeding behaviour and phylogeny: observations on early derivative Acari. *Experimental and Applied Acarology* **22**, 39–50.

Wang, H. & Nuttall, P. A. (1994). Excretion of host immunoglobulin in tick saliva and detection of IgG-binding proteins in tick haemolymph and salivary glands. *Parasitology* **109**, 525–30.

Waxman, L. & Connolly, T. M. (1993). Isolation of an inhibitor selective for collagen-stimulated platelet aggregation from the soft tick *Ornithodoros moubata*. *Journal of Biological Chemistry* **268**, 5445–5449.

Wikel, S. K. (1984). Tick and mite toxicoses and allergy. In *Handbook of Natural Toxins*, vol. 2, *Insect Poisons, Allergens and Other Invertebrate Venoms*, ed. Tu A. T., pp. 371–396. New York: Marcel Dekker.

Willadsen, P. & McKenna, R. V. (1983). Trypsin–chymotrypsin inhibitors from the tick, *Boophilus microplus*. *Australian Journal of Experimental Biology and Medical Science* **61**, 231–238.

Willadsen, P. & Riding, G. A. (1979). Characterization of a proteolytic-enzyme inhibitor with allergenic activity: multiple functions of a parasite-derived protein. *Biochemical Journal* **177**, 41–47.

Willadsen, P. & Riding, G. (1980). On the biological role of a proteolytic-enzyme inhibitor from the ectoparasitic tick *Boophilus microplus*. *Biochemical Journal* **189**, 295–303.

6 • Tick lectins and fibrinogen-related proteins

L. GRUBHOFFER, R. O. M. REGO, O. HAJDUŠEK, V. HYPŠA, V. KOVÁŘ,
N. RUDENKO AND J. H. OLIVER JR.

INTRODUCTION

Tissue-specific lectin/haemagglutinin activities have been investigated for both soft and hard ticks, although there are comparatively few papers published. Some tick lectins are proteins with binding affinity for sialic acid, various derivatives of hexosamines and different glycoconjugates. Most tick lectin/haemagglutinin activities are blood-meal enhanced, and could serve as molecular factors of self/non-self recognition in defence reactions against bacteria or fungi, as well as in pathogen/parasite transmission. Dorin M, the plasma lectin of *Ornithodoros moubata*, is the first tick lectin to be purified from tick haemolymph, and the first that has been fully characterized. Partial characterization of other tick lectins/haemagglutinins has been performed mainly with respect to their carbohydrate-binding specificities and immunochemical features. The main goal of this review is to provide an overview of our knowledge of lectins as tissue specific carbohydrate-binding proteins of ticks with emphasis on their structural properties and functional roles either in defence reactions or pathogen transmission. Other lectin reviews have been published dealing with tissue-specific lectins in blood-sucking arthropods (e.g. Ingram & Molyneux, 1991; Grubhoffer, Hypša & Volf, 1997; Grubhoffer & Jindrák, 1998). In addition, several publications have drawn attention to plant and animal lectins, and to lectins as tools in modern glycobiological research (e.g. Jacobson & Doyle, 1996; Rhodes & Milton, 1998).

Research on lectins began in 1888 with publication of the doctoral thesis of Herman Stilmark at the University of Dorpat, Estonia, on the agglutinins of the seeds of castor bean *Ricinus communis* (Sharon & Lis, 1988). Investigations of proteins able to agglutinate red blood cells (RBCs) have been improved significantly by modern biochemical and molecular biological techniques during the past few decades (Sharon & Lis, 2007; Slifkin & Doyle, 1990; Doyle & Slifkin, 1994; Jacobson & Doyle, 1996). The term lectin comes from the latin verb *legere* (to select) and reflects an ability of particular proteins (haemagglutinins) to aggregate different kinds of RBCs (Boyd & Shapleigh, 1954). Several authors improved the original definition of lectins as haemagglutinins as knowledge improved on the structural and functional properties of lectin molecules and their sugar-binding sites (Goldstein *et al.*, 1980; Kocourek & Hořejší, 1981; Barondes, 1988; Drickamer, 1988; Yoshizaki, 1990; Lee, 1992). According to Peumans & van Damme (1995), lectins are proteins or glycoproteins carrying at least one binding site that shows a reversible and specific interaction with a particular carbohydrate moiety. Lectins are ubiquitous proteins that are present most probably in all eukaryotes and many bacterial species, as well as in some viruses. This diverse group of proteins is of great importance for living organisms; in organisms that lack specific immunity some of them are considered to be functional analogues of immunoglobulins (Vasta & Marchalonis, 1984). Lectins as proteins with binding affinity for carbohydrate molecules take part mainly in protein–saccharide interactions. These interactions participate in both effector and regulatory processes in organisms, including those of disease-transmitting blood-sucking arthropods.

The list of identified lectins has grown rapidly in the last two decades with more than 10 000 publications dealing with these molecules. Based on their overall structure, three major types of lectins are distinguished: mero-, holo- and chimerolectins (Peumans & van Damme, 1995). Merolectins are proteins that have only a carbohydrate-binding domain. Hololectins comprise two or more identical or highly homologous domains. Chimerolectins possess a sugar-binding domain associated with an unrelated domain having different biological activity or another carbohydrate-binding specificity. Molecules with such combined functional activities have often been called lectinoids (Gilboa-Garber & Garber, 1989). Combining different binding activities in such molecules offers an effective way to function as regulatory tools for mechanisms of recognition and molecular identification by receptor/co-receptor systems. Proteins having an enzyme activity for carbohydrate substrates, regulatory proteins (e.g. repressor in the *lac* operon) or antibodies with idiotypic specificity to carbohydrates/glycoconjugates are not considered lectins (Kocourek & Hořejší, 1981).

Drickamer (1988) recognized lectins as both soluble and/or membrane-bound post-translationally modified proteins providing often multimolecular complexes built from a number of identical subunits. The author distinguished two major types of lectins based on their structure: S-type (recently called galectins) with thiol-group-dependent activity localized in both intra- and extracellular compartments, and C-type (some of them are called selectins) with Ca^{2+}-dependent activity in the extracellular space. The C-type lectins are more variable in structure and more specific to organs and tissues. However, their sequence data show a common and homologous region responsible for binding carbohydrates, the carbohydrate region domain (CRD) (Lee, 1992). Collectins represent an important subgroup of C-lectins. They have a collagen domain, a neck region and a globular C-type lectin-binding domain (Epstein *et al.*, 1996; Ezekowitz, Sastry & Reid, 1996).

An important function of lectins is their involvement in innate immunity of arthropods and vertebrates. They provide the primitive/ancient functions, such as aggregation of microorganisms and opsonizing them for phagocytosis (Ratcliffe & Rowley, 1987; Vasta, 1991; Vasta *et al.*, 1994). During evolution, lectins gained more sophisticated functions in defence against pathogens, which is evident from their relationship with the complement system (Matsushita, 1996; Vasta *et al.*, 1999; Matsushita *et al.*, 2001; Iwanaga, 2002). The lectin activation pathway of the complement system can be triggered not only by the mannose-binding lectin (MBL), a member of the collectins, but also by a group of other lectins called ficolins (Gadjeva, Thiel & Jesenius, 2001; Matsushita *et al.*, 2001). In the lectin activation pathway, an MBL with collagen-like and carbohydrate-recognition domains (CRD) (Taylor *et al.*, 1990) activates the complement cascade in cooperation with an MBL-associated serine protease (Hoffmann *et al.*, 1999). Ficolins occur in mammalian sera. Besides a collagen-like domain common in MBLs, ficolins contain a fibrinogen-like domain (Lu & Le, 1998). Recent characterization of ficolins present in human, mouse, and pig serum/plasma, as well as in the body fluids of ascidians, revealed their common binding specificity for GlcNAc. Human serum ficolins have been shown to function as opsonins (Matsushita *et al.*, 1996). Most probably, ficolins function as self/non-self recognition lectin molecules in haemolymph (Gokudan *et al.*, 1999). Members of the fibrinogen-related protein family have been identified both from vertebrates and invertebrates (Xu & Doolittle, 1990; Adema *et al.*, 1997; Kurachi *et al.*, 1998; Dimopoulos *et al.*, 2001; Kenjo *et al.*, 2001; Leonard *et al.*, 2001). Other recognized lectins are the I-type lectins (belonging to the immunoglobulin superfamily), pentraxins (lectin-like molecules, such as the C-reactive protein), and heparin-binding proteins (Vasta, Ahmed & Quesenberry, 1996). Lectins that recognize more complex structures at the cell surface, such as C-type lectins and galectins, are found in invertebrate organisms as well as in vertebrates, but the functions of these proteins have evolved separately in different animal lineages (Dodd & Drickamer, 2001).

Other proteins with carbohydrate-binding domains, which do not fall into the lectin family, can serve as scavenger receptors (Pearson, 1996). They occur on the surface of haemocytes carrying a binding specificity to anionic polysaccharides (e.g. fucoidan, dextran sulphate, chondroitin sulphate). Scavenger proteins as well as proteins with binding affinity for oligo/polynucleotides seem to play a significant role in the receptor functions over the whole evolution of innate immunity. Together with both lipopolysaccharide (LPS) and β-1,3 glucan-binding proteins, scavenger proteins constitute an important group of molecules engaged in non-self recognition in invertebrates.

ARTHROPOD LECTINS

Lectins may interact either with 'self' glycoconjugates or with glycosylated components of viral, bacterial, protozoan and metazoan pathogens. Soluble or membrane-bound invertebrate lectins take part in the processes of cell adhesion, opsonization, phagocytosis and cytolysis (Vasta & Marchalonis, 1983). They are clearly important molecules involved in humoral and cellular reactions for recognition and defence by opsonization of non-self objects within the body cavity of arthropods. Lectins may also play a part in the transmission of pathogens by vectors, for instance as receptors, homing factors or differentiation factors. A great number of lectins of various invertebrates have been described (Yeaton 1982*a*, *b*; Ratcliffe *et al.*, 1985). The best-known arthropod lectins are those from the horseshoe crabs *Limulus polyphemus* and *Tachypleus tridentatus*, and from the scorpions *Centruroides sculpturatus* and *Parauroctonus mesanensis* (Vasta & Marchalonis, 1984; see Olafsen 1986, 1996 for review). Almost all plasma lectins isolated from these chelicerates specifically bind *N*-acetyl neuraminic (sialic) acid and other *N*-acylamino-carbohydrates. Studies on invertebrate immunity show many molecular aspects common to vertebrate immune responses, such as similarities in the clotting mechanisms of haemolymph and blood. Remarkably, one of the four characterized haemolymph clotting factors of *Limulus*, factor C (serine protease zymogen), resembles the selectin family of cell-adhesion molecules. It belongs to the C-type (Ca^{2+} ion dependent) chimerolectins (for review see Iwanaga, 1993; Hoffmann, 1995).

Lectins purified from haemolymph of the Japanese horseshoe crab *T. tridentatus*, and called tachylectins 5A/5B (Tl-5A/5B), have been characterized and cloned (Gokudan *et al.*, 1999). Both possess a C-terminal fibrinogen (FBG) domain of high structural and sequence similarity to FBG domains of mammalian ficolins. However, Tl-5A/5B have no effector collagen domains and no lectin-associated serine proteases, as found in mammalian ficolins. Tl-5s, together with other defence molecules released from haemocytes, are employed in effective innate immunity against invading pathogens (Kawabata & Tsuda, 2002).

Lectins of disease vectors

Despite the great influence of vector-borne pathogens on humans and domestic animals, relatively little information is available on lectins of invertebrate vectors, with the exception of reduviid bugs, tsetse flies, mosquitoes and sandflies (Pereira, Andrade & Ribeiro, 1981; Wallbanks, Ingram & Molyneux, 1986; Maudlin & Welburn, 1988; Gomes, Furtado & Coelho, 1991; Volf, 1993; Hypša & Grubhoffer, 1995; Ratcliffe *et al.*, 1996; Chen & Billingsley, 1999).

The best-studied vector–parasite system involves lectins of various tsetse flies (*Glossina* spp.) (Maudlin & Welburn, 1988; Welburn & Maudlin, 1990; Ingram & Molyneux, 1991, Abubakar *et al.*, 2006). Some lectins in the gut and haemolymph of tsetse flies function as signalling factors for maturation of African trypanosome species. Other lectins act as lytic (killing) factors; an interaction between the gut lectin of the tsetse fly and symbiotic rickettsia-like organisms, or with products of their saccharide metabolism, affects the life cycle of trypanosomes in tsetse flies (Maudlin & Welburn, 1988). A similar interference between symbiotic rickettsiae and protozoa was thought to be involved in ticks although lectin activities were not considered at the time (Smith *et al.*, 1976). Sandflies, the vectors of leishmania, have also been well studied in relation to midgut lectins and their recep-

tor function in the process of leishmania transmission (Volf *et al.*, 1994; Palánová & Volf, 1997; Volf, Škařupová & Man, 2002).

Lectin–carbohydrate interactions participate in the vector–pathogen/parasite relationship. Tissue-specific lectins of blood-sucking arthropods interact with glycosylated molecules of receptors on the surfaces of pathogens or their cells within appropriate developmental stages and vice versa. For instance, bacteria or other intracellular parasites often employ their surface lectins to enter into phagocytes using a mechanism known as lectinophagocytosis (Ofek & Sharon, 1988).

Lectins of ticks (overview)

The lectins of ticks have not received the same level of attention as those of insects. A role for tick haemolymph lectins in transmission of pathogens by ticks has been suggested (Munderloh & Kurtti, 1995). Lectin/haemagglutinin activites in the haemolymph of four tick species (*Ixodes ricinus*, *Ornithodoros tartakovskyi*, *O. tholozani (papillipes)* and *Argas polonicus*) have been described (Vereš & Grubhoffer, 1990; Grubhoffer & Mat'ha, 1991; Grubhoffer, Vereš & Dusbábek, 1991). The partially characterized tick lectins have an affinity for sialic acid and *N*-acetyl-D-glucosamine but differ in their binding specificity to other sugars. It is postulated that these lectins interact with glycosylated structures of transmitted pathogens, but this has not yet been proved. The distribution of *Ixodes ricinus* haemolymph lectin in tick tissues supports the idea that the lectin is produced and/or stored in haemocytes and contributes to the immune system by recognition of foreign substances (Kuhn, Uhlíř & Grubhoffer, 1996). Like limulin, the sialic-acid-specific lectin of the horseshoe crab *Limulus polyphemus* and other sialic-acid-specific lectins of chelicerates, the haemolymph lectin of ixodid ticks may also recognize a wide range of Gram-negative bacteria owing to its site specificities for *N*-acetyl-D-glucosamine, D-galactose and 2-keto-3-deoxyoctonate acid (Vasta & Marchalonis, 1983; Grubhoffer *et al.*, 1991; Kuhn *et al.*, 1996). The haemocytes of *I. ricinus* phagocytose spirochaetes of *Borrelia burgdorferi* by the coiling method; coiling phagocytosis is thought to be a lectin-mediated process (Kuhn *et al.*, 1994).

Kamwendo *et al.* (1993) reported haemagglutination activities in the haemolymph, gut homogenates and salivary glands in the tick *Rhipicephalus appendiculatus*, the vector of East Coast fever. This is a serious tick-borne disease of cattle caused by the protozoan parasite *Theileria parva* (see

Chapter 14). Bovine RBCs were the most sensitive detection system for the tissue homogenates, whether from unfed or fed ticks (Kamwendo *et al.*, 1993). However, human erythrocytes of the ABO system and rabbit RBCs showed significantly higher values of haemagglutination titres in tissue samples from fed compared with unfed ticks, with the highest increase recorded in the gut extract. Later studies focused on the functional role of the tick salivary gland haemagglutinin in relation to the tick's ability to transmit *T. parva* parasites (Kamwendo *et al.*, 1995). *Rhipicephalus appendiculatus* gut lectin activity also influences transmission of the causative agent of theileriosis in Africa, causing agglutination of piroplasms (E. Sebitosi, personal communication). This led to the conclusion that the protein–carbohydrate interactions of piroplasms within the body cavity of the tick may be a target for the control of theileriosis.

Many functional and morphological aspects of tick tissues are involved in the process of parasite transmission (Friedhoff, 1990). However, little is known about the molecular aspects of cell communication and tick interactions with the surface components of transmitted pathogens. The following three sections are devoted to tissue specific lectins/agglutinins of the hard tick *Ixodes ricinus*, the soft tick *Ornithodoros moubata*, and FBG domain containing tick proteins that are similar to ficolins and tachylectins. Because of the lack of published data on tick lectins, most of the results are based on the research of the Tick Lectin Group at the Institute of Parasitology, České Budějovice, Czech Republic (see Table 6.1).

LECTINS OF THE IXODID TICK *IXODES RICINUS*

Gut

The sheep tick, *Ixodes ricinus*, which transmits tick-borne encephalitis (TBE) virus and the Lyme-disease-causing spirochaete (*Borrelia burgdorferi sensu lato*) is the most important tick vector in Europe. Gut haemagglutinating activity of *I. ricinus* has been partially characterized (Uhlíř, Grubhoffer & Volf, 1996). Native mouse erythrocytes were the most sensitive cell based detection system for a wide pH range (pH 6.5–8.0). Binding specificity analysis showed that simple sugars (*N*-acetyl-D-galactosamine, *N*-acetyl-D-glucosamine, rhamnose and dulcit) and glycoconjugates (fetuin, hyaluronic acid and laminarin (β-1,3-glucan)) and bacterial LPS strongly inhibit haemagglutination. Binding characteristics of the tick gut agglutinin are consistent with

LPS-binding agglutinins in the gut tissue of other blood-sucking arthropods (Grubhoffer *et al.*, 1997). It seems that the midgut agglutinin as a potential LPS-binding protein (LPS-BP) could (in cooperation with digestive enzymes) affect Gram-negative spirochaetes (e.g. *B. burgorferi*) passing through the gut epithelium. Midgut extracts from unfed ticks lack haemagglutinating activity, which is consistent with the observations of Kamwendo *et al.* (1993) for *R. appendiculatus*. Mouse polyclonal antibodies raised against midgut haemagglutinating activity (Yeaton, 1982*a*) recognized, by Western blotting, four proteins with molecular weights 37, 60, 65 and 73 kDa, as putative structural components of the lectin(s) (Uhlíř *et al.*, 1996). Whereas the 37- and 60-kDa proteins are glycoproteins modified by both high mannose and complex *N*-glycans, the 70-kDa subunit is modified only by a complex type of glycan (Uhlíř *et al.*, 1994). Using specific mouse polyclonal antibodies, the midgut haemagglutinin/lectin complex was immunohistochemically localized solely to the midgut cells, and was not detected in the gut contents or other tick tissues including haemolymph (Uhlíř *et al.*, 1996).

Two agglutinins/lectins of the gut haemagglutinating complex in *I. ricinus* were isolated from the gut homogenate after delipidation by acetone extraction (Durnová, 1998). Both agglutinins/lectins were purified by affinity chromatography either on immobilized bovine submaxillary mucine (BSM) or laminarin (β-1,3-glucan). Using SDS-PAGE, a 65-kDa protein was identified as the main agglutinin with binding affinity to BSM, and a 37-kDa protein showed strong binding specificity for laminarin (β-1,3-glucan). Although crude gut homogenate had no binding affinity for free sialic acid in the haemagglutination inhibition assay, the delipidated proteinaceous 'acetone powder' prepared from the crude gut homogenate showed a strong ability to bind sialic acid. It seems likely that lipids, including glycolipids abundant in the gut tissue, hide the lectin-binding site for sialic acid. The pure 65-kDa agglutinin/lectin also maintained affinity for free sialic acid. Mouse antibodies raised against the entire haemagglutinating activity of the crude gut homogenate recognized both proteins. Because the 37-kDa gut agglutinin/lectin functions as a β-1,3-glucan-binding protein, it might be part of the defence machinery protecting the tick against fungal infections.

Factors taking part in processes of pathogen/parasite transmission by ticks have been reviewed by Friedhoff (1990). This review proposed that the midgut agglutinin/LPS-binding protein may belong to factors associated with the development of pathogens/parasites

Table 6.1 *Tick lectins/agglutinins and their basic characteristics*

Tick species	Tissue	Binding specificity[a]	Agglutination of RBC	Structural subunit[b] (molecular size, kDa)	Reference
Ixodes ricinus	Haemolymph (plasma, haemocytes)	Sialic acid; N-acetyl-D-glucosamine; fetuin; asialofetuin; BSM	Mouse (pronase treated)	85	Kuhn, Uhlíř & Grubhoffer, 1996
	Gut	N-acetyl-galactosamine; N-acetyl-glucosamine; rhamnose; dulcit; fetuin; hyaluronic acid; laminarin; LPS	Mouse	37; 60; 65; 73 (haemagglutinating complex)	Uhlíř, Grubhoffer & Volf, 1996
	Gut (after acetone extraction of lipids)	Sialic acid; BSM; laminarin	Mouse	37 (laminarin[c]) 65 (BSM[c])	Durnová, 1998
	Salivary glands	Sialic acid; fructose; rhamnose; trehalose; fetuin; asialofetuin; BSM; laminarin; heparin	Mouse	70	Unpublished data
Rhipicephalus appendiculatus	Haemolymph Gut	n.d.	Bovine Human Rabbit	n.d.	Kamwendo *et al.*, 1993
	Salivary glands	Mannose[d]; turanose[d]	Bovine Human Rabbit	n.d.	Kamwendo *et al.*, 1995
Ornithodoros tartakovskyi	Haemolymph/ plasma	Sialic acid; N-acetyl-D-glucosamine; N-acetyl-D-galactosamine; fetuin; asialofetuin; BSM	Mouse (pronase treated)	30; 35	Grubhoffer, Vereš & Dusbábek, 1991
	Haemocytes			35	
Ornithodoros tholozani (papillipes)	Haemolymph/ plasma	Sialic acid; N-acetyl-D-glucosamine; D-galactose; fetuin; BSM	Mouse (pronase treated)	37; 40	Grubhoffer, Vereš & Dusbábek, 1991
Ornithodoros moubata	Haemolymph/ plasma	N-acetylneuraminyl-lactose; sialic acid	Mouse (pronase treated)	~ 37 Dorin M GenBank: AY 333989	Grubhoffer & Kovář, 1998 Kovář, Kopáček & Grubhoffer, 2000 Rego *et al.*, 2006
	Haemocytes, salivary glands	N-acetyl-D-galactosamine; N-acetyl-D-glucosamine; N-acetyl-D-mannosamine; fetuin; asialofetuin; BSM; PSM-I			
Argas polonicus	Haemolymph/ plasma	N-acetyl-D-glucosamine; D-galactose; fetuin; BSM	Mouse	n.d.	Grubhoffer, Vereš & Dusbábek, 1991

[a] Determined by the haemagglutination inhibition assay.

[b] Determined by SDS-PAGE and Western blotting.

[c] Ligands in affinity chromatography.

[d] Inhibition of *T. parva* transmission in vivo.

BSM, bovine submaxillary mucin; PSM-I, porcine submaxillary mucin; n.d., not determined.

in the digestive tract of the tick. Blood digestion in haematophagous insects proceeds rapidly and takes place in the lumen of the gut (Gooding, 1972). By contrast, digestion in ticks is intracellular and slow (Balashov, 1972; Akov *et al.*, 1976). In ixodid ticks, the blood meal is taken up by the digestive cells by means of fluid-phase endocytosis and via receptor-mediated endocytosis. Large particles such as whole cells are phagocytosed. The binding activity of the midgut agglutinin/LPS-BP is pH dependent. At pH 3.0, the optimum for gut proteases, the gut agglutinin is ineffective whereas at pH 6.5, the pH of the gut contents, haemagglutinating activity of this binding protein reaches its optimum (Uhlíř *et al.*, 1996). These observations suggest that the midgut agglutinin/LPS-BP may be involved in the preparatory phase of blood meal digestion, in addition to its role in defence.

Haemolymph (plasma and haemocytes)

Most lectins isolated from arthropods are constituents of the haemolymph, the extracellular circulating fluid that fills the body cavity or haemocoel. Haemolymph is physically isolated from direct contact with body tissues by the basal laminae. However, haemocytes within the haemolymph are in direct contact with this fluid (Mullins, 1985). Haemocytes are immunocytes of arthropods although they also have nutritional functions (Gupta, 1985, 1991). Haemolymph lectins and other humoral immune factors in the haemolymph are produced and secreted into the haemolymph mainly by haemocytes and fat body cells. Recently, Gudderra *et al.* (2002) published a comprehensive review on haemolymph proteins in ticks emphasizing the structural and functional features of vitellogenin and other tick proteins such as macroglobulins, haem lipoproteins, antimicrobial proteins and lectins.

Haemagglutinating activity in the haemolymph of *I. ricinus* is Ca^{2+} dependent, with a subunit molecular size of 85 kDa determined by non-reducing SDS-PAGE, and binding affinity for sialic acid, *N*-acetyl-D-glucosamine and D-galactose (Yeaton, 1982*a*; Grubhoffer *et al.*, 1991; Kuhn *et al.*, 1996, and unpublished observations). Immunoreactivity with poly- and monoclonal antibodies, prepared against haemolymph lectin (Yeaton, 1982*a*), was detected in the granules of both types of granular haemocytes, on the plasma membrane of haemocytes and on the basal laminae surrounding the haemocoel (Figs. 6.1 and 6.2). Immunoreactivity was also localized to cells attached to the midgut invaginations of

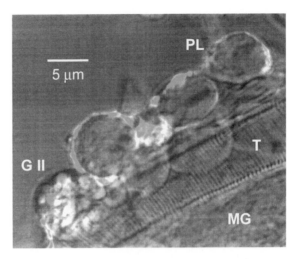

Fig. 6.1. Indirect immunofluorescence of the plasma lectin in *Ixodes ricinus* haemocytes detected using mouse monoclonal antibody 3/D3/C5 to the plasma lectin activity of *I. ricinus*. GII, type II granular haemocytes; PL, plasmatocyte; T, trachea; M, midgut. Laser confocal microscopy. (Courtesy Dr K.-H. Kuhn.)

Géné's organ and granular inclusions of nephrocytes, suggesting an association with eggs (Kuhn *et al.*, 1996).

The distribution of lectin immunoreactivity strengthens the assumption that the haemolymph lectin is produced and stored in haemocytes and contributes to the immune system of *I. ricinus*. Sialic-acid-binding lectins are common in chelicerates and recognize Gram-negative bacteria (Mandal & Mandal, 1990; Wagner, 1990; Zeng & Gabius, 1992). The haemolymph lectin of *I. ricinus* may also recognize a wide range of Gram-negative bacteria owing to its additional specificities for *N*-acetyl-D-glucosamine, D-galactose and unrelated molecules such as 2-keto-3-deoxyoctonate (Vasta & Marchalonis, 1983; Grubhoffer *et al.*, 1991). Johns *et al.* (2001*a*, *b*), studying differences in vector capability of various tick species for borrelia transmission, suggested a potential role of immune molecules. Several studies have shown that *B. burgdorferi* migrate from the midgut across the haemocoel to the salivary glands during tick feeding (Ribeiro *et al.*, 1987; Zung *et al.*, 1989). However, less than 5% of the midgut population of spirochaetes survive to reach the salivary glands (Coleman *et al.*, 1997). Moreover, in vitro assays have shown that *B. burgdorferi* is a target for the immune system of *I. ricinus* and is readily ingested by phagocytic haemocytes (Kuhn *et al.*, 1994). Additionally, Kurtti *et al.* (1988) reported that *B. burgdorferi* spirochaetes readily adhere to tick cells in tissue culture. Therefore, the lectin may mediate the interaction

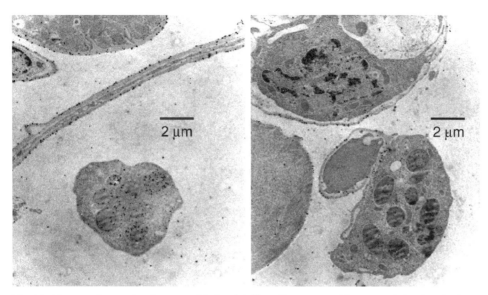

Fig. 6.2. Electron micrographs of immunogold–silver labelling of haemocytes with mouse polyclonal antibodies to the plasma lectin activity of *Ixodes ricinus*: granular inclusions of type II granular haemocytes. (Courtesy Dr K.-H. Kuhn.)

of pathogens like *B. burgdorferi* with the tissues of the vector. The main binding specificity of the haemolymph lectin is for sialic acid (Grubhoffer *et al.*, 1991), which is present in the wall of *B. burgdorferi* (Hulínská, Volf & Grubhoffer, 1992). In addition, the haemolymph lectin of the tick can bind complex oligosaccharides lacking sialic acid. The envelope glycoprotein of TBE virus exhibits biantennary complex oligosaccharides and can interact with the haemolymph lectin (Grubhoffer *et al.*, 1990)

Localization of lectin immunoreactivity in diverse tissues and cells of the tick does not imply that all labelled proteins are associated with the same function. Polyclonal antibodies most likely recognize common epitopes in functionally diverse proteins with multiple domains. Such crossreactivity has been shown for calcium-dependent globular lectin domains (C-type lectins) (Drickamer, 1993; Drickamer & Taylor, 1993) that are components of vertebrate haemostasis and complement systems (Thiel & Reid, 1989; Sastry & Ezekowitz, 1993; McEver, 1994) and a related protein, factor C (Muta *et al.*, 1991), and an immune protein of the complement-coagulation system of *L. polyphemus* (Iwanaga, 1993).

A cDNA library constructed from unfed *I. ricinus* females was screened using a probe obtained from the lectin gene of *B. burgdorferi* B31 (unpublished observations). Clones showing the strongest hybridization signals were selected and sequenced. The amino acid sequence of a putative carbo-

hydrate recognition domain of a recombinant clone (r58) from the *I. ricinus* cDNA library revealed strong homology (81%) with four lectins from *Periplaneta americana* cockroach haemolymph (Kawasaki, Kubo & Natori, 1996) (Fig. 6.3).

Salivary glands

Kamwendo *et al.* (1995) were the first to study the role of a tick salivary gland haemagglutinin in relation to pathogen transmission. They studied *Rhipicephalus appendiculatus*, the vector of *Theileria parva*, in relation to the mechanism of transmission of the protozoan parasite. A significant increase was observed in *T. parva* acinar infection rates in the salivary glands of *R. appendiculatus* fed on ears of rabbits infused with melibiose and raffinose (haemagglutination-inhibiting sugars). In contrast, mannose and turanose (non-inhibitory sugars) did not affect *T. parva* acinar infection rates. Haemagglutinating activity in *I. ricinus* salivary gland extracts was highest against mouse RBCs, and was inhibited by sialic acid, D-fructose, L-rhamnose, D-trehalose and several glycoconjugates (unpublished observations). Mouse polyclonal antibodies raised against the haemagglutinating activity identified a 70-kDa protein responsible for salivary gland haemagglutinating activity. The role of *I. ricinus* salivary gland lectins in the transmission of TBE virus and *B. burgdorferi* has yet to be determined.

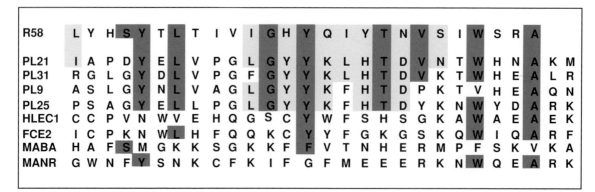

Fig. 6.3. Comparison of carbohydrate recognition domains of *I. ricinus* lectin (r58) with lectins of other animals: R58 (r58), clone from *I. ricinus* cDNA library (unpublished observations); PL21, 31, 9 and 25, four lectins from *Periplaneta americana* cockroach hemolymph (Kawasaki *et al.*, 1996); HLEC1, human hepatic lectin (Halberg *et al.*, 1987); FCE2, mucose immunoglobulin Fc receptor (Bettler *et al.*, 1989); MABA, rat mannose-binding protein A (Drickamer, Dordal & Reynolds, 1986); and MANR, human macrophage mannose receptor (Taylor *et al.*, 1990). Dark grey represents identical amino acid residues and light grey represents similar residues.

LECTINS OF THE ARGASID TICK *ORNITHODOROS MOUBATA*

Haemolymph (plasma and haemocytes)

The soft tick *Ornithodoros moubata* is a vector of African swine fever virus and *Borrelia duttoni*. Laboratory colonies of this species are valuable experimental models that have been employed to answer questions about the tick physiology and morphogenesis, such as vitellogenin synthesis (Chinzei, Chino & Takahashi, 1983; Chinzei & Yano, 1985; Chinzei, 1988; Gudderra *et al.*, 2002, for review), defensins as immunopeptides (Nakajima *et al.*, 2001) and peritrophic matrix synthesis (Grandjean, 1984). The immune response of *O. moubata* induced by infection with the filarial worm *Acanthocheilonema viteae* was described by Hutton, Reid & Towson (2000). A plasma/haemolymph lectin of *O. moubata* has been characterized and there is evidence of lectin activity in the salivary glands and gut (Grubhoffer & Kovář, 1998; Kovář, Kopáček & Grubhoffer, 2000, and unpublished observations)

Dorin M is a lectin with high haemagglutinating activity in the plasma of *O. moubata* (Grubhoffer & Kovář, 1998; Kovář *et al.*, 2000). Activity of the plasma lectin is inhibited by sialic acid, *N*-acetyl-D-hexosamines and acid (sialo)glycoproteins. It was purified to homogeneity using two different approaches: affinity chromatography on a column of bovine submaxillary mucin conjugated to Sepharose 4B and elution with *N*-acetyl-D-glucosamine (Grubhoffer &

Kovář, 1998), and chromatography on Blue Sepharose followed by anion-exchange FPLC on a MonoQ column (Kovář *et al.*, 2000). Dorin M is a glycoprotein with *N*-type glycosylation which, in the native state, forms aggregates with a molecular mass of about 640 kDa. Non-reducing SDS-PAGE revealed two non-covalently bound subunits migrating closely around 37 kDa. After chemical deglycosylation, only one band of about 32 kDa, without haemagglutinating activity, was detected. The amino acid sequence of Dorin M shows strong homology to other invertebrate lectins possessing a FBG domain (Rego *et al.*, 2006). With the burgeoning interest in such lectins of invertebrates, especially in disease vectors, their presence in both *O. moubata* and *I. ricinus* is discussed in a separate section below.

Research on tick haemolymph lectins was initiated more than 10 years ago with the study of *O. tartakovskyi* and *O. tholozani (papillipes)*, species closely realeated to *O. moubata* (Vereš & Grubhoffer, 1990; Grubhoffer *et al.*, 1991). Sialic-acid-specific lectins are present in plasma/haemolymph of these species, with minor differences in binding affinity to other free carbohydrates, such as *N*-acetyl-D-glucosmine, D-glucosamine and D-galactosamine. Two bands of 30 and 35 kDa in tick haemolymph and a single component of 35 kDa in the haemocyte lysate were distinguished by immunoblotting with polyclonal antibodies. In whole-body homogenates, a 150-kDa band was identified in addition to the 30-kDa protein. Both structural units of the haemolymph lectin activity were found to be glycoproteins by lectin affinoblotting. The

30-kDa protein was modified by the complex type of glycans with terminal-D galactose, whereas the 35-kDa protein complexed with high mannose type glycans (Grubhoffer et al., 1991).

A new galectin (OmGalec) from O. moubata was described by Huang et al. (2006). OmGalec consists of 333 amino acids with a predicted molecular weight of 77.4 kDa. It does not have a signal peptide or transmembrane domain, but it has tandem-repeated carbohydrate recognition domains. OmGalec was expressed at all stages of the life cycle and in multiple organs and was abundant in haemocytes, midgut and reproductive organs. Recombinant OmGalec showed significant binding affinity for lactosamine type disaccharides. Its functions in the development and immunity of ticks and/or tick–pathogen interactions are considered (Huang et al., 2006).

FIBRINOGEN-RELATED PROTEINS (FREPs) IN TICKS

The question of whether ticks possessed FBG domain containing lectins was answered with the purification and sequencing of Dorin M from O. moubata (Kovář et al., 2000, Rego et al., 2006). A similar protein has been identified in I. ricinus (Rego et al., 2005). Dorin M shows high sequence homology to the C-terminal fibrinogen domain of the tachylectins 5A and 5B from the horseshoe crab and to the sialic-acid-binding lectins from the slug Limax flavus and the snail Biomphalaria glabrata (Rego et al., 2006). All the aforementioned invertebrate lectins belong to a family of molecules containing a FBG domain but differing in their N-terminal regions. Lectins from this family induced by bacteria also include those in the ascidian Halocynthia roretzi (Kenjo et al., 2001), Anopheles gambiae (Dimopoulos et al., 2000) and Armigeres subalbatus (Wang et al., 2004). Their importance lies in recognizing pathogen-associated molecular patterns and hence they show similarities to lectins that play a large part in the innate immune system. Doolittle and colleagues (1997) suggest that the fibrinogen-related molecules evolved from a lectin-like molecule, which may have functioned as a non-self recognizing protein (Gokudan et al., 1999). At this point it is also important to note that in H. roretzi, a protochordate that is considered a phylogenetic intermediate between vertebrates and invertebrates, the ficolins having the FBG domain are associated with the lectin complement pathway and exhibit high homology to mammalian ficolins (Kenjo et al., 2001). Dorin M is localized within the haemocytes of O. moubata and other pub-

lished results clearly indicate that Dorin M is synthesized within haemocytes. (Rego et al., 2006). This is also the case for the Tl5s (Gokudan et al., 1999) and for Anopheles fibrinogens (Dimopoulos et al., 2001). The four conserved cysteine residues of Dorin M, also found in Tl5s, are likely to form disulphide bonds. Similarities are also seen in residues at positions that mediate binding between the carbohydrate and the protein (Rego et al., 2006).

OMFREP, isolated from O. moubata haemocyte cDNA, has a similar localization to Dorin M, in haemocytes and salivary glands but not in other tissues including the midgut and ovaries (Rego et al., 2005). Two proteins from I. ricinus, Ixoderin A and Ixoderin B, have sequence similarity to the O. moubata FREPs. Ixoderin A is detectable in haemocytes, salivary glands and midgut whereas Ixoderin B is expressed in the salivary glands. Differences in the localization and sequence of Ixoderin B suggest a function that is different from other tick FREPs (Rego et al., 2005). Knowledge that Anopheles mosquitoes, important disease vectors, may possess a family of FREPs as a result of the evolutionary pressures of haematophagy and exposure to parasites, led our group to search for more sequence homologues in ticks. Four different sequences were obtained from I. ricinus showing similarity to Ixoderin B, and two sequence homologues from O. moubata which are also more closely related to the Ixoderin B family than to Dorin M and OMFREP (unpublished results). In addition, a homologue of Dorin M and OMFREP was obtained from Amblyomma americanum and I. scapularis (unpublished results). Preliminary results of phylogenetic analysis (Fig. 6.4) show no correlation between tick phylogeny and sequence relatedness of tick FREPs, suggesting that individual forms of the tick FREP molecule are of ancient origin, arising before the evolutionary divergence of hard and soft ticks.

FUTURE DIRECTIONS

Protein–carbohydrate interactions are known to participate in vector–pathogen/parasite relationships. One of the most important issues implicating lectins in immune responses to pathogens is the inducibility/upregulation of their binding (haemagglutinating) activities after blood-feeding or pathogen invasion. To examine this question, subtractive cDNA libraries of I. ricinus ticks were analysed for the occurrence of upregulated clones and a gene with high homoalogy to jacalin, a plant lectin, was found (unpublished observations). Another important issue of pathogen/parasite transmission by ticks is the reciprocity of

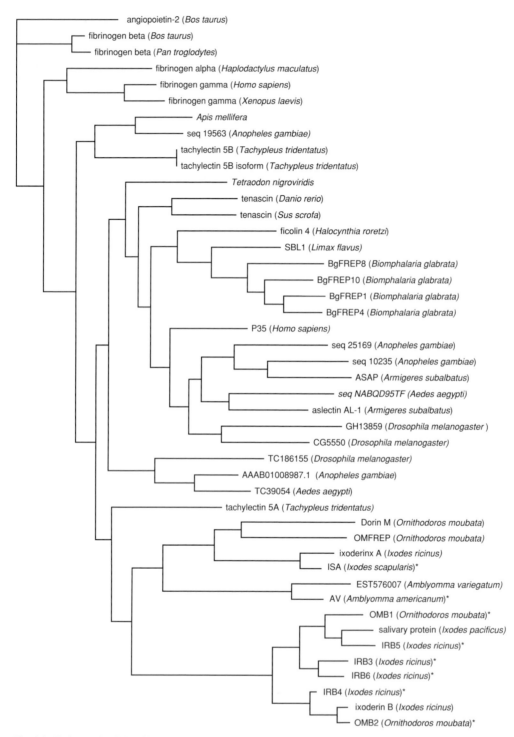

Fig. 6.4. Phylogenetic relationships among the fibrinogen-related molecules. The tree has been adapted from Rego *et al.* (2005) and modified with the additon of new sequences (marked with asterisks).

lectin–carbohydrate interactions. Here the lectins of trans-mitted pathogens/parasites (e.g. tick-borne viruses, borrelia spirochaetes, piroplasma, etc.) may recognize carbohydrate moieties of receptors on tick cells. This may facilitate binding of the pathogen to the tissues where they can evade antimi-crobial attacks by the host. A lectin/haemagglutinin from the relapsing fever spirochaete, *Borrelia recurrentis*, has been described as well as two different lectins/haemagglutinins from *B. burgdorferi* spirochaetes (Grubhoffer, Uhlíř & Volf, 1993; Leong *et al.*, 1995; Parveen & Leong, 2000; Rudenko, Golovchenko & Grubhoffer, 2000). With the increasing evi-dence of the existence of protein–carbohydrate interactions between tick tissues and pathogens, as discussed in this review, the next goal is to elucidate their biological roles and significance in tick–pathogen relationships and, ulti-mately, to design potential disease-transmission-blocking strategies based on our understanding of the tick or tick-borne pathogen lectins.

ACKNOWLEDGEMENTS

We are grateful to Dr K.-H. Kuhn (formerly the Uni-versity of Regensburg, Germany) for the fruitful collab-oration in the past, as well as to Eva Durnová, Vlad'ka Tlustá, Daniela Grubhofferová and Tomáš Klimpera for excellent technical help. This study was supported by the following grants: Z60220518 (Academy of Sciences of the Czech Republic); MSM 6007665801 and LC-06009 (Min-istry of Education, Youth and Sport of the Czech Republic); Grant Agency of the Academy of Sciences of the Czech Republic No. 62 252, No. A6022001; 524/06/1479 and 206/04/0520 (Grant Agency of the Czech Republic); Sci-ence and Technology collaboration between ASCR and NRC Canada (Z60220518/580-8580); and R37 AI-24899 from the US National Institutes of Health.

REFERENCES

Abubakar, L. U., Bulimo, W. D., Mulaa, F. J. & Osir, E. O. (2006). Molecular characterization of a tsetse fly midgut proteolytic lectin that mediates differentiation of African trypanosomes. *Insect Biochemistry and Molecular Biology* 36, 344–352.

Adema, C. M., Hertel, L. A., Miller, R. D. & Loker, E. S. (1997). A family of fibrinogen-related proteins that precipitates a parasite-derived molecule is produced by an invertebrate after infection. *Proceedings of the National Academy of Sciences of the USA* 94, 8691–8696.

Akov, S., Samish, M. & Galun, R. (1976). Protease activity in female *Ornithodoros tholozani* ticks. *Acta Tropica* 33, 36–52.

Balashov, Y. S. (1972). Bloodsucking ticks (Ixodidae): vector of disease of man and animals. *Miscellaneous Publications of the Entomological Society of America* 8, 161–362.

Barondes, S. H. (1988). Bifunctional properties of lectins: lectins redefined. *Trends in Biochemical Sciences* 12, 480–482.

Bettler, B., Hofstetter, H., Yokoyama, W. M., Kilchherr, F. & Conrad, D. H. (1989). Molecular structure and expression of the murine lymphocyte low-affinity receptor for IgE (Fc epsilon RII). *Proceedings of the National Academy of Sciences of the USA* 86, 7566–7570.

Boyd, W. C. & Shapleigh, E. (1954). Specific precipitating activity of plant agglutinins (lectins). *Science* 119, 419.

Chen, C. L. & Billingsley, P. F. (1999). Detection and characterization of a mannan-binding lectin from the mosquito, *Anopheles stephensi* (Liston). *European Journal of Biochemistry* 263, 360–366.

Chinzei, Y. (1988). A new method for determining vitellogenin in hemolymph of female *Ornithodoros moubata* (Acari: Argasidae). *Journal of Medical Entomology* 25, 548–550.

Chinzei, Y., Chino, H. & Takahashi, K. (1983). Purification and properties of vitellogenin and vitellin from a tick *Ornithodoros moubata*. *Journal of Comparative Physiology B* 152, 13–21.

Chinzei, Y. & Yano, I. (1985). Fat body is the site of vitellogenin synthesis in the soft tick, *Ornithodoros moubata*. *Journal of Comparative Physiology B* 155, 671–678.

Coleman, J. E., Gebbia, J. A., Piesman, J., *et al.* (1997). Plasminogen is required for efficient dissemination of *Borrelia burgdorferi* in ticks and for enhancement of spirochetemia in mice. *Cell* 89, 1111–1119.

Dimopoulos, G., Casavant, T. L., Chang, S. R., *et al.* (2000). *Anopheles gambiae* pilot gene discovery project: identification of mosquito innate immunity genes from expressed sequence tags generated from immune-competent cell lines. *Proceedings of the National Academy of Sciences of the USA* 97, 6619–6624.

Dodd, R. B. & Drickamer, K. (2001). Lectin-like proteins in model organisms: implications for evolution of carbohydrate-binding activity. *Glycobiology* 11, 71R–79R.

Doolittle, R. F., Spraggon, G. & Everse, S. J. (1997). Evaluation of vertebrate fibrin formation and the process of its dissolution. *Ciba Foundation Symposium* 212, 4–17.

Doyle, R. J. & Slifkin, M. (eds.) (1994). *Lectin–Microorganism Interactions*. New York: Marcel Dekker.

Drickamer, K. (1988). Two distinct classes of carbohydrate-recognition domains in animal lectins. *Journal of Biological Chemistry* **263**, 9557–9560.

Drickamer, K. (1993). Evolution of Ca^{2+}-dependent animal lectins. *Progress in Nucleic Acid Research Molecular Biology* **45**, 207–232.

Drickamer, K. & Taylor, M. E. (1993). Biology of animal lectins. *Annual Review of Cell Biology* **1993**, 236–264.

Drickamer, K., Dordal, M. S. & Reynolds (1986). Mannose-specific binding proteins isolated from rat liver contain carbohydrate-recognition domains linked to collagenous tails: complete primary structures and homology with pulmonary surfactant apoprotein. *Journal of Biological Chemistry* **261**, 6878–6887.

Durnová, E. (1998). Gut agglutinins (lectins) of common sheep tick, *Ixodes ricinus* L. (Ixodida, Ixodidae): isolation and partial characterization. Unpublished M.Sc. thesis, University of South Bohemia, České Budějovice (in Czech).

Epstein, J., Eichbaum, Q., Sheriff, S. & Ezekowitz, B. A. R. (1996). The collectins in inmate immunity. *Current Opinion in Immunology* **8**, 29–35.

Ezekowitz, R. A. B., Sastry, K. N. & Reid, K. B. M. (eds.) (1996). *Collectins and Innate Immunity*. Heidelberg, Germany: Springer-Verlag.

Friedhoff, K. T. (1990). Interaction between parasite and tick vector. *International Journal of Parasitology* **20**, 525–535.

Gadjeva, M., Thiel, S. & Jensenius, J. C. (2001). The mannan-binding lectin pathway of the innate immune response. *Current Opinion in Immunology* **13**, 74–78.

Gilboa-Garber, N. & Garber, N. (1989). Microbial lectin cofunction with lytic activities as a model for a general basic lectin role. *FEMS Microbiological Reviews* **63**, 211–222.

Gokudan, S., Muta, T., Tsuda, R., *et al.* (1999). Horseshoe crab acetyl group-recognizing lectins involved in innate immunity are structurally related to fibrinogen. *Proceedings of the National Academy of Sciences of the USA* **96**, 10086–10091.

Goldstein, I. J., Hughes, R. C., Monsigny, M., Osawa, T. & Sharon, N. (1980). What should be called a lectin? *Nature* **285**, 66.

Gomes, Y. M., Furtado, A. F. & Coelho, L. B. B. (1991). Partial purification and some properties of hemolymph lectin from *Panstrongylus megistus* (Heteroptera: Reduviidae). *Applied Biochemistry and Biotechnology* **31**, 97–107.

Gooding, R. H. (1972). Digestive processes of haematophagous insects. I. A literature review. *Quaestiones Entomologicae* **8**, 5–60.

Grandjean, O. (1984). Blood digestion in *Ornithodoros moubata* female. *Acarologia* **25**, 147–165.

Grubhoffer, L. & Jindrák, L. (1998). Lectin and tick-pathogen interactions: a minireview. *Folia Parasitologica* **45**, 9–13.

Grubhoffer, L. & Kovář, V. (1998). Arthropod lectins: affinity approaches in the analysis and preparation of carbohydrate binding proteins. In *Techniques in Insect Immunology FITC-5*, eds. Wiesner, A., Dunphy, G. B., Marmaras, V. J., *et al.*, pp. 47–57. Fair Haven, N.J.: SOS Publications.

Grubhoffer, L. & Maťha, V. (1991). New lectins of invertebrates. *Zoological Science (Tokyo)* **8**, 1001–1003.

Grubhoffer, L., Guirakhoo, F., Heinz, F. X. & Kunz, CH. (1990). Interaction of tick-borne encephalitis virus protein E with labelled lectins. In *Lectins: Biology, Biochemistry and Clinical Biochemistry*, vol. 7, eds. Kocourek, J. & Feed, D. L. J., pp. 313–319. St Louis, MO: Sigma Chemical Co.

Grubhoffer, L., Hypša V. & Volf, P. (1997). Lectins (agglutinins) in the gut of the important disease vectors. *Parasite* **4**, 203–216.

Grubhoffer, L., Uhlíř, J. & Volf, P. (1993). Functional and structural identification of a new lectin activity of *Borrelia recurrentis* spirochetes. *Comparative Biochemistry and Physiology B* **105**, 535–540.

Grubhoffer, L., Vereš, J. & Dusbábek, F. (1991). Lectins as molecular factors of recognition and defence reaction of ticks. In *Modern Acarology*, vol. 2, eds. Dusbábek, F. & Bukva, V., pp. 381–388. The Hague, the Netherlands: SPB Academic Publishing.

Gudderra, N. P., Sonenshine, D. E., Apperson, C. S. & Roe, R. M. (2002). Hemolymph proteins in ticks. *Journal of Insect Physiology* **48**, 269–278.

Gupta, A. P. (1985). Cellular elements in the hemolymph. In *Comprehensive Insect Physiology, Biochemistry and Pharmacology*, vol. 3, eds. Kerkut, G. A. & Gilbert L. I., pp. 401–451. Oxford, UK: Pergamon.

Gupta, A. P. (1991). Insect immunocytes and other hemocytes: roles in cellular and humoral immunity. In *Immunology of Insects and Other Arthropods*, ed. Gupta, A. P., pp. 19–118. Boca Raton, FL: CRC Press.

Halberg, D. F., Wager, R. E., Farell, D. C., *et al.* (1987). Major and minor forms of the rat liver asialoglycoprotein receptor are independent galactose-binding proteins: primary structure and glycosylation heterogeneity of minor receptor forms. *Journal of Biological Chemistry* **262**, 9828–9838.

Hoffmann, J. A. (1995). Innate immunity of insects. *Current Opinion in Immunology* **7**, 4–10.

Hoffmann, J. A., Kafatos, F. C., Janeway, C. A. & Ezekowitz, R. A. (1999). Phylogenetic perspectives in innate immunity. *Science* **284**, 1313–1318.

Huang, X., Tsuji, N., Miyoshi, T., *et al.* (2007). Molecular characterization and oligosaccharide binding properties of a galectin from the argasid tick *Ornithodoros moubata*. *Glycobiology* **17**, 313–323.

Hulínská, D., Volf, P. & Grubhoffer, L. (1992). Characterization of *Borrelia burgdorferi* glycoconjugates and surface carbohydrates. *Zentralblat für Bakteriologie, Microbiologie und Hygiene A* **276**, 473–480.

Hutton, D., Reid, A. P. & Towson, S. (2000). Immune responses of the argasid tick *Ornithodoros moubata* induced by infection with the filarial worm *Acanthocheilonema viteae*. *Journal of Helminthology* **74**, 233–239.

Hypša, V. & Grubhoffer, L. (1995). An LPS-binding hemagglutinin in the midgut of *Triatoma infestans*: partial characterization and tissue localization. *Archives of Insect Biochemistry and Physiology* **28**, 247–255.

Ingram, G. A. & Molyneux, D. H. (1991). Insect lectins: role in parasite-vector interactions. In *Lectin Reviews*, vol. 1, eds. Kilpatrick, D. C., Van Driessche, R. E. & Bøg-Hansen, T. C., pp. 103–127. St Louis, MO: Sigma Chemical Co.

Iwanaga, S. (1993). The *Limulus* clotting reaction. *Current Opinion in Immunology* **5**, 74–82.

Iwanaga, S. (2002). The molecular basis of innate immunity in the horseshoe crab. *Current Opinion in Immunology* **14**, 87–95.

Jacobson, R. L. & Doyle, R. J. (1996). Lectin–parasite interactions. *Parasitology Today* **12**, 55–60.

Johns, R., Ceraul, S., Sonenshine, D. E. & Hynes, W. L. (2001*a*). Tick immunity to microbial infections: control of representative bacteria in the hard tick *Dermacentor variabilis* (Acari: Ixodidae). In *Acarology: Proceedings of the 10th International Congress*, eds. Halliday, R. B., Walter, D. E., Proctor, H. C., Norton, R. A. & Colloff, M. J., pp. 638–644. Melbourne, Australia: CSIRO.

Johns, R., Ohnishi, J., Broadwater, A., *et al.* (2001*b*). Contrasts in tick innate immune response to *Borrelia burgdorferi* challenge: immunotolerance in *Ixodes scapularis* versus immunocompetence in *Dermacentor variabilis* (Acari: Ixodidae). *Journal of Medical Entomology* **38**, 99–107.

Kamwendo, S. P., Ingram, G. A., Musisi, F. L. & Molyneux, D. H. (1993). Haemagglutinin activity in tick (*Rhipicephalus appendiculatus*) haemolymph and extracts of gut and salivary glands. *Annals of Tropical Medicine and Parasitology* **87**, 303–305.

Kamwendo, S. P., Musis, F. L., Trees, A. J. & Molyneux, D. H. (1995). Effect of haemagglutinin (lectin) inhibitory sugars in *Theileria parva* infection in *Rhipicephalus appendiculatus*. *International Journal for Parasitology* **25**, 29–35.

Kawabata, S. & Tsuda, R. (2002). Molecular basis of non-self recognition by the horshoe crab tachylectins. *Biochimica et Biophysica Acta* **1572**, 414–421.

Kawasaki, K., Kubo, T. & Natori, S. (1996). Presence of the *Periplaneta* lectin-related protein family in the American cockroach *Periplaneta americana*. *Insect Biochemistry and Molecular Biology* **26**, 355–364.

Kenjo, A., Takahashi, M., Matsushita, M., *et al.* (2001). Cloning and characterization of novel ficolins from the solitary ascidian, *Halocynthia roretzi*. *Journal of Biological Chemistry* **276**, 19959–19965.

Kocourek, J. & Hořejší, V. (1981). Defining a lectin. *Nature* **290**, 188.

Kovář, V., Kopáček, P. & Grubhoffer, L. (2000). Isolation and characterization of Dorin M, a lectin from plasma of the soft tick *Ornithodoros moubata*. *Insect Biochemistry and Molecular Biology* **30**, 195–205.

Kuhn, K.-H., Ritting, M., Haupl, T. & Burmester, G. R. (1994). Haemocytes of the tick *Ixodes ricinus* express coiling phagocytosis of *Borrelia burgdorferi*. *Developmental and Comparative Immunology* **18**, S115.

Kuhn, K.-H., Uhlíř, J. & Grubhoffer, L. (1996). Ultrastructural localization of a sialic acid-specific hemolymph lectin in the hemocytes and other tissues of the hard tick *Ixodes ricinus* (Acari: Chelicerata). *Parasitology Research* **82**, 215–221.

Kurachi, S., Song, Z., Takagaki, M., *et al.* (1998). Sialic acid-binding lectin from the slug *Limax flavus*: cloning, expression of the polypeptide, and tissue localization. *European Journal of Biochemistry* **254**, 217–222.

Kurtti, T. J., Munderloh, U. G., Ahlstrand, G. G. & Johnson, R. C. (1988). *Borrelia burgdorferi* in tick cell culture: growth and cellular adherence. *Journal of Medical Entomology* **25**, 256–261.

Lee, Y. C. (1992). Biochemistry of carbohydrate–protein interaction. *FASEB Journal* **6**, 3193–3200.

Leonard, P. M., Adema, C. M., Zhang, S.-M. & Loker, E. S. (2001). Structure of two FREP genes that combine IgSF and fibrinogen domains, with comments on diversity of the FREP gene family in the snail *Biomphalaria glabrata*. *Gene* **269**, 155–165.

Leong, J. M., Morrissey, P. E., Ortega-Maria, E., Pereira, M. E. A. & Coburn, J. (1995). Hemagglutination and proteoglycan binding by the Lyme disease spirochete, *Borrelia burgdorferi*. *Infection and Immunity* **63**, 874–883.

Lu, J. & Le, Y. (1998). Ficolins and the fibrinogen-like domain. *Immunobiology* **199**, 190–199.

Mandal, C. & Mandal, C. (1990). Sialic acid-binding lectins. *Experientia* **46**, 433–441.

Matsushita, M. (1996). The lectin pathway of the complement system. *Microbiology and Immunology* **40**, 887–893.

Matsushita, M., Endo, Y., Nonaka, M. & Fujita, T. (2001). Activation of the lectin complement pathway by ficolins. *International Immunopharmacology* **1**, 359–363.

Matsushita, M., Endo, Y., Taira, S., *et al.* (1996). A novel human serum lectin with collagen- and fibrinogen- like domains that functions as an opsonin. *Journal of Biological Chemistry* **271**, 2448–2454.

Maudlin, I. & Welburn, S. C. (1988). The role of lectin and trypanosome genotype in the maturation of midgut infections in *Glossina morsitans*. *Tropical Medicine and Parasitology* **39**, 56–58.

McEver, R. P. (1994). Selectins. *Current Opinion in Immunology* **6**, 75–84.

Mullins, D. E. (1985). Chemistry and physiology of the hemolymph. In *Comprehensive Insect Physiology, Biochemistry and Pharmacology*, vol. 3, eds. Kerkut, G. A. & Gilbert, L. I., pp. 355–400. Oxford, UK: Pergamon.

Munderloh, U. G. & Kurtti, T. J. (1995). Cellular and molecular interrelationships between ticks and prokaryotic tick-borne pathogens. *Annual Review of Entomology* **40**, 221–243.

Muta, T., Miyata, T., Tokunaga, F., *et al.* (1991). *Limulus* factor-C: an endotoxin-sensitive serine protease zymogen with a mosaic structure of complement-like, epidermal growth factor-like, and lectin like domains. *Journal of Biological Chemistry* **266**, 6554–6561.

Nakajima, Y., van der Goes van Naters-Yasui, A., Taylor, D. & Yamakawa, M. (2001). Two isoforms of a member of the arthropod defensin family from the soft tick, *Ornithodoros moubata* (Acari: Argasidae). *Insect Biochemistry and Molecular Biology* **31**, 747–751.

Ofek, I. & Sharon, N. (1988). Lectinophagocytosis: a molecular mechanism of recognition between cell surface sugars and lectin in the phagocytosis of bacteria. *Infection and Immunity* **56**, 539–547.

Olafsen, J. A. (1986). Invertebrate lectins: biochemical heterogeneity as a possible key to their biological function. In *Immunity in Invertebrates*, ed. Bréhélin, M., pp. 95–111. Heidelberg, Germany: Springer-Verlag.

Olafsen, J. A. (1996). Lectins: model of natural and induced molecules in invertebrates. In *Advances in Comparative and Environmental Physiology*, vol. 24, ed. Cooper, E. L., pp. 49–76. Heidelberg, Germany: Springer-Verlag.

Palánová, L. & Volf, P. (1997). Carbohydrate-binding specificities and physico-chemical properties of lectins in various tissues of phlebotominae sandflies. *Folia Parasitologica* **44**, 71–76.

Parveen, N. & Leong, J. M. (2000). Identification of a candidate glycosaminoglycan-binding adhesin of the Lyme disease spirochete *Borrelia burgdorferi*. *Molecular Microbiology* **35**, 1220–1234.

Pearson, M. A. (1996). Scavenger receptors in innate immunity. *Current Opinion in Immunology* **8**, 20–28.

Pereira, M. A. E., Andrade, A. F. B. & Ribeiro, J. M. C. (1981). Lectins of distinct specificity in *Rhodnius prolixus* interact selectively with *Trypanosoma cruzi*. *Science* **211**, 597–600.

Peumans, W. J. & Van Damme, E. J. M. (1995). Lectins as plant defense proteins. *Plant Physiology* **109**, 347–352.

Ratcliffe, N. A., Nigam, Y., Mello, C. B., Garcia, E. S. & Azambuja, P. (1996). *Trypanosoma cruzi* and erythrocyte agglutinins: a comparative study of occurrence and properties in the gut and hemolymph of *Rhodnius prolixus*. *Experimental Parasitology* **83**, 83–93.

Ratcliffe, N. A. & Rowley, A. F. (1987). Insect responses to parasites and other pathogens. In *Immune Responses in Parasitic Infections*, vol. 4, ed. Soulsby, E. J. L., pp. 271–332. Boca Raton, FL: CRC Press.

Ratcliffe, N. A., Rowley, A. F., Fitzgerald, S. W. & Rhodes, C. P. (1985). Invertebrate immunity: basic concepts and recent advances. *International Review of Cytology* **97**, 183–349.

Rego, R. O. M., Hajdušek, O., Kovář, V., *et al.* (2005). Molecular cloning and comparative analysis of fibrinogen-related proteins from the soft tick *Ornithodoros moubata* and the hard tick *Ixodes ricinus*. *Insect Biochemistry and Molecular Biology* **35**, 991–1004.

Rego, R. O. M., Kovář, V., Kopáček, P., *et al.* (2006). The tick plasma lectin, Dorin M, is a fibrinogen-related molecule. *Insect Biochemistry and Molecular Biology* **36**, 291–299.

Ribeiro, J. M. C., Mather, T. N., Piesman, J. & Spielman, A. (1987). Dissemination and salivary delivery of Lyme disease spirochetes in vector ticks (Acari: Ixodiae). *Journal of Medical Entomology* **24**, 201–205.

Rhodes, J. M. & Milton, J. D. (1998). *Lectin Methods in Protocols*. Totowa, NJ: Humana Press.

Rudenko, N., Golovchenko, N. & Grubhoffer, L. (2000). Lectin-like sequences in genome of *Borrelia burgdorferi*. *Folia Parasitologica* **46**, 81–90.

Sastry, K. & Ezekowitz, R. A. (1993). Collectins: pattern recognition molecules involved in first line host defense. *Current Opinion in Immunology* **5**, 59–66.

Sharon, N. & Lis, H. (1988). A century of lectin research (1888–1988). *Trends in Biochemical Sciences* **12**, 488–491.

Sharon, N. & Lis, H. (2007). *Lectins*. Dordrecht: Springer-Verlag.

Slifkin, M. & Doyle, R. J. (1990). Lectins and their application to clinical microbiology. *Clinical Microbiology Reviews* **3**, 197–217.

Smiths, R. D., Sells, D. M., Stephenson, E. M., Ristic, M. & Huxdoll, D. L. (1976). Development of *Ehrlichia canis*, causative agent of canine ehrlichiosis, in the tick *Rhipicephalus sanguineus* and its differentiation from a symbiotic *Rickettsia*. *American Journal of Veterinary Research* **37**, 119–126.

Taylor, M. E., Conary, J. T., Lennertz, M. R., Stahl, P. D. & Drickamer, K. (1990). Primary structure of the mannose receptor contains multiple motifs resembling carbohydrate-recognition domains. *Journal of Biological Chemistry* **265**, 12156–12162.

Thiel, S. & Reid, K. B. M. (1989). Structures and functions associated with the group of mammalian lectins containing collagen-like sequences. *FEBS Letters* **250**, 78–84.

Uhlíř, J., Grubhoffer, L., Borský, I. & Dusbábek, F. (1994). Antigens and glycoproteins of larvae, nymphs and adults of the tick *Ixodes ricinus*. *Medical and Veterinary Entomology* **8**, 141–150.

Uhlíř, J., Grubhoffer, L. & Volf, P. (1996). Novel agglutinin in the midgut of the tick *Ixodes ricinus*. *Folia Parasitologica* **43**, 233–239.

Vasta, G. R. (1991). The multiple biological roles of invertebrate lectins: their participation in nonself recognition mechanisms. In *Phylogenesis of Immune Functions*, eds. Warr, G. W. & Cohen, N., pp. 183–199. Boca Raton, FL: CRC Press.

Vasta, G. R. & Marchalonis, J. J. (1983). Humoral recognition factors in the arthropoda: the specificity of chelicerate serum lectins. *American Zoology* **23**, 157–171.

Vasta, G. R. & Marchalonis, J. J. (1984). Summation: immunobiological significance of invertebrate lectins. *Progress in Clinical Biological Research* **154**, 177–191.

Vasta, G. R., Ahmed, H., Fink, N. E., *et al.* (1994). Animal lectins as self/non-self recognition molecules: biochemical and genetic approaches to understanding their roles and evolution. *Annals of the New York Academy of Sciences* **712**, 55–73.

Vasta, G. R., Ahmed, H. & Quesenberry, M. S. (1996). Invertebrate C-type lectins and pentraxins as non-self recognition molecules. In *New Directions in Invertebrate Immunology*, eds. Söderhäll, K., Iwanaga, S. & Vasta, G. R., pp. 189–227. Fair Haven, NJ: SOS Publications.

Vasta, G. R., Quesenberry, M. S., Ahmed, H. A. & O'Leary, N. (1999). C-type lectins and galectins mediate innate and adaptive immune functions: their roles in the complement activation pathway. *Developmental and Comparative Immunology* **23**, 401–420.

Vereš, J. & Grubhoffer, L. (1990). Detection and partial characterization of a new plasma lectin in the hemolymph of the tick *Ornithodoros tartakovskyi*. *Microbios Letters* **45**, 61–64.

Volf, P. (1993). Lectin activity in the gut extract of sandfly *Lutzomyia longipalpis*. *Folia Parasitologica* **40**, 155–156.

Volf, P., Killick-Kendrick, R., Bates, P. & Molyneux, D. H. (1994). Comparison of the haemagglutination activities in the gut and head extracts of various species and geographical populations of phlebotominae sandflies. *Annals of Tropical Medicine and Parasitology* **88**, 337–340.

Volf, P., Škařupová, S. & Man, P. (2002). Characterization of the lectin from females of *Phlebotomus duboscqi* sandflies. *European Journal of Biochemistry* **269**, 6294–6301.

Wagner, M. (1990). Sialic acid specific lectins. *Advances in Lectin Research* **3**, 36–82.

Wallbanks, K. R., Ingram, G. A. & Molyneux, D. H. (1986). The agglutination of erythrocytes and *Leishmania* parasites by sandfly gut extracts: evidence for lectin activity. *Tropical Medicine and Parasitology* **37**, 409–413.

Wang, X., Rocheleau, T. A., Fuchs, J. F., *et al.* (2004). A novel lectin with a fibrinogen-like domain and its potential involvement in the innate immune response of *Armigeres subalbatus* against bacteria. *Insect Molecular Biology* **133**, 273–282.

Welburn, S. C. & Maudlin, I. (1990). Haemolymph lectin and the maturation of trypanosome infections in tsetse. *Medical and Veterinary Entomology* **4**, 43–48.

Xu, X. & Doolittle, R. F. (1990). Presence of a vertebrate fibrinogen-like sequence in an echinoderm. *Proceedings of the National Academy of Sciences of the USA* **87**, 2097–2101.

Yeaton, R. W. (1982*a*). Invertebrate lectins. I. Occurrence. *Developmental and Comparative Immunology* **5**, 391–402.

Yeaton, R. W. (1982*b*). Invertebrate lectins. II. Diversity of specificity, biological synthesis and function in recognition. *Developmental and Comparative Immunology* **5**, 535–545.

Yoshizaki, N. (1990). Functions and properties of animal lectins. *Zoological Science (Tokyo)* **7**, 581–591.

Zeng, F. Y. & Gabius, H. J. (1992). Sialic acid-binding proteins: characterization, biological function and application. *Zeitschrift für Naturforschung C* **47**, 641–653.

Zung, J. L., Lewengrub, S., Rudzinska, M. A., *et al.* (1989). Fine structural evidence for the penetration of the Lyme disease spirochete *Borrelia burgdorferi* through the gut and salivary tissues of *Ixodes dammini*. *Canadian Journal of Zoology* **67**, 1737–1748.

7 • Endocrinology of tick development and reproduction

H. H. REES

INTRODUCTION

The developmental hormone systems of insects and crustaceans are probably best understood of all the arthropods (for reviews, see Gilbert, Iatrou & Gill, 2004; Wainwright & Rees, 2001). In other arthropod classes, information concerning the identification and functional significance of hormones is fragmentary or non-existent. The endocrine regulation of development and reproduction in ticks (acarines) has been reviewed (Oliver & Dotson, 1993; Lomas & Rees, 1998; Chang & Kaufman, 2004; Rees, 2004) and the reader is referred to these for further detail. However, there is a relative lack of new work in this field.

Blood meals are critical in ticks for triggering various events, including the endocrine system (see Chapter 8). In adult female ixodid ticks, the transition between the slow feeding phase and the rapid engorgement phase (that has been defined as the critical weight: Harris & Kaufman, 1984; Lindsay & Kaufman, 1988; Weiss & Kaufman, 2001) seems to be a critical control point for regulation of many endocrine events, including salivary gland degeneration, vitellogenesis and egg production (see Chapters 3 and 8). Thus, females prematurely removed from the host below the critical weight retain a host-seeking strategy and can reattach to a host if given the opportunity, do not undergo salivary gland degeneration and will not lay eggs. However, females prematurely removed above the critical weight are unable to reattach to a host, undergo salivary gland degeneration and will lay as many eggs as the acquired blood meal will support.

In this chapter, aspects of the endocrine regulation of development and reproduction will be considered. Various other factors involved in regulating reproduction are considered in detail in Chapter 8.

ECDYSTEROIDS

From evidence primarily based on a combination of the observed effects of exogenously administered moulting hormones (ecdysteroids) and the correlation of changes in ecdysteroid titres with physiological events, it appears that the functions of ecdysteroids in ticks are largely similar to those in insects and crustaceans. However, in ticks there is still a lack of information on direct effects of the hormones on individual gene transcription. The physiological roles of ecdysteroids in ticks have been reviewed (Diehl, Connat & Dotson, 1986; Sonenshine, 1991; Oliver & Dotson, 1993; Lomas & Rees, 1998).

Immature stages

MOULTING

There is a good body of evidence that ecdysteroids, particularly 20-hydroxyecdysone (Fig. 7.1), regulate moulting in ticks as they do in insects and crustaceans. It has been shown that feeding in vitro of nymphs of the argasid *Ornithodoros porcinus* with porcine blood containing 20-hydroxyecdysone accelerated moulting (cited in Solomon, Mango & Obenchain, 1982). Surprisingly, similar treatment of nymphs of *O. parkeri* with 20-hydroxyecdysone (Campbell & Oliver, 1984) or *O. moubata* with the incompletely hydroxylated ecdysteroid 22,25-dideoxyecdysone (Diehl *et al.*, 1986) did not affect the moulting period, although the ticks underwent additional moults without taking a further blood meal. However, as with *O. porcinus*, in nymphs of the ixodid *Hyalomma dromedarii*, topical application of 20-hydroxyecdysone speeded up moulting (Khalil *et al.*, 1984). As might be expected, the sensitivity of ticks is highly dependent upon the dose of ecdysteroid administered, the precise developmental time and the method of exposure

Ticks: Biology, Disease and Control, ed. Alan S. Bowman and Patricia A. Nuttall. Published by Cambridge University Press.
© Cambridge University Press 2008.

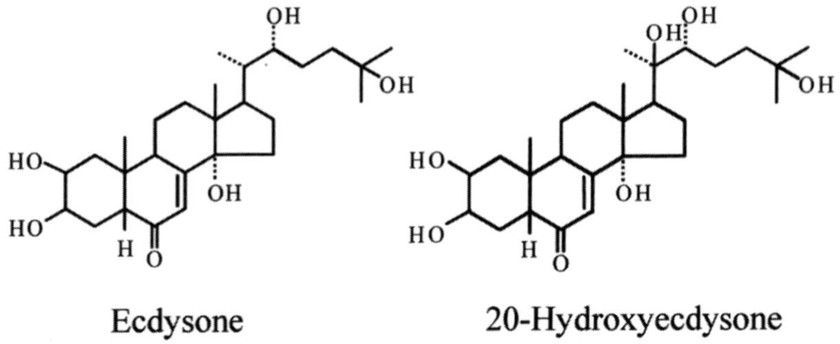

Ecdysone   20-Hydroxyecdysone

Fig. 7.1. Formulae of ecdysone and 20-hydroxyecdysone.

(Sonenshine, 1991). By injection of 20-hydroxyecdysone into female *O. parkeri*, a dose-dependent induction of apolysis and cuticle formation has been demonstrated (Pound, Oliver & Andrews, 1984).

There is also substantial evidence for a close temporal correlation between haemolymph ecdysteroid titres and the moulting process. Haemolymph and whole-body ecdysteroid titres in fifth-stage nymphs of the argasid *O. moubata* have been carefully correlated with the moulting processes which are induced by the blood meal (Germond, Diehl & Morici, 1982) (Fig. 7.2). Between days 2 and 3 post-feeding, when the ecdysteroid titres were low, a few procuticle lamellae were deposited and mitosis initiated. Ecdysteroid titres then began to increase (days 3–4), with the mitotic period ending on day 4. The hormone titres then rose sharply simultaneously with apolysis and the formation of the exuvial space (days 4–5), with highest titres being observed during deposition of the new epicuticle (days 5–6). The titres began to decrease concomitantly with the beginning of procuticle deposition and digestion of the old cuticle (day 6), and started to decrease to low values shortly before ecdysis (days 9–10). Similar correlations have been demonstrated in nymphs of *O. parkeri* (Zhu *et al.*, 1994) and in the ixodids *Amblyomma hebraeum* (Diehl, Germond & Morici, 1982) and *A. variegatum* (Stauffer & Connat, 1990), with probably a similar situation in larvae of *O. moubata* (cited in Diehl *et al.*, 1986; Dotson, Connat & Diehl, 1991).

Using well-synchronized *A. variegatum*, all the integumental events were realized along an anterior–posterior gradient. Of particular significance was the fact that a small peak of ecdysteroid preceded the major peak of hormone during this nymphal–adult moulting cycle. Many structural changes were observed in the integument, salivary glands and dermal glands during this moulting cycle, but additionally,

the muscles of the mouthparts and the pharynx underwent a temporary de-differentiation at the end of the first small peak of ecdysteroids. At the end of the second peak of hormone, when the ecdysteroid titre dropped to basal levels, these muscles reappeared in an adult differentiated state (Stauffer & Connat, 1990). This occurrence of two ecdysteroid peaks is reminiscent of the larval–pupal moult in holometabolous insect species.

In all tick species investigated, at high ecdysteroid titres, the principal hormone is 20-hydroxyecdysone, the presumed major active hormone, accompanied by ecdysone (see Delbecque, Diehl & O'Connor, 1978; Dees, Sonenshine & Breidling, 1984*a*). As in insects, the capacity to transform ecdysone into 20-hydroxyecdysone is widely distributed in tissues, including the midgut (cited in Diehl *et al.*, 1986) and fat body (Zhu, Oliver & Dotson, 1991*a*).

Certain argasids, but not ixodids, have been induced experimentally to supermoult by administration of ecdysteroids, although adult ticks do not normally moult (Diehl *et al.*, 1986). Apparently, sensitivity to induction of supermoulting depends on the species of tick, the hormone and the type and timing of application. Supermoulting in *O. moubata* and *O. parkeri* frequently resulted in failure of ecdysis, which rendered subsequent feeding impossible (Connat *et al.*, 1983*a*; Campbell & Oliver, 1984). However, in *O. porcinus*, supermoulting resulted in healthy specimens which could go on to feed, supermoult and oviposit (Mango, Odhiambo & Galun, 1976). Interestingly, the speculation was made in this case that supermoulting could occur in nature when the host feeds on ecdysteroid-rich plants.

When a recombinant *Autographa californica* multiple nuclear polyhedrosis virus (baculovirus) expressing a chitinase enzyme from the hard tick *Haemaphysalis longicornis* was used to express the chitinase gene in *Spodoptera*

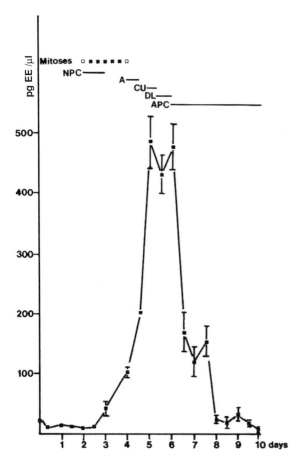

Fig. 7.2. Evolution of the integument structure and haemolymph ecdysteroid content after feeding. The concentration of radioimmunoassay-positive ecdysteroids is given in pg 20-hydroxyecdysone equivalents (EE)/μl. The main structural events of the integument are indicated. Mitoses (few, □; many, ■). NPC, deposition of additional nymphal procuticle lamellae. A, apolysis. Deposition of the cuticulin layer (CU) and of the dense layer (DL). APC, synthesis of the adult procuticle. Digestion of the nymphal cuticle, day 6 – moulting. All ticks from days 9, $9\frac{1}{2}$ and 10 had moulted. Reprinted from Germond, Diehl & Morici (1982) with permission from Elsevier.

TERMINATION OF LARVAL DIAPAUSE

It has been demonstrated that administration of ecdysteroids leads to termination of larval diapause in the ixodids *Dermacentor albipictus* and *Rhipicephalus sanguineus*, although the significance of this to the situation in vivo is unclear (Wright, 1969; Sannasi & Subramoniam, 1972).

ECDYSTEROID RECEPTORS

In insects, where the molecular action of ecdysteroids is best understood of the arthropods, this action is mediated by a heterodimeric receptor composed of two members of the nuclear superfamily of receptors, the Ecdysone Receptor (EcR) and Ultraspiracle (USP) (Henrich, Rybczynski & Gilbert, 1999; Koslova & Thummel, 2000; Riddiford, Cherbas & Truman, 2001; Henrich, 2004). USP is the insect homologue of the vertebrate Retinoid X Receptor (RXR) genes, which have a critical role in many nuclear signalling pathways (Mangelsdorf & Evans, 1995). In the case of numerous nuclear receptor ligands, such as ecdysteroids, they exert diverse effects on development, growth and physiological processes (Thummel, 2001, 2002). Such a diversity of effects is frequently due to the existence of multiple forms of the receptor protein that differentially transduce the ligand signals. Multiple forms of receptors can arise from expression of related gene families to yield receptor subtypes, from alternative RNA processing of a single gene to yield receptor isoforms, or from a combination of both mechanisms (Leid, Kastner & Chambon, 1992).

All nuclear receptors possess a characteristic modular structure which includes a variable N-terminal domain (A/B), a highly conserved DNA-binding domain (C), a variable hinge region (D), a conserved ligand-binding domain (E), together with a variable C-terminal region (F) of unknown significance that may be present in some but not all nuclear receptors (Fig. 7.3).

Complementary DNAs encoding three ecdysteroid receptor isoforms that have common DNA- and ligand-binding domains linked to distinct N-termini have been isolated from the ixodid tick *A. americanum* (Guo *et al.*, 1997). The DNA- and ligand-binding domains share an average of 86% and 64% identity, respectively, with such domains from insect EcR proteins. In these *A. americanum* EcRs, the N-termini are highly divergent and they also lack the 'F' C-terminal domain occurring in the insect EcRs. Analysis of the cDNAs showed that RNA processing is complex and in addition to producing transcripts with unique N-termini, produces EcR transcripts with different 5′ and 3′ untranslated regions, as well as splicing variants having incomplete

frugiperda (Sf9) insect cells, topical application of the supernatant from such cells on *H. longicornis* nymphs caused mortality (Assenga *et al.*, 2006). A synergistic effect was observed on tick killing when the recombinant-containing cell medium was used in combination with the recombinant virus. Interestingly, a mixture of recombinant virus and flumethrin halved the dose of the latter required.

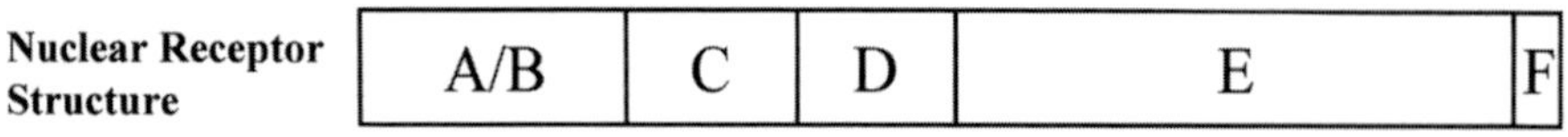

Nuclear Receptor Structure

A/B	C	D	E	F

Fig. 7.3. The modular structure characteristic of all nuclear receptors. A/B, variable N-terminal domain; C, a highly conserved DNA-binding domain; D, a variable hinge region; E, a conserved ligand-binding domain; F, a variable C-terminal region that is present in some but not all nuclear receptors.

open reading frames. Results of Northern blot analyses suggested both stage- and tissue-specific regulation of the EcR mRNA expression.

The vertebrate RXRs are encoded by a multigene family, whereas the insect RXR homologue, Ultraspiracle (USP), is encoded by a single gene. To determine the situation in acarines, cDNAs encoding two RXR homologues, Aam-RXR1 and AamRXR2, have been isolated from *A. americanum* (Guo et al., 1998). The DNA-binding domains share ~95% and 87% identity, respectively, with such domains from insect USP and vertebrate RXR proteins. Surprisingly, ligand-binding domains of the AamRXRs are more similar to such domains from vertebrate RXRs than to insect USP ligand-binding domains (~71% vs. ~52%, respectively). The biological significance of this is unclear. Furthermore, Northern blot and RT-PCR analysis revealed both unique and overlapping patterns of AamRXR1 and AamRXR2 expression during development. Both Aam-RXR1 and AamRXR2 proteins, when paired with any of the three AamEcR isoforms, or the *Drosophila* EcRB1 isoform, can activate transcription of ecdysone response element-containing reporters. Thus, despite differences in the ligand-binding domains of vertebrate RXRs, AamRXRs and insect USP proteins, all can pair with either insect or tick EcRs, yielding a functional ecdysteroid receptor in vitro in response to the ecdysteroid muristerone A (Guo et al., 1998). The foregoing studies on *A. americanum* EcRs and RXRs have been reviewed (Palmer, Harmon & Laudet, 1999).

As we have seen, the DNA-binding domains of arthropod USPs and their vertebrate homologues, the RXRs, are highly conserved. However, interestingly the ligand-binding domain sequences divide into two district groups: (1) sequences from members of the holometabolous higher insect orders, the Diptera and Lepidoptera, and (2) sequences from vertebrates, a fiddler crab (*Uca pugilator*), a tick (*A. americanum*) and a hemimetabolous orthopteran insect, the locust *Locusta migratoria* (Hayward et al., 1999). The reason for this evolutionary sharp divergence of the lepidopteran and dipteran sequences is unknown.

Determination of total body ecdysteroid titres during development in *A. americanum* revealed that one ecdys-

teroid peak was observed following larval apolysis. However, two distinct ecdysteroid peaks occurred in the nymphal moulting cycle, the first following apolysis and the second occurring at about the time of ecdysis. Determination of the whole-body profiles of the EcR and RXR mRNAs by RT-PCR showed that they were both correlated with the hormone titre (Palmer et al., 2002). However, using an electrophoretic gel mobility shift assay, it has been demonstrated that AamEcR–AamRXR1 (but not AamEcR–AamRXR2) exhibits broad DNA-binding specificity, suggesting that functional differences exist between the RXR1 and RXR2 proteins.

ECDYSTEROID INACTIVATION

Before considering ecdysteroid inactivation in ticks, it is germane to summarize such mechanisms in insects, since these are better understood. Ecdysteroid inactivation in insects may occur via several different routes, depending on species, stage in development and tissue (Lafont & Connat, 1989; Lafont et al., 2004) (Fig. 7.4). The major ecdysteroid inactivation pathways include: (1) conjugate (ester) formation, primarily polar phosphates or apolar fatty acyl derivatives. Such conjugates were originally identified as inactive maternal, storage ecdysteroid 22-esters in adult females/newly laid eggs for utilization in early stages of embryogenesis before *de novo* synthesis of hormone in embryos. However, analogous phosphate conjugates (at the C-2, C-3 or C-22 positions of the ecdysteroid) and fatty acyl esters (see Fig. 7.5) are formed in immature stages of insects; (2) an apparently fairly universal pathway of ecdysteroid inactivation involves ecdysteroid 26-oic acid formation via 26-hydroxyecdysteroid (see Fig. 7.6); (3) in many insect systems, irreversible 3-epiecdysteroid formation is emphasized. This occurs via ecdysone oxidase-catalysed 3-dehydroecdysteroid formation which then undergoes irreversible 3-dehydroecdysteroid 3α-reductase-catalysed reduction to 3-epiecdysteroid (see Fig. 7.7). Analogous reactions occur with both ecdysone and 20-hydroxyecdysone.

Most of the work on ecdysteroid inactivation/ metabolism in nymphs has been undertaken in *O. moubata*. In last-instar nymphs, injected [^{3}H]ecdysone was efficiently

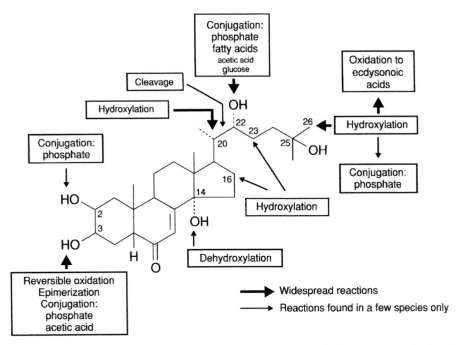

Fig. 7.4. Major reactions of ecdysone metabolism. Modified from Lafont *et al*. (2004) with permission from Elsevier.

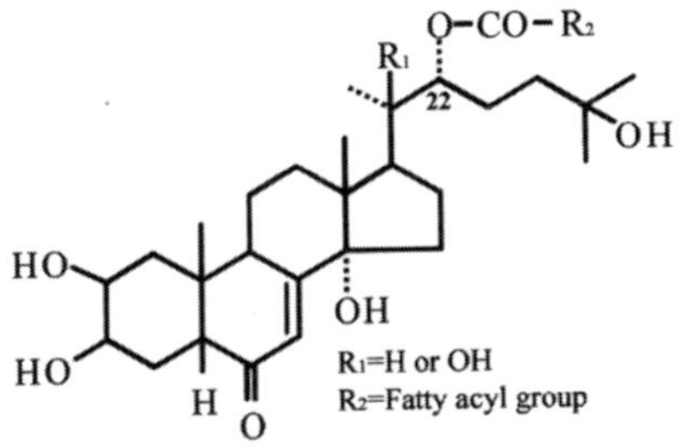

Ecdysteroid 22-long chain fatty acyl ester

Fig. 7.5. Formula of ecdysteroid 22-long chain fatty acyl esters.

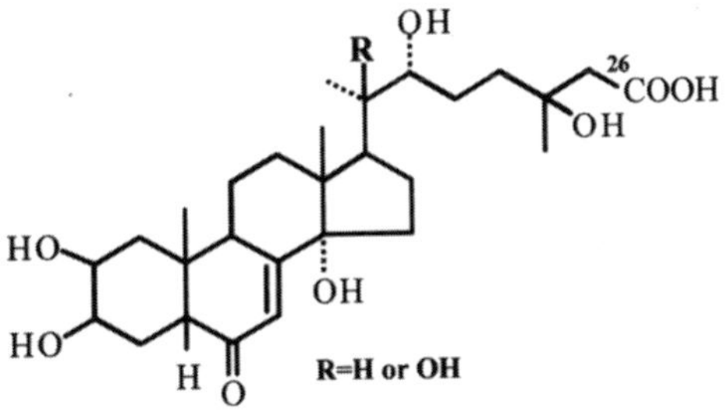

Ecdysteroid 26-oic acid

Fig. 7.6. Formula of ecdysteroid 26-oic acid.

converted into 20-hydroxyecdysone, the transformation having wide tissue distribution but being particularly active in the midgut (Bouvier, Diehl & Morici, 1982, cited in Diehl *et al.*, 1986). 20-Hydroxyecdysone is further transformed into putative 20,26-dihydroxyecdysone and all three ecdysteroids are subsequently converted into polar compounds of unknown structures. Significantly, this polar pathway was particularly active when the titre of endogenous ecdysteroids was decreasing, with accumulation of polar products in the digestive tract, thus suggesting its importance for inactivation of endogenous nymphal ecdysteroids. In contrast, it was

not operative with ingested ecdysteroids nor in adult females (see below).

However, apolar ecdysteroid 22-long chain fatty acyl esters are also quickly formed from injected ecdysteroids, but even more rapidly produced from ingested hormones (Diehl *et al.*, 1985; Connat, Diehl & Thompson, 1986*a*). Such apolar conjugates were then gradually converted into slightly more polar conjugates that are uncharacterized. Interestingly, most ingested ecdysteroid was retained in the digestive tract, where the midgut cells transformed it into apolar conjugates that accumulated in the midgut lumen (Diehl *et al.*, 1986).

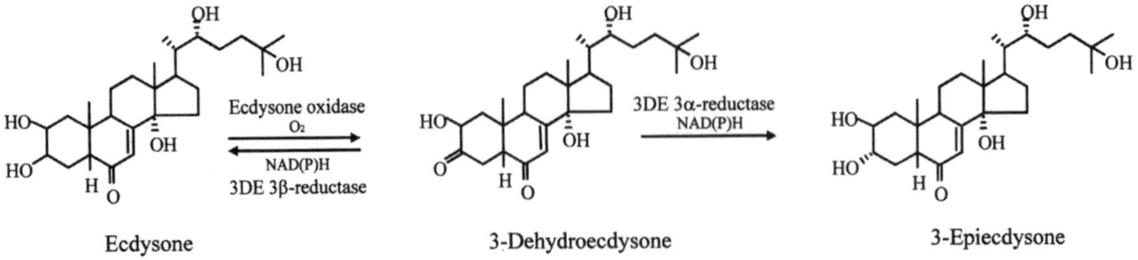

Fig. 7.7. Conversion of ecdysone into 3-epiecdysone.

It had been suggested that the apolar metabolic pathway might serve to inactivate ecdysteroids ingested in blood from hosts feeding on phytoecdysteroid-containing plants (Diehl *et al.*, 1985). Such apolar esters are important and predominant endogenous components in all nymphal stages of *O. moubata*, particularly in the fourth and fifth stages (Connat *et al.*, 1997). Interestingly, although 20-hydroxyecdysone predominated over ecdysone in all nymphal stages, the proportion of the latter was enhanced in the later three nymphal stages. The ecdysteroid metabolic pathways in fed nymphs of the ixodid *A. hebraeum* were similar to the ones in *O. moubata* (Diehl *et al.*, 1985).

Undoubtedly, formation of ecdysteroid 22-fatty acyl esters is a prominent metabolic route in ticks. Formation of such esters is widespread in arthropods (Connat & Diehl, 1986; Kubo *et al.*, 1987; Robinson *et al.*, 1987).

Adult males

SPERM MATURATION

In male ticks, sperm production and maturation has two phases (for review see Sonenshine, 1991; see also Chapter 8), spermatogenesis (the development of haploid spermatids by meiotic and mitotic division) and spermiogenesis (the growth and maturation of spermatids). Spermatogenesis begins during the nymphal–adult moult and in a few exceptional species proceeds to completion during that moult so that unfed males have mature prospermia and can mate (see Lomas & Rees, 1998). However, in most species, spermatogenesis stops after the development of secondary spermatogonia during adult ecdysis and remains arrested until the adult male begins feeding.

Although the mechanism of reinitiation of sperm development is uncertain, ecdysteroids appear to be involved. Ecdysteroids occur in males (Oliver, 1986*a*; Oliver & Dotson, 1993) and injection of 20-hydroxyecdysone into unfed males strongly stimulated germ cell DNA synthesis; similarly, injection of a crude synganglial extract could mimic this effect (Oliver, 1986*a*). It is significant that the accumulation of differentiated spermatocytes in *Dermacentor variabilis* nymphs is biphasic and parallels the ecdysteroid titres reported in *A. hebraeum* (Dumser & Oliver, 1981; Diehl *et al.*, 1982). There is evidence for ecdysteroid synthesis in insect testis under the influence of a brain peptide (Loeb *et al.*, 1987; Jarvis, Earley & Rees, 1994), but such steroid synthesis has not been demonstrated in testes of ticks.

The final growth and maturation of spermatids (spermiogenesis) can be divided further into a growth and elongation phase (resulting in prospermia) and a capacitation phase (sperm maturation: El-Said *et al.*, 1981). Spermatogenesis continues with sperm growth and elongation, resulting in mature prospermia throughout the testis after 5 days of feeding (Khalil, 1970). The final phase of sperm development, capacitation, requires 24 hours and is triggered by a low-molecular-weight protein(s) from the male accessory gland (El-Said *et al.*, 1981; Shepherd, Oliver & Hall, 1982).

For further discussion of various factors from male ticks involved in reproduction, see Chapter 8.

Adult females

REGULATION OF SALIVARY GLANDS

In female ixodid ticks, salivary glands perform several important functions relating to tick feeding and maintaining osmoregulation (for reviews see Kaufman, 1989; Sauer *et al.*, 1995; Sauer, Essenberg & Bowman, 2000). A major function is to secrete excess fluid from the blood meal back to the host, resulting in concentration of the blood meal and regulation of the haemolymph volume. A detailed account of tick salivary gland function and regulation is given in Chapter 3.

After the female tick has engorged, the salivary glands degenerate, an event characterized by the appearance of autophagic vacuoles (Harris & Kaufman, 1981). This autolytic degeneration of the salivary glands is triggered by 20-hydroxyecdysone (Harris & Kaufman, 1985; Kaufman, 1991). This hormone also triggers vitellogenesis (Sankhon et al., 1999) and, more recently, a role for ecdysteroid in inhibiting reattachment to the host has been demonstrated (Weiss & Kaufman, 2001).

The molecular action of ecdysteroid, particularly the role of the receptors EcR and USP, in salivary gland degeneration has been studied in *Amblyomma hebraeum*, where the glands degenerate within 4 days of engorgement. Extensive evidence has been furnished for the existence of a salivary gland EcR in *A. hebraeum*: (1) [^{3}H]ponasterone A binds to a salivary gland protein (Mao, McBlain & Kaufman, 1995); (2) the specific ponasterone A-binding protein binds to a *Drosophila* ecdysone response element, hsp27 EcRE, in a sequence-specific manner, the binding being enhanced by biologically active ecdysteroids; (3) monoclonal antibodies against *Drosophila* EcR and USP cross-react with counterparts in tick salivary gland extracts by Western blotting; and (4) the monoclonal antibody against USP retards the hsp27 EcRE–tick EcR complex in a gel mobility shift assay (Mao & Kaufman, 1998).

In *A. hebraeum*, the profile of the functional ecdysteroid receptor (EcR/USP) determined by [^{3}H]ponasterone A binding, gel mobility shift assays and Western blotting, has been correlated with the haemolymph ecdysteroid titre during the feeding period and six days post-engorgement (Mao & Kaufman, 1999). EcR was undetectable in unfed ticks, but following onset of feeding, specific ponasterone A binding and two major EcR bands detected by Western blots appeared. Both measurements were sustained throughout the feeding period, but declined after detachment when the salivary glands were degenerating. A discrete DNA-binding band, shown by gel mobility shift assay using *Drosophila* hsp27 EcRE as probe, intensified when haemolymph ecdysteroid titre reached its peak during the rapid phase of feeding, but declined along with decreasing EcR/USP levels and with specific ligand binding activity following engorgement. These results, taken together, substantiate a physiological role for ecdysteroid, acting via nuclear receptors, in salivary gland degeneration. Furthermore, the latter study suggested a role for the small haemolymph ecdysteroid peak during the rapid phase of feeding in initiating salivary gland degeneration, which may constitute a signal somewhat analogous to the 'commitment peak' of ecdysteroid in larval Lepidoptera and *Drosophila* that triggers the behavioural and molecular changes leading to metamorphosis (Mao & Kaufman, 1999).

REGULATION OF SEX PHEROMONE PRODUCTION

The various types of pheromone systems in ticks are considered in Chapter 21, with previous reviews also being available (Sonenshine, 1986, 1991; Hamilton, 1992). This short section will consider the involvement of ecdysteroids, particularly in regulation of sex pheromone production.

Evidence has been furnished that ecdysteroids may regulate aspects of sex pheromone production in some species. Production of the attractant sex pheromone 2,6-dichlorophenol by the foveal glands begins shortly after the nymphal–adult moult. An increase in 2,6-dichlorophenol production in unfed *Hyalomma dromedarii* females has been demonstrated when they were stimulated by exogenous 20-hydroxyecdysone prior to moulting (Dees, Sonenshine & Breidling, 1984b, 1985; also see Sonenshine, 1991). Furthermore, a similar increase was observed in unfed females and males when treated with 22,25-dideoxyecdysone (Jaffe et al., 1986).

In addition, ecdysteroids may also be components of pheromone systems per se. In *Dermacentor*, the genital sex pheromone consists of long-chain saturated fatty acids together with ecdysone and 20-hydroxyecdysone (Allan et al., 1988; Taylor, Sonenshine & Phillips, 1991c).

REGULATION OF OOGENESIS AND OVIPOSITION

In most ticks, feeding and mating are required for oogenesis and oviposition to occur. Ixodid females must copulate in order to complete the blood meal, whereas argasid females generally feed to repletion whether or not they have copulated. Whereas ixodid ticks undergo one gonotrophic cycle and die, argasids have several such cycles. It is clear that mating triggers factors regulating oogenesis and oviposition, with involvement of interactions between nervous, endocrine and reproductive systems (for earlier reviews, see Connat et al., 1986b; Oliver, 1986b; Oliver & Dotson, 1993; Lomas & Rees, 1998).

Egg maturation in all arthropods involves synthesis of the major egg yolk protein, vitellogenin (Vg) and its uptake in modified form, vitellin (Vn) by oocytes. In ticks, Vg is synthesized outside the ovary, released into the haemolymph and transported to the ovary, where it is selectively taken up as Vn (Chinzei & Yano, 1985; Rosell-Davis & Coons, 1989), which is a haemoglycolipoprotein (Chinzei, Chino & Takahashi, 1983; Rosell & Coons, 1991; James & Oliver, 1997).

In argasids, there is evidence that the fat body is a site of Vg production (e.g. Chinzei & Yano, 1985; for review see Chinzei, 1986). Clearly, vitellogenesis is under hormonal control in ticks, but there are appreciable conflicting data. In addition to neuropeptides, ecdysteroids and juvenile hormones are the main candidate hormones (see Oliver & Dotson, 1993; Lomas & Rees, 1998). However, the status of the occurrence of juvenile hormones in ticks is uncertain (see below).

In the argasid, *O. moubata*, Vg production is stimulated by a vitellogenesis inducing factor (VIF) released from the synganglion following the time of coxal fluid secretion during feeding and continues to be released for approximately 1 hour after feeding (Chinzei *et al.*, 1992). Although the mechanism of action of VIF is uncertain, it has been suggested to trigger the release of fat body stimulating factor (FSF), which in turn, stimulates vitellogenesis in the fat body (Chinzei & Taylor, 1990). Although the identity of the FSF is unknown (see Oliver & Dotson, 1993), some evidence exists that it may be a juvenile hormone (Pound & Oliver, 1979; Connat, Ducommun & Diehl, 1983*b*) with stronger evidence that it may be an ecdysteroid (Sankhon *et al.*, 1999; Taylor, Nakajima & Chinzei, 2000; Ogihara, 2003). Furthermore, evidence has been obtained in three other argasid species that an egg development stimulating factor (EDSF) from the synganglion may be involved in incorporation of Vg into oocytes (*Argas arboreus*: Shanbaky & Khalil, 1975; *Argas hermanni*: Shanbaky *et al.*, 1990; *O. parkeri*: Oliver *et al.*, 1992).

Administration of ecdysteroids to females of the argasid *O. moubata* not only induced supermoulting, but inhibited oogenesis and caused resorption of previously formed eggs (Diehl *et al.*, 1986). Such action of ecdysteroids has been incorporated into a working hypothesis for possible control of the gonotrophic cycle in *O. moubata* (Connat, Dotson & Diehl, 1987). Analogous inhibition of oogenesis by ecdysteroids has also been reported in ixodids, although there are conflicting results as well (see Diehl *et al.*, 1986). Although such inhibitory effects of ecdysteroids on oogenesis might at first sight appear to be difficult to reconcile with the notion of ecdysteroid being the vitellogenic hormone (FSF), the situation is not so different from that in *Drosophila* where 20-hydroxyecdysone may trigger oocyte resorption or vitellogenesis at certain different precise stages of oogenesis (Seller, Bownes & Kubli, 1999).

In ixodids, there is evidence that the prime sites of Vg synthesis are the fat body and midgut (Rosell & Coons, 1992; see Oliver & Dotson, 1993). Evidence has been furnished in *Hyalomma dromedarii* for the involvement in Vg production of a synganglion factor similar to VIF in argasids (Schriefer, 1991). Furthermore, in *Ixodes scapularis* and *D. varabilis*, a synganglion factor, analogous to EDSF in argasids, appears to stimulate Vg uptake by oocytes (Oliver & Dotson, 1993). Since high doses of 20-hydroxyecdysone also stimulate Vg uptake in *I. scapularis*, the possibility exists that the synganglion factor might be similar to the ecdysteroidotropic neurohormone (EtNH) identified in *A. hebraeum* synganglion (Lomas, Turner & Rees, 1997; see 'Neuroendocrine systems' section below).

For correlation of ecdysteroid titres with oocyte maturation, vitellogenesis and some other events in various species see Lomas & Rees (1998). Good evidence has been furnished for the regulation of vitellogenesis by ecdysteroid in two ixodid species. Using fat body organ cultures and backless explants of unfed female *D. variabilis*, it has been shown that 20-hydroxyecdysone stimulates vitellogenin production in the fat body trophocytes, whereas the juvenile hormone analogue methoprene was without significant effect (Sankhon *et al.*, 1999). Furthermore, injection of 20-hydroxyecdysone into partially fed female adults of this species initiated Vg synthesis, secretion and uptake by the ovary, whereas juvenile hormone III (JH-III) was again without effect (Thompson *et al.*, 2005). Similarly, it has been demonstrated in *A. hebraeum* that the natural rise in haemolymph Vg concentration lagged slightly behind the rise in haemolymph ecdysteroid and that fat body Vg synthesis and release was stimulated by injections of 20-hydroxyecdysone, but not of JH-III, into non-vitellogenic females (Friesen & Kaufman, 2002, 2004). However, the lack of significant Vg uptake by the oocytes was consistent with an earlier suggestion that such a process requires a distinct signal (Friesen & Kaufman, 2003). The pyrethroid insecticide cypermethrin stimulates vitellogenesis in *O. moubata* by stimulating release of the normal VIF from the synganglion, which subsequently triggers the release of FSF, which stimulates yolk synthesis in the fat body (Taylor *et al.*, 1991*a*). However, in *A. hebraeum*, cypermethrin inhibits vitellogenesis and egg development, possibly in part by inhibiting release of 20-hydroxyecdysone (Friesen & Kaufman, 2003). The reason for this discrepancy between these argasid and ixodid species is unclear. Interestingly the avermectin analogue Mk-243 injected into engorged females of *A. hebaeum* led to inhibition of egg development, possibly by inhibition of Vg uptake (Friesen, Suri & Kaufman, 2003). 20-Hydroxyecdysone had no demonstrable effect on Gené's organ growth or secretion of egg wax antimicrobial activity (Arrieta, Leskiw & Kaufman, 2006).

The transforming growth factor β (TGF-β) superfamily of proteins can be classified into three subgroups, viz. TGF-βs, bone morphogenetic proteins (BMPs) and activins. The protein follistatin is an activin-binding protein found in many vertebrates and inhibits secretion of a follicle-stimulating hormone. Furthermore, a structural homologue, follistatin-related protein (FRP), has been cloned from many vertebrate species and is suggested to play some roles in negative regulation of cell growth (Zwijsen *et al.*, 1994). Interestingly, a gene encoding an FRP has been isolated from an ovary cDNA library of *H. longicornis* (Zhou *et al.*, 2006). The gene was expressed throughout all the developmental stages, but principally in the ovary together with the fat body, haemocytes, salivary glands and midgut. The recombinant protein binds both activin A and bone morphogenetic protein-2 (BMP-2). Significantly, silencing of FRP by RNAi showed a decrease in tick oviposition, strongly indicating that the tick FRP has a role in tick oviposition.

ECDYSTEROID INACTIVATION

As alluded to above, ecdysteroid 22-fatty acyl esters are prominent metabolites in immature stages of ticks. The situation is similar in some cases in adult females. For example, in *R. appendiculatus*, apolar esters occur throughout oviposition and are incorporated, with free ecdysteroids, into the eggs (Magee, Jones & Rees, 1996). Incubation in vitro of ovaries of *O. moubata* with [^{3}H]ecdysone yielded 20-hydroxyecdysone and apolar 22-fatty acyl esters (Connat, Diehl & Morici, 1984). Furthermore, such esters accumulated in eggs of female *O. moubata* injected with [^{3}H]ecdysone or 20-hydroxyecdysone during vitellogenesis (Connat, Dotson & Diehl, 1988). The demonstration that such esters remained unchanged during embryonic development and during the moulting cycle of larvae led to the conclusion that in *O. moubata*, apolar ecdysteroid conjugates are inactivation metabolites that are not reutilized during development. However, in the ixodid species *Rhipicephalus* (*Boophilus*) *microplus*, the situation is apparently different (see 'Embryogenesis' section below). In this species, the tissue distribution of the production of fatty acyl esters from [^{3}H]ecdysone in vitro was widespread, including Malpighian tubules, ovaries, gut and fat body (Wigglesworth, Lewis & Rees, 1985). These esters were definitively identified as ecdysteroid 22-fatty acyl esters in eggs of *B. microplus* (Crosby *et al.*, 1986). The fact that these esters are utilized in early embryogenesis (see below) suggests that they are within the egg as opposed to being merely components of the egg wax. Similarly, in *A. hebraeum*, putative ecdysone 22-fatty acyl ester formation occurred rapidly in all tissues incubated in vitro with [^{3}H]ecdysone (Connat, Lafont & Diehl, 1986c). Malpighian tubules and gut had the highest 20-hydroxylase activity, although only low amounts of

20-hydroxyecdysone were formed. In this work, a massive conversion of ecdysone into 3-epiecdysone was observed in ovaries, together with detection of 3-dehydroecdysone, the expected intermediate (Fig. 7.7). In view of the large accumulation in the ovaries of only free endogenous hormones (ecdysone and 20-hydroxyecdysone: Connat *et al.*, 1985), and of incorporation of only free ecdysteroids into the newly laid eggs in labelling experiments in vivo (Connat *et al.*, 1987), the extensive formation of apolar esters in these ovary incubations is an enigma. As suggested for *R. appendiculatus* (Whitehead *et al.*, 1986), it was surmised (Connat *et al.*, 1987) that in vivo, the ecdysteroids may be bound to vitellogenins/vitellins and thus be protected from metabolism. Surprisingly, 3-epiecdysteroids were also not detected during [^{3}H]ecdysone metabolism in vivo (Connat *et al.*, 1987). It is conceivable that the epimerization enzymes occur in the oocytes for utilization in vivo during embryogenesis, when 3-epiecdysteroid conjugates are formed as hormone inactivation products (cited in Connat *et al.*, 1986c). The foregoing results demonstrating apolar ester and 3-epiecdysteroid formation in vitro but not in vivo demonstrate that experiments in vitro may not always accurately reflect the situation in vivo.

The foregoing results, taken together with a comparative study of ecdysteroids in newly laid eggs of various tick species (Connat & Dotson, 1988), indicate that ecdysteroids in eggs either occur in the free form (*A. hebraeum* and *A. variegatum*), or as polar fatty acyl esters (*O. moubata* and *B. microplus*), or a mixture of both forms (*R. appendiculatus* and *H. dromedarii*). In *Amblyomma* species, the significance of incorporation of the free ecdysteroids into the eggs is unclear.

After the identification of ecdysteroid 22-fatty acyl esters in newly laid eggs of ticks, such conjugates were also characterized in eggs of some insect species (for review, see Isaac & Slinger, 1989). Such esters may possibly serve as storage forms of maternal ecdysteroids, releasing free hormone during embryogenesis (Slinger & Isaac, 1988; Dinan, 1997), analogous to the situation for 22-phosphate conjugates in locusts (Rees & Isaac, 1984).

Embryogenesis

In the ixodid *B. microplus*, electron microscopic evidence was obtained during embryogenesis of the successive formation of three embryonic membranes/cuticles before larval cuticle deposition (T. Crosby, D. Lewis & H. H. Rees, unpublished data). It was difficult to correlate changes in free ecdysteroid titre during embryogenesis with specific cuticular events, probably due to the lack of synchronization in the developing embryos. However, the drastic reduction in the amounts of the maternal ecdysteroid 22-fatty acyl esters incorporated into the eggs in early embryogenesis suggests that they can at least serve as a source of free ecdysteroids until the embryos have differentiated the synthetic machinery (T. Crosby, H. H. Rees & L. O. Lomas, unpublished data). Obviously, this does not preclude the existence of other sources of hormone. During embryogenesis in *B. microplus* there is successive formation of ecdysteroid 26-oic acids together with a new class of apolar esters that are more polar on reversed-phase HPLC than the 22-fatty acyl esters, with both types of compound being presumed hormone inactivation products. The complete structure of the latter apolar esters is unclear, but evidence suggests that they may be fatty acyl derivatives of the 26-oic acids (Crosby *et al.*, 1986).

As alluded to earlier, in another ixodid, *A. hebraeum*, free maternal ecdysteroids are incorporated into the eggs. In this species, the embryonic cuticular events appear similar to *B. microplus*, with no significant peaks of hormone being detected (Dotson, Connat & Diehl, 1995). However, during embryogenesis, maternally incorporated [³H]ecdysteroid was converted into 22-fatty acyl esters and 3-epimers of the latter, both types of metabolites being regarded as hormone inactivation products. Significantly, the free ecdysteroids are apparently completely inactivated by the time of hatching. Various possible functions for the maternal free ecdysteroids have been suggested (Dotson *et al.*, 1995): they may be involved in oocyte maturation, such as a role in eggshell production; an involvement in production of the first two embryonic membranes/cuticles, before the free hormones are inactivated; and the ecdysteroids could act as a feeding deterrent to predators. In this respect, it is interesting that the sea spider *Pycnogonium littorale* has comparatively large concentrations of ecdysteroids in the integument, where they are believed to function as a feeding deterrent (Tomaschko, 1994). However, it is not clear how such ecdysteroids are prevented from causing developmental effects.

During a study of embryogenesis and larval development in the argasid species *O. moubata*, formation of three embryonic 'cuticles'/envelopes was observed, with the larval cuticulin commencing on day 8 of embryonic development and procuticle deposition continuing after hatching until apolysis of the larval cuticle (Dotson *et al.*, 1991). Whereas no distinct peaks in immunoreactive ecdysteroids were observed during deposition of the three envelopes, a very small peak occurred coincident with shortening of the germ band, with a second distinct peak coinciding with deposition of the larval epicuticle. As stated previously, maternal ecdysteroid 22-fatty acyl esters are deposited in newly laid eggs of *O. moubata*. However, in contrast to the situation in *B. microphus*, the embryos apparently did not hydrolyse the maternal ecdysteroid apolar esters during development, but apparently synthesized ecdysteroids at the times of the peaks, and these then appeared to be esterified (Dotson *et al.*, 1991). Furthermore, embryos could metabolize [³H]ecdysone in vitro into 22-fatty acyl esters, 20,26-dihydroxyecdysone (an intermediate in 26-oic acid formation) and 20-hydroxyecdysone. Significantly, 20-hydroxylation first became evident in 2-day-old embryos, with highest activity occurring during increasing endogenous ecdysteroid titres (Dotson, Connat & Diehl, 1993). Although fatty acylation occurred in all stages, it was more pronounced during periods of low endogenous ecdysteroid titres. It has been suggested that ecdysteroid fatty acyl ester formation may represent a detoxification mechanism in *O. moubata* eggs, although the true significance is uncertain (Dotson *et al.*, 1993).

Sites of ecdysteroid production

Although various organs had been proposed to synthesize ecdysteroids in ticks (for review see Oliver & Dotson, 1993), definitive evidence has been furnished for ecdysone synthesis in nymphal integument of *O. parkeri*, *D. variabilis* and *I. scapularis* (Zhu, Oliver & Dotson, 1991*b*). Significantly, when the integument of *O. parkeri* was incubated alone, only ecdysone was produced, whereas when both integument and fat body were incubated together, large amounts of 20-hydroxyecdysone together with a small amount of ecdysone were obtained, suggesting that the fat body is a site of 20-hydroxylation. Similarly, in adult female *A. hebraeum*, the integument is competent to synthesize ecdysteroids, but in this case synthesis is dependent on the presence of a synganglial peptide (Lomas, Turner & Rees, 1997; also see 'Neuroendocrine systems' section below). In immature stages of insects, the prothoracic glands are a major site of ecdysteroid synthesis, and it may be significant that both these glands and the epidermis are of ectodermal origin.

JUVENOIDS

Juvenile hormones (JHs; juvenoids) in insects function in immature stages in preventing metamorphosis (Riddiford, 1996) and in adults control aspects of reproduction (Goodman & Granger, 2004). In females, synthesis of vitellogenin,

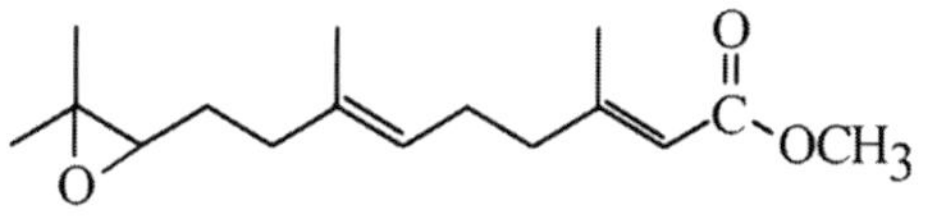

Juvenile hormone III

Fig. 7.8. Formula of juvenile hormone III.

its secretion into the haemolymph and uptake into the oocytes are controlled by JH and by ecdysteroid in some dipterans (Bellés, 1998). Outside the Lepidoptera, the most commonly occurring JH is JH-III (Fig. 7.8). In Crustacea, the juvenoid appears to be the un-epoxidized derivative of JH-III, methyl farnesoate (for review, see Wainwright & Rees, 2001).

In ticks, there is appreciable conflicting evidence regarding the occurrence and functioning of a juvenoid or JH-like compound. Although there is much indirect evidence, sometimes based on application of pharmacological doses of juvenoids, suggesting the functioning of such a compound, there are also contradictory reports. Such evidence has been thoroughly reviewed (Solomon et al., 1982; Sonenshine, 1991; Oliver & Dotson, 1993; Lomas & Rees, 1998).

It is difficult to assess the significance of conflicting reports of effects of juvenoids on moulting in ticks (Solomon et al., 1982). For example, JH-I delayed moulting in nymphs of O. porcinus (cited in Solomon et al., 1982) and H. dromedarii (Khalil et al., 1984). Furthermore, JH-III led to a dose-dependent reversal of 20-hydroxyecdysone-induced supermoulting in mated females of O. porcinus with promotion of normal oviposition (Obenchain & Mango, 1980).

There is appreciable evidence indicating that ticks possess a gonadotrophin having a somewhat similar function to insect JH. Administration of synganglion homogenate from fed, mated females induced oviposition in O. moubata (Aeschlimann, 1968), which is in accordance with ligation experiments implicating the synganglion as the source of gonadotrophin (Shanbaky & Khalil, 1975). Similarly, it has been suggested that the synganglion–lateral organ complex is the putative site of JH/gonadotrophin production (Binnington, 1981; Marzouk, Mohamed & Khalil, 1985). However, in some of these cases, neurosecretory products, in addition to hormones, may be involved in the regulatory processes being examined.

Experiments examining the effects of exogenous JHs have furnished indirect evidence suggesting a role for JH-like compounds in reproduction in ticks. Interestingly, topical application of a JH-mimic led to termination of reproductive

diapause in female A. arboreus (Bassal & Roshdy, 1974), with similar application of JH-I and JH-III inducing maturation and oviposition in fed, virgin female O. porcinus (Obenchain & Mango, 1980). However, a role of JH in egg development in the ixodid A. hebraeum could not be demonstrated (Lunke & Kaufman, 1993).

JH analogues, JH-I or JH-III induced vitellogenesis and oviposition in fed, virgin female O. moubata (Connat et al., 1983b). However, in contrast, in other experiments using unfed O. moubata or O. parkeri, vitellogenesis was not stimulated by JH and analogues (Chinzei, Taylor & Ando, 1991; Taylor et al., 1991b).

Interestingly, application of precocene-2, a compound which suppresses JH production in insect corpora allata, sterilized the argasid species Argas persicus and O. coriaceus and the ixodid R. sanguineus (Leahy & Booth, 1980). In O. parkeri, such precocene-2-induced sterility was partially reversed by JH-III application (Pound & Oliver, 1979). In contrast, clear-cut effects of lower doses of precocene analogues on fecundity in O. moubata were not observed (Connat & Nepa, 1990).

Many of the foregoing studies used fairly high doses of JHs and analogues and, thus, the effects observed may not necessarily be direct. However, collectively, they are suggestive of the occurrence of a JH-related hormone in ticks.

Various pieces of indirect biochemical evidence also support the occurrence of JH-like compounds in ixodid ticks. For example, HPLC fractionation of haemolymph from adult female D. variabilis and H. dromedarii with collection of fractions for JH-radioimmunoassay yielded peaks of immunoreactivity corresponding to JH-III (Sonenshine et al., 1989). However, as indicated by the authors, owing to probable cross-reactivity of the antiserum with other compounds related to JH-III, definitive conclusions could not be drawn regarding the identity of the immunoreactive material.

It is known that specific JH-binding proteins (JHPs) are involved in JH transport in insect haemolymph. When [^{3}H]EFDA, a photoreactive analogue of JH-III, was used, [^{3}H]EFDA-labelled proteins were demonstrated in haemolymph of mated, vitellogenic females, but not of virgin females of D. variabilis. The fact that JH-III displaced the [^{3}H]EFDA binding suggested that the binding proteins recognized JH-like substances in vitellogenic female ticks (Kulcsar, Prestwich & Sonenshine, 1989).

A major route of JH inactivation in insects involves hydrolysis by a haemolymph JH-specific esterase, with tissue epoxide hydrolase-catalysed conversion to JH-diol also occurring (Hammock, 1985; Chang & Kaufman, 2004). It has been shown in adult female *D. variabilis* that incubation of JHs with haemolymph yielded primarily JH-acid, demonstrating the occurrence of esterase activity, whereas with whole-body homogenates, appreciable epoxide hydrolase activity was observed as well (Venkatesh *et al.*, 1990). The fact that these enzymic activities were influenced by feeding and mating, with different profiles being observed for the JH ester hydrolysis and general esterase activity, is consistent with a conceivable role of these JH-metabolizing enzymes in the control of JH titre during development and reproduction.

An insect-type JH has never been identified in ticks. Recently, a thorough investigation by a variety of approaches did not provide direct evidence for the presence of such a JH or methyl farnesoate in either an ixodid (*D. variabilis*) or argasid (*O. parkeri*) tick species (Neese *et al.*, 2000). These authors provide strong evidence that indicates that ticks do not synthesize or have measurable amounts of farnesol, methyl farnesoate, JH-I, JH-II, JH-III or JH-III bisepoxide and do not appear to use these hormones to regulate nymphal–adult metamorphosis or vitellogenesis. Although based on negative evidence, this thorough study does strongly indicate that known juvenoids do not occur and function in ticks. Furthermore, tick egg and larval extracts had no apparent juvenilizing activity in insects. In contrast, in other work, tick extracts were reported to have juvenilizing activity on insects, although a known juvenoid could not be detected in haemolymph of *B. microplus* by gas chromatography-mass spectrometry (Connat, 1987). There are also further recent reports on the effects of a JH analogue, pyriproxyfen, on the lone star tick *A. americanum*, that include disruption of moulting (Donahue *et al.*, 1997; Strey, Teel & Langnecker, 2001).

NEUROENDOCRINE SYSTEMS

Nervous system

In ticks, there is a highly condensed, fused nerve mass where the cerebral ganglia and ventral nerve cord, with its associated segmental ganglia, have coalesced into a perioesophageal synganglion (Obenchain & Oliver, 1975; Binnington, 1987; Sonenshine, 1991). The whole synganglion is ensheathed by an acellular neurilemma and by perineural glial cells (Binnington, 1987). Peripheral nerves branch laterally from the synganglion and innervate all organs throughout the body. Paired discrete neurohaemal/endocrine organs of

unknown function, the lateral segmental organs and the retrocerebral organ complex, are associated with the synganglion and peripheral nerves. Unlike the situation in ticks, in other arthropods the body is segmented and discrete neuronal segments have been preserved in the form of segmental ganglia.

Neuroendocrine system

There is a paucity of information on the products of the neuroendocrine system in ticks (for reviews see Oliver & Dotson, 1993; Lomas & Rees, 1998). Neurosecretory centres within the synganglion have been identified primarily by paraldehyde fuchsin or vital dye (e.g. methylene blue) staining of histological sections. In the synganglion of *D. variabilis* and *O. parkeri*, 18 such centres have been identified (Pound & Oliver, 1982). It has been shown that a system of neurosecretory tracts extends between the foregoing identified neurosecretory cells, between the supraoesophageal and suboesophageal regions and between the retrocerebral organ complex (Obenchain & Oliver, 1975; see Sonenshine, 1991 for review). Furthermore, the neurilemma is innervated by axons from the neurosecretory cells that terminate in a diffuse network of neurosecretory terminals, suggesting a general release of neurohormones from the neurilemma. Interestingly, during periods of desiccation (unfed stages) and after engorgement, secretory products accumulate along the axonal pathways within the suboesophageal and supraoesophageal regions of the synganglion and extend to the perineural layers, including the retrocerebral organ complex (Obenchain & Oliver, 1975; Binnington & Oliver, 1982). However, the significance of the accumulated secretory products has not been elucidated. The involvement of neuropeptides in regulation of vitellogenesis and oogenesis has already been considered in the section 'Regulation of oogenesis and oviposition' (above).

Limited immunocytochemical studies have been undertaken to address the occurrence of possible neuropeptides in tick synganglia using antisera raised against non-tick species. It has been shown that three of the 18 paraldehyde fuchsin-positive synganglion regions of female *O. parkeri* (Pound & Oliver, 1982) reacted to an anti-bovine insulin antibody (Zhu *et al.*, 1991*a*). The occurrence of immunoreactivity as well in the extracellular surface of the neurilemma of the synganglion suggested a possible neurohaemal site. Similar, insulin-like, immunoreactivity has been demonstrated in the synganglion of nymphal and adult *D. variabilis* (Davis, Dotson & Oliver, 1994). The occurrence of such insulin-like immunoreactivity may be particularly significant, since bovine insulin and the neuropeptide bombyxin, a protho-racicotropic hormone (PTTH) that stimulates ecdysteroid synthesis in prothoracic glands from the silkworm *Bombyx mori*, have a high degree of amino acid sequence identity (Iwami *et al.*, 1989).

In another study, using a monoclonal antibody against the neuropeptide allatostatin I from the cockroach *Diploptera punctata*, immunoreactivity has been demonstrated in the synganglion of *D. variabilis* females (Zhu & Oliver, 2001). Immunoreactive cells were widely distributed within the synganglion. The observation that weak immunoreactiv-ity and fewer immunoreactive cells were apparent in newly moulted females compared to 1-month-old unfed females suggests that the immunoreactive products may be depleted during moulting and synthesized in females before feed-ing. Although the cockroach-type of allatostatins apparently only inhibit JH biosynthesis in cockroaches and crickets, they have additional physiological roles in certain insect species, such as antimyotropic effects and the inhibition of vitellogenin release by fat body (for reviews see Gade, Hoffmann & Spring, 1997; Hoffmann, Meyering-Vos & Lorenz, 1999). It may be that the wide distribution of allatostatin-like immunoreactive cells in the synganglion of *D. variabilis* females suggests that the secretory products of those cells may likewise have multiple functions.

In immature stages of insects, ecdysteroid synthesis in prothoracic glands is stimulated by PTTH produced by neu-rosecretory cells in the brain. Similarly, a synganglial pep-tide(s) (ecdysteroidotropic neurohormone, EtNH) has been identified in adult females of the ixodid tick *A. hebraeum*, that regulates ecdysteroid synthesis (Lomas, Turner & Rees, 1997). As previously demonstrated in immature stages of ixodid and argasid ticks (Zhu *et al.*, 1991*b*), the epidermal tissue of adult female *A. hebraeum* synthesizes ecdysteroids, but in this case, the synthesis is dependent on the presence of a synganglial peptide(s) in the incubation mixture (Lomas *et al.*, 1997). However, ligation experiments to examine the role of synganglial factors in stimulating ecdysteroid syn-thesis in nymphs of the argasid *O. parkeri* were inconclusive (see Oliver & Dotson, 1993).

The ecdysteroidotropic neurohormone (EtNH) from *A. hebraeum* synganglion referred to above stimulates ecdys-teroid synthesis in both a time- and dose-dependent man-ner. Furthermore, the effect of this peptide on epidermal ecdysteroidogenesis can be mimicked either by experimental elevation of endogenous cAMP concentration or by cAMP analogues, suggesting that cAMP may be involved in the action of the peptide (Lomas *et al.*, 1997), in an analogous

manner to the action of PTTH on prothoracic glands in insects (Smith *et al.*, 1996).

Significantly, two neuropeptides have been isolated recently from tick nervous system tissue. By a combination of immunocytochemistry and mass spectrometric analysis of single cells, a peptide (PALIPFPRV-NH$_2$; Ixori-PVK) was isolated from *I. ricinus* and *B. microplus* which shows high sequence homology to the insect periviscerokinin/CAP2b peptides (Neupert *et al.*, 2005). Although the function of the tick peptide is unknown, in view of the fact that such peptides in insects are involved in regulation of water balance, it is tempting to speculate that the function of Ixori-PVK is sim-ilar. In the other case, an opioid peptide (LVVYPWTKM), which shares similarity with vertebrate haemorphins, has been identified from synganglia of *A. testindiarium* (Liang *et al.*, 2005). This peptide displays an antinociceptive effect in mice and it has been speculated that it may be involved in escape from host immunosurveillance and inhibiting responses directed against the tick. Although the opioid peptide was purified from synganglia derived from unfed adults, the high similarity between the peptide and mam-malian haemorphins raises a question as to its exact origin (Liang *et al.*, 2005).

Particularly significant is the cloning and functional analysis of the first neuropeptide receptor from the Acari, the myokinin- or leucokinin-like peptide G-protein-coupled receptor (Holmes *et al.*, 2000, 2003). The leucokinins are a family of neuropeptides that have been demonstrated in several arthropod and invertebrate groups which have myotropic and diuretic activity in insects, and may also serve as neuromodulators of the central nervous system (Nässel, 1996). The cDNA encoding a leucokinin-like peptide recep-tor was cloned from larvae of *B. microplus* (Holmes *et al.*, 2000) and subsequently expressed in CHO-KI cells (Holmes *et al.*, 2003). Several myokinin peptides at nanomolar con-centrations specifically induced intracellular calcium release from intracellular stores in such transformed cells. Most leu-cokinins are characterized by the C-terminal pentapeptide FXSWGa (where X is F, H, S or Y) and this is the minimum sequence required for activity in this system, with the amino acids F and W being also essential (Taneja-Bageshwar *et al.*, 2006). The detection of receptor mRNA in all life stages of *B. microplus* indicates that myokinin peptides may have a critical role in the physiology of the tick (Holmes *et al.*, 2000). However, the receptor remains an orphan, since no endogenous ligand has been isolated as yet.

It is also significant that a factor, possibly a neuropep-tide, from the synganglion of *A. americanum* induces a phosphoinositide signalling pathway in salivary glands in vitro (McSwain *et al.*, 1989).

CONCLUDING REMARKS

It is evident that much of the literature cited is relatively old and there are major open questions in every aspect of the review; some of these are highlighted below. In the case of ecdysteroids, direct evidence for the functional significance of the peaks in haemolymph ecdysteroid titres during the life cycle is generally lacking. Understanding of the hormone biosynthetic and inactivation pathways (including their regulation) that are responsible for production of these mandatory changes in ecdysteroid profile is poor. There is evidence for the production of ecdysteroids in the 'integument', but it is not known whether this capacity is widespread in 'epidermal' cells or is more restricted to a discrete subpopulation of cells. Although ecdysteroid action in ticks is mediated by a heterodimeric receptor composed of the ecdysone receptor (EcR) and Ultraspiracle (USP) homologues, information on downstream events is practically lacking.

Current evidence strongly suggests that known insect-type juvenoids do not occur and function in ticks, but it is always difficult to draw firm conclusions from negative evidence. Presumably, it is still just conceivable that a novel (unique) type of juvenoid could exist and function in ticks.

As indicated in this review, uncharacterized neuropeptides/peptides appear to be involved in many different types of regulatory pathways. Importantly, the tick neural system is a potential acaricide target (Lees & Bowman, 2007). The continued improvement in techniques with enhanced sensitivity should make isolation and characterization of such peptides feasible. The ongoing projects to sequence the genomes of two representative tick species (the prostriate *I. scapularis* and, hopefully, the metastriate *B. microplus*) will be major advancements to furthering our understanding of the endocrinology in these important vectors of disease.

SUMMARY

Ecdysteroids (moulting hormones), juvenoids and neuropeptides have all been implicated in the reproduction and development of ticks. In immature stages of ticks, ecdysteroids have been shown to regulate moulting and to terminate larval diapause. Although there is a paucity of information on the molecular action of ecdysteroids in ticks, their action appears to be via a heterodimeric Ecdysone/Ultraspiracle Receptor, as in insects. Ecdysteroids are reported to play a role in sperm maturation in adult males. In females, ecdysteroids function in the regulation of salivary glands, of production of sex pheromones and of oogenesis and oviposition. There is evidence for ecdysteroid production in the integument and pathways of hormone inactivation are similar to those in insects. Ecdysteroids also function in embryogenesis. Although evidence for the occurrence and functioning of juvenile hormones in ticks has been contradictory, in recent thorough work, it has not been possible to detect known juvenile hormones in ticks, nor to demonstrate effects of extracts on insects. Factors (neuropeptides) from the synganglion affect physiological processes and limited immunocytochemical studies are reviewed. Significantly, a G-protein-coupled receptor has been cloned, expressed, and specifically responds to myokinins.

ACKNOWLEDGEMENTS

We thank The Wellcome Trust and the BBSRC for financial support of work from our laboratory and Dr Hajime Takeuchi for the figures.

REFERENCES

Aeschlimann, A. A. (1968). La ponte chez *Ornithodoros moubata* Murray (Ixodoidea: Argasidae). *Revue suisse de Zoologie* **75**, 1033–1039.

Allan, S. A., Phillips, J. S., Taylor, D. & Sonenshine, D. E. (1988). Genital sex pheromones of ixodid ticks: evidence for the role of fatty acids from the anterior reproductive tract in mating of *Dermacentor variabilis* and *Dermacentor andersoni*. *Journal of Insect Physiology* **34**, 315–323.

Arrieta, M. C., Leskiw, B. K. & Kaufman, W. R. (2006). Antimicrobial activity in the egg wax of the African cattle tick *Amblyomma hebraeum* (Acari: Ixodidae). *Experimental and Applied Acarology* **39**, 297–313.

Assenga, S. P., You, M., Shy, C. H., *et al.* (2006). The use of a recombinant baculovirus expressing a chitinase from the hard tick *Haemophysalis longicornis* and its potential application as a bioacaricide for tick control. *Parasitology Research* **98**, 111–118.

Bassal, T. T. M. & Roshdy, M. A. (1974). *Argas (Persicargas) arboreus*: juvenile hormone analog termination of diapause and oviposition control. *Experimental Parasitology* **36**, 34–39.

Bellés, X. (1998). Endocrine effectors in insect vitellogenesis. In *Recent Advances in Arthropod Endocrinology*, eds. Coast, G. M. & Webster, S. G., pp. 71–90. Cambridge, UK: Cambridge University Press.

Binnington, K. C. (1981). Ultrastructural evidence for the endocrine nature of the lateral organs of the cattle tick *Boophilus microplus*. *Tissue and Cell* **13**, 475–490.

Binnington, K. C. (1987). Histology and ultrastructure of the acarine synganglion. In *The Arthropod Brain: Its Evolution, Development, Structure, and Functions*, ed. Gupta, A. P., pp. 95–109. Oxford, UK: Pergamon Press.

Binnington, K. C. & Oliver, J. H. Jr (1982). Structure and function of the circulatory, nervous and neuroendocrine systems of ticks. In *Physiology of Ticks*, eds. Obenchain, F. D. & Galun, R., pp. 351–398. Oxford, UK: Pergamon Press.

Bouvier, J., Diehl, P. A. & Morici, M. (1982). Ecdysone metabolism in the tick *Ornithodoros moubata* (Argasidae, Ixodoidea). *Revue suisse de Zoologie* **89**, 967–976.

Campbell, J. D. & Oliver, J. H. Jr (1984). Membrane feeding and developmental effects of ingested β-ecdysone on *Ornithodoros parkeri* (Acari: Argasidae). In *Acarology VI*, vol. 1, eds. Griffiths, D. A. & Bowman, C. E., pp. 393–399. Chichester, UK: Ellis Horwood.

Chang, E. S. & Kaufman, W. R. (2004). Endocrinology of crustaceans and arachnids. In *Comprehensive Insect Science*, vol. 3, *Endocrinology*, eds. Gilbert, L. I., Iatrou, K. & Gill, S., pp. 805–842. Amsterdam: Elsevier.

Chinzei, Y. (1986). Vitellogenin biosynthesis and processing in a soft tick, *Ornithodoros moubata*. In *Host Regulated Development Mechanisms in Vector Arthropods*, eds. Borovsky, D. & Spielman, A., pp. 18–24. Vero Beach, FL: University of Florida Press.

Chinzei, Y. & Taylor, D. (1990). Regulation of vitellogenesis induction by engorgement in the soft tick (*Ornithodoros moubata*). *Advances in Invertebrate Reproduction* **5**, 565–570.

Chinzei, Y. & Yano, I. (1985). Fat body is the site of vitellogenin synthesis in the soft tick *Ornithodoros moubata*. *Journal of Comparative Physiology B* **155**, 671–678.

Chinzei, Y., Chino, H. & Takahashi, K. (1983). Purification and properties of vitellogenin and vitellin from a tick *Ornithodoros moubata*. *Journal of Comparative Physiology* **152**, 13–21.

Chinzei, Y., Taylor, D. & Ando, K. (1991). Effects of juvenile hormone and its analogs on vitellogenin synthesis and ovarian development in *Ornithodoros moubata* (Acari: argasidae). *Journal of Medical Entomology* **28**, 506–513.

Chinzei, Y., Taylor, D., Miura, K. & Ando, K. (1992). Vitellogenesis induction by synganglion factor in adult female tick, *Ornithodoros moubata* (Acari: Argasidae). *Journal of the Acarological Society of Japan* **1**, 15–26.

Connat, J.-L. (1987). Aspects endocrinologiques de la physiologie du développement et de la reproduction chez le tiques. Unpublished Ph.D. thesis, University of Bourgogne, France.

Connat, J.-L. & Diehl, P. A. (1986). Probable occurrence of ecdysteroid fatty acid esters in different classes of arthropods. *Insect Biochemistry* **16**, 91–97.

Connat, J.-L. & Dotson, E. M. (1988). Comparative investigation of the egg ecdysteroids of ticks using radioimmunoassay and metabolic studies. *Journal of Insect Physiology* **34**, 639–645.

Connat, J.-L. & Nepa, M.-C. (1990). Effects of different anti-juvenile hormone agents on the fecundity of the female tick *Ornithodoros moubata*. *Pesticide Biochemistry and Physiology* **37**, 266–274.

Connat, J.-L., Delbecque, J.-P., Alabouvette, J. & Pitoizet, N. (1997). Evolution of ecdysteroids and of their apolar conjugates during the post-embryonic development of the tick *Ornithodoros moubata*. *Archives of Insect Biochemistry and Physiology* **35**, 159–168.

Connat, J.-L., Diehl, P. A., Dumont, N., Carminati, S. & Thompson, M. J. (1983*a*). Effects of exogenous ecdysteroids on the female tick, *Ornithodoros moubata*: induction of supermolting and influence on oogenesis. *Zeitschrift für angewandte Entomologie* **96**, 520–530.

Connat, J.-L., Diehl, P. A., Gfeller, H. & Morici, M. (1985). Ecdysteroids in females and eggs of the ixodid tick *Amblyomma hebraeum*. *International Journal of Invertebrate Reproduction and Development* **8**, 103–116.

Connat, J.-L., Diehl, P. A. & Morici, M. (1984). Metabolism of ecdysteroids during the vitellogenesis of the tick *Ornithodoros moubata* (Ixodoidea: Argasidae): accumulation of apolar metabolites in the eggs. *General and Comparative Endocrinology* **56**, 100–110.

Connat, J.-L., Diehl, P. A. & Thompson, M. J. (1986*a*). Possible inactivation of ingested ecdysteroids by conjugation with long-chain fatty acids in the female tick *Ornithodoros moubata* (Acarina: Argasidae). *Archives of Insect Biochemistry and Physiology* **3**, 235–252.

Connat, J.-L., Dotson, E. M. & Diehl, P. A. (1987). Metabolism of ecdysteroids in the female tick *Amblyomma hebraeum* (Ixodoidea: Ixodidae): accumulation of free ecdysone and 20-hydroxyecdysone in the eggs. *Journal of Comparative Physiology B* **157**, 689–699.

Connat, J.-L., Dotson, E. M. & Diehl, P. A. (1988). Apolar conjugates of ecdysteroids are not used as a storage form of molting hormone in the argasid tick *Ornithodoros moubata*. *Archives of Insect Biochemistry and Physiology* **9**, 221–235.

Connat, J.-L., Ducommun, J. & Diehl, P. A. (1983*b*). Juvenile hormone-like substances can induce vitellogenesis in the tick

Ornithodoros moubata (Acarina: Argasidae). *International Journal of Invertebrate Reproduction* 6, 285–294.

Connat, J.-L., Ducommun, J., Diehl, P. A. & Aeschlimann, A. (1986b). Some aspects of the control of the gonotrophic cycle in the tick *Ornithodoros moubata* (Ixodoidea, Argasidae). In *Morphology, Physiology, and Behavioral Biology of Ticks*, eds. Sauer, J. R. & Hair, J. A., pp. 194–216. Chichester, UK: Ellis Horwood.

Connat, J.-L., Lafont, R. & Diehl, P. A. (1986c). Metabolism of [³H]ecdysone by isolated tissues of the female ixodid tick *Amblyomma hebraeum* (Ixodoidea: Ixodidae). *Molecular and Cellular Endocrinology* 47, 257–267.

Crosby, T., Evershed, R. P., Lewis, D., Wigglesworth, K. P. & Rees, H. H. (1986). Identification of ecdysone 22-long chain fatty acyl esters in newly laid eggs of the cattle tick *Boophilus microplus*. *Biochemical Journal* 240, 131–138.

Davis, H. H., Dotson, E. M. & Oliver, J. H. Jr (1994). Localization of insulin-like immunoreactivity in the synganglion of nymphal and adult *Dermacentor variabilis* (Acari: Ixodidae). *Experimental and Applied Acarology* 18, 111–122.

Dees, W. H., Sonenshine, D. E. & Breidling, E. (1984a). Ecdysteroids in the American dog tick, *Dermacentor variabilis* (Acari: Ixodidae), during different periods of tick development. *Journal of Medical Entomology* 21, 514–523.

Dees, W. H., Sonenshine, D. E. & Breidling, E. (1984b). Ecdysteroids in *Hyalomma dromedarii* and *Dermacentor variabilis* and their effects on sex pheromone activity. In *Acarology VI*, vol. 1, eds. Griffiths, D. A. & Bowman, C. E., pp. 405–413. Chichester, UK: Ellis Horwood.

Dees, W. H., Sonenshine, D. E. & Breidling, E. (1985). Ecdysteroids in the camel tick, *Hyalomma dromedarii* (Acari: Ixodidae) and comparison with sex pheromone activity. *Journal of Medical Entomology* 22, 22–27.

Delbecque, J. P., Diehl, P. A. & O'Connor, J. D. (1978). Presence of ecdysone and ecdysterone in the tick *Amblyomma hebraeum* Koch. *Experientia* 34, 1379–1381.

Diehl, P. A., Connat, J.-L. & Dotson, E. M. (1986). Chemistry, function, and metabolism of tick ecdysteroids. In *Morphology, Physiology and Behavioural Biology of Ticks*, eds. Sauer, J. R. & Hair, J. H., pp. 165–193. Chichester, UK: Ellis Horwood.

Diehl, P. A., Connat, J.-L., Girault, J. P. & Lafont, R. (1985). A new class of apolar ecdysteroid conjugates: esters of 20-hydroxy-ecdysone with long-chain fatty acids in ticks. *International Journal of Invertebrate Reproduction and Development* 8, 1–13.

Diehl, P. A., Germond, J. E. & Morici, M. (1982). Correlations between ecdysteroid titres and integument structure in nymphs of the tick, *Amblyomma hebraeum* Koch (Acarina: Ixodidae). *Revue suisse de Zoologie* 89, 859–868.

Dinan, L. (1997). Ecdysteroids in adults and eggs of the house cricket, *Acheta domesticus* (Orthoptera: Gryllidae). *Comparative Biochemistry and Physiology B* 116, 129–135.

Donahue, W. A., Teel, P. D., Strey, O. F. & Meola, R. W. (1997). Pyriproxyfen effects on newly engorged larvae and nymphs of the lone star tick (Acari: Ixodidae). *Journal of Medical Entomology* 34, 206–211.

Dotson, E. M., Connat, J.-L. & Diehl, P. A. (1991). Cuticle deposition and ecdysteroid titres during embryonic and larval development of the argasid tick *Ornithodoros moubata*. *General and Comparative Endocrinology* 82, 386–400.

Dotson, E. M., Connat, J.-L. & Diehl, P. A. (1993). Metabolism of [³H]ecdysone in embryos and larvae of the tick *Ornithodoros moubata*. *Archives of Insect Biochemisty and Physiology* 23, 67–78.

Dotson, E. M., Connat, J.-L. & Diehl, P. A. (1995). Ecdysteroid titre and metabolism and cuticle deposition during embryogenesis of the ixodid tick *Amblyomma hebraeum* (Koch). *Comparative Biochemistry and Physiology B* 110, 155–166.

Dumber, J. B. & Oliver, J. H. Jr (1981). Kinetics of spermatogenesis, cell cycle analysis, and testis development in nymphs of the tick *Dermacentor variabilis*. *Journal of Insect Physiology* 27, 743–753.

El-Said, A., Swiderski, Z., Aeschlimann, A. & Diehl, P. A. (1981). Fine structure of spermiogenesis in the tick *Amblyomma hebraeum* (Acari: Ixodidae): late stages of differentiation and structure of the mature spermatozoon. *Journal of Medical Entomology* 18, 464–476.

Friesen, K. J. & Kaufman, W. R. (2002). Quantification of vitellogenesis and its control by 20-hydroxyecdysone in the ixodid tick, *Amblyomma hebraeum*. *Journal of Insect Physiology* 48, 773–782.

Friesen, K. J. & Kaufman, W. R. (2003). Cypermethrin inhibits egg development in the ixodid tick, *Amblyomma hebraeum*. *Pesticide Biochemistry and Physiology* 76, 25–35.

Friesen, K. J. & Kaufman, W. R. (2004). Effects of 20-hydroxyecdysone and other hormones on egg development, and identification of a vitellin-binding protein in the ovary of the tick, *Amblyomma hebraeum*. *Journal of Insect Physiology* 50, 519–529.

Friesen, K. J., Suri, R. & Kaufman, W. R. (2003). Effects of the avermectin, MK-243, on ovary development and salivary gland degeneration in the ixodid tick, *Amblyomma hebraeum. Pesticide Biochemistry and Physiology* **76**, 82–90.

Gade, G., Hoffmann, K. H. & Spring, J. (1997). Hormonal regulation in insects: facts, gaps, and future directions. *Physiological Reviews* **77**, 963–1032.

Germond, J.-E., Diehl, P. A. & Morici, M. (1982). Correlations between integument structure and ecdysteroid titres in fifth-stage nymphs of the tick, *Ornithodoros moubata. General and Comparative Endocrinology* **46**, 255–266.

Gilbert, L. I., Iatrou, K. & Gill, S. (eds.) (2004). *Comprehensive Insect Science*, vol. 3, *Endocrinology*. Amsterdam: Elsevier.

Goodman, W. & Granger, N. (2004). The juvenile hormone. In *Comprehensive Insect Science*, vol. 3, *Endocrinology*, eds. Gilbert, L. I., Iatrou, K. & Gill, S., pp. 319–408. Amsterdam: Elsevier.

Guo, X., Harmon, M. A., Laudet, V., Mangelsdorf, D. J. & Palmer, M. J. (1997). Isolation of a functional ecdysteroid receptor homologue from the ixodid tick, *Amblyomma americanum* (L.). *Insect Biochemistry and Molecular Biology* **27**, 945–962.

Guo, X., Xu, Q., Harmon, M. A., *et al.* (1998). Isolation of two functional retinoid X receptor subtypes from the Ixodid tick, *Amblyomma americanum* (L.). *Molecular and Cellular Endocrinology* **139**, 45–60.

Hamilton, J. G. C. (1992). The role of pheromones in tick biology. *Parasitology Today* **8**, 130–133.

Hammock, B. D. (1985). Regulation of juvenile hormone titer: degradation. In *Comprehensive Insect Physiology, Biochemistry and Pharmacology*, vol. 7, eds. Kerkut, G. A. & Gilbert, L. I, pp. 431–472. Oxford, UK: Pergamon Press.

Harris, R A. & Kaufman, W. R. (1981). Hormonal control of salivary gland degeneration in the ixodid tick *Amblyomma hebraeum. Journal of Insect Physiology* **27**, 241–243.

Harris, R. A. & Kaufman, W. R. (1984). Neural involvement in the control of salivary gland degeneration in the ixodid tick, *Amblyomma hebraeum. Journal of Experimental Biology* **109**, 281–290.

Harris, R. A. & Kaufman, W. R. (1985). Ecdysteroids: possible candidates for the hormone which triggers salivary gland degeneration in the ixodid tick *Amblyomma hebraeum. Experientia* **41**, 740–742.

Hayward, D. C., Bastiani, M. J., Trueman, J. W. H., *et al.* (1999). The sequence of *Locusta* RXR, homologous to *Drosophila* Ultraspiracle, and its evolutionary implications. *Development Genes and Evolution* **209**, 564–571.

Henrich, V. (2004). The ecdysteroid receptor (EcR). In *Comprehensive Insect Science*, vol. 3, *Endocrinology*, eds.

Gilbert, L. I., Iatrou, K. & Gill, S., pp. 245–285. Amsterdam: Elsevier.

Henrich, V. C., Rybczynski, R. & Gilbert, L. I. (1999). Peptide hormones, and puffs: mechanisms and models in insect development. *Vitamins and Hormones – Advances in Research and Applications* **55**, 73–125.

Hoffmann, K. H., Meyering-Vos, M. & Lorenz, M. W. (1999). Allatostatins and allatotropins: is the regulation of corpora allata activity their primary function? *European Journal of Entomology* **96**, 255–266.

Holmes, S. P., Barhoumit, R., Nachman, R. J. & Pietrantonio, P. V. (2003). Functional analysis of a G protein-coupled receptor from the Southern cattle tick *Boophilus microplus* (Acari: Ixodidae) identifies it as the first arthropod myokinin receptor. *Insect Molecular Biology* **12**, 27–38.

Holmes, S. P., He, H., Chen, A. C., Ivie, G. W. & Pietrantonio, P. V. (2000). Cloning and transcriptional expression of a leucokinin-like peptide receptor from the Southern cattle tick, *Boophilus microplus* (Acari: Ixodidae). *Insect Molecular Biology* **9**, 457–465.

Isaac, R. E. & Slinger, A. J. (1989). Storage and excretion of ecdysteroids. In *Ecdysone*, ed. Koolman, J., pp. 250–253. Stuttgart, Germany: G. Thieme.

Iwami, M., Kawakami, A., Ishizaki, H., *et al.* (1989). Cloning of a gene encoding bombyxin, an insulin-like brain secretory peptide of the silkmoth *Bombyx mori* with prothoracicotropic activity. *Development Growth and Differentiation* **31**, 31–37.

Jaffe, H., Hayes, K. K., Sonenshine, D. E., *et al.* (1986). Controlled release reservoirs system for the delivery of insect steroid analogues against ticks. *Journal of Medical Entomology* **23**, 685–691.

James, A. M. & Oliver, J. H. Jr (1997). Purification and partial characterization of vitellin from the black-legged tick *Ixodes scapularis. Insect Biochemistry and Molecular Biology* **27**, 639–649.

Jarvis, T. D., Earley, F. G. & Rees, H. H. (1994). Ecdysteroid biosynthesis in larval testes of *Spodoptera littoralis. Insect Biochemistry and Molecular Biology* **24**, 531–537.

Kaufman, W. R. (1989). Tick–host interaction: a synthesis of current concepts. *Parasitology Today* **5**, 47–56.

Kaufman, W. R. (1991). Correlation between haemolymph ecdysteroid titre, salivary gland degeneration and ovarian development in the ixodid tick, *Amblyomma hebraeum* Koch. *Journal of Insect Physiology* **37**, 95–99.

Khalil, G. M. (1970). Biochemistry and physiological studies on certain ticks (Ixodoidea): gonad development and gametogenesis in *Hyalomma* (*H.*) *anatolicum excavatum*

Koch (Ixodidae). *Journal of Parasitology* **56**, 596–610.

Khalil, G. M., Sonenshine, D. E., Hanafy, H. A. & Abdelmonem, A. E. (1984). Juvenile hormone I effects on the camel tick, *Hyalomma dromedarii* (Acari: Ixodidae). *Journal of Medical Entomology* **21**, 561–566.

Kozlova, T. & Thummel, C. S. (2000). Steroid regulation of postembryonic development and reproduction in *Drosophila*. *Trends in Endocrinology and Metabolism* **11**, 276–280.

Kubo, I., Komatsu, S., Asaka, Y. & De Boer, G. (1987). Isolation and identification of apolar metabolites of ingested 20-hydroxyecdysone in frass of *Heliothis virescens* larvae. *Journal of Chemical Ecology* **13**, 785–794.

Kulcsar, P., Prestwich, G. G. & Sonenshine, D. E. (1989). Detection of binding proteins for juvenile hormone-like substances in ticks by photoaffinity labelling. In *Host Regulated Developmental Mechanisms in Vector Arthropods*, eds. Borovsky, D. & Spielman, A., pp. 18–23. Vero Beach, FL: University of Florida Press.

Lafont, R. & Connat, J.-L. (1989). Pathways of ecdysone metabolism. In *Ecdysone*, ed. Koolman, J., pp. 167–173. Stuttgart, Germany: G. Thieme.

Lafont, R., Dauphin-Villemant, C., Warren, J. & Rees, H. H. (2004). Ecdysteroid chemistry and biochemistry. In *Comprehensive Insect Science, vol. 3, Endocrinology*, eds. Gilbert, L. I., Iatrou, K. & Gill, S., pp. 125–195. Amsterdam: Elsevier.

Leahy, M. G. & Booth, K. S. (1980). Precocene induction of tick sterility and ecdysis failure. *Journal of Medical Entomology* **17**, 18–21.

Leid, M., Kastner, P. & Chambon, P. (1992). Multiplicity generates diversity in the retinoic acid signalling pathway. *Trends in Biochemical Sciences* **17**, 427–433.

Lees, K. & Bowman, A. S. (2007). Tick neurobiology: recent advances and the post-genomic era. *Invertebrate Neuroscience* **7**, 183–198.

Liang, J., Zhang, J., Lai, R. & Rees, H. H. (2005). An opioid peptide from synganglia of the tick, *Amblyomma testindinarium*. *Peptides* **26**, 603–606.

Lindsay, P. J. & Kaufman, W. R. (1988). Action of some steroids on salivary gland degeneration in the ixodid tick, *A. americanum*. *Journal of Insect Physiology* **34**, 351–359.

Loeb, M. J., Brandt, E. P., Woods, C. W. & Borkovec, A. B. (1987). An ecdysiotropic factor from brains *of Heliothis virescens* induces testes to produce immunodetectable ecdysteroid *in vitro*. *Journal of Experimental Zoology* **243**, 275–282.

Lomas, L. O. & Rees, H. H. (1998). Endocrine regulation of development and reproduction in acarines. In *Recent Advances in Arthropod Endocrinology*, eds. Coast, G. M. & Webster, S. G., pp. 91–124. Cambridge, UK: Cambridge University Press.

Lomas, L. O., Turner, P. C. & Rees, H. H. (1997). A novel neuropeptide–endocrine interaction controlling ecdysteroid production in ixodid ticks. *Proceedings of the Royal Society of London B* **264**, 589–596.

Lunke, M. D. & Kaufman, W. R. (1993). Hormonal control of ovarian development in the tick *Amblyomma hebraeum* Koch (Acari: Ixodidae). *Invertebrate Reproduction and Development* **23**, 25–38.

Magee, R. M., Jones, L. D. & Rees, H. H. (1996). Ecdysteroids in relation to adult development and reproduction in female *Rhipicephalus appendiculatus* (Acari: Ixodidae). *Archives of Insect Biochemistry and Physiology* **31**, 197–206.

Mangelsdorf, D. J. & Evans, R. M. (1995). The RXR heterodimers and orphan receptors. *Cell* **83**, 841–850.

Mango, C., Odhiambo, T. R. & Galun, R. (1976). Ecdysone and the super tick. *Nature* **260**, 318–319.

Mao, H. & Kaufman, W. R. (1998). DNA binding properties of the ecdysteroid receptor in the salivary gland of the female ixodid tick, *Amblyomma hebraeum*. *Insect Biochemistry and Molecular Biology* **28**, 947–957.

Mao, H. & Kaufman, W. R. (1999). Profile of the ecdysteroid hormone and its receptor in the salivary gland of the adult female tick, *Amblyomma hebraeum*. *Insect Biochemistry and Molecular Biology* **29**, 33–42.

Mao, H., McBlain, W. A. & Kaufman, W. R. (1995). Some properties of the ecdysteroid receptor in the salivary gland of the ixodid tick, *Amblyomma hebraeum*. *General and Comparative Endocrinology* **99**, 340–348.

Marzouk, A. S., Mohamed, F. S. A. & Khalil, G. M. (1985). Neurohemal–endocrine organs in the camel tick, *Hyalomma dromedarii* (Acari: Ixodoidea: Ixodidae). *Journal of Medical Entomology* **22**, 385–391.

McSwain, J. L., Tucker, J. S., Essenberg, R. C. & Sauer, J. R. (1989). Brain factor induced formation of inositol phosphates in tick salivary glands. *Insect Biochemistry* **19**, 343–349.

Nässel, D. R. (1996). Neuropeptides, amines, and amino acids in an elementary insect ganglion: functional and chemical anatomy of the unfused abdominal ganglion. *Progress in Neurobiology* **48**, 325–420.

Neese, P. A., Sonenshine, D. E., Kallapur, V. L., Apperson, C. S. & Roe, R. M. (2000). Absence of insect juvenile hormones in the American dog tick, *Dermacentor variabilis*

(Say) (Acari: Ixodidae), and in *Ornithodoros parkeri* Cooley (Acari: Argasidae). *Journal of Insect Physiology* **46**, 477–490.

Neupert, S., Predel, R., Russell, W. K., *et al.* (2005). Identification of tick periviscerokinin, the first neurohormone of Ixodidae: single cell analysis by means of MALDI-TOF/TOF mass spectrometry. *Biochemical and Biophysical Research Communications* **338**, 1860–1864.

Obenchain, F. D. & Mango, C. K. A. (1980). Effects of exogenous ecdysteroids and juvenile hormones on female reproductive development in *Ornithodoros p. porcinus*. *American Zoologist* **20**, Abstract No. 1192.

Obenchain, F. D. & Oliver, J. H. Jr (1975). Neurosecretory system of the American dog tick, *Dermacentor variabilis* (Acari: Ixodidae). II. Distribution of secretory cell types, axonal pathways and putative nerohemal–neuroendocrine associations: comparative histological and anatomical implications. *Journal of Morphology* **145**, 269–294.

Ogihara, K. (2003). Ecdysteroid hormone titer and expression of ecdysone receptor mRNA as related to vitellogenesis in the soft tick, *Ornothodoros moubata* (Acari: Argasidae). Unpublished M.Ag.Sci. thesis, University of Tsukuba, Japan.

Oliver, J. H. Jr (1986*a*). Relationship among feeding, gametogenesis, mating and syngamy in ticks. In *Host Regulated Development Mechanisms in Vector Arthropods*, eds. Borovsky, D. & Spielman, A., pp. 93–99. Vero Beach, FL: University of Florida Press.

Oliver, J. H. Jr (1986*b*). Induction of oogenesis and oviposition in ticks. In *Morphology, Physiology and Behavioural Biology of Ticks*, eds. Sauer, J. R. & Hair, J. A., pp. 233–247. Chichester, UK: Ellis Horwood.

Oliver, J. H. Jr & Dotson, E. M. (1993). Hormonal control of molting and reproduction in ticks. *American Zoologist* **33**, 384–396.

Oliver, J. H. Jr, Zhu, X. X., Vogel, G. N. & Dotson, E. M. (1992). Role of synganglion in oogenesis of the tick *Ornithodoros parkeri* (Acari: Argasidae). *Journal of Parasitology* **78**, 93–98.

Palmer, M. J., Harmon, M. A. & Laudet, V. (1999). Characterization of EcR and RXR homologues in the Ixodid tick, *Amblyomma americanum* (L.). *American Zoologist* **39**, 747–757.

Palmer, M. J., Warren, J. T., Jin, X., Guo, X. & Gilbert, L. I. (2002). Developmental profiles of ecdysteroids, ecdysteroid receptor mRNAs and DNA binding properties of ecdysteroid receptors in the ixodid tick *Amblyomma americanum* (L.). *Insect Biochemistry and Molecular Biology* **32**, 465–476.

Pound, J. M. & Oliver, J. H. Jr (1979). Juvenile hormone: evidence of its role in the reproduction of ticks. *Science* **206**, 355–357.

Pound, J. M. & Oliver, J. H. Jr (1982). Synganglial and neurosecretory morphology of female *Ornithodoros parkeri* (Cooley) (Acari: Argasidae). *Journal of Morphology* **173**, 159–177.

Pound, J. M., Oliver, J. H. Jr & Andrews, R. H. (1984). Induction of apolysis and cuticle formation in female *Ornithodoros parkeri* (Acari: Argasidae) by hemocoelic injections of β-ecdysone. *Journal of Medical Entomology* **21**, 612–614.

Rees, H. H. (2004). Hormonal control of tick development and reproduction. *Parasitology* **129**, S127–S143.

Rees, H. H. & Isaac, R. E. (1984). Biosynthesis of ovarian ecdysteroid phosphates and their metabolic fate during embryogenesis in *Schistocerca gregaria*. In *Biosynthesis, Metabolism and Mode of Action of Invertebrate Hormones*, eds. Hoffmann, J. & Porchet, M., pp. 181–195. Berlin: Springer-Verlag.

Riddiford, L. M. (1996). Juvenile hormone: the status of its 'status quo' action. *Archives of Insect Biochemistry and Physiology* **32**, 271–286.

Riddiford, L. M., Cherbas, P. & Truman, J. W. (2001). Ecdysone receptors and their biological actions. *Vitamins and Hormones – Advances in Research and Applications* **60**, 1–73.

Robinson, P. D., Morgan, E. D., Wilson, Y. D. & Lafont, R. (1987). The metabolism of ingested and injected [^{3}H]ecdysone by final instar larvae of *Heliothis armigera*. *Physiological Entomology* **12**, 321–330.

Rosell, R. & Coons, L. B. (1991). Purification and partial characterization of vitellin from the eggs of the hard tick *Dermacentor variabilis*. *Insect Biochemistry* **21**, 871–885.

Rosell, R. & Coons, L. B. (1992). The role of the fat body, midgut and ovary in vitellogenin production and vitellogenesis in the female tick *Dermacentor variabilis*. *International Journal for Parasitology* **22**, 341–349.

Rosell-Davis, R. & Coons, L. B. (1989). Relationship between feeding, mating, vitellogenin production and vitellogenesis in tick *Dermacentor variabilis*. *Experimental and Applied Acarology* **7**, 95–105.

Sankhon, N., Lockey, T., Rosell, R. C., Rothschild, M. & Coons, L. (1999). Effect of methoprene and 20-hydroxyecdysone on vitellogenin production in cultured

fat bodies and backless explants from unfed female
Dermacentor variabilis. *Journal of Insect Physiology* **45**,
755–761.

Sannasi, A. & Subramoniam, T. (1972). Hormonal rupture of
larval diapause in the tick *Rhipicephalus sanguineus* (Lat.).
Experientia **28**, 666–667.

Sauer, J. R., Essenberg, R. C. & Bowman, A. S. (2000).
Salivary glands in ixodid ticks: control and mechanism of
secretion. *Journal of Insect Physiology* **46**, 1069–1078.

Sauer, J. R., McSwain, J. L., Bowman, A. S. & Essenberg,
R. C. (1995). Tick salivary gland physiology. *Annual Review
of Entomology* **40**, 245–267.

Schriefer, M. E. (1991). Vitellogenesis in *Hyalomma dromedarii*
(Acari: Ixodidae): a model for analysis of endocrine
regulation in ixodid ticks. Unpublished Ph.D. thesis, Old
Dominion University & East Virginia Medical School,
Norfolk, VA.

Shanbaky, N. M. & Khalil, G. M. (1975). The sub-genus
Persicargus (Ixodoidea: Argasidae: *Argas*). XXII. The effect
of feeding on hormonal control of egg development in *Argas*
(*Persicargas*) *arboreus*. *Experimental Parasitology* **37**, 361–366.

Shanbaky, N. M., Mansour, M. M., Main, A. J., El-Said, A. &
Helmy, N. (1990). Hormonal control of vitellogenesis in
Argas (*Argas*) *hermanni* (Acari: Argasidae). *Journal of
Medical Entomology* **27**, 968–974.

Shepherd, J., Oliver, J. H. Jr & Hall, J. D. (1982). A
polypeptide from male accessory glands which triggers
maturation of tick spermatozoa. *International Journal of
Invertebrate Reproduction* **5**, 129–137.

Slinger, A. J. & Isaac, R. E. (1988). Ecdysteroid titers during
embryogenesis of the cockroach, *Periplaneta americana*.
Journal of Insect Physiology **34**, 1119–1125.

Smith, W. A., Varghese, A. H., Healy, M. S. & Lou, K. J.
(1996). Cyclic AMP is a prerequisite messenger in the action
of big PTTH in the prothoracic glands of pupal *Manduca
sexta*. *Insect Biochemistry and Molecular Biology* **26**, 161–170.

Soller, M., Bownes, M. & Kubli, E. (1999). Control of oocyte
maturation in sexually mature *Drosophila* females.
Developmental Biology **208**, 337–351.

Solomon, K. R., Mango, C. K. A. & Obenchain, F. D. (1982).
Endocrine mechanisms in ticks: effects of insect hormones
and their mimics on development and reproduction. In
Physiology of Ticks, eds. Obenchain, F. D. & Galun, R.,
pp. 399–438. Oxford, UK: Pergamon Press.

Sonenshine, D. E. (1986). Tick pheromones: an overview. In
Morphology, Physiology, and Behavioural Biology of Ticks,
eds. Sauer, J. R. & Hair, J. A., pp. 342–360. Chichester, UK:
Ellis Horwood.

Sonenshine, D. E. (1991). *Biology of Ticks*, vol. 1. Oxford, UK:
Oxford University Press.

Sonenshine, D. E., Roe, R. M., Venkatesh, K., *et al.* (1989).
Biochemical evidence of the occurrence of a juvenoid in
ixodid ticks. In *Host Regulated Developmental Mechanisms in
Vector Arthropods*, eds. Borovsky, D. & Spielman, A.,
pp. 9–17. Vero Beach, FL: University of Florida Press.

Stauffer, A. & Connat, J.-L. (1990). Anteroposterior gradient
during nymphal–adult moulting cycle of the tropical bont
tick, *Amblyomma variegatum* (Acarina: Ixodidae): correlations between ecdysteroid titers and integument structure.
Roux's Archives of Developmental Biology **198**, 309–321.

Strey, O. F., Teel, P. D. & Longnecker, M. T. (2001). Effects of
pyriproxyfen on off-host water-balance and survival of adult
lone star ticks (Acari: Ixodidae). *Journal of Medical
Entomology* **38**, 589–595.

Taneja-Bageshwar, S., Strey, A., Zubrzak, P., Pietrantonio,
P. V. & Nachman, R. J. (2006). Comparative
structure–activity analysis of insect kinin core analogs on
recombinant kinin receptors from southern cattle tick
Boophilus microplus (Acari: Ixodidae) and mosquito *Aedes
aegypti* (Diptera: Culicidae). *Archives of Insect Biochemistry
and Physiology* **62**, 128–140.

Taylor, D., Chinzei, Y., Ito, K., Higuchi, N. & Ando, K.
(1991*a*). Stimulation of vitellogenesis by pyrethroids in
mated and virgin female adults, and fourth instar females of
Ornithodoros moubata. *Journal of Medical Entomology* **28**,
322–329.

Taylor, D., Chinzei, Y., Miura, K. & Ando, K. (1991*b*).
Vitellogenin synthesis, processing and hormonal regulation
in the tick, *Ornithodoros parkeri* (Acari: Argasidae). *Insect
Biochemistry* **21**, 723–733.

Taylor, D., Nakajima, Y. & Chinzei, Y. (2000). Ecdysteroids
and vitellogenesis in the soft tick, *Ornithodoros moubata*
(Acari: Argasidae). In *Proceedings of the 3rd International
Conference on Ticks and Tick-Borne Pathogens: Into the 21st
Century*, eds. Kazimírova, M., Labuda, M. & Nuttall, P. A.,
pp. 223–227.

Taylor, D., Sonenshine, D. E. & Phillips, J. S. (1991*c*).
Ecdysteroids as a component of the genital sex pheromone
in two species of hard ticks, *Dermacentor variabilis* (Say) and
Dermacentor andersoni Stiles (Acari: Ixodidae). *Experimental
and Applied Acarology* **12**, 275–296.

Thompson, D. M., Khalil, S. M. S., Jeffers, L. A., *et al.*
(2005). *In vivo* role of 20-hydroxyecdysone in the regulation
of the vitellogenin mRNA and egg development in the
American dog tick, *Dermacentor variabilis* (Say). *Journal of
Insect Physiology* **51**, 1105–1116.

Thummel, C. S. (2001). Molecular mechanisms of developmental timing in C. *elegans* and *Drosophila*. *Developmental Cell* 1, 453–465.

Thummel, C. S. (2002). Ecdysone-regulated puff genes 2000. *Insect Biochemistry and Molecular Biology* 32, 113–120.

Tomaschko, K.-H. (1994). Ecdysteroids from *Pycnogonium litorale* (Arthropoda, Pantopoda) act as chemical defense against *Carcinus maenas* (Crustacea, Decapoda). *Journal of Chemical Ecology* 20, 1445–1455.

Venkatesh, K., Roe, R. M., Apperson, C. S., *et al.* (1990). Metabolism of juvenile hormone during adult development of *Dermacentor variabilis* (Acari: Ixodidae). *Journal of Medical Entomology* 27, 36–42.

Wainwright, G. & Rees, H. H. (2001). Hormonal regulation of reproductive development in crustaceans. In *Enviroment and Animal Development*, eds. Atkinson, D. & Thorndyke, M., pp. 71–84. Oxford, UK: Bios.

Weiss, B. L. & Kaufman, W. R. (2001). The relationship between 'critical weight' and 20-hydroxyecdysone in the female ixodid tick, *Amblyomma hebraeum*. *Journal of Insect Physiology* 47, 1261–1267.

Whitehead, D. L., Osir, E. W., Obenchain, F. D. & Thomas, L. S. (1986). Evidence for the presence of ecdysteroids and preliminary characterization of their carrier proteins in the eggs of the brown ear tick *Rhipicephalus appendiculatus* (Neumann). *Insect Biochemistry* 19, 112–133.

Wigglesworth, K. P., Lewis, D. & Rees, H. H. (1985). Ecdysteroid titre and metabolism of novel apolar derivatives

in adult female *Boophilus microplus* (Ixodidae). *Archives of Insect Biochemisty and Physiology* 2, 39–54.

Wright, J. E. (1969). Hormonal temination of larval diapause in *Dermacentor albipictus*. *Science* 163, 390–391.

Zhou, J., Liao, M., Hatta, T., *et al.* (2006). Identification of a follistatin-related protein from the tick *Haemophysalis longicornis* and its effect on tick oviposition. *Gene* 372, 191–198.

Zhu, X. X. & Oliver, J. H. Jr (2001). Cockroach allatostatin-like immunoreactivity in the synganglion (Acari: Ixodidae). *Experimental and Applied Acarology* 25, 1005–1013.

Zhu, X. X., Oliver, J. H. Jr & Dotson, E. M. (1991a). Immunocytochemical localization of an insulin-like substance in the synganglion of the tick, *Ornithodoros parkeri* (Acari: Argasidae). *Experimental and Applied Acarology* 13, 153–159.

Zhu, X. X., Oliver, J. H. Jr & Dotson, E. M. (1991b). Epidermis as the source of ecdysone in an argasid tick. *Proceedings of the National Academy of Sciences of the USA* 88, 3744–3747.

Zhu, X. X., Oliver, J. H. Jr, Dotson, E. M. & Ren, H. L. (1994). Correlation between ecdysteroids and cuticulogenesis in nymphs of the tick *Ornithodoros parkeri* (Acari: Argasidae). *Journal of Medical Entomology* 31, 479–485.

Zwijsen, A., Blockx, H., Van Arnhem, W., *et al.* (1994). Characterization of a rat C6 glioma-secreted follistatin-related protein (FRP): cloning and sequence of the human homologue. *European Journal of Biochemistry* 225, 937–946.

8 • Factors that determine sperm precedence in ticks, spiders and insects: a comparative study

W. R. KAUFMAN

INTRODUCTION

There is a broad tendency among female insects for fecundity to increase with mating frequency (Ridley, 1988). A female's eggs carry forward to the next generation her genes and those of her mate(s). Hence, one might guess that promiscuity should be of selective advantage for both sexes. On closer inspection it is not as simple as that. The survival of a female's offspring is as much a function of her mate's fitness as it is of her own, so she has some interest in determining the paternity of her offspring (Birkhead, 1998; Eberhard, 1998; Wedell & Karlsson, 2003). On the other hand, how can the male's paternity be assured if his spermatozoa have to compete with those of other males in the same arena? In this chapter some of the mechanisms used by male insects that assure their paternity in this 'promiscuous world' are reviewed. These will be compared to what is known from the less extensive literature on ticks, with the object of determining whether there are lessons from insects that might point our way to future enquiries in ticks.

SPERM PRECEDENCE AND PATERNITY

Because of the proverbial battle of the sexes, as well as that among competing males, one encounters numerous mating strategies among terrestrial arthropods, and a fascinating literature on sperm competition to match. The degree to which sperm competition occurs in any instance is determined by a number of factors (reviewed extensively by Simmons & Siva-Jothy, 1998, and by Parker, 1998): (1) How readily will a previously mated female accept further males (sperm preemption)? (2) How much mixing of sperm occurs in the spermathecae? (3) In copulating with a female, how readily can the male either flush out or otherwise remove previously deposited sperm, so as to improve the likelihood that his will be used for fertilization (sperm displacement)? (4)

Which of the following mechanisms are available to influence sperm precedence: sperm stratification (last sperm in has positional advantage), sperm loading (which male has provided the largest number of spermatozoa in the ejaculate) and sperm selection (non-random use of spermatozoa by the female)? (5) Are there substances in the male's ejaculate that can incapacitate the sperm of previous males? Finally, (6) what effort is made by the first male to avoid sperm competition altogether, such as by repeated copulation prior to oviposition, or by mate-guarding? Robust experiments have rarely been conducted in specific cases to distinguish among these theoretical mechanisms, even though they influence whether the first male's sperm or the nth male's sperm is more likely to fertilize the female (P1 and Pn, respectively).

Some studies have indicated that female mating strategy can vary even within a given species. In the green-veined white butterfly (*Pieris napi*), for example, some females are monandrous and others polyandrous, and there exists a genetic component to this variation in lifetime mating frequency (Wedell, Wilkund & Cook, 2001). The two conditions appear to coexist because of a balance between advantages and disadvantages associated with each condition. On the one hand, multiple spermatophores provide extra nutrition to the female, which increases longevity and lifetime fecundity. On the other hand, multiple mating incurs energetic costs, increased risks of predation/disease transmission, and a direct negative effect due to toxicity of excess seminal material (Wedell *et al.*, 2001). The latter authors discuss at length how the importance of these factors is influenced by environmental and other ecological conditions.

Ecological factors and the morphology of the female genital tract have a predominant influence on determining mating strategies (Thomas & Zeh, 1984). For example, some spiders possess conventional cul-de-sac spermathecae – simple diverticula of the common oviduct into which sperm enters for storage and leaves for fertilization via a common

Ticks: Biology, Disease and Control, ed. Alan S. Bowman and Patricia A. Nuttall. Published by Cambridge University Press.
© Cambridge University Press 2008.

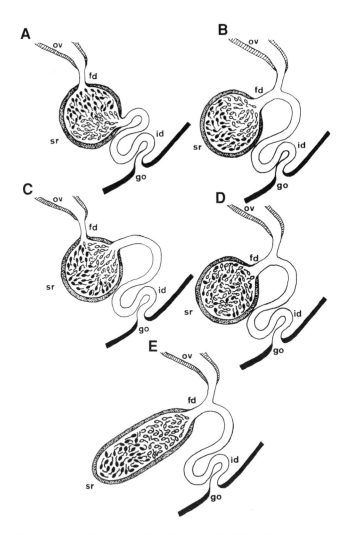

Fig. 8.1. Schematic representation of hypothetical relationships between the insemination duct, the seminal receptacle and the oviduct of various spiders. These depictions are designed to explain how morphology of the lower reproductive tract in the female can determine whether P1 or P2 paternity is favoured. (ov) oviposition duct, (fd) fertilization duct, (sr) seminal receptacle, (id) insemination duct, (go) genital opening. Black blobs in the seminal receptacle are spermatozoa from the first male to mate, and clear blobs are those from the second male. Cartoon A depicts a conduit-design spermathecal morphology that favours P1 paternity, because the first male's sperm encounter the ovulated eggs before those of the second male. The other cartoons (B to E) represent four distinct cul-de-sac design morphologies. In B, the sperm is stratified in the seminal receptacle such that P2 paternity is favoured. In C, the extent of stratification is ambiguous, so sperm precedence is not easily predicted. In D, sperm mixing occurs such that P1 and P2 are equally likely. In E, the ovoid seminal receptacle leads to sperm stratification and favours P2. Figure reprinted from Elgar (1998) with permission.

opening. Other spiders possess 'conduit' spermathecae – sperm enters for storage via one channel and leaves for fertilization via another (Elgar, 1998) (Fig. 8.1). Considering this diversity of spermathecal morphology in spiders, Austad (1984) predicted that first male sperm priority and the occurrence of mating plugs should be the dominant pattern among species with conduit spermathecae, while last male sperm priority or mixing of sperm should be most common in species with cul-de-sac spermathecae. Austad (1984) lists the spider families corresponding to the two generic spermathecal morphologies, and there is no overlap between these lists. But unfortunately, sperm priority has not been determined in more than a few species of spider, and for a number of reasons outlined by Austad, the data are not sufficiently reliable to test the hypothesis adequately. However, the bowl and doily spider (*Frontinella pyramitela*), with a conduit-type spermatheca, does show P1 precedence through a mechanism of remote mate-guarding. Subsequent males, arriving 24+ h after a female has first mated, readily engage in pre-insemination behaviour, but usually do not even attempt to transfer sperm (Austad, 1982). The nature of the signal conveyed by a mated female to discourage subsequent matings was not reported in this case. Austad (1982) speculated that sperm displacement would be difficult in spiders because of spermathecal morphology. In such a situation, insemination would be futile, and this might explain why males subsequent to the first one do not even attempt it. Huber (2005) has recently re-evaluated the extensive literature on sexual selection in spiders, with particular emphasis on the potential modalities for communication (visual, acoustic, chemical, vibratory and tactile signals).

There are a number of mechanisms whereby a male promotes his paternity. Among them are physiological effects of the reproductive accessory gland secretions.

REPRODUCTIVE ACCESSORY GLANDS

The female accessory glands

Some commonly stated functions of female accessory sex glands in insects are to synthesize material for the egg capsules, to produce the lipid waterproofing layer, to lubricate the egg while it moves through the ovipositor, and to digest the spermatophore by means of enzymes secreted into the reproductive tract (Gillott, 1988). In ticks, the female possesses three such glands. (1) A tubular accessory gland (argasid and ixodid ticks) produces a protein-rich secretion, the function of which is unknown (Diehl, Aeschlimann & Obenchain, 1982; Sonenshine, 1991). (2) A lobular accessory gland (ixodid ticks only). Lees & Beament (1948) demonstrated that it provides the initial waterproofing layer for the egg. Sonenshine (1991) suggests it is also the most likely source of the genital sex pheromone. (3) Gené's organ (found only in ticks) deposits wax on each egg as it is delivered from the vagina (Figs. 8.2 and 8.3). The wax coat waterproofs the eggs, causes them to stick together and probably protects them from soil fungi and microbes (Lees & Beament, 1948; Schöl *et al.*, 2001).

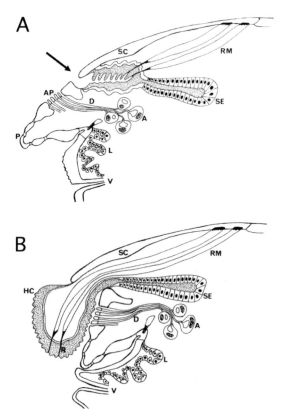

Fig. 8.2. Drawing of Gené's organ (sagittal section) in an engorged female *Boophilus microplus*, showing its relation to other anterior structures. (A) Normal appearance of Gené's organ (cuticular sac retracted). (B) Appearance of Gené's organ during oviposition (cuticular sac everted). The thick arrow points to the camerostomal fold (articulation between the capitulum and scutum: SC), the opening through which the cuticular sac everts. Other labels are: (A) accessory glands that secrete a component of the egg wax onto the external areae porosae (AP) via a number of ducts (D); (HC) horn cuticle of the cuticular sac that manipulates each egg as it is oviposited; (L) lobed gland cells which, according to Lees & Beament (1948), provide the initial waterproofing area of the egg; (P) palps; (R) cuticular rods that anchor the retractor muscles (RM) of Gené's organ to the horn cuticle; (SE) secretory epithelium of Gené's organ proper produces the bulk of the egg wax; (V) vagina. Figure reprinted from Booth *et al.* (1984), with permission from Elsevier.

The male accessory glands

Semen consists of spermatozoa bathed in a seminal fluid, the latter comprising secretions from the testis and secretions from the male accessory glands (MAGs). Commonly stated functions of MAG secretions in insects (reviewed by

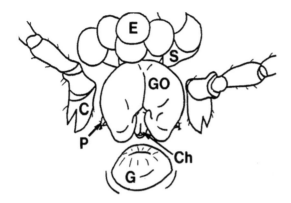

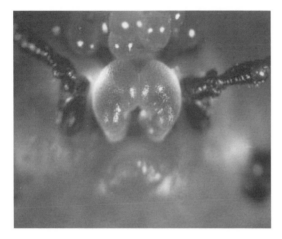

Fig. 8.3. Gené's organ in *Dermacentor reticulatus*; diagram above and photograph below. The figure shows the fully everted cuticular sac of Gené's organ (GO), showing the two prominent horns, just prior to receiving an egg about to be extruded via the gonopore (G). A few previously waxed eggs (E) can be seen above Gené's organ, lying on the scutum (S; part of the dorsal cuticle). The cheliceral digits (Ch) appear between the two horns of Gené's organ, and the tips of the pedipalps (P) can just be made out as a darkened area at the lower right of the coxa (C) of the first leg. Figure is taken from Sieberz & Gothe (2000), where the whole sequence of egg-laying can be seen in six similar figures. Figure reproduced with kind permission of Springer Science and Business Media.

Gillott, 1988, 1996, 2003) are: (1) to nourish/protect the spermatozoa within the male tract, (2) to secrete the spermatophore, (3) to alter the behaviour of the female in such ways as to (a) hasten oviposition and/or increase egg production and (b) attenuate mating behaviour in the female, (4) to secrete components of the mating plug, a coagulum which helps ensure P1 sperm precedence, (5) to secrete antibacte-

rial and antifungal agents which protect the gametes within the female reproductive tract (Lung, Kuo & Wolfner, 2001*b*), (6) to provide a nutritional supplement to the female (Friedel & Gillott, 1977), although Vahed (1998) comments that it is usually difficult to determine the extent to which spermatophore contents contribute to the overall nutritional budget of the female, and finally, (7) to extend the duration of sperm viability within the female genital tract (Tram & Wolfner, 1999; Chapman *et al.*, 2000), a function also suggested for ticks (Feldman-Muhsam, 1991).

Friedel & Gillott (1977) coined the terms 'fecundity-enhancing substances' (FES) and 'receptivity-inhibiting substances' (RIS) in reference to some MAG secretions. An FES is one that (directly or indirectly) increases egg production, oviposition or egg fertility. An RIS is one that renders the female less likely to copulate with further males, and thus constitutes the mechanism for remote mate-guarding. FESs and RISs are well documented for numerous insects (Gillott, 1988, 2003), and a few examples will be discussed in this chapter.

FECUNDITY-ENHANCING SUBSTANCES

The blood-sucking bug *Rhodnius prolixus*

In *Rhodnius*, transfer of a spermatophore influences fecundity in at least two ways: (1) by promoting migration of sperm to the spermathecae and (2) by stimulating ovulation and/or oviposition. Each of these effects is regulated by a distinct mechanism. (1) A spermatophore in the female's bursa copulatrix stimulates rhythmic contractions in the common oviduct, and within 5–10 min spermatozoa are found in the spermathecae (Fig. 8.4). The MAG comprises three transparent lobes (which produce the spermatophore) and one opaque lobe. Davey (1958) demonstrated that a secretion from the opaque lobe stimulates peristalsis in the common oviduct, thus effecting migration of sperm to the spermathecae (Fig 8.5). This MAG secretion is not itself a myotropin, but exerts its peristaltic effect on the oviduct via the nervous system (Davey, 1958). (2) The presence of sperm within the spermathecae leads to a doubling of the rate of oviposition compared to that of fed virgins (Davey, 1965*a*). Thus, if fed females are mated to males lacking either seminal vesicles (no sperm) or the opaque accessory gland lobes (sperm migration inhibited), the rate of oviposition equals that of normally fed virgins (Kuster & Davey, 1986). Likewise, the rate of oviposition in fed mated females surgically deprived of

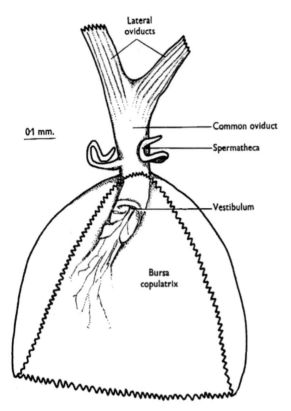

Fig. 8.4. Fig 8.4. Drawing of the lower female reproductive tract of *Rhodnius*. The bursa copulatrix (the functional vagina) receives the spermatophore. Spermatozoa released from the spermatophore are conveyed to the spermathecae, where they are stored, awaiting their turn to fertilize the eggs moving down the oviducts. Figure reproduced from Davey (1958) with the permission of the Company of Biologists, Cambridge, UK.

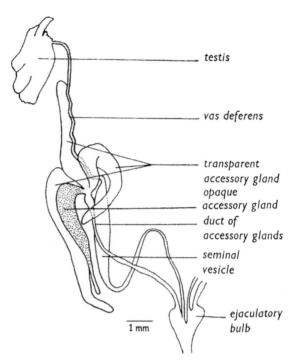

Fig. 8.5. Fig 8.5. The male reproductive tract of *Rhodnius*, showing the transparent and opaque lobes of the accessory gland. A secretion from the opaque lobe stimulates peristalsis in the common oviduct, thus effecting migration of sperm to the spermathecae. When they contain sperm, the spermathecae secrete a factor that stimulates ovulation and/or oviposition (Davey, 1965*a*; Kuster & Davey, 1986). Figure reproduced from Davey (1965*b*) with the permission of K. G. Davey and Pearson Education Ltd, Edinburgh, UK (1965).

their spermathecae is similar to that of normally fed virgins. Finally, fed virgins receiving implants of spermathecae from mated females lay significantly more eggs than fed virgins receiving implants of virgin spermathecae. Taken together, these experiments indicate that the spermathecae secrete a factor that stimulates ovulation and/or oviposition, but only when they contain sperm.

Further experiments (Davey, 1967) demonstrated that the spermathecal factor has its accelerating effect on oviposition by stimulating medial neurosecretory cells (mNSCs) in the brain to release a myotropin. This myotropin augments contractions in the ovarian sheath and lateral oviducts, and hence ovulation and oviposition follow (Kriger & Davey, 1982, 1983). Moreover, the release of the myotropin from

the mNSC. depends on the presence of ecdysteroids from the ovary (Ruegg *et al.*, 1981). Thus, the mNSCs require at least two inputs to release myotropin: a signal from mating (the spermathecal factor), and ovarian ecdysteroid. The myotropin of *Rhodnius* appears to be a peptide (~8.5 kDa) in the FMRF-amide family (Sevala *et al.* 1992).

Mosquitoes

Insemination produces profound effects in the female mosquito, including host-seeking, oviposition, receptivity and circadian flight activity (Klowden & Chambers, 1991). The MAGs of *Aedes aegypti* and other mosquitoes produce a secretion, originally named 'matrone' because, when injected into virgins, it elicits numerous characteristics of mated females (Fuchs, Craig & Hiss, 1968). According to earlier literature, matrone was composed of α- and β-fractions. The α-fraction alone stimulated oviposition, but both together were required to inhibit receptivity (Hiss & Fuchs, 1972; see further below). More recently, Klowden (1999) suggested that the term 'matrone' is now obsolete, since MAG extracts having matrone activity contain at least several distinct substances that may vary from one species to another. For example, heterologous MAG substances activate different subsets of behaviour in different species, and active material from different sources come in a wide range of molecular size.

A series of elegant experiments by Adlakha & Pillai (1975) demonstrated that MAG secretions are also necessary for normal fertilization to occur, perhaps by activating the sperm or by altering the egg surface so as to facilitate penetration of the sperm.

One important behavioural change attributed to MAG secretions in mosquitoes is a switch from swarming (= mating) to host-seeking flight behaviour (Gillott, 1988). The switch to host-seeking behaviour can be regarded ultimately as a fecundity-enhancing effect, because attaining a blood meal is required for producing a batch of eggs.

It has recently been shown in *Anopheles gambiae* that oviposition does not occur if the spermathecae of blood-fed females are removed either at emergence or after mating (Klowden, 2001). It is not known whether the spermathecal signal is neural or is a humoral factor, reminiscent of the system just described for *Rhodnius* (M. J. Klowden, personal communication). Whatever the nature of the spermathecal effect, it is not mimicked by implantation of a MAG (Klowden, 2001).

Drosophila

Most experimental studies on the MAGs of insects depend on delicate surgical techniques, and the implantation of tissues or injection of homogenates into the haemocoel (most often) or into the female genital tract (less often). Even though we have learned much from these studies, they obviously do not mimic the normal mating situation. In contrast, the relative ease of generating mutant strains of *Drosophila* which lack specific components of the MAG, or which do not produce sperm, enable an experimenter to study the effects of these genetic lesions under more natural conditions. Thus, it is not surprising that we know far more about the functions of MAG secretions in *Drosophila* than we do for any other insect. To date, no fewer than 83 gene products are predicted to be accessory gland proteins (Acps), all of them produced

in the 'main cells' (MC; 96% of total) of the MAG (Wolfner, 2002).

Extracts of the MAG, when injected into the haemocoel of virgin *Drosophila* females, markedly increase the number of eggs laid (Leahy & Lowe, 1967; Chen & Bühler, 1970). Kalb, DiBenedetto & Wolfner (1993) produced strains of flies in which males lacked spermatozoa or in which the MCs of the MAG were destroyed by directed cell ablation. They observed the effects on oviposition after selective cross-matings, and compared these to normal mated controls (average of 108 eggs in 5 days) and virgin controls (average of 12.2 eggs in 5 days). When mated with males producing neither spermatozoa nor MC secretions, females laid an average of 9.3 eggs in 5 days. When mated with males producing no spermatozoa (but having functional MCs), females laid an average of 40 eggs in 5 days. Thus, MC secretions cause only a partial, short-term fecundity-enhancing effect.

Acp62F shows sequence similarities to a neurotoxin (PhTx2-6) produced by a Brazilian spider, *Phoneutria nigriventer* (Wolfner *et al.*, 1997). PhTx2-6 prolongs action potentials and increases membrane excitability, suggesting that Acp62F may behave like the myotropin of *Rhodnius* and stimulate oviposition or regulation of sperm release from the seminal receptacle. Acp62F also has significant sequence homology to a class of serine protease inhibitors from the nematode *Ascaris* (Lung *et al.*, 2002).

RECEPTIVITY-INHIBITING SUBSTANCES

Mosquitoes

Female *Aedes aegypti* and other mosquitoes mate once only, and virgin females injected with MAG homogenates remain forever virgin (Craig, 1967). Extracts of *Aedes* MAG have at least some activity across 12 species of mosquitoes from three genera (Craig, 1967). In fact, even the sex peptide of *Drosophila* (Chen *et al.*, 1988) has biological activity in *Aedes* similar to that of α-matrone. Thus, injection into *Aedes* of extracts containing *Drosophila* sex peptide stimulates oviposition. In order to inhibit receptivity in *Aedes*, however, *Drosophila* sex peptide has to be injected together with β-matrone (Hiss & Fuchs, 1972). More recently, Klowden (2001) demonstrated that, in contrast to aedine and culicine mosquitoes, MAG substances from several anopheline mosquitoes do not control sexual receptivity as had been suggested earlier (e.g. Craig, 1967).

Drosophila

Whereas 95% of virgin females from a normal population will mate, only 3% of mated females will accept a male 1 day later (Kalb *et al.*, 1993). When mated with males producing neither spermatozoa nor MC secretions, 86% of the females mated again 1 day later. When mated with males producing MC secretions but no spermatozoa, 43% mated again 1 day later. These partially inhibitory effects were only short-lived, however, because 90% of females mated with sperm-less males remated 2–7 days later. Thus, stored spermatozoa are necessary for a long-term receptivity-inhibiting effect in *Drosophila*.

The grasshopper *Gomphocerus rufus*

The situation in *Gomphocerus* is unusual in that mating inhibits receptivity but does not increase the rate of egg production or oviposition (reviewed by Hartmann & Loher, 1999. A sexually immature female grasshopper will reject a male's attempt to copulate with her. This behaviour is called 'primary defence' (Hartmann & Loher, 1996). 'Secondary defence' is the rejection behaviour elicited by sexually mature females after they have mated (strong directed kicks at the male which make copulation impossible). Secondary defence endures until oviposition terminates, about 3–4 days following mating. Denervation of the spermathecae or severance of the ventral nerve cord immediately after mating abolishes secondary defence (Hartmann & Loher, 1996). Experiments by Hartmann & Loher (1996, 1999) have established that secondary defence is elicited by chemical and mechanical stimuli. The chemical stimulus comes from a specific MAG tubule (containing a white secretion), and the mechanical stimulus arises from stretch in the lateral oviducts as eggs move through them. Secondary defence in the grasshopper does not occur if the spermathecae are surgically removed prior to mating. This is because receptors for the white secretion (an identified group of sensory bristles) are found within the spermathecal bulb. In several insects discussed so far, on the other hand, the MAG secretion is effective even if injected into the haemocoel.

OTHER FUNCTIONS OF MAGs

Mating plugs

In numerous insects, a mating plug constitutes a physical barrier deposited in the female genital tract by the male. The evolutionary advantage of reduced competition from other males is clear. The spermatophore itself can serve as an effective mating plug, provided it remains in place until the first male's sperm has fertilized the eggs. But as Gillott (2003) points out, the spermatophore is frequently ejected or digested *in situ* within a few days. In ticks, the (ecto)spermatophore is not even internalized within the female, and the endospermatophore, when evaginating into the female, forms a capsule around the prospermia which follow closely behind (Feldman-Muhsam, 1986). These sperm capsules can remain intact for extended periods, at least in *Ornithodorus savignyi*, in which starved females still contained viable sperm within their capsules 1 year after mating (Feldman-Muhsam, 1986). However, it is doubtful that such sperm capsules are effective as mating plugs, because at least some ticks show P2 sperm preference (see below; Yuval & Spielman, 1990).

In many insects, the mating plug is a coagulum of the seminal fluid rather than the spermatophore. In *Drosophila*, a major component of the mating plug is a 38-kDa protein (PEB-me) from the ejaculatory bulb (Lung, Kuo & Wolfner, 2001*a*). The plug also contains at least one other Acp (Acp76A; see below).

Antimicrobial proteins

In numerous organisms, antibacterial and antifungal peptides constitute a major defence mechanism against pathogens. Male *Drosophila* transfer at least three antimicrobial peptides to the female during copulation, one of which, produced by the male ejaculatory duct, is the 28-kDa andropin. These peptides are believed to protect egg and sperm while in the female genital tract, for they are likely to be expelled during oviposition, several hours after mating (Lung *et al.*, 2001*b*). Although ticks do produce antimicrobial peptides (e.g. Johns, Sonenshine & Hynes, 1998, 2001; van der Goes van Naters-Yasui *et al.*, 2000), it is not known whether they are a component of the seminal fluid.

Protease inhibition and sperm storage

The seminal fluid of many organisms contains protease inhibitors, which affect male fertility (see Lung *et al.*, 2002, for references). Wolfner *et al.* (1997) identified genes encoding a dozen Acps in *Drosophila*, eight of which begin with putative secretion signals, and therefore are likely to be included in the semen. Acp76A, a 388-amino-acid proprotein, shows a sequence homology to the serpin class of protease inhibitors. Wolfner *et al.* (1997) hypothesized the following: Acp26Aa and Acp36DE may be activated in the female by means of regulated proteolytic action. It is conceivable that this proteolytic activation does not occur in the male genital tract because Acp76A inhibits proteolysis there. Acp76A itself may be inactivated following copulation, and hence female-derived proteases would be able to activate the other Acps. Acp76A may also play a role in forming the mating plug, because it can be recovered from plugs extruded by the female (Wolfner, 1997). Finally, Acps are also necessary for sperm to be stored in the spermathecae. Tram & Wolfner (1999) demonstrated that male *Drosophila*, deficient in MAG secretions through directed cell ablation, produced a normal amount of sperm and transferred sperm successfully. Females mated to these males, however, stored only 3–10% the amount of sperm after 6 h compared to those mated to control males. The specific peptide mediating this effect is Acp36DE (Chapman *et al.*, 2000). Sperm storage in *Rhodnius* is mediated by the spermathecal factor rather than a MAG secretion (Kuster & Davey, 1986). Ticks can also store sperm for extended periods: for a year or more in the case of *O. savignyi* (Feldman-Muhsam, 1986) and for at least several months in the case of *Amblyomma americanum* (Gladney & Drummond, 1971), although in neither case is the mechanism of long-term storage known.

With the foregoing as background, we now focus on more directed discussion of the feeding and mating strategies of ticks, and consider how information from various insect models might point the way to future studies.

There are two major families of ticks: Argasidae and Ixodidae. A third family (Nuttalliellidae) consists of a single species (*Nuttalliella namaqua*), about which very little is known, because only a handful of specimens have ever been found (in Namibia and Tanzania: Keirans *et al.*, 1976; Sonenshine, 1991). The Argasidae comprise five genera, in five corresponding subfamilies. The Ixodidae are divided into two phyletic lines: Prostriata and Metastriata, the former comprising a single genus (*Ixodes*), and the latter comprising 12 genera in four subfamilies (Sonenshine, 1991; and see Chapter 1). The major physiological difference between the prostriate and metastriate ticks, at least for our purposes, is that the former can mate off the host before taking a blood meal, whereas the latter mate only while on the host, and only after they have fed for at least a few days.

Feeding in ticks

In all haematophagous arthropods, the taking of a blood meal constitutes the signal for profound developmental and physiological changes. The blood-sucking insects *Rhodnius* and *Aedes* feed on a host for only a matter of minutes. The argasid tick *Ornithodorus moubata* feeds for about 30–60 min (Kaufman, Kaufman & Phillips, 1981, 1982). The latter species can lay a number of egg batches, each one requiring a blood meal. Ixodid ticks, on the other hand, remain attached to a host for 5–15 days, the specific duration depending on species, the developmental stage and other factors. Mating causes distinct behavioural and physiological changes in ixodid females (reviewed by Kaufman & Lomas, 1996; Lomas & Kaufman, 1999), and these will be considered further below.

The extended feeding period of female ixodid ticks is divided into three phases. (1) The preparatory feeding phase occupies the first day or two, when the female anchors itself in place by means of a cement-like substance, and the feeding lesion is formed. (2) The slow-feeding phase occurs over the subsequent 4–8 days, when the female mates, and increases its weight approximately 10-fold; mating normally occurs during this period. (3) The rapid-feeding phase occupies only about 24 h, during which the female increases its weight a further 10-fold and then detaches from the host.

The salivary glands (SGs) of *Amblyomma hebraeum* degenerate within 4 days following engorgement (Harris & Kaufman, 1981), a process controlled by 20-hydroxyecdysone (20E) (Harris & Kaufman, 1985; Kaufman, 1991; Mao & Kaufman, 1999). Oocyte maturation begins during the feeding period, and yolk synthesis (also controlled by 20E) and yolk uptake are under way by the 4th day post-engorgement. Oviposition begins around day 10, and continues for about 3 weeks, after which the female dies (Friesen & Kaufman, 2002).

The transition between the slow and rapid phases of feeding occurs at about 10-fold the unfed weight (named the 'critical weight' (CW) by Harris & Kaufman (1984)). Relative sizes of unfed, CW and fully engorged ticks are shown in Fig. 8.6. Females forcibly removed from the host below the CW do not lay eggs, but will reattach to a host if given the opportunity, and their SGs do not degenerate. Females forcibly removed from the host above the CW do lay eggs (the size of egg mass being proportional to the extent of engorgement), do not reattach to a host if given the opportunity, and their SGs degenerate (Kaufman & Lomas, 1996). Moreover, there are slightly different CWs for reattachment to the host, SG degeneration and egg development (Weiss & Kaufman, 2001).

Changes in sensory physiology as a result of feeding are probably responsible for the cessation of host-seeking behaviour in ticks above the CW. Anderson, Scrimgeour & Kaufman (1998) demonstrated that the ability of CO_2 to induce questing and walking in *A. hebraeum* changes

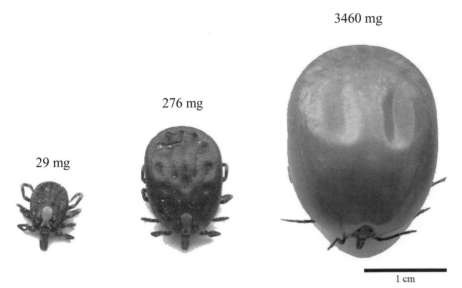

3460 mg

276 mg

29 mg

1 cm

Fig. 8.6. Photographs of *Amblyomma hebraeum* Koch showing three stages of feeding: unfed (29 mg); at approximately the CW (276 mg); and fully engorged (3460 mg). Scale bar = 1 cm. Photographs courtesy of Mr Alexander Smith, Department of Biological Sciences, University of Alberta, Canada.

dramatically with age, and feeding state. In general, ticks responded to CO_2 during periods in which they would normally be attracted to a host (e.g. when partially fed but below the CW), but were refractory to CO_2 at other times (e.g. when partially fed but above the CW, or when fully engorged).

MATING IN TICKS

The finding of a suitable mate involves a number of stereotyped behaviours that are triggered by pheromones. In ixodid ticks there are pheromones that promote assembly and aggregation–attachment, as well as a series of sex pheromones which cause attraction, mounting and probing the female genitals (see Chapter 21). Copulatory behaviour varies significantly with the group of ticks. The following very brief description is distilled from Feldman-Muhsam (1986) and Sonenshine (1991).

Sperm development

In the Metastriata, spermatogenesis (the production of haploid spermatids) begins in the nymph with the taking of a blood meal, and attains only the primary spermatocyte stage at the time of ecdysis to the adult.

When the adult feeds, spermatogenesis is completed and spermiogenesis follows. In ticks, the two phases of spermiogenesis are: (1) growth and elongation of spermatids to form

prospermia (the form in which sperm are packaged in the spermatophore and conveyed to the female), and (2) capacitation (precedes migration to, and fertilization of, the ova); capacitation occurs only after the prospermia are transferred to the female. Capacitation is triggered by a MAG substance which is discussed below in the section 'Mating factors in ticks'. After capacitation, the prospermium is called a spermiophore (synonymous with spermatozoon). A detailed description of spermiogenesis and spermatozoan structure is presented by El Said *et al.* (1981).

In contrast to the Metastriata, sperm development in the Argasidae and Prostriata advances to the prospermia stage by the time of adult ecdysis, thus accounting for these ticks being able to copulate prior to feeding.

Sperm transfer

The following pertains generally to argasid and ixodid ticks. Once the male and female are in the copulatory position, the male pushes its mouthparts through the female gonopore into the vagina, and may maintain this position for 1–2 h. At least one function for this probing behaviour is to detect the genital sex pheromone (Sonenshine, 1991; see Chapter 21). But perhaps this is not the only function because (1) the duration of the behaviour is so long and (2) probing occurs in both argasid and ixodid ticks, and only the latter are known to produce a genital sex pheromone. Whether or not the male can

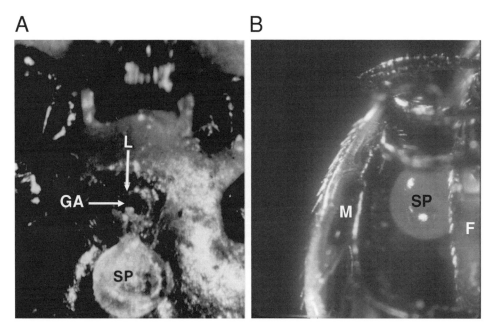

Fig. 8.7. (A) Ventral view of female *Dermacentor variabilis* with a spermatophore (SP) shown near the genital aperture (GA). The genital aperture of this experimental female had been occluded with a lacquer (L), which a male punctured with its chelicerae to insert the spermatophore. Figure reprinted from Sonenshine *et al.*, (1982) *Experimental Parasitology* **54**, 317–330, with permission from Elsevier. (B) Male and female *Ixodes ricinus in copulo*. The female (F) is shown with the lateral side facing the viewer and the male (M) is shown from a lateral–ventral perspective, with the spermatophore (SP) in between. The dark oval area (surrounding a white spot) situated immediately beneath the white 'M' is the male's right spiracle. Figure provided by courtesy of Univ.-Prof. Dr Franz-Rainer Matuschka, Charité–Universitätsmedizin Berlin. The two panels (A and B) are not magnified to the same scale.

use its mouthparts to remove sperm deposited by a previous male has not been reported, although the frequent occurrence of several spermiophore masses (from multiple males) in the female genital tract (see below under section 'Inhibition of receptivity in ticks') suggests that sperm removal does not occur. The spermatophore is then formed externally in less than a minute.

The ectospermatophore is a shell-like structure formed from a coagulum of mucopolysaccharides and proteins produced by the MAG (Feldman-Muhsam, 1986) (Fig. 8.7). Soon after the ectospermatophore is produced, the prospermia are ejaculated into it, followed immediately by a droplet containing yeast-like sperm symbionts (*Adlerocystis* spp.: Feldman-Muhsam, 1991). The last droplet secreted, the endospermatophore, seals off the ectospermatophore (Feldman-Muhsam, 1986).

Male ticks have no specialized copulatory organ, so their mouthparts push the tip of the spermatophore into the female's gonopore. While manipulating the spermatophore, the male also secretes saliva, which is believed to serve as a lubricant. Once the tip of the spermatophore has been inserted into the vagina, the endospermatophore evaginates into the female genital tract. There it forms capsules into which the prospermia and sperm symbionts are ultimately received. The generation of CO_2 within the ectospermatophore constitutes the force causing the evagination of the spermatophore contents into the female, although the mechanism of CO_2 production is not known (Feldman-Muhsam *et al.*, 1973).

MATING FACTORS OF MALE TICKS

A 'sperm capacitation factor' from argasid and ixodid ticks

As mentioned earlier, spermiogenesis comprises two phases: formation of the prospermia and capacitation (also called 'spermateleosis' by some authors), which takes place in the female. After copulation, an operculum at the end of each prospermium ruptures, and the prospermium evaginates to about double its original length (Sahli, Germond & Diehl, 1985) within about 60–90 min (Shepherd, Oliver & Hall, 1982). The latter authors studied capacitation of prospermia in hanging-drop cultures. They exposed prospermia of *Dermacentor variabilis* and *O. moubata* for 24 h to extracts of MAGs from fed males and found a detectable capacitation response with as little as 0.01–0.03 MAG per hanging drop (2–4 µl). The capacitation factor in both species was heat stable but destroyed by trypsin. Gel filtration revealed a single peak of activity at about 12.5 kDa for both species. In spite of the above similarities, the MAG extract of one species (tested at one concentration) was inactive on the prospermia of the other.

Because capacitation involves rupture of the operculum, Shepherd *et al.* (1982) hypothesized that the factor might be a proteolytic enzyme. However, treatment of prospermia of *O. moubata* with trypsin and pronase for several hours did not cause significantly more capacitation than observed in the controls. Moreover, these treated prospermia underwent normal capacitation when exposed to MAG extracts. They then exposed fresh prospermia to a MAG extract of *O. moubata* for 0–50 min and examined the cultures at 24 h. Percent elongation was then compared to that of similar cultures exposed to a MAG extract continuously for 24 h. As little as 2 min exposure to the MAG extract caused a significant degree of prospermial elongation (20%); elongation was 40% following 15 or 50 min exposure and was 97% after continuous exposure. Taken together, the results suggest that the capacitation factor acts more as a signalling molecule than as an enzyme. Only a brief exposure is necessary to trigger an effect which is not manifested until later. Capacitation of tick sperm is somewhat reminiscent of the acrosome reaction in sperm of vertebrates and marine invertebrates, which is triggered by Ca^{2+}, and Shepherd *et al.* (1982) speculated that a similar mechanism may apply to tick sperm.

A 'vitellogenesis-stimulating factor' (VSF) from argasid ticks

When virgin *O. moubata* feed, they undergo egg development for about a month, and then begin resorbing the yolk, a process called abortive vitellogenesis (Connat *et al.*, 1986). Within about 100 days after feeding, the ovary again resembles that of an unfed virgin (Sahli *et al.*, 1985; Connat *et al.*, 1986). If ticks are mated in the absence of a blood meal, oviposition does not occur. However, if a fed virgin is mated, even 4–5 months following the meal, eggs are produced and

laid; however, the egg mass is only about half that laid by females that mate at the time of feeding (Sahli *et al.*, 1985; Connat *et al.*, 1986).

Small metal beads inserted into the genital tracts of 'abortive virgins' stimulated a resumption of oogenesis, and many of these females oviposited an egg batch (only 30% smaller than normal) after a slightly prolonged pre-oviposition period (20–25 days: Connat *et al.*, 1986). Thus, in *O. moubata*, mechanical stimulation of the female's lower reproductive tract appears to be sufficient for a significant degree of egg development and oviposition.

The following summarizes the evidence suggesting that a chemical substance from the male (called VSF) also contributes to egg development. Homogenates of spermiophores taken from the seminal vesicles of *O. moubata* induce oviposition when injected into the haemocoels of abortive virgins (Ducommun, 1984); but homogenates of testis or MAGs do not have this effect (Connat *et al.*, 1986). Mature spermiophores, washed for 20 h in a physiological medium prior to injection, retain VSF activity, indicating that the factor is not a component of the testicular fluid. The wash medium, however, also contained some VSF activity.

Sahli *et al.* (1985) described the details of capacitation in *O. moubata* and proposed a potential secretory mechanism for VSF. Between 1 and 48 h after copulation, endospermatophores were dissected out of females, and the sperm masses were washed and incubated for 12 h in a physiological medium containing antibiotics. The sperm masses were centrifuged and the supernatants tested for VSF activity in a bioassay. No activity was found in the supernatant when the spermiophores were collected 1 h post-copulation; up to this time, evagination of the prospermia was incomplete. VSF was present in the supernatant 12–48 h post-copulation (capacitation complete), with the highest activity at 12 h. Also, VSF activity was destroyed by proteinase-K. Sahli *et al.* (1985) concluded that VSF is a protein synthesized by and contained within the prospermia while in the male, but liberated into the female genital tract during capacitation. The supernatants from the above experiments were subjected to electrophoresis; two proteins (~100 kDa and ~200 kDa) were particularly abundant in the 12-h supernatant. Sahli *et al.* (1985) proposed that these proteins constituted VSF.

A 'male factor' (MF) from ixodid ticks

The discovery of a protein (MF) from the male gonad of *Amblyomma hebraeum*, which exerts specific changes in female behaviour and physiology, arose from the following observations. Only a small minority of virgin females spontaneously feed beyond the CW, and none of these attain full engorgement even if left on a host for several weeks (Snow, 1969; Harris & Kaufman, 1984; Lomas & Kaufman, 1992a). SGs of mated or virgin females removed from the host below the CW do not degenerate within 14–21 days (Kaufman, 1983; Harris & Kaufman, 1984). SGs of mated females removed from the host above the CW degenerate within 4 days, whereas those of virgin females above the CW require 8 days to degenerate (Lomas & Kaufman, 1992a). Injecting an homogenate of male gonad (including or excluding the MAGs) into the haemocoel of virgin females above the CW induces SG degeneration within 4 (reduced from 8) days (Lomas & Kaufman, 1992a).

MF acts by stimulating the synthesis of 20E once females exceed the CW. Thus, injecting a male gonad homogenate into virgin females above the CW increases the haemolymph titre of 20E to levels characteristic of weight-matched, normal mated females (Lomas & Kaufman, 1992b). MF is a protein (heat labile and inactivated by proteinase-K: Lomas & Kaufman, 1992a), in the range 20 to 100 kDa (Kaufman & Lomas, 1996). MF is a testicular protein, because the MAG has no MF activity, and when partially purified homogenates of testis/vas deferens (T/VD) are layered on a discontinuous sucrose density gradient, MF activity is not associated with the spermatozoa, which sediment to the bottom (Lomas & Kaufman, 1992a). MF activity is found in abundance in the gonads of fed males, but is almost undetectable in the gonads of unfed males (Harris & Kaufman, 1984; Lomas & Kaufman, 1992a). MF activity is detectable in the haemolymph of mated females above and below the CW, but not in the haemolymph of virgin females of any size (Harris & Kaufman, 1984). It is not known whether the MF activity found in the haemolymph is due to MF itself or, analogous to the spermathecal factor of *Rhodnius* (Davey, 1967), to a substance secreted by the seminal receptacle after it receives a spermatophore.

An 'engorgement factor' (EF) from ixodid ticks

It has long been known that virgin females of ixodid ticks feed to only a fraction of the normal engorged weight range of mated females (e.g. Gregson, 1944; Sonenshine, 1967; Snow, 1969; Pappas & Oliver, 1971). Pappas and Oliver (1972) showed that female *D. variabilis* did not feed to repletion when exposed to males which had been irradiated or in which

the genital apertures had been blocked. These experiments suggested that an EF, produced by the male, promotes feeding to repletion.

The observations of Pappas & Oliver (1972) have been extended to *A. hebraeum* (Weiss *et al.*, 2002). We made a cDNA library from the T/VD portion of the male gonad and used a differential cross-screening approach to identify feeding-induced genes in this tissue. Thirty-five feeding-induced transcripts were identified, only two of which (AhT/VD16 and AhT/VD146) exhibited significant homologies to sequences in the GenBank. The predicted amino acid sequence of AhT/VD16 is 53% similar to human muscle-type acylphosphatase 2, and the predicted amino acid sequence of AhT/ VD146 is 44% similar to a 9-kDa basic protein (of unknown function) from *Drosophila*.

We have produced recombinant proteins (*rec*proteins) from those feeding-induced transcripts that had full open reading frames, and have identified EF among these *rec*proteins using a specific bioassay (Weiss & Kaufman, 2004); EF is neither AhT/VD16 nor AhT/VD146. Virgin females injected with EF engorged normally, achieved ovary weights significantly beyond those of control virgins (though less than those of normal mated females), and their SGs degenerated within 4 days. Virgin females injected with any other of the *rec*proteins did not feed beyond the CW, did not increase their ovary weights beyond those expected for small, partially fed virgins, and their SGs did not degenerate within 4 days.

On the following grounds, we hypothesize that MF and EF are the same protein and that they differ from the two other known male factors in ticks. (1) Both are found in homogenates of T/VD, whereas VSF is believed to be secreted by the prospermia and capacitation factor by the MAG. (2) Both are virtually undetectable in the gonads of unfed males, but are detectable at high levels in the gonads of fed males. (3) The molecular mass of MF (20–100 kDa) is different from that of the ixodid and argasid capacitation factors (12.5 kDa) and VSF (100–200 kDa). The molecular mass of natural EF has not yet been determined, but the combined molecular masses of *rec*AhEFα and *rec*AhEFβ is 27.7 kDa (Weiss & Kaufman, 2004). (4) As shown immediately above, *rec*EF (but none of the other *rec*proteins) also triggered SG degeneration and stimulated ovary growth – two known effects of MF. We proposed the name 'voraxin' for the natural EFs/MFs of ixodid ticks (from the Latin *vorax*, 'gluttonous', 'voracious').

Carvalho *et al.* (2006) demonstrated that meal size in *Drosophila* is also affected by a single MAG protein (Acp70, the 'sex peptide'); food intake in mated females is almost double that of virgins,

Site of action of the male factors

The sperm capacitation factor acts directly on the prospermia. The fact that VSF is effective when injected into the haemocoel suggests that the target tissue is outside the female genital tract. Whether VSF acts directly on the fat body (which synthesizes vitellogenin) or on the neuroendocrine system is not known. However, the synganglion contains at least several groups of NSCs, some of which appear to become active during egg development (Shanbaky, El-Said & Helmy, 1990). Also, homogenates of synganglion from mated fed females (but not from virgin fed or unfed) stimulate oogenesis, but not oviposition (Connat et al., 1986).

As MF in *A. hebraeum* stimulates ecdysteroid synthesis (Lomas & Kaufman, 1992b), and as ecdysteroid synthesis by the epidermis of *A. hebraeum* is controlled by a neuropeptide (Lomas, Turner & Rees, 1997), MF also may act via NSCs in the synganglion. Alternatively, MF might act directly on the epidermis to stimulate ecdysteroid synthesis. Finally, it is even possible that the EF effect of voraxin is mediated by NSCs and the MF effect via the epidermis.

DO PROSTAGLANDINS PLAY A ROLE IN TICK REPRODUCTION?

Prostaglandins (PGs) were originally detected in the semen of mammals, and are present there in astonishingly high concentration (20–25 µg/ml: Horrobin, 1978). Prostatic fluid stimulates sperm motility and sensitizes the uterine myometrium to seminal PGs in such a way as to cause the sperm to be sucked into the uterus (Mortimer, 1983). The pharmacological actions of PGs are far too diverse to consider here. Suffice it to say that they have direct actions on numerous smooth muscles, within and without the reproductive system, and modulate the effects of numerous other signalling molecules (Horrobin, 1978; Finn, 1983; Poyser, 1987). Thus PGs promote fecundity in mammals, including humans.

PGs also stimulate oviposition in the house cricket (*Acheta domesticus*) (Destephano, Brady & Farr, 1982). Inoculation of gravid, virgin females with PGE_2 or $PGF_{2\alpha}$ increased oviposition by the next day from 2 eggs/female (control) to 114 eggs/female (Destephano & Brady, 1977). Although the pharmacological effect of PGs implies that cricket ovaries can synthesize them, PG synthetase activity is absent from ovaries of virgin and mated females (Destephano & Brady, 1977). In contrast, mated female reproductive tracts show significant PG synthetase activity, as do testes, seminal vesicles/vas deferentia and spermatophores. The PG content of various tissues is even more revealing. The testes and spermatophores contain significant levels of PGs, as do the reproductive tracts of mated (but not virgin) females. PG titres in the mated female reproductive tract are much higher than can be accounted for by the contents of several spermatophores. These results indicate that mated females use the PG synthetase delivered by the spermatophore to synthesize PGs from arachidonic acid produced by the female. Support for this hypothesis comes from observations that egg-laying is inhibited by over 90% when acetaminophen (a PG synthetase inhibitor) is injected into normal mated females (Destephano & Brady, 1977).

The site of PG action (whether directly on the ovaries or via the neuroendocrine system) is not known. Destephano, Brady & Lovins (1974) suggested that PGs do not stimulate contractions in the ovarian sheath. However, the fact that as little as 2 ng PGE_2 injected directly into the common oviduct induces oviposition, whereas microgram quantities are required if injected into the haemocoel, suggests that PGs probably exert their effects within the female reproductive tract. If so, the site of action is not the spermatheca, because PGE_2 injected into the haemocoel still stimulates oviposition in spermathectomized females (Loher et al., 1981). More than one PG receptor probably exists in the oviduct of *Acheta*, because PGE_2 increases cyclic AMP production marginally, whereas $PGF_{2\alpha}$ inhibits octopamine-induced cyclic AMP production in the same tissue (A. Bowman, personal communication). The transfer of PG synthetase from male to female also occurs in the Australian field cricket (*Teleogryllus commodus*) (Loher et al., 1981).

Although a potential role for PGs in reproduction has been best elucidated in crickets, PGs have also been investigated in other insects (e.g. Yamaja Setty & Ramaiah, 1979, 1980; Lange, 1984; Wakayama, Dillwith & Blomquist, 1986; Brenner & Bernasconi, 1989; Medeiros et al., 2002).

Because PGs have pharmacological actions, all of which would promote the ability of ticks to feed (antihaemostatic, vasodilatory, immunosuppressive and anti-inflammatory), it is not surprising that most studies have concentrated on the role of PGs at the host–tick interface (Bowman, Dillwith & Sauer, 1996; Aljamali et al., 2002). PGs are synthesized in ticks from arachidonic acid imbibed in the blood meal, and are found in tick saliva of several species at concentrations 10–100-fold higher than in mammalian inflammatory exudates (Bowman *et al.*, 1996).

I am aware of only one study that reports the occurrence of PGs in the reproductive organs of ticks. Shemesh *et al.* (1979) measured the endogenous levels of PGF and PGE$_2$ in a number of tissues (testis, ovary and SGs from male and female *Hyalomma anatolicum excavatum*). Shemesh *et al.* (1979) also examined the ability of these tissues to synthesize PGs over 72 h using an organ-culture technique. Under these conditions, all tissues synthesized some PGF and PGE$_2$. The endogenous PG levels and amounts synthesized were similar in testis and ovary. However, whether PGs are incorporated within components of the spermatophore is not known (Sonenshine, 1991).

Oliver, Pound & Andrews (1984) demonstrated in male *Ornithodorus parkeri* that SG extracts, injected into the haemocoel of fed virgins, stimulated some of them to oviposit. When injected into the vagina, however, SG extracts stimulated some degree of oocyte maturation but not oviposition. As mentioned earlier, it has long been assumed that males salivate during copulation in order to lubricate the spermatophore (Feldman-Muhsam, 1986), but the results of Oliver *et al.* (1984) raise the possibility that, perhaps because of its PG content, male saliva may also play a role in oocyte development.

Pappas & Oliver (1972) blocked the gonopores of male *D. variabilis* with wax and put them with feeding virgins. Although the males could not produce spermatophores, they still probed the female gonopore with their mouthparts and presumably could still secrete saliva. The females did not engorge fully, but they fed to a weight (mean of 160 mg) significantly beyond that of control virgins not exposed to males (mean of 37 mg). They also eventually laid small, infertile egg batches. The extent to which this enhanced degree of feeding and egg development was due to the mechanical stimuli of male–female contact, or to the secretion of saliva into the female's reproductive tract, was not tested. Recall from above, however, that mechanical stimulation of the female genital tract of abortive virgin *O. moubata* is sufficient to trigger egg development and oviposition (Connat *et al.*, 1986). In *H. a. excavatum*, endogenous amounts of PGs in the SGs and the ability of these ticks to synthesize PGs were significantly higher in fed females than in fed males (Shemesh *et al.*, 1979). These observations are not inconsistent with a role for male-derived PGs in reproduction for the following reason: the latter authors neglected to normalize their PG values for amount of tissue. The SGs of fed females (*A. hebraeum*) are about 10 times heavier than those of males (Kaufman, 1976), so if a similar difference holds for *Hyalomma*, the PG concentration in male SGs would be significantly higher than that in female glands.

Inhibition of receptivity in ticks

The weight of current evidence indicates that inhibition of receptivity does not occur in mated female ticks. As early as 1948, Lees & Beament reported finding up to 12 sperm capsules in the genital tract of *O. moubata* females. When a number of virgin *O. moubata* were kept for 24 h with five males each, the females engaged in multiple copulations, in most cases with several partners, and *Ornithodorus tholozani* females are said to be able to copulate even while ovipositing (Feldman-Muhsam, 1986). Most argasid ticks require a meal in order to produce each batch of eggs. They will readily copulate before or after each blood meal, although only the initial copulation is necessary to produce several batches of fertile eggs (Aeschlimann & Grandjean, 1973; Connat *et al.*, 1986).

Ixodid females produce only a single large batch of eggs and then die, so they probably require only a single copulation. Although the seminal receptacle may contain multiple sperm capsules (Feldman-Muhsam, 1986), it is not clear whether these come from the same or different males. The following study, however, demonstrates that ixodid females can also mate with multiple partners.

Sperm precedence in ticks

Yuval & Spielman (1990), taking advantage of the fact that prostriate ticks are able to mate prior to feeding (preprandial) as well as during feeding (perprandial), conducted a study on sperm precedence in *Ixodes dammini* (more commonly now named as *I. scapularis*). A group of males were sterilized by exposure to cobalt-60 irradiation. When 22 such males were fed in the company of females, none of the laid eggs hatched. One group of virgin females was then kept (off the host) with irradiated males and another group with normal males for 2 weeks. Following this, the females were placed on a host, but this time in the company of the opposite group of males. Thus, females mated at least twice, each time with a different male. The results indicated that mated females remain receptive to subsequent males, and that sperm from a second mating takes precedence over that of the first. Thus, when the first male was irradiated and the second normal, more than 75% of the eggs of 78% of the ovipositing females hatched, whereas fewer than 25% of the eggs of the remaining females hatched. However, when the first male was normal and the second irradiated, fewer than 25% of the eggs of 86% of the ovipositing females hatched, and all of the eggs of the remaining females hatched.

Yuval & Spielman (1990) also showed that irradiated male *I. scapularis* can copulate normally. Twenty-five virgin females were placed on each ear of a rabbit with 20 normal plus 20 irradiated males. Of the egg batches laid by the engorged females, 49% showed no hatching, in 43% the hatch was greater than 75% and in the remaining 8% the hatch was 25–50%, results to be expected if all males were equally competitive. Irradiated males of *Argas persicus* likewise compete successfully for females (Sternberg, Peleg & Galun, 1973).

The results of Yuval & Spielman (1990) suggest that a male *I. scapularis* engaging in preprandial mating will be successful in passing his genes to the next generation only if the female, on finding a host, does not encounter another male. Otherwise, his paternity will pass to only a minority of her eggs. (By the same token, the female benefits from preprandial mating whether or not she encounters another male on the host.) Also, the viability of sperm in the female reproductive tract falls progressively over 8 weeks (Gladney & Drummond, 1971). The foregoing arguments perhaps explain Yuval & Spielman's (1990) observation that the incidence of insemination in female *I. scapularis* collected from vegetation in the wild, though significant (30–72% had copulated preprandially), was somewhat lower than that in females collected from deer (90–100% had copulated perprandially). These results also explain the male's tendency to remain in close contact with the female throughout the feeding period (an example of proximate mate-guarding): his physical proximity would preclude the advances of other males (Kiszewski, Matuschka & Spielman, 2002). A similar situation occurs in the male desert locust which, during the gregarious phase, breeds under crowded conditions. The male remains on the female throughout the period of oviposition, for if the pair is separated before then, the female will accept another male immediately (Seidelmann & Ferenz, 2002).

But subsequent work has shown that proximate mate-guarding is not the only strategy available to male *I. scapularis*. Males that mount previously inseminated females attempt to copulate with them for much less time than they attempt with virgin females, irrespective of whether the males and females encounter each other in vials or on rabbits. Moreover, successful insemination is much less frequent when the female has been inseminated previously (Kiszewski & Spielman, 2002). These inhibitory effects lasted for at least 60 days, though they did wane gradually. However, the inhibitory effect of prior insemination on male copulatory attempts was markedly reduced in partially fed females, effectively disappearing after about 5 days of feeding. In contrast, inhibition of spermatophore transfer by such males did not wane as a function of female feeding duration (Kiszewski & Spielman, 2002). The latter authors also demonstrated a potential reproductive advantage of preprandial insemination. Preprandially inseminated females, when subsequently fed in the company of males, contained double the amount of sperm after full engorgement compared to females not inseminated preprandially.

The mechanism(s) behind these inhibitory effects have not been determined, although Kiszewski & Spielman (2002) offer some speculations. Males do not appear to alter their behaviour until after inserting their mouthparts into the female gonopore; perhaps it is the detection of a spermatophore that causes them to disengage more rapidly. Or prior insemination may attenuate secretion of the genital sex pheromone, or of some other substance detected by the cheliceral digits of the male as they probe the female gonopore. The latter possibilities are not mutually exclusive, of course. Although most attention has been paid to the male's behaviour, it is certainly possible that disengagement is favoured by some change in the female's mating behaviour. For example, more than twice as many inseminated females, observed in vials, seemed to avoid copulations of any duration compared to uninseminated females (Kiszewski & Spielman, 2002).

Voraxin (EF/MF) is clearly an FES, because egg production is much reduced in females that exceed the CW but that do not engorge (Kaufman & Lomas, 1996). Voraxin and the VSF in argasid ticks stimulate vitellogenesis (Connat *et al.*, 1986; Weiss & Kaufman, 2004), and following stimulation, some days are required for oviposition to occur. Most of the insect FESs described earlier stimulate ovulation and/or oviposition, and so their effects are often manifested within only a matter of hours. A notable exception was the effect of MAG extracts on mosquitoes, which also act at the level of vitellogenesis (Borovsky, 1985; Klowden & Chambers, 1991). Does this suggest that the blood-sucking lifestyle may be a factor in determining the level (vitellogenesis vs. ovulation or oviposition) at which mating factors act? A few prominent exceptions suggest otherwise. We have already seen that in *Rhodnius*, the male's influence is not manifested at the level of vitellogenesis but at ovulation and/or oviposition

(Davey, 1967; Kriger & Davey, 1982, 1983). Also, there are at least a few instances among non-haematophagous insects from several orders, including *Drosophila*, in which MAG secretions promote egg development, apparently by stimulating juvenile hormone synthesis by the corpora allata (reviewed by Gillott, 2003).

FUTURE DIRECTIONS

Although P2 precedence has been demonstrated in an argasid (Sternberg *et al.*, 1973) and an ixodid tick (Yuval & Spielman, 1990), the mechanism (sperm stratification, sperm removal by probing mouthparts or sperm incapacitation) has not been explored. In at least one tick (*A. americanum*), sperm from fed males remained viable within the unfed female for up to 8 weeks, although viability declined progressively over that time (Gladney & Drummond, 1971). The mechanisms for sperm storage in ticks are unknown, and the viability of such sperm in partially fed females is not known. But consider that a mated female can be dislodged from a host in the partially fed state by grooming or by a variety of other means. If it is below the CW after removal, it will be able to seek another host to complete feeding (Kaufman & Lomas, 1996; Lomas & Kaufman, 1999). For how long can the sperm already received remain viable in relation to the time probably needed to find another host? As mentioned earlier, Feldman-Muhsam (1986) reported seeing mobile sperm in *O. savignyi* dissected 1 year following copulation, so sperm probably remain viable for extended periods.

Recall that the spermathecal factor of *Rhodnius* (Kuster & Davey, 1986) and Acp36DE of *Drosophila* (Chapman *et al.*, 2000) promote sperm viability within the spermatheca. The yeast-like sperm symbiont *Adlerocystis* spp. potentially has a similar role in ticks (Feldman-Muhsam, 1991). *Adlerocystis* develops in the posterior lobes of the MAG and is packaged in the spermatophore, separate from the sperm. The symbiont encounters the sperm only after copulation, and in argasid species (though not in ixodid species) the symbiont attaches to the spermiophores over the next day or so (Feldman-Muhsam, 1991). A unique observation of two sperm capsules within a single female *O. tholozani*, one with dead sperm (and lacking *Adlerocystis*), and the other with live sperm (containing *Adlerocystis*) suggested that the symbiont might promote sperm longevity in the female (Feldman-Muhsam, 1991). Most of these observations were made in the 1960s and 1970s, and should be extended.

It is not known whether the seminal fluid of ixodid males contains a myotropin, analogous to that of *Rhodnius*, which triggers ovulation and/or oviposition. That seems unlikely, however, because in ticks, copulation normally occurs long before vitellogenesis and oviposition. For example, Mao & Kaufman (1999) demonstrated that 69% of female *A. hebraeum* had mated by the 5th day of feeding. In this population of ticks, the normal time to engorgement was about day 12, the initiation of vitellogenesis was about day 16 and the beginning of oviposition was about day 22. However, tick semen may contain a myotropic peptide for some other function (e.g. for the transport of sperm within the genital tract, as occurs *in Rhodnius*: Davey, 1958).

The role for NSCs in the control of egg development has been demonstrated in several ticks, and the stimulation of specific NSCs by feeding has been well documented (reviewed by Kaufman, 1997; Chang & Kaufman, 2005). It is less clear, however, whether mating also plays a role in NSC activation, because in most previous studies, the synganglia extracts tested for stimulating egg development and/or oviposition were taken from fed, mated females.

The role for PG in reproduction, irrespective of its source (the male's saliva or its spermatophore, or endogenous to the female), has not yet been adequately explored in ticks. Kaufman & Lomas (1996) considered the possibility that MF (voraxin) might be a PG or PG synthetase. But PGs infused into the haemocoel of virgin females above the CW did not reduce salivary fluid secretion, as would have been expected for a substance having MF activity. Quite the contrary, $PGF_{2\alpha}$ and PGE_2 both significantly increased salivary fluid secretion, which is not surprising considering that PGE_2 also stimulates protein secretion by SGs (Sauer, Essenberg & Bowman, 2000). Even though voraxin is unlikely to be a PG synthetase, PGs may play other roles in reproduction.

Does the seminal fluid of ticks contain antimicrobial substances? Tick haemolymph contains defensin (Johns *et al.*, 1998, 2001; van der Goes van Naters-Yasui *et al.*, 2000) but the extent to which it appears in various secretions is not known. A substance in tick egg wax inhibits bacterial growth (Arrieta, Leskiw & Kaufman, 2006). This substance might be secreted by Gené's organ (the major source of the egg wax), although it remains to be seen whether it is a defensin (peptide) or a sterol amide-like substance similar to boophiline (Potterat *et al.*, 1997).

The morphology and histology of the MAGs have been described for several ticks (e.g. Khalil, 1970). The MAG is probably the largest gland in the male body, and consists of

14 lobes in *Ornithodorus* (*Pavlovskyella*) *erraticus*, showing characteristic histological differences (El Shoura, 1987) but, as outlined in this review, very little is known about the variety of substances produced by tick MAGs.

Proximate mate-guarding has been well established in ticks, and this seems understandable considering the crowded conditions under which they often feed. Moreover, Wang *et al.* (1998) demonstrated that male *Rhipicephalus appendiculatus* secrete an immunoglobulin-binding protein (IgBP) into the feeding site shared with the female. This IgBP leads to a 24% enhancement of female engorged weight, which in turn would lead to a larger egg batch carrying his genes. But the existence of mating plugs (unlikely), or of putative pheromones inhibiting rival males, has not been adequately investigated.

The VSF of *Ornithodoros* and the voraxin of *A. hebraeum* both stimulate fecundity. However, the differences between them (voraxin acts by stimulating engorgement, VSF does not; voraxin is in the testicular fluid, VSF is produced in the prospermia; their molecular sizes are different) may not be as substantial as appear at first sight. Both may be transported from the seminal receptacle to the haemolymph, and a reasonable hypothesis for both is that they act at the synganglion. Molecular homology between the two factors should be investigated.

In conclusion, this comparative view of male factors that affect female reproductive behaviour and physiology indicates that less is known about the male tick's influences on the female than is known for several insects. Many of the most recent citations on male tick reproductive physiology referred to here are 10 or more years old, a sad reflection of how little attention has been devoted to this topic in recent years. However, we know enough already to formulate some interesting questions that can be easily answered with current biochemical and molecular biological techniques. Let us hope that more experimenters will answer the call.

ACKNOWLEDGEMENTS

I am most grateful to Dr Cedric Gillott, Department of Biology, University of Saskatchewan, Canada, who thoroughly reviewed an early draft of the manuscript and recommended numerous, substantial revisions. I also thank Dr Brian Weiss, Department of Epidemiology and Public Health, Yale School of Medicine, Dr A. S. Bowman, Department of Zoology, University of Aberdeen, UK and Dr P. A. Diehl, Institut de Zoologie, Université de Neuchâtel, Switzerland, for their valuable insights on various sections of the manuscript. I am most grateful to Dr Tim Booth, Director, Viral Diseases Division, National Microbiology Laboratory, Public Health Agency of Canada, Winnipeg, Manitoba, who provided me with the scanned images shown in Fig. 8.2, Mr Alexander Smith, Department of Biological Sciences, University of Alberta, Canada, who provided me with the images shown in Fig. 8.6, and Univ.-Prof. Dr Franz-Rainer Matuschka, Charité–Universitätsmedizin Berlin, Germany, for providing me with the image shown in Fig. 8.7B. Research in my laboratory is generously supported by the Natural Sciences and Engineering Research Council (NSERC) of Canada.

REFERENCES

Adlakha, V. & Pillai, M. K. K. (1975). Involvement of male accessory gland substances in the fertility of mosquitoes. *Journal of Insect Physiology* **21**, 1453–1455.

Aeschlimann, A. A. & Grandjean, O. (1973). Observations on fecundity in *Ornithodoros moubata* Murray (Ixodoidea: Argasidae): relationships between mating and oviposition. *Acarologia* **15**, 206–217.

Aljamali, M., Bowman, A. S., Dillwith, J. W., *et al.* (2002). Identity and synthesis of prostaglandins in the lone star tick, *Amblyomma americanum* (L.), as assayed by radio-immunoassay and gas chromatography/mass spectrometry. *Insect Biochemistry and Molecular Biology* **32**, 331–341.

Anderson, R. B., Scrimgeour, G. J. & Kaufman, W. R. (1998). Responses of the tick, *Amblyomma hebraeum* (Acari: Ixodidae), to carbon dioxide. *Experimental and Applied Acarology* **22**, 667–681.

Arrieta, M. C., Leskiw, B. K. & Kaufman, W. R. (2006). Antimicrobial activity in the egg wax of the African cattle tick *Amblyomma hebraeum* (Acari: Ixodidae). *Experimental and Applied Acarology* **39**, 297–313.

Austad, S. N. (1982). First male sperm priority in the bowl and doily spider, *Frontinella pyramitela* (Walckenaer). *Evolution* **36**, 777–785.

Austad, S. N. (1984). Evolution of sperm priority patterns in spiders. In *Sperm Competition and the Evolution of Animal Mating Systems*, ed. Smith, R. L., pp. 223–249. San Diego, CA: Academic Press.

Birkhead, T. R. (1998). Cryptic female choice: criteria for establishing female sperm choice. *Evolution* **52**, 1212–1218.

Booth, T. F., Beadle, D. J. & Hart, R. J. (1984) Ultrastructure of the accessory glands of Gené's organ in the cattle tick, *Boophilus*. *Tissue and Cell* **16**, 589–599.

Borovsky, D. (1985). The role of male accessory gland fluid in stimulating vitellogenesis in *Aedes taeniorhynchus*. *Archives of Insect Biochemistry and Physiology* **2**, 405–413.

Bowman, A. S., Dillwith, J. W. & Sauer, J. R. (1996). Tick salivary prostaglandins: presence, origin and significance. *Parasitology Today* **12**, 388–396.

Brenner, R. R. & Bernasconi, A. (1989). Prostaglandin biosynthesis in the gonads of the hematophagus insect *Triatoma infestans*. *Comparative Biochemistry and Physiology B* **93**, 1–4.

Carvalho, G. B., Kapahi, P., Anderson, D. J. & Benzer, S. (2006). Allocrine modulation of appetite by the sex peptide of *Drosophila*. *Current Biology* **16**, 692–696.

Chang, E. S. & Kaufman, W. R. (2005). *Endocrinology* of Crustacea and Chelicerata. In *Comprehensive Molecular Insect Science*, vol 3, *Endocrinology*, eds. Gilbert, L. I., Iatrou, K. & Gill, S. S., pp. 805–842. Amsterdam: Elsevier.

Chapman, T., Neubaum, D. M., Wolfner, M. F. & Partridge, L. (2000). The role of male accessory gland protein Acp36DE in sperm competition in *Drosophila melanogaster*. *Proceedings of the Royal Society of London B* **267**, 1097–1105.

Chen, P. S. & Bühler, R. (1970). Paragonial substance (sex peptide) and other free ninhydrin-positive components in male and female adults of *Drosophila melanogaster*. *Journal of Insect Physiology* **16**, 615–627.

Chen, P. S., Stumm-Zollinger, E., Aigaki, T. *et al.* (1988). A male accessory gland peptide that regulates reproductive behaviour of female *D. melanogaster*. *Cell* **54**, 291–298.

Connat, J.-L., Ducommun, J., Diehl, P.-A. & Aeschlimann, A. A. (1986). Some aspects of the control of the gonotrophic cycle in the tick, *Ornithodoros moubata* (Ixodoidea, Argasidae). In *Morphology, Physiology and Behavioural Biology of Ticks*, eds. Sauer, J. R. & Hair, J. A., pp. 194–216. Chichester, UK: Ellis Horwood.

Craig, G. B. Jr (1967). Mosquitoes: female monogamy induced by male accessory gland substance. *Science* **156**, 1499–1501.

Davey, K. G. (1958). The migration of spermatozoa in the female of *Rhodnius prolixus* Stahl. *Journal of Experimental Biology* **35**, 694–701.

Davey, K. G. (1965*a*). Copulation and egg production in *Rhodnius prolixus*: the role of the spermathecae. *Journal of Experimental Biology* **42**, 373–378.

Davey, K. G. (1965*b*) *Reproduction in the Insects*. Edinburgh, UK: Oliver & Boyd.

Davey, K. G. (1967). Some consequences of copulation in *Rhodnius prolixus*. *Journal of Insect Physiology* **13**, 1629–1636.

Destephano, D. B. & Brady, U. E. (1977). Prostaglandin and prostaglandin synthetase in the cricket *Acheta domesticus*. *Journal of Insect Physiology* **23**, 905–911.

Destephano, D. B., Brady, U. E. & Farr, C. A. (1982). Factors influencing oviposition behaviour in the cricket *Acheta domesticus*. *Annals of the Entomological Society of America* **75**, 111–114.

Destephano, D. B., Brady, U. E. & Lovins, R. E. (1974). Synthesis of prostaglandins by reproductive tissue of the house cricket, *Acheta domesticus*. *Prostaglandins* **6**, 71–79.

Diehl, P. A., Aeschlimann, A. A. & Obenchain, F. D. (1982). Tick reproduction: oogenesis and oviposition. In *Physiology of Ticks*, eds. Obenchain, F. D. & Galun, R., pp. 277–350. Oxford, UK: Pergamon Press.

Ducommun, J. (1984). Contribution à la connaissance de la reproduction chez la tique *Ornithodoros moubata*, Murray, 1877; *sensu* Walton, 1962 (Ixodoidea: Argasidae). Unpublished Ph.D. thesis, University of Neuchatel, Switzerland.

Eberhard, W. G. (1998). Female roles in sperm competition. In *Sperm Competition and Sexual Selection*, eds. Birkhead, T. R. & Moller, A. P., pp. 91–116. San Diego, CA: Academic Press.

Elgar, M. A. (1998). Sperm competition and sexual selection in spiders and other arachnids. In *Sperm Competition and Sexual Selection*, eds. Birkhead, T. R. & Moller, A. P., pp. 307–339. San Diego, CA: Academic Press.

El Said, A., Swiderski, Z., Aeschlimann, A. & Diehl, P. A. (1981). Fine structure of spermiogenesis in the tick *Amblyomma hebraeum* (Acari: Ixodidae): late stages of differentiation and structure of the mature spermatozoon. *Journal of Medical Entomology* **18**, 464–476.

El Shoura, S. (1987). Fine structure of the vasa deferentia, seminal vesicle, ejaculatory duct and accessory gland of male *Ornithodoros* (*Pavlovskyella*) *erraticus* (Acari: Ixodoidea: Argasidae). *Journal of Medical Entomology* **24**, 235–242.

Feldman-Muhsam, B. (1986). Observations on the mating behaviour of ticks. In *Morphology, Physiology and Behavioural Biology of Ticks*, eds. Sauer, J. R. & Hair, J. A., pp. 217–232. Chichester, UK: Ellis Horwood.

Feldman-Muhsam, B. (1991). The role of *Adlerocystis* sp. in the reproduction of argasid ticks. In *The Acari: Reproduction, Development and Life-History Strategies*, eds. Schuster, R. & Murphy, P. W., pp. 179–190. London: Chapman & Hall.

Feldman-Muhsam, B., Borut, S., Saliternik-Givant, S. & Eden, C. (1973). On the evacuation of sperm from the spermatophore of the tick *Ornithodoros savignyi*. *Journal of Insect Physiology* **19**, 951–962.

Finn, C. A. (ed.) (1983). *Oxford Reviews of Reproductive Biology*, vol. 5. Oxford UK: Clarendon Press.

Friedel, T. & Gillott, C. (1977). Contribution of male-produced proteins to vitellogenesis in *Melanoplus sanguinipes*. *Journal of Insect Physiology* 23, 145–151.

Friesen, K. & Kaufman, W. R. (2002). Quantification of vitellogenesis and its control by 20-hydroxyecdysone in the ixodid tick, *Amblyomma hebraeum*. *Journal of Insect Physiology* 48, 773–782.

Fuchs, M. S., Craig, G. B. Jr & Hiss, E. A. (1968). The biochemical basis of female monogamy in mosquitoes. I. Extraction of the active principle from *Aedes aegypti*. *Life Sciences* 7, 835–839.

Gillott, C. (1988). Arthropoda–Insecta. In *Reproductive Biology of Invertebrates*, vol. 3, *Accessory Sex Glands*, eds. Adiyodi, K. G. & Adiyodi, R. G., pp. 319–471. New York: John Wiley.

Gillott, C. (1996). Male insect accessory glands: functions and control of secretory activity. *Invertebrate Reproduction and Development* 30, 199–205.

Gillott, C. (2003). Male accessory gland secretions: modulators of female reproductive physiology and behavior. *Annual Reveiw of Entomology* 48, 163–184.

Gladney, W. J. & Drummond, R. O. (1971). Spermatophore transfer of lone star ticks off the host. *Annals of the Entomological Society of America* 64, 379–381.

Gregson, J. D. (1944). The influence of fertility on the feeding rate of the female of the Wood Tick, *Dermacentor andersoni* Stiles. *Proceedings of the Entomological Society of Ontario* 74, 46–47.

Harris, R. A. & Kaufman, W. R. (1981). Hormonal control of salivary gland degeneration in the ixodod tick *Amblyomma hebraeum*. *Journal of Insect Physiology* 27, 241–248.

Harris, R. A. & Kaufman, W. R. (1984). Neural involvement in the control of salivary gland degeneration in the ixodid tick *Amblyomma hebraeum*. *Journal of Experimental Biology* 109, 281–290.

Harris, R. A. & Kaufman, W. R. (1985). Ecdysteroids: possible candidates for the hormone which triggers salivary gland degeneration in the ixodid tick *Amblyomma hebraeum*. *Experientia* 41, 740–742.

Hartmann, R. & Loher, W. (1996). Control mechanisms of the behavior 'secondary defense' in the grasshopper *Gomphocerus rufus* L. (Gomphocerinae: Orthoptera). *Journal of Comparative Physiology A* 178, 329–336.

Hartmann, R. & Loher, W. (1999). Post-mating effects in the grasshopper, *Gomphocerus rufus* L. mediated by the spermatheca. *Journal of Comparative Physiology A* 184, 325–332.

Hiss, E. A. & Fuchs, M. S. (1972). The effect of matrone on oviposition in the mosquito, *Aedes aegypti*. *Journal of Insect Physiology* 18, 2217–2225.

Horrobin, D. F. (1978). *Prostaglandins: Physiology, Pharmacology and Clinical Significance*. Montreal, Canada: Eden Press.

Huber, B. A. (2005) Sexual selection research on spiders: progress and biases. *Biological Reviews* 80, 363–385.

Johns, R., Sonenshine, D. E. & Hynes, W. L. (1998). Control of bacterial infections in the hard tick *Dermacentor variabilis* (Acari: Ixodidae): evidence for the existence of antimicrobial proteins in tick hemolymph. *Journal of Medical Entomology* 35, 458–464.

Johns, R., Sonenshine, D. E. & Hynes, W. L. (2001). Identification of a defensin from the hemolymph of the American dog tick, *Dermacentor variabilis*. *Insect Biochemistry and Molecular Biology* 31, 857–865.

Kalb, J. M., Delbenedetto, A. J. & Wolfner, M. F. (1993). Probing the function of *Drosophila melanogaster* accessory glands by directed cell ablation. *Proceedings of the National Academy of Sciences of the USA* 90, 8093–8097.

Kaufman, W. R. (1976). The influence of various factors on fluid secretion by *in vitro* salivary glands of ixodid ticks. *Journal of Experimental Biology* 64, 727–742.

Kaufman, W. R. (1983). The function of tick salivary glands. *Current Topics in Vector Research* 1, 215–247.

Kaufman, W. R. (1991). Correlation between haemolymph ecdysteroid titre, salivary gland degeneration and ovarian development in the ixodid tick, *Amblyomma hebraeum* Koch. *Journal of Insect Physiology* 37, 95–99.

Kaufman, W. R. (1997). Arthropoda–Chelicerata. In *Reproductive Biology of Invertebrates*, vol. 8, Part A, *Progress in Reproductive Endocrinology*, ed. Adams, T. S., pp. 211–245. New York: John Wiley.

Kaufman, W. R. & Lomas, L. O. (1996). 'Male factors' in ticks: their role in feeding and egg development. *Invertebrate Reproduction and Development* 30, 191–198.

Kaufman, S. E., Kaufman, W. R. & Phillips, J. E. (1981). Fluid balance in the argasid tick, *Ornithodorus moubata*, fed on modified blood meals. *Journal of Experimental Biology* 93, 225–242.

Kaufman, S. E., Kaufman, W. R. & Phillips, J. E. (1982). Mechanism and characteristics of coxal fluid execretion in the argasid tick *Ornithodorus moubata*. *Journal of Experimental Biology* 98, 343–352.

Keirans, J. E., Clifford, C. M., Hoogstraal, H. & Easton, E. R. (1976). Discovery of *Nuttalliella namaqua* Bedford (Acarina: Ixodoidea: Nutalliellidae) in Tanzania and redescription of the female based on scanning electron microscopy. *Annals of the Entomological Society of America* **69**, 926–932.

Khalil, G. M. (1970). Biochemical and physiological studies of certain ticks (Ixodoidea): gonad development and gametogenesis in *Hyalomma* (*H.*) *anatolicum excavatum* Koch (Ixodidae). *Journal of Parasitology* **56**, 596–610.

Kiszewski, A. E., Matuschka, F.-R. & Spielman, A. (2002). Mating strategies and spermiogenesis in ixodid ticks. *Annual Review of Entomology* **46**, 167–182.

Kiszewski, A. E. & Spielman, A. (2002). Preprandial inhibition of re-mating in *Ixodes* ticks (Acari: Ixodidae). *Journal of Medical Entomology* **39**, 847–853

Klowden, M. J. (1999). The check is in the male: male mosquitoes affect female physiology and behavior. *Journal of the American Mosquito Control Association* **15**, 213–220.

Klowden, M. J. (2001). Sexual receptivity in *Anopheles gambiae* mosquitoes: absence of control by male accessory gland substances. *Journal of Insect Physiology* **47**, 661–666.

Klowden, M. J. & Chambers, G. M. (1991). Male accessory gland substances activate egg development in nutritionally stressed *Aedes aegypti* mosquitoes. *Journal of Insect Physiology* **37**, 721–726.

Kriger, F. L. & Davey, K. G. (1982). Ovarian motility in mated *Rhodnius prolixus* requires an intact cerebral neurosecretory system. *General and Comparative Endocrinology* **48**, 130–134.

Kriger, F. L. & Davey, K. G. (1983). Ovulation in *Rhodnius prolixus* Stal is induced by an extract of neurosecretory cells. *Canadian Journal of Zoology* **61**, 684–686.

Kuster, J. E. & Davey, K. G. (1986). Mode of action of cerebral neurosecretory cells on the function of the spermatheca in *Rhodnius prolixus*. *International Journal of Invertebrate Reproduction and Development* **10**, 59–69.

Lange, A. B. (1984). The transfer of prostaglandin-synthesizing activity during mating in *Locusta migratoria*. *Insect Biochemistry* **14**, 551–556.

Leahy, M. G. & Lowe, M. L. (1967). Purification of the male factor increasing egg deposition in *D. melanogaster*. *Life Sciences* **6**, 151–156.

Lees, A. D. & Beament, J. W. L. (1948). An egg-waxing organ in ticks. *Quarterly Journal of Microscopical Science* **89**, 291–232.

Loher, W., Ganjian, I., Kubo, I., Stanley-Samuelson, D. & Tobe, S. S. (1981). Prostaglandins: their role in egg-laying of the cricket *Teleogryllus commodus*. *Proceedings of the National Academy of Sciences of the USA* **78**, 7835–7838.

Lomas, L. O. & Kaufman, W. R. (1992a). The influence of a factor from the male genital tract on salivary gland degeneration in the female ixodid tick *Amblyomma hebraeum*. *Journal of Insect Physiology* **38**, 595–601.

Lomas, L. O. & Kaufman, W. R. (1992b). An indirect mechanism by which a protein from the male gonad hastens salivary gland degeneration in the female ixodid tick *Amblyomma hebraeum*. *Archives of Insect Biochemistry and Physiology* **21**, 169–178.

Lomas, L. O. & Kaufman, W. R. (1999). What is the meaning of 'critical weight' to female ixodid ticks?: a 'grand unification theory'! In *Acarology IX*, vol. 2, *Symposia*, eds. Needham, G. R., Mitchell, R., Horn, D. J. & Welbourn, W. C., pp. 481–485. Columbus, OH: Ohio Biological Survey.

Lomas, L. O., Turner, P. C. & Rees, H. H. (1997). A novel neuropeptide–endocrine interaction controlling ecdysteroid production in ixodid ticks. *Proceedings of the Royal Society of London B* **264**, 589–596.

Lung, O., Kuo, M. F. & Wolfner, M. F. (2001a). Identification and characterization of the major *Drosophila melanogaster* mating plug protein. *Insect Biochemistry and Molecular Biology* **31**, 543–551.

Lung, O., Kuo, M. F. & Wolfner, M. F. (2001b). *Drosophila* males transfer antibacterial proteins from their accessory gland and ejaculatory duct to their mates. *Journal of Insect Physiology* **47**, 617–622.

Lung, O., Tram, U., Finnerty, C. M., *et al.* (2002). The *Drosophila melanogaster* seminal fluid protein Acp62F is a protease inhibitor that is toxic upon ectopic expression. *Genetics* **160**, 211–224.

Mao, H. & Kaufman, W. R. (1999). Profile of the ecdysteroid hormone and its receptor in the salivary gland of the adult female tick, *Amblyomma hebraeum*. *Insect Biochemistry and Molecular Biology* **29**, 33–42.

Medeiros, M. N., Oliveira, D. N. P., Paiva-Silva, G. O., *et al.* (2002). The role of eicosanoids on *Rhodnius* heme-binding protein (RHBP) endoscytosis by *Rhodnius prolixus* ovaries. *Insect Biochemistry and Molecular Biology* **32**, 537–545.

Mortimer, D. (1983). Sperm transfer in the human female reproductive tract. In *Oxford Reviews of Reproductive Biology*, vol. 5, ed. Finn, C. A., pp. 30–61. Oxford, UK: Clarendon Press.

Oliver, J. H. Jr, Pound, J. M. & Andrews, R. H. (1984). Induction of egg maturation and oviposition in the tick, *Ornithodoros parkeri* (Acari: Argasidae). *Journal of Parasitology* **70**, 337–342.

Pappas, P. J. & Oliver, J. H. Jr (1971). Mating necessary for complete feeding of female *Dermacentor variabilis* (Acari: Ixodidae). *Journal of the Georgia Entomological Society* **6**, 122–124.

Pappas, P. J. & Oliver, J. H. Jr (1972). Reproduction in ticks (Acari: Ixodoidea). II. Analysis of the stimulus for rapid and complete feeding of female *Dermacentor variabilis* (Say). *Journal of Medical Entomology* **9**, 47–50.

Parker, G. A. (1998). Sperm competition and the evolution of ejaculates: towards a theory base. In *Sperm Competition and Sexual Selection*, eds. Birkhead, T. R. & Moller, A. P., pp. 3–54. San Diego, CA: Academic Press.

Potterat, O., Hostettmann, K., Höltzel, A., *et al.* (1997). Boophiline, an antimicrobial sterol amide from the cattle tick *Boophilus microplus*. *Helvetica Chimica Acta* **80**, 2066–2072.

Poyser, N. L. (1987) Prostaglandins and ovarian function. In *Eicosanoids and Reproduction*, ed. Hillier, K., pp. 1–29. Lancaster, UK: MTP Press.

Ridley, M. (1988). Mating frequency and fecundity in insects. *Biological Reviews* **63**, 509–549.

Ruegg, R. P., Kriger, F. L., Davey, K. G. & Steel, C. G. H. (1981). Ovarian ecdysone elicits release of a myotropic ovulation hormone in *Rhodnius* (Insecta: Hemiptera). *International Journal of Invertebrate Reproduction* **3**, 357–361.

Sahli, R., Germond, J. E. & Diehl, P. A. (1985). *Ornithodoros moubata*: spermateleosis and secretory activity of the sperm. *Experimental Parasitology* **60**, 383–395.

Sauer, J. R., Essenberg, R. C. & Bowman, A. S. (2000). Salivary glands in ixodid ticks: control and mechanism of secretion. *Journal of Insect Physiology* **46**, 1069–1078.

Schöl, H., Sieberz, J., Göbel, E. & Gothe, R. (2001). Morphology and structural organization of Gené's organ in *Dermacentor reticulatus* (Acari: Ixodidae). *Experimental and Applied Acarology* **25**, 327–352.

Seidelmann, K. & Ferenz, H.-J. (2002). Courtship inhibition pheromone in desert locusts, *Schistocerca gregaria*. *Journal of Insect Physiology* **48**, 991–996.

Sevala, V. L., Sevala, V. M., Davey, K. G. & Loughton, B. G. (1992). A FMRFamide-like peptide is associated with the myotropic ovulation hormone in *Rhodnius prolixus*. *Archives of Insect Biochemistry and Physiology* **20**, 193–203.

Shanbaky, N. M., El-Said, A. & Helmy, N. (1990). Changes in neurosecretory cell activity in female *Argas* (*Argas*) *hermanni* (Acari: Argasidae). *Journal of Medical Entomology* **27**, 975–981.

Shemesh, M., Hadani, A., Shklar, A., Shore, L. S. & Meleguir, F. (1979). Prostaglandins in the salivary glands and reproductive organs of *Hyalomma anatolicum excavatum* Koch (Acari: Ixodidae). *Bulletin of Entomological Research* **69**, 381–385.

Shepherd, J., Oliver, J. H. Jr & Hall, J. D. (1982). A polypeptide from male accessory glands which triggers maturation of tick spermatozoa. *International Journal of Invertebrate Reproduction* **5**, 129–137.

Sieberz, J. & Gothe, R. (2000). Modus operandi of oviposition in *Dermacentor reticulates* (Acari: Ixodidae). *Experimental and Applied Acarology* **24**, 63–76.

Simmons, L. W. & Siva-Jothy, M. T. (1998). Sperm competition in insects: mechanisms and the potential for selection. In *Sperm Competition and Sexual Selection*, eds. Birkhead, T. R. & Moller, A. P., pp. 341–434. San Diego, CA: Academic Press.

Snow, K. R. (1969). Egg maturation in *Hyalomma anatolicum anatolicum* Koch, 1844 (Ixodoidea, Ixodidae). In *Proceedings of the 2nd International Congress of Acarology 1967*, ed. Owen Evans, G., pp. 349–355.

Sonenshine, D. E. (1967). Feeding time and oviposition of *Dermacentor variabilis* (Acarina: Ixodidae) as affected by delayed mating. *Annals of the Entomological Society of America* **60**, 489–490.

Sonenshine, D. E. (1991). *Biology of Ticks*, vol. 1, Oxford, UK: Oxford University Press.

Sonenshine, D. E., Khalil, G. M., Homsher, P. J. & Mason, S. N. (1982). *Dermacentor variabilis* and *Dermacentor andersoni*: genital sex pheromones. *Experimental Parasitology* **54**, 317–330.

Sternberg, S., Peleg, B. A. & Galun, R. (1973). Effect of irradiation on mating competitiveness of the male tick, *Argas persicus* (Oken). *Journal of Medical Entomology* **10**, 137–142.

Thomas, R. W. & Zeh, D. W. (1984). Sperm transfer and utilization strategies in arachnids: ecological and morphological constraints. In *Sperm Competition and the Evolution of Animal Mating Systems*, ed. Smith, R. L., pp. 179–221. San Diego, CA: Academic Press.

Tram, U. & Wolfner, M. F. (1999). Male seminal fluid proteins are essential for sperm storage in *Drosophila melanogaster*. *Genetics* **153**, 837–844.

Vahed, K. (1998). The function of nuptial feeding in insects: a review of empirical studies. *Biological Reviews* **73**, 43–78.

Van der Goes van Naters-Yasui, A., Taylor, D., Shono, T. & Yamakawa, M. (2000). Purification and partial amino acid sequence of antibacterial peptides from the hemolymph of the soft tick, *Ornithodoros moubata*, (Acari: Argasidae). In *Proceedings of the 3rd International Conference Ticks and Tick-Borne Pathogens: Into the 21st Century*, eds.

Kazimirova, M., Labuda, M. & Nuttall, P. A., pp. 189–194.

Wakayama, E. J., Dillwith, J. W. & Blomquist, G. J. (1986). Occurrence and metabolism of prostaglandins in the housefly, *Musca domestica* (L.). *Insect Biochemistry* **16**, 895–902.

Wang, H., Paesen, G. C., Nuttall, P. A. & Barbour, A. G. (1998). Male ticks help their mates to feed. *Nature* **391**, 753–754.

Wedell, N. & Karlsson, B. (2003) Paternal investment directly affects female reproductive effort in an insect. *Proceedings of the Royal Society of London B* **270**, 2065–2071.

Wedell, N., Wiklund, C. & Cook, P. A. (2001). Monandry and polyandry as alternative lifestyles in a butterfly. *Behavioral Ecology* **13**, 450–455.

Weiss, B. L. & Kaufman, W. R. (2001). The relationship between 'critical weight' and 20-hydroxyecdysone in the female ixodid tick, *Amblyomma hebraeum. Journal of Insect Physiology* **47**, 1261–1267.

Weiss, B. L. & Kaufman, W. R. (2004). Two feeding-induced proteins from the male gonad trigger engorgement of the female tick *Amblyomma hebraeum. Proceedings of the National Academy of Science of the USA* **101**, 5874–5879.

Weiss, B. L., Stepczynski, J. M., Wong, P. & Kaufman, W. R. (2002). Identification and characterization of genes differentially expressed in the testis/vas deferens of the fed male tick, *Amblyomma hebraeum. Insect Biochemistry and Molecular Biology* **32**, 785–793.

Wolfner, M. F. (1997). Tokens of love: functions and regulation of *Drosophila* male accessory gland products. *Insect Biochemistry and Molecular Biology* **27**, 179–192.

Wolfner, M. F. (2002). The gifts that keep on giving: physiological functions and evolutionary dynamics of male seminal proteins in *Drosophila. Heredity* **88**, 85–93.

Wolfner, M. F., Harada, H. A., Bertram, M. J., *et al.* (1997). New genes for male accessory gland proteins in *Drosophila melanogaster. Insect Biochemistry and Molecular Biology* **10**, 825–834.

Yamaja Setty, B. N. & Ramaiah, T. R. (1979). Isolation and identification of prostaglandins from the reproductive organs of male silkmoth, *Bombyx mori* L. *Insect Biochemistry* **9**, 613–617.

Yamaja Setty, B. N. & Ramaiah, T. R. (1980). Effect of prostaglandins and inhibitors of prostaglandin biosynthesis on oviposition in the siltmoth, *Bombyx mori. Indian Journal of Experimental Biology* **18**, 539–541.

Yuval, B. & Spielman, A. (1990). Sperm precedence in the deer tick *Ixodes dammini. Physiological Entomology* **15**, 123–128.

9 • Tick immunobiology

M. BROSSARD AND S. K. WIKEL

INTRODUCTION

Understanding of the tick–host–pathogen interface has increased dramatically in recent years and will continue to benefit from tick salivary gland transcriptome analysis, proteomics, functional genomics and the first tick genome project. Ticks modulate host haemostasis, pain/itch responses, wound healing and immune defences (Wikel, 1996, 1999; Schoeler & Wikel, 2001; Brossard & Wikel, 2004; Nuttall & Labuda, 2004). Salivary gland genes responsible for these activities are being identified (Valenzuela, 2004; Ribeiro et al., 2006; Alarcon-Chaidez, Sun & Wikel, 2007). Pharmacologically active compounds have been identified in insect and tick salivary glands and their biological activities established (Ribeiro, 1995a, b, 2004; Steen, Barker & Alewood, 2006). In addition to facilitating blood-feeding, these molecules are increasingly recognized as important factors in transmission and establishment of tick-borne infectious agents (Schoeler & Wikel, 2001; Brossard & Wikel, 2004). The biological activities of these molecules can be exploited for development of novel vector and transmission blocking vaccines. Characterizing the immunobiology of the dynamic interactions of the tick–host–pathogen interface is important for understanding infectious agent transmission, innate and specific acquired immune responses developed to the vector and to the pathogen, and for vaccine design. In the following sections, we address current knowledge in key aspects of tick immunobiology.

VECTOR SALIVA POTENTIATES PATHOGEN TRANSMISSION

Vector saliva enhances transmission of a variety of infectious agents transmitted by haematophagous insects and ticks. Mechanisms of heightened infectivity and the responsible saliva molecules have not been definitively identified; however, modulators of host innate immune defences could

be key elements (Schoeler & Wikel, 2001). Examples among insects started with the observations that salivary gland homogenate of the sandfly *Lutzomyia longipalpis* increased *Leishmania major* in cutaneous lesions up to 5000 times and the size of lesions up to 10 times upon injection of a mixture of salivary gland homogenate and promastigotes into mouse foot pads (Titus & Ribeiro, 1988). Feeding of *Aedes aegypti*, *Aedes triseriatus* or *Culex pipiens* within 4 h of the inoculation of Cache Valley virus into the feeding sites enhanced infection (Edwards et al., 1998). *Plasmodium berghei* sporozoites inoculated by *Anopheles stephensi* bites were more infectious than intravenously inoculated sporozoites (Vaughan et al., 1999). Inoculation of West Nile virus at the site where *A. aegypti* had fed 30 minutes earlier resulted in higher viraemia and accelerated neuroinvasion than injection of a similar dose of virus in the absence of mosquito feeding (Schneider et al., 2006). Five-day-old chickens infected with West Nile virus by *Culex pipiens* or *Culex tarsalis* had similar infection profiles of higher viraemia and greater viral shedding than chickens needle-inoculated with virus (Styer, Bernard & Kramer, 2006).

Direct and indirect evidence that tick saliva potentiates pathogen transmission has been reported for at least 11 tick-borne agents (see Chapter 10). Most of the early reports were for tick-borne viruses and doubts were expressed that the phenomenon applied to non-viral pathogens such as *Borrelia burgdorferi*, the bacterium that causes Lyme disease. However, there is now extensive evidence that potentiation of transmission and/or infection also affects tick-borne bacteria. For example, *Ixodes scapularis* salivary gland lysate enhanced the amount of *Borrelia burgdorferi* in host tissues when the two were inoculated together, but infection with the Portuguese strain *Borrelia lusitaniae* was not enhanced by *I. scapularis* salivary gland lysate (Zeidner et al., 2002). *Ixodes ricinus* salivary gland extract (SGE) increased the number of *B. burgdorferi* spirochaetes found in the skin 1 day after both

Ticks: Biology, Disease and Control, ed. Alan S. Bowman and Patricia A. Nuttall. Published by Cambridge University Press.
© Cambridge University Press 2008.

were inoculated together (Machackova, Obornik & Kopecky, 2006). Co-inoculation of *Francisella tularensis* with *I. ricinus* SGE resulted in increased numbers of bacteria at the inoculation site as well as in draining lymph nodes and spleen (Krocova *et al.*, 2003).

Exposure to feeding by pathogen-free ticks induces host responses that provide protection against subsequent transmission of infectious agents by the sensitizing tick species. The first report of this phenomenon was for rabbits infested with uninfected *Dermacentor andersoni*, which induced resistance to subsequent transmission of fully virulent *F. tularensis* by *D. andersoni* nymphs (Bell *et al.*, 1979). When Thogoto virus infected *Rhipicephalus appendiculatus* adults were fed on tick resistant guinea pigs, significantly fewer co-feeding uninfected nymphs became infected with Thogoto virus (Jones & Nuttall, 1990). Although immunity to tick bite did not develop, mice infested four times with pathogen-free *I. scapularis* nymphs were significantly resistant to acquisition of *B. burgdorferi* infection when subsequently exposed to infected nymphs (Wikel *et al.*, 1997). In contrast, repeated infestations of guinea pigs with uninfected *I. scapularis* nymphs induced resistance to both tick feeding and transmission of *B. burgdorferi* by infected nymphs (Nazario *et al.*, 1998). Protection against *Leishmania major* infection occurs due to a cutaneous delayed-type hypersensitivity response induced by sandfly *Phlebotomus papatasi* saliva (Kamhawi *et al.*, 2000).

IMMUNE RESPONSES TO TICK FEEDING

Some tick–host relationships are characterized by the acquisition of resistance to tick feeding which develops as a result of repeated infestations (Willadsen, 1980; Wikel, 1996). Acquired resistance to ticks is expressed as reduced engorgement weight, increased duration of feeding, decreased numbers of ova, reduced viability of ova, inhibition of moulting and death of engorging ticks (Bowessidjaou, Brossard & Aeschlimann, 1977; Wikel, 1996). *Ixodes ricinus* females feeding on immune rabbits digest haemoglobin with difficulty. Indeed an analysis of multiple regression showed that haemoglobin concentration of tick midgut correlates only with the time after a first infestation and shows no correlation with time after subsequent infestations. In ticks fed on immune rabbits, haemoglobin concentration and body weight were dependent on the quantity of midgut C3 (Papatheodorou & Brossard, 1987).

As early as 1939, Trager observed that guinea pigs become strongly immune to *Dermacentor variabilis* larvae (Trager,

1939). Since then, immunity to ticks has been studied most extensively in laboratory animals (guinea pigs, rabbits or mice) and bovines (Willadsen, 1980; Brossard & Wikel, 1997; Willadsen & Jongejan, 1999). Adaptive anti-tick immunity in laboratory animals is often more intense than that observed for natural hosts (Ribeiro, 1989). Laboratory mice are exceptions as they do not acquire resistance to immature stages of *I. ricinus* (Mbow *et al.*, 1994), *I. scapularis*, *I. pacificus* (Schoeler, Manweiler & Wikel, 2000) or *Rhipicephalus sanguineus* (Ferreira & Silva, 1999). In contrast, reductions in numbers of engorged ticks and in weights of fed larvae are observed on BALB/c mice repeatedly infested with *D. variabilis* (denHollander & Allen, 1985). Wild rodents vary between those that do acquire immunologically based resistance to ticks (e.g. *Clethrionomys glareolus*) and those that do not (e.g. *Apodemus sylvaticus*) (Dizij & Kurtenbach, 1995).

It has long been recognized that some breeds of cattle develop immunity after repeated infestations with ticks (reviewed in Willadsen, 1980, 1987; and see Chapter 19). The host–parasite relationship of bovines with *Rhipicephalus* (*Boophilus*) *microplus* (hereafter referred to as *B. microplus*) is the most extensively studied from an immunological perspective. Acquired resistance or susceptibility of bovines is also described for infestations with *Hyalomma anatolicum anatolicum* and *Rhipicephalus evertsi* (Latif, 1984), *R. appendiculatus* (Latif *et al.*, 1991), and *Hyalomma* spp. (Sahibi *et al.*, 1997).

Tick feeding induces a complex array of host immune responses involving antigen-presenting cells (APCs), T-cells, B-cells, antibodies, cytokines, complement, basophils, mast cells, eosinophils and a number of bioactive molecules (reviewed in Wikel, 1996; Brossard & Wikel, 1997, 2004). These complex interactions can be viewed as a balance between host defences raised against the parasite and tick evasion strategies, facilitating feeding and transmission of pathogens. Circulating immunoglobulin G (IgG) antibodies to saliva antigens, which are induced by tick feeding, are detected in different host animals (Brossard, 1976; Wikel, 1996). By passively transferring immune serum to naive laboratory animals, it was shown that humoral factors are involved in the acquisition of immunity against ticks (Brossard & Girardin, 1979). Immunity against *B. microplus*, even though weakened, was also transmitted passively to cattle (Roberts & Kerr, 1976).

Antibodies are not the only effector elements of the immune system contributing to acquired resistance. The role of lymphocytes in expression of acquired resistance to *D. andersoni* larvae was proven by adoptively transferring lymph node cells from resistant guinea pigs into naive animals (Wikel & Allen, 1976). Involvement of T-cells was also established by applying the immunosuppressor cyclosporin A to rabbits infested with *I. ricinus* adults (Girardin & Brossard, 1989, 1990). T-cells are key elements in regulation and effector functions of the immune system, including antibody production and cell mediated immunity. Lymphocytes derived from infested laboratory animals (Wikel, Graham & Allen, 1978; Schorderet & Brossard, 1994; Ganapamo, Rutti & Brossard, 1997) and cattle (George, Osburn & Wikel, 1985) undergo in vitro blastogenesis when cultured in the presence of SGE. Reactivity is generally more intense with cells obtained during repeated exposures. In BALB/c mice repeatedly infested with nymphs of *I. ricinus*, lymphocytes from lymph nodes draining the tick attachment site produce significant levels of tumour necrosis factor alpha (TNF-α) and granulocyte/monocyte colony stimulating factor (GM-CSF) when stimulated in vitro with concanavalin A (ConA) or anti-CD3 antibodies (Ganapamo, Rutti & Brossard, 1996*a*). GM-CSF induces the maturation of Langerhans cells (LCs) in vitro by maintaining their viability and inducing high expression of Ia molecules (Berthier *et al.*, 2000). LC migration from the skin to the regional lymph nodes and their accumulation in these secondary lymphoid organs is controlled by TNF-α (Cumberbatch & Kimber, 1995). LCs in the skin trap antigens, as well as interacting with a variety of cell types through an array of cytokines (Salmon, Armstrong & Ansel, 1994). They are one of the first cells to be exposed to tick immunogens in the skin, from where they migrate to draining lymph nodes (Allen, Khalil & Wikel, 1979). In the paracortical area of the lymph nodes, LCs transform into dendritic cells (DCs) and function there as APCs for T-cells (Nithiuthai & Allen, 1984*a*). Ultraviolet (UV) radiation treatment of guinea pig ears before primary infestation with *D. andersoni* larvae reduced the number of LCs and acquired resistance (Nithiuthai & Allen, 1984*b*). When UV was applied to resistant animals a marked diminution of resistance during reinfestation was observed. Short-wavelength UVC was found to be more effective than mid-wavelength UVB in depleting LCs (Nithiuthai & Allen, 1984*b*). Functional LCs are necessary to initiate adaptive immunity and to stimulate a secondary immune response. The ability of T-lymphocytes from tick infested BALB/c mice to respond in vitro to *I. ricinus* SGE was demonstrated after spleen accessory cell antigen processing (Ganapamo, Rutti & Brossard, 1995*b*).

The importance of spleen DCs to induce an anti-tick immune response was further studied. Whole spleen DCs pulsed in vitro with tick saliva upregulated mostly B7-2 costimulator expression and were able to prime naive splenic

cells. Using RT-PCR, elevated amounts of interleukin-4 (IL-4) mRNA compared to interferon-gamma (IFN-γ) mRNA were detected in proliferating cells, demonstrating establishment of a primary T-helper 2 (Th2) immune response in vitro (Mejri & Brossard, 2007). Murine spleen DCs were purified and separated into different subsets: CD8α+ and CD8α− (CD4+ or CD4−). Each subset was pulsed in vitro with *I. ricinus* female saliva and injected into naive BALB/c mice. The whole population of DCs and CD8α− DCs triggered a Th2 immune response as shown by measuring the IL-4/IFN-γ ratio for lymphocytes restimulated in vitro by SGE. In contrast, CD8α+ DCs seem to polarize the response towards Th1.

The complement system is also involved in development of immunity against ticks. Acquired resistance to *D. andersoni* larvae in guinea pigs was inhibited by dramatically lowering the levels of C3 with cobra venom factor (Wikel & Allen, 1977). C4-deficient guinea pigs acquired resistance showing that the alternate pathway of complement activation was important in the expression of this immunity (Wikel, 1979). C3 was deposited in the dermal–epithelial junction near the location of the tick bite (Allen *et al.*, 1979). Activation of the complement system in the vicinity of the bite site may result in generation of mediators of inflammation, chemotaxis and opsonins, and could locally attract basophils and other cells linked to host resistance.

Cutaneous reactions at tick attachment sites on cattle and laboratory animals expressing acquired immunity contain infiltrates of basophils and eosinophils (Allen, 1973; Allen, Doube & Kemp, 1977; Brossard & Fivaz, 1982; Steeves & Allen, 1991). This type of reaction is termed cutaneous basophil hypersensitivity (CBH), which is a form of delayed-type hypersensitivity (Askenase *et al.*, 1978). Infiltration of basophils was more pronounced in *D. andersoni*-infested guinea pigs (Allen, 1973) than in *I. ricinus*-infested rabbits (Brossard & Fivaz, 1982). CBH also developed after the infestation of rabbits or cattle with *Hyalomma anatolicum anatolicum* (Gill & Walker, 1985; Gill, 1986) and guinea pigs or cattle infested with *Ixodes holocyclus* (Allen, Doube & Kemp, 1977) or *Amblyomma americanum* (Brown & Askenase, 1981; Brown, Barker & Askenase, 1984). Saliva antigens complex with homocytotropic antibodies bound to Fc receptors on mast cells and basophils. Those cells are activated and release bioactive molecules, including histamine, leukotrienes, prostaglandins and enzymes, at the bite site, which contribute to expression of acquired immunity (Brossard, Monneron & Papatheodorou, 1982; Wikel, 1996). A greater number of degranulated cells was found during reinfestation (Schleger *et al.*, 1976; Brossard & Fivaz, 1982).

In cattle infested with *B. microplus*, release of histamine causes skin irritation which leads to increased host grooming. In this way, some ectoparasites are actively removed from the host (Koudstaal, Kemp & Kerr, 1978). A positive correlation was established between skin histamine concentration and the degree of resistance acquired by cattle against *B. microplus* (Willadsen, Wood & Riding, 1979). In guinea pigs infested with *D. andersoni*, histamine-rich basophils concentrated in the area around the attachment site (Wikel, 1982). The very high level of resistance observed after a single infestation is partially broken down by treating the host with histamine antagonists (Brossard, 1982; Wikel, 1982).

PHARMACOLOGY OF TICK SALIVA

Tick saliva is a complex mixture of several hundred secreted proteins differentially expressed during the course of feeding (Ribeiro *et al.*, 2006). Redundancy and antigenic variants appear to be common among salivary gland gene families (Ribeiro *et al.*, 2006). Distinct differences occur in the composition of pharmacologically active saliva molecules for different tick species (Ribeiro, 1995b; Steen, Barker & Alewood, 2006). As one would expect in any randomly bred population, variations occur in the salivary secretions of individual ticks within a species (Wang, Kaufman & Nuttall, 1999). Diverse patterns of expression and complex mixtures of saliva molecules that target similar host defence pathways provide a biological advantage when blood meals are obtained from a phylogenetically diverse group of hosts. Redundancy and antigenic variation, in addition to differential expression of proteins during the course of engorgement, add mechanisms by which the tick reduces the likelihood that host immune responses will neutralize biological activities of the saliva that are essential for successful blood-feeding.

The pace of discovery of pharmacologically active molecules in insect and tick saliva is dramatically increasing due to transcriptome analysis of salivary glands (see Chapter 4). Describing all pharmacologically active molecules reported for tick salivary glands is beyond the scope of this chapter. Selected examples are provided of inhibitors of host haemostasis, angiogenesis, inflammation and innate/specific acquired immune defences. Reviews of this topic provide an underpinning for both current and forthcoming reported studies (Ribeiro, 1989; Titus & Ribeiro, 1990; Champagne, 1994; Champagne & Valenzuela, 1996; Valenzuela, 2004; Andrade *et al.*, 2005; Steen, Barker & Alewood, 2006).

Antihaemostasis factors are a popular focus of studies. Important background information on haemostasis can be obtained from the excellent reviews of the molecular basis of blood coagulation (Furie & Furie, 2005) and platelet function (Plow & Abrams, 2005). Saliva modulation of host blood coagulation is relevant to host immunobiology, since there is considerable cross-talk between the host inflammatory/innate immune responses and coagulation (Esmon, 2004). *Ixodes scapularis* salivary glands produce Kunitz-domain-containing proteins, serine protease inhibitors with one, two or five domains (Ribeiro *et al.*, 2006). A two-Kunitz-domain protein, ixolaris, shares extensive sequence homology to tissue factor pathway inhibitor. It binds factors X or Xa scaffolds to inhibit the extrinsic coagulation pathway, and inhibits thrombus formation upon in vivo administration (Nazareth *et al.*, 2006). Penthalaris is an inhibitor of factor VIIa/tissue factor induced factor X activation, containing five tandem Kunitz domains, found in the saliva of *I. scapularis*, which may act together with ixolaris to facilitate blood-feeding (Francischetti, Mather & Ribeiro, 2004). These factors are examples of redundant targeting of a host defence pathway by two different, yet related, molecules. Iris, a serine protease inhibitor secreted with *I. ricinus* saliva, disrupts coagulation and fibrinolysis via an antiproteolytic reactive centre loop (RCL) domain (Prévôt *et al.*, 2006). As shown with mutants, one or more other domains could be responsible for primary haemostasis inhibition.

Genes encoding putative metalloproteases are expressed in salivary glands of *I. scapularis* (Ribeiro *et al.*, 2006), *D. andersoni* (Alarcon-Chaidez, Sun & Wikel, 2007) and *I. ricinus* (Decrem *et al.*, 2008) which may modify host extracellular matrix and haemostasis. A metalloprotease detected in *I. scapularis* saliva is related to the reprolysin family and similar to snake venom haemorrhagic proteases (Francischetti, Mather & Ribeiro, 2003). The cysteine-rich domain of the metalloprotease may facilitate binding to host extracellular matrix and contribute to the feeding site cavity by cleaving fibrinogen (Francischetti, Mather & Ribeiro, 2003). The metalloproteinase of *I. scapularis* may facilitate *B. burgdorferi* dissemination (Francischetti, Mather & Ribeiro, 2003), since spirochaetes upregulate and activate matrix metalloproteinase gelatinase B and collagenase to penetrate across host extracellular matrix (Gebbia, Coleman & Benach, 2001). Tick remodelling of extracellular matrix impacts host inflammatory and immune responses to blood-feeding, since extracellular matrix interactions with leukocytes are essential for cellular transit through extravascular tissues (Nabi, 1999; Young, 1999).

Tick salivary gland molecules modulate expression of skin endothelial cell adhesion molecules, which has implications for impaired inflammatory and immune cell movement from the vascular compartment and accumulation at tick bite sites (Maxwell *et al.*, 2005). *Ixodes scapularis* SGE significantly reduces expression of P-selectin and vascular cell adhesion molecule-1 (VCAM-1) while *D. andersoni* SGE down regulates intercellular adhesion molecule-1 (ICAM-1). Reduced endothelial cell expression of ICAM-1 has implications for the immune response to endothelial cells infected with spotted fever group rickettsiae. Infected endothelial cells present antigen in association with increased ICAM-1 expression, in the absence of chemokines, to anti-rickettsial T-cells (Valbuena & Walker, 2004). *Dermacentor andersoni* is a competent vector of Rocky Mountain spotted fever rickettsiae (Walker, 1998). ICAM-1 is an important component of the immunological synapse contributing to adhesion between dendritic cells and T-lymphocytes, a factor controlling T-cell activation (Grakoui *et al.*, 1999; van Gisbergen *et al.*, 2005). *Ixodes scapularis* saliva also contains a potent inhibitor of both endothelial cell proliferation and angiogenesis (Francischetti, Mather & Ribeiro, 2005).

An increasing number of tick salivary gland molecules with properties that impact host immune responses are being identified since the initial discovery of *A. americanum* calreticulin (Jaworski *et al.*, 1995). Besides *A. americanum*, calreticulin has been cloned and sequenced from *B. microplus* (Ferreira *et al.*, 2002) and four other ixodid species, including *I. scapularis* (Xu *et al.*, 2004). A chromatographic fraction enriched with a 65-kDa protein isolated from salivary glands of partially fed female *I. ricinus* induces in vitro a specific CD4+T-cell proliferation of lymph node cells from infested mice (Ganapamo, Rutti & Brossard, 1997). As shown by peptide fragmentation sequencing, this immunogen has sequence homology with *I. scapularis* calreticulin (N. Mejri and M. Brossard, personal communication). Immunoglobulin G-binding proteins are present in SGEs of *Amblyomma variegatum*, *I. hexagonus* and *R. appendiculatus* (Wang & Nuttall, 1995).

Other pharmacologically active tick saliva molecules include a number of cytokine-binding proteins (Gillespie *et al.*, 2001; Hajnicka *et al.*, 2001), histamine- and serotonin-binding proteins (Paesen *et al.*, 1999; Sangamnatdej *et al.*, 2002) and complement inhibitors (Valenzuela *et al.*, 2000; Couvreur *et al.*, 2008) (see Chapter 4). Tick-induced cutaneous responses stimulate grooming that results in tick removal (Snowball, 1956). Host itch response would be inhibited by histamine-binding proteins and a kininase in the saliva of *I. scapularis* (Ribeiro & Mather, 1998).

Homologues of macrophage migration inhibition factor occur in the salivary glands of *A. americanum* (Jaworski

et al., 2001) and *Haemaphysalis longicornis* (Umemiya *et al.*, 2007), which may inhibit macrophage accumulation at the tick feeding site. A 36-kDa protein in the saliva of *D. andersoni* inhibits T-lymphocyte proliferation (Bergman, Ramachandra & Wikel, 1998; Bergman *et al.*, 2000), and genes encoding sequence similarity were found in both *A. variegatum* (Nene *et al.*, 2002) and *R. appendiculatus* (Nene *et al.*, 2004). Salp15 is an inhibitor of host T-lymphocyte activation in *I. scapularis* saliva that acts by depressing T-cell receptor initiated calcium-dependent signalling (Anguita *et al.*, 2002). Tick saliva contains abundant prostaglandins (Bowman, Dillwith & Sauer, 1996), which are known to possess immunosuppressive properties (Harris *et al.*, 2002).

B-lymphocyte inhibitory proteins occur in the saliva of *I. ricinus* (Hannier *et al.*, 2004) and *Hyalomma asiaticum asiaticum* (Yu *et al.*, 2006). The *I. ricinus* protein greatly inhibited *B. burgdorferi* OspA and OspC outer surface lipoprotein induced proliferation of B-lymphocytes in vitro, suggesting a possible role in modulating host responses to tick transmitted spirochaetes. The *H. asiaticum* protein possessed no in silico similarity to known proteins. A histamine-releasing factor homologue occurs in *D. variabilis* saliva (Mulenga & Azad, 2005). The recombinant form of this protein induced basophil release of histamine (Mulenga *et al.*, 2003), which is a dichotomy relative to histamine-binding proteins in tick saliva. A cystatin found in *I. scapularis* saliva has anti-inflammatory and immunosuppressive properties as indicated by inhibition of proliferation of the IL-2-dependent cytotoxic T-cell lymphoma line CTLL-2 (Kotsyfakis *et al.*, 2006).

As will become evident in the next section, salivary gland transcriptome analysis is increasing the pace of discovery of modulators of host inflammation and innate and specific acquired immune response factors. In addition to enhancing understanding of tick–host–pathogen interactions, the therapeutic potential of many of these molecules will be explored.

TICK MODULATION AND DEVIATION OF THE HOST IMMUNE DEFENCES

Tick countermeasures against host immune defences target pathways important in acquisition and expression of acquired immunity (Wikel, Ramachandra & Bergman, 1994; Wikel, 1996). Modulation of host immune defences not only promotes successful blood-feeding but it also enhances the ability of tick-borne pathogens to establish effectively in the host (Wikel, 1996, 1999; Schoeler & Wikel, 2001). An overview of tick modulation of host immune defences is provided in Table 9.1.

Table 9.1 *Overview of cells and molecules of the cost immune system and tick countermeasures*

Cell or molecule	Role in immune defence	Tick countermeasure	Possible significance
Complement (general)	Classical, alternative, mannose-binding pathways to generate inflammatory mediators and opsonins, C3 pivotal molecule for all three pathways	C3b deposition inhibited by *I. scapularis* saliva (Ribeiro, 1987)	Reduce inflammatory response and lysis of microbes
Alternative complement pathway	Antibody independent, mediates inflammation and direct lysis of microbes	Role in acquired resistance to *D. andersoni* (Wikel, 1979). Inhibited by SGEs of *I. ricinus*, *I. hexagonus*, *I. uriae* (Lawrie et al., 1999). Binding of C3B and uncoupling Bb by *I. scapularis* 18.5-kDa protein (Valenzuela et al., 2000). Two homologues of Isac, Irac 1 and 2 in *I. ricinus* (Daix et al., 2007)	Reduce inflammatory response to tick and tick-transmitted microbes
Anaphylatoxin	Complement components C3a, C4a, C5a bind to cells and promote acute inflammation, neutrophil chemotaxis, mast cell activation	*I. scapularis* saliva antagonist of anaphylatoxins (Ribeiro & Spielman, 1986)	Reduce cellular infiltrate and release of mast cell mediators (histamine, leukotrienes, cytokines)
NK cells	Lymphocytes (neither B nor T) that directly kill microbe infected cells and secrete IFN-γ	SGE of *D. reticulatus* effector function of NK cells from healthy humans, affecting effector–target cell interaction (Kubes et al., 2002). Similar, but lesser activity in SGEs of *A. variegatum* and *H. inermis* (Kubes et al., 2002)	Reduce killing of microbe-infected cells and suppress IFN-γ, which activates macrophages and can polarize acquired immune response to Th1 profile
Chemokines	Cytokines that stimulate leukocyte migration (chemotaxis) from blood into tissues. Numerous different chemokines identified, including IL-8	Anti-IL-8 activity in SGEs of *A. variegatum*, *D. reticulatus*, *H. inermis*, *I. ricinus* and *R. appendiculatus* (Hajnická et al., 2001).	IL-8 is chemotactic for neutrophils, basophils and T-lymphocytes. All of these cells have roles in host immune responses to limit tick feeding. Blocking IL-8 confers a survival advantage for the tick
Adhesion molecules	Cell surface molecules that interact with other cells or extracellular matrix to promote cell migration and activation (also in adaptive immunity)	Infestation of BALB/c mice with *D. andersoni* nymphs reduced lymphocyte expression of the lymphocyte integrins LFA-1 and VLA-4, which respectively bind to the adhesion molecules ICAM-1 and 2 and VCAM-1 on activated endothelium (Macaluso & Wikel, 2001)	Modulation of host leukocyte migration to tick bite site and reduced immune response to infestation

(cont.)

Table 9.1 (*Cont.*)

Cell or molecule	Role in immune defence	Tick countermeasure	Possible significance
Monocyte/macrophage function	Blood monocytes become tissue macrophages, which are phagocytic and activated by IFN-γ. Activated macrophages kill engulfed microbes, present antigens to CD4 T-lymphocytes and secrete pro-inflammatory cytokines (TNF-α, IL-1β). Macrophage-derived IL-12 promotes IFN-γ from NK and T-cells to activate macrophages. NO is involved in intracellular killing of microbes by macrophages	Macrophage IL-1β and TNF-α production reduced by *D. andersoni* SGE (Ramachandra & Wikel, 1992). Saliva of *I. scapularis* suppressed macrophage NO production (Urioste *et al.*, 1994). Saliva of *R. sanguineus* inhibits IFN-γ macrophage activation and reduced NO production (Ferreira & Silva, 1998). SGE of *I. ricinus* reduced production of NO (Kopecky & Kuthejlova, 1998). SGE of *I ricinus* inhibits macrophage killing of *Borrelia afzelii* (Kuthejlova *et al.*, 2001). SGE of *R. appendiculatus* inhibits in vitro transcription and secretion of IL-1α, TNF-α, IL-10 and production of NO by murine macrophage cell line (Gwakisa *et al.*, 2001). Macrophage migration inhibition factor in salivary glands of *A. americanum* (Jaworski *et al.*, 2001) has reported roles inhibiting NK-cell-mediated lysis and delayed-type hypersensitivity responses	Prevent inflammatory responses, reduce cytokines that could orchestrate anti-tick responses, reduce ability to clear microbes, and possibly alter antigen presentation
Histamine-binding proteins and serotonin-binding proteins	Histamine and serotonin are mediators of the itch response. Histamine increases vascular permeability and it is a mediator of inflammation. Concurrent blocking of type-1 and type-2 histamine receptors reduced acquired resistance to *D. andersoni* (Wikel, 1982). Histamine and serotonin reduce sucking and salivation of *D. andersoni* (Paine, Kemp & Allen, 1983)	Histamine blocking by SGE of *R. sanguineus* (Chinery & Ayitey-Smith, 1977). Histamine-binding proteins in SGE of *R. appendiculatus* (Paesen *et al.*, 1999). Lipocalin of *D. reticulatus* binds histamine and serotonin (Sangamnatdej *et al.*, 2002)	Reduce inflammation and itch response at bite site. Reduce direct impact of histamine and serotonin on tick feeding

Table 9.1 (*Cont.*)

Cell or molecule	Role in immune defence	Tick countermeasure	Possible significance
Lymphocyte cytokine modulation	Cytokines mediate differentiation/activation of lymphocytes and other cells and orchestrate innate and adaptive immune responses. IL-2: T-cell, NK cell and B-cell proliferation and antibody synthesis. IL-4: Th2 polarization/differentiation and proliferation, inhibits macrophage activation, and mast cell proliferation. IL-10: inhibitor of IL-12, alters expression of major histocompatibility complex (MHC) class II and costimulatory molecules, inhibits activated dendritic cells/macrophages. IFN-γ: activates macrophages and polarizes toward Th1 response	Influence of ticks on host cytokines have been the focus of numerous studies. A general pattern that has emerged is downregulation of Th1 responses and polarization toward Th2 cytokine profiles (see text and review by Schoeler & Wikel, 2001; Brossard & Wikel, 2004)	Modulation or deviation of T-lymphocyte cytokines can reduce immune responses to tick feeding and in turn facilitate transmission/establishment of microbes
Lymphocyte proliferation	Proliferation of T- and B-lymphocytes is essential for increased population and to perform effector functions. Numerous factors and receptor–ligand interactions influence lymphocyte proliferation. Cell-mediated immune responses linked to acquired resistance	Several investigators reported tick suppression of in vitro T-cell proliferation as a result of infestation or exposure of lymphocytes in vitro to saliva, salivary gland derived molecules, or recombinant salivary gland proteins (see text and review by Schoeler & Wikel, 2001; Brossard & Wikel, 2004)	Reduce immune responses to tick saliva proteins, which also reduce host immunity to tick-transmitted microbes
Antibody responses	Antibodies carry out a variety of functions including neutralization, complement fixation (classical pathway) and facilitating leukocyte interaction with target cells (including microbes). Antibodies linked to expression of acquired resistance	Tick infestation reduces antibody responses to heterologous antigens, but the basis for this suppression is unknown (Wikel, 1985; Fivaz, 1989)	Reduce the likelihood of developing antibodies that could neutralize tick saliva proteins essential for successful blood-feeding

Antigen-specific T-cell activation triggers cytokine signalling pathways, cell differentiation and proliferation (Janeway *et al.*, 2005). Examination of lymphocyte-elaborated cytokines provides insights into helper functions and cell-mediated immune responses to infestation (Mosmann & Coffman, 1989). Tick infestation is often characterized by a reduction in the response of host lymphocytes to ConA and phytohaemagglutin (PHA) stimulation, which is generally interpreted as an immunosuppressive phenomenon (Brossard & Wikel, 1997). The reduced response may be the result of decreased IL-2 production caused by tick saliva components like prostaglandin E_2 (PGE_2) (Ribeiro *et al.*, 1985; Inokuma, Kemp & Willadsen, 1994) or IL-2 binding proteins (Gillespie *et al.*, 2001). PGE_2 primes naive T-cells for production of high levels of IL-4, IL-10 and IL-13, and low levels of IL-2, IFN-γ, TNF-α and TNF-α (Demeure *et al.*, 1997; Harris *et al.*, 2002). Peripheral blood lymphocytes of uninfested pure-bred *Bos indicus* and *Bos taurus* were reduced in responsiveness to ConA when cultured in the presence of *D. andersoni* SGE (Ramachandra & Wikel, 1995). SGE or saliva of *I. ricinus* females inhibited proliferation (Mejri, Rutti & Brossard, 2002) and in vitro production of IL-2 by spleen lymphocytes from naive BALB/c mice stimulated with ConA (Ganapamo, Rutti & Brossard, 1996*a*). Those effects may be due to saliva immunosuppressive molecules and IL-10 production by T-cells (Ganapamo, Rutti & Brossard, 1996*b*). The saliva of *I. scapularis* inhibited splenic T-cell proliferation and IL-2 secretion in response to ConA or PHA in a dose-dependent manner, while nitric oxide production by macrophages was diminished after stimulation by lipopolysaccharide (LPS) (Urioste *et al.*, 1994). A saliva protein of 5 kDa molecular weight or higher was responsible for the effects.

Modulation of host cytokine and T-cell proliferation has distinct advantages for the tick, circumventing responses that reduce the ability to obtain a blood meal. Altered host immune defences also provide an environment that is favourable for establishment of tick-transmitted infectious agents.

Guinea pigs infested with *D. andersoni* larvae were evaluated for their ability to develop a primary IgM antibody response to immunization with a thymic-dependent antigen, sheep red blood cells (SRBC), during tick exposure (Wikel, 1985). Antibody production was determined by the direct haemolytic plaque-forming cell assay. Infested animals produced significantly fewer anti-SRBC antibodies than uninfested controls. Production of anti-SRBC antibodies returned to normal levels when animals were immunized with SRBC 4 days after termination of blood-feeding. Similar results were obtained with mice infested with *I. rici-*

nus and immunized with SRBC (Mejri, Rutti & Brossard, 2002). Rabbits infested with adult *R. appendiculatus* were suppressed in their ability to develop an antibody response to bovine serum albumin when immunized during the peak of tick feeding (Fivaz, 1989). Decreasing the host antibody response during infestation reduces the potential both for damage to the tick from anti-tick immunoglobulins and neutralization of tick saliva molecules needed for blood feeding.

Expression of IL-4 and IFN-γ mRNA in skin, draining lymph nodes and spleen was measured by competitive quantitative RT-PCR after infestation of BALB/c mice with *I. ricinus* larvae (Lorimier, 2003). Peak IL-4 mRNA levels were detected in the epidermis at 18 hours after infestation, followed by an increase of IL-4 mRNA in the dermis. This early production of IL-4 could determine the development of an anti-tick Th2 response. An increase of IL-4 mRNA was concomitantly observed in spleen, and 24 hours later in lymph nodes draining the tick feeding site. IFN-γ mRNA consistently remained at low levels. Lymphocytes collected from draining lymph nodes of BALB/c mice infested with *I. ricinus* larvae, nymphs or adults produced high levels of IL-4 and low levels of IFN-γ after in vitro stimulation with tick saliva (Mejri *et al.*, 2001) or with ConA (Ganapamo, Rutti & Brossard, 1995*a*). These observations confirm a Th2 polarization of the immune response. Infested mice developed IgE antibodies and intense immediate-type hypersensitivity against tick antigens (Mbow *et al.*, 1994). An increase of IL-5 and IL-10 was also observed in cultured CD4+ T-cells stimulated with ConA (Ganapamo, Rutti & Brossard, 1996*b*). Draining lymph node cells from mice of different genetic backgrounds, DBA (H-2d), C57BL/6 (H-2b), CBA (H-2k), C3H (H-2k), SJL (H-2s) and FVB (H-2q), and infested with nymphs of *I. ricinus*, produced more IL-4 than IFN-γ when stimulated in vitro with ConA (Christe, Rutti & Brossard, 1999). Lymph node cells from C3H/HeJ mice infested with *R. sanguineus* also developed a Th2 cytokine profile, represented by augmented IL-4, IL-10 and transforming growth factor beta (TGF-β) and inhibited production of IL-2, IFN-γ and IL-12 (Ferreira & Silva, 1999). C3H/HeJ mice did not develop protection against *R. sanguineus*, as shown for BALB/c, C57BL/6 and C3H mice infested with *I. ricinus* nymphs (Christe, Rutti & Brossard, 2000). Likewise, C3H/HeN mice did not acquire resistance against nymphs of *I. scapularis* and *I. pacificus*; cytokine production was also biased toward a Th2 profile, with suppression of pro-inflammatory Th1 cytokines (Schoeler *et al.*, 2000). Thus saliva of several ixodid tick species appears highly potent in polarizing mice to a Th2 immune response and preventing development of anti-tick immunity.

Salivary gland extracts of female *D. andersoni* collected daily over the course of feeding were assessed for their ability to alter the elaboration of cytokines by normal murine macrophages and T-cells stimulated with mitogens (Ramachandra & Wikel, 1992). Macrophage cytokine IL-1 elaboration was significantly suppressed by SGE. Production of TNF-α was also reduced. In another experiment, the effects on lymphocyte proliferation, cytokine production and expression of adhesion molecules were studied following repeated infestations of BALB/c mice with *D. andersoni* nymphs (Macaluso & Wikel, 2001). After two infestations, production of IL-2 was decreased but that of IFN-γ remained unchanged while the production of IL-4 and IL-10 was significantly enhanced. Macrophages collected from naive pure-bred *B. indicus* and *B. taurus* were also suppressed in their ability to elaborate IL-1 and TNF-α in vitro (Ramachandra & Wikel, 1995). T-cell elaboration of IL-2 and IFN-γ was inhibited and IL-4 production was unchanged in the presence of *D. andersoni* SGE. Expression of IFN-α, TNF-α, IL-1α, IL-1β, IL-5, IL-6, IL-7 and IL-8 mRNA by human peripheral blood leukocytes was reduced when treated with a mixture of LPS and *R. appendiculatus* SGE (Fuchsberger *et al.*, 1995). Exposure of a macrophage-like cell line to LPS and SGE of *R. appendiculatus* reduced mRNA and secreted IL-1β, IL-10 and TNF-α (Gwakisa *et al.*, 2001). B-cell functions are also influenced by tick salivary gland molecules. A 18-kDa protein in *I. ricinus* SGE inhibits B-cell production of IL-10 and expression of the early activation marker CD69 on both B-cells and T-cells (Hannier *et al.*, 2003, 2004). Expression of leukocyte function-associated antigen-1 (LFA-1) and very late antigen-1 (VLA-1) by lymphocytes was suppressed. LFA-1 is particularly important in T-cell adhesion to endothelial cells and to APCs (Janeway *et al.*, 2005). VLA-4 is upregulated following T-cell activation and is important for recruiting effector T-cells into sites of infection. Infestation of BALB/c mice with *D. andersoni* nymphs reduced lymphocyte expression of the lymphocyte integrins LFA-1 and VLA-4, which respectively bind to the adhesion molecules ICAM-1 and ICAM-2, and VCAM-1, on activated endothelium (Macaluso & Wikel, 2001). Tick modulation of lymphocyte trafficking is potentially advantageous in reducing immune responses that negatively impact blood-feeding.

As previously mentioned, genes are induced during the feeding process that result in the secretion of proteins involved in modulation of host immune and haemostatic responses (Leboulle *et al.*, 2002b; Valenzuela *et al.*, 2002). The properties of an immunomodulatory salivary gland protein induced during the *I. ricinus* feeding process, which

is called Iris for '*I. ricinus* immunosuppressor', have been characterized. Immunoblot and confocal microscopy, using a specific antiserum, were used to demonstrate that Iris is secreted in saliva during blood feeding by female *I. ricinus* with an increasing expression of Iris from day 3 to day 5 of engorgement (Leboulle *et al.*, 2002a). The effect of the corresponding recombinant protein (rIris) was studied in the context of human peripheral blood mononuclear cells (PBMCs) cytokine production, using T-cells (PHA, ConA, CD3/CD28 and PMA/CD28), macrophages (LPS) and antigen-presenting cell (PPD) activators. ELISA and ELISPOT assays showed that the rIris suppressed the production of IFN-γ by human T-cells and APCs, while IL-5 and IL-10 levels remained unchanged. In contrast, rIris enhanced expression of IL-10 by macrophages. The expression of the pro-inflammatory cytokines IL-6 and TNF-α by macrophages, T-cells and APCs was inhibited, while IL-10 expression remained unaffected. By showing that rIris protein inhibited IFN-γ production by T-cells, it was clearly established that the recombinant protein contributed to the observed immunomodulation by inducing a Th2 type immune response. In addition, Iris modulated innate immune defences by inhibiting the production of pro-inflammatory cytokines (IL-6 and TNF-α). Moreover, as mentioned above, Iris also interferes with coagulation pathways, fibrinolysis and platelet adhesion (Prévôt *et al.*, 2006).

Modulation of host immunity by tick saliva is of major importance both for successful blood-feeding and for transmission of tick-borne pathogens (see Chapter 10). In addition, modulation of host immunity by tick-borne pathogens may benefit tick feeding. For example, transmission of *B. burgdorferi* and *Anaplasma phagocytophylum* by *I. scapularis* synergized to suppress splenic IL-2 production and diminish IFN-γ production (Zeidner *et al.*, 2000). Splenic IL-4 production was increased after infestation of mice with co-infected ticks and levels of this cytokine were significantly higher than those induced by transmission of either pathogen individually.

GENOMIC, PROTEOMIC AND FUNCTIONAL GENOMIC ANALYSES OF TICK–HOST–PATHOGEN INTERACTIONS

A widely used and powerful method for identification of genes expressed in blood feeding arthropod tissues is the expressed sequence tag (EST), which is a partial nucleic acid sequence derived from clones selected at random from a cDNA library prepared from mRNA of a tissue of interest (Adams *et al.*, 1991). A sequenced clone is compared with EST gene sequences, translated amino acid sequences and motifs recorded in public databases. Studies examining salivary gland transcriptomes of blood-feeding arthropods have steadily increased in number and scope during the past decade. The complexity and diversity of the molecules found in haematophagous insect and tick salivary glands are remarkable, with some molecules common to many species while others are limited to a single species (Ribeiro & Francischetti, 2003). Genes encoding secreted products with no known function are frequently identified; most probably they are specific to the blood-feeding process. Differential expression of tick salivary gland genes occurs along with preferential expression of individual members of gene families at different times during the days of blood-feeding (Ribeiro *et al.*, 2006). New insights are emerging into how blood-feeding insects and ticks modulate host haemostasis, immune defences, angiogenesis, tissue remodelling and wound healing, and facilitate the transmission and establishment of infectious agents they transmit.

The first tick EST published study described a cDNA library prepared from whole, unfed larvae of *B. microplus* (Crampton *et al.*, 1998). Of the 234 unique ESTs identified, database matches were not found for 39% of them. Whole *A. americanum* larvae and adults were used to construct cDNA libraries, and 1462 adult and 480 larval ESTs were sequenced with 56% having limited or no similarity to previously identified encoded proteins (Hill & Gutierrez, 2000). A far more extensive analysis of *B. microplus* gene expression was based on a normalized cDNA library prepared with RNA pooled from larvae exposed to different stimuli (acaricides, heat, cold, host odour), infected with *Babesia bovis*, as well as ova, nymphs, adults and dissected tissues (Guerrero *et al.*, 2005). Of 11 520 clones sequenced, 8270 unique sequences were identified with 44% of sequences sharing significant similarity to sequences in public databases. Specific changes in transcripts due to infection could not be identified in this study.

Tick salivary gland transcripts were characterized for unfed and 5-day-fed *I. ricinus* (Leboulle *et al.*, 2002*b*); 7- to 14-day-fed male *A. americanum* and *D. andersoni* (Bior, Essenberg & Sauer, 2002); 5-day-fed female *A. variegatum* (Nene *et al.*, 2002); female *I. pacificus* (Francischetti *et al.*, 2005); unfed female *I. scapularis* or females fed for 6–12 hours, 18–24 hours, 3 to 4 days and unfed or fed nymphs (Ribeiro *et al.*, 2006); and 18–24-hour-fed female *D. andersoni* (Alarcon-Chaidez, Sun & Wikel, 2007). *Ixodes scapularis*

salivary gland transcripts are the most extensively analysed to date over the course of feeding (Ribeiro *et al.*, 2006). Secreted proteins comprised 49% with 15% of the transcripts having no in silico matches with any known gene. Approximately 0.5% of transcripts represented transposable elements, indicating the ongoing potential for gene rearrangement. Over-expressed variants were not detected in the salivary glands of 6–12-hour-fed females; the highest number occurred in salivary glands of 18–24-hour-fed females. Biologically relevant genes are likely to be redundant and encode antigenic variants. Identified gene families whose products are potentially involved in antihaemostasis are listed in Table 4.2 (Chapter 4). Other gene groups include those for: lipocalins; short-coding cysteine-rich peptides that potentially cleave cartilage and inhibit angiogenesis; proline and glycine-rich peptides that might inhibit angiogenesis or contribute to attachment cement; small peptides that resemble scorpion toxins; anti-protease peptides that include cystatins; and mucins that may contribute to mouthpart structure.

Dermacentor andersoni is a member of the Metastriata while *I. scapularis* is a member of the evolutionarily earlier phyletic line, the Prostriata (Sonenshine, 1991; see Chapter 1). Although many similar predicted genes are found in both lineages, differences exist in the salivary gland transcripts identified in these lineages. One example is that both *A. variegatum* (Nene *et al.*, 2002) and *R. appendiculatus* (Nene *et al.*, 2004) have genes encoding products showing sequence similarity with an immunosuppressant protein first described in *D. andersoni* salivary glands (Bergman *et al.*, 2000), but not detected in the *I. scapularis* salivary gland transcriptome. Approximately 26% of *D. andersoni* salivary gland transcripts identified to date have no in silico matches with any known gene in public databases; 5% of transcripts encode putative proteases; 4% of genes are related to possible attachment cement components; and 2% encoded proteins potentially link to modulation of host immune defenses (Alarcon-Chaidez, Sun & Wikel, 2007).

Infection with a tick-borne pathogen has surprisingly little impact on tick gene expression according to the few studies reported to date. No major differences were detected in abundantly expressed transcripts when 9162 ESTs from uninfected *R. appendiculatus* salivary glands were compared with 9844 ESTs prepared from salivary glands of the same tick species infected with *Theileria parva* (Nene *et al.*, 2004). Fully engorged female *I. ricinus* were held for 7 days after detachment prior to isolation of RNA and preparation of cDNA libraries from whole ticks that fed on uninfected or *B. burgdorferi*-infected-guinea pigs (Rudenko *et al.*, 2005).

Consumption of an infected blood meal resulted in induction of a defensin-like gene in the midgut and two thioredoxin peroxidase genes in haemolymph, midgut and salivary glands. Significant differential representation occurred for 10 EST clusters of secreted proteins when salivary glands of *B. burgdorferi*-infected *I. scapularis* nymphs were compared to ESTs from salivary glands of uninfected fed nymphs (Ribeiro *et al.*, 2006).

In addition to EST analyses of salivary glands and other tissues, whole genomes are being sequenced for a number of important arthropod vectors. The results are relevant to all areas of vector biology, host–vector–pathogen relationships, and vector control research. The first vector genome sequenced was that of *Anopheles gambiae* (Holt *et al.*, 2002) and the first release of the *A. aegypti* genome sequence occurred on 14 June 2006 (http://msc.tigr.org/aedes/release.shtml). Genomes of blood-feeding arthropods being sequenced as of October 2006 included the body louse (*Pediculus humanus humanus* www.genome.gov/10002154), mosquito (*Culex pipiens quinquefasciatus* www.broad.mit.edu/), tsetse fly (*Glossina morsitans morsitans* www.sanger.ac.uk/), sandflies (*Lutziomiya longipalpis* and *Phlebotomus papatasi* www.genome.gov/10002154), and the tick *I. scapularis* (Hill and Wikel, 2005). Additional tick genomes will be sequenced in the future. Comparison of a member of the Metastriata with *I. scapularis* genome is a logical next step. A compelling argument has been made for a genome project for the southern cattle tick, *B. microplus* (Guerrero *et al.*, 2006). However, the large size of tick genomes is constraining progress. The *I. scapularis* genome is approximately 2.1 Gbp and that of *B. microplus* 7.1 Gbp (Ullmann *et al.*, 2005), more than twice the size of the 3.2-Gbp human genome (Venter *et al.*, 2001). Both of these tick genomes are larger than insect genomes (Ullmann *et al.*, 2005).

powerful tools for understanding the biology, evolutionary relationships and host–vector–pathogen interactions of blood-feeding arthropods. The next steps are to take the information gained from the use of these fundamental tools of modern biology and address the functional genomics of gene transcription, gene translation profiles, protein–protein interactions and functions of encoded molecules. As would be expected, these types of studies are beginning to be reported. High-throughput approaches are needed to determine predicted functional activities adapted to cloning, screening and DNA vaccination (see Chapter 4). Protein analysis by two-dimensional gel electrophoresis and mass spectrometry provide tools for moving to protein analysis in the post-genomic phase of analysis (Untalan *et al.*, 2005; Vennestrom & Jensen, 2007). RNA interference (RNAi) is increasingly used to assess tick gene function. For example, capillary feeding was used to deliver double-stranded RNA (dsRNA) to silence an anti-complement gene of *I. scapularis*, which reduced tick feeding weight but also affected expression of several salivary gland and other proteins, suggesting that more than the target tick gene was silenced by RNAi (Soares *et al.*, 2005). This observation points out a potential problem with RNAi, since transcriptome studies indicate the presence of gene families characterized by antigenic variation. Silencing one gene might not result in a phenotype or it might cause the silencing of related and unrelated genes.

These tools will result in an accelerated pace of discovery, particularly in previously difficult to study aspects of tick biology. The challenge to researchers is how to use genomic, proteomic and functional genomic data to address practical problems of disease and vector control to improve the health of humans, livestock, companion animals and wildlife.

CONCLUSIONS

Ticks are of vast medical and veterinary public health importance due to direct damage caused by feeding and their roles in transmitting well-known and emerging infectious agents. Ticks and tick-borne pathogens stimulate the immune system of the host. Those immune interactions are of importance in tick biology, pathogen transmission and the control of ticks and tick-borne diseases. Both innate and specific acquired immune defences are involved in the responses of vertebrate hosts to infestation. Ticks have evolved countermeasures to circumvent host immune defences. Transcriptome analysis and whole-genome sequencing are

ACKNOWLEDGEMENTS

The research of S.K.W. is supported in part by National Institutes of Health Grant 1R01-AI062735 and the United States Army Medical Research and Materiel Command award DAMD 17-03-1-0075.

REFERENCES

Adams, M. D., Kelley, J. M., Gocayne, J. D., *et al.* (1991). Complementary DNA sequencing: expressed sequence tags and human genome project. *Science* **252**, 1651–1656.

Alarcon-Chaidez, F. J., Sun, J. & Wikel, S. K. (2007). Transcriptome analysis of the salivary glands of the tick *Dermacentor andersoni* Stiles (Acari: Ixodidae). *Insect Biochemistry and Molecular Biology* **37**, 48–71.

Allen, J. R. (1973). Tick resistance: basophils in skin reactions of resistant guinea pigs. *International Journal for Parasitology* **3**, 195–200.

Allen, J. R., Doube, B. M. & Kemp, D. H. (1977). Histology of bovine skin reactions to *Ixodes holocyclus* Neuman. *Canadian Journal of Comparative Medicine* **41**, 26–35.

Allen, J. R., Khalil, H. M. & Wikel, S. K. (1979). Langerhans cells trap tick salivary gland antigens in tick-resistant guinea pigs. *Journal of Immunology* **122**, 563–565.

Andrade, B. B., Teixeira, C. R., Barral, A. & Barral-Netto, M. (2005). Haematophagous arthropod saliva and host defense system: a tale of tear and blood. *Annals of the Brazilian Academy of Sciences* **77**, 665–693.

Anguita, J., Ramamoorthi, N., Hovius, J. W. R., *et al.* (2002). Salp15, and *Ixodes scapularis* salivary protein, inhibits CD4+ T cell activation. *Immunity* **16**, 849–859.

Askenase, P. W., Debernardo, R., Tauben, D. & Kashgarian, M. (1978). Cutaneous basophil anaphylaxis: immediate vasopermeability increases and anaphylactic degranulation of basophils at delayed hypersensitivity reactions challenged with additional antigen. *Immunology* **35**, 741–755.

Bell, J. F., Stewart, S. J. & Wikel, S. K. (1979). Resistance to tick-borne *Francisella tularensis* by tick-sensitized rabbits: allergic klendusity. *American Journal of Tropical Medicine and Hygiene* **28**, 876–880.

Bergman, D. K., Palmer, M. J., Caimano, M. J., Radolf, J. D. & Wikel, S. K. (2000). Isolation and cloning of a secreted immunosuppressant protein from *Dermacentor andersoni* salivary gland. *Journal of Parasitology* **86**, 516–525.

Bergman, D. K., Ramachandra, R. N. & Wikel, S. K. (1998). Characterization of an immunosuppressant protein from *Dermacentor andersoni* (Acari: Ixodidae) salivary glands. *Journal of Medical Entomology* **35**, 505–509.

Berthier, R., Martinon-Ego, C., Laharie, A. M. & Marche, P. N. (2000). A two-step culture method starting with early growth factors permits enhanced production of functional dendritic cells from murine splenocytes. *Journal of Immunological Methods* **239**, 95–107.

Bior, A. D., Essenberg, R. C. & Sauer, J. R. (2002). Comparison of differentially expressed genes in the salivary glands of male ticks, *Amblyomma americanum* and *Dermacentor andersoni*. *Insect Biochemistry and Molecular Biology* **32**, 645–655.

Bowessidjaou, J., Brossard, M. & Aeschlimann, A. (1977). Effects and duration of resistance acquired by rabbits on feeding and egg laying in *Ixodes ricinus* L. *Experientia* **33**, 548–550.

Bowman, A. S., Dillwith, J. W. & Sauer, J. R. (1996). Tick salivary prostaglandins: presence, origin and significance. *Parasitology Today* **12**, 388–396.

Brossard, M. (1976). Relations immunologiques entre Bovins et Tiques, plus particulièrement entre Bovins et *Boophilus microplus*. *Acta Tropica* **33**, 15–36.

Brossard, M. (1982). Rabbits infested with adult *Ixodes ricinus* L.: effects of mepyramine on acquired resistance. *Experientia* **38**, 702–704.

Brossard, M. & Fivaz, V. (1982). *Ixodes ricinus* L.: mast cells, basophils and eosinophils in the sequence of cellular events in the skin of infested or reinfested rabbits. *Parasitology* **85**, 583–592.

Brossard, M. & Girardin, P. (1979). Passive transfer of resistance in rabbits infested with adult *Ixodes ricinus* L: Humoral factors influence feeding and egg laying. *Experientia* **35**, 1395–1396.

Brossard, M. & Wikel, S. K. (1997). Immunology of interactions between ticks and hosts. *Medical and Veterinary Entomology* **11**, 270–276.

Brossard, M. & Wikel, S. K. (2004). Tick immunobiology. *Parasitology* **129**, S161–S176.

Brossard, M., Monneron, J. P. & Papatheodorou, V. (1982). Progressive sensitization of circulating basophils against *Ixodes ricinus* L. antigens during repeated infestations of rabbits. *Parasite Immunology* **4**, 355–361.

Brown, S. J. & Askenase, P. W. (1981). Cutaneous basophil responses and immune resistance of guinea pigs to ticks: passive transfer with peritoneal exudate cells or serum. *Journal of Immunology* **127**, 2163–2167.

Brown, S. J., Barker, R. W. & Askenase, P. W. (1984). Bovine resistance to *Amblyomma americanum* ticks: an acquired immune response characterized by cutaneous basophil infiltrates. *Veterinary Parasitology* **16**, 147–165.

Champagne, D. E. (1994). The role of salivary vasodilators in blood-feeding and parasite transmission. *Parasitology Today* **10**, 430–433.

Champagne, D. E. & Valenzuela, J. G. (1996). Pharmacology of haematophagous arthropod saliva. In *The Immunology of Host–Ectoparasitic Arthropod Relationships*, ed. Wikel, S. K., pp. 85–106. Wallingford, UK: CAB International.

Chinery, W. A. & Ayitey-Smith, E. (1977). Histamine blocking agent in the salivary gland homogenate of the tick, *Rhipicephalus sanguineus*. *Nature* **265**, 366–367.

Christe, M., Rutti, B. & Brossard, M. (1999). Influence of genetic background and parasite load of mice on immune response developed against nymphs of *Ixodes ricinus*. *Parasitology Research* **85**, 557–561.

Christe, M., Rutti, B. & Brossard, M. (2000). Cytokines (IL-4 and IFN-γ) and antibodies (IgE and IgG2a) produced in mice infected with *Borrelia burgdorferi* sensu stricto via nymphs of *Ixodes ricinus* ticks or syringe inoculations. *Parasitology Research* **86**, 491–496.

Couvreur, B., Beaufays, J., Charon, C., *et al.* (2008). Variability and action mechanism of a family of anticomplement proteins in *Ixodes ricinus*. *PLoS ONE* 3(1): e1400 doi:10.1371/journal.pone.0001400.

Crampton, A. L., Miller, C., Baxter, G. D. & Barker, S. C. (1998). Expressed sequence tags and new genes from the cattle tick, *Boophilus microplus*. *Experimental and Applied Acarology* **22**, 177–186.

Cumberbatch, M. & Kimber, I. (1995). Tumour necrosis factor-α is required for accumulation of dendritic cells in draining lymph nodes and for optimal contact sensitization. *Immunology* **84**, 31–35.

Daix, V., Schroeder, H., Praet, N., *et al.* (2007). *Ixodes* ticks belonging to the *Ixodes ricinus* complex encode a family of anti-complement proteins. *Insect Molecular Biology* **16**, 155–166.

Decrem, Y., Beaufays, J., Blasioli, V., *et al.* (2008). A family of putative metalloproteases in the salivary glands of the tick *Ixodes ricinus*. *FEBS Journal* **275**, 1485–1489.

Demeure, C. E., Yang, L. P., Desjardins, C., Raynauld, P. & Delespesse, G. (1997). Prostaglandin E₂ primes naive T cells for the production of anti-inflammatory cytokines. *European Journal of Immunology* **27**, 3526–3531.

DenHollander, N. & Allen, J. R. (1985). *Dermacentor variabilis*: acquired resistance to ticks in BALB/c mice. *Experimental Parasitology* **59**, 118–129.

Dizij, A. & Kurtenbach, K. (1995). *Clethrionomys glareolus*, but not *Apodemus flavicollis*, acquired resistance to *Ixodes ricinus* L., the main European vector of *Borrelia burgdorferi*. *Parasite Immunology* **17**, 177–183.

Edwards, J. F., Higgs, S. & Beaty, B. J. (1998). Mosquito feeding-induced enhancement of Cache Valley virus (Bunyaviridae) infection in mice. *Journal of Medical Entomology* **35**, 261–265.

Esmon, C. T. (2004). Interactions between the innate immune and blood coagulation systems. *Trends in Immunology* **25**, 536–542.

Ferreira, B. R. & Silva, J. S. (1998). Saliva of *Rhipicephalus sanguineus* tick impairs T cell proliferation and

IFN-gamma-induced macrophage microbicidal activity. *Veterinary Immunology and Immunopathology* **31**, 279–293.

Ferreira, B. R. & Silva, J. S. (1999). Successive tick infestations selectively promote a T-helper 2 cytokine profile in mice. *Immunology* **96**, 434–439.

Ferreira, C. A. S., Vaz, I. D. S. Jr, Da Silva, S. S., *et al.* (2002). Cloning and partial characterization of a *Boophilus microplus* (Acari: Ixodidae) calreticulin. *Experimental Parasitology* **101**, 25–34.

Fivaz, B. H. (1989). Immune suppression induced by the brown ear tick *Rhipicephalus appendiculatus* Neumann, 1901. *Journal of Parasitology* **75**, 946–952.

Francischetti, I. M., Mather, T. N. & Ribeiro, J. M. (2003). Cloning of a salivary gland metalloprotease and characterization of gelatinase and fibrin(ogen)lytic activities in the saliva of the Lyme disease tick vector *Ixodes scapularis*. *Biochemical and Biophysical Research Communications* **305**, 869–875.

Francischetti, I. M., Mather, T. N. & Ribeiro, J. M. (2004). Penthalaris, a novel recombinant five-Kunitz tissue factor pathway inhibitor (TFPI) from the salivary gland of the tick vector of Lyme disease, *Ixodes scapularis*. *Thrombosis and Haemostasis* **91**, 886–98.

Francischetti, I. M., Mather, T. N. & Ribeiro, J. M. (2005). Tick saliva is a potent inhibitor of endothelial cell proliferation and angiogenesis. *Thrombosis and Haemostasis* **94**, 167–174.

Francischetti, I. M. B., Pham, V. M., Mans, B. J., *et al.* (2005). The transcriptome of the salivary glands of the female western black-legged tick *Ixodes pacificus* (Acari: Ixodidae). *Insect Biochemistry and Molecular Biology* **35**, 1142–1161.

Fuchsberger, N., Kita, M., Hajnicka, V., *et al.* (1995). Ixodid tick salivary gland extracts inhibit production of lipopolysaccharide-induced mRNA of several different human cytokines. *Experimental and Applied Acarology* **19**, 671–676.

Furie, B. & Furie, B. C. (2005). Molecular basis of blood coagulation. In *Hematology: Basic Principles and Practice*, 4th edn, eds. Hoffman, R., Benz, E. J. Jr, Shatti, S. J., *et al.*, pp. 1931–1953. Philadelphia, PA: Elsevier.

Ganapamo, F., Rutti, B. & Brossard, M. (1995a). *In vitro* production of interleukin-4 and interferon-gamma by lymph node cells from BALB/c mice infested with nymphal *Ixodes ricinus* ticks. *Immunology* **85**, 120–124.

Ganapamo, F., Rutti, B. & Brossard, M. (1995b). Spleen accessory cells antigen processing and *in vitro* induction of specific lymphocyte proliferation in BALB/c mice infested with nymphal *Ixodes ricinus* ticks. In *Dendritic Cells in Fundamental and Clinical Immunology*, vol. 2, eds. Banchereau, J. & Schmitt, D., pp. 195–197. New York: Plenum Press.

Ganapamo, F., Rutti, B. & Brossard, M. (1996*a*). Cytokine production by lymph node cells from mice infested with *Ixodes ricinus* ticks and the effect of tick salivary gland extracts on IL-2 production. *Scandinavian Journal of Immunology* **44**, 388–393.

Ganapamo, F., Rutti, B. & Brossard, M. (1996*b*). Immunosuppression and cytokine production in mice infested with *Ixodes ricinus* ticks: a possible role of laminin and interleukin-10 on the *in vitro* responsiveness of lymphocytes to mitogens. *Immunology* **87**, 259–263.

Ganapamo, F., Rutti, B. & Brossard, M. (1997). Identification of an *Ixodes ricinus* salivary gland fraction through its ability to stimulate CD4 T cells present in BALB/c mice lymph nodes draining the tick fixation site. *Parasitology* **115**, 91–96.

Gebbia, J. A., Coleman, J. L. & Benach, J. L. (2001). *Borrelia* spirochetes upregulate release and activation of matrix metalloproteinase gelatinase B (MMP-9) and collagenase (MMP-1) in human cells. *Infection and Immunity* **69**, 456–462.

George, J. E., Osburn, R. L. & Wikel, S. K. (1985). Acquisition and expression of resistance by *Bos indicus* and *Bos indicus* × *Bos taurus* calves to *Amblyomma americanum* infestation. *Journal of Parasitology* **71**, 174–182.

Gill, H. S. (1986). Kinetics of mast cell, basophil and eosinophil populations at *Hyalomma anatolicum anatolicum* feeding sites on cattle and the acquisition of resistance. *Parasitology* **93**, 305–315.

Gill, H. S. & Walker, A. R. (1985). Differential cellular responses at *Hyalomma anatolicum anatolicum* feeding sites on susceptible and tick-resistant rabbits. *Parasitology* **91**, 591–607.

Gillespie, R. D., Dolan, M. C., Piesman, J. & Titus, R. G. (2001). Identification of an IL-2 binding protein in the saliva of the Lyme disease vector tick, *Ixodes scapularis*. *Journal of Immunology* **166**, 4319–4327.

Girardin, P. & Brossard, M. (1989). Effects of cyclosporin-A on the humoral immunity to ticks, and on the cutaneous immediate (type I) and delayed (type IV) hypersensitivity reactions to *Ixodes ricinus*, L. salivary gland antigens in re-infested rabbits. *Parasitology Research* **75**, 657–662.

Girardin, P. & Brossard, M. (1990). Rabbits infested with *Ixodes ricinus* L. adults: effects of treatment with cyclosporin A on the biology of ticks fed on resistant or naive hosts. *Annales de Parasitologie Humain et Comparative* **65**, 262–266.

Grakoui, A., Bromley, S. K., Sumen, C., *et al.* (1999). The immunological synapse: a molecular machine controlling T cell activation. *Science* **285**, 221–227.

Guerrero, F. D., Miller, R. J., Rousseau, M.-E., *et al.* (2005). BmiGI: a database of cDNAs expressed in *Boophilus microplus*, the tropical/southern cattle tick. *Insect Biochemistry and Molecular Biology* **35**, 585–595.

Guerrero, F. D., Nene, V. M., George, J. E., Barker, S. C. & Willadsen, P. (2006). Sequencing a new target genome: the *Boophilus microplus* (Acari: Ixodidae) genome project. *Journal of Medical Entomology* **43**, 9–16.

Gwakisa, P., Yoshihara, K., To, T. L., *et al.* (2001). Salivary gland extract of *Rhipicephalus appendiculatus* ticks inhibits *in vitro* transcription and secretion of cytokines and production of nitric oxide by LPS-stimulated JA-4 cells. *Veterinary Parasitology* **99**, 53–61.

Hajnická, V., Kocakova, P., Slavikova, M., *et al.* (2001). Anti-interleukin-8 activity of tick salivary gland extracts. *Parasite Immunology* **23**, 483–489.

Hannier, S., Liversidge, J., Sternberg, J. M. & Bowman, A. S. (2003). *Ixodes ricinus* tick salivary gland extract inhibits IL-10 secretion and CD69 expression by mitogen-stimulated murine splenocytes and induces hyporesponsiveness in B lymphocytes. *Parasite Immunology* **25**, 27–37.

Hannier, S., Liversidge, J., Sternberg, J. M. & Bowman, A. S. (2004). Characterization of the B-cell inhibitory protein factor in *Ixodes ricinus* tick saliva: a potential role in enhanced *Borrelia burgdorferi* transmission. *Immunology* **113**, 401–408.

Harris, S. G., Padilla, J., Koumas, L., Ray, D. & Phipps, R. P. (2002). Prostaglandins as modulators of immunity. *Trends in Immunology* **23**, 144–150.

Hill, C. A. & Gutierrez, J. A. (2000). Analysis of the expressed genome of the lone star tick, *Amblyomma americanum* (Acari: Ixodidae) using expressed sequence tag approach. *Microbial and Comparative Genomics* **5**, 89–101.

Hill, C. A. & Wikel, S. K. (2005). The *Ixodes scapularis* genome project: an opportunity for advancing tick research. *Trends in Parasitology* **21**, 151–153.

Holt, R. A., Subramanian, G. M., Halpern, A., *et al.* (2002). The genome sequence of the malaria mosquito *Anopheles gambiae*. *Science* **298**, 129–149.

Inokuma, H., Kemp, D. H. & Willadsen, P. (1994). Prostaglandin E2 production by the cattle tick (*Boophilus microplus*) into feeding sites and its effect on the response of bovine mononuclear cells to mitogen. *Veterinary Parasitology* **53**, 293–299.

Janeway, C. A. Jr, Travers, P., Walport, M. & Shlomchik, M. J. (2005). *Immunobiology*, 6th edn. New York: Garland.

Jaworski, D. C., Jasinskas, A., Metz, C. N., Bucala, R. & Barbour, A. G. (2001). Identification and characterization of a homologue of the pro-inflammatory cytokine macrophage migration inhibitory factor in the tick, *Amblyomma americanum*. *Insect Molecular Biology* **10**, 323–331.

Jaworski, D. C., Simmen, F. A., Lamoreaux, W., *et al.* (1995). A secreted calreticulin protein in ixodid tick (*Amblyomma americanum*) saliva. *Journal of Insect Physiology* **41**, 369–375.

Jones, L. D. & Nuttall, P. A. (1990). The effect of host resistance to tick infestation on the transmission of Thogoto virus by ticks. *Journal of General Virology* **71**, 1039–1043.

Kamhawi, S., Belkaid, Y., Modi, G., Rowton, E. & Sacks, D. (2000). Protection against cutaneous leishmaniasis resulting from bites of uninfected sandflies. *Science* **290**, 1351–1354.

Kopecky, J. & Kuthejlova, M. (1998). Suppressive effect of *Ixodes ricinus* salivary gland extract on mechanisms of natural immunity *in vitro*. *Parasite Immunology* **20**, 169–174.

Kotsyfakis, M., Sa-Nuñes, A., Francischetti, I. M. B., *et al.* (2006). Anti-inflammatory and immunosuppressive activity of sialostatin L, a salivary cystatin from the tick, *Ixodes scapularis*. *Journal of Biological Chemistry* **36**, 26298–26307.

Koudstaal, D., Kemp, D. H. & Kerr, J. D. (1978). *Boophilus microplus*: rejection of larvae from British breed cattle. *Parasitology* **76**, 379–386.

Krocova, Z., Macela, A., Hernychova, L., *et al.* (2003). Tick salivary gland extract accelerates proliferation of *Francisella tularensis* in the host. *Journal of Parasitology* **89**, 14–20.

Kubes, M., Kocakova, P., Slovak, M., *et al.* (2002). Heterogeneity in the effect of different ixodid tick species on natural killer cell activity. *Parasite Immunology* **24**, 23–28.

Kuthejlova, M., Kopecky, J., Stepanova, G. & Macela, A. (2001). Tick salivary gland extract inhibits killing of *Borrelia afzelli* spirochetes by mouse macrophages. *Infection and Immunity* **69**, 575–578.

Latif, A. A. (1984). Resistance to *Hyalomma anatolicum anatolicum* Koch (1844) and *Rhipicephalus evertsi* Neumann (1897) (Ixodoidea: Ixodidae) by cattle in Sudan. *Insect Science Applications* **5**, 509–511.

Latif, A. A., Punyua, D. K., Capstick, P. B. & Newson, R. M. (1991). Tick infestations on Zebu cattle in Western Kenya: host resistance to *Rhipicephalus appendiculatus* (Acari: Ixodidae). *Journal of Medical Entomology* **28**, 127–132.

Lawrie, C. H., Randolph, S. E. & Nuttall, P. A. (1999). *Ixodes* ticks: serum species sensitivity of anti-complement activity. *Experimental Parasitology* **93**, 207–214.

Leboulle, G., Crippa, M., Decrem, Y., *et al.* (2002*a*). Characterization of a novel salivary immunosuppressive protein from *Ixodes ricinus* ticks. *Journal of Biological Chemistry* **277**, 10083–10089.

Leboulle, G., Rochez, C., Louahed, J., *et al.* (2002*b*). Isolation of *Ixodes ricinus* salivary gland mRNA encoding factors induced during blood feeding. *American Journal of Tropical Medicine and Hygiene* **66**, 225–233.

Lorimier, Y. (2003). Réponse immunitaire de souris à la salive de tiques *Ixodes ricinus*: influence des cellules dendritiques et Tγδ. Ph.D. thesis, University of Neuchâtel, Switzerland. Available at http://doc.rero.ch/search.py?recid=2625&ln=fr.

Macaluso, K. R. & Wikel, S. K. (2001). *Dermacentor andersoni*: effects of repeated infestations on lymphocyte proliferation, cytokine production, and adhesion-molecule expression by BALB/c mice. *Annals of Tropical Medicine and Parasitology* **95**, 413–427.

Machackova, M., Obornik, M. & Kopecky, J. (2006). Effect of salivary gland extract from *Ixodes ricinus* ticks on the proliferation of *Borrelia burgdorferi* sensu stricto *in vivo*. *Folia Parasitologia* **53**, 153–158.

Maxwell, S. S., Stoklasek, T. A., Dash, Y., Macaluso, K. R. & Wikel, S. K. (2005). Tick modulation of the in-vitro expression of adhesion molecules by skin-derived endothelial cells. *Annals of Tropical Medicine and Parasitology* **99**, 661–672.

Mbow, M. L., Christe, M., Rutti, B. & Brossard, M. (1994). Absence of acquired resistance to nymphal *Ixodes ricinus* ticks in BALB/c mice developing cutaneous reactions. *Journal of Parasitology* **80**, 81–87.

Mejri, N. & Brossard, M. (2007). Splenic dendritic cells pulsed with *Ixodes ricinus* tick saliva prime naive CD4[+] T to induce Th2 cell differentiation *in vitro* and *in vivo*. *International Immunology* **19**, 535–543.

Mejri, N., Franscini, N., Rutti, B. & Brossard, M. (2001). Th2 polarization of the immune response of BALB/c mice to *Ixodes ricinus* instars, importance of several antigens in activation of specific Th2 subpopulations. *Parasite Immunology* **23**, 61–69.

Mejri, N., Rutti, B. & Brossard, M. (2002). Immunosuppressive effects of *Ixodes ricinus* tick saliva or salivary gland extracts on innate and acquired immune response of BALB/c mice. *Parasitology Research* **88**, 192–197.

Mosmann, T. R. & Coffman, R. L. (1989). Th1 and Th2 cells: different patterns of lymphokine secretion lead to different functional properties. *Annual Review of Immunology* 7, 145–173.

Mulenga, A. & Azad, A. F. (2005). The molecular and biological analysis of ixodid ticks histamine releasing factors. *Experimental and Applied Acarology* 37, 215–229.

Mulenga, A., Macaluso, K. R., Simser, J. A. & Azad, A. F. (2003). The American dog tick encodes a functional histamine release factor homolog. *Insect Biochemistry and Molecular Biology* 33, 911–919.

Nabi, I. R. (1999). The polarization of the motile cell. *Journal of Cell Science* 112, 1803–1811.

Nazareth, R. A., Tomaz, L. S., Ortiz-Costa, S., *et al.* (2006). Antithrombic properties of ixolaris, a potent inhibitor of the extrinsic pathway of the coagulation cascade. *Thrombosis and Haemostasis* 96, 7–13.

Nazario, S., Das, S., De Silva, A. M., *et al.* (1998). Prevention of *Borrelia burgdorferi* transmission in guinea pigs by tick immunity. *American Journal of Tropical Medicine and Hygiene* 58, 780–785.

Nene, V., Lee, D., Kang'A, S., *et al.* (2004). Genes transcribed in the salivary glands of female *Rhipicephalus appendiculatus* ticks infected with *Theileria parva*. *Insect Biochemistry and Molecular Biology* 34, 1117–1128.

Nene, V., Lee, D., Quackenbush, J., *et al.* (2002). AvGI, and index of genes transcribed in the salivary glands of the ixodid tick *Amblyomma variegatum*. *International Journal for Parasitology* 32, 1447–1456.

Nithiuthai, S. & Allen, J. R. (1984a). Significant changes in epidermal Langerhans cells of guinea pigs infested with ticks (*Dermacentor andersoni*). *Immunology* 51, 133–141.

Nithiuthai, S. & Allen, J. R. (1984b). Effects of ultraviolet irradiation on the acquisition and expression of tick resistance in guinea pigs. *Immunology* 51, 153–159.

Nuttall, P. A. & Labuda, M. (2004). Tick–host interactions: saliva-activated transmission. *Parasitology* 129, S177–S189.

Paesen, G. C., Adams, P. L., Harlos, K., Nuttall, P. A. & Stuart, D. I. (1999). Tick histamine-binding proteins: isolation, cloning, and three-dimensional structure. *Molecular Cell* 3, 661–671.

Paine, S. H., Kemp, D. H. & Allen, J. R. (1983). *In vitro* feeding of *Dermacentor andersoni* (Stiles): effects of histamine and other mediators. *Parasitology* 86, 419–428.

Papatheodorou, V. & Brossard, M. (1987). C3 levels in the sera of rabbits infested and reinfested with *Ixodes ricinus* L. and in midguts of fed ticks. *Experimental and Applied Acarology* 3, 53–59.

Plow, E. F. & Abrams, C. S. (2005). The molecular basis of platelet function. In *Hematology: Basic Principles and Practice*, 4th edn, eds. Hoffman, R., Benz, E. J. Jr, Shatti, S. J., *et al.*, pp. 1881–1897. Philadelphia, PA: Elsevier.

Prévôt, P. P., Adam, B., Boudjeltia, K. Z., *et al.* (2006). Anti-hemostatic effects of a serpin from the saliva of the tick *Ixodes ricinus*. *Journal of Biological Chemistry* 281, 26361–26369.

Ramachandra, R. N. & Wikel, S. K. (1992). Modulation of host-immune responses by ticks (Acari: Ixodidae): effect of salivary gland extracts on host macrophages and lymphocyte cytokine production. *Journal of Medical Entomology* 29, 818–826.

Ramachandra, R. N. & Wikel, S. K. (1995). Effects of *Dermacentor andersoni* (Acari: Ixodidae) salivary gland extracts on *Bos indicus* and *B. taurus* lymphocytes and macrophages: *in vitro* cytokine elaboration and lymphocyte blastogenesis. *Journal of Medical Entomology* 32, 338–345.

Ribeiro, J. M. C. (1987). Role of saliva in blood-feeding by arthropods. *Annual Review of Entomology* 32, 463–478.

Ribeiro, J. M. C. (1989). Role of saliva in tick/host interactions. *Experimental and Applied Acarology* 7, 15–20.

Ribeiro, J. M. C. (1995a). Blood-feeding arthropods: live syringes or invertebrate pharmacologists? *Infectious Agents and Disease* 4, 143–152.

Ribeiro, J. M. C. (1995b). How ticks make a living. *Parasitology Today* 11, 91–93.

Ribeiro, J. M. C. (2004). Bugs, blood, and blisters. *Journal of Investigative Dermatology* 123, xvi.

Ribeiro, J. M. C. & Francischetti, I. M. B. (2003). Role of arthropod saliva in blood feeding: sialome and post-sialome perspectives. *Annual Review of Entomology* 48, 73–88.

Ribeiro, J. M. C. & Mather, T. N. (1998). *Ixodes scapularis*: salivary kininase activity is a metallo dipeptidyl carboxypeptidase. *Experimental Parasitology* 89, 213–221.

Ribeiro, J. M. C. & Spielman, A. (1986). *Ixodes dammini*: salivary anaphylatoxin inactivating activity. *Experimental Parasitology* 62, 292–297.

Ribeiro, J., Alarcon-Chaidez, F., Francischetti, I. M. B., *et al.* (2006). An annotated catalog of salivary gland transcripts from *Ixodes scapularis* ticks. *Insect Biochemistry and Molecular Biology* 36, 111–129.

Ribeiro, J. M. C., Makoul, G. T., Levine, J., Robinson, D. R. & Spielman, A. (1985). Antihemostatic, antiinflammatory, and immunosuppressive properties of the saliva of a tick, *Ixodes dammini*. *Journal of Experimental Medicine* 161, 332–344.

Roberts, J. A. & Kerr, J. D. (1976). *Boophilus microplus*: passive transfer of resistance in cattle. *Journal of Parasitology* **62**, 485–488.

Rudenko, N., Golovchenko, M., Edwards & Grubhoffer, L. (2005). Differential expression of *Ixodes ricinus* tick genes induced by blood feeding or *Borrelia burgdorferi* infection. *Journal of Medical Entomology* **42**, 36–41.

Sahibi, H. F., Rhalem, A., Tikki, N., Ben Kouka, F. & Barriga, O. (1997). *Hyalomma* ticks: bovine resistance under field conditions as related to host age and breed. *Parasite* **4**, 159–165.

Salmon, J. K., Armstrong, C. A. & Ansel, J. C. (1994). The skin as an immune organ. *Western Journal of Medicine* **160**, 146–152.

Sangamnatdej, S., Paesen, G. C., Slovak, M. & Nuttall, P. A. (2002). A high affinity serotonin- and histamine-binding lipocalin from tick saliva. *Insect Molecular Biology* **11**, 79–86.

Schleger, A. V., Lincoln, D. T., Mckenna, R. V., Kemp, D. H. & Roberts, J. A. (1976). *Boophilus microplus*: cellular responses to larval attachment and their relationship to host resistance. *Australian Journal of Biological Science* **29**, 499–512.

Schneider, B. S., Soong, L., Girard, Y. A., et al. (2006). Potentiation of West Nile encephalitis by mosquito feeding. *Viral Immunology* **19**, 74–82.

Schoeler, G. B. & Wikel, S. K. (2001). Modulation of host immunity by haematophagous arthropods. *Annals of Tropical Medicine and Parasitology* **95**, 755–771.

Schoeler, G. B., Manweiler, S. A. & Wikel, S. K. (2000). Cytokine responses of C3H/HeN mice infested with *Ixodes scapularis* or *Ixodes pacificus* nymphs. *Parasite Immunology* **22**, 31–40.

Schorderet, S. & Brossard, M. (1994). Effects of human recombinant interleukin-2 on the resistance, and on the humoral and cellular response of rabbits infested with adult *Ixodes ricinus* ticks. *Veterinary Parasitology* **54**, 375–387.

Snowball, G. J. (1956). The effect of self-licking by cattle on infestation of cattle tick, *Boophilus microplus* (Canestrini). *Australian Journal of Agricultural Research* **7**, 227–232.

Soares, C. A. G., Lima, C. M. R., Dolan, M. C., et al. (2005). Capillary feeding of specific dsRNA induces silencing of the *isac* gene in nymphal *Ixodes scapularis* ticks. *Insect Molecular Biology* **14**, 443–452.

Sonenshine, D. E. (1991). *Biology of Ticks*, vol. 1. Oxford, UK: Oxford University Press.

Steen, N. A., Barker, S. C. & Alewood, P. F. (2006). Proteins in the saliva of the Ixodidae (ticks): pharmacological features and biological significance. *Toxicon* **47**, 1–20.

Steeves, E. B. & Allen, J. R. (1991). Tick resistance in mast cell-deficient mice: histological studies. *International Journal for Parasitology* **21**, 265–268.

Styer, L. M., Bernard, K. A. & Kramer, L. D. (2006). Enhanced early West Nile virus infection in young chickens infected by mosquito bite: effect of viral dose. *American Journal of Tropical Medicine and Hygiene* **75**, 337–345.

Titus, R. G. & Ribeiro, J. M. C. (1988). Salivary gland lysates from the sand fly, *Lutzomyia longipalpis*, enhance *Leishmania* infectivity. *Science* **239**, 1306–1308.

Titus, R. G. & Ribeiro, J. M. C. (1990). The role of vector saliva in transmission of arthropod-borne disease. *Parasitology Today* **6**, 157–160.

Trager, W. (1939). Acquired immunity to ticks. *Journal of Parasitology* **25**, 57–81.

Ullmann, A. J., Lima, C. M. R., Guerrero, F. D., Piesman, J. & Black, W. C. IV (2005). Genome size and organization in the blacklegged tick, *Ixodes scapularis*, and the Southern cattle tick, *Boophilus microplus*. *Insect Molecular Biology* **14**, 217–222.

Umemiya, R., Hatta, T., Liao, M., et al. (2007). *Haemaphysalis longicornis*: molecular characterization of a homologue of the macrophage migration inhibitory factor from the partially fed ticks. *Experimental Parasitology* **115**, 135–142.

Untalan, P. M., Guerrero, F. D., Haines, L. R. & Pearson, T. W. (2005). Proteome analysis of abundantly expressed proteins from unfed larvae of the cattle tick, *Boophilus microplus*. *Insect Biochemistry and Molecular Biology* **35**, 141–151.

Urioste, S., Hall, L. R., Telford, S. R. III & Titus, R. G. (1994). Saliva of the Lyme disease vector, *Ixodes dammini*, blocks cell activation by a nonprostaglandin E2-dependent mechanism. *Journal of Experimental Medicine* **180**, 1077–1085.

Valbuena, G. & Walker, D. H. (2004). Effect of blocking the CXCL9/10-CXCR3 chemokine system on the outcome of endothelial-target rickettsial infections. *American Journal of Tropical Medicine and Hygiene* **71**, 393–399.

Valenzuela, J. G. (2004). Exploring tick saliva: from biochemistry to 'sialomes' and functional genomics. *Parasitology* **129**, S83–S94.

Valenzuela, J. G., Charlab, R., Mather, T. N. & Ribeiro, J. M. C. (2000). Purification, cloning, and expression of a novel salivary anticomplement protein from the tick, *Ixodes scapularis*. *Journal of Biological Chemistry* **275**, 18717–18723.

Valenzuela, J. G., Francischetti, I. M. B., Pham, V. M., et al. (2002). Exploring the sialome of the tick *Ixodes scapularis*. *Journal of Experimental Biology* **205**, 2843–2864.

Van Gisbergen, K. P. J. M., Paessens, L. C., Geijtenbeek, T. B. H. & Van Kooyk, Y. (2005). Molecular mechanisms that set the stage for DC-T cell engagement. *Immunology Letters* 97, 199–208.

Vaughan, J. A., Scheller, L. F., Wirtz, R. A. & Azad, A. F. (1999). Infectivity of *Plasmodium berghei* sporozoites delivered by intravenous inoculation versus mosquito bite: implications for sporozoite vaccine trials. *Infection and Immunity* 67, 4285–4289.

Vennestrom, J. & Jensin, P. M. (2007). *Ixodes ricinus*: the potential of two-dimensional gel electrophoresis as a tool for studying host–vector–pathogen interactions. *Experimental Parasitology* 115, 53–58.

Venter, J. C., Adams, M. D., Myers, E. W., *et al.* (2001). The sequence of the human genome. *Science* 291, 1304–1351.

Walker, D. H. (1998). Tick-transmitted infectious diseases in the United States. *Annual Review of Public Health* 19, 237–269.

Wang, H. & Nuttall, P. A. (1995). Immunoglobulin-G binding proteins in the ixodid ticks, *Rhipicephalus appendiculatus*, *Amblyomma variegatum* and *Ixodes hexagonus*. *Parasitology* 111, 161–165.

Wang, H., Kaufman, W. R. & Nuttall, P. A. (1999). Molecular individuality: polymorphism of salivary gland proteins in three species of ixodid tick. *Experimental and Applied Acarology* 23, 969–975.

Wikel, S. K. (1979). Acquired resistance to ticks: expression of resistance by C4-deficient guinea pigs. *American Journal of Tropical Medicine and Hygiene* 28, 586–590.

Wikel, S. K. (1982). Histamine content of tick attachment sites and the effects of H1 and H2 histamine antagonists on the expression of resistance. *Annals of Tropical Medicine and Parasitology* 76, 179–185.

Wikel, S. K. (1985). Effects of tick infestation on the plaque-forming cell response to a thymic dependent antigen. *Annals of Tropical Medicine and Parasitology* 79, 195–198.

Wikel, S. K. (1996). Host immunity to ticks. *Annual Review of Entomology* 41, 1–22.

Wikel, S. K. (1999). Tick modulation of host immunity: an important factor in pathogen transmission. *International Journal for Parasitology* 29, 851–859.

Wikel, S. K. & Allen, J. R. (1976). Acquired resistance to ticks. I. Passive transfer of resistance. *Immunology* 30, 311–316.

Wikel, S. K. & Allen, J. R. (1977). Acquired resistance to ticks. III. Cobra venom factor and the resistance response. *Immunology* 32, 457–465.

Wikel, S. K., Graham, J. E. & Allen, J. R. (1978). Acquired resistance to ticks. IV. Skin reactivity and *in vitro* lymphocyte responsiveness to salivary gland antigen. *Immunology* 34, 257–263.

Wikel, S. K., Ramachandra, R. N. & Bergman, D. K. (1994). Tick-induced modulation of the host immune response. *International Journal for Parasitology* 24, 59–66.

Wikel, S. K., Ramachandra, R. N., Bergman, D. K., Burkot, T. R. & Piesman, J. (1997). Infestation with pathogen-free nymphs of the tick *Ixodes scapularis* induces host resistance to transmission of *Borrelia burgdorferi* by ticks. *Infection and Immunity* 65, 335–338.

Willadsen, P. (1980). Immunity to ticks. *Advances in Parasitology* 18, 293–313.

Willadsen, P. (1987). Immunological approaches to the control of ticks. *International Journal for Parasitology* 17, 671–677.

Willadsen, P. & Jongejan, F. (1999). Immunology of the tick–host interaction and the control of ticks and tick-borne diseases. *Parasitology Today* 15, 258–262.

Willadsen, P., Wood, G. M. & Riding, G. A. (1979). The relation between skin histamine concentration, histamine sensitivity, and the resistance of cattle to the tick, *Boophilus microplus*. *Zeitschrift für Parasitenkunde* 59, 87–93.

Xu, G., Fang, Q. Q., Keirans, J. E. & Durden, L. A. (2004). Cloning and sequencing of putative calreticulin complementary DNAs from four hard tick species. *Journal of Parasitology* 90, 73–78.

Young, A. J. (1999). The physiology of lymphocyte migration through the single lymph node *in vivo*. *Seminars in Immunology* 11, 73–83.

Yu, D., Liang, J., Yu, H., *et al.* (2006). A tick B-cell inhibitory protein from salivary glands of the hard tick, *Hyalomma asiaticum asiaticum*. *Biochemical and Biophysical Research Communications* 343, 585–590.

Zeidner, N., Dolan, M., Massung, R., Piesman, J. & Fish, D. (2000). Coinfection with *Borrelia burgdorferi* and the agent of human granulocytic ehrlichiosis suppresses IL-2 and IFN-gamma production and promotes an IL-4 response in C3H/HeJ mice. *Parasite Immunology* 22, 581–588.

Zeidner, N. S., Schneider, B. S., Nuncio, M. S., Gern, L. & Piesman, J. (2002). Coinoculation of *Borrelia* spp. with tick salivary gland lysate enhances spirochete load in mice and is tick species-specific. *Journal of Parasitology* 88, 1276–1278.

10 • Saliva-assisted transmission of tick-borne pathogens

P. A. NUTTALL AND M. LABUDA

INTRODUCTION

Saliva-assisted transmission (SAT) is the indirect promotion of arthropod-borne pathogen transmission via the actions of arthropod saliva molecules on the vertebrate host. This phenomenon has been reported for most blood-feeding arthropods that transmit disease causing agents via their saliva, but the greatest number of examples has been recorded in ticks. The skin site where ticks feed is highly modified by the pharmacologically active molecules secreted in tick saliva. For pathogens, it is an ecologically privileged niche they can exploit. Here we review evidence for SAT and consider candidates for SAT factors among the tick pharmacopoeia of antihaemostatic, anti-inflammatory and immunomodulatory molecules. SAT factors appear to differ for different pathogens and tick vector species, and possibly even depend on the vertebrate host species on which a tick feeds. Most probably, SAT is mediated by a suite of molecules that act together to overcome the redundancy in host response mechanisms. The quest to identify the tick molecules that mediate SAT is an exciting one, offering new insights into host inflammatory and immune mechanisms, and novel ways of controlling ticks and tick-borne diseases.

TICK–HOST–PATHOGEN INTERACTIONS

The relationships between tick-borne pathogens, their tick vectors and diverse vertebrate hosts, can be represented by a triangle of parasitic interactions (Fig. 10.1). The interactions are between (i) pathogen–tick, (ii) pathogen–host and (iii) tick–host. In (i) the pathogen interacts with its vector, infecting and replicating within tick cells or extracellular spaces (including those of the gut, haemocoel and salivary glands). The ability of a particular tick species to act as a vector depends on whether the pathogen can survive and overcome several 'barriers' within the tick, e.g. the environment within the midgut where the pathogen is initially taken up in the blood meal, the gut infection barrier and the salivary gland infection barrier. The role of the tick immune system in controlling infections within the tick is largely unknown. For example, does RNA interference help ticks control virus infections (Garcia et al., 2005)? Another unanswered question is the nature of any conspecific or heterospecific interactions between strains or genotypes of tick-borne pathogens during multiple infections of an individual tick.

Interaction (ii) also involves infection: the pathogen interacts with its vertebrate host, infecting and replicating extracellularly or within cells. The outcome of infection depends on the tropism of the pathogen for specific host cell or tissue types, or locations (e.g. nervous tissue, joints), and its pathogenic phenotype, and on the age and immune status of the host, and genetic background.

The third component of the triangle (Fig. 10.1, iii) is the interaction between the tick and its host. Although this is the non-infective face of the triangle, the skin site at which ticks attach to feed is the crossroads for pathogens. It is the initial site at which pathogens gain access to either their host or vector: the tick–host–pathogen interface. This crossroads is particularly busy for tick-borne pathogens because blood-feeding of ticks is such a complex and comparatively long and highly ordered process, especially in ixodid species. Attachment and feeding takes several days to complete (minutes to hours for argasids), and involves sawing through the epidermis by means of toothed chelicerae, inserting the mouthparts (barbed hypostome and chelicerae) into the resulting wound site which are then (for ixodids) cemented in place, followed by the formation of a feeding pool resulting from tick and host activities. For ixodid species, the greater part of the blood meal is not taken up until the last day of attachment (Kemp, Stone & Binnington, 1982). Such a profound physical and chemical assault on the host should provoke strong

Ticks: Biology, Disease and Control, ed. Alan S. Bowman and Patricia A. Nuttall. Published by Cambridge University Press.

205

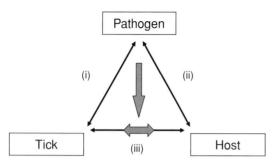

Fig. 10.1. The triangle of tick–host–pathogen interactions:
(i) pathogen–tick interactions, (ii) pathogen–host interactions and
(iii) tick–host interactions. The vertical arrow represents
interactions of a pathogen with tick–host interactions at the skin
site of tick attachment and blood-feeding (the tick–host–pathogen
interface) and the double-headed arrow represents the crossroads
for pathogens at which they pass from infected tick vector to
vertebrate host, or from infected host to tick vector.

haemostatic, inflammatory and immune responses. However, despite the host's armoury of rejection mechanisms, the tick manages to remain attached and achieve engorgement. The successful feeding of ticks relies on a pharmacy of chemicals located in their complex salivary glands and secreted, in tick saliva, into the feeding pool (see Chapters 4 and 9).

Increasing evidence indicates that the survival of tick-borne pathogens depends on their ability to exploit the pharmacological activities of tick saliva molecules. This is depicted in Fig. 10.1 by the broad vertical arrow representing the interaction of the pathogen with the modified tick–host interface. Why do we use such a forceful representation? Because it demonstrates our view of pathogen adaptation to the very specific environment created by the intimate and dramatic interplay between host and vector, and concentrated in the tick feeding site. By exploiting this unique ecological niche – the feeding site – tick-borne pathogens are making an easier living for themselves, increasing their chances of survival in nature.

SALIVA-ASSISTED TRANSMISSION (SAT)

Promotion of tick-borne pathogen transmission, via the actions of tick saliva components on the host, has been termed saliva-activated transmission. As we understand more about the mechanisms underlying this phenomenon, the term saliva-assisted transmission (SAT) seems a more accurate descriptor. It represents the pathogen's quest for

saliva-mediated host modulation, represented by the large vertical arrow in Fig. 10.1. 'Saliva-activated transmission' was first used to describe the promotion of Thogoto virus transmission by salivary gland extract (SGE) of *Rhipicephalus appendiculatus* (Nuttall & Jones, 1991). Since then, the phenomenon has been demonstrated for a number of tick-borne pathogens (Tables 10.1 and 10.2). Direct and indirect evidence of SAT has also been reported for several other vector-borne pathogens (summarized in Nuttall & Labuda, 2004).

Experimentally, SAT is demonstrated by enhanced transmission of infectivity when the pathogen plus saliva or SGE is syringe-inoculated into a host, compared with the level of infectivity when the pathogen alone is injected. For example, comparison of pooled experimental data for Thogoto virus infection reveals the attenuating effect of SGE (Table 10.3). Experimental inoculation of guinea pigs with a mixture of the virus plus SGE of partially fed *R. appendiculatus* or *Amblyomma variegatum* female ticks increased the number of nymphs that became infected approximately 10-fold compared with inoculation with virus alone. Similar direct evidence of SAT has been reported for tick-borne encephalitis virus (TBEV), the Lyme disease spirochaetes *Borrelia afzelii*, *B. burgdorferi* sensu stricto and *B. lusitaniae*, and *Franciscella tularensis* (Table 10.1). Real-time PCR demonstrated the effect of SGE on spirochaete proliferation within the inoculated host (Machackova *et al.*, 2006). Identification of a saliva protein of *Ixodes scapularis*, Salp15, that promotes transmission of *B. burgdorferi* s.s., enabled direct demonstration of SAT by co-inoculation of mice with the recombinant Salp15 protein and the spirochaete, and by use of the RNA-mediated interference technique (Ramamoorthi *et al.*, 2005).

Indirect evidence of SAT is provided by observations of efficient non-systemic transmission between infected and uninfected ticks co-feeding on the same host (Table 10.2). The first unequivocal example was demonstrated by the non-viraemic transmission of Thogoto virus (Jones *et al.*, 1987). Remarkably, virus transmission from infected to uninfected ticks co-feeding on non-viraemic guinea pigs was more efficient than transmission on hamsters that exhibited high levels of viraemia. Non-viraemic transmission has been shown for several other tick-borne viruses (Table 10.2). A possible example exists for Kyasanur forest disease virus transmitted by *Argas persicus* feeding on a domestic chick (Singh, Goverdhan & Bhat, 1971). Comparable non-systemic transmission has been reported for *B. burgdorferi* s.s., *B. garinii* and *B. afzelii* (Table 10.2). Transmission of *Ehrlichia ruminantium*, the causative agent of heartwater in cattle, between infected

Table 10.1 *Tick-borne pathogens showing direct evidence of saliva-assisted transmission*

Pathogen	Tick species source of SGE	Reference
Viruses:		
Tick-borne encephalitis virus	*Ixodes ricinus, Dermacentor reticulatus, D. marginatus, Rhipicephalus appendiculatus*	Alekseev *et al.*, 1991; Labuda *et al.*, 1993*b*
Thogoto virus	*Rhipicephalus appendiculatus, R. evertsi, Amblyomma cajennense, A. hebraeum, A. variegatum, Boophilus microplus, Hyalomma dromedarii, H. marginatum rufipes*	Jones *et al.*, 1989, 1992*a, b, c*
Bacteria:		
Borrelia afzelii	*Ixodes ricinus*	Pechova *et al.*, 2002
Borrelia burgdorferi sensu stricto	*Ixodes scapularis*	Zeidner *et al.*, 2002; Ramamoorthi *et al.*, 2005; Machackova, Obornik & Kopecky, 2006
Borrelia lusitaniae	*Ixodes ricinus*	Zeidner *et al.*, 2002
Franciscella tularensis	*Ixodes ricinus*	Krocova *et al.*, 2003

Table 10.2 *Tick-borne pathogens showing indirect evidence of saliva-assisted transmission (non-systemic transmission)*

Pathogen	Tick species involved	Reference
Viruses:		
Bhanja virus	*Dermacentor marginatus, Rhipicephalus appendiculatus*	Labuda *et al.*, 1997*a*
Crimean–Congo haemorrhagic fever virus	*Hyalomma marginatum*	Gordon, Linthicum & Moulton, 1993; Zeller *et al.*, 1994
Louping ill virus	*Ixodes ricinus*	Jones *et al.*, 1997
Tick-borne encephalitis virus	*Ixodes persulcatus, I. ricinus, Dermacentor marginatum, D. reticulatus, Rhipicephalus appendiculatus*	Alekseev & Chunikhin, 1990; Labuda *et al.*, 1993*a, c*
Thogoto virus	*Rhipicephalus appendiculatus, Amblyomma variegatum*	Jones *et al.*, 1987, 1990*b*
Palma virus	*R. sanguineus, Dermacentor reticulatus, D. marginatus*	Labuda *et al.*, 1997*a*
West Nile virus	*Ornithodoros moubata*	Lawrie *et al.*, 2004
Bacteria:		
Borrelia afzelii	*Ixodes ricinus*	Richter *et al.*, 2002; Severinova, *et al.*, 2005
Borrelia burgdorferi s.s.	*Ixodes ricinus, I. scapularis*	Gem & Rais, 1996; Patrican, 1997
Borrelia garinii	*Ixodes persulcatus*	Sato & Nakao, 1997

Table 10.3 *Thogoto virus infection of ticks via different*
transmission routes

Transmission route	% ticks infected[a] (no. of experiments)
Syringe inoculation: virus	6% (6)
Syringe inoculation: virus + SGE	58% (7)
Co-feeding with infected ticks	85% (11)

[a] Infection of *Rhipicephalus appendiculatus* nymphs feeding on guinea pigs that were either inoculated with virus ± SGE or co-infested with Thogoto virus infected ticks.
Source: Modified from Nuttall (1998).

and uninfected ticks co-feeding on tortoises, is another possible example of non-systemic transmission (Bezuidenhout, 1987). By contrast, *Anaplasma marginale*, the tick-borne rickettsial pathogen of cattle, and the agent of human granulocytic ehrlichiosis (*Anaplasma phagocytophilum*), do not appear to utilize co-feeding non-systemic transmission (Hodzic *et al.*, 2001; Kocan & de la Fuente, 2003). However, *A. phagocytophilum* induces salivary gland expression of the *I. scapularis salp16* gene during blood-feeding, which appears to be required for successful infection of the salivary glands by the rickettsia-like pathogen (Sukumaran *et al.*, 2006).

Results of co-feeding experiments using pathogen-immune natural rodent hosts also support the concept of SAT. These animals supported TBEV transmission in the presence of virus-specific neutralizing antibodies (Labuda *et al.*, 1997*b*). In sharp contrast, immunization with specific tick-derived antigens significantly diminished TBEV transmission and, surprisingly, also increased the survival of mice following an otherwise lethal infective *I. ricinus* tick bite (Labuda *et al.*, 2006). The protective effect of anti-tick immunity against TBEV infection (which has been observed with other pathogen–vector systems) underpins the concept of transmission-blocking vaccines (see Chapter 19).

Further indirect evidence of SAT is based on in vitro studies in which SGE is shown to promote infections by tick-borne pathogens. For the Lyme disease spirochaete, SGE of unfed female *I. ricinus* stimulated growth of *B. garinii* cultures in vitro while SGE of unfed female *D. reticulatus*, a non-competent vector of the Lyme disease spirochaete, did not (Rudolf & Hubalek, 2003). In vitro studies with murine macrophages showed that killing of *B. afzelii* spirochaetes was inhibited by SGE of *I. ricinus* females fed for 5 days (Kuthejlova *et al.*, 2001). As macrophages represent one

of the first lines of host defence against this spirochaete, suppression of their antimicrobial activity may well contribute to SAT. A stimulatory chemotactic effect of SGE from *I. scapularis* on *B. burgdorferi* s.s. was observed using a modified U-tube chemotaxis assay (Shih, Chao & Yu, 2002). This apparent chemotactic potential of SGE may contribute to the acquisition of Lyme disease spirochaetes by feeding ticks, which also can be considered a form of SAT. Thus the effect of saliva at the feeding site of uninfected ticks may act as a magnet, drawing spirochaetes to the site. The mechanism for this, if it occurs, is unknown.

In vitro treatment of bovine lymphocytes with SGE of *R. appendiculatus* enhanced their susceptibility to infection by *Theileria parva* sporozoites (Shaw, Tilney & McKeever, 1993). Curiously, the effect was observed with SGE from unfed ticks as well as ticks that had fed for 4 days, and it occurred within only 90 min of treatment. Studies are needed to determine whether the apparent stimulating effect on *T. parva* invasion of host cells reflects an ability of the tick vector to promote transmission of this important pathogen. Tick SGE has even been shown to accelerate replication of vesicular stomatitis virus, an arthropod-borne virus (arbovirus) not transmitted by ticks (see section 'Cytokine inhibitors', below).

An interesting approach to investigating whether tick-induced immunomodulation affects tick-borne pathogen infection was reported for *A. phagocytophilum*, the aetiological agent of human granulocytic ehrlichiosis (Borjesson *et al.*, 2002). Mice were infected via syringe inoculation and then infested with uninfected *I. scapularis* nymphs. At the same time as the infestation, a suture was placed through the skin, distant from the ticks. The suture served as a non-specific inflammatory source and acted as a positive control for the specific inflammation induced by the feeding ticks. A marked increase in bacteraemia and rate of *A. phagocytophilum* infection of neutrophils were observed following tick feeding. The increased bacteraemia was not explained in terms of a specific tick-induced modulatory mechanism; nevertheless, this approach warrants further investigation.

All the direct and indirect evidence for SAT involves ixodid tick species, with one clear exception (Table 10.2). Whether this indicates a greater capacity of ixodid species for SAT remains to be determined, as comparatively few studies have been reported with argasid species. The one exception is the reported non-viraemic transmission of West Nile virus (WNV) between co-feeding infected and uninfected *Ornithodoros moubata* (Lawrie *et al.*, 2004). This report is unusual because WNV is regarded as a mosquito-borne

Table 10.4 *Tick–pathogen specificity for SAT*

Tick-borne pathogen	Tick species		Reference
	SAT positive	SAT negative	
Thogoto virus	*Rhipicephalus appendiculatus,* *R. evertsi,* *Amblyomma cajennense,* *A. hebraeum,* *A. variegatum,* *Hyalomma dromedarii,* *H. marginatum rufipes*	*Ixodes ricinus,* *I. hexagonus,* *Carios maritimus*	Jones *et al.*, 1989, 1992*a*
TBE virus	*I. ricinus,* *Dermacentor reticulatus,* *R. appendiculatus*	none identified	Labuda *et al.*, 1993*b*
B. burgdorferi s.s.	*I. scapularis*	*I. ricinus*	Zeidner *et al.*, 2002
B. lusitaniae	*I. ricinus*	*I. scapularis*	

virus although epidemiological evidence indicates it can be maintained in nature by argasid species (see Chapter 12).

MECHANISM OF SAT

The first SAT factor to be identified from ticks is Salp15, a 15-kDa saliva protein from *I. scapularis* that inhibits CD4[+] T cell activation (Anguita *et al.*, 2002). Expression of the encoding gene, *salp15*, in *I. scapularis* salivary glands is enhanced by infection with *B. burgdorferi* s.s. The Salp15 protein binds to the outer surface protein (OspC) of the spirochaete, protecting it from antibody-mediated killing in vitro and facilitating replication of the spirochaete in vivo even in hosts apparently immune to the spirochaete (Ramamoorthi *et al.*, 2005).

For tick-borne viruses, SAT factors appear to promote transmission through their activity on the host rather than by a direct effect on or interaction with the virus. However, currently no SAT factors of tick-borne viruses have been characterized. When Thogoto virus was mixed with SAT-active SGE and then assayed in cell culture and mice, viral infectivity was unchanged (Jones, Hodgson & Nuttall, 1989, 1990*a*). This suggests that the enhancing factor is not a proteolytic enzyme that cleaves a viral surface protein to expose a more infectious virus particle, as occurs with some insect-borne viruses (Borucki *et al.*, 2002; Takamatsu *et al.*, 2003). More compelling is the observation that SAT was observed

when SGE and Thogoto virus were inoculated at different times into guinea pigs (Jones, Kaufman & Nuttall, 1992*b*). The key to successful enhancement was the inoculation of virus and SGE into the same skin site; when inoculated at different sites, the numbers of infected nymphs fell to levels observed when guinea pigs were inoculated with virus alone (Jones *et al.*, 1989). Immunomodulatory effects of tick saliva that may mediate SAT and explain these observations are considered in the next section.

Tick-borne transmission studies undertaken with laboratory animals suggest a correlation between pathogen, vector competence and SAT (Table 10.4). For example, SAT of Thogoto virus was demonstrated with SGE or saliva of competent vector species but not with tick species such as *I. ricinus* that were unable to transmit Thogoto virus following *per os* infection. However, SAT was demonstrated with TBEV and SGE from *I. ricinus*, its natural vector. Thus the SAT factor that promotes TBEV transmission does not affect Thogoto virus transmission. Comparable observations have been made for the Lyme disease spirochaete (Table 10.4). The observation that the gene encoding Salp15, the SAT factor of *B. burgdorferi*, is upregulated in *I. scapularis* infected with *B. burgdorferi* s.s. but not by *A. phagocytophilum* suggests pathogen specificity of SAT even in a competent vector species (Ramamoorthi *et al.*, 2005). Similarly, the apparent correlation between SAT and vector competence is contradicted by the fact that tick SGE accelerates the replication in vitro of vesicular stomatitis virus, which is not tick-transmitted (Hajnická et al., 1998).

Together, these observations on the mechanism of SAT demonstrate that the interface between vector and host, with which the pathogen interacts (indicated by the vertical arrow, Fig. 10.1), is highly complex, remarkably specific and considerably variable. Both the vector and the vertebrate host species of the pathogen, and possibly even the strain of the pathogen, have roles to play in determining whether SAT occurs. The implications of SAT and non-systemic transmission for tick-borne pathogen survival in nature have been discussed in other publications (see Randolph et al., 1996, 1999; Nuttall & Labuda, 2003; Norman et al., 2004).

SAT FACTORS

To date, a SAT factor has been identified for only one of the pathogens (B. burgdorferi s.s.) listed in Tables 10.1 and 10.2. However, it is not known whether SAT of a particular tick-borne pathogen involves one or more than one saliva molecule. Expression of the gene encoding Salp15, the SAT factor of B. burgdorferi s.s., is selectively increased in infected nymphal ticks during engorgement (Ramamoorthi et al., 2005). Studies with Thogoto virus demonstrated that the SAT factor is not present in the salivary glands of unfed ticks but that it accumulates in the salivary glands and is secreted in saliva as feeding progresses. Maximal SAT activity was shown by SGE and saliva at 5 to 8 days feeding for uninfected female A. variegatum and 6 days for adult female R. appendiculatus, whereas for adult female R. (Boophilus) microplus, SGE collected at a feeding weight range of 3 to 250 mg all showed SAT and with no obvious peak (Jones et al., 1989, 1992b; Jones, Matthewson & Nuttall, 1992c). Whatever the one-host tick species B. microplus produces that promotes Thogoto virus transmission appears to be present for most, if not all, the adult female feeding period. The dynamics of SAT activity for Thogoto virus and its tick vector, R. appendiculatus, suggest that the active saliva ingredient is probably not an antihaemostatic, anti-inflammatory or anti-complement factor, as these activities are expressed early during the feeding period, in parallel with activation of the matching host responses. Physicochemical analysis indicates that the SAT factor for Thogoto virus is one or more proteins or peptides (Jones et al., 1990a).

As mentioned in the preceding section, the limited evidence to date indicates that the SAT factor differs for different pathogens and vector species. The SAT factor may

even differ with different vertebrate host species, a possibility that has not been explored. Nevertheless, all the evidence for SAT factors points to the bioactive molecules present in the salivary glands and secreted in saliva. These saliva molecules are responsible for countering host haemostatic, inflammatory and immune responses. Although their precise role and mode of action in tick-borne pathogen transmission are unknown, a number of these bioactive molecules are likely to affect the host in a way that is inadvertently beneficial to pathogen transmission. Their isolation and activities are described in Chapters 4 and 9, and in recent reviews (Steen, Barker & Alewood, 2006; Titus, Bishop & Mejia, 2006). Here we consider certain tick saliva molecules and immunomodulatory activities as SAT factor candidates for promoting tick-borne pathogen transmission.

Antihaemostatic molecules

The interface between coagulation and inflammation (involving serine protease receptors) suggests opportunities exist for SAT (Cirino et al., 2000; see also Chapter 9). Indeed, a tick serpin has been identified that interferes with both haemostasis and the immune response (Prévôt et al., 2006) A study of the acquisition of Anaplasma phagocytophilum by Ixodes scapularis nymphs feeding on infected mice identified a role for the formation of small haemorrhages within the feeding pool (Borjesson et al., 2002).

Anti-inflammatory molecules

HISTAMINE-BINDING PROTEINS
Histamine is a key mediator of inflammation and also affects immune functions. It is secreted by mast cells, basophils and (in some species) blood platelets, and is believed to function as a primitive means of controlling parasites (Stebbings, 1974). The importance of histamine as a mediator in successful anti-tick responses, particularly in previously exposed hosts that have developed an immune (anamnestic) response, has been demonstrated in several host species, including cattle, rabbits and guinea pigs (Willadsen, Wood & Riding, 1979; Brossard, 1982; Wikel, 1982).

Ticks have adopted a unique strategy to control the effects of histamine: they produce saliva molecules that bind histamine directly (Paesen et al., 1999, 2000). Tick histamine-binding proteins (HBPs) are structurally related to a large family of barrel-shaped ligand-binding proteins known as lipocalins. Their high affinity for histamine enables them to outcompete histamine receptors and thereby prevent a histamine-mediated inflammatory response. HBPs have been isolated from several ixodid tick species including *Rhipicephalus appendiculatus* and *Ixodes hexagonus*. Homologues of the *R. appendiculatus* HBPs have been identified in the salivary glands of *I. scapularis* (see Chapters 4 and 9) but it remains to be determined whether they bind histamine. In addition, a histamine release factor homologue is secreted by at least one ixodid tick species, suggesting ticks have a multifaceted control mechanism for histamine (Mulenga *et al.*, 2003).

A site of inflammation is a hostile environment for invading pathogens. Histamine upregulates certain cytokines such as tumour necrosis factor alpha (TNF-α), a suppressor of *B. burgdorferi* infections, and activates natural killer (NK) cells, which have antiviral activity. Through the H_4 receptor, histamine is involved in leukocyte trafficking, including chemotaxis of eosinophils and mast cells, and recruitment of neutrophils (Takeshita *et al.*, 2003). In mouse models of allergic asthma and acute respiratory distress syndrome, an HBP of *R. appendiculatus* largely abrogated these inflammatory diseases (Couillin *et al.*, 2004; Ryffel *et al.*, 2005). Manipulation of the many functions of histamine, by ticks, is likely to benefit all tick-borne pathogens to some degree although there is no evidence to date that HBPs act as SAT factors.

COMPLEMENT INHIBITORS

The complement system is the principal effector arm of the humoral immune system in vertebrates, involved in inflammation and innate immunity. The three activation pathways (classical, alternative and lectin) comprise parallel cascades that converge on the C3 protein resulting in complement activation and release of the anaphylatoxins C3a and C5a (Law & Reid, 1995; Gadjeva *et al.*, 2001).

Activation of host complement occurs when ticks feed, contributing to resistance to tick infestation (Wikel & Allen, 1977; Allen, Khalil & Graham, 1979). Not surprisingly, ticks have evolved complement inhibitors. Saliva or SGE of several *Ixodes* species (*I. scapularis*, *I. ricinus*, *I. hexagonus* and *I. uriae*) inhibit the alternative complement pathway (Ribeiro & Spielman, 1986; Ribeiro, 1987; Astigarraga *et al.*, 1997; Lawrie, Randolph & Nuttall, 1999; Valenzuela *et al.*, 2000; Lawrie, Sim & Nuttall, 2004; Ribeiro *et al.*, 2006; Daix *et al.*, 2007) while *Ornithodoros* species target both the classical and alternative pathways (Astigarraga *et al.*, 1997; Nunn *et al.*, 2005) (see also Chapter 4).

For *I. ricinus*, anti-complement activity is demonstrated by SGE from unfed as well as feeding ticks (Lawrie *et al.*, 1999). This reflects the fact that activation of the alterna-

tive complement cascade is one of the first events following tick attachment, and a constant threat throughout tick feeding. However, the existence of significant anti-complement activity in unfed *I. ricinus* SGE indicates the complement inhibitor is not the SAT factor of TBEV, as SGE from unfed adult female *I. ricinus* does not demonstrate SAT activity (Labuda *et al.*, 1993*b*). Nevertheless, complement inhibitors may promote transmission of Lyme disease spirochaetes (Kyckova & Kopecky, 2006). Complement sensitivity is a major influence in Lyme disease ecology and pathogenicity (Kurtenbach *et al.*, 1998; van Dam, 2002; Ullmann *et al.*, 2003; see also Chapter 11). The borreliacidal effect is due to the activity of the alternative complement pathway, and specific mechanisms appear to mediate resistance (Kurtenbach *et al.*, 1998; Hellwage *et al.*, 2001; Pausa *et al.*, 2003).

CYSTATINS

Cystatins are reversible, tight-binding inhibitors of cysteine proteases that play a role in inflammation and defence against parasitic and microbial infections (Abrahamson, Alvarez-Fernandez & Nathanson, 2003). Ticks secrete cystatins to aid blood-feeding (Karim *et al.*, 2005). Sialostatin L, the cystatin secreted by *I. scapularis*, targets cathepsin L (Kotsyfakis *et al.*, 2006). It exhibits an anti-inflammatory role and inhibits proliferation of cytotoxic T-lymphocytes, activities that may facilite pathogen transmission.

Immunomodulators

IMMUNOGLOBULIN-BINDING PROTEINS

When ticks feed, a small proportion of host plasma proteins escape digestion and pass through the gut wall, into the haemocoel. These host proteins include immunoglobulin G (IgG) molecules, some of which may be pathogen-specific antibodies (Fujisaki, Kamio & Kitaoka, 1984; Jasinskas, Jaworski & Barbour, 2000). The fate of host immunoglobulins that enter the tick haemocoel was unknown until the discovery that adult female *R. appendiculatus* excrete intact IgG in their saliva (Wang & Nuttall, 1994). Further investigations revealed a family of immunoglobulin G-binding proteins (IGBPs) in the haemolymph and salivary glands of adult *R. appendiculatus*, and proteins with similar activity in other ixodid species, including *I. ricinus* and *Amblyomma variegatum*, and a sequence homologue in *I. scapularis* (Wang & Nuttall, 1999; Packila & Guilfoile, 2002). Although IGBPs have not been reported for argasid tick species, circumstantial evidence suggests they may occur (Minoura, Chinzei & Kitamura, 1985).

The prevalence and abundance of IGBPs indicate that they play an important role in blood-feeding. Possibly they provide a tick immunoglobulin excretion system (TIES) that enables ticks to ferry potentially damaging antibodies safely through the haemocoel to their salivary glands from where they are excreted (Wang & Nuttall, 1999). Such a system may benefit tick-borne pathogens by protecting them (and possibly the infected tick) from pathogen-specific antibodies taken up in the blood meal, as shown for Thogoto virus (Jones & Nuttall, 1989). In addition, male *R. appendiculatus* ticks seem to use the host's immunoglobulins to induce local immunosuppression for the benefit of the female, a novel form of mate-guarding (Wang *et al.*, 1998). Whether such an effect aids in the transmission or acquisition of tick-borne pathogens by feeding ticks remains to be explored.

CYTOKINE INHIBITORS

Cytokines are the chemical mediators of inflammation and immunity. The type 1 interferons (IFNs) are a cytokine superfamily comprising four subfamilies of which IFN-α and IFN-β are induced by viral infections. Most cell types produce them (Stark *et al.*, 1998; Goodbourn, Didcock & Randall, 2000). Arboviruses are not generally recognized as strong inducers of IFNα/β and there have been few (if any) studies of IFN induction following vector-borne virus transmission. A notable exception as a strong IFN inducer is vesicular stomatitis virus, an insect-borne rhabdovirus. Although it is not tick-borne, SGE from partly fed adult *R. appendiculatus* or *D. reticulatus* increased viral yields by 100- to 1000-fold in mouse cell cultures, the first published evidence that tick SGE can promote virus replication *in vitro* (Hajnická *et al.*, 1998). The effect appeared to result from inhibition of the antiviral effect of IFN by salivary gland products, possibly acting through the IFNα/β receptor rather than directly affecting IFN (Hajnická *et al.*, 2000).

Interferon α/β-induced viral resistance is mediated by antiviral factors such as Mx gene products that are active against viruses of several different families including the Orthomyxoviridae, Bunyaviridae and Togaviridae. The interferon-induced mouse Mx1 protein has intrinsic antiviral activity against influenza A and B viruses, and the tick-borne orthomyxovirus, Thogoto virus (Haller *et al.*, 1995). Mice carrying the *Mx1* gene are resistant to Thogoto virus infection by needle injection. However, they are susceptible to tick-borne virus challenge (non-viraemic transmission from infected to uninfected co-feeding ticks) and, to a lesser degree, injection of virus mixed with tick SGE (Dessens & Nuttall, 1998). These data are consistent with the ability of

tick SGE to interfere with the antiviral action of IFNα/β. Inhibition of IFNα/β appears to be a good candidate for the SAT factor of Thogoto virus.

In addition to the action on IFN (described above), several studies have reported effects of tick feeding or tick salivary gland products on cytokine expression or activities (see Chapter 9). Many of the effects on cytokine activities appear to be due to a wealth of cytokine-binding molecules produced by ixodid tick species and secreted in their saliva (Hajnická *et al.*, 2000, 2005; Gillespie *et al.*, 2001; Power, Proudfoot & Fraunenschuh, 2005). Cytokine binders react with interleukins IL-2, IL-4, IL-8, MCP-1, MIP-1α, RANTES and eotaxin, and probably more. The IL-2-binding protein of *I. scapularis* provides a mechanism for suppressing T-cell proliferation and other IL-2-stimulated immune responses (Gillespie *et al.*, 2001). The IL-8 binder of *D. reticulatus* outcompetes IL-8 receptors on neutrophils, inhibiting IL-8-induced chemotaxis of neutrophils (Hajnická *et al.*, 2001). ChBP, a 20-kDa chemokine binding protein from *Rhipicephalus sanguineus*, binds the C-C chemokines, MCP-1, MIP-1α and RANTES, although it appears to have highest affinity for MIP-1α (Power *et al.*, 2005). The apparent strategy of manipulating the cytokine network most likely overcomes redundancy in this innate immune system, and should greatly facilitate blood-feeding.

By binding different cytokines, ixodid ticks provide a potential gateway for tick-borne pathogens. To what extent this gateway is exploited remains to be determined. The only indication to date is that inoculation of C3H/HeJ mice with a mixture of TNF-α, IFN-γ and IL-2 at the time of tick feeding suppressed *B. burgdorferi* transmission by *I. scapularis*, suggesting that cytokine manipulation by ticks might aid borrelia transmission (Zeidner *et al.*, 1996). Anti-TNF-α activity has been identified in the saliva of *I. ricinus* (Konik *et al.*, 2006). Clearly, the extent to which ticks manipulate the cytokine network, and the consequences for tick-borne pathogen transmission, need to be explored.

LEUKOCYTE MODULATORS

Numerous effects of tick saliva or SGE on T-cell, B-cell and macrophage function have been described in vitro (see Chapter 9). Many of them are mediated through the effects on cytokines. The potential benefits to tick-borne pathogens are obvious, and have been demonstrated for Salp15, an inhibitor of CD4+ T cell activation (Anguita *et al.*, 2002; Garg *et al.*, 2006). Neutrophils phagocytize *B. burgdorferi* hence the ability of their tick vector's saliva to inhibit neutrophil function could be highly significant (Ribeiro, Weis &

Telford, 1990; Suhonen, Hartiala & Viljanen, 1998; Montgomery *et al.*, 2004). Langerhans cells (dendritic cells) are thought to play a key role in non-viraemic transmission of TBEV (see 'The "red herring" hypothesis' in Nuttall & Labuda, 2003). Tick saliva inhibits the differentiation, maturation, and function of murine bone marrow-derived dendritic cells (Cavassani *et al.*, 2005). Obviously the effect of SGE on Langerhans cells needs to be examined in the search for the SAT factor of TBEV. The function of the homologue of macrophage migration inhibitory factor (MIF, a mammalian pro-inflammatory cytokine) is unknown. It has been detected in both salivary glands and midgut tissues of unfed and feeding *A. americanum* adult females and, in vitro, inhibited macrophage migration (Jaworski *et al.*, 2001). Potentially it may reduce macrophage microbicidal activity.

Tick infestation is characterized by polarization towards a Th2 response. Iris, a 43-kDa protein from *I. ricinus*, appears to be one of the tick saliva molecules involved (Leboulle *et al.*, 2002). Th2 polarization benefits tick-borne pathogens such as *B. burgdorferi* and *Babesia bovis* (Zeidner *et al.*, 1996, 1997; Christe, Rutti & Brossard, 2000; Goff *et al.*, 2003). Indirect evidence of the potential benefit for TBEV was demonstrated by the reduction of co-feeding transmission in laboratory mice in which a Th1-like response was induced by immunization with a recombinant tick salivary gland protein (M. Labuda and M. Lickova, unpublished data). Mice repeatedly infested with *I. scapularis* nymphs, and showing a Th2-polarized response, became resistant to *B. burgdorferi* transmission indicating that the benefits of such polarization may be overcome by the host (Wikel *et al.*, 1997).

Direct immunosuppression of B-cells by *I. ricinus* SGE (Hannier *et al.*, 2003) may explain the reduced ability of tick-infested hosts to produce antibodies (Wikel, 1985; Christe, Rutti & Brossard, 2000). Such immunosuppression is also likely to affect B-cell production of immune regulatory cytokines such as IFN-γ. B-cells play a crucial role in antimicrobial immunity (Ochsenbein *et al.*, 1999; Baumgarth, 2000). The 18-kDa B-cell inhibitory protein (BIP) of *I. ricinus* inhibits both Osp A- and Osp C-induced proliferation of naive murine B-lymphocytes (Hannier, *et al.*, 2004). Potentially it may enhance *B. burgdorferi* transmission by preventing T-independent B-cell activation during the initial stage of host infection. Sequencing of the BIP of *Hyalomma asiaticum* revealed a novel sequence predicted to encode a protein of 13 kDa (Yu *et al.*, 2006).

The role of natural killer (NK) cells includes control of viral infections through the killing of cells that express viral antigens on their surface. They may even kill extracellular bacteria. NK cells secrete cytokines, providing an important means by which innate immunity communicates with the acquired immune system (Lanier, 2000). Direct evidence of a role for NK cells during tick feeding is lacking, nevertheless tick salivary glands contain a protein that suppresses NK cell activity (Kubeš *et al.*, 1994). Such activity was demonstrated with SGE from partially fed *D. reticulatus*, *A. variegatum* and *Haemaphysalis inermis*, but not from *I. ricinus* or *R. appendiculatus* (Kubeš *et al.*, 2002). The apparent absence of activity for the latter two important vectors suggests that control of NK cells does not play an important role in promoting tick-borne pathogen transmission, at least for these two tick species. However, a suppressive effect on NK cell activity was observed in a mouse model with SGE from partially fed female *I. ricinus* (Kopecky & Kuthejlova, 1998). Interactions between dendritic cells and NK cells, during the early stages of *B. burgdorferi* infection, influence development of a protective humoral response in mice (Mbow *et al.*, 2001). Hence suppression of NK cell activity may affect *B. burgdorferi*, although possibly at the infection stage following SAT.

CONCLUSIONS

The hunt for tick saliva molecules that promote pathogen transmission (the SAT factors) continues. To date, only one SAT factor has been identified (Salp15). However, evidence suggests we are searching for a suite of molecules that act cooperatively. For ticks, cooperative salivary activity is the only way they can overcome redundancy in host protective systems to a degree that allows them to complete their sumptuous meal. Cooperation balances the need, for example, to control blood flow and cell recruitment so that ticks have plenty to feed on but they do not have excessive host factors to fight against. Tick-borne pathogens apparently have evolved to exploit a combination of immunomodulatory activities in order to establish a crucial toehold in the site of tick feeding. Anti-tick vaccines may have to neutralize the suite of activities to be effective in blocking tick-borne pathogen transmission.

One specific difficulty in the quest for SAT factors is that, in most cases, SGE or saliva is taken from adult females. However, for many tick-borne pathogens, epidemiological data implicate nymphs as the key transmitters of these pathogens, and larvae as the acquirers of infections. The epidemiological role of adult females is to lay eggs and produce more acquirers (larvae) and transmitters (nymphs) of tick-borne pathogens. Thus more studies are needed on immature stages, to determine how their pharmacological prowess compares with that of their mothers. Identification of candidate SAT factors, and testing their effects on pathogen transmission (singly or as cocktails of different molecules), should help identify the key ingredients in the tick pharmacopoeia that promote pathogen transmission. The challenge then will be to determine how this information can be used to control ticks and tick-borne diseases.

ACKNOWLEDGEMENTS

The work of ML was supported by the Slovak Research and Development Agency under contract No. APVV-51-004505.

REFERENCES

Abrahamson, M., Alvarez-Fernandez, M. & Nathanson, C. M. (2003). Cystatins. *Biochemistry Society Symposium* **70**, 179–199.

Alekseev, A. N. & Chunikhin, S. P. (1990). Exchange of tick-borne encephalitis virus between Ixodidae simultaneously feeding on animals with subthreshold levels of viraemia. *Meditsinskaya Parazitologiya i Parazitarnye Bolezni* **2**, 48–50.

Alekseev, A. N., Chunikhin, S. P., Rukhkyan, M. Y. & Stefutkina, L. F. (1991). Possible role of Ixodidae salivary gland substrate as an adjuvant enhancing arbovirus transmission. *Meditsinskaya Parazitologiya i Parazitarnye Bolezni* **1**, 28–31.

Allen, J. R., Khalil, H. A. & Graham, J. E. (1979). The location of tick salivary antigens, complement and immunoglobulin in the skin of guinea-pigs infested with *Dermacentor andersoni* larvae. *Immunology* **38**, 467–472.

Anguita, J., Ramamoorthi, N., Das, G., *et al.* (2002). Salp15, an *Ixodes scapularis* saliva protein, inhibits CD4$^+$ T cell activation. *Immunity* **16**, 849–859.

Astigarraga, A., Oleaga-Perez, A., Perez-Sanchez, R., Baranda, J. A. & Encinas-Grandes, A. (1997). Host immune response evasion strategies in *Ornithodoros erraticus* and *O. moubata* and their relationship to the development of an antiargasid vaccine. *Parasite Immunology* **19**, 401–410.

Baumgarth, N. (2000). A two-phase model of B-cell activation. *Immunology Reviews* **176**, 171–180.

Bezuidenhout, J. D. (1987). Natural transmission of heartwater. *Onderstepoort Journal of Veterinary Medicine* **54**, 349–351.

Borjesson, D. L., Simon, S. I., Hodzic, E., *et al.* (2002). Roles of neutrophil β2 integrins in kinetics of bacteremia, extravasation, and tick acquisition of *Anaplasma phagocytophila* in mice. *Blood* **101**, 3257–3264.

Borucki, M. K., Kempf, B. J., Blitvich, B. J., Blair, C. D. & Beaty, B. J. (2002). La Crosse virus: replication in vertebrate and invertebrate hosts. *Microbes and Infection* **4**, 341–350.

Brossard, M. (1982). Rabbits infested with adult *Ixodes ricinus* L.: effects of mepyramine on acquired resistance. *Experientia* **38**, 702–704.

Cavassani, K. A., Aliberti, J. C., Dias, A. R., Silva, J. S. & Ferreira, B. R. (2005). Tick saliva inhibits differentiation, maturation and function of murine bone-marrow-derived dendritic cells. *Immunology* **114**, 235–245.

Christe, M., Rutti, B. & Brossard, M. (2000). Cytokines (IL-4 and IFN-gamma) and antibodies (IgE and IgG2a) produced in mice infected with *Borrelia burgdorferi* sensu stricto via nymphs of *Ixodes ricinus* ticks or syringe inoculations. *Parasitology Research* **86**, 491–496.

Cirino, G., Napoli, C., Bucci, M. & Cicala, C. (2000). Inflammation–coagulation network: are serine protease receptors the knot? *Trends in Pharmacological Science* **21**, 170–172.

Couillin, I., Maillet, I., Jacobs, M., *et al.* (2004). Arthropod-derived histamine binding protein prevents murine allergic asthma. *Journal of Immunology* **173**, 3281–3286.

Daix, V., Schroeder, N., Praet, N., *et al.*, (2007). *Ixodes* ticks belonging to the *Ixodes ricinus* complex encode a family of anti-complement proteins. *Insect Biochemistry and Molecular Biology* **16**, 155–166.

Dessens, J. T. & Nuttall, P. A. (1998). Mx1-based resistance to Thogoto virus in A2G mice is bypassed in tick-mediated virus delivery. *Journal of Virology* **72**, 8362–8364.

Fujisaki, K., Kamio, T. & Kitaoka, S. (1984). Passage of host serum components, including antibodies specific for *Theileria sergenti*, across the digestive tract of argasid and ixodid ticks. *Annals of Tropical Medicine and Parasitology* **78**, 449–450.

Gadjeva, M., Thiel, S. & Jensenius, J. C. (2001). The mannan-binding-lectin pathway of the innate immune system. *Current Opinions in Immunology* **13**, 74–78.

Garcia, S., Billecocq, A., Crance, J. M., *et al.* (2005). Nairovirus RNA sequences expressed by a Semliki Forest virus replicon induce RNA interference in tick cells. *Journal of Virology* **79**, 8942–8947.

Garg, R., I. Juncadella, I. J., Ramamoorthi, N., *et al.*, (2006). Cutting edge: CD4 is the receptor for the tick saliva immunosuppressor, Salp15. *Journal of Immunology* **177**, 6579–6583.

Gern, L. & Rais, O. (1996). Efficient transmission of *Borrelia burgdorferi* between cofeeding *Ixodes ricinus* ticks (Acari: Ixodidae). *Journal of Medical Entomology* **33**, 189–192.

Gillespie, R. D., Dolan, M. C., Piesman, J. & Titus, R. G. (2001). Identification of an IL-2 binding protein in the saliva of the Lyme disease vector tick, *Ixodes scapularis*. *Journal of Immunology* **166**, 4319–4327.

Goff, W., Johnson, W., Horn, R., Barrington, G. & Knowles, D. (2003). The innate response in calves to *Boophilus microplus* tick transmitted *Babesia bovis* involves type-1 cytokine induction and NK-like cells in the spleen. *Parasite Immunology* **25**, 185–188.

Goodbourn, S., Didcock, L. & Randall, R. E. (2000). Interferons: cell signalling, immune modulation, antiviral responses and virus countermeasures. *Journal of General Virology* **81**, 2341–2364.

Gordon, S. W., Linthicum, K. J. & Moulton, J. R. (1993). Transmission of Crimean–Congo hemorrhagic fever virus in two species of *Hyalomma* ticks from infected adults to cofeeding immature forms. *American Journal of Tropical Medicine and Hygiene* **48**, 576–580.

Hajnická, V., Fuchsberger, N., Slovak, M., *et al.* (1998). Tick salivary gland extracts promote virus growth *in vitro*. *Parasitology* **116**, 533–538.

Hajnická, V., Kocáková, P., Sláviková, M., *et al.* (2001). Anti-interleukin-8 activity of tick salivary gland extracts. *Parasite Immunology* **23**, 483–489.

Hajnická, V., Kocaková, P., Slovák, M., *et al.* (2000). Inhibition of the antiviral action of interferon by tick salivary gland extract. *Parasite Immunology* **22**, 201–206.

Hajnická, V., Vancova, I., Kocakova, P., *et al.* (2005). Manipulation of host cytokine network by ticks: a potential gateway for pathogen transmission. *Parasitology* **130**, 333–342.

Haller, O., Frese, M., Rost, D., Nuttall, P. A. & Kochs, G. (1995). Tick-borne Thogoto virus infection in mice is inhibited by the orthomyxovirus resistance gene product mx I. *Journal of Virology* **69**, 2596–2601.

Hannier, S., Liversidge, J., Sternberg, J. M. & Bowman, A. S. (2003). *Ixodes ricinus* tick salivary gland extract inhibits IL-10 secretion and CD69 expression by mitogen-stimulated murine splenocytes and induces hyporesponsiveness in B lymphocytes. *Parasite Immunology* **25**, 27–37.

Hannier, S., Liversidge, J., Sternberg, J. M. & Bowman, A. S. (2004). Characterization of the B-cell inhibitory protein factor in *Ixodes ricinus* tick saliva: a potential role in enhanced *Borrelia burgdorferi* transmission. *Immunology* **113**, 401–408.

Hellwage, J., Meri, T., Heikkila, T., *et al.* (2001). The complement regulator factor H binds to the surface protein OspE of *Borrelia burgdorferi*. *Journal of Biological Chemistry* **276**, 8427–8435.

Hodzic, E., Borjesson, D. L., Feng, S. & Barthold, S. W. (2001). Acquisition dynamics *of Borrelia burgdorferi* and the agent of human granulocytic ehrlichiosis at the host–vector interface. *Vector Borne Zoonotic Disease* **1**, 149–158.

Jasinskas, A., Jaworski, D. C. & Barbour, A. G. (2000). *Amblyomma americanum*: specific uptake of immunoglobulins into tick hemolymph during feeding. *Experimental Parasitology* **96**, 213–221.

Jaworski, D. C., Jasinskas, A., Metz, C. N., Bucala, R. & Barbour, A. G. (2001). Identification and characterization of a homologue of the pro-inflammatory cytokine Macrophage Migration Inhibitory Factor in the tick, *Amblyomma americanum*. *Insect Molecular Biology* **10**, 323–331.

Jones, L. D. & Nuttall, P. A. (1989). The effect of virus-immune hosts on Thogoto virus infection of the tick, *Rhipicephalus appendiculatus*. *Virus Research* **14**, 129–140.

Jones, L. D., Davies, C. R., Steele, G. M. & Nuttall, P. A. (1987). A novel mode of arbovirus transmission involving a nonviraemic host. *Science* **237**, 775–777.

Jones, L. D., Davies, C. R., Williams, T., Cory, J. & Nuttall, P. A. (1990*b*). Non-viraemic transmission of Thogoto virus: vector efficiency of *Rhipicephalus appendiculatus* and *Amblyomma variegatum*. *Transactions of the Royal Society of Tropical Medicine and Hygiene* **84**, 846–848.

Jones, L. D., Gaunt, M., Hails, R. S., *et al.* (1997). Transmission of louping-ill virus between infected and uninfected ticks co-feeding on mountain hares. *Medical and Veterinary Entomology* **11**, 172–176.

Jones, L. D., Hodgson, E. & Nuttall, P. A. (1989). Enhancement of virus transmission by tick salivary glands. *Journal of General Virology* **70**, 1895–1898.

Jones, L. D., Hodgson, E. & Nuttall, P. A. (1990*a*). Characterization of tick salivary gland factor(s) that enhance Thogoto virus transmission. *Archives of Virology* (Suppl.) **1**, 227–234.

Jones, L. D., Hodgson, E., Williams, T., Higgs, S. & Nuttall, P. A. (1992*a*). Saliva activated transmission (SAT) of Thogoto virus: relationship with vector potential of different haematophagous arthropods. *Medical and Veterinary Entomology* **6**, 261–265.

Jones, L. D., Kaufman, W. R. & Nuttall, P. A. (1992*b*). Modification of the skin feeding site by tick saliva mediates virus transmission. *Experientia* **48**, 779–782.

Jones, L. D., Matthewson, M. & Nuttall, P. A. (1992*c*). Saliva-activated transmission (SAT) of Thogoto virus: dynamics of SAT activity in the salivary glands of *Rhipicephalus appendiculatus, Amblyomma variegatum*, and *Boophilus microplus. Experimental and Applied Acarology* 13, 241–248.

Karim, S., Miller, N. J., Valenzuela, J. G. Sauer, J. R. & Mather, T. N. (2005). RNAi-mediated gene silencing to assess the role of synaptobrevin and cystatin in tick blood feeding. *Biochemical and Biophysical Research Communications* 334, 1336–1342.

Kemp, D. H., Stone, B. F. & Binnington, K. C. (1982). Tick attachment and feeding: role of the mouthparts, feeding apparatus, salivary gland secretions and host response. In *Physiology of Ticks*, eds. Obenchain, F. D. & Galun, R., pp. 119–168. Oxford, UK: Pergamon Press.

Kocan, K. M. & de la Fuente, J. (2003). Co-feeding studies of ticks infected with *Anaplasma marginale. Veterinary Parasitology* 112, 295–305.

Konik, P., Slavikova, V., Salat, J., *et al.* (2006). Anti-tumour necrosis factor-alpha activity in *Ixodes ricinus. Parasite Immunology* 28, 649–656.

Kopecky, J. & Kuthejlova, M. (1998). Suppressive effect of *Ixodes ricinus* salivary gland extract on mechanisms of natural immunity *in vitro. Parasite Immunology* 20, 169–174.

Kotsyfakis, M., Sa-Nuñes, A., Francischetti, I. M. B., *et al.* (2006). Antiinflammatory and immunosuppressive activity of sialostatin L, a salivary cystatin from the tick *Ixodes scapularis. Journal of Biological Chemistry* 281, 26298–26307.

Krocova, Z., Macela, A., Hernychova, L., *et al.* (2003). Tick salivary gland extract accelerates proliferation of *Franciscella tularensis* in the host. *Journal of Parasitology* 89, 14–20.

Kubeš, M., Fuchsberger, N., Labuda, M., Zuffova, E. & Nuttall, P. A. (1994). Salivary gland extracts of partially fed *Dermacentor reticulatus* ticks decrease natural killer cell activity *in vitro. Immunology* 82, 113–116.

Kubeš, M., Kocáková, P., Slovák, M., *et al.* (2002). Hetero-genity in the effect of different ixodid tick species on human natural killer cell activity. *Parasite Immunology* 24, 23–28.

Kurtenbach, K., Sewell, H., Ogden, N., Randolph, S. E. & Nuttall, P. A. (1998). Serum complement sensitivity as a key factor in Lyme disease ecology. *Infection and Immunity* 66, 1248–1251.

Kuthejlova, M., Kopecky, J., Stepanova, G. & Macela, A. (2001). Tick salivary gland extract inhibits killing of *Borrelia afzelii* spirochaetes by mouse macrophages. *Infection and Immunity* 69, 575–578.

Kyckova, K. & Kopecky, J (2006). Effect of tick saliva on mechanisms of innate immune response against *Borrelia afzelii. Journal of Medical Entomology* 43, 1208–1214.

Labuda, M., Alves, M. J., Eleckova, E., Kozuch, O. & Filipe, A. R. (1997*a*). Transmission of tick-borne bunyaviruses by cofeeding ixodid ticks. *Acta Virologica* 41, 325–328.

Labuda, M., Jones, L. D., Williams, T., Danielova, D. & Nuttall, P. A. (1993*a*). Efficient transmission of tick-borne encephalitis virus between cofeeding ticks. *Journal of Medical Entomology* 30, 295–299.

Labuda, M., Jones, L. D., Williams, T. & Nuttall, P. A. (1993*b*). Enhancement of tick-borne encephalitis virus transmission by tick salivary gland extracts. *Medical and Veterinary Entomology* 7, 193–196.

Labuda, M., Kozuch, O., Zuffova, E., Eleckova, E., Hails, R. S. & Nuttall, P. A. (1997*b*). Tick-borne encephalitis virus transmission between ticks co-feeding on specific immune natural rodent hosts. *Virology* 235, 138–143.

Labuda, M., Nuttall, P. A., Kozuch, O., *et al.* (1993*c*). Non-viraemic transmission of tick-borne encephalitis virus: a mechanism for arbovirus survival in nature. *Experientia* 49, 802–805.

Labuda, M., Trimnell, A. R., Lickova, M., *et al.* (2006). An antivector vaccine protects against a lethal vector-borne pathogen. *PLoS Pathogens* 2, e27.

Lanier, L. (2000). The origin and functions of natural killer cells. *Clinical Immunology* 95, S14–S18.

Law, S. K. & Reid, K. B. M. (1995). *Complement*. New York: Oxford University Press.

Lawrie, C. H., Randolph, S. E. & Nuttall, P. A. (1999). *Ixodes* ticks: serum species sensitivity of anti-complement activity. *Experimental Parasitology* 93, 207–214.

Lawrie, C. H., Sim, R. B. & Nuttall, P. S. (2004). Investigation of the mechanisms of anti-complement activity in *Ixodes* ticks. *Molecular Immunology* 42, 31–38.

Lawrie, C. H., Uzcategui, N. Y., Gould, E. A. & Nuttall, P. A. (2004). Ixodid and argasid ticks and West Nile virus. *Emerging Infectious Diseases* 10, 653–657.

Leboulle, G., Crippa, M., Decrem, Y., *et al.* (2002). Characterization of a novel salivary immunosuppressive protein from *Ixodes ricinus* ticks. *Journal of Biological Chemistry* 277, 10083–10089.

Machackova, M., Obornik, M. & Kopecky, J. (2006). Effect of salivary gland extract from *Ixodes ricinus* ticks on the proliferation of *Borrelia burgdorferi* sensu stricto *in vivo. Folia Parasitologica* 53, 153–158.

Mbow, M. L., Zeidner, N. S., Gilmore, R. D. J., *et al.* (2001). Major histocompatibility complex class 11-independent generation of neutralizing antibodies against T-cell-dependent *Borrelia burgdorferi* antigens presented by dendritic cells: regulation by NK and γδ T cells. *Infection and Immunity* **69**, 2407–2425.

Minoura, H., Chinzei, Y. & Kitamura, S. (1985). *Ornithodoros moubata*: host immunoglobulin G in tick haemolymph. *Experimental Parasitology* **60**, 355–363.

Montgomery, R. R., Lusitani, D., De Boisfleury Chevance, A. & Malawista, S. E. (2004). Tick saliva reduces adherence and area of human neutrophils. *Infection and Immunity* **72**, 2989–2994.

Mulenga, A., Macaluso, K. R., Simser, J. A. & Azad, A. F. (2003). The American dog tick, *Dermacentor variabilis*, encodes a functional histamine release factor homolog. *Insect Biochemistry and Molecular Biology* **33**, 911–919.

Norman, R., David, D., Laurenson, M. K. & Hudson, P. J. (2004). The role of non-viraemic transmission on the persistence and dynamics of a tick borne virus: louping ill in red grouse (*Lagopus lagopus scoticus*) and mountain hares (*Lepus timidus*). *Journal of Mathematical Biology* **48**, 119–134.

Nunn, M. A., Sharma, A., Paesen, G. C., *et al.* (2005). Complement inhibitor of C5 activation from the soft tick *Ornithodoros moubata*. *Journal of Immunology* **174**, 2084–2091.

Nuttall, P. A. (1998). Displaced tick–parasite interactions at the host interface. *Parasitology* **116** (Suppl.), S65–S72.

Nuttall, P. A. & Jones, L. D. (1991). Non-viraemic tick-borne virus transmission: mechanism and significance. In *Modern Acarology*, eds. Dusbabek, F. & Bukva, V., pp. 3–6. The Hague, Netherlands: SPB Academic.

Nuttall, P. A. & Labuda, M. (2003). Dynamics of infection in tick vectors and at the tick–host interface. *Advances in Virus Research* **60**, 233–272.

Nuttall, P. A. & Labuda, M. (2004). Tick–host interactions: saliva-activated transmission. *Parasitology* **129**, S177–S190.

Ochsenbein, A. F., Fehr, T., Lutz, C., *et al.* (1999). Control of early viral and bacterial distribution and disease by natural antibodies. *Science* **286**, 2156–2159.

Packila, M. & Guilfoile, P. G. (2002). Mating male *Ixodes scapularis* express several genes including those with sequence similarity to immunoglobulin-binding proteins and met alloproteases. *Experimental and Applied Acarology* **27**, 151–160.

Paesen, G. C., Adams, P. L., Harlos, K., Nuttall, P. A. & Stuart, D. I. (1999). Tick histamine-binding proteins: isolation, cloning, and three-dimensional structure. *Molecular Cell* **3**, 661–671.

Paesen, G. C., Adams, P. L., Nuttall, P. A. & Stuart, D. L. (2000). Tick histamine-binding proteins: lipocalins with a second binding cavity. *Biochimica et Biophysica Acta* **1482**, 92–101.

Patrican, L. (1997). Acquisition of Lyme disease spirochetes by cofeeding *Ixodes scapularis* ticks. *American Journal of Tropical Medicine and Hygiene* **57**, 589–593.

Pausa, M. P. V., Cinco, M., Giulianini, P. G., *et al.* (2003). Serum-resistant strains of *Borrelia burgdorferi* evade complement-mediated killing by expressing a CD59-like complement inhibitory molecule. *Journal of Immunology* **170**, 3214–3222.

Pechova, J., Stepanova, G., Kovar, L. & Kopecky, J. (2002). Tick salivary gland extract-activated transmission of *Borrelia afzelii* spirochaetes. *Folia Parasitologica* **49**, 153–159.

Power, C., Proudfoot, A. & Frauenschuh, A. (2005). CC-chemokine-binding tick proteins. Patent WO/2005/063812.

Prévôt, P., Adam, B., Boudjeltia, K. Z., *et al.* (2006). Anti-hemostatic effects of a serpin from the saliva of the tick *Ixodes ricinus*. *Journal of Biological Chemistry* **281**, 26361–26369.

Ramamoorthi, N., Narasimhan, S., Pal, U., *et al.* (2005). The Lyme disease agent exploits a tick protein to infect the mammalian host. *Nature* **436**, 573–577.

Randolph, S. E., Gern, L. & Nuttall, P. A. (1996). Co-feeding ticks: epidemiological significance for tick-borne pathogen transmission. *Parasitology Today* **12**, 472–479.

Randolph, S. E., Miklisova, D., Lysy, J., Rogers, D. J. & Labuda, M. (1999). Incidence from coincidence: patterns of tick infestations on rodents facilitate transmission of tick-borne encephalitis virus. *Parasitology* **118**, 177–186.

Ribeiro, J. M. C. (1987). *Ixodes dammini*: salivary anti-complement activity. *Experimental Parasitology* **64**, 347–353.

Ribeiro, J. & Spielman, A. (1986). *Ixodes dammini*: salivary anaphylatoxin inactivating activity. *Experimental Parasitology* **62**, 292–297.

Ribeiro, J. M. C., Alarcon-Chaidez, F., Francischetti, I. M. B., *et al.* (2006). An annotated catalog of salivary gland transcripts from *Ixodes scapularis* ticks. *Insect Biochemistry and Molecular Biology* **36**, 111–129.

Ribeiro, J. M. C., Weiss, J. J. & Telford, S. R. III (1990). Saliva of the tick *Ixodes dammini* inhibits neutrophil function. *Experimental Parasitology* **70**, 382–388.

Richter, D., Allgower, R. & Matuschka, F.-R. (2002). Co-feeding transmission and its contribution to the perpetuation of the Lyme disease spirochaete *Borrelia afzelii*. *Emerging Infectious Diseases* 8, 1421–1425.

Rudolf, I. & Hubalek, Z. (2003). Effect of the salivary gland and midgut extracts from *Ixodes ricinus* and *Dermacentor reticulatus* (Acari: Ixodidae) on the growth of *Borrelia garinii in vitro*. *Folia Parasitologica* 50, 159–160.

Ryffel, B., Couillin, I., Maillet, I., *et al.* (2005). Histamine scavenging attenuates endotoxin-induced acute lung injury. *Annals of the New York Academy of Science* 1056, 197–205.

Sato, Y. & Nakao, M. (1997). Transmission of the Lyme disease spirochaete, *Borrelia garinii*, between infected and uninfected *Ixodes persulcatus* during cofeeding on *mice*. *Journal of Parasitology* 83, 547–550.

Severinova, J., Salat, J., Krocova, Z., *et al.* (2005). Co-inoculation of *Borrelia afzelii* with tick salivary gland extract influences distribution of immunocompetent cells in the skin and lymph nodes of mice. *Folia Microbiologica* 50, 457–463.

Shaw, M. K., Tilney, L. G. & Mckeever, D. J. (1993). Tick salivary gland extract and interleukin-2 stimulation enhance susceptibility of lymphocytes to infection by *Theileria parva* sporozoites. *Infection and Immunity* 61, 1486–1495.

Shih, C.-M., Chao, L. L. & Yu, C. P. (2002). Chemotactic migration of the Lyme disease spirochete (*Borrelia burgdorferi*) to salivary gland extracts of vector ticks. *American Journal of Tropical Medicine and Hygiene* 66, 616–621.

Singh, K. R. P., Goverdhan, M. K. & Bhat, U. K. M. (1971). Transmission of Kyasanur forest disease virus by soft tick, *Argas persicus* (Ixodoidea: Argasidae). *Indian Journal of Medical Research* 59, 213–218.

Stark, G. R., Kerr, I. M., Williams, B. R., Silverman, R. H. & Schreiber, R. D. (1998). How cells respond to interferons. *Annual Review of Biochemistry* 67, 227–254.

Stebbings, J. H. J. (1974). Immediate hypersensitivity: a defense against arthropods? *Perspectives in Biology and Medicine* 17, 233–239.

Steen, N. A., Barker, S. C., & Alewood, P. F. (2006). Proteins in the saliva of the Ixodida (ticks): pharmacological features and biological significance. *Toxicon* 47, 1–20.

Suhonen, J., Hartiala, K. & Viljanen, M. K. (1998). Tube phagocytosis, a novel way for neutrophils to phagocytose *Borrelia burgdorferi*. *Infection and Immunity* 66, 3433–3435.

Sukumaran, B., Narasimhan, S., Anderson, J. F., *et al.* (2006). An *Ixodes scapularis* protein required for survival of *Anaplasma phagocytophilum* in tick salivary glands. *Journal of Experimental Medicine* 203, 1507–1517.

Takamatsu, H., Mellor, P. S., Mertens, P. P., *et al.* (2003). A possible overwintering mechanism for bluetongue virus in the absence of the insect vector. *Journal of General Virology* 84, 227–235.

Takeshita, K., Sakai, K., Bacon, K. B. & Gantner, F. (2003). Critical role of histamine H4 receptor in leukotriene B4 production and mast cell-dependent neutrophil recruitment inuced by zymosan *in vivo*. *Journal of Pharmacology and Experimental Therapeutics* 307, 1–7.

Titus, R. G., Bishop, J. V. & Mejia, J. S. (2006). The immunomodulatory factors of arthropod saliva and the potential for these factors to serve as vaccine targets to prevent pathogen transmission. *Parasite Immunology* 28, 131–141.

Ullmann, A. J., Lane, R. S., Kurtenbach, K., *et al.* (2003). Bacteriolytic activity of selected vertebrate sera for *Borrelia burgdorferi* sensu stricto and *Borrelia bissettii*. *Journal of Parasitology* 89, 1256–1257.

Valenzuela, J. G., Charlab, R., Mather, T. N. & Ribeiro, J. M. C. (2000). Purification, cloning, and expression of a novel salivary anticomplement protein from the tick, *Ixodes scapularis*. *Journal of Biological Chemistry* 275, 18717–18723.

Van Dam, A. P. (2002). Diversity of *Ixodes*-borne *Borrelia* species: clinical, pathogenetic, and diagnostic implications and impact on vaccine development. *Vector-Borne Zoonotic Disease* 2, 249–254.

Wang, H. & Nuttall, P. A. (1994). Excretion of host immunoglobulin in tick saliva and detection of IgG-binding proteins in tick haemolymph and salivary glands. *Parasitology* 109, 525–530.

Wang, H. & Nuttall, P. A. (1999). Immunoglobulin binding proteins in ticks: new target for vaccine development against a blood-feeding parasite. *Cellular and Molecular Life Sciences* 56, 286–295.

Wang, H., Paesen, G. C., Nuttall, P. A. & Barbour, A. G. (1998). Male ticks help their mates to feed. *Nature* 391, 753–754.

Wikel, S. K. (1982). Histamine content of tick attachment sites and the effects of HI and H2 histamine antagonists on the expression of resistance. *Annals of Tropical Medicine and Parasitology* 76, 179–185.

Wikel, S. K. (1985). Effect of tick infestation on the plaque-forming cell response to a thymic dependent antigen. *Annals of Tropical Medicine and Parasitology* 79, 195–198.

Wikel, S. K. & Allen, J. R. (1977). Acquired resistance to ticks. III. Cobra venom factor and the resistance response. *Immunology* 32, 457–465.

Wikel, S. K., Ramachandra, R. N., Bergman, D. K., Burkot, T. R. & Piesman, J. (1997). Infestation with pathogen-free nymphs of the tick *Ixodes scapularis* induces host resistance to transmission of *Borrelia burgdorferi* by ticks. *Infection and Immunity* **65**, 335–338.

Willadsen, P., Wood, G. M. & Riding, G. A. (1979). The relation between skin histamine concentration, histamine sensitivity and the resistance of cattle to the tick *Boophilus microplus*. *Zeitschrift für Parasitenkunde* **59**, 87–93.

Yu, D., Liang, J., Yu, C., *et al.* (2006). A tick B-cell inhibitory protein from salivary glands of the hard tick, *Hyalomma asiaticum asiaticum*. *Biochemical and Biophysical Research Communications* **343**, 585–590.

Zeidner, N., Dreitz, M., Belasco, W. & Fish, D. (1996). Suppression of acute *Ixodes scapularis*-induced *Borrelia burgdorferi* infection using tumour necrosis factor-alpha, interleukin-2 and interferon-gamma. *Journal of Infectious Diseases* **173**, 187–195.

Zeidner, N., Mbow, M., Dolan, M., *et al.* (1997). Effects of *Ixodes scapularis* and *Borrelia burgdorferi* on modulation of the host immune response: induction of the Th2 cytokine response in Lyme disease-susceptible (C3H/HeJ) mice but not in disease-resistant (BALB/c) mice. *Infection and Immunity* **65**, 3100–3106.

Zeidner, N. S., Schneider, B. S., Nuncio, M. S., Gern, L. & Piesman, J. (2002). Coinoculation of *Borrelia* spp. with tick salivary gland lysate enhances spirochaete load in mice and is tick species-specific. *Journal of Parasitology* **88**, 1276–1278.

Zeller, H. G., Cornet, J.-P. & Camicas, J.-L. (1994). Experimental transmission of Crimean–Congo hemorrhagic fever virus by West African wild ground-feeding birds to *Hyalomma marginatum rufipes* ticks. *American Journal of Tropical Medicine and Hygiene* **50**, 676–681.

11 • Lyme borreliosis in Europe and North America

J. PIESMAN AND L. GERN

INTRODUCTION

Arthropod-borne spirochaetes have long caused human suffering and disease. Louse-borne relapsing fever (LBRF), caused by *Borrelia recurrentis* and transmitted by the human body louse (*Pediculus humanus*), was once widespread in the extensive areas where human body lice were found. Today, LBRF is reported mainly from northeastern and central Africa including the countries of Ethiopia, Somalia and Sudan, in discrete foci where human body lice remain prevalent (Porcella *et al.*, 2000). Tick-borne relapsing fever (TBRF) was first described in Africa where the argasid (soft) tick *Ornithodoros moubata* was found to transmit *Borrelia duttoni* (see historical review by Burgdorfer, 2001). Isolated endemic cycles of TBRF caused by individual species of relapsing fever spirochaetes and their matching argasid vector species have been described in Asia, Europe and the Americas (Felsenfeld, 1979). Recent reports detailing the epidemiology and biology of relapsing fever include studies in Tanzania, where *B. duttoni* frequently causes human disease (Melkert & Stel, 1991; Fukunaga *et al.*, 2001), as well as studies in North America where *Borrelia hermsii* is the primary aetiologic agent of relapsing fever (Dworkin *et al.*, 2002). Although *Borrelia* spp. were known to cause human disease in isolated pockets, scant attention was directed towards the study of these organisms in the latter half of the twentieth century until an epidemic of arthritis was described in Lyme, Connecticut (Steere *et al.*, 1977*b*). In sequential fashion, this condition was associated with a typical rash previously described in Europe as erythema chronicum migrans (later shortened to erythema migrans or EM) and the bite of the blacklegged tick, *Ixodes scapularis* (Steere, Broderick & Malawista, 1978; Steere & Malawista, 1979). A significant breakthrough occurred in 1982 when Burgdorfer *et al.* (1982) reported the discovery of a spirochaete in *I. scapularis*, and a few months later in *Ixodes ricinus* (Burgdorfer *et al.*, 1983), which proved to be the aetiologic agent of Lyme disease or Lyme borreliosis. This spirochaete was subsequently named *Borrelia burgdorferi* (Johnson *et al.*, 1984). It seems appropriate to review at this time (two decades following the discovery of *B. burgdorferi*), the large body of knowledge accumulated concerning the ecology, entomology, epidemiology, microbiology and prevention of Lyme borreliosis in the two areas of the world where the most human cases have been described: Europe and North America.

By necessity, this review must focus solely on Lyme borreliosis in Europe and North America since the topic is extensive and the literature vast on this subject alone. The subject of Lyme borreliosis in Asia, where *Ixodes persulcatus* is the primary vector, has recently been reviewed by Miyamoto & Masuzawa (2002), and Korenberg, Gorelova & Kovalevskii (2002). Moreover, the focus of this review is placed on the aspects of Lyme borreliosis that principally affect human health. An extensive review of Lyme borreliosis in livestock, companion animals and wildlife is beyond the scope of the current review. In the veterinary literature, the most comprehensive body of knowledge for disease in animals has been developed through the use of a canine model (Appel *et al.*, 1993). Initial studies on developing an equine model of infection have also been reported (Chang *et al.*, 2000).

Although it has been over two decades since the discovery of the Lyme disease spirochaete, Lyme borreliosis is an expanding public health problem that has defied our attempts to control it. By comparing the accumulated experience of investigators in North America and Europe, where the disease is most frequently reported, we hope to advance the cause of developing novel approaches to combat Lyme borreliosis.

Ticks: Biology, Disease and Control, ed. Alan S. Bowman and Patricia A. Nuttall. Published by Cambridge University Press.
© Cambridge University Press 2008.

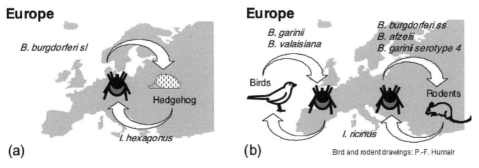

Fig. 11.1. Transmission cycle of *Borrelia burgdorferi* in Europe. (a) Cycle involving *Ixodes hexagonus* and hedgehogs. (b) Cycles involving *I. ricinus*, various genospecies of *B. burgdorferi* sl, and birds and rodents.

BORRELIA BURGDORFERI SENSU LATO IN EUROPE

In Europe, *B. burgdorferi* sensu lato (sl) has been reported from 26 countries (Hubálek & Halouzka, 1997). The reported mean rates of *B. burgdorferi* infection in unfed *I. ricinus* ticks vary from 0% to 11% (mean 2%) for larvae, from 2% to 43% (mean 11%) for nymphs and from 3% to 58% (mean 17%) for adults (Hubálek & Halouzka, 1998). Occasionally, higher infection rates have been reported, mainly using PCR, as for example in Portugal where *B. burgdorferi* DNA was detected in 75% of *I. ricinus* ticks (de Michelis et al., 2000).

Seven *Borrelia* genospecies have been found associated with *I. ricinus*: *B. burgdorferi* sensu stricto (ss) (Johnson et al., 1984), *B. garinii* (Baranton et al., 1992), *B. afzelii* (Canica et al., 1993), *B. valaisiana* (Wang et al., 1997), *B. lusitaniae* (Le Fleche et al., 1997) and *B. spielmanii* (Richter et al., 2006) (Fig. 11.1). Only recently, *B. bissettii* was detected in ticks in Europe. In fact, a single *I. ricinus* from Slovakia was found to be reactive with probes specific for *B. bissettii* (Hanincová et al., 2003b); the fact that this tick was also reactive with probes for two other genospecies of *B. burgdorferi* complicated the specific identification of the spirochaetes. Therefore, the presence of this *Borrelia* species in *I. ricinus* has to be confirmed.

Soon after the discovery of *B. burgdorferi*, phenotyping of *Borrelia* isolates showed that the protein profiles of *B. burgdorferi* sl isolates are heterogeneous (Barbour, Heiland & Howe, 1985). A few years later, outer surface protein A (OspA) and outer surface protein C (OspC) serotyping of isolates was established using sets of monoclonal antibodies (Wilske et al., 1993, 1995, 1996). Eight OspA serotypes of *B. burgdorferi* sl have been defined (Wilske et al., 1993, 1996). These serotypes correlate well with the delineated three most frequent genospecies: serotype 1 corresponds to *B. burgdorferi* ss, serotype 2 to *B. afzelii*, and serotypes 3 to 8 correspond to *B. garinii*. The heterogeneity among *B. garinii* isolates was confirmed on a genetic basis (Will et al., 1995). Strikingly, *B. garinii* serotype 4 isolates have been cultivated from cerebrospinal fluid (CSF) from patients from Germany, the Netherlands, Denmark and Slovenia, and have been more frequently cultivated from CSF than other serotypes (Wilske et al., 1993, 1996; Van Dam et al., 1997) but were only recently shown to be transmitted by *I. ricinus* ticks (Hu et al., 2001). A recent meta-analysis showed that *B. garinii* and *B. afzelii* are the most frequent and most widely distributed species whereas *B. burgdorferi* ss and *B. valaisiana* are less common (Rauter & Hartung, 2005).

Borrelia lusitaniae was first isolated from *I. ricinus* ticks in Portugal (Nuncio et al., 1993) and has subsequently been reported in the Czech Republic, Moldavia, Ukraine (Postic et al., 1997), Slovakia (Gern et al., 1999), Tunisia (Zhioua et al., 1999; Younsi et al., 2001; Dsouli et al., 2006), Morocco (Sarih et al., 2003), Poland (Mizak & Krol, 2000), Spain (Escudero et al., 2000; Barral et al., 2002) and Switzerland (Jouda et al., 2003, 2004a, b; Poupon et al., 2006). Interestingly, in Portugal (de Michelis et al., 2000), Tunisia (Zhioua et al., 1999; Younsi et al., 2001; Dsouli et al., 2006) and Morocco (Sarih et al., 2003), *B. lusitaniae* is common and greatly exceeds the other genospecies in *I. ricinus* ticks whereas *B. lusitaniae* is only sporadically reported in ticks from other areas. Although *B. lusitaniae* can be found outside its well-defined foci in southern Europe, the dominance of *B. lusitaniae* in *I. ricinus* ticks in Portugal, Tunisia and Morocco indicates that the genospecies diversity of *B. burgdorferi* sl decreases towards the southern margin of its European distribution. The distribution of *Borrelia* spp. in different parts of Europe has been recorded by Rauter & Hartung (2005).

Many *Borrelia* species may circulate in an endemic area, resulting in mixed infections. Such infections are reported less frequently than single infections and are often detected by PCR methods. Analysis of data collected throughout Europe identified 13% mixed infections in ticks (Rauter & Hartung, 2005). Mixed infections in ticks may result from the feeding of ticks on a host infected by multiple *Borrelia* species or from infected ticks feeding simultaneously on a host and exchanging *Borrelia* species through co-feeding transmission (Gern & Rais, 1996; Randolph, Gern & Nuttall, 1996; Hu *et al.*, 2003). Moreover, ticks may acquire various *Borrelia* species through their successive blood meals on different hosts, and maintain the infection to the subsequent stage via trans-stadial transmission. Infections by multiple *B. burgdorferi* sl genospecies have been observed in ticks in many parts of Europe, including the Netherlands (Rijpkema *et al.*, 1995), Croatia (Rijpkema *et al.*, 1996), Switzerland (Leuba-Garcia *et al.*, 1994; Jouda *et al.*, 2003, 2004*a, b*), France (Pichon *et al.*, 1995; Ferquel *et al.*, 2006), Austria (Stunzner *et al.*, 1998), Belgium (Misonne, Van Impe & Hoet, 1998), Estonia, Kirghizia, Moldavia, Russia and Ukraine (Postic *et al.*, 1997), Ireland (Kirstein *et al.*, 1997), Italy (Cinco *et al.*, 1998), Germany (Liebisch *et al.*, 1998*b*; Hu *et al.*, 2001; Kurtenbach *et al.*, 2001; Kampen *et al.*, 2004), Latvia, United Kingdom and Slovakia (Kurtenbach *et al.*, 2001), Norway (Jenkins *et al.*, 2001), Finland (Junttila *et al.*, 1999), Czech Republic (Basta *et al.*, 1999) and Poland (Stanczak *et al.*, 2000). Different combinations of mixed infections with two or three genospecies have been detected in *I. ricinus*. *Borrelia garinii* and *B. valaisiana* constitute the majority of mixed infections followed by mixed infections with *B. garinii* and *B. afzelii*.

BORRELIA BURGDORFERI SENSU LATO IN NORTH AMERICA

Spirochaete diversity was thought to be much greater in Europe than in North America until close examination of spirochaete populations across the Atlantic was initiated during the 1990s. A landmark study involved molecular characterization of a total of 186 strains from throughout the United States (Mathiesen *et al.*, 1997). These strains fell into two major groups: a fairly uniform B31 division and a more heterogeneous division from more moderate climates, resembling the well-characterized 25015 strain. A smaller group included several isolates from *Ixodes dentatus* ticks in

Missouri. All the strains examined that were human-derived fell within the B31 group.

Three formal genospecies have now been well defined in North America. The predominant one is the former B31 division, corresponding to the genospecies *B. burgdorferi* ss according to molecular criteria (Baranton *et al.*, 1992; Postic *et al.*, 1994). This is the only genospecies that has been demonstrated to infect humans in North America. It is ubiquitous in *I. scapularis* ticks in hyperendemic regions of the northeastern United States (Seinost *et al.*, 1999). The 25015 division (also called the DN127 group) was defined as a unique genospecies named *B. bissettii* by Postic *et al.* (1998). Several *B. bissettii* strains were described from *I. pacificus* ticks collected in California in this original description. In addition, a large number of strains isolated from an enzootic cycle involving woodrats and *I. spinipalpis* ticks in Colorado were found to be *B. bissettii* (Schneider *et al.*, 2000) (Fig. 11.2). *Borrelia bissettii* has also been isolated from a variety of rodents and ticks in the southern United States (Lin, Oliver & Gao, 2002), and from rodents in the metropolitan Chicago area (Picken & Picken, 2000). The third recognized genospecies in North America has been isolated from rabbits and from a tick associated with rabbits, *I. dentatus*. These spirochaetes were formally described as a new genospecies (*B. andersonii*) by Marconi, Liveris & Schwartz (1995) (Fig. 11.2). Spirochaetes that appear to fit the definition of *B. burgdorferi* sl but that are distinctly different from any well-described genospecies have also been detected in California (Postic *et al.*, 1998; Brown, Peot & Lane, 2006) and Florida (Lin *et al.*, 2002). Although the known diversity of *B. burgdorferi* sl in North America is likely to expand, it must be stressed that human-derived culture confirmed isolates have all been *B. burgdorferi* ss. Additional attempts to make human-derived *Borrelia* isolates in culture medium from various geographical locations in North America are urgently needed.

A group of spirochaetes quite separate and distinct from *B. burgdorferi* sl have been reported to infect hard ticks in North America (see Chapter 16). These include *B. lonestari* from the lone star tick, *Amblyomma americanum* (Barbour *et al.*, 1996), 'Novel *Borrelia*-MP2000' from *I. scapularis* (Scoles *et al.*, 2001) and *B. theileri* from *Boophilus microplus* and *Rhipicephalus* spp. (Rich *et al.*, 2001). Recently, *B. lonestari* has been cultured in the presence of tick cells (Varela *et al.*, 2004). Based on molecular analysis this group of spirochaetes is more closely related to relapsing fever spirochaetes than to *B. burgdorferi* sl. They have been informally called 'hard-tick relapsing fever spirochaetes'.

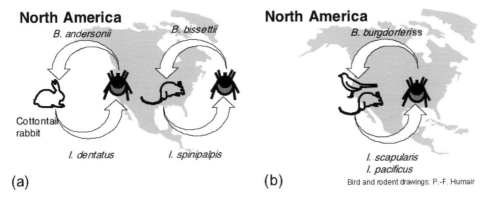

Fig. 11.2. Transmission cycle of *B. burgdorferi* in North America. (a) Cycles involving *B. andersonii*, *I. dentatus* and rabbits; and *B. bissettii*, *I. spinipalpis* and rodents. (b) Cycle involving *I. scapularis* or *I. pacificus*, *B. burgdorferi* ss and birds and rodents.

Moreover, they are very closely related to a spirochaete, *B. miyamotoi*, isolated from *I. persulcatus* in Japan (Fukunaga *et al.*, 1995), and a spirochaete from *I. ricinus* in Sweden (Fraenkel, Garpmo & Berglund, 2002), and in Germany and France (Richter *et al.*, 2003). Although *B. lonestari* DNA has been detected in an erythema migrans lesion and an associated *A. americanum* (James *et al.*, 2001), the pathogenic potential of all of these hard tick relapsing fever spirochaetes is still undetermined. Nevertheless, these spirochaetes are important because they cause confusion when surveying tick populations for *B. burgdorferi* sl. Specific tools to differentiate the hard tick relapsing fever spirochaetes from *B. burgdorferi* sl are needed to survey ticks for *Borrelia* in North America (Ullmann *et al.*, 2005). In both North America and Europe, special care is required to distinguish spirochaetes infecting unfed larvae in estimates of transovarial transmission rates, owing to the possibility that hard tick relapsing fever spirochaetes are transovarially transmitted, as suggested by Rich *et al.* (2001).

CLINICAL ASPECTS, DIAGNOSIS AND TREATMENT OF LYME DISEASE IN EUROPE

The first clinical case of what is now known as Lyme borreliosis was reported in Europe at the end of the nineteenth century (Buchwald, 1883). In the following years, erythema migrans (EM), lymphadenosis benigna cutis, acrodermatitis chronica atrophicans (ACA) and meningopolyneuritis were described (Weber & Pfister, 1993).

In Europe, the clinical case definition described by the European Union Concerted Action on Risk Assessment (Stanek *et al.*, 1996) serves as a guideline for clinical diagnosis of the disease. Two different aspects can be distinguished in the development of the infection: (i) localized infection, and (ii) disseminated infection.

Erythema migrans is the hallmark of Lyme borreliosis. The erythema begins as a red macule or papule, often with a central clearing at the site of the tick bite, appearing a few days to 1 month after the tick bite. At this stage the infection is localized.

The disseminated form of the disease appears a few days or weeks after the tick bite. Manifestations of early neurological involvement including meningitis, unilateral facial palsy, other cranial neuritis and radiculitis may occur. Chronic involvement of the central nervous system includes encephalomyelitis and chronic meningitis; however, these manifestations are rare. Lyme arthritis includes brief attacks of joint swelling with occasional persistence of synovitis. Cardiac involvement appears as an acute onset of disturbances in atrioventricular conduction. Endomyocarditis, pericarditis and rhythm disturbances have also been reported (Stanek *et al.*, 1996).

The EM is similar in the whole geographical distribution of Lyme borreliosis whereas there appear to be differences in the manifestations of the disease as well as the frequency and severity of the disease. This may be related to the distribution of the various pathogenic genospecies and their prevalences. In Europe, where more pathogenic genospecies have been described than in North America, the disease expresses itself under a wider range of manifestations (Stanek *et al.*, 1996, 2002). It is believed that this may reflect regional differences in the distribution and frequency of the different *Borrelia* genospecies. Currently, three *Borrelia* species (*B. burgdorferi*

Table 11.1 *Characteristics of Lyme borreliosis in North America and Eurasia*

Characteristic	North America	Eurasia
Vector	*I. scapularis, I. pacificus*	*I. ricinus, I. persulcatus*
Aetiological agents	*B. burgdorferi* ss	*B. burgdorferi* ss, *B. afzelii*,[a] *B. garinii*[b]
Clinical features	Erythema migrans, arthritis, facial palsy, meningitis, peripheral radiculoneuropathy, atrioventricular block	Erythema migrans, acrodermatitis chronica atrophicans,[a] lymphocytoma, arthritis, facial palsy,[b] meningitis,[b] peripheral radiculoneuropathy,[b] atrioventricular block

[a] *B. afzelii* is associated with skin disease, including specifically acrodermititis chronica atrophicans.
[b] *B. garinii* is associated with neurological disease, including facial palsy, meningitis and peripheral radiculoneuropathy.

ss, *B. garinii* and *B. afzelii*) have been frequently isolated from patients suffering from Lyme borreliosis. The status of *B. valaisiana* as a pathogen has yet to be confirmed (Wang *et al.*, 1999a, 1999b). A *B. lusitaniae* strain has been isolated from a Portuguese patient (Collares-Pereira *et al.*, 2004) indicating this species has pathogenic potential, suggested by its pathogenicity for laboratory mice (Zeidner *et al.*, 2001). *Borrelia spielmanii* has been isolated or detected from patient skin in the Netherlands (Van Dam *et al.*, 1993), Germany (Michel *et al.*, 2004) and Hungary (Földvari *et al.*, 2005).

Various studies suggest an association between clinical manifestations and *Borrelia* spp. in Europe. The three pathogenic species (*B. burgdorferi* ss, *B. garinii* and *B. afzelii*) differ in their organo-tropism and generally cause different clinical manifestations (Assous *et al.*, 1993; Van Dam *et al.*, 1993; Dressler, Ackermann & Steere, 1994; Balmelli & Piffaretti, 1995; Busch *et al.*, 1996a, b; Eiffert *et al.*, 1998; Picken *et al.*, 1998; Jaulhac *et al.*, 2000) (Table 11.1). *Borrelia afzelii* predominates in human skin isolates and *B. garinii* in CSF isolates (Wilske *et al.*, 1993, 1996) whereas considerable heterogeneity has been described for *Borrelia* species detected in synovial fluid (Vasiliu *et al.*, 1998). Currently, the situation appears to be even more complicated. Recent studies reported that a few groups of *Borrelia* within the three pathogenic species are responsible for the disseminated form of the disease (Seinost *et al.*, 1999; Baranton *et al.*, 2001; Lagal *et al.*, 2003). In fact, the authors of these studies showed that 69 OspC groups could be defined within the three pathogenic species based on OspC sequence analysis of various *Borrelia* isolates obtained from ticks and patients, and that all isolates from patients with disseminated forms of Lyme borreliosis belonged to only 24 of the OspC groups. All tick and EM isolates belonged to the other groups. This suggests that the OspC gene is involved in invasiveness, leading

to either localized infections due to non-invasive clones, or to the disseminated form of the disease due to a few clones that are invasive. The geographical distribution and frequency of these various OspC groups are unknown.

Stanek *et al.* (1996), in their paper describing clinical manifestations of Lyme borreliosis in Europe, also reported on laboratory evidence supporting the clinical findings. Diagnosis of Lyme borreliosis by serological testing is difficult, and additionally complicated in Europe owing to the presence of at least three different pathogenic species (Dressler *et al.*, 1994; Hauser *et al.*, 1998). A European multicentre study on immunoblotting showed that it would be difficult to have a standardized immunoblotting method because it would require agreement on the strains used as antigens (Robertson *et al.*, 2000). Moreover, this approach appears unworkable because of the local distribution of species and strains of *B. burgdorferi* sl and the heterogeneity within the strains. A new test developed in the United States (Liang *et al.*, 1999) (see next section) based on the vlsE protein of *B. burgdorferi* may help in the future to improve serological testing in Europe. However, it is clear from all accumulated studies on Lyme borreliosis serology that serological testing should be used to support a clinical diagnosis rather than as confirmation.

Treatment practices in Europe and North America are fairly similar. They are described in detail in the following section.

CLINICAL ASPECTS, DIAGNOSIS AND TREATMENT OF LYME DISEASE IN NORTH AMERICA

Lyme disease in North America was first described as a distinct clinical entity in Lyme, Connecticut among a population of children believed to have juvenile rheumatoid arthritis (Steere *et al.*, 1977*b*). In rapid succession, the skin (Steere *et al.*, 1977*a*), neurological (Reik *et al.*, 1979) and cardiac (Steere *et al.*, 1980) manifestations of Lyme disease were brilliantly elucidated. The clinical manifestations of Lyme disease in North America can be broken down into an acute and a chronic phase. The earliest stage of the disease, often called localized early infection, usually starts as a macule or papule at the site of a tick bite, 3 to 32 days following exposure. This spreads into a large annular lesion, most often with a bright red border and partial central clearing (Steere, 1994). This so-called bull's-eye or target lesion was originally called erythema chronicum migrans (ECM), as per the older literature in Europe. This was shortened to erythema migrans (EM), in part because these lesions proved not to be chronic in North American as they can be in European patients. A large-scale multicentre study that examined 10 936 participants described 118 patients with microbiologically confirmed erythema migrans; curiously, most of these patients had fairly homogenous EM lesions, with only 9% demonstrating classical bull's-eye lesions with a central clearing (Smith *et al.*, 2002). The reason for the lack of bull's-eye rashes was thought to be the short duration between onset of symptoms and presentation at the clinics (mean = 3 days) for diagnosis and treatment when compared to previous studies (Nadelman & Wormser, 2002).

The next stage of the disease has been called early disseminated infection. This stage follows the original EM lesion and may include systemic symptoms, e.g. severe headache, mild neck stiffness, fever, chills, migratory musculoskeletal pain, arthralgias and profound malaise and fatigue (Steere, 1994). Another key characteristic of this stage is the presence of secondary EM lesions at sites remote from the original lesion. These lesions may reflect the haematogenous spread of spirochaetes from the original tick bite site. Interestingly, in a study of patients in a clinic in Westchester County, New York, patient-derived isolates of *B. burgdorferi* ss fell into three distinct genetic subtypes based on restriction fragment length polymorphism (RFLP Types 1, 2, 3). RFLP Type 1 strains were found in the blood of patients with disseminated disease, as compared to skin lesion biopsies from patients with localized disease (predominantly Type 2 and 3) (Wormser *et al.*, 1999). In an elegant series of studies, Seinost *et al.* (1999) and Qiu *et al.* (2002) demonstrated that at least 21 major clonal groups of *B. burgdorferi* ss (as defined by OspA and OspC haplotypes) have been isolated from *I. scapularis* ticks along the eastern seaboard; 15 of these groups have been isolated from primary EM lesions. However, only

four of the clonal groups (A, B, I, K) have been isolated from secondary sites of infection (e.g. blood and CSF). Groups A, B and K are three of the most common haplotypes; the type strain of *B. burgdorferi* ss (B31) is a group A strain. The proportion of Lyme disease patients that present with an EM lesion is estimated at between 80% (Steere, 2001) and 90% (Nadelman & Wormser, 2002). Early symptoms seem to disappear within several weeks.

Prior to the association of Lyme disease with a specific bacterial aetiology, many cases of Lyme disease in North America were not treated with antibiotics. This permitted the natural course of the disease to evolve in patients and observation by physicians. Several months after the acute disease, approximately 60% of patients begin intermittent attacks of joint swelling and pain, particularly in the large joints; the knee is the joint most commonly affected, but not exclusively so (Steere, 1989, 1994). This pattern can best be described as an oligoarticular arthritis. Although the arthritis can move from joint to joint, in a small number of patients the lesions in one or both knees may become chronic with actual erosion of cartilage and bone. These types of severe chronic arthritic lesions are rare in North America today owing to prompt recognition and treatment of the disease in its early stages.

Several weeks after the onset of illness, about 5% of untreated patients in North America develop cardiac involvement (Steere, 2001). The most common cardiac abnormality is an atrioventricular block of fluctuating degrees. In some cases, more diffuse cardiac involvement occurs, including acute myopericarditis, left ventricular dysfunction, cardiomegaly or pancarditis (Steere, 1994). Symptoms may include lightheadedness, palpitations and chest pains.

The most complex manifestations of Lyme disease involve neurological disease. This occurs in about 15% of untreated patients in North America (Steere, 1989, 1994). Neurological abnormalities include meningitis, subtle encephalitic signs, cranial neuritis, bilateral facial palsy, motor or sensory radiculoneuropathy, mononeuritis multiplex, chorea or myelitis. The usual pattern is fluctuating symptoms of meningitis accompanied by facial palsy and peripheral radiculoneuropathy. The CSF may show a lymphocytic pleiocytosis at this point of about 100 cells per ml. Although these symptoms may resolve even in untreated patients, and respond well to treatment, a small minority of patients in North America may develop a late neurological syndrome called 'Lyme encephalopathy' manifested by subtle cognitive disturbances (Halperin *et al.*, 1989;

Logigian, Kaplan & Steere, 1990). The frequency and severity of these cognitive disturbances appear to be the source
of much controversy in the United States. Severe neurological consequences of Lyme disease in the United States are
rare in the present day. However, at least one case of permanent bilateral blindness in a child due to increased cranial
pressure has been reported (Rothermel, Hedges & Steere,
2001). The comparative clinical aspects of Lyme disease in
United States and Europe (Table 11.1) have been succinctly
reviewed by Steere (2001).

Like many bacterial diseases, the ideal basis for diagnosis of Lyme disease is isolation of the aetiologic agent,
Borrelia burgdorferi. The standard culture medium is called
Barbour–Stoenner–Kelly medium or BSK. Tissue samples
are generally surface disinfected, minced, placed in BSK,
and incubated at 33–34 °C. Successful culture of frank EM
lesions has been achieved on a routine basis in research set-
tings in highly endemic regions of the United States through
biopsy and culture of the skin at the affected site (Schwartz
et al., 1992). Recently, quantitative PCR techniques have
proved quite successful in the detection of spirochaetes in
EM lesions, with detection of *Borrelia* DNA in up to 80%
of the lesions tested (Nowakowski *et al.*, 2001; Liveris *et al.*,
2002). Large-volume blood cultures yielded positives in 44%
of early-stage Lyme disease patients (Wormser *et al.*, 2001);
the yield of spirochaetes or DNA in late-stage Lyme disease
from blood, CSF or synovial fluid is, however, much less successful (Nocton *et al.*, 1994; Steere, 2001). Unfortunately, in
the United States, Lyme disease has become a potential diagnosis in an extremely large number of patients lacking a frank
EM and presenting with a complex of symptoms including
fatigue and vague feelings of ill health. Although diagnosis of
Lyme disease is fundamentally based on a clinical evaluation
of the patient, in practice diagnosis is often based upon serology. In fact, a market analysis predicted that approximately
2.8 million serological tests for Lyme disease were performed
in the United States during 1995 (Johnson *et al.*, 1996). The
majority of those tested do not have Lyme disease. Thus,
serological diagnosis of Lyme disease in the United States
has become an area of current controversy. In general, a two-
tiered testing regime that involves an ELISA screening test
and a confirmatory Western blot test produces reliable results
(Dressler *et al.*, 1993; Centers for Disease Control and Prevention, 1995; Johnson *et al.*, 1996). Due to the large number
of serological samples that are tested each year, however, the
specificity of this testing regime is not robust enough to eliminate completely the problem of false positive results. This is
a particularly acute problem with immunoglobulin M (IgM)

blots conducted after the first month of illness. Thus, only
IgG results should be used to support the diagnosis of Lyme
disease after the first month of infection (Steere, 2001). A
new test, based on a recombinant protein of a portion (C6) of
the vlsE protein of *B. burgdorferi* shows promise for improving the sensitivity and specificity of the serological diagnosis
of Lyme disease in the future (Liang *et al.*, 1999; Philipp
et al., 2001).

Practice guidelines for the treatment of Lyme disease
have been issued by the Infectious Diseases Society of America (Wormser *et al.*, 2000, 2006), and are summarized by
Piesman & Gern (2004). Clinical trials that have attempted
to enrol patients with 'chronic Lyme disease' or 'post-Lyme
disease syndrome' have generally been terminated early due
to a lack of enrollees with objective evidence of Lyme disease. In these limited trials, however, no benefit of continued
antibiotic treatment was found (Klempner *et al.*, 2001).

The practice of prophylactic treatment of tick bite in
Lyme disease endemic areas has generated much public discussion. A cost–benefit analysis concluded that prophylactic treatment should only be considered in areas with an
extremely high risk of Lyme disease transmission (Magid
et al., 1992). In a recent clinical trial, Nadelman *et al.* (2001)
found that prophylactic treatment of bites by partially fed
nymphal *I. scapularis* with a single dose of doxycycline in
Westchester County, New York was 87% effective in preventing Lyme disease. Experimental work with rodents has
demonstrated that transmission of *B. burgdorferi* ss becomes
efficient after nymphal *I. scapularis* are attached for >48 h
(des Vignes *et al.*, 2001). Prophylactic treatment of only those
patients exposed to infected *I. scapularis* that have fed for
more than 2 days would be ideal. It remains to be seen
whether this ideal plan can be put into widespread clinical practice, since it would involve rapid testing of ticks for
infection and estimation of the duration of attachment based
on a scutal index of tick engorgement.

LYME DISEASE VECTOR ECOLOGY IN EUROPE

In Europe, three tick species are considered vectors of *B.
burgdorferi* sl: *Ixodes ricinus*, *I. hexagonus* and *I. uriae*. These
three species have very different ecologies, but they all are
three-host ticks with each parasitic stage (larva, nymph and
adult female) feeding on different hosts. Although adult
male *Ixodes* sometimes ingest fluids from hosts, they do not
ingest significant amounts of blood and their role as vectors is probably insignificant but has not been thoroughly

evaluated. These three species, however, have a rather different host range and different biology. Vector biology is an important factor since it dictates much of the epidemiology of the diseases. According to their habitats, the three recognized European vectors of *B. burgdorferi* sl can be divided into non-nidicolous ticks, *I. ricinus*, and nidicolous ticks, *I. uriae* and *I. hexagonus*. Non-nidicolous ticks occupy open habitats whereas nidicolous ticks live in caves, burrows, or nests of their hosts. The differences concerning behaviour and physiology between nidicolous and non-nidicolous ticks are enormous, especially in their host-finding behaviours. The non-nidicolous tick *I. ricinus* awaits a host on the vegetation. Nidicolous tick species, like *I. uriae* and *I. hexagonus*, have closer contact with their host by living in their nests or in their close environment. This implies that contact between non-nidicolous ticks and humans is more frequent than between nidicolous ticks and humans, illustrating the importance of vector biology.

The common European tick species, *I. ricinus*, is the main vector of *B. burgdorferi* sl. This tick species has a wide geographical distribution throughout Europe. It has been described within the latitudes 65° and 39°, from Portugal into Russia (Gern & Humair, 2002), and also in North Africa (Tunisia, Algeria and Morocco) (Dsouli *et al.*, 2006). This wide geographical distribution of *I. ricinus* implies that this tick survives under various environmental conditions. *Ixodes ricinus* prefers deciduous woodlands and mixed forests. High humidity is a prerequisite for tick survival as *I. ricinus* is susceptible to desiccation when questing for hosts on vegetation. High humidity is found at the base of vegetation in the leaf litter where ticks periodically return to take up atmospheric water. Hence *I. ricinus* survives only where the relative humidity of the microenvironment is higher than 80% (Kahl & Knülle, 1988; Randolph *et al.*, 2000). If saturation deficit (a measure of the drying power of the air) is above approximately 4 mm Hg (calculated according to Randolph & Storey, 1999), *I. ricinus* shows positive geotropism (McLeod, 1935). In nature, abrupt declines in questing tick density have been reported to coincide with abrupt increases in saturation deficit (Perret *et al.*, 2000, 2004; Randolph *et al.*, 2002). Temperature, which is known to have an effect on tick questing activity and on tick development rates, varies throughout the geographical distribution of *I. ricinus*. Specific dynamics of seasonal activity have been demonstrated under different climatic conditions (Steele & Randolph, 1985; Tälleklint & Jaenson, 1996; Korenberg, 2000; Perret *et al.*, 2000; Randolph *et al.*, 2002; Dsouli *et al.*, 2006). The seasonal activity pattern is either unimodal or bimodal. Data and model

predictions from Randolph *et al.* (2002) suggest a simple life cycle for *I. ricinus* with a single cohort of each stage starting in the autumn, and not two separate cohorts as previously thought (Lees & Milne, 1951; Donnelly, 1976; Gray, 1982, 1985, 1991; Walker, 2001). The height at which *I. ricinus* ticks quest on vegetation depends on each stage and on the vegetation structure, and it influences host encounter (Mejlon & Jaenson, 1997). *Ixodes ricinus* feeds on an extraordinarily broad array of hosts, from small, medium and large-sized mammals to birds and reptiles (Anderson, 1991), and is the tick species which most frequently bites humans in Europe (for more information see Chapter 2).

Ixodes hexagonus is an endophilous nidicole tick and is one of the most widespread tick species in Europe (Morel, 1965). It has been reported from Northern Europe (Jaenson *et al.*, 1994) to the north of Africa (Bailly-Choumara, Morel & Rageau, 1974). This tick species lives in the nest and burrow of its hosts, an environment which provides a suitable microclimate for tick survival; hence, the geographical distribution of *I. hexagonus* is less dependent on meso- and microclimatic conditions than that of *I. ricinus*. Temperature also influences duration of *I. hexagonus* development (Toutoungi, Aeschlimann & Gern, 1993). This suggests that the duration of tick development may be longer during the winter months than in spring or summer. In view of its habitats, *I. hexagonus* rarely comes in contact with humans. Nevertheless, humans can be bitten occasionally, particularly when they handle nests of hedgehogs when gardening. Hedgehogs are a frequent host of *I. hexagonus* and have surface nests that are common in gardens in Europe. *Ixodes hexagonus* parasitizes primarily Mustelidae (e.g. *Meles meles* – European badger, *Martes fouina* – beach marten, *Mustela putorius* – European polecat, *Mustela ermina* – ermine) and hedgehogs (*Erinaceus europaeus*). In Switzerland, *I. hexagonus* has been collected from 15 animal species, especially from foxes and Mustelidae, but also from domestic animals like dogs and cats (Toutoungi *et al.*, 1991). *Ixodes hexagonus* may also occasionally infest birds (*Pica pica* – magpie, *Falco tinnunculus* – kestrel) and deer (*Capreolus capreolus* – roe deer) (Hubbard, Baker & Cann, 1998; Toutoungi *et al.*, 1991). An additional tick species that is widely distributed throughout Europe and may serve a secondary role as an enzootic vector of *B. burgdorferi* is *I. trianguliceps*, a tick that feeds predominantly on rodents and has been found to be infected with *B. burgdorferi* (Gorelova *et al.*, 1996).

The third known vector of *B. burgdorferi* sl in Europe, *I. uriae*, is a tick species parasitizing seabirds, which has a three-stage life cycle that usually corresponds to one stage each year. Each stage attaches to the host for a single, long blood meal and then returns to the host nesting substrate to overwinter (Eveleigh & Threlfall, 1974). It has been reported that the prevalence and abundance of *I. uriae* are correlated in both space and time at the scale of the host-breeding cliff (Danchin, Boulinier & Massot, 1998; McCoy *et al.*, 1999). The seabird tick, *I. uriae*, has a distribution area covering coasts situated at high latitudes both in northern and southern hemispheres. In Europe, this includes the coasts of Ireland, Iceland, Norway, Sweden, Denmark, UK and France (Olsen *et al.*, 1995a; Hillyard, 1996). Although *I. uriae* infests principally seabirds, it has occasionally been reported on mammals such as seals, river otters and humans (Olsen, 1995).

LYME DISEASE VECTOR ECOLOGY IN NORTH AMERICA

The two principal vectors of Lyme disease in North America are the blacklegged tick (*Ixodes scapularis*) in the eastern half of the continent and the western blacklegged tick (*I. pacificus*) in the western half of the continent (Fig. 11.2). The vast majority of Lyme disease infections in North America are acquired through the bites of nymphal *I. scapularis*. Both biotic and abiotic factors control the distribution of these fascinating and nefarious ticks. One key component that controls *I. scapularis* distribution is humidity. These ticks are very susceptible to desiccation. In Westchester County, New York researchers found that the distribution of *I. scapularis* was positively associated with a remotely sensed greenness–wetness index (Dister *et al.*, 1997). In the north-central United States, the presence of *I. scapularis* was positively associated with deciduous, dry to mesic forests and alfisol-type soils of sandy or loam–sand textures overlying sedimentary rock (Guerra *et al.*, 2002). Ticks were found in moist soils, but not in poorly drained soils where standing water occurred.

There seems little doubt that *I. scapularis* ticks are forest inhabitants. An analysis of *I. scapularis* populations in suburban landscapes in Westchester County, New York demonstrated that tick populations were highest in the woods, intermediate in ecotonal vegetation and sparse in ornamental planting and lawns (Maupin *et al.*, 1991). In Long Point (Ontario, Canada), *I. scapularis* was most abundant in maple forests, followed by oak savannah; ticks were rare in white pine forest and cottonwood dunes (Lindsay *et al.*, 1999). In general, *I. scapularis* is found in hardwood forests where abundant leaf litter provides ample cover from desiccation

and protective cover during snowfall. The importance of leaf litter was demonstrated in an experiment where leaf litter was actually removed from a forested plot; *I. scapularis* populations decreased by 72–100% as a result of litter removal (Schulze, Jordan & Hung, 1995). Although populations of *I. scapularis* are mainly associated with mature oak–maple forests in the northeastern United States, they have the flexibility to inhabit diverse habitats. In coastal regions and islands, ticks can be found in extremely dense shrub-like habitat that contains bayberry, rose and scrub oak as predominant vegetation (Piesman & Spielman, 1979). Another interesting observation has been the influence of masting in oak trees (wherein a massive crop of acorns is produced every 2–5 years) on the density of ticks within forests and the annual variation in populations of *I. scapularis* (Jones *et al.*, 1998). Although populations of *I. scapularis* are mainly associated with hardwood forests, they can also be found in pine forests that are surrounded by hardwoods (Schulze, Jordan & Hung, 1998). The minimal amount of ground cover and hardwood leaf litter that allows *I. scapularis* to thrive in a conifer dominated forest has not been established.

In eastern North America, *I. scapularis* essentially takes 2 years to go through its life cycle (Fig. 11.3). Setting Year 0 as the first year of active questing, larvae generally feed in August–September (Piesman & Spielman, 1979). The majority of larvae moult to nymphs that overwinter as unfed nymphs. Nymphs feed the following year (Year 1) in May–July. These nymphs then moult to adults that begin questing that same year in the fall. Adults quest from October of Year 1 until April of the next year (Year 2). In areas with cold winter temperatures, adult feeding ceases in the middle of winter, but in southern areas adult questing may actually peak in February (Goddard, 1992). In Year 2, replete females lay eggs in May and June; larvae hatch by August and begin questing thus completing the 2-year life cycle. This standard life cycle has been elegantly described by Yuval & Spielman (1990). Interestingly, in some colder climates this idealized life cycle may not take place. Lindsay *et al.* (1995) discovered that in locations in northern Ontario that lacked sufficient degree–days >11 °C, larvae did not emerge from eggs laid in May or June during that same year. Although some eggs could survive the winter and hatch the next year, the overall survivability of these eggs was quite low in this extreme environment. The key factor was not just latitude, since one area (Kenora) that was north and west of other areas had sufficient degree–days, while other areas to the south and east did not have sufficient degree–days during most years

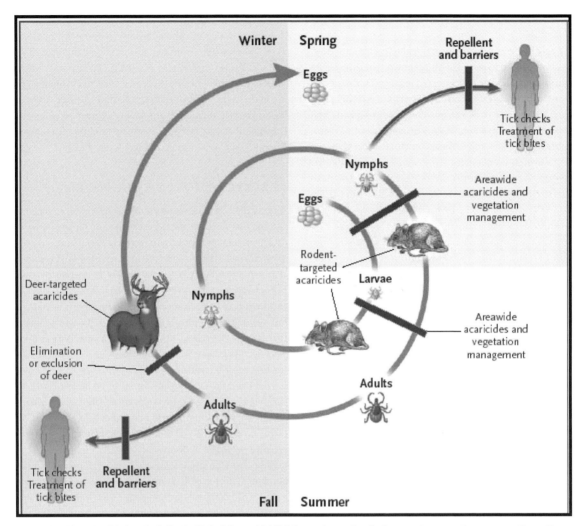

Fig. 11.3. Life cycle of *B. burgdorferi* in the United States, highlighting various points for intervention strategies to prevent Lyme disease. (Reprinted with permission from the Massachusetts Medical Society; Hayes & Piesman, 2003.)

(Kapsuskasing, Geraldton, Thunder Bay). Sufficient thermal warming during the summer months may be required for *I. scapularis* to complete its life cycle efficiently. This key factor may be much more important than the mean or minimum winter temperatures, when the ticks may be insulated by snow cover and well protected. Thus, there may be thermal limits on the areas where *I. scapularis* can efficiently go through its life cycle in North America.

The distribution of *I. scapularis* in the United States was surveyed in 1991, 1994 and 1997 via questionnaires delivered to entomologists and public health officials. A county-by-county map was produced showing records of *I. scapu-*

laris in 952 of the 3141 counties within the United States (Dennis *et al.*, 1998). This tick is distributed along the eastern seaboard from Florida to Maine and as far west as central Texas. Interestingly, the distribution is not continuous, with a gap in the distribution seen in central states (e.g. West Virginia, Ohio, Kentucky, Tennessee). The possibility that *I. scapularis* was once extant over the entire eastern United States, then reduced to a few refugia due to human encroachment and host reduction, and now slowly but surely regaining its former range, is intriguing; this hypothesis will be difficult to test. A 4-year prospective ecological Lyme disease risk survey of the entire eastern United States based on populations of nymphal *I. scapularis* is currently ongoing (Diuk-Wasser *et al.*, 2006).

There has been debate concerning the species status of *I. scapularis*. Spielman *et al.* (1979) proposed that this species be divided into a southern species (*I. scapularis*) and northern species (*I. dammini*). In contrast, Oliver *et al.* (1993*b*) challenged the validity of *I. dammini* as a species separate and distinct from *I. scapularis*. Molecular studies on the mitochondrial genes 16S and 12S suggest two distinct clades of *I. scapularis*: (i) the southern clade, stretching from Florida to North Carolina, and (ii) the northern or all-American clade stretching from Massachusetts to Mississippi (Rich *et al.*, 1995; Norris *et al.*, 1996). The southern clade was considered the basal group of *I. scapularis* by Norris *et al.* (1996). A recent phylogenetic analysis of *I. scapularis* from South Carolina to Massachusetts confirmed that two mitochondrial clades of *I. scapularis* exist (Qiu *et al.*, 2002). These researchers found that the northern clade (Clade A) had a low within-population sequence divergence, while the southern clade (Clade B) had a much higher diversity. Qiu *et al.* (2002) suggested that the northern and southern clades have separate and different evolutionary histories, perhaps influenced by separation during the last glacial period, a maximum of 18 000 years ago. The hypothesis that the northern clade of *I. scapularis* results from several refugia established during the last Ice Age and that this population is now expanding across the eastern United States, with these ticks being the principal ticks responsible for transmitting Lyme disease in North America, is worthy of further objective study. However, the complexity and diversity of host populations (see section 'Reservoir hosts in North America' below) present in different geographical areas may play a predominant role in determining the local risk of Lyme disease transmission, in addition to the genetic make-up of local *I. scapularis* populations.

The western blacklegged tick, *I. pacificus*, is found along the Pacific Coast from British Columbia to Baja California Norte, Mexico (Kain, Sperling & Lane, 1997). Although populations of *I. pacificus* are principally coastal, isolated inland populations have been described in such arid overall climates as Mohave County, Arizona (Olson *et al.*, 1992) and the southwestern corner of Utah (Kain *et al.*, 1997). There is no evidence of genetic isolation among these diverse populations of *I. pacificus* (Kain *et al.*, 1999). Habitats where adult *I. pacificus* can be collected are extremely varied, from redwood or Douglas fir forest, to more open habitats such as chaparral and open grasslands. When collecting adult *I. pacificus* along trails established by people or animals, several researchers have found that the majority of these ticks can be found on the uphill side of the trail. Like *I. scapularis*, *I. pacificus* has been collected mainly at lower altitudes, but *I. pacificus* has been collected, in at least one instance, at an altitude of 2345 m (Olson *et al.*, 1992). The habitats where questing nymphal *I. pacificus* can be collected in abundance are more restricted than comparable adult habitat. Areas of high nymphal abundance are mainly characterized by the presence of mature trees and abundant leaf litter (Tälleklint-Eisen & Lane, 1999; Eisen, Eisen & Lane, 2006).

RESERVOIR HOSTS IN EUROPE

Among the main tick vectors of *B. burgdorferi* sl, *I. ricinus* is the one that feeds on the largest variety of vertebrate hosts (>300 vertebrate species) (Anderson, 1991). However, only a few dozen of these vertebrate host species have been currently identified as reservoir hosts for *B. burgdorferi* sl in Europe (Gern *et al.*, 1998). Thus, little information is available on the real significance of most animal hosts as sources for infecting ticks with *B. burgdorferi* sl. Of the few tick hosts that have been studied up to now, some act as reservoirs whereas others appear to be refractory to infection. A distinction must be made between animals that serve as hosts for ticks and are occasionally found to be infected with *B. burgdorferi* versus true reservoirs of infection that infect a significant proportion of the immature ticks that feed on them. A careful analysis of the term reservoir capacity was recently published by Kahl *et al.* (2002).

The enzootic cycle involves larval and nymphal ticks becoming infected with *B. burgdorferi* while feeding on their hosts. Small mammals are frequent hosts of these developmental stages and are certainly the group that has been the most extensively investigated in Europe and North America. Currently, several species of mice, voles, rats and shrews are recognized as competent reservoirs of *B. burgdorferi* sl in Europe (Gern *et al.*, 1998). Evidence that mice (*Apodemus flavicollis*, *A. sylvaticus*, *A. agrarius*) and the vole *Clethrionomys glareolus* act as reservoirs for *B. burgdorferi* sl has been obtained in many European countries (Aeschlimann *et al.*, 1986; Matuschka *et al.*, 1992; de Boer *et al.*, 1993; Humair *et al.*, 1993*a*; Gern *et al.*, 1994; Kurtenbach *et al.*, 1994, 1995, 1998*b*; Tälleklint & Jaenson, 1994; Randolph & Craine, 1995; Hu *et al.*, 1997; Humair, Rais & Gern, 1999; Richter *et al.*, 1999; Hanincová *et al.*, 2003*a*). Once infected, *Apodemus* spp. remain persistently infectious for ticks (Gern *et al.*, 1994); and since small rodents are frequently parasitized by nymphal and larval *I. ricinus* (Humair *et al.*, 1993*a*; Randolph *et al.*, 2000), they are potent reservoir hosts. However, a study highlighted different transmission patterns in nature between *Apodemus* spp. and ticks, and *C. glareolus* and ticks (Humair *et al.*, 1999). The authors of this study observed that each host species seems to have developed different strategies towards tick infestation and *Borrelia* infection. *Borrelia* infection in *Apodemus* spp. is rarely detected by *Borrelia* isolation; this may be related to the fact that *Apodemus* spp. appear to maintain a low level of *Borrelia* infection (Kurtenbach *et al.*, 1994). However, *Borrelia* is efficiently transmitted from *Apodemus* spp. to ticks (Humair *et al.*, 1999). By contrast, in *C. glareolus*, *Borrelia* infection is easily detectable by isolation and spirochaetes are readily transmitted to ticks; however, most ticks do not feed completely or do not moult (Humair *et al.*, 1999). This is in line with the observation that *C. glareolus* develops an immune response to ticks that prevents ticks from engorging and moulting successfully (Kurtenbach *et al.*, 1994; Dizij & Kurtenbach, 1995). Consequently, the reservoir competence of *Apodemus* spp. and *C. glareolus* is modulated by their immune responses against both the pathogen and the tick.

More limited information has been obtained on the implications of other small mammals in the maintenance cycles of *Borrelia* spp. in nature. Nevertheless, another species of vole, *Microtus agrestis*, in Sweden (Tälleklint & Jaenson, 1994), and black rats (*Rattus rattus*) and Norway rats (*R. norvegicus*) in urbanized environments in Germany (Matuschka *et al.*, 1996, 1997) and in Madeira (Matuschka *et al.*, 1994a) may serve to infect *I. ricinus* ticks. Similarly, only a few studies mention *B. burgdorferi* sl in shrews (*Sorex minutus, S. araneus* and *Neomys foediens*) or in ticks attached on them (Humair *et al.*, 1993a; Tälleklint & Jaenson, 1994). Observations in endemic areas in Germany and in France showed that edible dormice (*Glis glis*) (Matuschka *et al.*, 1994b) and garden dormice (*Eliomys quercinus*) (Matuschka *et al.*, 1999; Richter *et al.*, 2004) are reservoir hosts for *Borrelia* spp. In Germany, edible dormice were frequently parasitized by subadult ticks, and infected around 95% of larvae feeding on them (Matuschka *et al.*, 1994b).

Additional rodents, such as grey squirrels (*Sciurus carolinensis*) in the UK (Craine *et al.*, 1997b) and red squirrels (*S. vulgaris*) in Switzerland (Humair & Gern, 1998), also contribute to the amplification of *Borrelia* spp. in the tick population. Observations indicate that red and grey squirrels are heavily infested with ticks; one study reported a high prevalence of infection (69%) in ticks feeding on red squirrels (Humair & Gern, 1998).

Several researchers demonstrated that the European hedgehog (*Erinaceus europaeus*) also perpetuates *B. burgdorferi* sl in Ireland (Gray *et al.*, 1994), Germany (Liebisch,

Finkbeiner-Weber & Liebisch, 1996), and Switzerland (Gern *et al.*, 1997b) (Fig. 11.1). In Switzerland, an enzootic transmission cycle of *B. burgdorferi* sl involving hedgehogs and another tick vector, *I. hexagonus*, has been described in an urban environment (Gern *et al.*, 1997b).

Another group of animals, lagomorphs (*Lepus europaeus, L. timidus* and *Oryctolagus cuniculus*), plays a role in the support of the enzootic cycle of *B. burgdorferi* sl (Tälleklint & Jaenson, 1993, 1994; Jaenson & Tälleklint, 1996; Matuschka *et al.*, 2000).

Among larger mammals, the red fox (*Vulpes vulpes*) is implicated in the maintenance of *Borrelia* spp. in nature as described in two studies in Germany (Kahl & Geue, 1998; Liebisch *et al.*, 1998a). These animals did not appear to be potent reservoirs since spirochaetes were poorly transmitted to ticks.

Not all tick hosts are competent to serve as reservoirs. This is the case for cervids in general which act primarily as sources of blood for ticks. In fact, studies on roe deer (*Capreolus capreolus*) (Jaenson & Tälleklint, 1992), moose (*Alces alces*) (Tälleklint & Jaenson, 1994), red deer (*Cervus elaphus*) (Gray *et al.*, 1995), and fallow deer (*Dama dama*) (Gray *et al.*, 1992) suggest that these species do not infect feeding ticks with *B. burgdorferi*. Interestingly, sheep have been found to be reservoirs of *B. burgdorferi* in areas of the UK, but the principal mechanism by which they infect *I. ricinus* is co-feeding (Ogden, Nuttall & Randolph, 1997). This is a phenomenon in which the host does not necessarily become infected, but neighbouring ticks serve to infect each other while feeding on the host.

After a long period of controversy, the role of birds in the maintenance of *B. burgdorferi* sl in endemic areas is now recognized (Humair, 2002). The first report in Europe of *B. burgdorferi* sl in *I. ricinus* ticks feeding on birds dates back to 1993 (Humair *et al.*, 1993b). The same year, Olsen *et al.* (1993) demonstrated the existence of a transmission cycle of *B. burgdorferi* sl in seabird colonies among razorbills (*Alca torda*) and *I. uriae* on a Swedish island. Later, spirochaetes were reported in *I. ricinus* ticks collected from migratory birds in Sweden (Olsen, Jaenson & Bergström, 1995b) and Switzerland (Poupon *et al.*, 2006), and in ticks feeding on birds captured in endemic areas in the Czech Republic (Hubálek *et al.*, 1996) and the UK (Craine *et al.*, 1997). In 1998, two studies clearly defined the reservoir role of birds, one on a passerine bird, the blackbird (*Turdus merula*) (Humair *et al.*, 1998), the other on a gallinaceous species, the pheasant (*Phasianus colchicus*) (Kurtenbach *et al.*, 1998a). Both studies demonstrated the reservoir role of these avian species using xenodiagnosis, providing evidence that birds contribute to the circulation of *Borrelia* in endemic areas. The involvement of seabirds and *I. uriae* in the marine environment was also confirmed by additional studies in both the northern and the southern hemispheres (Olsen *et al.*, 1995*a*; Gylfe *et al.*, 1999).

In endemic areas in Europe, at least seven *Borrelia* genospecies may circulate between vertebrate hosts and ticks. The first findings on host specificity of *Borrelia* spp. came from a study conducted in Switzerland (Humair *et al.*, 1995). In this study, it was shown that *Borrelia* spp. isolated from *Apodemus* spp. captured in two different endemic sites all belonged to *B. afzelii*, whereas genospecies diversity in ticks collected by flagging vegetation in these sites displayed heterogeneity. Later, it was shown that small rodents of the genus *Apodemus*, such as woodmice (*A. sylvaticus*) and yellow-necked mice (*A. flavicollis*), and of the genus *Clethrionomys* as well as red squirrels and grey squirrels, are usually infected by *B. afzelii* and less frequently by *B. burgdorferi* ss; moreover, these hosts transmit these *Borrelia* spp. to ticks feeding on them (Craine *et al.*, 1997; Hu *et al.*, 1997; Humair *et al.*, 1999; Kurtenbach *et al.*, 1998*b*). An increasing body of evidence first showed that *B. garinii* is mostly associated with migratory birds (Olsen *et al.*, 1995*b*), and later that *B. garinii* and *B. valaisiana* are associated with blackbirds and pheasants (Humair *et al.*, 1998; Kurtenbach *et al.*, 1998*a*, *b*). *Borrelia garinii* was also described as the species involved in marine environments, in seabird colonies located on the northern and southern hemispheres (Olsen *et al.*, 1995*a*; Gylfe *et al.*, 1999). This gives us the opportunity to reiterate that Olsen and colleagues, in their 1995*a* study, detected the presence of *B. garinii* DNA in North America (Alaska). In fact, amplified flagellin gene fragments from positive *I. uriae* ticks collected from fork-tailed storm petrels (*Oceanodroma furcata*) on Egg Island (Alaska) subjected to DNA sequencing showed that they were closely related to the *fla* gene of *B. garinii* suggesting the presence of *B. garinii* in North America, at least in a very specific area (marine enzootic cycles).

Borrelia garinii has occasionally been described associated with rodents in Austria, Germany and Russia (Khanakah *et al.*, 1994; Gorelova *et al.*, 1995; Richter *et al.*, 1999). The serotype of the incriminated *B. garinii* is unknown. However, in view of recent findings showing that laboratory mice challenged with nymphs collected in nature were able to transmit *B. garinii* OspA serotype 4 to xenodiagnostic ticks (Hu *et al.*, 2001) and that *Apodemus* spp. captured in Switzerland transmitted *B. garinii* OspA serotype 4 to xenodiagnostic ticks (Huegli *et al.*, 2002), rodents (at least in some well-specified areas) can be reservoir hosts for *B. garinii* (Huegli *et al.*, 2002).

In summary, rodents are mainly associated with *B. afzelii*, and to a lesser degree with *B. burgdorferi* ss and *B. garinii* OspA serotype 4; other *B. garinii* serotypes are associated with birds (Kurtenbach *et al.*, 2002*b*). *Borrelia valaisiana* has been described only in birds and never in rodents and therefore appears to be specific to birds (Olsen *et al.*, 1993, 1995*a, b*; Olsen, 1995; Humair *et al.*, 1998; Kurtenbach *et al.*, 1998*b*; Gylfe *et al.*, 2000) (Fig. 11.1).

Although *B. lusitaniae* may be common in *I. ricinus* ticks in some areas of North Africa (Younsi *et al.*, 2001; Sarih *et al.*, 2003) and Portugal (de Michelis *et al.*, 2000), and infections in ticks often show large numbers of spirochaetes (authors' unpublished data), its reservoir hosts were identified only recently. Poupon *et al.* (2006) frequently observed *B. lusitaniae* in *I. ricinus* larvae feeding on birds migrating between southwestern Europe/North Africa and northwestern Europe. These authors considered that migratory birds play a role in the dispersal of *B. lusitaniae*. Dsouli *et al.* (2006) demonstrated the reservoir role of the lizard *Psammodronus algirus* for *B. lusitaniae* in Tunisia. Another recent study identified *E. quercinus* as one of the reservoirs of *B. spielmanii* in Central Europe (Richter *et al.*, 2004).

The host complement system appears to be a major determinant of host-specificity of *B. burgdorferi* sl in Europe (Kurtenbach *et al.*, 1998*c*, 2002*a*). In vitro, *Borrelia* spp. show different patterns of resistance or sensitivity to the serum of different host species, which corresponds to their host specificity observed in nature (Kurtenbach *et al.*, 1998*c*, 2002*a*). The specificity of complement lysis of genospecies of *B. burgdorferi* is apparently mediated through expression of the *erp* gene loci. These gene loci encode for the so-called CRASPs (complement regulatory-acquiring surface proteins) which bind differentially to complement inhibitors (e.g. factor H). Research is ongoing into the specific receptors involved in this interaction between host and pathogen (Kraiczy *et al.*, 2001; Stevenson *et al.*, 2002). The evolutionary consequences of this system are an intriguing subject for further study.

RESERVOIR HOSTS IN NORTH AMERICA

There is little doubt that white-tailed deer (*Odocoileus virginianus*) play a key role in the Lyme disease enzootic cycle in North America because they serve as the principal hosts for the adult stage of the principal tick vector. Early studies demonstrated that white-tailed deer support large numbers of adult *I. scapularis* (Piesman *et al.*, 1979; Spielman *et al.*, 1985). In addition, populations of white-tailed deer have exploded during the latter half of the twentieth century, exactly coinciding with the time period of the Lyme disease epidemic in North America. Observations on islands, with and without deer, added to the impression that these tick populations were dependent on the presence of deer (Wilson, Adler & Spielman, 1985; Anderson *et al.*, 1987). Although other animals such as medium-sized mammals (Fish & Dowler, 1989), dogs and cats are often infested with adult *I. scapularis*, these hosts generally do not support the large numbers needed to support populations of *I. scapularis*.

Despite the importance of white-tailed deer as hosts for the adult stage of the principal vector of Lyme disease spirochaetes in North America, these hosts are not an important reservoir of *B. burgdorferi* due to the lytic properties of the complement contained in deer sera. Telford *et al.* (1988) demonstrated that larval *I. scapularis* dropping off deer carcasses were not infected with *B. burgdorferi* in an area of Massachusetts highly endemic for this aetiologic agent. European researchers have now demonstrated that complement contained in deer sera is highly lytic to a wide variety of *B. burgdorferi* (Kurtenbach *et al.*, 2002*b*) and this observation has been confirmed using a North American strain of *B. burgdorferi* ss (Nelson *et al.*, 2000). Thus, white-tailed deer are a doubled-edged sword in the maintenance of the Lyme disease spirochaete enzootic cycle in eastern North America; deer are needed to support large populations of vector ticks, but these hosts also serve to decrease the proportion of the tick population that becomes infected with spirochaetes. Columbian black-tailed deer (*O. hemionus columbianus*), and other so-called 'mule deer' play a similar role in support of *I. pacificus* populations (Westrom, Lane & Anderson, 1985) in western North America.

Another group of animals that serve as important hosts for *I. scapularis* and *I. pacificus* but apparently do not serve as reservoirs for spirochaetal infection are lizards. Spielman *et al.* (1985) originally coined the phrase 'zooprophylaxis' for hosts that fail to infect the ticks that infest them. Lizards appear to be important zooprophylactic hosts, serving to drive down the *B. burgdorferi* infection rates of questing ticks in key areas. In the southern United States, lizards serve as hosts to the majority of immature *I. scapularis* (Apperson *et al.*, 1993; Oliver, Cummins & Joiner, 1993*a*). Similarly, lizards also serve as important hosts for immature *I. pacificus* (Lane & Loye, 1989). Several researchers speculated that lizards were incompetent hosts for *B. burgdorferi* and that nymphs and adult ticks that had previously fed on these hosts were not infected with *B. burgdorferi* (Spielman, 1988). Recently, Kuo, Lane & Gicias (2000) demonstrated that complement from *Sceloporus* and *Elgaria* lizards was highly lytic for *B. burgdorferi* ss, thus supporting the rationale that lizards are zooprophylactic hosts. A note of caution must be sounded, however, based on the observations of Levin *et al.* (1996). These researchers found that lizards in the genus *Eumeces* and *Anolis* could serve to infect *I. scapularis* with spirochaetes under experimental conditions. The activity of the complement of a wide range of lizards that serve as hosts for *Ixodes* ticks should be studied before general conclusions are made about the reservoir competence of reptilian hosts.

Birds play two roles in the support of the enzootic cycle of *B. burgdorferi* in North America. Migrating birds may serve to move immature *I. scapularis* and *I. pacificus* into new locations (Spielman, 1988), and some birds may serve as reservoir hosts infecting the ticks that feed on them (Fig. 11.2). The fact that a variety of avian species serve as hosts for immature *I. scapularis* has been documented numerous times (Battaly & Fish, 1993; Stafford, Bladen & Magnarelli, 1995). In general, ground-feeding and ground-nesting birds are the most heavily infested (Weisbrod & Johnson, 1989). Numerous isolates of *B. burgdorferi* have been obtained from birds in North America (Anderson, Magnarelli & Stafford, 1990; McLean *et al.*, 1993), but the reservoir competence or ability of birds to infect larval *I. scapularis* feeding on them has been controversial. The fact that larval *I. scapularis* removed from many avian species in the wild are infected with *B. burgdorferi* ss has been repeatedly documented (Weisbrod & Johnson, 1989; Stafford *et al.*, 1995). The first experiments attempting to infect xenodiagnostic larval *I. scapularis* by feeding them on grey catbirds (*Dumatella carolinensis*) indicated that these birds could not serve as reservoirs of *B. burgdorferi* (Mather *et al.*, 1989*a*). In contrast, American robins (*Turdus migratorius*) efficiently infected larval *I. scapularis* that fed upon them (Richter *et al.*, 2000). The degree to which various avian species serve to infect *I. scapularis* with *B. burgdorferi* in the eastern United States needs further research. In addition, the relationship between birds and *I. pacificus* seems to vary from site to site, with birds in some western regions carrying extremely light burdens of ticks (Manweiler *et al.*, 1990), and birds in other regions being heavily infested (Wright *et al.*, 2000). The observation that unidentified spirochaetes were found in blood smears from birds in Placer County, California (Wright *et al.*, 2000) needs further evaluation.

Rodents are clearly the primary reservoir hosts of *B. burgdorferi* ss in the regions most highly endemic for Lyme disease in North America (Fig. 11.2). Initial studies on Nantucket Island, Massachusetts pointed toward the white-footed mouse (*Peromyscus leucopus*) as the primary host for larval and nymphal *I. scapularis*. Spielman, Levine & Wilson (1984) estimated that 91% of larval and nymphal *I. scapularis* fed on *P. leucopus*; these hosts infected as many as 76% of larval ticks feeding on them (Mather *et al.*, 1989*a*). Clearly, the contribution of *P. leucopus* as a reservoir host serving to infect ticks with *B. burgdorferi* ss in coastal New England is substantial. A study on Monhegan Island (Maine) demonstrated that on an island where *P. leucopus* is absent, other rodents such as Norway rats could serve as efficient substitute reservoir hosts for *B. burgdorferi* (Smith *et al.*, 1993). Short-tailed shrews (*Blarina brevicauda*) have also been mentioned as effective reservoirs of *B. burgdorferi* (Telford *et al.*, 1990; Brisson & Dykhuisen, 2006). Interestingly, the reservoir potential of eastern chipmunks (*Tamias striatus*) appears to differ from region to region. In coastal Massachusetts, white-footed mice were found to infect many more immature *I. scapularis* with *B. burgdorferi* compared to chipmunks (Mather *et al.*, 1989*b*), but in the midwestern state of Illinois, chipmunks may be more important than white-footed mice as reservoirs of *B. burgdorferi* (Slajchert *et al.*, 1997). Local variation in the importance of various reservoir hosts in the enzootic cycle of *B. burgdorferi* in eastern North America is an important factor that must be taken into account when designing control strategies for Lyme disease spirochaetes transmitted by *I. scapularis*. A simple mathematical model has been proposed to predict the importance of specific hosts as reservoirs of select OspC groups of *B. burgdorferi* ss that most commonly cause disseminated Lyme disease in humans (Brisson & Dykhuizen, 2006). The role of host diversity in promoting *B. burgdorferi* diversity is an interesting area for further study (Hanincová *et al.*, 2006; Kurtenbach *et al.*, 2006).

The importance of rodents as reservoirs of *B. burgdorferi* in areas of western North America, where *I. pacificus* is the principal vector, is a complex subject. A study in Oregon demonstrated that rodents such as *Neotoma fuscipes*, *Peromyscus maniculatus* and *P. boylii* were infected with *B. burgdorferi* and infested with both *I. pacificus* and *I. spinipalpis* (Burkot *et al.*, 1999). In northern California, various rodents were infected with *B. burgdorferi* and infested with *I. spinipalpis* (Peavey, Lane & Kleinjan, 1997). A possible scenario exists wherein *I. spinipalpis* serves as the principal enzootic vector of *B. burgdorferi* in western North America, transmitting the pathogen from rodent to rodent; people are at risk of acquiring these rodent-derived strains only when

I. pacificus acquires an infection from rodents and subsequently transmits the spirochaetes to humans. This hypothesis warrants further investigation.

TRANSMISSION DYNAMICS IN EUROPE

Borrelia burgdorferi sl is transmitted orally while ticks are feeding on hosts. Indeed, it is currently well established for North American and European tick vectors that *B. burgdorferi* sl is transmitted to the host via infected saliva during the blood meal. Only a few studies in Europe have investigated the transmission dynamics of *B. burgdorferi* sl by *I. ricinus*. In most infected unfed *I. ricinus* nymphs and adults, spirochaetes are present in the midgut and migrate during blood-feeding to the salivary glands from where they are transmitted to the host via saliva (Gern, Zhu & Aeschlimann, 1990; Gern, Lebet & Moret, 1996; Zhu, 1998). However, microscopic examination of unfed nymphal and adult *I. ricinus* collected in endemic areas in Switzerland demonstrated that spirochaetes may infect salivary glands even before any blood uptake (Lebet & Gern, 1994; Leuba-Garcia *et al.*, 1994; Zhu, 1998). These systemic or generalized infections may be more common than what has been described for *I. scapularis* in North America. When unfed *I. ricinus* attaches to a vertebrate host, *Borrelia* transmission does not occur at the beginning of the blood uptake but later, and transmission efficiency increases with the duration of engorgement (Kahl *et al.*, 1998). In a laboratory study, early transmission of borreliae with high efficiency was described for *I. ricinus*. In fact, Kahl *et al.* (1998) reported that 50% of laboratory animals were infected by *B. burgdorferi* sl after only 16.7 hours of tick attachment. The observations of high infection rates in salivary glands of unfed *I. ricinus* suggest that systemically infected ticks may transmit *Borrelia* early after attachment to hosts (Lebet & Gern, 1994; Leuba-Garcia *et al.*, 1994). This may be a factor that influences the duration of the delay in transmission after the attachment of infected ticks to uninfected hosts. This delay may also be influenced by the *Borrelia* spp. infecting the ticks (Crippa, Rais & Gern, 2002). In fact, earlier transmission by *I. ricinus* may occur when ticks are infected by *B. afzelii* rather than *B. burgdorferi* ss. Crippa *et al.* (2002) noted that during the first 48 h of attachment to the host, *B. burgdorferi* ss-infected ticks did not infect 18 exposed experimental mice whereas *B. afzelii*-infected ticks transmitted the infection to 33% of mice. This study showed that *I. ricinus* transmits *B. afzelii* earlier than *B. burgdorferi* ss, and also that *I. ricinus* is a more efficient vector for *B. afzelii* than for *B. burgdorferi* ss.

It is well known from studies on *I. scapularis* that spirochaetes express OspA in the tick midgut and that during blood feeding, OspA synthesis is repressed and OspC synthesis is induced (Schwan & Piesman, 2002). In *I. ricinus*, few studies have addressed this point. Leuba-Garcia, Martinez & Gern (1998) observed that *B. afzelii* spirochaetes expressing OspA and OspC were present in the midgut of unfed ticks and that spirochaetes expressing OspA were not detected in ticks attached to the host for more than 24 hours. In salivary glands of engorged ticks, *B. afzelii* spirochaetes expressed OspC. This study also reported that in the skin of mice infected by *B. afzelii*-infected nymphs, borreliae expressed OspC. Later Fingerle *et al.* (2002), using different *B. afzelii* and *B. garinii* strains, demonstrated that in capillary-infected *I. ricinus* ticks OspA was expressed in the tick midgut and that the proportion of OspC-positive borreliae was usually greater when the borreliae reached the salivary glands. In this study, a *B. afzelii* strain unable to produce OspC failed to disseminate and infect the salivary glands, showing the role of OspC in *Borrelia* dissemination in *I. ricinus*. The influence of strain specificity on the dynamics of Osp expression and the dissemination of spirochaetes in the vector is an interesting topic. Interactions of the various *Borrelia* species and strains with *I. ricinus* are clearly extremely complex.

TRANSMISSION DYNAMICS IN NORTH AMERICA

A recent review by Schwan & Piesman (2002) summarized the intricate relationship between *Borrelia* and their tick vectors. One of the systems that has received the most research attention is the *B. burgdorferi–I. scapularis* interaction during spirochaete transmission. Transmission by nymphal *I. scapularis*, in particular, has received close scrutiny since the vast majority of Lyme disease cases in North America acquire infection from nymphal ticks (Piesman *et al.*, 1987a; Piesman, 1989; Falco *et al.*, 1999). A key factor in the transmission dynamics of *B. burgdorferi* is the duration of attachment required for efficient transmission of spirochaetes to the host. Several animal studies with laboratory infected and/or field collected nymphal *I. scapularis* have demonstrated that nymphs must be attached to hosts for >48 h in order for *B. burgdorferi* ss to be efficiently transmitted (Piesman *et al.*, 1987b; des Vignes *et al.*, 2001). Observations with patients in Lyme disease endemic regions also support the concept that only those ticks feeding for > 48 h transmit an infectious dose of spirochaetes (Sood *et al.*, 1997; Nadelman *et al.*, 2001). Although the smaller nymphs often

escape detection and feed for a sufficient interval to transmit spirochaetes, the larger adult female *I. scapularis* is more routinely detected and removed before feeding long enough to transmit spirochaetes (Piesman *et al.*, 1991; Falco, Fish & Piesman, 1996) (Fig. 11.4). Thus, Lyme disease cases occur in North America virtually exclusively when nymphal *I. scapularis* are active (May–July) as opposed to when adults are active (October–April).

The underlying reasons for the delay between tick attachment and spirochaete transmission have been described (Ohnishi, Piesman & de Silva, 2001; Schwan & Piesman, 2002). Spirochaetes in unfed nymphs are restricted mainly to the tick midgut; most of these spirochaetes express OspA. When feeding commences, rapid multiplication of the spirochaetes occurs; OspA is downregulated and a proportion of the population now expresses OspC. The downregulation of OspA may allow the spirochaetes to leave the midgut since OspA apparently binds to uncharacterized tick midgut proteins (Pal *et al.*, 2000). Spirochaete populations increase in tick salivary glands during the feeding process and are eventually transmitted to the skin adjacent to the feeding site (Ohnishi *et al.*, 2001; Piesman, Schneider & Zeidner, 2001). A tick salivary gland protein, Salp15, plays an intriguing role in binding to OspC and facilitating transmission of the spirochaete and initial survival in the vertebrate host (Ramamoorthi *et al.*, 2005; see Chapter 10). Transmission dynamics of genospecies other than *B. burgdorferi* ss may differ dramatically (Crippa *et al.*, 2002). The requirement for ticks to be attached for a period >48 h in order to transmit *B. burgdorferi* ss efficiently also seems to hold true for *I. pacificus* (Peavey & Lane, 1995).

PREVENTION IN NORTH AMERICA AND IN EUROPE

The first line of defence against Lyme disease in North America and in Europe is clearly personal protection. This involves avoidance of tick-infested habitat, wearing protective clothing, the prudent use of tick repellents, and daily tick checks to detect and promptly remove attached ticks. In order to be effective, personal protection requires that residents of highly endemic regions have a fairly sophisticated knowledge of the enzootic cycle of Lyme disease and the biology of infected ticks.

In North America, the knowledge, attitudes and practice of Lyme disease prevention differ from region to region (Herrington *et al.*, 1997). Certainly, people living in the endemic regions of the northeastern United States have an

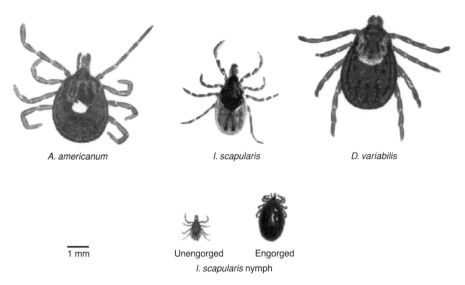

A. americanum   I. scapularis   D. variabilis

1 mm   Unengorged   Engorged
I. scapularis nymph

Fig. 11.4. The principal ticks found commonly biting people in eastern North America, including the principal vector of Lyme disease, *Ixodes scapularis*. (Reprinted with permission from the Massachusetts Medical Society; Hayes & Piesman, 2003.)

in-depth understanding of the biology of Lyme disease; however, prevention practices are not routinely employed even in places like Nantucket Island, Massachusetts that have been dealing with the Lyme disease epidemic for decades (Phillips *et al.*, 2001). Intense education campaigns are currently under way in select communities in Massachusetts, Connecticut, New York and New Jersey to see if the community education approach to Lyme disease prevention can be optimized.

Tick control methods are also an essential part of Lyme disease prevention in North America (Fig. 11.3). A single, well-timed area-wide application of acaricides such as carbaryl, cyflutherin or deltamethrin to vegetation can reduce populations of questing nymphs by 68–100% (Stafford, 1991; Curran, Fish & Piesman, 1993; Schulze *et al.*, 2001). Although many homeowners do allow licensed pest control operators to apply acaricides to their properties in endemic areas (Stafford, 1997), the majority of homeowners have reservations about using this approach due to concerns about the effect of these chemicals on non-target species and human toxicity. It is therefore incumbent upon the public health community to present homeowners with alternative means for tick control.

Vegetation management is one alternative means of tick control. Burning (Stafford, Ward & Magnarelli, 1998), brush removal (Wilson, 1986), leaf litter removal (Schulze *et al.*, 1995) and the establishment of woodchip barriers have all been tested as means of reducing contact with ticks. In addi-

tion, treatment of vegetation with soaps and desiccants holds promise for tick reduction (Allan & Patrican, 1995). The use of biological control agents such as parasitoid wasps, parasitic nematodes and fungi is under evaluation (Zhioua *et al.*, 1995; Stafford, Denicola & Magnarelli, 1996; Benjamin, Zhioua & Ostfeld, 2002; see also Chapter 20). Vegetation management and biological control agents are well received by the community owing to their reputation as non-toxic environmentally sound control methodologies. In general, however, these 'bio-friendly' methods are less consistently effective than chemical acaricidal agents. Host-targeted control methods have also been tested for their efficacy in reducing populations of *I. scapularis*. Some of these methods aim at overall population reduction and others aim at specifically reducing the number of questing nymphs infected with *B. burgdorferi*. Due to their role as primary hosts for the adult stage of *I. scapularis*, white-tailed deer have been targeted for various intervention strategies to control ticks. These include deer eradication (Wilson *et al.*, 1988), deer reduction (Deblinger *et al.*, 1993), fencing (Stafford, 1993; Daniels & Fish, 1995), and the application of acaricides to deer via bait stations charged with chemicals on paint-roller delivery systems. This latter method is called the 'four-poster' method; it has been shown to be effective against *A. americanum* (Pound, Miller & George, 2000) and is currently under evaluation for the control of *I. scapularis*. Although deer-target methods for the control of *I. scapularis* hold great promise for the future, they have yet to be put into common practice in

Lyme disease endemic regions in North America, with the possible exception of fencing of individual properties.

Rodents have also been the target of tick control intervention efforts in North America. A commercial product that utilizes permethrin-treated cotton balls collected specifically by white-footed mice has been developed. It appears to have reduced the population of questing *I. scapularis* in some trials, but not in others (Mather, Ribeiro & Spielman, 1987; Daniels, Fish & Falco, 1991; Deblinger & Rimmer, 1991; Stafford, 1992). One of the reasons suggested for its varied success is the importance of white-footed mice versus other reservoir hosts in different locations. Permethrin-treated cotton balls were also ineffective when tested against *I. pacificus* (Leprince & Lane, 1996). Other rodent-targeted methods involve the use of bait stations for applying acaricides to various rodent species (Sonenshine & Haines, 1985; Gage *et al.*, 1997; Lane *et al.*, 1998). Trials are currently being conducted using bait boxes treated with fipronil, a promising new topical acaricide for the control of *I. scapularis* infected with *B. burgdorferi* (Dolan *et al.*, 2004). Significant decreases in ticks on rodents and questing nymphs have been observed where these bait boxes are undergoing field trials. Systemic treatment of deer and rodents with ivermectin or closely related compounds is also a future avenue for research.

The current status of tick control methods is in a rapid state of change. No one single method holds promise as the 'magic bullet' for tick control. Certainly, an integrated pest management approach (IPM) that utilizes various methods in diverse situations will be the most effective response to prevent Lyme disease in the future.

Although much research on control strategies against *I. scapularis* in North America has been carried out, less has been done in Europe to control *I. ricinus*. This is probably essentially due to different risk exposures. In the highly endemic areas of the northeastern United States, people have close contact with *I. scapularis* and residential exposure dominates. In Europe, the majority of residential properties do not have borders with woodlands or forests and most people contract Lyme borreliosis while visiting tick-endemic areas. Tick bites are primarily due to occupational or recreational exposure.

A study undertaken in Sweden evaluated permethrin-treated rodent nesting materials similar to those used in North America (Mejlon, Jaenson & Mather, 1995). The authors concluded that application of these methods in natural ecosystems would not represent a practical and economic method to reduce the risk to humans of developing Lyme borreliosis mainly because host reservoirs for

B. burgdorferi in Europe are diverse and are not restricted to rodents. A review of various potential methods to reduce *I. ricinus* in Sweden similarly concluded that most methods were not appropriate to reduce Lyme borreliosis cases (Tälleklint & Jaenson, 1995). Therefore, in Europe personal protection is favoured. Simple measures include use of tick repellents before entering a tick-infested area, avoidance of (or minimization of) exposure to tick-infested areas, and thorough examination of clothes and skin after exposure. Protective clothing, particularly boots, long trousers tucked into boots and shirts tucked into trousers, are recommended. Light-coloured clothes favour detection of ticks crawling on humans before they find a suitable place to attach to skin. Attached ticks should be removed as soon as possible because *I. ricinus* may transmit *B. burgdorferi* sl quite early after tick attachment (Kahl *et al.*, 1998; Crippa *et al.*, 2002). Therefore, regular checks for ticks on clothes and the body immediately when one leaves the area where ticks are present are highly recommended. A survey conducted in various countries of Europe on Lyme borreliosis awareness showed that awareness was greater in countries showing a high prevalence of Lyme borreliosis like Germany, Russia, Sweden and the Czech Republic whereas it was lower in countries with low prevalence such as the UK and Ireland (Gray *et al.*, 1998). Interestingly, the media were the main source of information for most people.

In the United States, a Lyme disease vaccine based on a recombinant OspA protein was tested (Sigal *et al.*, 1998; Steere *et al.*, 1998) and licensed for use. Despite the fact that it was reasonably efficacious, and an adverse events registry reported only minor side effects associated with the vaccine (Lathrop *et al.*, 2002), the vaccine was withdrawn from the market essentially due to lack of demand. Problems associated with the vaccine also included the need for a series of three shots and an undetermined need to get booster injections, high cost and a theoretical but widespread concern that the vaccine could cause serious autoimmune reactions (Gross *et al.*, 1998). The possibility that an OspA vaccine could be used in rodent populations to decrease the number of questing nymphs infected with *B. burgdorferi* requires further field testing but holds potential for the future (Tsao *et al.*, 2004).

In Europe, where at least three pathogenic species are infecting *I. ricinus* ticks, the variability in OspA expression among *B. burgdorferi* sl isolates was often used as an argument against the development of an OspA vaccine. However, a study showed that a vaccine compatible with human use, containing multiple OspA antigens, was highly effective at protecting mice against the three causative agents of Lyme borreliosis in Europe, *B. burgdorferi* ss, *B. garinii*, and *B. afzelii*, using the natural mode of transmission (Gern *et al.*, 1997*a*). However, after the licensed vaccine in North America was withdrawn from the market, the development programme of this OspA vaccine for Europe was stopped.

As discussed in the 'Treatment' section of this review, prophylactic treatment of tick bites in highly endemic regions of North America remains an option (Nadelman *et al.*, 2001). In Europe, prophylactic treatment after a tick bite is usually not recommended (Stanek & Kahl, 1999). In the end, the most effective prevention campaign against Lyme disease may involve education of the community to practise personal protection measures, IPM campaigns for tick control, and an alert and dedicated public health infrastructure to combat Lyme disease.

REFERENCES

Aeschlimann, A., Chamot, E., Gigon, F., *et al.* (1986). *B. burgdorferi* in Switzerland. *Zentralblatt für Bakteriologie, Mikrobiologie und Hygiene A* **263**, 450–458.

Allan, S. A. & Patrican, L. A. (1995). Reduction of immature *Ixodes scapularis* (Acari: Ixodidae) in woodlots by application of desiccant and insecticidal soap formulations. *Journal of Medical Entomology* **32**, 16–20.

Anderson, J. F. (1991). Epizootiology of Lyme borreliosis. *Scandinavian Journal of Infectious Diseases Supplement* **77**, 23–34.

Anderson, J. F., Johnson, R. C., Magnarelli, L. A., Hyde, F. W. & Myers, J. E. (1987). Prevalence of *Borrelia burgdorferi* and *Babesia microti* in mice on islands inhabited by white-tailed deer. *Applied and Environmental Microbiology* **53**, 892–894.

Anderson, J. F., Magnarelli, L. A. & Stafford, K. C. III (1990). Bird-feeding ticks transstadially transmit *Borrelia burgdorferi* that infect Syrian hamsters. *Journal of Wildlife Diseases* **26**, 1–10.

Appel, M. J., Allan, S., Jacobson, R. H., *et al.* (1993). Experimental Lyme disease in dogs produces arthritis and persistent infection. *Journal of Infectious Diseases* **167**, 651–664.

Apperson, C. S., Levine, J. F., Evans, T. L., Braswell, A. & Heller, J. (1993). Relative utilization of reptiles and rodents as hosts by immature *Ixodes scapularis* (Acari: Ixodidae) in the coastal plain of North Carolina, USA. *Experimental and Applied Acarology* **17**, 719–731.

Assous, M. V. D., Postic, D., Paul, G., Nevot, P. & Baranton, G. (1993). Western blot analysis of sera from Lyme borreliosis patients according to the genomic species of the *Borrelia* strains used as antigens. *European Journal of Clinical Microbiology and Infectious Diseases* **12**, 261–268.

Bailly-Choumara, H., Morel, P. C. & Rageau, J. (1974). Première contribution au catalogue des tiques du Maroc (Acari: Ixodidae). *Bulletin de la Société des Sciences Naturelles du Maroc* **54**, 71–80.

Balmelli, T. & Piffaretti, J. C. (1995). Association between different clinical manifestations of Lyme disease and different species of *Borrelia burgdorferi* sensu lato. *Research in Microbiology* **146**, 329–340.

Baranton, G., Postic, D., Saint Girons, I., *et al.* (1992). Delineation of *Borrelia burgdorferi* sensu stricto, *Borrelia garinii* sp. nov.; and group VS461 associated with Lyme borreliosis. *International Journal of Systematic Bacteriology* **42**, 378–383.

Baranton, G., Seinost, G., Theodore, G., Postic, D. & Dykhuisen, D. (2001). Distinct levels of genetic diversity of *Borrelia burgdorferi* are associated with different aspects of pathogenicity. *Research in Microbiology* **152**, 149–156.

Barbour, A. G., Heiland, R. A. & Howe, T. R. (1985). Heterogeneity of major proteins in Lyme disease borreliae: a molecular analysis of North American and European isolates. *Journal of Infectious Diseases* **152**, 478–484.

Barbour, A. G., Maupin, G. O., Teltow, G. J., Carter, C. J. & Piesman, J. (1996). Identification of an uncultivable *Borrelia* species in the hard tick *Amblyomma americanum*: possible agent of a Lyme disease-like illness. *Journal of Infectious Diseases* **173**, 403–409.

Barral, M., Garcia-Perez, A. L., Juste, R. A., *et al.* (2002). Distribution of *Borrelia burgdorferi* sensu lato in *Ixodes ricinus* (Acari: Ixodidae) ticks from the Basque Country, Spain. *Journal of Medical Entomology* **39**, 177–184.

Basta, J., Plch, J., Hulinska, D. & Daniel, M. (1999). Incidence of *Borrelia garinii* and *Borrelia afzelii* in *Ixodes ricinus* ticks in an urban environment, Prague, Czech Republic, between 1995 and 1998. *European Journal of Clinical Microbiology and Infectious Diseases* **18**, 515–517.

Battaly, G. R. & Fish, D. (1993). Relative importance of bird species as hosts for immature *Ixodes dammini* (Acari: Ixodidae) in a suburban residential landscape of southern New York State. *Journal of Medical Entomology* **30**, 740–747.

Benjamin, M. A., Zhioua, E. & Ostfeld, R. S. (2002). Laboratory and field evaluation of the entomopathogenic fungus *Metarhizium anisopliae* (Deuteromycetes) for controlling questing adult *Ixodes scapularis* (Acari: Ixodidae). *Journal of Medical Entomology* **39**, 723–728.

Brisson, D. & Dykhuizen, D. E. (2006). A modest model explains the distribution and abundance of *Borrelia burgdorferi* strains. *American Journal of Tropical Medicine and Hygiene* **74**, 615–622.

Brown, R. N., Peot, M. A. & Lane, R. S. (2006). Sylvatic maintenance of *Borrelia burgdorferi* (Spirochaetales) in Northern California: untangling the web of transmission. *Journal of Medical Entomology* **43**, 743–751.

Buchwald, A. (1883). Ein Fall von diffuser idiopathischer Hautatrophie. *Vierteljahrschrift für Dermatologie* **15**, 553–556.

Burgdorfer, W. (2001). Arthropod-borne spirochetoses: a historical perspective. *European Journal of Clinical Microbiology and Infectious Disease* **20**, 1–5.

Burgdorfer, W., Barbour, A. G., Hayes, S. F., *et al.* (1982). Lyme disease: a tick-borne spirochetosis? *Science* **216**, 1317–1319.

Burgdorfer, W., Barbour, A. G., Hayes, S. F., Péter, O. & Aeschlimann, A. (1983). Erythema migrans: a tick-borne spirochetosis. *Acta Tropica* **40**, 79–83.

Burkot, T. R., Clover, J. R., Happ, C. M., Debess, E. & Maupin, G. O. (1999). Isolation of *Borrelia burgdorferi* from *Neotoma fuscipes*, *Peromyscus maniculatus*, *Peromyscus boylii*, and *Ixodes pacificus* in Oregon. *American Journal of Tropical Medicine and Hygiene* **60**, 453–457.

Busch, U., Hizo-Teufel, C., Boehmer, R., *et al.* (1996*a*). Three species of *Borrelia burgdorferi* sensu lato (*B. burgdorferi* sensu stricto, *B. afzelii* and *B. garinii*) identified from cerebrospinal fluid isolates by pulsed-field gel electrophoresis and PCR. *Journal of Clinical Microbiology* **34**, 1072–1078.

Busch, U., Hizo-Teufel, C., Boehmer, R., *et al.* (1996*b*). *Borrelia burgdorferi* sensu lato strains isolated from cutaneous Lyme borreliosis biopsies differentiated by pulsed-field gel electrophoresis. *Scandinavian Journal of Infectious Diseases* **28**, 583–589.

Canica, M. M., Nato, F., Du Merle, L., *et al.* (1993). Monoclonal antibodies for identification of *Borrelia afzelii* sp. nov. associated with late cutaneous manifestations of Lyme borreliosis. *Scandinavian Journal of Infectious Diseases* **25**, 441–448.

Centers for Disease Control and Prevention (1995). Recommendations for test performance and interpretation from the Second National Conference on Serologic Diagnosis of Lyme Disease. *Morbidity and Mortality Weekly Report* **44**, 590–591.

Chang, Y. F., Novosol, V., McDonough, S. P., *et al.* (2000). Experimental infection of ponies with *Borrelia burgdorferi* by exposure to Ixodid ticks. *Veterinary Pathology* **37**, 68–76.

Cinco, M., Padovan, D., Murgia, R., *et al.* (1998). Rate of infection of *Ixodes ricinus* ticks with *Borrelia burgdorferi* sensu stricto, *Borrelia garinii*, *Borrelia afzelii* and group VS116 in an endemic focus of Lyme disease in Italy. *European Journal of Clinical Microbiology and Infectious Diseases* **17**, 90–94.

Collares-Pereira, M., Couceiro, S., Franca, I., *et al.* (2004). First isolation of *Borrelia lusitaniae* from a patient. *Journal of Clinical Microbiology* **42**, 1216–1318.

Craine, N. G., Nuttall, P. A., Marriott, A. C. & Randolph, S. E. (1997). Role of grey squirrels and pheasants in the transmission of *Borrelia burgdorferi* sensu lato, the Lyme disease spirochaete, in the UK. *Folia Parasitologica* **44**, 155–160.

Crippa, M., Rais, O. & Gern, L. (2002). Investigations on the mode and dynamics of transmission and infectivity of *Borrelia burgdorferi* sensu stricto and *Borrelia afzelii* in *Ixodes ricinus* ticks. *Vector Borne and Zoonotic Diseases* **2**, 3–9.

Curran, K. L., Fish, D. & Piesman, J. (1993). Reduction of nymphal *Ixodes dammini* (Acari: Ixodidae) in a residential suburban landscape by area application of insecticides. *Journal of Medical Entomology* **30**, 107–113.

Danchin, E., Boulinier, T. & Massot, M. (1998). Conspecific reproductive success and breeding habitat selection: implications for the study of coloniality. *Ecology* **79**, 2415–2428.

Daniels, T. J. & Fish, D. (1995). Effect of deer exclusion on the abundance of immature *Ixodes scapularis* (Acari: Ixodidae) parasitizing small and medium-sized mammals. *Journal of Medical Entomology* **32**, 5–11.

Daniels, T. J., Fish, D. & Falco, R. C. (1991). Evaluation of host-targeted acaricide for reducing risk of Lyme disease in southern New York State. *Journal of Medical Entomology* **28**, 537–543.

Deblinger, R. D. & Rimmer, D. W. (1991). Efficacy of a permethrin-based acaricide to reduce the abundance of *Ixodes dammini* (Acari: Ixodidae). *Journal of Medical Entomology* **28**, 708–711.

Deblinger, R. D., Wilson, M. L., Rimmer, D. W. & Spielman, A. (1993). Reduced abundance of immature *Ixodes dammini* (Acari: Ixodidae) following incremental removal of deer. *Journal of Medical Entomology* **30**, 144–150.

De Boer, R., Hovius, K. E., Nohlmans, M. K. E. & Gray, J. S. (1993). The woodmouse (*Apodemus sylvaticus*) as a reservoir of tick-transmitted spirochaetes (*Borrelia burgdorferi*) in the Netherlands. *Zentralblatt für Bakteriologie* **279**, 404–416.

De Michelis, S., Sewell, H. S., Collares-Pereira, M., *et al.* (2000). Genetic diversity of *Borrelia burgdorferi* sensu lato in ticks from mainland Portugal. *Journal of Clinical Microbiology* 38, 2118–2133.

Dennis, D., Nekomoto, T. S., Victor, J. C., Paul, W. S. & Piesman, J. (1998). Reported distribution of *Ixodes scapularis* and *Ixodes pacificus* (Acari: Ixodidae) in the United States. *Journal of Medical Entomology* 35, 629–638.

Des Vignes, F., Piesman, J., Heffernan, R., *et al.* (2001). Effect of tick removal on transmission of *Borrelia burgdorferi* and *Ehrlichia phagocytophila* by *Ixodes scapularis* nymphs. *Journal of Infectious Diseases* 183, 773–778.

Dister, S. W., Fish, D., Bros, S. M., Frank, D. H. & Wood, B. L. (1997). Landscape characterization of peridomestic risk for Lyme disease using satellite imagery. *American Journal of Tropical Medicine and Hygiene* 57, 687–692.

Diuk-Wasser, M. A., Gatewood, A. G., Cortinas, M. R., *et al.* (2006). Spatiotemporal patterns of host-seeking *Ixodes scapularis* nymphs (Acari: Ixodidae) in the United States. *Journal of Medical Entomology* 43, 166–176.

Dizij, A. & Kurtenbach, K. (1995). *Clethrionomys glareolus*, but not *Apodemus flavicollis*, acquires resistance to *Ixodes ricinus* L., the main European vector of *Borrelia burgdorferi*. *Parasite Immunology* 17, 177–183.

Dolan, M. C., Maupin, G. O., Schneider, B. S., *et al.* (2004). Control of immature *Ixodes scapularis* (Acari: Ixodidae) on rodent reservoirs of *Borrelia burgdorferi* in a residential community of southeastern Connecticut. *Journal of Medical Entomology* 41, 1043–1054.

Donnelly, J. (1976). The life cycle of *Ixodes ricinus* L. based on recent published findings. In *Tick-Borne Diseases and their Vectors*, ed. Wilde, J. K. H., pp. 56–60. Edinburgh, UK: CTVM, University of Edinburgh.

Dressler, F., Ackermann, R. & Steere, A. C. (1994). Antibody responses to the three genomic groups of *Borrelia burgdorferi* in European Lyme borreliosis. *Journal of Infectious Diseases* 169, 313–318.

Dressler, F., Whalen, J. A., Reinhardt, B. N. & Steere, A. C. (1993). Western blotting in the serodiagnosis of Lyme disease. *Journal of Infectious Diseases* 167, 392–400.

Dsouli, N., Younsi-Kabachi, H., Postic, D., *et al.* (2006). Reservoir role of the lizard, *Psammodromus algirus*, in the transmission cycle of *Borrelia burgdorferi* sensu lato (Spirochaetacea) in Tunisia. *Journal of Medical Entomology* 43, 737–742.

Dworkin, M. S., Shoemaker, P. C., Fritz, C. L., Dowell, M. E. & Anderson, D. E. Jr (2002). The epidemiology of tick-borne relapsing fever in the United States. *American Journal of Tropical Medicine and Hygiene* 66, 753–758.

Eiffert, H., Karsten, A., Thomssen, R. & Christen, H. J. (1998). Characterization of *Borrelia burgdorferi* strains in Lyme arthritis. *Scandinavian Journal of Infectious Diseases* 30, 265–268.

Eisen, R. J., Eisen, L. & Lane, R. S. (2006). Predicting density of *Ixodes pacificus* nymphs in dense woodlands in Mendocino County, California, based on geographic information systems and remote sensing versus field-derived data. *American Journal of Tropical Medicine and Hygiene* 74, 632–640.

Escudero, R., Barral, M., Perez, A., *et al.* (2000). Molecular and pathogenic characterization of *Borrelia burgdorferi* sensu lato isolates from Spain. *Journal of Clinical Microbiology* 38, 4026–4033.

Eveleigh, E. S. & Threlfall, W. (1974). The biology of *Ixodes* (*Ceratoixodes*) *uriae* White, 1852 in Newfoundland. *Acarologia* 16, 621–635.

Falco, R. C., Fish, D. & Piesman, J. (1996). Duration of tick bites in a Lyme disease-endemic area. *American Journal of Epidemiology* 143, 187–192.

Falco, R. C., McKenna, D. F., Daniels, T. J., *et al.* (1999). Temporal relationship between *Ixodes scapularis* abundance and risk for Lyme disease associated with erythema migrans. *American Journal of Epidemiology* 149, 771–776.

Felsenfeld, O. (1979). Borrelia. In *CRC Handbook Series in Zoonoses*, Section A, *Bacterial, Rickettsial, and Mycotic Diseases*, vol. 1, ed. Steele, J. H., pp. 79–96. Boca Raton, FL: CRC Press.

Ferquel, E., Garnier, M., Marie, J., *et al.* (2006). Prevalence of *Borrelia burgdorferi* sensu lato and *Anaplasmataceae* members in *Ixodes ricinus* ticks in Alsace, a focus of Lyme borreliosis endemicity in France. *Applied and Environmental Microbiology* 72, 3074–3078.

Fingerle, V., Rauser, S., Hammer, B., *et al.* (2002). Dynamics of dissemination and outer surface protein expression of different *Borrelia burgdorferi* s.l. strains in artificially infected *Ixodes ricinus* nymphs. *Journal of Clinical Microbiology* 40, 1456–1463.

Fish, D. & Dowler, R. C. (1989). Host associations of ticks (Acari: Ixodidae) parasitizing medium-sized mammals in a Lyme disease endemic area of southern New York. *Journal of Medical Entomology* 26, 200–209.

Földvari, G., Farkas, R. & Lakos, A. (2005). *Borrelia spielmanii* erythema migrans, Hungary. *Emerging Infectious Diseases* 11, 1794–1795.

Fraenkel, C.-J., Garpmo, U. & Berglund, J. (2002). Determination of novel *Borrelia* genospecies in Swedish

Ixodes ricinus ticks. *Journal of Clinical Microbiology* **40**, 3308–3312.

Fukunaga, M., Takahashi, Y., Tsuruta, Y., *et al.* (1995). Genetic and phenotypic analysis of *Borrelia miyamotoi* sp. nov., isolated from the ixodid tick *Ixodes persulcatus*, the vector for Lyme disease in Japan. *International Journal of Systematic Bacteriology* **45**, 804–810.

Fukunaga, M., Ushijima, Y., Aoki, L. Y. & Talbert, A. (2001). Detection of *Borrelia duttonii*, a tick-borne relapsing fever agent in central Tanzania, within ticks by flagellin gene-based nested polymerase chain reaction. *Vector Borne Zoonotic Diseases* **1**, 331–338.

Gage, K. L., Maupin, G. O., Montieneri, J., *et al.* (1997). Flea (Siphonaptera: Ceratophyllidae, Hystrichopsyllidae) and tick (Acarina: Ixodidae) control on wood rats using host-targeted liquid permethrin in bait tubes. *Journal of Medical Entomology* **34**, 46–51.

Gern, L. & Humair, P. F. (2002). Ecology of *Borrelia burgdorferi* sensu lato in Europe. In *Lyme Borreliosis: Biology, Epidemiology and Control*, eds. Gray, J., Kahl, O., Lane, R. S. & Stanek, G., pp. 149–174. Wallingford, UK: CAB International.

Gern, L. & Rais, O. (1996). Efficient transmission of *Borrelia burgdorferi* between co-feeding *Ixodes ricinus* ticks (Acari: Ixodidae). *Journal of Medical Entomology* **33**, 189–192.

Gern, L., Estrada-Peña, A., Frandsen, F., *et al.* (1998). European reservoir hosts of *Borrelia burgdorferi* sensu lato. *Zentralblatt für Bakteriologie* **287**, 196–204.

Gern, L., Hu, C. M., Kocianova, E., Vyrostekova, V. & Rehacek, J. (1999). Genetic diversity of *Borrelia burgdorferi* sensu lato isolates obtained from *Ixodes ricinus* ticks collected in Slovakia. *European Journal of Epidemiology* **15**, 665–669.

Gern, L., Hu, C. M., Voet, P., Hauser, P. & Lobet, Y. (1997*a*). Immunization with a polyvalent OspA vaccine protects mice against *Ixodes ricinus* tick bites infected by *Borrelia burgdorferi* ss, *B. garinii* and *B. afzelii*. *Vaccine* **15**, 1551–1557.

Gern, L., Lebet, N. & Moret, J. (1996). Dynamics of *Borrelia burgdorferi* infection in nymphal *Ixodes ricinus* ticks during feeding. *Experimental and Applied Acarology* **20**, 649–658.

Gern, L., Rouvinez, E., Toutoungi, L. N. & Godfroid, E. (1997*b*). Transmission cycles of *Borrelia burgdorferi* sensu lato involving *Ixodes ricinus* and/or *I. hexagonus* ticks and the European hedgehog, *Erinaceus europaeus*, in suburban and urban areas in Switzerland. *Folia Parasitologica* **44**, 309–314.

Gern, L., Siegenthaler, M., Hu, C. M., *et al.* (1994). *Borrelia burgdorferi* in rodents (*Apodemus flavicollis* and *A. sylvaticus*):

duration and enhancement of infectivity for *Ixodes ricinus* ticks. *European Journal of Epidemiology* **10**, 75–80.

Gern, L., Zhu, Z. & Aeschlimann, A. (1990). Development of *Borrelia burgdorferi* in *Ixodes ricinus* females during blood feeding. *Annales de Parasitologie Humaine et Comparée* **65**, 89–94.

Goddard, J. (1992). Ecological studies of adult *Ixodes scapularis* in central Mississippi: questing activity in relation to time of year, vegetation type, and meteorologic conditions. *Journal of Medical Entomology* **29**, 501–506.

Gorelova, N. B., Korenberg, E. I., Kovalevskii, Y. V., Postic, D. & Baranton, G. (1996). [The isolation of *Borrelia* from the tick *Ixodes trianguliceps* (Ixodidae) and the possible significance of this species in the epizootiology of ixodid tick-borne borrelioses.] (In Russian) *Parazitologiia* **30**, 13–18.

Gorelova, N. B., Korenberg, E. I., Kovalevskii, Y. V. & Shcherbakov, S. V. (1995). Small mammals as reservoir hosts for *Borrelia* in Russia. *Zentralblatt für Bakteriologie* **282**, 315–322.

Gray, J. S. (1982). The development and questing activity of *Ixodes ricinus* (L.) (Acari: Ixodidae) under field conditions in Ireland. *Bulletin of Entomological Research* **72**, 263–270.

Gray, J. S. (1985). Studies on the larval activity of the tick *Ixodes ricinus* (L.) in Co. Wicklow, Ireland. *Experimental and Applied Acarology* **1**, 307–316.

Gray, J. S. (1991). The development and seasonal activity of the tick *Ixodes ricinus* (L.) a vector of Lyme borreliosis. *Review of Medical and Veterinary Entomology* **79**, 323–333.

Gray, J. S., Granström, M., Cimmino, M., *et al.* (1998). Lyme borreliosis awareness. *Zentralblatt für Bakteriologie* **287**, 253–265.

Gray, J. S., Kahl, O., Janetzki, C. & Stein, J. (1992). Studies on the ecology of Lyme disease in a deer forest in County Galway, Ireland. *Journal of Medical Entomology* **29**, 915–920.

Gray, J. S., Kahl, O., Janetzki-Mittman, C., Stein, J. & Guy, E. (1994). Acquisition of *Borrelia burgdorferi* by *Ixodes ricinus* ticks fed on the European hedgehog, *Erinaceus europaeus* L. *Experimental and Applied Acarology* **18**, 485–491.

Gray, J. S., Kahl, O., Janetzki, C., Stein, J. & Guy, E. (1995). The spatial distribution of *Borrelia burgdorferi*-infected *Ixodes ricinus* in the Connemara region of Co. Galway, Ireland. *Experimental and Applied Acarology* **18**, 485–491.

Gross, D. M., Forsthuber, T., Tary-Lehmann, M., *et al.* (1998). Identification of Lfa-1 as a candidate autoantigen in treatment-resistant Lyme arthritis. *Science* **281**, 703–706.

Guerra, M., Walker, E., Jones, C., *et al.* (2002). Predicting the risk of Lyme disease: habitat suitability for *Ixodes scapularis* in the north central United States. *Emerging Infectious Diseases* **8**, 289–297.

Gylfe, Å., Bergström, S., Lunström, J. & Olsen, B. (2000). Epidemiology: reactivation of *Borrelia* infection in birds. *Nature* **403**, 724–725.

Gylfe, Å., Olsen, B., Strasevicius, D., *et al.* (1999). Isolation of Lyme disease *Borrelia* from puffins (*Fratercula arctica*) and seabird ticks (*Ixodes uriae*) on the Faroe Islands. *Journal of Clinical Microbiology* **37**, 890–896.

Halperin, J. J., Luft, B. J., Anand, A. K., *et al.* (1989). Lyme neuroborreliosis: central nervous system manifestations. *Neurology* **39**, 753–759.

Hanincová, K., Kurtenbach, K., Diuk-Wasser, M., Brei, B. & Fish, D. (2006). Epidemic spread of Lyme borreliosis, northeast United States. *Emerging Infectious Diseases* **12**, 604–611.

Hanincová, K., Schäfer, S. M., Etti, S., *et al.* (2003a). Association of *Borrelia afzelii* with rodents in Europe. *Parasitology* **126**, 11–20.

Hanincová, K., Taragelová, V., Koci, J., *et al.* (2003b). Association of *Borrelia garinii* and *B. valaisiana* with songbirds in Slovakia. *Applied and Environmental Microbiology* **69**, 2825–2830.

Hauser, U., Krahl, H., Peters, H., Fingerle, V. & Wilske, B. (1998). Impact of strain heterogeneity on Lyme disease serology in Europe: comparison of enzyme-linked immunosorbent assays using different species of *Borrelia burgdorferi* sensu lato. *Journal of Clinical Microbiology* **36**, 427–436.

Hayes, E. B. & Piesman, J. (2003). How can we prevent Lyme disease? *New England Journal of Medicine* **348**, 2424–2430.

Herrington, J. E. Jr, Campbell, G. L., Bailey, R. E., *et al.* (1997). Predisposing factors for individuals' Lyme disease prevention practices: Connecticut, Maine, and Montana. *American Journal of Public Health* **87**, 2035–2038.

Hillyard, P. D. (1996). *Ticks of North-West Europe*, eds. Barnes, R. S. K. & Crothers, J. H., vol. 52. Shrewsbury, UK: Field Studies Council.

Hu, C. M., Cheminade, Y., Perret, J.-L., *et al.* (2003). Early detection of *Borrelia burgdorferi* sensu lato infection in BALB/c mice by co-feeding *Ixodes ricinus* ticks. *International Journal of Medical Microbiology* **293**, 421–426.

Hu, C. M., Humair, P. F., Wallich, R. & Gern, L. (1997). *Apodemus* sp. rodents, reservoir hosts for *Borrelia afzelii* in an endemic area in Switzerland. *Zentralblatt für Bakteriologie* **285**, 558–564.

Hu, C. M., Wilske, B., Fingerle, V., Lobet, Y. & Gern, L. (2001). Transmission of *Borrelia garinii* OspA serotype 4 to BALB/c mice by *Ixodes ricinus* ticks collected in the field. *Journal of Clinical Microbiology* **39**, 1169–1171.

Hubálek, Z. & Halouzka, J. (1997). Distribution of *Borrelia burgdorferi* sensu lato genomic groups in Europe: a review. *European Journal of Epidemiology* **13**, 951–957.

Hubálek, Z. & Halouzka, J. (1998). Prevalence rates of *Borrelia burgdorferi* sensu lato in host-seeking *Ixodes ricinus* ticks in Europe. *Parasitology Research* **84**, 167–172.

Hubálek, Z., Anderson, J. F., Halouzka, J. & Hájek, V. (1996). Borreliae in immature *Ixodes ricinus* (Acari: Ixodidae) ticks parasitizing birds in the Czech Republic. *Journal of Medical Entomology* **33**, 766–771.

Hubbard, M. J., Baker, A. S. & Cann, K. J. (1998). Distribution of *Borrelia burgdorferi* s.l. spirochaete DNA in British ticks (Argasidae and Ixodidae) since the nineteenth century, assessed by PCR. *Medical and Veterinary Entomology* **12**, 89–97.

Huegli, D., Hu, C. M., Humair, P.-F., Wilske, B. & Gern, L. (2002). *Apodemus* species mice, reservoir hosts of *Borrelia garinii* OspA serotype 4 in Switzerland. *Journal of Clinical Microbiology* **40**, 4735–4737.

Humair, P. F. (2002). Birds and *Borrelia*. *International Journal of Medical Microbiology* **291**, 70–74.

Humair, P. F. & Gern, L. (1998). Relationship between *Borrelia burgdorferi* sensu lato species, red squirrels (*Sciurus vulgaris*) and *Ixodes ricinus* in enzootic areas in Switzerland. *Acta Tropica* **69**, 213–227.

Humair, P. F., Péter, O., Wallich, R. & Gern, L. (1995). Strain variation of Lyme disease spirochetes isolated from *Ixodes ricinus* ticks and rodents collected in two endemic areas in Switzerland. *Journal of Medical Entomology* **32**, 433–438.

Humair, P. F., Postic, D., Wallich, R. & Gern, L. (1998). An avian reservoir (*Turdus merula*) of the Lyme borreliosis spirochaetes. *Zentralblatt für Bakteriologie* **287**, 521–538.

Humair, P. F., Rais, O. & Gern, L. (1999). Transmission of *Borrelia afzelii* from *Apodemus* mice and *Clethrionomys* voles to *Ixodes ricinus* ticks: differential transmission pattern and overwintering maintenance. *Parasitology* **118**, 33–42.

Humair, P. F., Turrian, N., Aeschlimann, A. & Gern, L. (1993a). *Borrelia burgdorferi* in a focus of Lyme borreliosis: epizootiologic contribution of small mammals. *Folia Parasitologica* **40**, 65–70.

Humair, P. F., Turrian, N., Aeschlimann, A. & Gern, L. (1993b). *Ixodes ricinus* immatures on birds in a focus of Lyme borreliosis. *Folia Parasitologica* **40**, 237–242.

Jaenson, T. G. T. & Tälleklint, L. (1992). Incompetence of roe deer as reservoirs of the Lyme borreliosis spirochete. *Journal of Medical Entomology* **29**, 813–817.

Jaenson, T. G. T. & Tälleklint, L. (1996). Lyme borreliosis spirochetes in *Ixodes ricinus* (Acari: Ixodidae) and the varying hare on isolated islands in the Baltic sea. *Journal of Medical Entomology* **33**, 339–343.

Jaenson, T. G. T., Tälleklint, L., Lundqvist, L., *et al.* (1994). Geographical distribution, host associations and vector roles of ticks (Acari: Ixodidae, Argasidae) in Sweden. *Journal of Medical Entomology* **31**, 240–256.

James, A. M., Liveris, D., Wormser, G. P., *et al.* (2001). *Borrelia lonestari* infection after a bite by an *Amblyomma americanum* tick. *Journal of Infectious Diseases* **183**, 1810–1814.

Jaulhac, B., Heller, R., Limbach, F. X., *et al.* (2000). Direct molecular typing of *Borrelia burgdorferi* sensu lato species in synovial samples from patients with Lyme arthritis. *Journal of Clinical Microbiology* **38**, 1895–1900.

Jenkins, A., Kristiansen, B. E., Allum, A. G., *et al.* (2001). *Borrelia burgdorferi* sensu lato and *Ehrlichia* spp. in *Ixodes* ticks from southern Norway. *Journal of Clinical Microbiology* **39**, 3666–3671.

Johnson, B. J., Robbins, K. E., Bailey, R. E., *et al.* (1996). Serodiagnosis of Lyme disease: accuracy of a two-step approach using a flagella-based ELISA and immunoblotting. *Journal of Infectious Diseases* **174**, 346–353.

Johnson, R. C., Schmid, G. P., Hyde, F. W., Stiegerwalt, A. G. & Brenner, D. J. (1984). *Borrelia burgdorferi* sp. nov.: etiologic agent of Lyme disease. *International Journal of Systematic Bacteriology* **34**, 496–497.

Jones, C. G., Ostfeld, R. S., Richard, M. P., Schauber, E. M. & Wolff, J. O. (1998). Chain reaction linking acorns to gypsy moth outbreaks and Lyme disease risk. *Science* **279**, 1023–1026.

Jouda, F., Crippa, M., Perret, J.-L. & Gern, L. (2003). Distribution and prevalence of *Borrelia burgdorferi* sensu lato in *Ixodes ricinus* ticks of Canton Ticino (Switzerland). *European Journal of Epidemiology* **18**, 907–912.

Jouda, F., Perret, J.-L. & Gern, L. (2004a). *Ixodes ricinus* density, and distribution and prevalence of *Borrelia burgdorferi* sensu lato infection along an altitudinal gradient. *Journal of Medical Entomology* **41**, 162–170.

Jouda, F., Perret, J.-L. & Gern, L. (2004b). Density of questing *Ixodes ricinus* nymphs and adults infected by *Borrelia burgdorferi* sensu lato in Switzerland: spatio-temporal pattern at a regional scale. *Vector Borne and Zoonotic Diseases* **4**, 23–32.

Junttila, J., Peltomaa, M., Soini, H., Marjamäki, M. & Viljamnen, M. K. (1999). Prevalence of *Borrelia burgdorferi* in *Ixodes ricinus* ticks in urban recreational areas of Helsinki. *Journal of Clinical Microbiology* **37**, 1361–1365.

Kahl, O. & Geue, L. (1998). Laboratory study on the possible role of the European fox, *Vulpes vulpes* as a potential reservoir of *Borrelia burgdorferi* s.l. In *2nd International Conference on Tick-Borne Pathogens at the Host–Vector Interface: A Global Perspective*, eds. Coons, L. & Rothschild, M., p. 29.

Kahl, O. & Knülle, W. (1988). Water vapour uptake from subsaturated atmospheres by engorged immature ixodid ticks. *Experimental and Applied Acarology* **4**, 73–83.

Kahl, O., Janetzki-Mittmann, C., Gray, J. S., *et al.* (1998). Risk of infection with *Borrelia burgdorferi* sensu lato for a host in relation to the duration of nymphal *Ixodes ricinus* feeding and the method of tick removal. *Zentralblatt für Bakteriologie* **287**, 41–52.

Kahl, O., Gern, L., Eisen, L. & Lane, R. S. (2002). Ecological research on *Borrelia burgdorferi* sensu lato: terminology and some methodological pitfalls. In *Lyme Borreliosis: Biology, Epidemiology and Control*, eds. Gray, J., Kahl, O., Lane, R. S. & Stanek, G., pp. 29–46. Wallingford, UK: CAB International.

Kain, D. E., Sperling, F. A. H., Daly, H. V. & Lane, R. S. (1999). Mitochondrial DNA sequence variation in *Ixodes pacificus* (Acari: Ixodidae). *Heredity* **83**, 378–386.

Kain, D. E., Sperling, F. A. H. & Lane, R. S. (1997). Population genetic structure of *Ixodes pacificus* (Acari: Ixodidae) using allozymes. *Journal of Medical Entomology* **34**, 441–450.

Kampen, H., Rötzel, D. C., Kurtenbach, K., Maier, W. A. & Seitz, H. M. (2004). Substantial rise in the prevalence of Lyme borreliosis spirochetes in a region of Western Germany over a 10-year period. *Applied and Environmental Microbiology* **70**, 1576–1582.

Khanakah, G., Kmety, E., Radda, A. & Stanek, G. (1994). Micromammals as reservoir of *Borrelia burgdorferi* in Austria. In *Abstract Book of the 6th Conference on Lyme Borreliosis*, eds. Cevenini, R., Sambri, V. & La Girons, M., Abstract P077W.

Kirstein, F., Rijpkema, S., Molkenboer, M. & Gray, J. S. (1997). The distribution and prevalence of *Borrelia burgdorferi* genomospecies in *Ixodes ricinus* ticks in Ireland. *European Journal of Epidemiology* **13**, 67–72.

Klempner, M. S., Hu, L. T., Evans, J., *et al.* (2001). Two controlled trials of antibiotic treatment in patients with persistent symptoms and a history of Lyme disease. *New England Journal of Medicine* **345**, 85–92.

Korenberg, E. I. (2000). Seasonal population dynamics of *Ixodes* ticks and tick-borne encephalitis virus. *Experimental and Applied Acarology* **24**, 665–681.

Korenberg, E. I., Gorelova, N. B. & Kovalevskii, Y. V. (2002). Ecology of *Borrelia burgdorferi* sensu lato in Russia. In *Lyme Borreliosis: Biology, Epidemiology and Control*, eds. Gray, J., Kahl, O., Lane, R. S. & Stanek, G., pp. 175–200. Wallingford, UK: CAB International.

Kraiczy, P., Skerka, C., Brade, V. & Zipfel, P. F. (2001). Further characterization of complement regulator-acquiring surface proteins of *Borrelia burgdorferi*. *Infection and Immunity* **69**, 7800–7809.

Kuo, M. M., Lane, R. S. & Gicias, P. C. (2000). A comparative study of mammalian and reptilian alternative pathway of complement-mediated killing of the Lyme disease spirochete (*Borrelia burgdorferi*). *Journal of Parasitology* **86**, 1223–1228.

Kurtenbach, K., Carey, D., Hoodless, A. N., Nuttall, P. A. & Randolph, S. E. (1998a). Competence of pheasants as reservoirs for Lyme disease spirochetes. *Journal of Medical Entomology* **35**, 77–81.

Kurtenbach, K., De Michelis, S., Etti, S., *et al.* (2002a). Host association of *Borrelia burgdorferi* sensu lato: the key role of host complement. *Trends in Microbiology* **10**, 74–79.

Kurtenbach, K., De Michelis, S., Sewell, H. S., *et al.* (2001). Distinct combinations of *Borrelia burgdorferi* sensu lato genospecies found in individual questing ticks from Europe. *Applied and Environmental Microbiology* **67**, 4926–4929.

Kurtenbach, K., Dizij, A., Seitz, H. M., *et al.* (1994). Differential immune responses to *Borrelia burgdorferi* in European wild rodent species influence spirochete transmission to *Ixodes ricinus* L. (Acari: Ixodidae). *Infection and Immunity* **62**, 5344–5352.

Kurtenbach, K., Hanincová, K., Tsao, J. I., *et al.* (2006). Fundamental processes in the evolutionary ecology of Lyme borreliosis. *Nature Reviews in Microbiology* **4**, 660–669.

Kurtenbach, K., Kampen, H., Dizij, A., *et al.* (1995). Infestation of rodents with larval *Ixodes ricinus* (Acari: Ixodidae) is an important factor in the transmission cycle of *Borrelia burgdorferi* s.l. in German woodlands. *Journal of Medical Entomology* **32**, 807–817.

Kurtenbach, K., Peacey, M., Rijpkema, S. G. T., *et al.* (1998b). Differential transmission of the genospecies of *Borrelia burgdorferi* sensu lato by game birds and small rodents in England. *Applied and Environmental Microbiology* **64**, 1169–1174.

Kurtenbach, K., Schäfer, S. M., Sewell, H. S., *et al.* (2002b). Differential survival of Lyme borreliosis spirochetes in ticks that feed on birds. *Infection and Immunity* **70**, 5893–5895.

Kurtenbach, K., Sewell, H. S., Ogden, N. H., Randolph, S. E. & Nuttall, P. A. (1998c). Serum complement sensitivity as a key factor in Lyme disease ecology. *Infection and Immunity* **66**, 1248–1251.

Lagal, V., Postic, D., Ruzic-Sabljic, E. & Baranton, G. (2003). Genetic diversity among *Borrelia* strains determined by single-strand conformation polymorphism analysis of the *ospC* gene and its association with invasiveness. *Journal of Clinical Microbiology* **41**, 5059–5065.

Lane, R. S. & Loye, J. E. (1989). Lyme disease in California: interrelationship of *Ixodes pacificus* (Acari: Ixodidae), the western fence lizard (*Sceloporus occidentalis*), and *Borrelia burgdorferi*. *Journal of Medical Entomology* **26**, 272–278.

Lane, R. S., Casher, L. E., Peavey, C. A. & Piesman, J. (1998). Modified bait tube controls disease-carrying ticks and fleas. *California Agriculture* **52**, 43–48.

Lathrop, S. L., Ball, R., Haber, P., *et al.* (2002). Adverse events reports following vaccination for Lyme disease: December 1998–July 2000. *Vaccine* **20**, 1603–1608.

Lebet, N. & Gern, L. (1994). Histological examination of *Borrelia burgdorferi* infection in unfed *Ixodes ricinus* nymphs. *Experimental and Applied Acarology* **18**, 177–183.

Lees, A. D. & Milne, A. (1951). The seasonal and diurnal activities of individual sheep ticks (*Ixodes ricinus*). *Parasitology* **41**, 180–209.

Le Fleche, A., Postic, D., Girardet, K., Péter, O. & Baranton, G. (1997). Characterization of *Borrelia lusitaniae* sp. nov. by 16S ribosomal DNA sequence analysis. *International Journal of Systematic Bacteriology* **47**, 921–925.

Leprince, D. J. & Lane, R. S. (1996). Evaluation of permethrin-impregnated cotton balls as potential nesting material to control ectoparasites of woodrats in California. *Journal of Medical Entomology* **33**, 355–360.

Leuba-Garcia, S., Kramer, M. D., Wallich, R. & Gern, L. (1994). Characterization of *Borrelia burgdorferi* isolated from different organs of *Ixodes ricinus* ticks collected in nature. *Zentralblatt für Bakteriologie* **280**, 468–475.

Leuba-Garcia, S., Martinez, R. & Gern, L. (1998). Expression of outer surface proteins A and C of *Borrelia afzelii* in *Ixodes ricinus* ticks and in the skin of mice. *Zentralblatt für Bakteriologie und Hygiene* **287**, 475–484.

Levin, M., Levine, J. F., Yang, S., Howard, P. & Apperson, C. S. (1996). Reservoir competence of the southeastern five-line skink (*Eumeces inexpectatus*) and the green anole

(*Anolis carolinensis*) for *Borrelia burgdorferi*. *American Journal of Tropical Medicine and Hygiene* **54**, 92–97.

Liang, F. T., Steere, A. C., Marques, A. R., *et al.* (1999). Sensitive and specific serodiagnosis of Lyme disease by enzyme-linked immunosorbent assay with a peptide based on an immunodominant conserved region of *Borrelia burgdorferi* vlsE. *Journal of Clinical Microbiology* **37**, 3990–3996.

Liebisch, G., Dimpfl, B., Finkbeiner-Weber, B., Liebisch, A. & Frosch, M. (1998*a*). The red fox (*Vulpes vulpes*) a reservoir competent host for *Borrelia burgdorferi* sensu lato. In *2nd International Conference on Tick-Borne Pathogens at the Host–Vector Interface: A Global Perspective*, eds. Coons, L. & Rothschild, M., p. 238.

Liebisch, G., Finkbeiner-Weber, B. & Liebisch, A. (1996). The infection with *Borrelia burgdorferi* s.l. in the European hedgehog (*Erinaceus europaeus*) and its ticks. *Parassitologia* **38**, 385.

Liebisch, G., Sihns, B. & Bautsch, W. (1998*b*). Detection and typing of *Borrelia burgdorferi* sensu lato in *Ixodes ricinus* ticks attached to human skin by PCR. *Journal of Clinical Microbiology* **36**, 3355–3358.

Lin, T., Oliver, J. H. Jr & Gao, L. (2002). Genetic diversity of the outer surface protein C gene of southern *Borrelia* isolates and its possible epidemiological, clinical, and pathogenetic implications. *Journal of Clinical Microbiology* **40**, 2572–2583.

Lindsay, L. R., Barker, I. K., Surgeoner, G. A., *et al.* (1995). Survival and development of *Ixodes scapularis* (Acari: Ixodidae) under various climatic conditions in Ontario, Canada. *Journal of Medical Entomology* **32**, 143–152.

Lindsay, L. R., Mathison, S. W., Barker, I. K., Mcewen, S. A. & Surgeoner, G. A. (1999). Abundance of *Ixodes scapularis* (Acari: Ixodidae) larvae and nymphs in relation to host density and habitat on Long Point, Ontario. *Journal of Medical Entomology* **36**, 243–254.

Liveris, D., Wang, G., Girao, G., *et al.* (2002). Quantitative detection of *Borrelia burgdorferi* in 2-millimeter skin samples of erythema migrans lesions: correlation of results with clinical laboratory findings. *Journal of Clinical Microbiology* **40**, 1249–1253.

Logigian, E. L., Kaplan, R. F. & Steere, A. C. (1990). Chronic neurologic manifestations of Lyme disease. *New England Journal of Medicine* **323**, 1438–1444.

Magid, D., Schwartz, B., Craft, J. & Schwartz, J. S. (1992). Prevention of Lyme disease after tick bites: a cost effectiveness analysis. *New England Journal of Medicine* **327**, 534–541.

Manweiler, S. A., Lane, R. S., Block, W. M. & Morrison, M. L. (1990). Survey of birds and lizards for ixodid ticks (Acari) and spirochetal infection in northern California. *Journal of Medical Entomology* **27**, 1011–1015.

Marconi, R. T., Liveris, D. & Schwartz, I. (1995). Identification of novel insertion elements, restriction fragment length polymorphism patterns, and discontinuous 23S rRNA in Lyme disease spirochetes: phylogenetic analyses of rRNA genes and their intergenic spacers in *Borrelia japonica* sp. nov. and genomic group 21038 (*Borrelia andersonii* sp. nov.) isolates. *Journal of Clinical Microbiology* **33**, 2427–2434.

Mather, T. N., Ribeiro, J. M. & Spielman, A. (1987). Lyme disease and babesiosis: acaricide focused on potentially infected ticks. *American Journal of Tropical Medicine and Hygiene* **36**, 609–614.

Mather, T. N., Telford, S. R. III, Maclachlan, A. B. & Spielman, A. (1989*a*). Incompetence of catbirds as reservoirs for the Lyme disease spirochete (*Borrelia burgdorferi*). *Journal of Parasitology* **75**, 66–69.

Mather, T. N., Wilson, M. L., Moore, S. I., Ribeiro, J. M. & Spielman, A. (1989*b*). Comparing the relative potential of rodents as reservoirs of the Lyme disease spirochete (*Borrelia burgdorferi*). *American Journal of Epidemiology* **130**, 143–150.

Mathiesen, D. A., Oliver, J. H. Jr, Kolbert, C. P., *et al.* (1997). Genetic heterogeneity of *Borrelia burgdorferi* in the United States. *Journal of Infectious Diseases* **175**, 98–107.

Matuschka, F. R., Allgöwer, R., Spielman, A. & Richter, D. (1999). Characteristics of garden dormice that contribute to their capacity as reservoirs for Lyme disease spirochetes. *Applied and Environmental Microbiology* **65**, 707–711.

Matuschka, F. R., Eiffert, H., Ohlenbusch, A., *et al.* (1994*a*). Transmission of the agent of Lyme disease on a subtropical island. *Tropical Medicine and Parasitology* **45**, 39–44.

Matuschka, F. R., Eiffert, H., Ohlenbusch, A. & Spielman, A. (1994*b*). Amplifying role of edible dormice in Lyme disease transmission in Central Europe. *Journal of Infectious Diseases* **170**, 122–127.

Matuschka, F. R., Endepols, S., Richter, D., *et al.* (1996). Risk of urban Lyme disease enhanced by the presence of rats. *Journal of Infectious Diseases* **174**, 1108–1111.

Matuschka, F. R., Endepols, S., Richter, D. & Spielman, A. (1997). Competence of urban rats as reservoir hosts for Lyme disease spirochetes. *Journal of Medical Entomology* **34**, 489–493.

Matuschka, F. R., Fischer, P., Heiler, M., Richter, D. & Spielman, A. (1992). Capacity of European animals as reservoir hosts for the Lyme disease spirochete. *Journal of Infectious Diseases* **165**, 479–483.

Matuschka, F. R., Schinkel, T. W., Klug, B., Spielman, A. & Richter, D. (2000). Relative incompetence of European rabbits for Lyme disease spirochaetes. *Parasitology* **121**, 297–302.

Maupin, G. O., Fish, D., Zultowsky, J., Campos, E. G. & Piesman, J. (1991). Landscape ecology of Lyme disease in a residential area of Westchester County, New York. *American Journal of Epidemiology* **133**, 1105–1113.

McCoy, K. D., Boulinier, T., Chardine, J. W., Danchin, E. & Michalakis, Y. (1999). Dispersal and distribution of the tick *Ixodes uriae* within and among seabird populations: the need for a population genetic approach. *Journal of Parasitology* **85**, 196–202.

McLean, R. G., Ubico, S. R., Hughes, C. A., Engstrom, S. M. & Johnson, R. C. (1993). Isolation and characterization of *Borrelia burgdorferi* from blood of a bird captured in the Saint Croix River Valley. *Journal of Clinical Microbiology* **31**, 2038–2043.

McLeod, J. (1935). *Ixodes ricinus* in relation to its physical environment. IV. An analysis of the ecological complexes controlling distribution and activities. *Parasitology* **28**, 295–319.

Mejlon, H. A. & Jaenson, T. G. T. (1997). Questing behaviour of *Ixodes ricinus* ticks (Acari: Ixodidae). *Experimental and Applied Acarology* **21**, 747–754.

Mejlon, H., Jaenson, T. G. T. & Mather, T. N. (1995). Evaluation of host-targeted applications of permethrin for control of *Borrelia*-infected *Ixodes ricinus* (Acari: Ixodidae). *Medical and Veterinary Entomology* **9**, 207–210.

Melkert, P. W. & Stel, H. V. (1991). Neonatal *Borrelia* infections (relapsing fever): report of five cases and review of the literature. *East African Medical Journal* **68**, 999–1005.

Michel, H., Wilske, B., Hettche, G., *et al.* (2004). An OspA-polymerase chain reaction/restriction fragment length polymorphism-based method for sensitive detection and reliable differentiation of all European *Borrelia burgdorferi* sensu lato species and OspA types. *Medical Microbiology and Immunology* **193**, 219–226.

Misonne, M. C., Van Impe, G. & Hoet, P. P. (1998). Genetic heterogeneity of *Borrelia burgdorferi* sensu lato in *Ixodes ricinus* ticks collected in Belgium. *Journal of Clinical Microbiology* **36**, 3352–3354.

Miyamoto, K. & Masuzawa, T. (2002). Ecology of *Borrelia burgdorferi* sensu lato in Japan and East Asia. In *Lyme Borreliosis: Biology, Epidemiology and Control*, eds. Gray, J.,

Kahl, O., Lane, R. S. & Stanek, G., pp. 201–222. Wallingford, UK: CAB International.

Mizak, B. & Krol, J. (2000). Analysis of Polish isolates of *Borrelia burgdorferi* by amplification of (5S)-rrl (23S) intergenic spacer. *Bulletin of the Veterinary Institute in Pulawy* **44**, 147–154.

Morel, P. C. (1965). Les tiques d'Afrique et du bassin méditerranéen. Maison Alfort (I.E.M.T.Y.). Photocopied document.

Nadelman, R. B. & Wormser, G. P. (2002). Recognition and treatment of erythema migrans: are we off target? *Annals of Internal Medicine* **136**, 477–479.

Nadelman, R. B., Nowakowski, J., Fish, D., *et al.* (2001). Prophylaxis with single-dose doxycycline for the prevention of Lyme disease after an *Ixodes scapularis* tick bite. *New England Journal of Medicine* **345**, 79–84.

Nelson, D. R., Rooney, S., Miller, N. J. & Mather, T. R. (2000). Complement-mediated killing of *Borrelia burgdorferi* by nonimmune sera from sika deer. *Journal of Parasitology* **86**, 1232–1238.

Nocton, J. J., Dressler, F., Rutledge, B. J., *et al.* (1994). Detection of *Borrelia burgdorferi* DNA by polymerase chain reaction in synovial fluid from patients with Lyme arthritis. *New England Journal of Medicine* **330**, 229–234.

Norris, D. E., Klompen, J. S. H., Keirans, J. E., *et al.* (1996). Population genetics of *Ixodes scapularis* (Acari: Ixodidae) based on mitochondrial 16S and 12S genes. *Journal of Medical Entomology* **33**, 78–89.

Nowakowski, J., Schwartz, I., Liveris, D., *et al.* (2001). Laboratory diagnostic techniques for patients with early Lyme disease associated with erythema migrans: a comparison of different techniques. *Clinical Infectious Diseases* **33**, 2023–2037.

Nuncio, M. S., Péter, O., Alves, M. J., Bacellar, F. & Filipe, A. R. (1993). Isolamento e caracterizacão de borrélias de *Ixodes ricinus* L. em Portugal. *Revista Portuguesa Doencas Infecciosas* **16**, 175–179.

Ogden, N. H., Nuttall, P. A. & Randolph, S. E. (1997). Natural Lyme disease cycles maintained via sheep by co-feeding ticks. *Parasitology* **115**, 591–599.

Ohnishi, J., Piesman, J. & De Silva, A. M. (2001). Antigenic and genetic heterogeneity of *Borrelia burgdorferi* populations transmitted by ticks. *Proceedings of the National Academy of Sciences of the USA* **98**, 670–675.

Oliver, J. H. Jr, Cummins, G. A. & Joiner, M. S. (1993a). Immature *Ixodes scapularis* (Acari: Ixodidae) parasitizing lizards from the southeastern USA. *Journal of Parasitology* **79**, 684–689.

Oliver, J. H. Jr, Owsley, M. R., Hutcheson, H. J., *et al.* (1993*b*). Conspecificity of the ticks *Ixodes scapularis* and *I. dammini* (Acari: Ixodidae). *Journal of Medical Entomology* **30**, 54–63.

Olsen, B. (1995). Birds and *Borrelia*. Unpublished Ph.D. thesis, Umeå University, Umeå, Sweden.

Olsen, B., Duffy, D. C., Jaenson, T. G. T., *et al.* (1995*a*). Transhemispheric exchange of Lyme disease spirochetes by seabirds. *Journal of Clinical Microbiology* **33**, 3270–3274.

Olsen, B., Jaenson, T. G. T., Noppa, L., Bunikis, J. & Bergström, S. (1993). A Lyme borreliosis cycle in seabirds and *Ixodes uriae* ticks. *Nature* **362**, 340–342.

Olsen, B., Jaenson, T. G. T. & Bergström, S. (1995*b*). Prevalence of *Borrelia burgdorferi* sensu lato-infected ticks on migrating birds. *Applied and Environmental Microbiology* **61**, 3082–3087.

Olson, C. A., Cupp, E. W., Luckhart, S., Ribeiro, J. M. C. & Levy, C. (1992). Occurrence of *Ixodes pacificus* (Parasitoformes: Ixodidae) in Arizona. *Journal of Medical Entomology* **29**, 1060–1062.

Pal, U., de Silva, A. M., Montgomery, R. R., *et al.* (2000). Attachment of *Borrelia burgdorferi* within *Ixodes scapularis* mediated by outer surface protein A. *Journal of Clinical Investigation* **106**, 561–569.

Peavey, C. A. & Lane, R. S. (1995). Transmission of *Borrelia burgdorferi* by *Ixodes pacificus* nymphs and reservoir competence of deer mice (*Peromyscus maniculatus*) infected by tick-bite. *Journal of Parasitology* **81**, 175–178.

Peavey, C. A., Lane, R. S. & Kleinjan, J. E. (1997). Role of small mammals in the ecology of *Borrelia burgdorferi* in a peri-urban park in north coastal California. *Experimental and Applied Acarology* **21**, 569–584.

Perret, J. L., Guigoz, E., Rais, O. & Gern, L. (2000). Influence of saturation deficit and temperature on *Ixodes ricinus* tick questing activity in a Lyme borreliosis-endemic area (Switzerland). *Parasitology Research* **86**, 554–557.

Perret, J. L., Rais, O. & Gern, L. (2004). Influence of climate on the proportion of *Ixodes ricinus* nymphs and adults questing in a tick population. *Journal of Medical Entomology* **41**, 361–365.

Philipp, M. T., Bowers, L. C., Fawcett, P. T., *et al.* (2001). Antibody response to IR6, a conserved immunodominant region of the vlsE lipoprotein, wanes rapidly after antibiotic treatment of *Borrelia burgdorferi* infection in experimental animals and in humans. *Journal of Infectious Diseases* **184**, 870–878.

Phillips, C. B., Liang, M. H., Sangha, O., *et al.* (2001). Lyme disease and preventive behaviors in residents of Nantucket Island, Massachusetts. *American Journal of Preventive Medicine* **20**, 219–224.

Pichon, B., Godfroid, E., Hoyois, B., *et al.* (1995). Simultaneous infection of *Ixodes ricinus* by two *Borrelia burgdorferi* sensu lato species: possible implications for clinical manifestations. *Emerging Infectious Diseases* **1**, 89–90.

Picken, R. N. & Picken, M. M. (2000). Molecular characterization of *Borrelia* spp. isolates from greater metropolitan Chicago reveals the presence of *Borrelia bissettii*: preliminary report. *Journal of Molecular Microbiology and Biotechnology* **2**, 505–507.

Picken, R. N., Strle, F., Picken, M. M., *et al.* (1998). Identification of three species of *Borrelia burgdorferi* sensu lato (*B. burgdorferi* sensu stricto, *B. garinii*, and *B. afzelii*) among isolates from acrodermatitis chronica atrophicans lesions. *Journal of Investigative Dermatology* **110**, 211–214.

Piesman, J. (1989). Transmission of Lyme disease spirochetes (*Borrelia burgdorferi*). *Experimental and Applied Acarology* **7**, 71–80.

Piesman, J. & Gern, L. (2004). Lyme borreliosis in Europe and North America. *Parasitology* **129** (Suppl.), S191–S220.

Piesman, J. & Spielman, A. (1979). Host-associations and seasonal abundance of immature *Ixodes dammini* in southeastern Massachusetts. *Annals of the Entomological Society of America* **72**, 829–832.

Piesman, J., Mather, T. N., Dammin, G. J., *et al.* (1987*a*). Seasonal variation of transmission risk: Lyme disease and human babesiosis. *American Journal of Epidemiology* **126**, 1187–1189.

Piesman, J., Mather, T. N., Sinsky, R. J. & Spielman, A. (1987*b*). Duration of tick attachment and *Borrelia burgdorferi* transmission. *Journal of Clinical Microbiology* **25**, 557–558.

Piesman, J., Maupin, G. O., Campos, E. G. & Happ, C. M. (1991). Duration of adult female *Ixodes dammini* attachment and transmission of *Borrelia burgdorferi*, with description of a needle aspiration isolation method. *Journal of Infectious Diseases* **163**, 895–897.

Piesman, J., Schneider, B. S. & Zeidner, N. S. (2001). Use of quantitative PCR to measure the density of *Borrelia burgdorferi* in the midgut and salivary glands of feeding tick vectors. *Journal of Clinical Microbiology* **39**, 4145–4148.

Piesman, J., Spielman, A., Etkind, P., Ruebush, T. K. II & Juranek, D. D. (1979). Role of deer in the epizootiology of *Babesia microti* in Massachusetts, USA. *Journal of Medical Entomology* **15**, 437–440.

Porcella, S. F., Raffel, S. J., Schrumpf, M. E., *et al.* (2000). Serodiagnosis of louse-borne relapsing fever with glycerophosphodiester phosphodiesterase (GlpQ) from

Borrelia recurrentis. Journal of Clinical Microbiology **38**, 3561–3571.

Postic, D., Assous, M. V., Grimont, P. A. & Baranton, G. (1994). Diversity of *Borrelia burgdorferi* sensu lato evidenced by restriction fragment length polymorphism of rrf (5S)–rrl (23S) intergenic spacer amplicons. *International Journal of Systematic Bacteriology* **44**, 743–752.

Postic, D., Korenberg, E., Gorelova, N., *et al.* (1997). *Borrelia burgdorferi* sensu lato in Russia and neighbouring countries: high incidence of mixed isolates. *Research in Microbiology* **148**, 691–702.

Postic, D., Ras, N. M., Lane, R. S., Hendson, M. & Baranton, G. (1998). Expanded diversity among California *Borrelia* isolates and description of *Borrelia bissettii* sp. nov. (formerly *Borrelia* group DN127). *Journal of Clinical Microbiology* **36**, 3497–3504.

Pound, J. M., Miller, J. A. & George, J. E. (2000). Efficacy of amitraz applied to white-tailed deer by the 'four-poster' topical treatment device in controlling free-living lone star ticks (Acari: Ixodidae). *Journal of Medical Entomology* **37**, 878–884.

Poupon, M.-A., Lommano, E., Humair, P.-F., *et al.* (2006). Prevalence of *Borrelia burgdorferi* sensu lato in ticks collected from migratory birds in Switzerland. *Applied and Environmental Microbiology* **72**, 976–979.

Qiu, W. G., Dykhuizen, D. E., Acosta, M. S. & Luft, B. J. (2002). Geographic uniformity of the Lyme disease spirochete (*Borrelia burgdorferi*) and its shared history with tick vector (*Ixodes scapularis*) in the northeastern United States. *Genetics* **160**, 833–849.

Ramamoorthi, N., Narasimhan, S., Pal, U., *et al.* (2005). The Lyme disease agent exploits a tick protein to infect the mammalian host. *Nature* **436**, 573–577.

Randolph, S. E. & Craine, N. G. (1995). General framework for comparative quantitative studies on transmission of tick-borne diseases using Lyme borreliosis in Europe as an example. *Journal of Medical Entomology* **32**, 765–777.

Randolph, S. E. & Storey, K. (1999). Impact of microclimate on immature tick-rodent host interactions (Acari: Ixodidae): implications for parasite transmission. *Journal of Medical Entomology* **36**, 741–748.

Randolph, S. E., Gern, L. & Nuttall, P. A. (1996). Co-feeding ticks: epidemiological significance for tick-borne pathogen transmission. *Parasitology Today* **12**, 472–479.

Randolph, S. E., Green, R. M., Hoodless, A. N. & Peacey, M. F. (2002). An empirical quantitative framework for the seasonal population dynamics of the tick *Ixodes ricinus*. *International Journal for Parasitology* **32**, 979–989.

Randolph, S. E., Green, R. M., Peacey, M. F. & Rogers, D. J. (2000). Seasonal synchrony: the key to tick-borne encephalitis foci identified by satellite data. *Parasitology* **121**, 15–23.

Rauter, C. & Hartung, T. (2005). Prevalence of *Borrelia burgdorferi* sensu lato genospecies in *Ixodes ricinus* ticks in Europe: a metaanalysis. *Applied and Environmental Microbiology* **71**, 7203–7216.

Reik, L., Steere, A. C., Bartenhagen, N. H., Shope, R. E. & Malawista, S. E. (1979). Neurologic abnormalities of Lyme disease. *Medicine (Baltimore)* **58**, 281–294.

Rich, S. M., Armstrong, P. M., Smith, R. D. & Telford, S. R. III (2001). Lone star tick-infecting borreliae are most closely related to the agent of bovine borreliosis. *Journal of Clinical Microbiology* **39**, 494–497.

Rich, S. M., Caporale, D. A., Telford, S. R., *et al.* (1995). Distribution of the *Ixodes ricinus*-like ticks of eastern North America. *Proceedings of the National Academy of Sciences of the USA* **92**, 6284–6288.

Richter, D., Endepols, S., Ohlenbusch, A., *et al.* (1999). Genospecies diversity of Lyme disease spirochetes in rodent reservoirs. *Emerging Infectious Diseases* **5**, 291–296.

Richter, D., Postic, D., Sertour, N., *et al.* (2006). Delineation of *Borrelia burgdorferi* sensu lato species by multilocus sequence analysis and confirmation of the delineation of *B. spielmanii* sp. nov. *International Journal of Systematic and Evolutionary Microbiology* **56**, 873–881.

Richter, D., Schlee, D. B., Allgover, R. & Matuschka, F. R. (2004). Relationships of a novel Lyme disease spirochete, *Borrelia spielmanii* sp. nov., with its hosts in Central Europe. *Applied and Environmental Microbiology* **70**, 6414–6419.

Richter, D., Schlee, D. B. & Matuschka, F. R. (2003). Relapsing fever-like spirochetes infecting European vector tick of Lyme disease agent. *Emerging Infectious Diseases* **9**, 697–701.

Richter, D., Spielman, A., Komar, N. & Matuschka, F. R. (2000). Competence of American robins as reservoir hosts for Lyme disease spirochetes. *Emerging Infectious Diseases* **6**, 133–138.

Rijpkema, S. G. T., Golubic, D., Molkenboer, M., Verbeeek-De Kruif, N. & Schellenkens, J. F. P. (1996). Identification of four genomic groups of *Borrelia burgdorferi* sensu lato in *Ixodes ricinus* ticks collected in a Lyme borreliosis endemic region of northern Croatia. *Experimental and Applied Acarology* **20**, 23–30.

Rijpkema, S. G. T., Molkenboer, M. J. C. H., Schouls, L. M., Jongejan, F. & Schellenkens, J. P. P. (1995). Simultaneous detection and genotyping of three genomic groups of *Borrelia burgdorferi* sensu lato in Dutch *Ixodes ricinus* ticks by characterization of the amplified intergenic spacer region between 5S and 23 S rRNA genes. *Journal of Clinical Microbiology* 33, 3091–3095.

Robertson, J., Guy, E., Andrews, N., *et al.* (2000). A European multicenter study of immunoblotting in serodiagnosis of Lyme borreliosis. *Journal of Clinical Microbiology* 38, 2097–2102.

Rothermel, H., Hedges, T. R. III & Steere, A. C. (2001). Optic neuropathy in children with Lyme disease. *Pediatrics* 108, 477–481.

Sarih, M., Jouda, F., Gern, L. & Postic, D. (2003). First isolation of *Borrelia burgdorferi* sensu lato from *Ixodes ricinus* ticks in Morocco. *Vector Borne and Zoonotic Diseases* 3, 133–139.

Schneider, B. S., Zeidner, N. S., Burkot, T. R., Maupin, G. O. & Piesman, J. (2000). *Borrelia* isolates in northern Colorado identified as *Borrelia bissettii*. *Journal of Clinical Microbiology* 38, 3103–3105.

Schulze, T. L., Jordan, R. A. & Hung, R. W. (1995). Suppression of subadult *Ixodes scapularis* (Acari: Ixodidae) following removal of leaf litter. *Journal of Medical Entomology* 32, 730–733.

Schulze, T. L., Jordan, R. A. & Hung, R. W. (1998). Comparison of *Ixodes scapularis* (Acari: Ixodidae) populations and their habitats in established and emerging Lyme disease areas in New Jersey. *Journal of Medical Entomology* 35, 64–70.

Schulze, T. L., Jordan, R. A., Hung, R. W., *et al.* (2001). Efficacy of granular deltamethrin against *Ixodes scapularis* and *Amblyomma americanum* (Acari: Ixodidae) nymphs. *Journal of Medical Entomology* 38, 344–346.

Schwan, T. G. & Piesman, J. (2002). Vector interactions and molecular adaptations of Lyme disease and relapsing fever spirochetes associated with transmission by ticks (Perspectives). *Emerging Infectious Diseases* 8, 115–121.

Schwartz, I., Wormser, G. P., Schwartz, J. J., *et al.* (1992). Diagnosis of early Lyme disease by polymerase chain reaction amplification and culture of skin biopsies from erythema migrans lesions. *Journal of Clinical Microbiology* 30, 3082–3088.

Scoles, G. A., Papero, M., Beati, L. & Fish, D. (2001). A relapsing fever group spirochete transmitted by *Ixodes scapularis* ticks. *Vector Borne and Zoonotic Diseases* 1, 21–34.

Seinost, G., Dykhuizen, D. E., Dattwyler, R. J., *et al.* (1999). Four clones of *Borrelia burgdorferi* sensu stricto cause invasive infection in humans. *Infection and Immunity* 67, 3518–3524.

Sigal, L. H., Zahradnik, J. M., Lavin, P., *et al.* (1998). A vaccine consisting of recombinant *Borrelia burgdorferi* outer-surface protein A to prevent Lyme disease: recombinant outer surface protein A Lyme disease vaccine study consortium. *New England Journal of Medicine* 339, 216–222.

Slajchert, T., Kitron, U. D., Jones, C. J. & Mannelli, A. (1997). Role of the eastern chipmunk (*Tamias striatus*) in the epizootiology of Lyme borreliosis in northwestern Illinois, USA. *Journal of Wildlife Diseases* 33, 40–46.

Smith, R. P. Jr, Rand, P. W., Lacombe, E. H., *et al.* (1993). Norway rats as reservoir hosts for Lyme disease spirochetes on Monhegan Island, Maine. *Journal of Infectious Diseases* 168, 687–691.

Smith, R. P., Schoen, R. T., Rahn, D. W., *et al.* (2002). Clinical characteristics and treatment outcome of early Lyme disease in patients with microbiologically confirmed erythema migrans. *Annals of Internal Medicine* 136, 421–428.

Sonenshine, D. E. & Haines, G. (1985). A convenient method for controlling populations of the American dog tick, *Dermacentor variabilis* (Acari: Ixodidae), in the natural environment. *Journal of Medical Entomology* 22, 577–583.

Sood, S. K., Salzman, M. B., Johnson, B. J. B., *et al.* (1997). Duration of tick attachment as a predictor of the risk of Lyme disease in an area in which Lyme disease is endemic. *Journal of Infectious Diseases* 175, 996–999.

Spielman, A. (1988). Lyme disease and human babesiosis: evidence incriminating vector and reservoir hosts. In *Biology of Parasitism*, eds. Englund, P. T. & Sher, A., pp. 147–165. New York: Alan R. Liss.

Spielman, A., Clifford, C. M., Piesman, J. & Corwin, M. D. (1979). Human babesiosis on Nantucket Island: description of the vector, *Ixodes (Ixodes) dammini*, n. sp. (Acarina: Ixodidae). *Journal of Medical Entomology* 15, 218–234.

Spielman, A., Levine, J. F. & Wilson, M. L. (1984). Vectorial capacity of North American *Ixodes* ticks. *Yale Journal of Biology and Medicine* 57, 507–513.

Spielman, A., Wilson, M. L., Levine, J. L. & Piesman, J. (1985). Ecology of *Ixodes dammini*-borne human babesiosis and Lyme disease. *Annual Review of Entomology* 30, 439–460.

Stafford, K. C. III (1991). Effectiveness of carbaryl applications for the control of *Ixodes dammini* (Acari: Ixodidae) nymphs in an endemic residential area. *Journal of Medical Entomology* 28, 32–36.

Stafford, K. C. III (1992). Third year evaluation of host-targeted permethrin for the control of *Ixodes dammini*

(Acari: Ixodidae) in southeastern Connecticut. *Journal of Medical Entomology* **29**, 717–720.

Stafford, K. C. III (1993). Reduced abundance *of Ixodes scapularis* (Acari: Ixodidae) with exclusion of deer by electric fencing. *Journal of Medical Entomology* **30**, 986–996.

Stafford, K. C. III (1997). Pesticide use by licensed applicators for the control of *Ixodes scapularis* (Acari: Ixodidae) in Connecticut. *Journal of Medical Entomology* **34**, 552–558.

Stafford, K. C. III, Bladen, V. C. & Magnarelli, L. A. (1995). Ticks (Acari: Ixodidae) infesting wild birds (Aves) and white-footed mice in Lyme, CT. *Journal of Medical Entomology* **32**, 453–466.

Stafford, K. C. III, Denicola, A. J. & Magnarelli, L. A. (1996). Presence of *Ixodiphagus hookeri* (Hymenoptera: Encyrtidae) in two populations of *Ixodes scapularis* (Acari: Ixodidae). *Journal of Medical Entomology* **33**, 183–188.

Stafford, K. C. III, Ward, J. S. & Magnarelli, L. A. (1998). Impact of controlled burns on the abundance of *Ixodes scapularis* (Acari: Ixodidae). *Journal of Medical Entomology* **35**, 510–513.

Stanczak, J., Kubica-Biernat, B., Racewicz, M., Kruminis-Lozowska, W. & Kur, J. (2000). Detection of three genospecies of *Borrelia burgdorferi* sensu lato in *Ixodes ricinus* ticks collected in different regions of Poland. *International Journal of Microbiology* **290**, 559–566.

Stanek, G. & Kahl, O. (1999). Chemoprophylaxis for Lyme borreliosis? *Zentralblatt für Bakteriologie* **289**, 655–695.

Stanek, G., O'Connell, S., Cimmino, M., *et al.* (1996). European Union concerted action on risk assessment in Lyme borreliosis: clinical case definitions for Lyme borreliosis. *Wiener klinische Wochenschrift* **108**, 741–747.

Stanek, G., Strle, J., Gray, J. & Wormser, G. P. (2002). History and characteristics of Lyme borreliosis. In *Lyme Borreliosis: Biology, Epidemiology and Control*, eds. Gray, J., Kahl, O., Lane, R. S. & Stanek, G., pp. 1–28. Wallingford, UK: CAB International.

Steele, G. M. & Randolph, S. E. (1985). An experimental evaluation of conventional control measures against the sheep tick, *Ixodes ricinus* (L.) (Acari: Ixodidae). I. A unimodal seasonal activity pattern. *Bulletin of Entomological Research* **75**, 489–499.

Steere, A. C. (1989). Lyme disease. *New England Journal of Medicine* **308**, 586–596.

Steere, A. C. (1994). Lyme borreliosis. In *Thirteenth Edition Harrison's Principles of Internal Medicine*, vol. 1, eds. Isselbacher, I., Braunwald, E., Wilson, J. D. *et al.*, pp. 745–747. New York: McGraw-Hill.

Steere, A. C. (2001). Lyme disease. *New England Journal of Medicine* **345**, 115–125.

Steere, A. C. & Malawista, S. E. (1979). Cases of Lyme disease in the United States: locations correlated with distribution of *Ixodes dammini*. *Annals of Internal Medicine* **91**, 730–733.

Steere, A. C., Batsford, W. P., Weinberg, M., *et al.* (1980). Lyme carditis: cardiac abnormalities of Lyme disease. *Annals of Internal Medicine* **93**, 8–16.

Steere, A. C., Broderick, T. F. & Malawista, S. E. (1978). Erythema chronicum migrans and Lyme arthritis: epidemiologic evidence for a tick vector. *American Journal of Epidemiology* **108**, 312–321.

Steere, A. C., Malawista, S. E., Hardin, J. A., *et al.* (1977a). Erythema chronicum migrans and Lyme arthritis: the enlarging clinical spectrum. *Annals of Internal Medicine* **86**, 685–698.

Steere, A. C., Malawista, S. E., Snydman, D. R., *et al.* (1977b). Lyme arthritis: an epidemic of oligoarticular arthritis in children and adults in three Connecticut communities. *Arthritis and Rheumatology* **20**, 7–17.

Steere, A. C., Sikand, V. K., Meurice, F., *et al.* (1998). Vaccination against Lyme disease with recombinant *Borrelia burgdorferi* outer-surface lipoprotein A with adjuvant: Lyme disease vaccine study group. *New England Journal of Medicine* **339**, 209–215.

Stevenson, B., El-Hage, N., Mines, M. A., Miller, J. C. & Babb, K. (2002). Differential binding of host complement inhibition factor H by *Borrelia burgdorferi* erp surface proteins: a possible mechanism underlying the expansive host range of Lyme disease spirochetes. *Infection and Immunity* **70**, 491–497.

Stunzner, D., Hubalek, Z., Halouzka, J., *et al.* (1998). Prevalence of *Borrelia burgdorferi* s.l. in *Ixodes ricinus* ticks from Styria (Austria) and species identification by PCR-RFLP analysis. *Zentralblatt für Bakteriologie* **228**, 471–478.

Tälleklint, L. & Jaenson, T. G. T. (1993). Maintenance by hares of European *Borrelia burgdorferi* in ecosystems without rodents. *Journal of Medical Entomology* **30**, 273–276.

Tälleklint, L. & Jaenson, T. G. T. (1994). Transmission of *Borrelia burgdorferi* s.l. from mammal reservoirs to the primary vector of Lyme borreliosis, *Ixodes ricinus* (Acari: Ixodidae), in Sweden. *Journal of Medical Entomology* **31**, 880–886.

Tälleklint, L. & Jaenson, T. G. T. (1995). Control of Lyme borreliosis in Sweden by reduction of tick vectors: an impossible task? *International Journal of Angiology* **4**, 34–37.

Tälleklint, L. & Jaenson, T. G. T. (1996). Seasonal variations in density of questing *Ixodes ricinus* (Acari: Ixodidae) nymphs and prevalence of infection with *Borrelia burgdorferi* sl in south central Sweden. *Journal of Medical Entomology* 33, 592–597.

Tälleklint-Eisen, L. & Lane, R. S. (1999). Variation in the density of questing *Ixodes pacificus* (Acari: Ixodidae) nymphs infected with *Borrelia burgdorferi* at different spatial scales in California. *Journal of Parasitology* 85, 824–831.

Telford, S. R. III, Mather, T. N., Adler, G. H. & Spielman, A. (1990). Short-tailed shrews as reservoirs of the agents of Lyme disease and human babesiosis. *Journal of Parasitology* 76, 681–683.

Telford, S. R. III, Mather, T. N., Moore, S. L., Wilson, M. L. & Spielman, A. (1988). Incompetence of deer as reservoirs of the Lyme disease spirochete. *American Journal of Tropical Medicine and Hygiene* 39, 105–109.

Toutoungi, L. N., Gern, L., Aeschlimann, A. & Debrot, S. (1991). A propos du genre *Pholeoixodes*, parasite des carnivores en Suisse. *Acarologia* 32, 311–328.

Toutoungi, L., Aeschlimann, A. & Gern, L. (1993). Biology of immature stages of *Pholeoixodes hexagonus* under laboratory conditions. *Experimental and Applied Acarology* 17, 655–662.

Tsao, J. I., Wootton, J. R., Bunikis, J., *et al.* (2004). An ecological approach to preventing human infection: vaccinating wild mouse reservoirs intervenes in the Lyme disease cycle. *Proceedings of the National Academy of Sciences of the USA* 101, 18159–18164.

Ullmann, A. J., Gabitzsch, E. S., Schulze, T. L., Zeidner, N. S. & Piesman, J. (2005). Three multiplex assays for detection of *Borrelia burgdorferi* sensu lato and *Borrelia miyamotoi* sensu lato in field-collected *Ixodes* nymphs in North America. *Journal of Medical Entomology* 42, 1057–1062.

Van Dam, A. P., Kuiper, H., Vos, K., *et al.* (1993). Different genospecies of *Borrelia burgdorferi* are associated with distinct clinical manifestations of Lyme borreliosis. *Clinical Infectious Disease* 17, 708–717.

Van Dam, A. P., Oei, A., Jaspars, R., *et al.* (1997). Complement-mediated serum sensitivity among spirochetes that cause Lyme disease. *Infection and Immunity* 65, 1228–1236.

Varela, A. S., Luttrell, M. P., Howerth, E. W., *et al.* (2004). First culture isolation of *Borrelia lonestari*, putative agent of southern tick-associated rash illness. *Journal of Clinical Microbiology* 42, 1163–1169.

Vasiliu, V., Herzer, P., Roessler, D., Lehnert, G. & Wilske, B. (1998). Heterogeneity of *Borrelia burgdorferi* sensu lato demonstrated by an *ospA*-type specific PCR in synovial fluid from patients with Lyme arthritis. *Medical Microbiology and Immunology* 187, 97–102.

Walker, A. R. (2001). Age structure of a population of *Ixodes ricinus* (Acari: Ixodidae) in relation to its seasonal questing. *Bulletin of Entomological Research* 91, 69–78.

Wang, G., Van Dam, A. P. & Dankert, J. (1999*a*). Phenotypic and genetic characterization of a novel *Borrelia burgdorferi* sensu lato isolate from a patient with Lyme borreliosis. *Journal of Clinical Microbiology* 37, 3025–3028.

Wang, G., Van Dam, A. P., Le Fleche, A., *et al.* (1997). Genetic and phenotypic analysis of *Borrelia valaisiana* sp. nov. (*Borrelia* genomic groups VS116 and Ml9). *International Journal of Systematic Bacteriology* 47, 926–932.

Wang, G., Van Dam, A. P., Schwartz, I. & Dankert, J. (1999*b*). Molecular typing of *Borrelia burgdorferi* sensu lato: taxonomic, epidemiological and clinical implications. *Clinical Microbiology Review* 12, 633–635.

Weber, K. & Pfister, H.-W. (1993). History of Lyme borreliosis in Europe. In *Aspects of Lyme Borreliosis*, eds. Weber, K. & Burgdorfer, W., pp. 1–20. Heidelberg, Germany: Springer-Verlag.

Weisbrod, A. R. & Johnson, R. C. (1989). Lyme disease and migrating birds in the Saint Croix River Valley. *Applied and Environmental Microbiology* 55, 1921–1924.

Westrom, D. R., Lane, R. S. & Anderson, J. R. (1985). *Ixodes pacificus* (Acari: Ixodidae): population dynamics and distribution on Columbian black-tailed deer (*Odocoileus hemionus columbianus*). *Journal of Medical Entomology* 22, 507–511.

Will, G., Jauris-Heipke, S., Schwab, E., *et al.* (1995). Sequence analysis of *ospA* genes shows homogeneity within *Borrelia burgdorferi* sensu stricto and *Borrelia afzelii* strains but reveals major subgroups within the *Borrelia garinii* species. *Medical Microbiology and Immunology* 184, 73–80.

Wilske, B., Busch, U., Eiffert, H., *et al.* (1996). Diversity of OspA and OspC among cerebrospinal fluid isolates of *Borrelia burgdorferi* sensu lato from patients with neuroborreliosis in Germany. *Medical Microbiology and Immunology* 184, 195–201.

Wilske, B., Jauris-Heipke, S., Lobentanzer, R., *et al.* (1995). Phenotypic analysis of outer surface protein C (OspC) of *Borrelia burgdorferi* sensu lato by monoclonal antibodies: relationship to genospecies and OspA serotype. *Journal of Clinical Microbiolology* 33, 103–109.

Wilske, B., Preac-Mursic, V., Göbel, B. U., *et al.* (1993). An OspA serotyping system for *Borrelia burgdorferi* based on reactivity with monoclonal antibodies and OspA sequence analysis. *Journal of Clinical Microbiolology* 31, 340–350.

Wilson, M. L. (1986). Reduced abundance of adult *Ixodes dammini* (Acari: Ixodidae) following destruction of vegetation. *Journal of Economic Entomology* **79**, 693–696.

Wilson, M. L., Adler, G. H. & Spielman, A. (1985). Correlation between deer abundance and that of the deer tick *Ixodes dammini* (Acari: Ixodidae). *Annals of the Entomological Society of America* **78**, 172–176.

Wilson, M. L., Telford, S. R. III, Piesman, J. & Spielman, A. (1988). Reduced abundance of immature *Ixodes dammini* (Acari: Ixodidae) following elimination of deer. *Journal of Medical Entomology* **25**, 224–228.

Wormser, G. P., Bittker, S., Cooper, D., *et al.* (2001). Yield of large-volume blood cultures in patients with early Lyme disease. *Journal of Infectious Diseases* **184**, 1070–1072.

Wormser, G. P., Dattwyler, R. J., Shapiro, E. D., *et al.* (2006). The clinical assessment, treatment, and prevention of Lyme disease, human granulocytic anaplasmosis, and babesiosis: clinical practice guidelines by the Infectious Diseases Society of America. *Clinical Infectious Diseases* **43**, 1089–1134.

Wormser, G. P., Liveris, D., Nowakowski, J., *et al.* (1999). Association of specific subtypes of *Borrelia burgdorferi* with hematogenous dissemination of early Lyme disease. *Journal of Infectious Diseases* **180**, 720–725.

Wormser, G. P., Nadelman, R. D., Dattwyler, R. J., *et al.* (2000). Practice guidelines for the treatment of Lyme disease. *Clinical Infectious Diseases* **31** (Suppl.), S1–S14.

Wright, S. A., Thompson, M. A., Miller, M. J., *et al.* (2000). Ecology of *Borrelia burgdorferi* in ticks (Acari: Ixodidae), rodents, and birds in the Sierra Nevada foothills, Placer County, California. *Journal of Medical Entomology* **37**, 909–918.

Younsi, H., Postic, D., Baranton, G. & Bouattour, A. (2001). High prevalence of *Borrelia lusitaniae* in *Ixodes ricinus* ticks in Tunisia. *European Journal of Epidemiology* **17**, 53–56.

Yuval, B. & Spielman, A. (1990). Duration and regulation of the development cycle of *Ixodes dammini* (Acari: Ixodidae). *Journal of Medical Entomology* **27**, 196–201.

Zeidner, N. S., Nuncio, M. S., Schneider, B. S., *et al.* (2001). A Portuguese isolate of *Borrelia lusitaniae* induces disease in C3H/HeN mice. *Journal of Medical Microbiology* **50**, 1055–1060.

Zhioua, E., Bouattour, A., Hu, C. M., *et al.* (1999). Infections of *Ixodes ricinus* (Acari: Ixodidae) by *Borrelia burgdorferi* sensu lato in North Africa (Tunisia). *Journal of Medical Entomology* **36**, 216–218.

Zhioua, E., Lebrun, R. A., Ginsberg, H. S. & Aeschlimann, A. (1995). Pathogenicity of *Steinernema carpocapsae* and *S. glaseri* (Nematoda: Steinernematidae) to *Ixodes scapularis* (Acari: Ixodidae). *Journal of Medical Entomology* **32**, 900–905.

Zhu, Z. (1998). Histological observations on *Borrelia burgdorferi* growth in naturally infected female *Ixodes ricinus*. *Acarologia* **39**, 11–22.

12 • Viruses transmitted by ticks

M. LABUDA AND P. A. NUTTALL

INTRODUCTION

Ticks transmit a wide variety of arboviruses (**arthropod-borne** viruses). Tick-borne viruses are found in six different viral families (Asfarviridae, Reoviridae, Rhabdoviridae, Orthomyxoviridae, Bunyaviridae, Flaviviridae) and at least nine genera. Some as yet unassigned tick-borne viruses may belong to a seventh family, the Arenaviridae. With only one exception (African swine fever virus) all tick-borne viruses (as well as all other arboviruses) are RNA viruses. Some tick-borne viruses pose a significant threat to the health of humans (tick-borne encephalitis virus, Crimean–Congo haemorrhagic fever virus) or livestock (African swine fever virus, Nairobi sheep disease virus). This chapter first considers the characteristics of ticks important in virus transmission and then presents an overview of the tick-borne members of different virus families.

TICKS AS VECTORS OF ARBOVIRUSES

Ticks are not insects. The significance of this statement is considered in a review of the marked contrasts between the biology of ticks and that of insects, and the consequences for their potential to transmit micro-organisms (Randolph, 1998). Interestingly, tick-borne viruses are found in all the RNA virus families in which insect-borne members are found, with the exception of the family Togaviridae. Virus–tick–vertebrate host relationships are highly specific, and fewer than 10% of all tick species (Argasidae and Ixodidae) are known to play a role as vectors of arboviruses. However, a few tick species transmit several (e.g. *Ixodes ricinus, Amblyomma variegatum*) or many (*I. uriae*) tick-borne viruses. Transmission is 'biological', the virus replicating within the tick prior to transmission to a vertebrate host; the tick acts as a vector. Many unique features of ticks contribute to their remarkable success as virus vectors. These features are summarized below.

Tick life cycle and longevity

One of the most outstanding features of ticks is their remarkable longevity. The complete life cycle of ticks is usually measured in years, and individual stages can survive long periods without a blood meal (Sonenshine, 1991). Experimental data indicate that virus infections persist in ticks for the duration of the ticks' lifespan (Rehacek, 1965; Davies, Jones & Nuttall, 1986). Ecological and epidemiological data also support the observation that tick-borne virus survival is greatly dependent on persistent infections in tick populations (Blaskovic & Nosek, 1972).

Tick life stages (eggs, larvae, nymphs or adults) readily survive from one year to the next. Ixodid tick species show marked differences in the number of generations completed within a year. The survival strategies of different tick species inhabiting temperate climates loosely reflect the prevailing conditions in which the ticks are found. For example, *Rhipicephalus sanguineus*, which prefers the warmer environment of southern Europe, can have up to two generations in one year and will remain active over winter. In contrast, *I. ricinus*, which has a more northerly distribution and is active in colder climates, generally has two peaks of activity, in the spring and autumn. However, generally in each year *I. ricinus* only feeds once and only passes through one developmental stage, thus taking at least 3 years to complete its life cycle. Ticks that do not find a host in the autumnal activity period will overwinter to become active again the following spring, hence the life cycle can take up to 6 years to complete. This may be regarded as a survival strategy to meet the demands of a harsh climate. Some tick species, such as

Ticks: Biology, Disease and Control, ed. Alan S. Bowman and Patricia A. Nuttall. Published by Cambridge University Press.
© Cambridge University Press 2008.

Dermacentor spp., appear to show an intermediate strategy, with the adults overwintering and one generation being completed in each year.

The survival strategy of ticks is important for the survival of the viruses they transmit. Because of the exceptional longevity of ticks, they can carry tick-borne viruses over prolonged periods of time. As a result, ticks are not only vectors but also excellent reservoir hosts for the viruses they carry.

If the long-term survival of viruses depends on their tick vector, selection must favour infections that have no detrimental effect on the tick. Generally this appears to be the case, although few studies have been published concerning arboviral effects on tick vectors. Differences were not detected in the reproductive output, moulting success and survival of uninfected *R. appendiculatus*, compared with ticks of the same population that were infected with Thogoto virus (THOV) at the larval stage (L. D. Jones, personal communication). However, the salivary glands of partially fed adult female *R. appendiculatus* ticks infected with THOV secreted fluid in vitro at about 75% the rate of controls (Kaufman, Bowman & Nuttall, 2002). The significance of this observation is unknown. Detrimental effects of infection with African swine fever virus (ASFV) on adult *Ornithodoros moubata* ticks have been reported. A significant increase in mortality rates was observed amongst the adult ticks that fed on an infective compared with a normal (uninfective) blood meal (Rennie, Wilkinson & Mellor, 2000). Such reports of adverse effects are exceptional whereas there are several reports that insect-borne viruses can adversely affect their vectors (Turell, 1988).

A specific mode of arboviral persistence in the vector population is via vertical transmission in which virus from the infected female is transmitted via the egg to the succeeding generation. Although evidence from experimental studies of vertical transmission has been recorded for numerous tick-borne viruses, the levels of vertical transmission and filial infection in nature are usually low. Certainly, the high levels of vertical transmission of certain insect-borne viruses associated with stabilized infections of their vectors (Turell, 1988) have not been recorded for any tick-borne viruses. If, as discussed above, tick-borne viruses rely on their vectors for persistence, then any deleterious effects of vertical transmission on ticks may outweigh the advantages to the virus. The balancing of costs and benefits of vertical transmission, together with the gains from co-feeding and non-viraemic transmission (see Chapter 10) may explain why vertical transmission is common among tick-borne viruses but occurs at a low level.

Non-viraemic transmission has another important implication for vertical transmission. As mentioned above, vertical transmission is common among tick-borne viruses but occurs at an apparently low level. This has led to claims that vertical transmission is not a significant factor in the ecology and epidemiology of tick-borne viruses (Rehacek, 1965). A laboratory study of low-level tick-borne encephalitis virus (TBEV) infection in a population of larval ticks (detectable only by polymerase chain reaction) demonstrated that the infection was amplified by non-viraemic co-feeding to yield a significant number of infected nymphal ticks (Labuda *et al.*, 1993). Opportunities for such amplification of vertically transmitted infections occur in the field where a low prevalence of TBEV infection in *I. ricinus* larvae has been documented (Danielova *et al.*, 2002). Similar results have been reported for Colorado tick fever virus (CTFV) (Calisher, 2001) and Crimean–Congo haemorrhagic fever virus (CCHFV) (Gordon, Linthicum & Moulton, 1993), whereas a higher filial infection prevalence was reported for ASFV (Rennie, Wilkinson & Mellor, 2001). Because larvae that hatch from the same egg mass often quest in clusters, several of them may attach to the same individual host. This behaviour provides many opportunities for amplification of vertically acquired tick-borne virus infections, in the vector population, through non-viraemic transmission between co-feeding larvae (Labuda *et al.*, 1993; Danielova *et al.*, 2002). As a result of such amplification, vertical transmission might be the difference between survival and extinction of certain tick-borne viruses in nature.

Host-finding and host preferences in the virus transmission cycle

The most important requirement of ticks to accomplish their life cycle is to find a suitable vertebrate host. Some tick species prefer to feed on a particular vertebrate host species, whereas others feed on a range of hosts. As a rule, a successful tick vector species has a wide host range but there are notable exceptions, e.g. populations of the seabird tick *I. uriae* often feed year after year on the same seabird species. The number of hosts on which a tick feeds during its lifetime varies depending on the species of tick and its preferences and also on whether it is a one-, two- or three-host tick. In Europe, most tick species are three-host (e.g. *Ixodes* spp. *Rhipicephalus* spp., *Dermacentor* spp. and *Amblyomma* spp.), of which the larva, nymph and adult each feed on a separate host and are free-living between feeding periods. Some species (e.g. *Hyalomma* spp.) are described as two-host ticks, where the larva and nymph feed on the same host, but the adult feeds on a different one. Thus, there is not only the possibility for transmission of viruses between hosts of the same species, but because of the range of potential hosts, there is also the important possibility of disease transfer between vertebrate species including humans.

Owing to their feeding preferences, ticks restrict the potential range of hosts for a virus. For example, *I. ricinus* has a very wide host range, including many species of mammals, birds and even lizards. In spite of such a variety of hosts, the majority of *I. ricinus* ticks feed on only a few mammalian species and tick infestation is frequently limited to only a part of the host population. Typically, the overdispersed distribution of ticks on their hosts results in a significant proportion of the hosts carrying large numbers of ticks that feed together.

Overdispersion arises from the non-random distribution of questing ticks and host genetic, behavioural and immunological heterogeneities. These factors determine the differential probabilities of an individual host picking up ticks. In Central Europe, the most abundant rodent hosts of immature *I. ricinus* are frequently yellow-necked mice (*Apodemus flavicollis*) and bank voles (*Clethrionomys glareolus*). Coincident aggregated distributions of *I. ricinus* larvae and nymphs on these species, in western Slovakia, resulted in 20% of the animals feeding about three-quarters of both the larval and nymphal populations (Randolph *et al.*, 1999). As a result, the number of larvae exposed to infection by feeding alongside potentially infected nymphs was doubled compared with the null hypothesis of independent distribution of these two tick stages. The observed pattern of co-feeding is typical for *I. ricinus* larvae and nymphs and not for other tick species occurring in western Slovakia. Overall, only 3% of *I. ricinus* nymphs were recorded on hosts that were not carrying at least one larva. In contrast, as many as 28% of *D. reticulatus* nymphs were found on hosts (from the same area and at the same time) that were not feeding with larvae of the same species. As both species are 'competent' vectors of TBEV in the laboratory, the virus can potentially be exchanged between ticks of each species where they coexist. However, these tick species make differential use of voles and mice as hosts. More *D. reticulatus* have been recorded on *C. glareolus* than on *A. flavicollis*, while *I. ricinus* showed the reverse host association (Randolph *et al.*, 1999). These particular patterns of tick infestation on transmission-competent rodent hosts help provide a quantitative explanation for the focality

of TBEV, as described by Randolph (see Chapter 2). The concept was first expounded by Pavlovsky as the 'nidality' of TBEV in Euroasia (Zilber & Soloviev, 1946). Focality and nidality reflect the fact that TBEV survival in nature results from the critically balanced relationships within the virus–vector–host triangle in the given environment.

Primary and secondary vector species

Certain tick species are crucial for the maintenance of virus transmission cycles and are considered primary vectors. All the ecological and physiological characteristics of such tick species appear to be well suited for maintaining particular tick-borne viruses in nature. Other species may be involved as secondary vectors. For example, TBEV is maintained in Europe primarily by *I. ricinus* ticks and in Asia by *I. persulcatus*. However, it seems that all competent tick species occurring in sufficiently high numbers and having a sympatric distribution with the primary tick vector species may become infected and subsequently transmit TBEV. Although experimental studies have shown that numerous tick species are competent vectors, their ecological roles vary. For example, the vector competence of *I. hexagonus* for TBEV has been demonstrated in the laboratory, including transmission of TBEV to hedgehogs, the principal host of this tick species, and TBEV has been isolated from field-collected *I. hexagonus*. *Ixodes arboricola*, a bird tick, was shown to be a competent vector in the laboratory. Similarly, *Haemaphysalis concinna*, *H. inermis* and *H. punctata* are competent vectors, and TBEV has been isolated from field-collected specimens (Gresikova & Calisher, 1988). Yet, in the natural situation only the two primary vectors are able to perpetuate efficiently TBEV transmission cycles as documented many times over the huge territory in which TBEV is endemic. In contrast to the many tick species capable of transmitting TBEV, the epizootology of louping ill virus (LIV) implicates *I. ricinus* as the exclusive vector even though *D. reticulatus* and *H. punctata* are present in the UK where the virus is endemic.

In the endemic area of Kyasanur forest disease, 36 tick species have been recorded. Of the 15 species of *Haemaphysalis* present in the area, Kyasanur forest disease virus (KFDV) has been isolated from *H. spinigera* and eight other species which paints a picture even more complicated than that for TBEV. However, the record for the highest number of infected tick species goes to CCHFV, isolated from at least 31 different species and sub-species.

Taking a blood meal

Ticks are pool feeders, yet another important feature distinguishing them from many blood-feeding insects. To create a specific feeding pool in the dermis, ticks attach themselves to the host skin using their chelicerae and toothed hypostome. Ixodid ticks may feed for a few days or up to 2 weeks, cementing their mouthparts into the skin. Only during the last day of attachment is the majority of the blood meal taken up (Sonenshine, 1991). Such a profound physical and chemical assault on the host provokes the host's haemostatic, inflammatory and immune responses. Despite the massive armoury of rejection mechanisms, the tick manages to remain attached and achieve engorgement. The success of the tick is based upon a pharmacy of chemicals located in its complex salivary glands and secreted, in tick saliva, into the feeding pool (Nuttall, 1999; see also Chapter 4). The main route of virus transmission by infected ticks is via saliva secreted during feeding (Kaufman & Nuttall, 1996). Virus transmitted by this route enters a skin site that is profoundly altered by the effects of tick saliva (Titus & Ribeiro, 1990). Tick saliva possesses pharmacologically active substances that have antihaemostatic, vasodilatory, anti-inflammatory, antinociceptive and immunosuppressive activities (see Chapter 9). Modulation of host responses at the feeding site not only allows the attached ticks to feed but also increases the rate at which other co-feeding ticks acquire infection from the same host. Thus tick blood-feeding inadvertently provides an advantage to tick-borne viruses in that modulation of the site of infection promotes virus transmission and survival (see Chapter 10).

Tick competence, digestion and moulting

In virus–vector systems other than those mentioned above, we do not observe a broad range of vector competence among tick species. For example, experiments comparing different methods of infecting ticks with Dhori virus and Dugbe virus have demonstrated the presence of a 'gut barrier' to virus infection. Dhori virus and Dugbe virus replicated in *R. appendiculatus* after inoculation into the tick haemocoel, a route of infection bypassing the gut. However, neither virus established an infection when the ticks were fed on an infective blood meal by the per-oral route of infection (Steele & Nuttall, 1989). The presence of a gut barrier in ticks indicates that there is a specific interaction between virus (imbibed in the blood meal) and midgut cells. Although the nature of the gut barrier has not been determined, it appears to vary for different virus–tick systems.

The susceptibility of arthropod midgut cells to virus infection is one of the most important determinants of vector competence. Understanding the determinants of vector competence is important in explaining why certain tick-borne viruses have many tick vectors (e.g. CCHFV) whereas others have few (e.g. NSDV).

The initial stages of virus infection are likely to differ markedly for ticks and insects, and may be the principal reason why tick-borne viruses are rarely, if ever, transmitted by insects. Thus, viruses entering the tick midgut are exposed to different environmental conditions compared with those existing in, for example, the mosquito midgut. This is because ticks are heterophagous, i.e. blood meal digestion is primarily an intracellular process occurring within midgut cells (Sonenshine, 1991). In contrast, the blood meal of insect vectors is digested extracellularly (within the midgut lumen). Studies with mosquitoes and La Crosse virus (Bunyaviridae, Bunyavirus) indicate that cleavage of a protein exposed on the surface of virus particles (virions) is necessary to initiate vector infection (Ludwig et al., 1991). The necessary proteolytic conditions apparently occur in the midgut of mosquitoes, but such conditions may not be present in the midgut of heterophagic ticks.

If the method of blood meal digestion in ticks exerts a strong selective pressure on arboviruses, the structure of the outer surface of tick-borne viruses is likely to differ significantly from that of related insect-borne viruses (given that virion–cell surface interactions are the first phase of infection). Presently, this hypothesis cannot be tested as there are insufficient data, for arboviruses, on the three-dimensional structure of virions and the nature of virus receptors. However, comparative sequence data have revealed significant differences in the virion surface proteins of midge-transmitted orbiviruses (e.g. bluetongue virus) and the tick-transmitted orbivirus Broadhaven virus (Iwata, Yamagawa & Roy, 1992); and in the surface envelope protein of tick-borne flaviviruses (e.g. TBEV) which contains a unique region of six continuous amino acids not found in the envelope protein of mosquito-borne flaviviruses, e.g. yellow fever virus (Shiu et al., 1991). Recently, much attention has been given to the interaction of viral surface proteins with glycosaminoglycans, which are largely distributed on cell surfaces but vary with respect to their composition and quantity. For TBEV, it has been proposed that the affinity of the viral surface for glycosaminoglycan molecules such as heparin sulphate may be an important determinant of tissue tropism (Mandl et al., 2001). Attention has been also given to vector molecules. A lectin, named Dorin M, has been identified in the haemocytes and plasma of *O. moubata* ticks. Since these lectin types were reported to function as non-self recognizing molecules, Dorin M may play a role in innate immunity and pathogen transmission (Kovar, Kopacek & Grubhoffer, 2000; and see Chapter 6). It remains to be determined whether and to what degree such molecules govern the specific adaptations of arboviruses to either tick or insect vectors.

As a result of the feeding behaviour of ticks, viruses must persist from one instar to the next in order to be transmitted to a vertebrate host. This means that the 'extrinsic incubation period', which is so important in determining the transmission dynamics of insect-borne viruses (Turell, 1988), is not significant for virus transmission by ixodid ticks because it is unlikely to exceed the comparatively long moulting period. However, the extrinsic incubation period is important in terms of virus survival, and in the rare cases of interrupted feeding by ticks (see next section).

In relation to virus survival during the extrinsic incubation period, the histolytic enzymes and tissue replacement associated with moulting provide a potentially hostile environment (Balashov, 1972). Several authors have suggested that the dynamics of viral replication within the tick reflect these events: a fall in virus titre, followed by an increase in the titre as the virus infects and replicates in replacement tick tissues (Rehacek, 1965; Burgdorfer & Varma, 1967). However, the replication of some viruses (e.g. THOV in *R. appendiculatus*, Langat virus in *I. ricinus*) is not obviously correlated with any particular stage of the moulting period (Varma & Smith, 1972; Davies *et al.*, 1986). These conflicting observations can be explained by the variety of infection strategies adopted by tick-borne viruses. The apparent targeting of specific cell types, tissues or organs may reflect mechanisms by which different tick-borne viruses have adapted to survive the moulting period, viz. by establishing an infection in at least one cell type not involved in histolysis.

An additional factor bearing on the extrinsic incubation period is the resorption and regeneration of salivary glands during moulting (see Chapter 3). Hence virus infection of the salivary glands is likely to be a relatively late event in the infection cycle within ticks. A few reports describe virus in the salivary glands but the timing of infection varies. TBEV and Powassan virus infect the salivary glands prior to feeding; presumably they can be transmitted to the vertebrate host as soon as feeding is initiated (Rehacek, 1965; Chernesky & McLean, 1969). In contrast, THOV and Dugbe virus accumulate in the salivary glands after feeding commences (Booth *et al.*, 1989, 1991), although in ticks infected in the preceding stadium, THOV is present in the salivary glands prior to blood-feeding (Kaufman & Nuttall, 2003).

Interrupted feeding

The duration of the extrinsic incubation period (see preceding section) is also important when ticks are interrupted in their feeding on a host. For example, a host may be killed or it may die from a virulent tick-borne virus infection. Partially fed infected ticks can detach from their deceased host and may successfully reattach and feed on a new host. The consequences of interrupted feeding were investigated experimentally with THOV which kills hamsters before its nymphal or adult vector (*R. appendiculatus*) has completed engorgement (Wang & Nuttall, 2001). Ticks that had partially fed on infected hamsters were able to transmit the infection to new uninfected hosts on which they completed engorgement. The periods between feeds varied from 7 to 28 days, presumably within the extrinsic incubation period.

Although interrupted feeding of infected ticks is comparatively rare in nature, it may contribute to outbreaks of rapid and fatal tick-borne viral diseases such as Kyasanur forest disease in monkeys (Sreenivasan, Bhat & Rajagopalan, 1979). Additionally interrupted feeding provides an increased risk of transmission to humans and domestic animals during slaughter and game hunting.

TICK-BORNE ARBOVIRUSES

Arboviruses are taxonomically a heterogenous group of vertebrate viruses unified by a unique ecological feature, namely transmission by haematophagous arthropods in which they replicate. Intriguingly, all arboviruses (with one exception) are RNA viruses. The only DNA arbovirus (African swine fever virus, see next section) is transmitted by argasid ticks (*Ornithodoros* spp.). At least 500 arboviruses are registered (Karabatsos, 1985). About half of them are mosquito-borne and approximately one-third are transmitted by ticks. Tick-borne viruses belong to six virus families (Tables 12.1–12.5). Each family is characterized by a unique genome organization and replication strategy. Thus tick-borne virus transmission has evolved independently at least six times during the phylogenetic period that can be traced today. Among virus families containing arboviruses, only the Togaviridae (genus *Alphavirus*) does not contain any tick-borne members, although some mosquito-borne alphaviruses (e.g. Sindbis virus) have occasionally been isolated from ticks (Gresikova *et al.*, 1978).

Table 12.1 *Tick-borne viruses of the families Asfarviridae, Rhabdoviridae and Orthomyxoviridae*

Virus	Tick	
Family, genus, species and strain[a]	Main vector species[b]	Geographical distribution
FAMILY Asfarviridae		
Genus *Asfivirus*		
African swine fever virus (3)		
African swine fever virus – Ba71V, LIL20/1, LIS57	*O. moubata, O. erraticus*	Sub-Saharan Africa, southern Europe[c], South America[c]
FAMILY Orthomyxoviridae		
Genus *Thogotovirus*		
Thogoto virus (2)		
Thogoto virus	*Rhipicephalus, R.* (*Boophilus*), *Hyalomma* spp.; *A. variegatum*	Central and East Africa, southern Europe
Araguari virus	unknown	South America
Dhori virus (2)		
Batken virus	*Hy. marginatum*	Kyrgyzstan
Dhori virus	*Hy. dromedarii, Hy. marginatum*	India, eastern Russia, Egypt, southern Portugal
FAMILY Rhabdoviridae		
Genus *Vesiculovirus*		
Isfahan virus		
Isfahan virus	*Hy. asciaticum*	Turkmenistan
Unassigned family members:		
Kern Canyon group (4)		
Barur virus	*H. intermedia*	India, Kenya, Somalia
Fukuoka virus	unknown	Japan
Kern Canyon virus	unknown	USA (California)
Nkolbisson virus	unknown	Cameroon
Sawgrass virus group (3)		
Connecticut virus	unknown	USA (Connecticut)
New Minto virus	unknown	USA (Alaska)
Sawgrass virus	*D. variabilis, H. leporispalustris*	USA (Florida)

[a] Virus species name in italics corresponds to that of the prototype virus of the species; number in parenthesis indicates number of viruses (serotypes/strains/subtypes) if more than one (listed in succeeding rows). Virus named in parenthesis is considered synonymous.

[b] Argasid species: *Argas, Carios* (C.), *Ornithodoros* (O.). Ixodid species: *Amblyomma* (A.), *Dermacentor* (D.), *Haemaphysalis* (H.), *Hyalomma* (Hy.), *Ixodes* (I.), *Rhipicephalus* (R.).

[c] Epizootics followed by eradication.

Source: After Fauquet *et al.* (2005) unless otherwise stated.

Table 12.2 *Tick-borne viruses of the family Reoviridae*

Virus	Tick	
Genus, species and strain	Main vector species	Geographical distribution
Genus *Orbivirus*		
Chenuda virus (7)		
Baku virus	*C. maritimus*	Caspian Sea, Uzbekistan
Chenuda virus	*Argas hermanni*	Egypt, Uzbekistan
Essaouria virus	*C. maritimus*	Morocco
Great Saltee Island virus – GS80-5, -6	*C. maritimus*	Ireland (south-east)
Kala Iris virus	*C. maritimus*	Morocco
Moellez virus	*C. maritimus*	France
Chobar Gorge virus (2)		
Chobar Gorge virus	*Carios* spp.	Nepal
Fomede virus	unknown (from bats)	Guinea
Great Island virus (33)		
Above Maiden virus	*I. uriae*	Scotland (Isle of May)
Aniva	*I. uriae*	Russia (north-east)
Arbroath virus	*I. uriae*	Scotland
Bauline virus	*I. uriae*	Canada (Newfoundland)
Broadhaven virus	*I. uriae*	Scotland (St Abb's Head)
Cape Wrath virus	*I. uriae*	Scotland
Colony virus	*I. uriae*	Scotland (Isle of May)
Colony B North virus	*I. uriae*	Scotland (Isle of May)
Ellidaey virus	*I. uriae*	Iceland
Foula virus	*I. uriae*	Scotland (Shetland Islands)
Great Island virus	*I. uriae*	Canada (Newfoundland)
Great Saltee Island virus – GS80-4, -7, -8	*I. uriae*	Ireland (southeast)
Grimsey virus	*I. uriae*	Iceland
Inner Fame virus	*I. uriae*	England (Fame Islands)
Kenai virus	*I. signatus*	USA (Alaska)
Lundy virus	*I. uriae*	England
Maiden virus	*I. uriae*	Scotland (Isle of May)
Mill Door virus	*I. uriae*	Scotland (Isle of May)
Mykines virus	*I. uriae*	Denmark (Faroe Islands)
North Clett virus	*I. uriae*	Scotland (Isle of May)
North End virus	*I. uriae*	England (Lundy)
Nugget virus	*I. uriae*	Australia (Macquarie Island)
Okhotskiy virus	*I. uriae, I. signatus*	Russia (east, north-west), USA (Alaska)
Poovoot virus	*I. uriae*	USA (Alaska)
Røst Islands virus	*I. uriae*	Norway (Lofoten)
St Abb's Head virus	*I. uriae*	Scotland
Shiant Islands virus	*I. uriae*	Scotland
Thormódseyjarklettur virus	*I. uriae*	Iceland
Tillamook virus	*I. uriae*	USA (California, Oregon)
Tindholmur virus	*I. uriae*	Denmark (Faroe Islands)

(cont.)

Table 12.2 (*cont.*)

Virus	Tick	
Genus, species and strain	Main vector species	Geographical distribution
Vaerøy virus	*I. uriae*	Norway (Lofoten)
Wexford virus	*I. uriae*	Ireland (south-east)
Yaquina Head virus	*I. uriae*	USA (Alaska, Oregon)
Kemerovo virus[a] (4)		
Kemerovo virus	*I. persulcatus, I. ricinus*	Russia, Slovakia
Kharagysh virus	*I. ricinus*	Moldova
Lipovnik virus	*I. ricinus*	Slovakia, Czech Republic
Tribeč virus	*I. ricinus, H. punctata*	Slovakia, Italy, Belorussia
Mono Lake virus[b] (3)		
Huacho virus	*C. amblus*	Peru
Mono Lake virus	*Argas cooleyi*	USA (California)
Sixgun City virus	*Argas cooleyi*	USA (Colorado, Texas)
Wad Medani virus (2)		
Seletar virus	*R. (Boophilus) microplus*	Malaysia, Singapore
Wad Medani virus	*R. sanguineus, Hyalomma* spp.	East Africa, Asia, Jamaica
Lake Clarendon virus[c]	*Argas robertsi*	Australia (Queensland)
Matucare virus[c]	*O. boliviensis*	Bolivia
St Croix River virus[c]	*I. scapularis*	USA
Genus *Coltivirus*		
Colorado tick fever virus (2)		
California hare virus – California S6-14-03	unknown (from a hare)	USA (California)
Colorado tick fever virus – Florio	*D. andersoni, D. occidentalis, D. albipictus*	USA
Eyach virus (3)		
Eyach virus – France 577	*I. ricinus, I. ventalloi*	France
Eyach virus – France 578		France
Eyach virus – Germany	*I. ricinus*	Germany

See Table 12.1 legend.

[a] Listed under *Great Island virus* by Fauquet *et al.* (2005).

[b] Listed under *Chenuda viru*s by Fauquet *et al.* (2005) .

[c] Tentative species.

Source: After Fauquet *et al.* (2005).

Given that arboviruses represent the largest biological group of vertebrate viruses, it is reasonable to assume that their lifestyle is a successful one. This is despite the fact that arboviruses, during their evolution, have solved a very specific problem: how to replicate successfully in two phylogenetically distinct systems, swapping between an arthropod cell and a vertebrate cell. Arboviruses have achieved this irrespective of whether they have a RNA genome that is double-stranded or single-stranded, segmented or non-segmented, or of positive or negative polarity.

Arbovirus groups having insect-borne members frequently also include tick-borne viruses. Insect-borne arboviruses outnumber tick-borne viruses in the Bunyaviridae and Flaviviridae families, but tick-borne viruses are exclusive to the nairoviruses, Uukuniemi virus group of phleboviruses, and coltiviruses. Interestingly, of

Table 12.3 *Tick-borne viruses of the family Bunyaviridae*

Virus	Tick	
Genus, species and strain	Main vector species	Geographical distribution
Genus *Orthobunyavirus*		
Estero Real virus		
Estero Real virus – K329	*O. tadaridae*	Cuba
Tete virus (6)		
Bahig virus	*Hyalomma* spp.	Egypt, Cyprus, Italy
Batama	unknown	Africa
Matruh virus	*Hy. marginatum*	Egypt, Italy
Tete virus SAAn 3518	unknown	South Africa, Nigeria
Tsuruse virus	unknown	Japan
Weldona virus	(*Culicoides* spp.)	USA (Colorado)
Genus *Nairovirus*		
Crimean–Congo haemorrhagic fever virus (4)		
Crimean–Congo haemorrhagic fever virus – AP92	*R. bursa*	Greece
Crimean-Congo haemorrhagic fever virus – C68031	*Hy. marginatum*; isolated from many ixodid spp.	Many countries in Asia and Africa; parts of Europe (e.g. Albania, Bulgaria)
Hazara virus	*I. redikorzevi*	Pakistan
Khasan virus	*H. longicornis*	Russia
Dera Ghazi Khan virus (6)		
Abu Hammad virus	*Argas hermanni*	Egypt, Iran
Abu Mina virus	unknown	Egypt
Dera Ghazi Khan virus – JD254	*Hy. dromedarii*	Pakistan
Kao Shuan virus	*Argas robertsi*	Taiwan, Australia, Java
Pathum Thani virus	*Argas robertsi*	Thailand
Pretoria virus	*Argas africolumbae*	South Africa
Dugbe virus (2)		
Dugbe virus	*A. variegatum*	sub-Saharan Africa
Nairobi sheep disease virus (Ganjam virus)	*R. appendiculatus, H. intermedia* (*Culicoides* spp., mosquitoes)	East and Central Africa, India
Hughes virus (10)		
Farallon virus	*C. denmarki*	USA (California, Oregon)
Fraser Point virus		
Great Saltee virus	*C. maritimus, I. uriae*	Ireland
Hughes virus	*C. denmarki*	USA (Florida), Trinidad, Venezuela, Cuba
Puffin Island virus	*C. maritimus*	Wales
Punta Salinas virus	*C. amblus, Argas arboreus*	Peru
Raza virus	*C. denmarki*	USA, Mexico
Sapphire II virus	*Argas cooleyi*	USA (Montana)
Soldado virus	*C. capensis, C. denmarki, C. maritimus, A. loculosum*	Trinidad, Hawaii, Seychelles, Ethiopia, South Africa, Morocco, France, Wales
Zirqa virus	*C. muesebecki*	Persian Gulf
Qalyub virus (4)		
Bakel virus	unknown	Senegal
Bandia virus	*O. sonrai*	Senegal
Omo virus	unknown (from *Mastomys*)	Ethiopia
Qalyub virus	*O. erraticus*	Egypt

(cont.)

Table 12.3 (*cont.*)

Virus	Tick	
Genus, species and strain	Main vector species	Geographical distribution
Sakhalin virus (6)		
Avalon virus (Paramushir virus)	*I. uriae, I. signatus*	Canada (Newfoundland), Russia (east)
Clo Mor virus	*I. uriae*	Scotland
Kachemak Bay virus	*I. signatus*	USA (Alaska)
Sakhalin virus	*I. uriae, I. signatus*	Russia (northeast, east)
Taggert virus	*I. uriae*	Australia (Macquarie Island)
Tillamook virus	*I. uriae*	USA (California, Oregon)
Genus *Phlebovirus*		
Uukuniemi virus (13)		
EgAn 1825-61 virus	unknown	Africa
Fin V 707 virus	unknown	Europe
Grand Arbaud virus	*Argas reflexus, Argas hermanni*	France
Manawa virus	*Argas abdussalami, Rhipicephalus* spp.	Pakistan
Murre virus	unknown	
Oceanside virus	*I. uriae*	USA (California, Oregon)
Ponteves virus	*Argas reflexus*	France
Precarious Point virus	*I. uriae*	Australia (Macquarie Island)
RML 105355 virus	*I. uriae*	USA (Pribilof Islands)
Saint Abb's Head virus	*I. uriae*	Scotland
Tunis virus	*A. reflexus*	Tunisia
Uukuniemi virus – S 23	*I. ricinus, I. uriae*	Europe, Lithuania
Zaliv Terpeniya virus	*I. uriae, I. signatus*	Russia (northwest, east), France (Brittany), England (Lundy), USA (California, Oregon)
Unassigned family members:		
Bhanja group (3)		
Bhanja virus	*H. punctata* + many others	Africa, Asia, southern Europe
Forecariah virus	*R. (Boophilus) geigyi*	Guinea
Kismayo virus	*R. pulchellus*	Somalia
Kaisodi group (3)		
Kaisodi virus	*H. spinigera*	India
Lanjan virus	*Dermacentor, Haemaphysalis* spp.	Malaysia
Silverwater virus	*H. leporispalustris*	Canada (Ontario)
Upolu group (2)		
Aransas Bay virus	*C. capensis*	USA (Texas)
Upolu virus	*C. capensis*	Australia
Ungrouped viruses (7)		
Chim virus	*R. turanicus*	Uzbekistan
Issyk-Kul virus (Keterah virus)	*Argas* spp.	Kyrgyzstan, Malaysia
Lone Star virus	*A. americanum*	USA (Kentucky)
Razdan virus	*D. marginatus*	Armenia
Sunday Canyon virus	*Argas cooleyi*	USA (Texas)
Tamdy	*Hy. asiaticum*	Turkmenistan, Uzbekistan
Wanowrie virus	*Hyalomma* spp.	Egypt, India, Iran, Sri Lanka

See Table 12.1 legend.
Source: After Fauquet *et al.* (2005).

Table 12.4 *Tick-borne viruses of the family Flaviviridae, genus* Flavivirus

Virus	Tick	
Group and species	Main vector species	Geographical distribution
Gadgets Gully virus		
Gadgets Gully virus	*I. uriae*	Australia (Macquarie Island)
Kadam virus		
Kadam virus	*R. pravus*	Saudi Arabia, Uganda
Karshi virus		
Karshi virus	*O. tholozani, Hy. anatolicum*	Kazakhstan, Uzbekistan
Kyasanur Forest disease virus (2)		
Alkhurma virus	unknown	Saudi Arabia
Kyasanur Forest disease virus	*H. spinigera*[‡]	India
Langat virus		
Langat virus	*I. granulatus*	Malaysia
Meaban virus		
Meaban virus	*C. maritimus*	France
Omsk haemorrhagic fever virus		
Omsk haemorrhagic fever virus	*D. reticulatus*[‡]	Western Siberia
Powassan virus (2)		
Deer tick virus	*I. scapularis*	USA (northeast), Canada (Ontario)
Powassan virus	*I. cookei*	Canada, USA, Russia
Royal Farm virus		
Royal Farm virus	*Argas hermanni*	Afghanistan
Saumarez Reef virus		
Saumarez Reef virus	*C. capensis, I. eudyptidis*	Australia (Tasmania)
Tick-borne encephalitis virus (4)		
Eastern tick-borne encephalitis virus		
– Far Eastern subtype	*I. persulcatus*[‡]	Northern Asia
– Siberian subtype	*I. persulcatus*	Siberia
Louping ill virus[†]		
– British subtype	*I. ricinus*	England, Scotland, Wales, Norway
– Irish subtype	*I. ricinus*	Ireland
– Spanish subtype	*I. ricinus*	Spain
Turkish encephalitis virus		
– Greek subtype	*I. ricinus*	Greece
– Turkish subtype	*I. ricinus*	Turkey
Western tick-borne encephalitis virus	*I. ricinus*[‡]	Northern Europe
Tyuleniy virus		
Tyuleniy virus	*I. uriae*	Russia (east, northwest), USA (Alaska, Oregon), Norway (Røst Islands)
West Nile virus[‡‡]		
West Nile virus	*C. maritimus, Argas hermanni, Hyalomma* spp.	Azerbaijan, Russia, Turkmenistan
Unassigned family member:		
Ngoye virus	*R. evertsi*	Senegal

See Table 12.1 legend.

[†] Negishi virus is now recognized as a laboratory contaminant of louping ill virus.

[‡] Numerous other tick species possibly involved; however, the given species is the main vector.

[‡‡] Mosquito-borne virus.

Source: After Fauqust *et al.* (2005), modified according to Grard *et al.* (2006).

Table 12.5 *Unassigned tick-borne viruses*

Virus	Tick	
Group, virus name	Main vector species	Geographical distribution
Nyaminini virus group (3)		
Hirota virus	*C. capensis*	Japan (Aomatsushima Island)
Midway virus	*C. capensis, C. denmarki*	Hawaii
Nyaminini virus	*Argas arboreus, Argas walkerae*	Egypt, Nigeria, South Africa
Quaranfil virus group (2)		
Johnston Atoll virus	*C. capensis, C. denmarki*	Central Pacific Islands, Australia, New Zealand, southwest Africa
Quaranfil virus	*Argas* spp.	Egypt, Nigeria, South Africa, Afghanistan, Iran
Ungrouped (7)[a]		
Aride virus	*A. loculosum*	Seychelles
Caspiy virus	*C. maritimus*	Azerbaijan, Turkmenistan, Uzbekistan
Jos virus	*A. variegatum, R. (Boophilus) decoloratus*	Nigeria, Senegal
Mayes virus	*I. uriae*	Southern Ocean (Kerguelen Archipelago)
Røst Islands virus – NorArV-958	*I. uriae*	Norway
Runde virus – Ru E81, RuE85	*I. uriae*	Norway
Slovakia virus	*Argas persicus*	Slovakia

See Table 12.1 legend.
[a] Not listed in Fauquet *et al.* (2005).
Source: After Fauquet *et al.* (2005).

approximately 200 named tick-borne viruses, 80% are members of the *Orbivirus, Nairovirus, Phlebovirus* and *Flavivirus* genera.

In general, the association between the arthropod vector and the transmitted virus is very intimate and highly specific. Comparatively few arthropod species act as vectors. In fact, fewer than 10% of the known tick species are incriminated as virus vectors and they are mostly found in large tick genera. For argasid ticks these are *Ornithodoros, Carios* and *Argas* and, among ixodid ticks, virus vectors are found mostly in the genera *Ixodes, Haemaphysalis, Hyalomma, Amblyomma, Dermacentor* and *Rhipicephalus* and the subgenus *Rhipicephalus (Boophilus)*.

Some tick vector species transmit one or two viral species and a few transmit several species; *I. ricinus* is a good example of the latter. This cosmopolitan tick is widespread across most of the European continent reaching northern parts of Africa. In many forested areas it is the most abun-

dant tick species with a very broad vertebrate host range. All these features make it a highly efficient vector of several arboviruses and also other pathogens (e.g. Lyme disease borreliae). Indeed, *I. ricinus* is the main vector of viruses from three different virus families, e.g. TBEV and LIV of the Flaviviridae, Tribeč virus and Eyach virus of the Reoviridae and Uukuniemi virus of the Bunyaviridae. Another example of a wide virus range transmitted by a single tick species is that of *Ixodes uriae*. It is a specific tick of seabird colonies with a circumpolar distribution along the sea shore. At least six virus species of the genus *Orbivirus, Nairovirus, Phlebovirus* and *Flavivirus* have been isolated from this tick species. Interestingly, viruses of the same genus or even of the same virus group are vectored by both *I. ricinus* and *I. uriae*. These viruses are orbiviruses of Great Island virus (some 43 serotypes) transmitted by *I. uriae* and Kemerovo virus (all four serotypes) transmitted by *I. ricinus*; flaviviruses Tyuleniy virus and Gadgets Gully virus vectored by *I. uriae* and TBEV and LIV by *I. ricinus*; the phleboviruses Zaliv Terpeniya virus (a serotype of Uukuniemi virus) transmitted by *I. uriae* and Uukuniemi virus by both *I. ricinus* and *I. uriae*. The implications for the spread, vector specificity and evolution of these viruses are unknown. We can only speculate that tick genetics, physiology, ecology and life cycle are probably the key factors allowing both vector species to be involved in the maintenance of so many virus species in nature.

Returning to the basic question: how do arboviruses switch between replicating in vertebrate and invertebrate cells? Very little is known about the molecular mechanisms governing the relationship of arboviruses with ticks versus vertebrate hosts. The following characteristics of tick-borne members of different virus families and genera provide some clues and demonstrate the remarkable variety of viruses that have adapted to a life style using ticks as vectors.

Family Asfarviridae

Only a single genus, *Asfivirus*, is currently recognized within the Asfarviridae family and there is a single species, *African swine fever virus* (ASFV) (Table 12.1). The name of the family is derived from **A**frican **s**wine **f**ever **a**nd **r**elated viruses. ASFV represents the only known DNA arbovirus and is transmitted by tick bite or by a direct oral route (Vinuela, 1985; Dixon *et al.*, 2000).

Virions of ASFV have a nucleoprotein core structure, 70–100 nm in diameter, within an icosahedral capsid, 170–190 nm in diameter, which is surrounded by internal lipid layers. Extracellular virions with an external lipid-containing envelope have a diameter of 175–215 nm. The viral genome comprises a single molecule of linear, covalently closed, double-stranded DNA, 170–190 kbp in size. The end sequences are present as two flip-flop forms that are inverted and complementary with respect to each other; adjacent to both termini are identical tandem repeat arrays about 2.1 kbp long. Whether these reflect adaptation to infect tick or vertebrate host cells, or both, is not known. The complete nucleotide sequence of the tissue culture-adapted isolate has been published (Yanez *et al.*, 1995) as well as partial sequences of two other virulent isolates. The genome comprises about 150 open reading frames, which are closely spaced (intergenic distances are generally less than 200 bp) and are read from both DNA strands. A few intergenic regions contain short tandem repeat arrays of unknown significance for virus virulence (Dixon *et al.*, 2000).

Virions contain more than 50 proteins including a number of enzymes involved in nucleotide metabolism, DNA replication and repair or transcription, post-translational protein modification, an enzyme involved in synthesis of isoprenoid compounds, and factors needed for early mRNA transcription and processing. Enzymes packaged into virions further include RNA polymerase, poly(A) polymerase, guanyltransferase and protein kinase. There are at least eight characterized major structural proteins, and a further two DNA-binding proteins and seven proteins with putative transmembrane regions found in virions. In addition, 26 proteins with predicted transmembrane domains are encoded with some known to modify host cell function. Proteins that may modulate the host response to virus infection are also present. They include proteins similar to the T-cell surface protein CD2, IkB, the apoptosis inhibitors Bcl2 and IAP, a protein similar to a *Herpes simplex* virus encoded neurovirulence factor ICP34.5, a myeloid differentiation antigen and the gadd34 protein. Large length variations are observed between genomes of different isolates, resulting from the gain or loss of members of five multigene families found in the genomic regions close to the termini, but such variations do not appear to result in any significant change in virus properties (Yozawa *et al.*, 1994).

ASFV infects warthogs and bushpigs without any apparent ill effects. By contrast it causes severe disease, characterized by haemorrhage, in domestic and wild swine. The virus replicates in cells of the mononuclear phagocytic system and reticuloendothelial cells in lymphoid tissues and organs. Virus enters cells by receptor-mediated endocytosis and early mRNA synthesis begins in the cytoplasm immediately following entry. Virus DNA replication and assembly takes place in perinuclear areas ('virus factories') with peak DNA replication about 8 hours post infection. DNA replication may proceed by a self-priming mechanism. The cell nucleus is required for productive infection. Genes are expressed in an ordered cascade. Early genes are expressed prior to DNA replication; expression of late genes is dependent on the onset of DNA replication. Expression of some early genes continues throughout infection. Intermediate genes are expressed late but their expression does not depend on the onset of DNA replication. Virus morphogenesis takes place in the virus factories. Two layers of membrane, derived from the endoplasmic reticulum, are incorporated as internal lipid membranes. Formation of the icosahedral capsid is thought to occur on these membranes. The virus genome and enzymes are packaged into a nucleoprotein core. Extracellular virus has a loose-fitting external lipid envelope possibly derived by budding through the plasma membrane.

Soft (argasid) ticks of the genus *Ornithodoros* are the main vectors of ASFV. In parts of Africa south of the Sahara, *O. moubata* acts as the vector; it is replaced by *O. erraticus* in southern Europe. ASFV replication in the tick midgut epithelium is required for infection of ticks (Kleiboeker *et al.*, 1999). The virus can be transmitted in ticks transstadially, transovarially and sexually. Transovarial transmission of ASFV in experimentally infected *O. moubata* ranged in filial infection prevalence from 1.2% to 35.5%. Immunohistochemistry showed that virus replicated in the developing larval cells and not in the yolk sac cells or within the outer layer of the eggs (Rennie *et al.*, 2001). It appears that there is an increased mortality among adult *Ornithodoros* ticks infected by ASFV (Rennie *et al.*, 2000).

Ticks transmit ASFV to warthogs, bushpigs and swine. Transmission between domestic swine can also occur by direct contact, ingestion of infected meat and fomites, or mechanically by biting flies. Disease is endemic in domestic swine in many African countries and in Sardinia. The virus was first introduced into Europe via Portugal in 1957 and became endemic in parts of the Iberian Peninsula from 1960 until 1995. Sporadic outbreaks that were successfully controlled have occurred in Belgium, Brazil, Cuba, the Dominican Republic, France, Haiti, the Netherlands and Malta. Using restriction endonuclease analysis of genomic DNA, ASFV isolates can be distinguished into five groups (Gibbs, 2001). European and American isolates fall into one group and African isolates into the remainder, reflecting the probable African origin of the virus. American isolates appear to have originated in Europe.

Virus isolates differ in virulence and may produce a variety of signs ranging from acute to chronic, or they may produce inapparent infections. Virulent isolates can cause 100% mortality in 7–10 days. Less virulent isolates may produce a mild disease from which a number of infected swine recover and become carriers (Vinuela, 1985; Salas, 1994). A striking feature of ASFV infections is the absence of neutralizing antibody production. This has severely hampered attempts to produce an effective vaccine although the use of gene-deleted virus strains has shown promise in protecting against virulent strains (Gibbs, 2001).

Family Orthomyxoviridae

The family contains three genera of influenza viruses (*Influenzavirus A, B* and *C*). In addition to these well-known viruses causing respiratory diseases of humans, the fourth genus in the family (*Thogotovirus*) comprises tick-borne viruses (Table 12.1).

GENUS *THOGOTOVIRUS*

The morphology and morphogenesis of these viruses show similarities with the influenza viruses. Virions are spherical or pleomorphic, 80–120 nm in diameter, and filamentous forms occur. The virion envelope is derived from cell membrane lipids and bears surface glycoprotein projections, 10–14 nm in length and 4–6 nm in diameter. Virions contain six or seven segments of linear, negative-sense single-stranded RNA. Total genomic size is about 10 kbp. Like the influenza viruses, each viral RNA segment possesses conserved regions of semicomplementary nucleotides at the 3′ and 5′ termini and mRNA synthesis is primed by host-derived cap structures. Both the 3′ and 5′ sequences of virion RNA are required for viral RNA promoter activity and the cap-snatching mechanism appears unique (Leahy, Dessens & Nuttall, 1997).

The type species, *Thogoto virus* (THOV), contains six single-stranded RNA segments. Four of them encode gene products that correspond to the viral polymerase (PB1, PB2 and PA) and nucleocapsid protein (NP) of influenza viruses (Weber *et al.*, 1998). However, the fourth largest segment encodes a surface glycoprotein that is unrelated to any influenza viral protein but instead shows striking sequence homology to a baculovirus surface glycoprotein (Morse, Marriott & Nuttall, 1992). The same is true for *Dhori virus* (DHOV) (Freedman-Faulstich & Fuller, 1990). This unique glycoprotein is probably the key to the ability of members of the *Thogotovirus* genus to infect ticks (Nuttall *et al.*, 1995). Influenza viruses use sialic acid residues on the surface of vertebrate cell membranes as receptors for infecting cells. Sialylation in invertebrates is somewhat controversial although sialylated glycoconjugates have been detected in the salivary glands of female *I. ricinus* (Vancova *et al.*, 2006). Clearly *Thogotovirus* members have circumvented this problem by evolving a different mechanism of cell infection to that of influenza viruses. Comparison of the glycoprotein sequences of eight THOV isolates, two DHOV isolates and one Batken virus isolate with the glycoprotein sequence of 10 nucleopolyhedrosis viruses (insect baculoviruses) suggests that the sequence similarity may represent convergent evolution rather than a common ancestry for the encoding gene (S. Turner, personal communication).

THOV has been isolated from *Boophilus* and *Rhipicephalus* spp. ticks in Africa (Kenya) and Europe (Sicily) as well as from *Amblyomma variegatum* ticks in Nigeria and *Hyalomma* spp. ticks in Nigeria and Egypt. THOV has also been isolated in the Central African Republic, Cameroon, Uganda and Ethiopia, and in several countries of southern Europe. Infections in sheep have been associated with high levels of abortion. THOV also infects cattle, goats and mongoose, and there are two reported cases of infections in humans associated with clinical conditions (Woodall, 2001). DHOV has a somewhat different but overlapping geographical distribution to that of THOV, occurring in India, eastern Russia, as well as Egypt and southern Portugal. DHOV has been isolated from *Hyalomma* spp. ticks. There is no detectable serological reactivity between THOV and DHOV and the structural differences (THOV has six RNA segments and DHOV has seven) and sequence diversity of 37% and 31% in the nucleoprotein and the envelope protein, respectively, support their separate species status. Batken virus, isolated from mosquitoes and ticks from Russia, cross-reacts serologically with DHOV and shares 98% identity in a portion of the nucleoprotein and 90% identity in a portion of the envelope protein. These and other data suggest that Batken virus is closely related to DHOV (Frese *et al.*, 1997). Araguari virus, isolated from an opossum in Brazil, is proposed as a member of the *Thogotovirus* genus, most closely related to THOV (Da Silva *et al.*, 2005). None of the viral proteins of members of the *Thogotovirus* genus are related antigenically to those of influenza viruses.

THOV has been used extensively in experimental studies of saliva-assisted transmission in which THOV has been shown to exploit the pharmacological properties of its vector's saliva (see Chapter 10). It has also been used to investigate reassortment using temperature-sensitive mutants to follow the exchange of genomic segments between viruses. The ability of THOV to reassort has been demonstrated in both ticks and a vertebrate host (Davies *et al.*, 1987; Jones *et al.*, 1987). However, the significance of such genetic exchange in the epidemiology of this virus is unknown.

Family Rhabdoviridae

The family Rhabdoviridae contains six genera, at least six other serogroups not assigned to any genus, and a number of individual unassigned viral species. The most important and best known representative is rabies virus (genus *Lyssavirus*), one of the most deadly of all human pathogens. Typical rhabdoviruses infecting vertebrates have virions character-

istically bullet-shaped, 100–430 nm long and 45–100 nm in diameter. The outer surface of virions is covered with projections (peplomers) comprising trimers of the viral glycoprotein. The viral genome is a single molecule of linear, negative-sense single-stranded RNA and contains at least five open reading frames. Viruses generally have five structural polypeptides.

Many viruses of the genus *Vesiculovirus* are typical arboviruses replicating in both arthropods and vertebrates, such as vesicular stomatitis virus (VSV). Phlebotomine sandflies are incriminated as vectors although recent data indicate that black flies (*Simulium vittatum*) transmit VSV by co-feeding on non-viraemic hosts (Mead *et al.*, 2000). *Isfahan virus* has been isolated from sandflies and also from *Hyalomma asiaticum* ticks in Turkmenia (Karabatsos, 1985). Antibody prevalence in humans is comparatively high in endemic regions but there are no reports of associated disease. The only other rhabdoviruses isolated from ticks are currently unassigned. These are Barur virus and Sawgrass virus of the Kern Canyon and Sawgrass virus groups, respectively (Table 12.1). Further studies are needed to confirm that ticks transmit these three rhabdoviruses biologically, and to determine whether other members of the Kern Canyon and Sawgrass virus groups are tick-borne.

Family Reoviridae

Reoviridae is a large family containing viruses with very diverse biological properties. The virions have icosahedral symmetry but may appear spherical in shape. Each virion has a capsid, which comprises concentric protein layers organized as one, two or three distinct capsid shells, having an overall diameter of 60–80 nm. Virions contain 10, 11 or 12 segments of linear double-stranded RNA, depending on the genus. The 12 genera of the family can be divided into two groups. One group contains those viruses in which intact virus particles or cores have relatively large 'spikes' or 'turrets' situated at the 12 vertices of the icosahedron (members of the genera *Orthoreovirus*, *Cypovirus*, *Aquareovirus*, *Fijivirus*, *Oryzavirus*, *Idnoreovirus* and *Mycoreovirus* as well as most of the unclassified or unassigned reoviruses from invertebrates). The second group of viruses has relatively smooth surfaces without prominent surface projections at their fivefold axes (members of the *Orbivirus*, *Rotavirus*, *Coltivirus*, *Phytoreovirus* and *Seadornavirus* genera). Orbiviruses and coltiviruses are arboviruses and include tick-borne viruses (Table 12.2).

Virions are spherical in appearance but have icosahedral symmetry (Roy, 2001). Although no lipid envelope is present on mature virions, they can leave the host cell by budding through the cell plasma membrane. During this process they transiently acquire an unstable membrane envelope. Unpurified virus is often associated with cellular membranes. The outer capsid has an ordered structure with icosahedral symmetry. The surface layer of the core particle is composed of capsomeres arranged as hexameric rings. These rings give rise to the name of the genus. The core particle also contains a complete inner capsid shell surrounding the 10 segments of double-stranded RNA that comprise the viral genome. Minor core proteins are attached to the inner surface of the subcore at the fivefold symmetry axes. Replication is characterized by production of viral 'tubules' and viral inclusion bodies and may be accompanied by formation of flat hexagonal crystals of the major outer core protein in the cytoplasm of infected cells. Orbiviruses are transmitted between vertebrate hosts by a variety of haematophagous arthropods. They do not appear to cause disease in their natural hosts. Most orbiviruses associated with birds have been isolated from arthropod vectors that feed on birds, but evidence of infection of birds has been obtained for only a few orbiviruses (Nuttall, 1993).

The type species of the *Orbivirus*, *Bluetongue virus* (BTV), causes an economically important disease of sheep and is transmitted by midges of the genus *Culicoides*. It is one of the best-characterized arboviruses. Approximately 60 tick-borne orbiviruses have been identified (Table 12.2). Most of these are variants (serotypes or topotypes) of *Great Island virus* (formerly in the Kemerovo serogroup) transmitted by ixodid ticks that parasitize seabirds (guillemots or murres, puffins, penguins, gannets, gulls etc.); at least 40 viruses have been isolated from the common seabird tick, *Ixodes uriae*. The distribution of *Great Island virus* reflects the bipolar distribution of *I. uriae* (Main *et al*., 1973; Nuttall, 1984; Chastel, 1988). Although these viruses are confined to seabird colonies, the presence of antibodies in various mammals including humans living in close proximity to the colonies (Lvov *et al*., 1972), and the recovery of *I. uriae* from 'land' birds (Arthur, 1963), indicate the possibility of extension of their normal range.

The four viruses of the species *Kemerovo virus* are maintained in the Palearctic region among small mammals and birds by two related species of ixodid ticks: Kemerovo virus by *I. persulcatus* in western Siberia (Libikova *et al*., 1964), Tribeč virus and Lipovnik virus by *I. ricinus* in central

Europe (Gresikova *et al*., 1965) and Kharagysh virus by *I. ricinus* in Moldova (Skofertsa *et al*., 1972). This group of viruses is listed under *Great Island virus* by the International Committee on Taxonomy of Viruses (Fauquet *et al*., 2005). We consider *Kemerovo virus* a distinct species because it shows limited genome segment reassortment with three representatives of *Great Island virus*, and none with *Chenuda virus*, Essaouira virus or Mono Lake virus (Nuttall & Moss, 1989). Speciation of *Kemerovo virus* and *Great Island virus* may be at a transitional stage in which ancestral links can be detected under highly selective experimental conditions. Presumably genetic exchange between these two species does not occur in nature, particularly given their different ecologies.

The species *Chenuda virus* includes seven different serotypes from soft ticks of the genera *Argas*, *Carios* and *Ornithodoros* parasitizing birds (Hoogstraal, 1973). Chenuda virus was originally isolated from *Argas reflexus hermanni* collected from a pigeon house in Egypt (Taylor *et al*., 1966). Related viruses have been isolated from *Carios* (formerly *Ornithodoros*) *coniceps* from pigeon nests in the Chatkal mountains and *C. maritimus* ticks infesting gulls (*Larus argentatus*) on an island in the Caspian Sea (Baku virus), gull colonies in Morocco (Essaouira virus and Kala Iris virus) and gull and shag (*Phalacrocorax aristotelis*) colonies in Brittany (Moellez virus). Two isolates of Great Saltee Island virus (GS80-5, GS80-6) were from eggs and adult male *C. maritimus* feeding in a breeding colony of shags on a small island off the southeast coast of Ireland. This virus is classed as a possible member of the species based on its geographical location, and vector and host associations.

Huacho virus, Mono Lake virus and Sixgun City virus are classified by the International Committee on Taxonomy of Viruses as belonging to the species *Chenuda virus* (Fauquet *et al*., 2005). Members of an orbivirus species are distinguished by their ability to exchange genomic segments. Based on this definition, we recognize a distinct species, *Mono Lake virus*. The basis for this speciation is the inability of Mono Lake virus to demonstrate genome segment reassortment when tested experimentally with representatives of *Chenuda virus*, *Great Island virus* and *Kemerovo virus* (Nuttall & Moss, 1989). Mono Lake virus was originally isolated from a pool of 10 adult *Argas cooleyi* ticks collected in 1966 from the nest of the California gull (*L. californicus*) on an island in Mono Lake, California. Other species members are recognized by their antigenic cross-reactivity with Mono Lake virus. These are Huacho virus, isolated from *C. amblus* nymphs collected from rocks of a guano seabird colony at Punta Salinas, and

Sixgun City virus from *Argas cooleyi* collected in the nests of cliff swallows (*Petrochelidon pyrrhonota*) in Texas.

Chobar Gorge virus has two serotypes, Chobar Gorge virus and Fomede virus. They are associated with bats. Wad Medani virus has been isolated from sheep and from ixodid ticks of the genera *Boophilus*, *Hyalomma*, *Ambylomma* and *Rhipicephalus* in Sudan, West Pakistan, India, Russia and Jamaica. The closely related Seletar virus was isolated from *B. microplus* in Singapore and Malaysia.

The Great Island orbiviruses associated with seabird colonies provide a fascinating model in the study of virus evolution. Main *et al.* (1973) found two distinct but closely related serotypes, Great Island virus and Bauline virus, among puffins and petrels on Great Island, with little or no indication of other serotypes. They suggested that the antigenic identity of serotypes may be maintained by the isolation of the primary hosts of the vector and that each serotype developed in separate demes on the island. Yunker (1975) suggested that each virus within defined complexes has been influenced by common patterns of geography, ecology and behaviour. These include a discontinuous distribution, the nidicolous activity of the ticks that maintain them and the homing–colonial instinct of the bird hosts. Isolation produces a large number of strong insular variants, which in some cases develop into separate serotypes. Experimental studies have shown that Great Island viruses from ticks collected in seabird colonies in Scotland, Ireland, Iceland, Newfoundland and Macquarie Island in the sub-Antarctic can swap RNA segments. Thus, despite their extensive geographical distribution, this group of viruses constitutes a single gene pool clearly representing one virus species (Nuttall & Moss, 1989).

Structural and genetic studies have revealed significant differences between the tick-borne orbiviruses and *Bluetongue virus* (BTV, the type species of the genus) that may provide clues to their different ecologies. The most notable is in the structure and function of the two outer capsid proteins. In BTV and other insect-borne orbiviruses, VP2 and VP5 are the protein components of the outer capsid. VP2, the major determinant of the virus serotype, carries neutralizing epitopes and the binding site for the vertebrate cell receptor. In contrast, the major determinant of serotype in the tick-borne orbiviruses is VP5 while VP4 is the second component of the outer capsid. VP4 probably bears the binding site for the vertebrate cell receptor as it determines neurovirulence. Interaction of VP4 and VP5 of *Great Island virus* (GIV) modulates viral pathogenicity. Thus, by reassortment, new viruses can be produced experimentally that are more

or less virulent than either of their parental viruses (Nuttall *et al.*, 1992). The significance of these studies for the natural ecology of GIV is unknown, particularly as the virus does not cause overt disease in its seabird hosts.

Comparison of three-dimensional models of BTV and Broadhaven virus (*Great Island virus*) indicate remarkable similarity except for differences in accessibility of the outer shell proteins (Schoehn *et al.*, 1997). This may reflect the need to access and cleave VP2 of BTV within the midgut of its *Culicoides* vector. Cleavage of VP2 exposes the core protein, VP7, that bears the Arg–Gly–Asp (RGD) motif involved in insect cell infection. In contrast to insects, blood meal digestion in ticks occurs intracellularly. Furthermore, none of the viral proteins of tick-borne orbiviruses has been reported to carry an RGD motif. Hence tick-borne viruses most likely have evolved a mechanism of tick cell infection that does not rely on proteolysis in the midgut, unlike their insect-borne relatives.

GENUS *COLTIVIRUS*

The most important feature distinguishing coltiviruses from the other members of the Reoviridae family is a genome comprising 12 segments of double-stranded RNA. During replication, viruses are found in the cell cytoplasm but immunofluorescent staining also reveals nucleolar fluorescence. Coltiviruses are transmitted to vertebrate hosts by ticks.

Coltivirus particles are 60–80 nm in diameter having two concentric capsid shells with a core that is about 50 nm in diameter. Electron microscope studies using negative staining reveal particles that have a relatively smooth surface capsomeric structure and icosahedral symmetry. Particles are found intimately associated with filamentous structures and granular matrices in the cytoplasm. The majority of the viral particles are non-enveloped, but a few acquire an envelope structure during passage through the endoplasmic reticulum.

Coltiviruses have been isolated from several mammalian species (including humans) and from ticks. *Colorado tick fever virus* (CTFV), the type species of the genus, causes an acute febrile illness in humans. There appears to be a single serotype involved in human infections. Because of the ease of isolation of the virus from patients and ticks, the geographical distribution has been determined as essentially that of the major vector, *Dermacentor andersoni*, in mountainous northwestern USA and Canada. Other tick species from which CTFV has been isolated include *D. occidentalis*, *D. albipictus*, *D. parumapertus*, *Haemaphysalis leporispalustris* and a species of *Otobius*. Vertebrate reservoirs include rodents, ground squirrels, pine squirrels, chipmunks, meadow voles and porcupines (Burgdorfer, 1977).

CTFV persists within erythrocytes in human patients and in experimentally infected animals (Emmons *et al.*, 1972; Hughes, Casper & Clifford, 1974). Ticks become infected with CTFV on ingestion of a blood meal from an infected vertebrate host. Both adult and nymphal ticks become persistently infected and provide an overwintering mechanism for the virus. CTFV is transmitted trans-stadially and also transovarially as demonstrated in *D. andersoni* ticks experimentally, and by the finding of related viruses from field-collected *Ixodes* larvae in France and Germany (Calisher, 2001). Some rodent species have prolonged viraemias (more than 5 months), which may also facilitate overwintering, and virus persistence. Humans usually become infected when bitten by adult *D. andersoni* ticks but probably do not act as a source of reinfection for other ticks. Transmission from person to person has been recorded as the result of blood transfusion. The prolonged viraemia observed in humans and rodents is thought to be due to the intra-erythrocytic location of virions, protecting them from immune clearance (Emmons *et al.*, 1972). The isolate from *Lepus californicus* is known as California hare virus (California s6-14-03) and listed under *Colorado tick fever virus* (Fauquet *et al.*, 2005).

In 1972, a second serotype of CTFV was isolated from *I. ricinus* ticks collected near Tübingen, West Germany (Rehse-Kupper *et al.*, 1976). This virus is now recognized as a distinct species, *Eyach virus*. The virus has been associated with meningo-encephalitis in humans based on detection of specific immunoglobulins IgM and IgG in patients with the disease. Two distinct serotypes of Eyach virus have been isolated in France (Chastel *et al.*, 1984).

Family Bunyaviridae

The family Bunyaviridae (named after *Bunyamwera virus*, the prototype virus) consists of almost 400 named viruses making it the largest animal RNA virus family. Biologically, these viruses encompass a wide range of characteristics. Five genera are presently recognized within the family (*Orthobunyavirus*, *Hantavirus*, *Nairovirus*, *Phlebovirus* and *Tospovirus*). No arthropod vector has been demonstrated for rodent-associated hantaviruses transmitted among hosts by direct contact and aerosol-borne infections. Tospoviruses are transmitted by thrips between plants and are able to replicate in both thrips and plants. Viruses in the other three genera are typical arboviruses capable of replicating alternately in ver-

tebrates and arthropods. Different viruses are transmitted by different species of a large variety of haematophagous arthropods. Some viruses of the family display a very narrow host range, especially for arthropod vectors. Tick-borne viruses are found in each of the three (*Orthobunyavirus*, *Nairovirus* and *Phlebovirus*) arbovirus genera (Table 12.3).

Viruses of the family Bunyaviridae have similar morphological features. The virions range between 80 and 120 nm in size, and display short (5–10 nm) surface glycoprotein projections embedded in a lipid bilayer envelope. The viral particles appear spherical or pleomorphic, depending on the method used for fixation for electron microscopic examination. An envelope, derived usually from cellular Golgi membranes, surrounds a core consisting of the genome and its associated proteins. All bunyaviruses have two glycoproteins, designated G_c and G_n.

Members of the family Bunyaviridae contain three unique molecules of single-stranded RNA of negative or ambisense polarity designated large (L), medium (M) and small (S) with a total size of 11–19 kbp. The terminal nucleotides of each genomic RNA segment are base-paired forming non-covalently closed, circular RNAs and ribonucleocapsids. The terminal nucleotide sequences of genome segments are conserved among viruses in each genus but are different from those viruses in other genera. Genomic segments from different but closely related viruses can reassort when cells are co-infected with two viruses, but reassortment is limited to closely related viruses (a complex or a serogroup). All members of the family appear to utilize the same coding strategy, with the L RNA segment encoding a large (L) protein, the polymerase, the M RNA segment coding for the viral glycoproteins (G_c or G_1 and G_n or G_2), and the S RNA segment encoding a nucleocapsid protein (N). A number of non-structural proteins have also been described and assigned to specific segments. Each virion contains three internal nucleocapsids comprising genomic RNA associated with many copies of the N protein and a few copies of the L protein. Each nucleocapsid is arranged in a non-covalently closed circle which may be visualized in electron micrographs (Schmaljohn & Hooper, 2001).

Unlike most other enveloped viruses, viruses of the Bunyaviridae (with some exceptions) do not bud from the plasma membrane of infected cells. Rather, they mature by budding into intracytoplasmic vesicles associated with the Golgi apparatus. Virus release may occur through cell death and rupture or by transport of vesicles containing assembled virions to the cell surface. The glycoproteins are associated with the Golgi apparatus even when expressed independently of the other viral proteins indicating that the glycoproteins have specific processing and transport signals (Elliott, 1996).

GENUS *ORTHOBUNYAVIRUS*

Classification of viruses belonging to the family Bunyaviridae was originally based on their serological relationships. These data have now been supplemented and largely supported by biochemical and genetic analyses. Viruses of the genus *Orthobunyavirus* are serologically unrelated to members of other genera. At the genome level, the most important distinguishing feature is the consensus terminal nucleotide sequence of the L, M and S genome segments. The N and NSs proteins are encoded in overlapping reading frames by the S RNA and are translated from the same complementary mRNA by alternative AUG initiation codon usage.

The genus *Orthobunyavirus* comprises over 150 viruses representing nearly 50 viral species. Most bunyaviruses are mosquito-borne and some have been isolated from *Culicoides* midges; only a few tick-borne viruses have been recorded (Table 12.3). Isolates of Bahig virus and Matruh virus (representatives of *Tete virus*) originate from Africa (Egypt, South Africa, Nigeria), Europe (Italy), Cyprus and Japan. Most isolates are from migratory birds whereas a few isolates are from ticks, yet ticks are considered the main vectors. *Estero Real virus* is the only other orthobunyavirus believed to be tick-borne.

GENUS *NAIROVIRUS*

The genus *Nairovirus* is named after Nairobi sheep disease virus (NSDV) and contains some of the most important tick-borne pathogens. It was divided into seven serogroups: Crimean–Congo haemorrhagic fever, Dera Ghazi Khan, Hughes, Nairobi sheep disease (NSD), Qalyub, Sakhalin and Thiafora groups (Clerx, Casals & Bishop, 1981; Zeller *et al.*, 1989*a*). These have now been replaced by seven viral species with *Dugbe virus* replacing NSDV. All but one species (*Thiafora virus*) are tick-borne viruses (Table 12.3). The high genetic variation of viruses of the genus *Nairovirus* reflects the diversity of their predominant tick vectors and their genomic plasticity in being able to both reassort genomic segments and undergo genetic recombination (Honig, Osborne & Nichol, 2004; Devde *et al.*, 2006).

Virions are morphologically similar to other members of the family with very small surface units, which appear as a peripheral fringe. The L RNA segment is considerably larger than the L segments of other members of the family (Marriott & Nuttall, 1996), and the S segment does not

encode a non-structural protein. The M segment encodes a single gene product, which is processed in a complex and poorly defined manner to yield the structural glycoproteins and at least three non-structural proteins. Processing of G_n is dependent on subtilase SKI-1, a cellular protease, whereas G_c processing is not. Determination of glycoprotein processing in ticks may provide important insights into nairovirus replication in the vector. Recent progress in applying reverse genetic technology to CCHFV offers the real prospect of understanding the biology of this important virus genus.

The most distinctive biological feature of nairoviruses is that they are all tick-borne, with relatively few isolates also reported from mosquitoes (Dugbe virus and Ganjam virus) or *Culicoides* midges, although insect-borne transmission is unlikely. CCHFV is the most medically important member of the genus and the best-studied representative. This virus was originally isolated independently in two distant parts of the world (Democratic Republic of Congo and the Crimea); subsequent laboratory comparison demonstrated the two viruses to be identical (Hoogstraal, 1979). It has one of the widest geographical distributions of the medically important arboviruses (Nuttall, 2001). Infections in humans result in haemorrhagic fever with severe typhoid-like symptoms and mortality rates up to 50% (Watts *et al.*, 1989). Livestock movements and migratory birds play an important role in the transport of infected ticks. Although CCHFV has been isolated from at least 31 different tick species and subspecies (including two argasid species), the principal vectors are ixodid ticks of the genus *Hyalomma* (such as *H. marginatum*, *H. rufipes*, *H. anatolicum* and *H. asiaticum*). In contrast to ixodid ticks, CCHFV does not replicate in argasid ticks indicating that CCHFV isolated from argasid ticks in nature merely represents virus survival in the recently ingested blood meal (Shepherd *et al.*, 1989). The distribution of CCHFV in scattered enzootic foci in sub-Saharan Africa, the Middle East and southern Eurasia falls within the limits of the distribution of the genus *Hyalomma* and human infections have been associated principally with bites by ticks of this genus (Swanepoel *et al.*, 1987). Direct transmission from human to human is a frequent cause of nosocomial infections. Hares (*Lepus* spp.), hedgehogs (*Erinaceus* and *Hemiechinus* spp.) and cattle are probably important amplifying hosts in nature, although virus screening of wild and domestic animals has failed to identify viraemic host species. Transmission from infected to uninfected ticks co-feeding on non-viraemic hosts (e.g. sheep and ground-feeding birds) may be an effective transmission route (Nuttall, 2001).

The International Committee on Taxonomy of Viruses lists Kodzha virus as a variant of CCHFV and cites under this virus two strains, C68031 and AP92. Strain AP92 is from Greece where evidence of the disease in humans has not been recorded. The Greek strain is phylogenetically distinct from all other CCHFV strains examined to date. Other viruses related to CCHFV, Hazara virus and Khasan virus, likewise have not been associated with disease.

Dera Ghazi Khan virus (DGKV) was isolated originally from *Hyalomma dromedarii* collected from a camel in western Pakistan (reference strain JD 254). At least five other related viruses are known. Relatives of DGKV have been isolated from argasid ticks, except for Abu Mina virus for which the tick vector is unknown.

Until recently, Dugbe virus was classified in the NSD serogroup but the serogroup now takes its name from *Dugbe virus*. It is antigenically and genetically related to CCHFV but of comparatively low pathogenicity, which makes it a good model for experimental studies. Dugbe virus replicates and persists trans-stadially in orally infected *A. variegatum* ticks and can subsequently be transmitted by tick feeding to a vertebrate host (Steele & Nuttall, 1989). The virus has been isolated from numerous other tick species (e.g. *H. truncatum*, *B. decoloratus* and *R. appendiculatus*) but has a much more restricted geographical distribution compared with CCHFV, possibly because it is not associated with birds.

Nairobi sheep disease virus (NSDV) is highly pathogenic for sheep and goats, and causes disease in humans. Some of the earliest work on tick-borne virus transmission is recorded for NSDV (Daubney & Hudson, 1931). More recent studies have been limited by the high degree of containment needed to handle this virus. Genetic and serological data indicate that Ganjam virus is an Asian variant of NSDV.

The largest group of nairoviruses representing a distinct species is that of *Hughes virus*. Hughes group viruses are transmitted by *Carios* (formerly *Ornithodoros*) ticks of the 'capensis' complex associated with seabirds, such as *C. maritimus* and *C. denmarki*, and the ixodid seabird tick *I. uriae*. Hughes virus was first isolated from *C. denmarki* ticks collected in 1962 on Bush Key, Dry Tortugas, Florida and subsequently from the same tick species from Soldado Rock, off the southwest corner of Trinidad. The virus was also isolated from the blood of nestling terns (*Sterna fuscata*) captured on Soldado Rock. The number and geographical range of Hughes group viruses is comparable to that of orbiviruses of the Great Island group, which also circulate in seabird colonies (Table 12.2). The abundance of these two virus groups may simply reflect the level of activity in collecting ticks and isolating viruses from the attractive habitats of seabirds. Alternatively, evolutionary pressures resulting from unique features of their ecology may have given rise to a large number of distinguishable virus variants.

Qalyub virus was originally isolated from *Ornithodoros erraticus* in Egypt and many subsequent isolates were obtained from the same tick species (Darwish & Hoogstraal, 1981). The related Bandia virus was isolated from *O. sonrai*, a member of the *O. erraticus* group, and from rodents in Senegal. The capacity of viruses in this group to serve as human pathogens is unknown.

GENUS *PHLEBOVIRUS*

The surface morphology of phleboviruses is distinct in having small round subunits with a central hole. The S RNA exhibits an ambisense coding strategy. It is transcribed by the virion RNA polymerase to a subgenomic virus-complementary sense mRNA that encodes the N protein and, from full-length antigenome S RNA, to a subgenomic virus sense mRNA that encodes a non-structural protein, NSs. The M segment gene order and coding strategy vary. Tick-borne phleboviruses (e.g. Uukuniemi virus) encode only G_c and G_n glycoproteins whereas the mosquito- and sandfly-vectored phleboviruses also have coding information for a non-structural protein, NSm (Schmaljohn & Hooper, 2001).

The *Phlebovirus* genus comprises some 30 viruses transmitted by phlebotomine sandflies, mosquitoes, or ceratopogonids of the genus *Culicoides*. The type species is *Rift Valley fever virus* transmitted by mosquitoes. At least 10 viruses transmitted by ticks were formerly assigned to an independent *Uukuvirus* genus but are now included within the *Phlebovirus* genus based on similarities in their coding strategies (Simons, Hellman & Pettersson, 1990). Uukuniemi virus was originally isolated in southeast Finland from *I. ricinus* ticks (reference strain S 23). Subsequently, isolates have been recovered in many other European countries, predominantly from *I. ricinus* ticks. This virus appears to straddle terrestrial and marine biotopes, having been isolated from *I. uriae* collected in seabird colonies on the island of Runde in Norway (strain Ru E82). The other representatives of the species have been isolated repeatedly from passerines and seabirds or from their ticks, particularly in Scandinavia and in other parts of Palaearctic and Nearctic regions. They include Grand Arbaud virus (from *Argas reflexus* ticks in the Rhone delta of France), Manawa virus (from *A. abdussalami* in West Pakistan), Oceanside virus (from *I. uriae* collected at

Three Arch Rocks and Yaquina Head, Oregon and Flat Iron Rock, California), Ponteves virus (from *A. reflexus*, France), Precarious Point virus (from *I. uriae* collected in penguin rookeries on Macquarie Island in the sub-Antarctic) and the closely related Murre virus and RML 105355 virus, St Abbs Head virus (from *I. uriae*, east Scotland), and Zaliv Terpeniya virus (from *I. uriae* collected in Sakhalin and Kamchatka regions, Far-eastern Russia and the Murmansk region in the European part of Russia) (Table 12.3).

In addition to the orthobunya-, nairo- and phlebovirus tick-borne members of the Bunyaviridae, there are at least 15 other tick-borne viruses that have yet to be assigned to a particular genus. The best known of these is Bhanja virus isolated from ticks of the genera *Haemaphysalis*, *Boophilus*, *Amblyomma* and *Hyalomma* in India, Africa (Nigeria, Cameroon, Senegal) and Europe (Italy, Balkan states). Palma virus is an isolate from Portugal closely related to Bhanja virus. Others include Silverwater virus from *Haemaphysalis leporispalustris* collected from snowshoe hare in Ontario and Upolu virus from *Carios* ticks collected in a tern colony (*Sterna* sp.) on a coral atoll in the Great Barrier Reef. Issyk-Kul virus appears to be transmissible by both mosquitoes and ticks, and has been associated with febrile illness in humans in Tajikistan.

Family Flaviviridae

Viruses within the family Flaviviridae are classified into three different genera (*Flavivirus*, *Pestivirus*, *Hepacivirus*) and exhibit very different biological characteristics. Many of the viruses in the genus *Flavivirus* are arboviruses; viruses in the other two genera only infect mammals. Indeed, on ecological grounds the composition of the Flaviviridae is questionable. It remains to be seen whether phylogenetic analyses support the taxonomic structure of the family.

GENUS *FLAVIVIRUS*

Most flaviviruses (at least 50 species) are transmitted to vertebrate hosts by mosquitoes. They include such medically important pathogens as *Yellow fever virus*, *Japanese encephalitis virus*, *West Nile virus* (WNV), the four serotypes of *Dengue virus* and several others. Tick-borne viruses currently comprise 12 species and there are at least 14 species that are zoonotic viruses transmitted between rodents or bats with no known arthropod vector.

Flavivirus virions are enveloped, roughly spherical, with a diameter of 40–60 nm. The virion contains a nucleocapsid core of 20–30 nm composed of a single capsid protein.

The envelope contains two virus-encoded proteins (envelope and membrane proteins). Immature, intracellular virions contain a precursor membrane protein, which is proteolytically cleaved during virus maturation. The viral genome is a single molecule of single-stranded RNA of approximately 11 kbp which is positive-sense and infectious. The virion RNA appears to be identical to the mRNA. Three structural proteins and seven non-structural proteins are encoded from one long open reading frame flanked by terminal non-coding regions that form specific secondary structures required for genome replication, translation or packaging. Viral proteins are synthesized as part of a polyprotein of more than 3000 amino acids, which is co- and post-translationally cleaved by viral and cellular proteases (Lindenbach & Rice, 2001). The envelope glycoprotein (E protein) is the major structural protein and plays an important role in membrane binding and inducing a protective immune response following virus infection. It carries epitopes detected by neutralization and haemagglutination-inhibition tests that have been used to identify different flaviviral subgroups and species.

Determination of the crystallographic structure of a soluble form of the E protein of a TBEV revealed that, unlike the spikes seen on many viruses, the flavivirus envelope protein is situated parallel to the virion surface in the form of head-to-tail homodimeric rods. Residues that influence binding of monoclonal antibodies occur on the outward facing surface of the protein (Rey et al., 1995). Interestingly, some mosquito-borne flaviviruses carry a putative integrin-binding motif Arg–Gly–Asp which is not found in the E protein of TBEV. There is an obvious analogy with the orbiviruses (see above).

All flaviviruses are serologically related, as demonstrated in binding assays such as ELISA and by haemagglutination-inhibition using polyclonal and monoclonal antibodies. Neutralization assays are more discriminating and have been used to define several serocomplexes within the genus *Flavivirus*. Comparative analyses of the nucleotide and amino acid sequence of the E gene have been used to investigate the phylogeny of flaviviruses (Marin et al., 1995). Cell fusing agent virus (CFAV) (tentatively placed in the genus *Flavivirus*) E gene was used as the outgroup based on the distant relationship of CFAV to tick- and mosquito-borne flaviviruses. The analyses showed that tick- and mosquito-borne flaviviruses diverged as two distinct and major genetic lineages. Interestingly, analyses of the E and non-structural NS5 gene sequences to compare mutational regimes indicated that mosquito-borne flaviviruses are evolving almost twice as fast as tick-borne flaviviruses (Marin et al., 1995; Gould, Zanotto & Holmes, 1997).

The tick-borne flaviviruses are currently classified into two groups: the mammalian tick-borne virus group and the seabird tick-borne virus group (Fauquet *et al.*, 2005). A third group, Kadam virus group, has been proposed based on phylogenetic analysis (Grard *et al.*, 2007). In addition, *Karshi virus* is proposed as a new species, and tick-borne encephalitis and louping ill viruses are considered a unique species, *Tick-borne encephalitis virus*. As the classification of tick-borne flaviviruses is still evolving, we have simply listed the viruses in alphabetical order (Table 12.4).

Medically, by far the most important tick-borne flaviviruses are those classically grouped into the TBEV serogroup, the most important being TBEV. Tick-borne encephalitis was recognized clinically in the Far East of the former USSR (now Russia) in the early 1930s. Severe cases of encephalitis were observed in humans residing in formerly unhabited areas. A special expedition was organized in 1937 by L. A. Zilber to determine the cause of the disease. Virus isolates were obtained from the blood of patients and from *I. persulcatus* ticks. The disease has several names, including Russian spring–summer encephalitis (RSSE), Far Eastern encephalitis and forest spring encephalitis (Zilber & Soloviev, 1946). In 1948, a similar though less severe form of encephalitis affected humans residing in central Bohemia, Czech Republic. The virus recovered from the blood of a patient and from *I. ricinus* ticks was related to the isolates from RSSE cases (Rampas & Gallia, 1949). Similar or milder forms of the disease, called bi-phasic meningoencephalitis, were observed in other central and eastern European countries. TBEV is endemic over a wide area, covering Europe, northern Asia and China. Several thousand human cases are recorded annually, with considerable variation from year to year. In general, the Far Eastern subtypes of TBEV cause severe disease in humans with a mortality that can be 50% in some outbreaks; disease associated with the European subtype is less severe and mortality is usually under 5% (Gresiková & Calisher, 1988). Based on genome analyses, recently a third Siberian subtype has been established. Members of this subtype are known to cause chronic infections in humans.

Louping ill virus (LIV) derives its name from the disease of sheep (louping ill), which has been recognized in southern Scotland for at least two centuries (Smith & Varma, 1981). 'Louping' refers to the characteristic behaviour of 'leaping' shown by sheep infected with LIV. The disease is found throughout much of the upland sheep farming areas of Scotland, northern and southwestern England, Ireland and Wales affecting sheep and red grouse (*Lagopus scoticus*), with other species of domestic animals and humans affected less frequently. The main vector is *I. ricinus* (Reid, 1984). Two subtypes of LIV are recognized in the British Isles (Irish and British subtypes), as well as in Norway. The virus can cause severe encephalitis in humans; however, there are only about 35 recorded cases and they have been confined mostly to laboratory workers, veterinary surgeons, farmers and abattoir workers (Smith & Varma, 1981).

Powassan virus (POWV) is the North American member of the TBE subgroup although there are several records of POWV from ticks and mosquitoes collected also in Russia. The virus was originally isolated from a pool of *D. andersoni* ticks collected in 1952 in Colorado (Thomas, Kennedy & Eklund, 1960) and derives its name from the town in Northern Ontario where the first fatal case of the encephalitic disease was recognized. About 20 cases of disease associated with POWV infection have been reported in North America and there was one fatal case recorded in Russia (Artsob, 1988). Deer tick virus, a relative of POWV isolated from *Ixodes* and *Dermacentor* spp., and from human brain (Telford *et al.*, 1997), is not listed by the International Committee on the Taxonomy of Viruses (Fauquet *et al.*, 2005).

Omsk haemorrhagic fever was first reported in 1941 when physicians recorded sporadic cases in the forest steppes of the Omsk Region of Siberia among muskrat hunters. Though closely related to TBEV, *Omsk haemorrhagic fever virus* (OHFV) is unique with respect to both the clinical features of the disease (haemorrhagic symptoms compared with encephalitis) and the ecology of the virus. The primary tick vector of OHFV is *D. reticulatus* and the hosts are voles, particularly the narrow-skulled vole (*Microtus gregalis*). Another tick species, *I. apronophorus*, plays a role together with its main host, the water vole (*Arvicola terrestris*). *Dermacentor marginatus* and *I. persulcatus* ticks play secondary roles in the transmission cycles of OHFV (Lvov, 1988).

Kyasanur Forest disease virus (KFDV) is another highly pathogenic member of the tick-borne flaviviruses producing haemorrhagic disease in infected humans. Since the first record of the disease in 1957 in Karnataka State, India, outbreaks have occurred every year in the affected region. The highest recorded incidence was in 1983 with 1555 cases and 150 deaths. Preceding the first human epidemic, a large number of sick and dead monkeys had been noticed in the nearby forest area. KFDV was isolated from dead monkeys, sick patients and from *Haemaphysalis* and other tick species. Clinically, the disease is particularly interesting because, in contrast to most other tick-borne flaviviruses that cause encephalitis, KFDV causes a haemorrhagic disease even more severe than that associated with OHFV

(Banerjee, 1988). Alkhurma virus, a close relative of KFDV, was first isolated in 1995 from patients with haemorrhagic fever in Saudi Arabia (Zaki, 1997; Charrel *et al.*, 2001). Little is known of the ecology and epidemiology of this virus.

Langat virus, which is transmitted by *I. granulatus*, is the least pathogenic representative of the tick-borne flaviviruses that infect mammals. This virus was isolated in Malaysia (Karabatsos, 1985). Because of its naturally low virulence and serological cross-reactivity with TBEV it was assessed as a candidate vaccine against TBE in Russia with disastrous consequences.

The remaining members of the group (Kadam virus, Karshi virus, Royal Farm virus and Gadgets Gully virus) are not considered human pathogens and consequently knowledge of their ecology and epidemiology is limited (Karabatsos, 1985). Similarly, little is known about Ngoye virus, which appears to represent a separate lineage within the *Flavivirus* genus (Grard *et al.*, 2006).

Ticks are potential reservoirs of at least two mosquito-borne flaviviruses. Yellow fever virus (YFV) has been isolated from the eggs and emergent larvae of an ixodid tick species collected in the field (Germain *et al.*, 1979); there is no epidemiological evidence that ticks are vectors of YFV. West Nile virus (WNV), although clearly a mosquito-borne virus, is included in Table 12.4 because it has been isolated repeatedly from bird-infesting argasid and ixodid ticks in arid regions of Russia and from *H. detritum* in a desert region of Turkmenistan (Lvov *et al.*, 1975). The five isolates of WNV from *C. maritimus* ticks collected in the nest site of gulls (*Larus argentatus*) were from Glinyanyi Island in the Caspian Sea, where mosquitoes are reported to be absent. For WNV, long-term survival and co-feeding transmission of the virus has been demonstrated in laboratory studies with *Ornithodoros* ticks (Vermeil, Lavillaureix & Reeb, 1959; Lawrie *et al.*, 2004).

The surprisingly broad spectrum of ecological conditions occupied by tick-borne flaviviruses is worthy of further comment. As we have seen, they range from the seabird colonies occupied by *I. uriae* as the dominant tick species, to the typically forested habitats of the Northern Hemisphere with *I. ricinus* and *I. persulcatus* as the dominant tick species. In addition, representatives of the TBEV serogroup have dispersed across Asia and Europe in a cline (continuous or gradual evolution across a geographical area). The most divergent lineages (often incorrectly referred to as the most ancient viruses) appeared in the east and the more recent lineages emerged in the west, with LIV at the extreme west of this geographical distribution (Zanotto *et al.*, 1995).

Unassigned viruses

A significant number of viruses have not yet been assigned to a recognized virus family or genus. Some of these appear to be tick-borne viruses (Table 12.5).

Nyamanini virus was first isolated from cattle egrets (*Bulbulcus ibis*) collected in 1957 in South Africa. Subsequent isolations have been from cattle egrets and ticks in Africa. It is antigenically related to (though clearly distinct from) Midway virus, which has been isolated from ticks collected in seabird colonies. Virions are enveloped and contain RNA.

Quaranfil virus was isolated from *Argas arboreus* ticks collected near Cairo, Egypt in 1953. The virus has also been isolated from humans with febrile illness, birds (ibis and pigeons) and ticks. Little more is known about the related Johnston Atoll virus, which was isolated from seabird ticks. This virus has an envelope together with three major and two minor proteins, two of which appear to be glycoproteins. Both Quaranfil virus and Johnston Atoll virus have morphological and morphogenetic characteristics similar to members of the Arenaviridae (Zeller *et al.*, 1989*b*). No members of this family are currently recognized as arboviruses.

Six of the seven ungrouped viruses listed in Table 12.5 have been isolated from bird-feeding ticks. Aride virus was recovered from ticks collected from dead roseate terns (*Sterna dougallii*) found on Bird Island in the Seychelles. The birds breed on nearby Aride Island, hence the virus name. Caspiy virus is found in seabird colonies of the Caspian Sea basin. One isolate was from a clinically sick gull (*Larus* sp.). Slovakia virus was isolated from a single pool of engorged female ticks collected during an investigation of the cause of sickness and deaths among domestic chickens heavily infested with *Argas persicus*.

Runde virus was isolated from two pools of *I. uriae* collected from a puffin (*Fratercula arctica*) colony on the island of Runde, Norway, and antibodies were detected (Traavik, 1979). The two isolates appeared to be antigenically identical but were unrelated to any major arbovirus group. Electron microscopy revealed morphological features similar to those of members of the Coronaviridae, a virus family that has no recognized arbovirus members to date. However, no antigenic relationship to infectious bronchitis virus (an avian coronavirus) was detected. The virus appears to have a single-stranded RNA genome. Røst Islands virus (NorAr V-958) showed a similar coronavirus-like morphology when examined by electron microscopy.

Jos virus was originally isolated in 1967 from the blood of a cow during slaughter at the abattoir in Jos, north central

Nigeria (Lee *et al.*, 1974). The 42 additional isolates were from ticks collected at cattle markets in southern Nigeria. The virus has also been isolated from *A. variegatum* collected in Senegal and Ethiopia. Although the virus replicated in Vero cell culture, attempts to characterize the virus by electron microscopy were unsuccessful (Zeller *et al.*, 1989*a*).

CONCLUSIONS

Most tick-borne viruses were isolated during the 1950s, 1960s and 1970s when many American, Russian and European scientists went out into the field, in many parts of the world, to collect specimens that were then screened for arboviruses in specialist laboratories. Inevitably, the distribution of these viruses reflects the scientists' activities and the accessibility of their published results. For example, many viruses have been isolated from ticks collected in seabird colonies where the permanence of the breeding sites, and the near guarantee of a blood meal, provide an ideal habitat for ticks and an attraction for tick collectors. Comparatively few of the viruses listed in Tables 12.1 to 12.5 are from China, which is most probably a misrepresentation of the true picture. There are surely many more tick-borne viruses yet to be discovered. Likewise, there are exciting discoveries to be made about how tick-borne viruses infect and replicate in both vertebrate and tick cells, and what determines whether the infection is pathogenic or benign. Advances in applying reverse genetics, together with site-directed mutagenesis and RNA interference techniques, will provide insights into the molecular basis of the tick-borne virus life cycle. However, there will still be the need for field studies to identify the biotic and abiotic factors that determine tick-borne virus survival in nature.

ACKNOWLEDGEMENTS

The work of ML was supported by the Slovak Research and Development Agency under contract No. APVV-51-004505.

REFERENCES

Arthur, D. R. (1963). *British Ticks*. London: Butterworths.

Artsob, H. (1988). Powassan encephalitis. In *The Arboviruses: Epidemiology and Ecology*, vol. 4, ed. Monath, T. P., pp. 29–49. Boca Raton, FL: CRC Press.

Balashov, Y. S. (1972). Bloodsucking ticks (Ixodoidea): vectors of diseases of man and animals. *Miscellaneous Publications of the Entomological Society of America* 8, 160–376.

Banerjee, K. (1988). Kyasanur Forest Disease. In *The Arboviruses: Epidemiology and Ecology*, vol. 3, ed. Monath, T. P., pp. 93–116. Boca Raton, FL: CRC Press.

Blaskovic, D. & Nosek, J. (1972). The ecological approach to the study of tick-borne encephalitis. *Progress in Medical Virology* 14, 275–320.

Booth, T. F., Davies, C. R., Jones, L. D., Staunton, D. & Nuttall, P. A. (1989). Anatomical basis of Thogoto virus infection in BHK cell culture and in the ixodid tick vector, *Rhipicephalus appendiculatus*. *Journal of General Virology* 70, 1093–1104.

Booth, T. F., Steele, G. M., Marriott, A. C. & Nuttall, P. A. (1991). Dissemination, replication, and trans-stadial persistence of Dugbe virus (Nairovirus, Bunyaviridae) in the tick vector *Amblyomma variegatum*. *American Journal of Tropical Medicine and Hygiene* 45, 146–157.

Burgdorfer, W. (1977). Tick-borne diseases in the United States: Rocky Mountain spotted fever and Colorado tick fever. *Acta Tropica* 34, 103–112.

Burgdorfer, W. & Varma, M. G. R. (1967). Trans-stadial and transovarial development of disease agents in arthropods. *Annual Review of Entomology* 12, 347–376.

Calisher, C. H. (2001). Colorado tick fever. In *The Encyclopedia of Arthropod-Transmitted Infections*, ed. Service, M. W., pp. 121–126. Wallingford, UK: CAB International.

Charrel, R. N., Zaki, A. M., Attoui, H. *et al.* (2001). Complete coding sequence of the Alkhurma virus, a tick-borne flavivirus causing severe haemorrhagic fever in humans in Saudi Arabia. *Biochemical and Biophysical Research Communications* 287, 455–461.

Chastel, C. (1988). Tick-borne virus infections of marine birds. *Advances in Disease Vector Research* 5, 25–60.

Chastel, C., Main, A. J., Couatarmanac'h, A., *et al.* (1984). Isolation of Eyach virus (Reoviridae, Colorado tick fever group) from *Ixodes ricinus* and *I. ventalloi* ticks in France. *Archives of Virology* 82, 161–171.

Chernesky, M. A. & McLean, D. M. (1969). Localization of Powassan virus in *Dermacentor andersoni* ticks by immunofluorescence. *Canadian Journal of Microbiology* 15, 1399–1408.

Clerx, J. P. M., Casals, J. & Bishop, D. H. L. (1981). Structural characteristics of nairoviruses (genus *Nairovirus*, Bunyaviridae). *Journal of General Virology* 55, 165–178.

Da Silva, E. V., Da Rosa, A. P., Nunes, M., *et al.* (2005). Araguari virus, a new member of the family

Orthomyxoviridae: serologic, ultrastructural, and molecular characterization. *American Journal of Tropical Medicine and Hygiene* **73**, 1050–1058.

Danielova, V., Holubova, J., Pejcoch, M. & Daniel, M. (2002). Potential significance of transovarial transmission in the circulation of tick-borne encephalitis virus. *Folia Parasitologica* **49**, 323–325.

Darwish, M. & Hoogstraal, H. (1981). Arboviruses infecting humans and lower animals in Egypt: a review of thirty years of research. *Journal of the Egyptian Public Health Association* **61**, 1–112.

Daubney, R. & Hudson, J. R. (1931). Nairobi sheep disease. *Parasitology* **23**, 507–524.

Davies, C. R., Jones, L. D., Green, B. M. & Nuttall, P. A. (1987). *In vivo* reassortment of Thogoto virus (a tick-borne influenza-like virus) following oral infection of *Rhipicephalus appendiculatus* ticks. *Journal of General Virology* **68**, 2331–2338.

Davies, C. R., Jones, L. D. & Nuttall, P. A. (1986). Experimental studies on the transmission cycle of Thogoto virus, a candidate orthomyxovirus, in *Rhipicephalus appendiculatus* ticks. *American Journal of Tropical Medicine and Hygiene* **35**, 1256–1262.

Devde, V. M., Khristova, M. L., Rollin, P. E. & Ksiazek, T. G. (2006). Crimean–Congo hemorrhagic fever virus genomics and global diversity. *Journal of Virology* **80**, 8834–8842.

Dixon, L. K., Costa, J. V., Escribano, J. M., *et al.* (2000). Family Asfarviridae. In *Virus Taxonomy*, 7th Report of the International Committee on Taxonomy of Viruses, eds. van Regenmortel, M. H. V., Fauquet, C. M., Bishop, D. H. L., *et al.*, pp. 159–165. San Diego, CA: Academic Press.

Emmons, R. W., Oshiro, L. S., Johnson, H. N. & Lennette, E. H. (1972). Intraerythrocytic location of Colorado tick fever. *Journal of General Virology* **17**, 185–195.

Elliott, R. M. (ed.) (1996). *The Bunyaviridae*. New York: Plenum Press.

Fauquet, C. M., Mayo, M. A., Maniloff, J., Desselberger, U. & Ball, L. A. (eds.) (2005). *Virus Taxonomy: Classification and Nomenclature of Viruses*, 8th Report of the International Committee on Taxonomy of Viruses. San Diego, CA: Academic Press.

Freedman-Faulstich, E. Z. & Fuller, F. J. (1990). Nucleotide sequence of the tick-borne, orthomyxo-like Dhori/Indian/1313/61 virus envelope gene. *Virology* **175**, 10–18.

Frese, M., Weeber, M., Weber, F., Speth, V. & Haller, O. (1997). MX1 sensitivity: Batken virus is an orthomyxovirus

closely related to Dhori virus. *Journal of General Virology* **78**, 2453–2458.

Germain, M., Saluzzo, J. F., Cornet, J. P., *et al.* (1979). Isolement du virus de la fièvre jaune à partir de la ponte et de larves d'une tique *Amblyomma variegatum*. *Comptes Rendus des Séances de l'Academie des Sciences D* **289**, 635–637.

Gibbs, E. P. J. (2001). African swine fever. In *The Encyclopedia of Arthropod-Transmitted Infections*, ed. Service, M. W., pp. 7–13. Wallingford, UK: CAB International.

Gordon, S. W., Linthicum, K. J. & Moulton, J. R. (1993). Transmission of Crimean–Congo hemorrhagic fever virus in two species of *Hyalomma* ticks from infected adults to cofeeding immature forms. *American Journal of Tropical Medicine and Hygiene* **48**, 576–580.

Gould, E. A., Zanotto, P. M. de A. & Holmes, E. C. (1997). The genetic evolution of flaviviruses. In *Factors in the Emergence of Arbovirus Diseases*, eds. Saluzzo, J. F. & Dodet, B., pp. 51–63. Paris: Elsevier.

Grard, G., Lemasson, J.-J., Sylla, M., *et al.* (2006). Ngoye virus: a novel evolutionary lineage within the genus Flavivirus. *Journal of General Virology* **87**, 3273–3277.

Grard, G., Moureau, G., Charrel, R. N., *et al.* (2007). Genetic characterization of tick-borne flaviviruses: new insights into evolution, pathogenetic determinants and taxonomy. *Virology* **361**, 80–92.

Gresikova, M. & Calisher, C. H. (1988). Tick-borne encephalitis. In *The Arboviruses: Epidemiology and Ecology*, vol. 4, ed. Monath, T. P., pp. 177–202. Boca Raton, FL: CRC Press.

Gresikova, M., Nosek, J., Kozuch, O., Ernek, E. & Lichard, M. (1965). Study on the ecology of Tribeč virus. *Acta Virologica* **9**, 83–89.

Gresikova, M., Sekeyova, M., Tempera, G., Guglielmino, S. & Castro, A. (1978). Identification of Sindbis virus strain isolated from *Hyalomma marginatum* ticks in Sicily. *Acta Virologica* **22**, 231–232.

Honig, J. E., Osborne, J. C. & Nichol, S. T. (2004). The high genetic variation of viruses of the genus Nairovirus reflects the diversity of their predominant tick hosts. *Virology* **318**, 10–16.

Hoogstraal, H. (1973). Viruses and ticks. In *Viruses and Invertebrates*, ed. Gibbs, A. J., pp. 349–390. Amsterdam: North-Holland.

Hoogstraal, H. (1979). The epidemiology of tick-borne Crimean–Congo hemorrhagic fever in Asia, Europe, and Africa. *Journal of Medical Entomology* **15**, 307–417.

Hughes, L. E., Casper, E. A. & Clifford, C. M. (1974). Persistence of Colorado tick fever in red blood cells.

American Journal of Tropical Medicine and Hygiene **23**, 530–532.

Iwata, H., Yamagawa, M. & Roy, P. (1992). Evolutionary relationships among the gnat-transmitted orbiviruses that cause African horse sickness, bluetongue, and epizootic hemorrhagic disease as evidenced by their capsid protein sequences. *Virology* **191**, 251–261.

Jones, L. D., Davies, C. R., Green, B. M. & Nuttall, P. A. (1987). Reassortment of Thogoto virus (a tick-borne influenza-like virus) in a vertebrate host. *Journal of General Virology* **68**, 1299–1306.

Karabatsos, N. (1985). *International Catalogue of Arboviruses Including Certain Other Viruses of Vertebrates*, 3rd edn. San Antonio, TX: American Society of Tropical Medicine and Hygiene.

Kaufman, W. R., Bowman, A. S. & Nuttall, P. A. (2002). Salivary fluid secretion in the ixodid tick *Rhipicephalus appendiculatus* is inhibited by Thogoto virus infection. *Experimental and Applied Acarology* **25**, 661–674.

Kaufman, W. R. & Nuttall, P. A. (1996). *Amblyomma variegatum* (Acari: Ixodidae): mechanism and control of arbovirus secretion in tick saliva. *Experimental Parasitology* **82**, 316–323.

Kaufman, W. R. & Nuttall, P. A. (2003). *Rhipicephalus appendiculatus* (Acari: Ixodidae): dynamics of Thogoto virus infection in female ticks during feeding on guinea pigs. *Experimental Parasitology* **104**, 20–25.

Kleiboeker, S. B., Scoles, G. A., Burrage, T. G. & Sur, J. H. (1999). African swine fever virus replication in the midgut epithelium is required for infection of *Ornithodoros* ticks. *Journal of Virology* **73**, 8587–8598.

Kovar, V., Kopacek, P. & Grubhoffer, L. (2000). Isolation and characterization of Dorin M, a lectin from plasma of the soft tick *Ornithodoros moubata*. *Insect Biochemistry and Molecular Biology* **30**, 195–205.

Labuda, M., Danielova, V., Jones, L. D. & Nuttall, P. A. (1993). Amplification of tick-borne encephalitis virus infection during co-feeding of ticks. *Medical and Veterinary Entomology* **7**, 339–342.

Lawrie, C. H., Uzcategui, N. Y., Gould, E. A. & Nuttall, P. A. (2004). Ixodid and argasid tick species and West Nile virus. *Emerging Infectious Diseases* **10**, 653–657.

Leahy, M. B., Dessens, J. T. & Nuttall, P. A. (1997). *In vitro* polymerase activity of Thogoto virus: evidence for a unique cap snatching mechanism in a tick-borne orthomyxovirus. *Journal of Virology* **71**, 8347–8351.

Lee, V. H., Kemp, G. E., Madbouly, M. H., *et al.* (1974). Jos, a new tick-borne virus from Nigeria. *American Journal of Veterinary Research* **35**, 1165–1167.

Libikova, H., Mayer, V., Kozuch, O., *et al.* (1964). Isolation from *Ixodes persulcatus* ticks of cytopathic agents (Kemerovo virus) differing from tick-borne encephalitis virus and some of their properties. *Acta Virologica* **8**, 289–301.

Lindenbach, B. D. & Rice, C. M. (2001). *Flaviviridae*: the viruses and their replication. In *Fields' Virology*, 4th edn, eds. Knipe, D. M., Howley, P. M. *et al.*, pp. 991–1041. Philadelphia, PA: Lippincott Williams & Wilkins.

Ludwig, G. V., Israel, B. A., Christensen, B. M., Yuill, T. M. & Schultz, K. T. (1991). Role of La Crosse virus glycoproteins in attachment of virus to host cells. *Virology* **181**, 564–571.

Lvov, D. K. (1988). Omsk haemorrhagic fever. In *The Arboviruses: Epidemiology and Ecology*, vol. 3, ed. Monath, T. P., pp. 205–216. Boca Raton, FL: CRC Press.

Lvov, D. K., Chervonski, V. I., Gostinshchikova, I. N., *et al.* (1972). Isolation of Tyuleniy virus from ticks *Ixodes* (*Ceratixodes*) *putus* Pick.-Camb. 1878 collected on Commodore Islands. *Archiv für gesamte Virusforschung* **38**, 139–142.

Lvov, D. K., Timopheeva, A. A., Smirnov, V. A., *et al.* (1975). Ecology of tick-borne viruses in colonies of birds in the USSR. *Medical Biology* **53**, 325–330.

Main, A. J., Downs, W. G., Shope, R. E. & Wallis, R. C. (1973). Great Island and Bauline: two new Kemerovo group orbiviruses from *Ixodes uriae* in eastern Canada. *Journal of Medical Entomology* **10**, 229–235.

Mandl, C. W., Kroschewski, H., Allison, S. L., *et al.* (2001). Adaptation of tick-borne encephalitis virus to BHK-21 cells results in the formation of multiple heparin sulfate binding sites in the envelope protein and attenuation *in vivo*. *Journal of Virology* **75**, 5627–5637.

Marin, M. S., Zanotto, P. M. de A., Gritsun, T. S. & Gould, E. A. (1995). Phylogeny of Tyu, SRE and CFA virus: different evolutionary rates in the genus *Flavivirus*. *Virology* **206**, 1133–1139.

Marriott, A. C. & Nuttall, P. A. (1996). Large RNA segment of Dugbe nairovirus encodes the putative RNA polymerase. *Journal of General Virology* **77**, 1775–1780.

Mead, D. G., Ramberg, F. B., Besselsen, D. G. & Mare, C. J. (2000). Transmission of vesicular stomatitis virus from infected to noninfected black flies co-feeding on nonviremic deer mice. *Science* **287**, 485–487.

Morse, M. A., Marriott, A. C. & Nuttall, P. A. (1992). The glycoprotein of Thogoto virus (a tick-borne orthomyxo-like virus) is related to the baculovirus glycoprotein gp64. *Virology* **186**, 640–646.

Nuttall, P. A. (1984). Tick-borne viruses in seabird colonies. *Seabird* **1**, 31–41.

Nuttall, P. A. (1993). Orbiviruses associated with birds. In *Virus Infections of Vertebrates*, vol. 4, *Virus Infections of Birds*, eds. McFerran J. B. & McNulty, M. S., pp. 195–198. Amsterdam: Elsevier.

Nuttall, P. A. (1999). Pathogen–tick–host interactions: *Borrelia burgdorferi* and TBE virus. *Zentralblatt für Bakteriologie* **289**, 492–505.

Nuttall, P. A. (2001). Crimean–Congo haemorrhagic fever. In *The Encyclopedia of Arthropod-Transmitted Infections*, ed. Service, M. W., pp. 126–132. Wallingford, UK: CAB International.

Nuttall, P. A. & Moss, S. R. (1989). Genetic reassortment indicates a new grouping for tick-borne orbiviruses. *Virology* **171**, 156–161.

Nuttall, P. A., Jacobs, S. C., Jones, L. D., Carey, D. & Moss, S. R. (1992). Enhanced neurovirulence of tick-borne orbiviruses resulting from genetic modulation. *Virology* **187**, 407–412.

Nuttall, P. A., Morse, M. A., Jones, L. D. & Portela, A. (1995). Adaptation of members of the Orthomyxoviridae family to transmission by ticks. In *Molecular Basis of Virus Evolution*, eds. Gibbs, A. J., Calisher, C. H. & García-Arenal, F., pp. 416–425. Cambridge, UK: Cambridge University Press.

Rampas, J. & Gallia, F. (1949). [Isolation of encephalitis virus from *Ixodes ricinus* ticks.] (in Czech) *Casopis Lekaru Ceskych* **88**, 1179–1180.

Randolph, S. E. (1998). Ticks are not insects: consequences of contrasting vector biology for transmission potential. *Parasitology Today* **14**, 186–192.

Randolph, S. E., Miklisova, D., Lysy, J., Rogers, D. J. & Labuda, M. (1999). Incidence from coincidence: patterns of tick infestations on rodents facilitate transmission of tick-borne encephalitis virus. *Parasitology* **118**, 177–186.

Rehacek, J. (1965). Development of animal viruses and rickettsiae in ticks and mites. *Annual Review of Entomology* **10**, 1–24.

Rehse-Kupper, B., Casals, J., Rehse, E. & Ackermann, R. (1976). Eyach: an arthropod-borne virus related to Colorado tick fever virus in the Federal Republic of Germany. *Acta Virologica* **20**, 339–342.

Reid, H. W. (1984). Epidemiology of louping-ill. In *Vector Biology*, eds. Mayo, M. A. & Harrap, K. A., pp. 161–178. London: Academic Press.

Rennie, L., Wilkinson, P. J. & Mellor, P. S. (2000). Effects of infection of the tick *Ornithodoros moubata* with African swine fever virus. *Medical and Veterinary Entomology* **14**, 355–360.

Rennie, L., Wilkinson, P. J. & Mellor, P. S. (2001). Transovarial transmission of African swine fever virus in the argasid tick *Ornithodoros moubata*. *Medical and Veterinary Entomology* **15**, 140–146.

Rey, F. A., Heinz, F. X., Mandl, C., Kunz, C. & Harrison, S. C. (1995). The envelope glycoprotein from tick-borne encephalitis virus at 2 Å resolution. *Nature* **375**, 291–298.

Roy, P. (2001). Orbiviruses. In *Fields' Virology*, 4th edn, eds. Knipe, D. M., Howley, P. M. *et al.*, pp. 1835–1869. Philadelphia, PA: Lippincott, Williams & Wilkins.

Salas, M. L. (1994). African swine fever virus. In *Encyclopedia of Virology*, vol. 1, eds. Webster, R. G. & Granoff, A., pp. 1–29. London: Academic Press.

Schmaljohn, C. S. & Hooper, J. W. (2001). *Bunyaviridae*: the viruses and their replication. In *Fields' Virology*, 4th edn, eds. Knipe, D. M., Howley, P. M. *et al.*, pp. 1581–1602. Philadelphia, PA: Lippincott, Williams & Wilkins.

Schoehn, G., Moss, S. R., Nuttall, P. A. & Hewat, E. A. (1997). Structure of Broadhaven virus by cryo-electron microscopy: correlation of structural and phenotypic properties of Broadhaven virus and bluetongue virus outer capsid proteins. *Virology* **235**, 191–200.

Shepherd, A. J., Swanepoel, R., Cornel, A. J. & Mathee, O. (1989). Experimental studies on the replication and transmission of Crimean–Congo hemorrhagic fever virus in some African tick species. *American Journal of Tropical Medicine and Hygiene* **40**, 326–331.

Shiu, S. Y. W., Ayres, M. D. & Gould, E. A. (1991). Economic sequence of the structural proteins of louping ill virus: comparative analysis with tick-borne encephalitis virus. *Virology* **180**, 411–415.

Simons, J. F., Hellman, U. & Pettersson, R. F. (1990). Uukuniemi virus S RNA segment: ambisense coding strategy, packaging of complementary strands into virions, and homology to members of the genus *Phlebovirus*. *Journal of Virology* **64**, 247–255.

Skofertsa, P. G., Gaidamovich, S. I. A., Obukhova, V. R., Korchmar' Nd, I. A. & Rovoi, P. I. (1972). Isolation of Kemerovo group Kharagysh virus on the territory of the Moldavian SSR. *Voprosy Virusologii* **17**, 709–711.

Smith, C. E. G. & Varma, M. G. R. (1981). Louping ill. In *CRC Handbook Series of Zoonoses*, Section B, *Viral Diseases*, vol. 1, ed. Beran, G. W., pp. 191–200. Boca Raton, FL: CRC Press.

Sonenshine, D. E. (1991). *Biology of Ticks*, vol. 1. Oxford, UK: Oxford University Press.

Sreenivasan, M. A., Bhat, H. R. & Rajagopalan, P. K. (1979). Studies on the transmission of Kyasanur forest disease virus by partly fed ixodid ticks. *Indian Journal of Medical Research* **69**, 708–713.

Steele, G. M. & Nuttall, P. A. (1989). Difference in vector competence of two species of sympatric ticks, *Amblyomma variegatum* and *Rhipicephalus appendiculatus* for Dugbe virus (*Nairovirus*: Bunyaviridae). *Virus Research* **14**, 73–84.

Swanepoel, R., Shepherd, A. J., Leman, P. A., *et al.* (1987). Epidemiological and clinical features of Crimean–Congo hemorrhagic fever in southern Africa. *American Journal of Tropical Medicine and Hygiene* **36**, 120–132.

Taylor, R. M., Hurlbut, H. S., Work, T. H. & Kingston, J. R. (1966). Arboviruses isolated from *Argas* ticks in Egypt: Quaranfil, Chenuda and Nyaminini. *American Journal of Tropical Medicine and Hygiene* **15**, 76–86.

Telford, S. R., Armstrong, P., Katavolos, P., *et al.* (1997). A new tick-borne encephalitis-like virus infecting New England deer ticks, *Ixodes dammini. Emerging Infectious Diseases* **3**, 165–170.

Thomas, L. A., Kennedy, R. C. & Eklund, C. M. (1960). Isolation of a virus closely related to Powassan virus from *Dermacentor andersoni* collected along north Cache la Poudre River, Colo. *Proceedings of the Society of Experimental Biology and Medicine* **104**, 355–359.

Titus, R. G. & Ribeiro, J. M. C. (1990). The role of vector saliva in transmission of arthropod-borne diseases. *Parasitology Today* **6**, 157–160.

Traavik, T. (1979). Arboviruses in Norway. In *Arctic and Tropical Arboviruses*, ed. Kurstak, E., pp. 67–81. New York: Academic Press.

Turell, M. J. (1988). Horizontal and vertical transmission of viruses by insect and tick vectors. In *The Arboviruses: Ecology and Epidemiology*, vol. 1, ed. Monath, T. P., pp. 127–152. Boca Raton, FL: CRC Press.

Vancova, M., Zacharovova, K., Grubhoffer, L. & Nebesarova, J. (2006). Ultrastructure and lectin characterization of granular salivary cells from *Ixodes ricinus* females. *Journal of Parasitology* **92**, 431–440.

Varma, M. G. R. & Smith, C. E. G. (1972). Multiplication of Langat virus in the tick *Ixodes ricinus. Acta Virologica* **16**, 159–167.

Vermeil, C., Lavillaureix, J. & Reeb, E. (1959). Infection et transmission expérimentales du virus West Nile par *Ornithodoros coniceps* (Canestrini) de souche Tunisienne. *Bulletin de la Société de Pathologie Exotique* **53**, 489–495.

Vinuela, E. (1985). African swine fever. *Current Topics in Microbiology and Immunology* **116**, 151–170.

Wang, H. & Nuttall, P. A. (2001). Intra-stadial tick-borne Thogoto virus (Orthomyxoviridae) transmission:

accelerated arbovirus transmission triggered by host death. *Parasitology* **122**, 439–446.

Watts, D. M., Ksiazek, T. G., Lithicum, K. J. & Hoogstraal, H. (1989). Crimean–Congo hemorrhagic fever. In *The Arboviruses: Epidemiology and Ecology*, vol. 2, ed. Monath, T. P., pp. 177–260. Boca Raton, FL: CRC Press.

Weber, F., Jambrina, E., Gonzalez, S., *et al.* (1998). *In vivo* reconstitution of active Thogoto virus polymerase: assays for the compatibility with other orthomyxovirus core proteins and template RNAs. *Virus Research* **58**, 13–20.

Woodall, J. (2001). Thogoto virus. In *The Encyclopedia of Arthropod-Transmitted Infections*, ed. Service, M. W., pp. 504–506. Wallingford, UK: CAB International.

Yanez, R. J., Rodriguez, J. M., Nogal, M. L., *et al.* (1995). Analysis of the complete nucleotide sequence of African swine fever virus. *Virology* **208**, 249–278.

Yozawa, T., Kutish, G. F., Afonso, C. L., Lu, Z. & Rock, D. L. (1994). Two novel multigene families, 530 and 300, in the terminal variable regions of African swine fever virus genome. *Virology* **202**, 997–1002.

Yunker, C. E. (1975). Tick-borne viruses associated with seabirds in North America and related islands. *Medical Biology* **53**, 302–311.

Zaki, A. M. (1997). Isolation of a flavivirus related to the tick-borne encephalitis complex from human cases in Saudi Arabia. *Transactions of the Royal Society of Tropical Medicine and Hygiene* **91**, 179–181.

Zanotto, P. M. de A., Gao, G. F., Gritsun, T., *et al.* (1995). An arbovirus cline across the northern hemisphere. *Virology* **210**, 152–159.

Zeller, H. G., Karabatsos, N., Calisher, C. H., *et al.* (1989*a*). Electron microscopic and antigenic studies of uncharacterized viruses. I. Evidence suggesting placement of viruses in the families *Arenaviridae, Paramyxoviridae,* or *Poxviridae. Archives of Virology* **108**, 191–209.

Zeller, H. G., Karabatsos, N., Calisher, C. H., *et al.* (1989*b*). Electron microscopic and antigenic studies of uncharacterized viruses. II. Evidence suggesting placement of viruses in the family *Bunyaviridae. Archives of Virology* **108**, 211–227.

Zilber, L. A. & Soloviev, V. D. (1946). Far Eastern tick-borne spring-summer (spring) encephalitis. *American Review of Soviet Medicine* (Special Suppl.), 1–80.

13 • Babesiosis of cattle

R. E. BOCK, L. A. JACKSON, A. J. DE VOS AND W. K. JORGENSEN

INTRODUCTION

Babesiosis (also known as tick fever or cattle fever) is caused by intraerythrocytic protozoan parasites of the genus *Babesia* that infect a wide range of domestic and wild animals, and occasionally humans. The disease is tick-transmitted and distributed worldwide. Economically, tick fever is the most important arthropod-borne disease of cattle, with vast areas of Australia, Africa, South and Central America and the United States continuously under threat. Tick fever was the first disease for which transmission by an arthropod to a mammal was implicated at the turn of the twentieth century, and is the first disease to be eradicated from a continent (North America). This review describes the biology of *Babesia* spp. in the host and the tick, the scale of the problem to the cattle industry, the various components of control programmes, epidemiology, pathogenesis, immunity, vaccination and future research. The emphasis is on *Babesia bovis* and *Babesia bigemina*, the two most important species infecting cattle.

Babes (1888) investigated disease outbreaks causing haemoglobinuria in cattle in Romania in 1888 and was the first to describe piroplasms in the blood of cattle. He believed it to be a bacterium and named it *Haematococcus bovis* (Angus, 1996). Shortly afterwards investigations by Smith and Kilborne (1893) in the United States of America demonstrated the causative organism of 'Texas Fever' (babesiosis), which they called *Pyrosoma bigeminum* (= *Babesia bigemina*). They were the first to demonstrate transmission of a disease organism from an arthropod to a mammalian host when they showed the organism was transmitted by *Boophilus annulatus* to cattle. Starcovici in 1893 was the first to apply the genus name *Babesia* to this group of parasites (Uilenberg, 2006). After the publication of this work, babesiosis was discovered in various parts of the world. Dr Sidney Hunt confirmed that bovine 'redwater' in Australia was identical in aetiology to Texas Fever (Angus, 1996). In Argentina,

Lignières (1903) described two forms of 'Tristeza' (babesiosis) 'forme A' and 'forme C'. These were later known as *Babesia bigemina* and *Babesia argentina* (= *B. bovis*). From descriptions of these parasites, it is now believed that both *Babesia bigemina* and *Babesia bovis* were also present in cases of babesiosis in Australia and USA in the late 1890s and early 1900s (Angus, 1996). The four species of bovine babesia that are now recognized are *Babesia bovis* (= *B. argentina; B. berbera; B. colchica*), *Babesia bigemina*, *Babesia divergens* (= *B. caucasica; B. occidentalis; B. karelica*) and *B. major* (Angus, 1996).

PARASITES AND DISTRIBUTION

The genus *Babesia* belongs to the phylum Apicomplexa, class Sporozoasida, order Eucoccidiorida, suborder Piroplasmorina and family Babesiidae (Allsopp *et al.*, 1994; Levine, 1971, 1985). Criado-Fornelio *et al.* (2003) used the 18S rRNA gene for phylogenetic analysis and divided piroplasmids into five proposed clades: (1) the *B. microti* group with *B. rodhaini, B. felis, B. leo, B. microti* and *Theileria annae* (Archaeopiroplasmids); (2) the theileriid-like group of small piroplasmids of cervids and dogs from USA (prototheileriids); (3) the *Theileria* group, containing all *Theileria* species from Bovinae and including *Theileria (ex Babesia) equi* (theileriids); (4) a group of *Babesia* species including *B. canis* and *B. gibsoni* from canids together with *B. divergens* and *B. odocoilei* (babesiids); and (5) a group comprising mainly *Babesia* species from ungulates: *B. caballi, B. bigemina, B. ovis, B. bovis* and *Babesia* spp. from cattle (ungulibabesiids). By definition, babesia parasites do not form pigment in parasitized cells, and replicate only in erythrocytes; infection is passed transovarially in the tick vector (Uilenberg, 2006). Therefore only the babesiids and ungulibabesiids of Criado-Fornelio *et al.* (2003) are true babesia parasites.

Ticks: Biology, Disease and Control, ed. Alan S. Bowman and Patricia A. Nuttall. Published by Cambridge University Press.

Table 13.1 *Major* Babesia *species infective to domestic animals, their ixodid tick vectors and geographical distribution as adapted from Uilenberg (2006)*

Babesia species	Major ixodid vectors	Known distribution	Domestic species affected
Babesia bigemina	*Boophilus microplus* *Boophilus decoloratus* *Boophilus annulatus* *Boophilus geigyi* *Rhipicephalus evertsi*	Africa, Asia, Australia, Central and South America and Southern Europe	Cattle, buffalo
Babesia bovis	*Boophilus microplus* *Boophilus annulatus* *Boophilus geigyi*	As for *Babesia bigemina*, but less widespread in Africa due to *Boophilus microplus* competition with *B. decoloratus*	Cattle, buffalo
Babesia divergens	*Ixodes ricinus* *Ixodes persulcatus*	Northwest Europe, Spain, Great Britain, Ireland	Cattle
Babesia major	*Haemaphysalis punctata*	Europe, Northwest Africa, Asia	Cattle
Babesia ovata	*Haemaphysalis longicornis*	Eastern Asia	Cattle
Babesia orientalis	*Rhipicephalus haemaphysaloides*	Asia	Buffalo
Babesia ovis	*Rhipicephalus bursa*	Southeastern Europe, North Africa, Middle East, Asia	Sheep and goat
Babesia motasi	*Haemaphysalis punctata*	Southeastern Europe, North Africa and Asia	Sheep and goat
Babesia caballi	*Dermacentor* spp. *Hyalomma marginatus* *Hyalomma truncatum* *Rhipicephalus evertsi evertsi*	Africa, South and Central America and southern USA, Europe, Asia	Horses, donkey, mule
Babesia canis	*Rhipicephalus sanguineus* *Dermacentor* spp. *Haemaphysalis* spp. *Hyalomma* spp.	Europe	Dog
Babesia rossi	*Haemaphysalis leachi*	Africa	Dog
Babesia vogeli	*Rhipicephalus sanguineus*	Africa, Asia, Europe, Australia, America	Dog
Babesia gibsoni	*Haemaphysalis* spp. *Rhipicephalus sanguineus*	Africa, Asia, Europe, North America	Dog
Babesia trautmanni	*Rhipicephalus* spp.	Southern Europe, former USSR, Africa	Pig

The major *Babesia* spp. known to infect domestic animals, and their proven vectors, are listed in Table 13.1. *Babesia bovis* and *B. bigemina* are present in many countries between 40° N and 32° S (McCosker, 1981). The genus *Rhipicephalus* is considered paraphyletic with respect to the genus *Boophilus* and thus *Boophilus* has become a subgenus of the genus *Rhipicephalus* (Barker & Murrell, 2002, 2004). However, there is a wealth of literature on *Boophilus* species

and a large part of the tick, babesia and livestock community uses this name on a regular basis. Also the application of polyphasic taxonomy argues for the retention of *Boophilus* as a separate genus (Uilenberg, Thiaucourt & Jongejan, 2004). Therefore throughout this review we have retained the *Boophilus* name.

Boophilus microplus is the most important and widespread vector of *B. bovis* and *B. bigemina*, but in southern Africa a closely related tick, *Boophilus decoloratus*, interferes with its spread in drier and colder areas. Interbreeding between the two tick species produces sterile progeny which creates a zone through which *Boophilus microplus* has difficulty passing (Sutherst, 1987). Generally *B. bovis* and *B. bigemina* have the same distribution, but in Africa *B. bigemina* is more widespread than *B. bovis* because of the ability of *Boophilus decoloratus* and *Rhipicephalus evertsi* to act as vectors for *B. bigemina* (Friedhoff, 1988). *Boophilus annulatus* is the principal vector of *B. bovis* and *B. bigemina* in northern Africa (Bouattour, Darghouth & Daoud, 1999; Ndi *et al.*, 1991; Sahibi *et al.*, 1998), the Middle East (Pipano, 1997), Turkey (Sayin *et al.*, 1996) and in some areas of southern Europe (Caeiro, 1999).

Babesia divergens is transmitted almost exclusively by *Ixodes ricinus* in northern Europe (Friedhoff, 1988) and this probably explains its limited distribution. As this review concentrates on *B. bovis* and *B. bigemina*, a recent review by Zintl *et al.* (2003) should be referred to for a comprehensive summary of *B. divergens* biology, including its life cycle, host specificity and morphology, and the current state of knowledge about both human and bovine infections.

ECONOMIC IMPACT

Most of the 1.2 billion cattle in the world are exposed to babesiosis but this figure is not a true reflection of the number at risk of disease (McCosker, 1981). Breeds of cattle that are indigenous to babesia endemic regions often have a certain degree of natural resistance to these diseases and the consequences of infection are not as serious as when exotic *Bos taurus* breeds are involved. In addition, in tropical areas with a high vector population, natural exposure can occur at an early age and cattle are therefore immune to subsequent challenge as adults (see section 'Endemic stability' below).

Costs due to babesiosis are incurred not only from mortality, ill-thrift, abortions, loss of milk/meat production and draught power, and from control measures (such as acaricide treatments, purchase of vaccines and therapeutics), but also through its impact on international cattle trade. McLeod & Kristjanson (1999) developed a spreadsheet model (Tick Cost) to assess the overall impact of ticks and tick-borne diseases. They calculated that losses due to babesiosis and anaplasmosis and control measures cost the Australian cattle industry US$16.9 m per annum with 'tick worry' adding a further US$6.4 m to annual losses. The model further estimated the cost in Kenya, Zimbabwe, Tanzania, South Africa, China, India, Indonesia and Philippines to be US $5.1, $5.4, $6.8, $21.6, $19.4, $57.2, $3.1 and $0.6 million annually, respectively.

LIFE CYCLE STAGES AND DEVELOPMENT OF *BABESIA*

The development of *Babesia* spp. in ticks was reviewed by Friedhoff (1988). Despite many detailed studies, our understanding of the life cycles of *Babesia* spp. is still incomplete.

In electron microscope studies, Rudzinska *et al.* (1983) showed sexual reproduction in *B. microti*. More recently, DNA measurements showed that sexual reproduction does occur for *B. divergens* (Mackenstedt *et al.*, 1990), and *B. bigemina* and *B. canis* (Mackenstedt *et al.*, 1995), and is therefore likely in the other species. Mackenstedt *et al.* (1995) also revealed important differences in the life cycle of the *Babesia* species they studied, indicating that the genus does not have a characteristic life cycle. Therefore, the emphasis here will be on *B. bovis* and *B. bigemina* with particular reference to the latter as more information is available on this species.

The development of *B. bovis* and *B. bigemina* follow similar patterns in adult *Boophilus* spp. (Friedhoff, 1988; Mehlhorn & Shein, 1984; Potgieter, 1977; Potgieter & Els, 1976). The life cycle (Fig. 13.1) as currently understood is outlined below, and alternatives are given in parentheses to avoid confusion about the terminology used in the literature.

Babesia spp. do not parasitize any vertebrate host cell other than erythrocytes (Friedhoff, 1988). During an asexual growth cycle of babesia parasites in a natural host, each extracellular sporozoite (merozoite) invades a host erythrocyte via multiple adhesive interactions of several protozoan ligands with the target receptors on the host cell surface (Yokoyama, Okamura & Igarashi, 2006). Once inside, it transforms into a trophozoite from which two merozoites develop by a process of merogony (binary fission) (Friedhoff, 1988; Potgieter & Els, 1977a; Potgieter & Els, 1979). In *B. bigemina* Mackenstedt *et al.* (1995) identified an ovoid type of merozoite they called a gamont precursor, which unlike other piroplasms studied, is diploid. These gamont precursors do not develop further until they are taken up by the tick.

Changes experienced in passage from host blood to the midgut of the tick vector stimulate the development of two populations of ray bodies (strahlenkörper) from the gamonts (gametocytes) (Gough, Jorgensen & Kemp, 1998). The ray bodies undergo further multiplication within the erythrocyte that continues after they have emerged. Large aggregations of multinucleated ray bodies form. Once division is complete, single-nucleated ray bodies that are now

LIFE CYCLE FOR *BABESIA BIGEMINA*

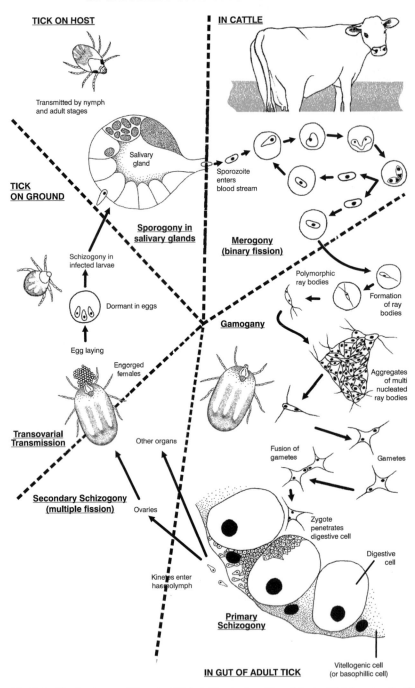

Fig. 13.1. The development life cycle of *Babesia bigemina* in cattle and the ixodid tick vector *Boophilus microplus* as currently understood and adapted from Gough, Jorgensen & Kemp (1998), Mackenstedt *et al.* (1995) and Mehlhorn & Shein (1984).

Fig. 13.2. Dorsal and ventral view of *Boophilus microplus* ticks and their egg mass. *Babesia* parasites develop in oocytes of *Boophilus microplus* resulting in transovarial transmission, an essential component in babesiosis epidemiology as *Boophilus* spp. are one-host ticks.

haploid and assumed to be gametes (Mackenstedt *et al.*, 1995) emerge from the aggregates. These then fuse together in pairs (syngamy) (Gough *et al.*, 1998) to form a spherical cell (zygote) (Friedhoff, 1988). The zygotes selectively infect the digestive cells of the tick gut where they probably multiply and then infect the basophilic cells (vitellogenin synthesizing cells). Further multiplication occurs in the basophilic cells with development to kinetes (vermicules) that escape into the tick haemolymph (Agbede, Kemp & Hoyte, 1986).

Babesia bigemina undergoes a one-step meiosis to form a haploid zygote at some stage in development in the gut (Mackenstedt *et al.*, 1995). In the gut cells (probably vitellogenic cells), schizogony (multiple fission, sporogony) occurs with the formation of polyploid kinetes (vermicules, sporokinetes, large merozoites) (Mackenstedt *et al.*, 1995). These motile club-shaped kinetes then escape into the haemolymph and infect a variety of cell types and tissues, including the oocytes where successive cycles of secondary schizogony take place. Thus, transovarial transmission occurs with further development taking place in the larval stage.

This is an important life cycle adaptation as the *Boophilus* species vectors are one-host ticks (Fig. 13.2). Kinetes enter the salivary glands and are transformed into multinucleated stages (sporogony) and these then break up to form sporozoites (small haploid merozoites) (Mackenstedt *et al.*, 1995). In all species, sporozoite development usually only begins when the infected tick attaches to the vertebrate host. In *B. bigemina*, some development takes place in the feeding larvae, but infective sporozoites take about 9 days to appear

and therefore only occur in the nymphal and adult stages of the tick (Hoyte, 1961; Potgieter & Els, 1977*b*). Transmission can occur throughout the rest of the nymphal stage and by adult females and males (Callow & Hoyte, 1961; Dalgliesh, Stewart & Callow, 1978; Riek, 1964). In the case of *B. bovis*, the formation of infective sporozoites usually occurs within 2 to 3 days of larval tick attachment (Riek, 1966). *Babesia bovis* does not persist in an infective form in the tick beyond the larval stage (Mahoney & Mirre, 1979).

EPIDEMIOLOGY

The usual prepatent period for *B. bigemina* is 12–18 days after tick attachment (Callow, 1984). However, despite its classification as a one-host tick, *Boophilus microplus* (particularly males) can transfer amongst cattle in close proximity. This can lead to a much-shortened prepatent period of 6–12 days (Callow, 1979; Callow & Hoyte, 1961). The prepatent period for *B. bovis* is generally 6–12 days and peak parasitaemias are reached about 3–5 days after that (Callow, 1984). Heat stimulation of larval ticks prior to attachment (37 °C for 3 days or 30 °C for 8 days) enables transmission of *B. bovis* immediately the larvae attach; this can lead to naturally shortened prepatent periods, particularly in summer (Dalgliesh & Stewart, 1982).

Endemic stability

Endemic stability in a population occurs when clinical disease is scarce despite a high frequency of infection. For babesiosis, passively acquired resistance from colostrum lasts about 2 months and is followed by innate immunity (as defined in section 'Immunity') from 3 to 9 months of age (Mahoney, 1974; Mahoney & Ross, 1972). As a result, calves infected with *Babesia* spp. during the first 6 to 9 months rarely show clinical symptoms and develop a solid long-lasting immunity (Dalgliesh, 1993). Mahoney (1974) estimated that if at least 75% of calves are exposed to *B. bovis* infection by 6 to 9 months of age the disease incidence will be very low and a state of natural endemic stability will exist.

One infected tick is sufficient to transmit *B. bovis*. However, tick infection rates can be low and the rate of transmission to cattle correspondingly low. Consequently, under conditions of endemic instability, some animals fail to become infected for a considerable period after birth but develop severe, life-threatening disease if infected later in life (Callow, 1984). In an Australian field study, only 0.04% of larval ticks were found to be infected with *B. bovis* in paddock surveys where *Bos taurus* cattle were involved and even fewer in paddocks stocking *Bos indicus* cattle (Mahoney & Mirre, 1971; Mahoney *et al.*, 1981). A higher tick infection prevalence for *B. bigemina* (0.23%) was found in the same study (Mahoney & Mirre, 1971). Consequently, transmission rates for *B. bigemina* were higher than *B. bovis*, and endemic stability was more likely to arise for *B. bigemina* than *B. bovis*.

Breed resistance to babesiosis

Bos indicus breeds almost invariably exhibit milder clinical response to primary *B. bovis* infections than *Bos taurus* breeds (Bock *et al.*, 1997*a*; Bock, Kingston & de Vos, 1999*a*; Callow, 1984; Lohr, 1973). This phenomenon is thought to be a result of the evolutionary relationship between *Bos indicus* cattle, *Boophilus* spp. and babesia (Dalgliesh, 1993).

Mahoney *et al.* (1981) compared transmission rates of *B. bovis* in *Bos taurus* and in three-eighths to half *Bos indicus* crossbred cattle in Australia. They concluded that, in an environment unfavourable for tick survival, stocking with *Bos indicus* or *Bos indicus* crossbred cattle will, over several seasons, almost lead to the disappearance of the ticks. However, Bock *et al.* (1999*a*) showed that if *Bos indicus* cattle that have not been previously exposed to *Boophilus microplus* are moved to a paddock with a high resident *Boophilus microplus* infestation, babesia transmission rates can initially be very high.

In Australia, *B. bigemina* is usually of lower pathogenicity than *B. bovis* (Callow, 1979). There is some ambiguity in the literature concerning the relative susceptibility of *Bos indicus* and *Bos taurus* breeds to *B. bigemina*, but most studies conclude that *Bos indicus* are much more resistant (Bock *et al.*, 1997*a*; 1999*b*; Lohr, 1973; Parker *et al.*, 1985).

IMMUNITY

The immune response of cattle to infection with *B. bovis* or *B. bigemina* involves both innate and acquired immune mechanisms.

Innate immune mechanisms

Innate immunity is non-specific and includes factors such as host–parasite specificity, genetic factors, age of the host and the response of host cells (such as the mononuclear phagocyte system and polymorphonuclear leukocytes).

Most *Babesia* spp. are highly host specific and often splenectomy is needed to establish an infection in an unnatural host (Mahoney, 1972). Cattle with long-standing infections of *Babesia* spp. can relapse to the infections if splenectomized (Callow, 1977) and susceptible cattle that have been splenectomized develop much higher parasitaemias to primary *Babesia* spp. infections than do intact cohorts (Callow, 1984). These observations imply that the spleen plays a major role in the immune response to *Babesia* spp.

Different breeds of cattle are known to have different susceptibilities to infection with *B. bovis* and *B. bigemina* (see section 'Breed resistance to babesiosis'). The genetically determined factors that might control the susceptibility of different cattle breeds have not yet been identified.

There is an age-related immunity to primary infection of cattle with *B. bovis* and *B. bigemina* (Riek, 1963), with young calves exhibiting a strong innate immunity compared to adult cattle (Goff *et al.*, 2001; Trueman & Blight, 1978). Initially this innate immunity was thought to be due to the passive transfer of protective antibodies in the colostrum of immune dams (Mahoney, 1967*b*), but calves from non-immune mothers exhibit resistance to *B. bovis* and *B. bigemina* (Riek, 1963). Young calves control acute *B. bovis* infections with spleen-dependent innate immune mechanisms that involve the early induction of interleukin (IL)-12 that stimulates natural killer (NK) cells to produce interferon gamma (IFN-γ) which in turn activates macrophages to produce nitric oxide (Goff *et al.*, 2001, 2006). Nitric oxide kills the parasites early enough to prevent pathological consequences of infection, but there is sufficient exposure to ensure an acquired immune response. The nitric oxide response decreases to pre-infection levels by day 10 post-infection. In contrast in *B. bovis* infection of adult cattle, IL-12 and IFN-γ are induced later and nitric oxide is not produced early in the infection (Goff *et al.*, 2001).

Activated monocytes, macrophages and neutrophils provide the first line of defence during an infection with *Babesia* spp., employing antimicrobial agents, including reactive nitrogen intermediates (RNI), reactive oxygen intermediates (ROI) and phagocytosis (Court, Jackson & Lee, 2001). In addition, these cells secrete regulators of the inflammatory response such as cytokines IL-1β, IL-12, and tumour necrosis factor alpha (TNF-α), as well as nitric oxide (Court *et al.*, 2001; Shoda *et al.*, 2000).

Increased phagocytic activity of macrophages has been proposed as a mechanism for the elimination of *B. bovis* in cattle (Jacobson *et al.*, 1993; Mahoney, 1972), with specific antibodies playing an important role as opsonins (Jacobson *et al.*, 1993). During a primary *B. bovis* infection in cattle, peripheral blood monocytes display a suppressed phagocytic ability and neutrophils exhibit an increased phagocytic ability that coincides with peak parasitaemia (Court *et al.*, 2001).

Nitric oxide, an RNI, is produced by macrophages, monocytes, neutrophils and endothelial cells during acute infections. In vitro experiments indicated that nitric oxide reduces the viability of *B. bovis* (Johnson *et al.*, 1996) and that both intact *B. bovis* merozoites and fractionated merozoite antigens induce the production of nitric oxide by monocytes/macrophages in the presence of IFN-γ and TNF-α (Goff *et al.*, 2002; Stich *et al.*, 1998). Indirect in vivo evidence suggests that nitric oxide has a role in the pathology caused by *B. bovis* infections (Gale *et al.*, 1998). Gale *et al.* (1998) administered aminoguanidine, an inhibitor of inducible nitric oxide synthase, to cattle during a virulent *B. bovis* infection and caused a reduction in parasitaemia and amelioration of anaemia and pyrexia.

Phagocytosis is closely linked with an oxidative burst and the release of ROI (superoxide anion, hydrogen peroxide and hydroxyl radicals) within the phagocyte (Auger & Ross, 1992). The oxidative burst reaction is stimulated by cross-linking of receptors on the phagocyte surface, such as Fc receptors for immunoglobulin G (IgG); ROIs are released within the phagosome and into the extracellular environment. In vitro experiments demonstrated that ROIs (superoxide anion and hydroxyl radicals, but not hydrogen peroxide) produced by activated macrophages are babesicidal (Johnson *et al.*, 1996). Furthermore, ex vivo derived monocytes exhibit increased, and neutrophils display decreased oxidative activity during a primary *B. bovis* infection (Court *et al.*, 2001).

Acquired immune mechanisms

Hyperimmune serum (from cattle infected with *B. bovis* many times) or a mixture of IgG1 and IgG2 prepared from hyperimmune serum can be used to passively protect naive calves against *B. bovis* infection (Mahoney *et al.*, 1979a). However, this protection is strain specific. Splenectomized calves given hyperimmune serum and challenged with *B. bovis* recovered as effectively as similarly challenged naive intact calves. Serum collected after a primary infection was found to be less effective than hyperimmune serum for protecting naive calves and it was thought that this could be related to the isotype of antibody present in the serum (Goodger, Wright & Waltisbuhl, 1985). Antibodies probably act as opsonins for increased phagocytosis rather than having a direct effect on the parasite's viability (Mahoney *et al.*, 1979a) and are thought to be important effectors during a secondary infection (Mahoney, 1986). Immune complexes of

babesial antigen, bovine immunoglobulin and C3 of the complement cascade form following *B. bovis* infection (Goodger, Wright & Mahoney, 1981). The major immunoglobulin in the complexes is IgM but IgG1 and IgG2 are also present in lower concentrations. Both bovine IgG1 and IgG2 are capable of fixing complement; bovine IgG2 is the superior opsonizing antibody subclass (McGuire, Musoke & Kurtti, 1979). Following *B. bovis* infection, IgG1 and IgM (but not IgG2) complement-fixing antibodies have been detected (Goff *et al.*, 1982). Antibody-dependent cell-mediated cytotoxicity may be involved in the resolution of *B. bovis* infection in cattle (Goff, Wagner & Craig, 1984).

Adoptive transfer studies have not yet been performed with cattle to investigate the definitive role of T-cells in the immune response to *B. bovis* and *B. bigemina*. However, in vitro experiments performed with T-cell lines and clones from immune cattle have demonstrated a role for T cells. Peripheral blood mononuclear leukocytes and CD4$^+$ T-cell clones from immune cattle show proliferation in response to *B. bovis* antigens (Brown & Logan, 1992; Brown *et al.*, 1991). The cytokine response by these CD4$^+$ helper T-cell clones was either a Th1 (IL-2, IFN-γ and TNF-α) or Th0 (IL-2, IL-4, IFN-γ and TNF-α) response with no Th2 clones detected (Brown *et al.*, 1993). Furthermore, Th1 clones can promote IgG2 production by autologous B-cells (Rodriguez *et al.*, 1996) and produce IFN-γ and TNF-α that contribute to the induction of macrophages to produce nitric oxide (Stich *et al.*, 1998).

The model of acquired immunity to *B. bovis* and *B. bigemina* relies on the rapid activation of memory and effector CD4$^+$ helper T-cells that secrete IFN-γ, which activates phagocytic cells and enhances antibody production by B-cells (Brown *et al.*, 2006b).

Babesis bovis and *B. bigemina* cross-protection

Protective cross-species immunity against infection cannot be induced with *B. bovis* or *B. bigemina* (Smith *et al.*, 1980). This is in agreement with Legg (1935) who found that *B. bovis* infection does not protect against *B. bigemina* challenge. However, *B. bigemina* immunity does give some cross-protection against *B. bovis* (Wright *et al.*, 1987). The cross-protection was exploited in Australia for a short period in the 1930s, but was found to be unreliable (Callow, 1984).

Duration of immunity

In *Bos taurus* cattle, Mahoney, Wright & Mirre (1973) found that tick-transmitted *B. bovis* infection persisted for at least

4 years but *B. bigemina* infection usually lasted less than 6 months (Mahoney *et al.*, 1973). Immunity to both parasites persisted for at least 4 years. Tjornehoj *et al.* (1996) placed great importance on the correlation between the persistence of antibodies to *B. bovis* and the duration of immunity. However, there is ample evidence in the literature suggesting the presence of antibodies is not necessarily an indication of immunity nor is absence of detectable antibodies necessarily an indication of a lack of immunity. Mahoney *et al.* (1979*a*) clearly showed that while antibodies alone provided good homologous protection, the same did not apply in cross-protection between strains.

Johnston, Leatch & Jones (1978) showed that immunity of *Bos indicus*-cross cattle to *B. bovis* lasted at least 3 years despite the fact that most of the cattle eliminated the infection during that time. Unfortunately, Johnston and co-workers did not monitor antibody levels in the cattle. In a different study, however, Callow *et al.* (1974*a*, 1974*b*) showed that sera from vaccinated cattle sterilized with imidocarb were generally negative within 6 months of treatment, but this decrease in indirect fluorescent antibody test (IFAT) titre was not associated with a loss of immunity. In a long-term study of immunity in *Bos taurus* cattle to *B. bovis*, Mahoney, Wright & Goodger (1979*b*) found that 30% to 50% of the vaccinated or naturally infected trial cattle became seronegative in an indirect haemagglutination test during the trial period. However, these cattle were still immune 4 years after initial vaccination or exposure. Therefore, it appears that a persistent, detectable antibody titre is not a prerequisite for immunity, but it is a very effective indicator of recent natural infection or vaccination.

Strain variation and persistence of infection

Isolates and strains of *B. bovis* and *B. bigemina* differ antigenically (Dalgliesh, 1993). Cross-immunity experiments have shown that recovered cattle are more resistant to challenge with the same (homologous) isolates than with different (heterologous) ones (Dalgliesh, 1993; de Vos, Dalgliesh & Callow, 1987; Shkap *et al.*, 1994).

Antigenic variation is also known to occur during *B. bovis* infections with recovered cattle retaining a latent infection for from 6 months to several years. Episodes of detectable recrudescence of parasitaemia occur at irregular intervals during the latent phase of the infection (Mahoney & Goodger, 1969). *Babesia bovis* parasites are able to establish infections of long duration in immune hosts by adhering to endothelial cells to avoid splenic clearance, and by antigenic variation (Allred & Al-Khedery, 2004). Allred (2003) suggested that antigenic variation, cytoadhesion/ sequestration, host-protein binding, and induction of immunosuppression all facilitate persistence in the individual immune host. He also suggested that monoallelic expression of different members of a multigene family might facilitate multiple infections of immune hosts, and population dispersal in endemic areas (Allred, 2003).

PATHOGENESIS, CLINICAL SIGNS AND PATHOLOGY

Babesia bovis infection

Cytokines and other pharmacologically active agents have an important function in the immune response to *Babesia* spp. The outcome is related to the timing and the quantity produced. For example, the severe pathogenesis of *B. bovis* infections has been related to the overproduction of IFN-γ, TNF-α and nitric oxide (Brown *et al.*, 2006*a*) causing vasodilation, hypotension, increased capillary permeability, oedema, vascular collapse, coagulation disorders, endothelial damage and circulatory stasis (Ahmed, 2002; Brown *et al.*, 2006*a*, Wright *et al.*, 1989).

Although stasis is induced in the microcirculation by aggregation of infected erythrocytes in all capillary beds, probably the most deleterious pathophysiological lesions occur in the brain and lung. This can result in cerebral babesiosis and a respiratory distress syndrome associated with infiltration of neutrophils, vascular permeability and oedema (Brown & Palmer, 1999; Wright & Goodger, 1988).

Progressive haemolytic anaemia develops during the course of *B. bovis* infections. While this is not a major factor during the acute phase of the disease, it contributes to the disease process in more protracted cases.

Acute babesiosis generally runs a course of 3 to 7 days and fever ($>40\,°C$) is usually present for several days before other signs become obvious. This is followed by inappetence, depression, increased respiratory rate, weakness and a reluctance to move. A hypotensive shock syndrome develops with intravascular stasis and accumulation of parasitized red blood cells in the peripheral circulation, and a consequential drop in packed cell volume. Intravascular haemolysis with haemoglobinuria is often present later in the clinical phase; hence, the disease is known as redwater in some countries. Anaemia and jaundice occur, especially in more protracted cases. Muscle wasting, tremors and recumbency develop in advanced cases, followed terminally by coma (de Vos, De Waal & Jackson, 2004). The fever during infections may cause pregnant cattle to abort (Callow, 1984) and bulls to show reduced fertility lasting 6 to 8 weeks (Singleton, 1974). Cerebral babesiosis manifests itself as a variety of signs of central nervous system involvement, and is almost invariably fatal (de Vos *et al.*, 2004).

Haematological, biochemical and histopathological changes are described by deVos *et al.* (2004). Clinical pathology centres on a haemolytic anaemia, which is characteristically macrocytic and hypochromic. Dead animals appear anaemic with haemoglobinuria, excess thick granular bile, ecchymotic haemorrhages of the epicardium and endocardium, and congestion of the brain and visceral organs. A cherry-pink discolouration of the cerebral cortex is characteristic of acute *B. bovis* infections. The spleen is markedly swollen with bulging of the contents from the cut surface. When the disease is more protracted the carcass may be pale with slight jaundice, the kidneys and brain moderately congested or even pale, the spleen and liver only slightly enlarged, and the heart may present a more haemorrhagic appearance than in acute cases. Thick, granular bile is a constant finding.

Non-fatal cases may take several weeks to regain condition but recovery is usually complete. In subacute infections, clinical signs are less pronounced and sometimes difficult to detect. Calves that become infected before they attain 9 months of age often develop subclinical infections only (Callow, 1984). Recovered cases remain symptomless carriers for a number of years with the duration of infection being breed dependent (Johnston *et al.*, 1978; Mahoney, 1969).

Babesia bigemina infection

The pathogenesis of *B. bigemina* infection is almost entirely related to parasite-induced intravascular haemolysis. Anaemia, jaundice, haemoglobinuria and excess thick granular bile are commonly seen in fatal cases, but not congestion of the brain and visceral organs. Cardiac haemorrhages and splenic enlargement are not as marked as after *B. bovis* infection, but pulmonary oedema is a more regular feature (de Vos *et al.*, 2004). Haemoglobinuria is present earlier and more consistently than in *B. bovis* infections, but fever is less of a feature. There is no cerebral involvement and recovery in non-fatal cases is usually rapid and complete. However, in some cases the disease can develop very rapidly with sudden and severe anaemia, jaundice and death, which may occur with little warning. Animals that recover from

B. bigemina remain infective for ticks for 4 to 7 weeks and carriers for only a few months (Johnston *et al.*, 1978; Mahoney, 1969).

DIAGNOSIS

Babesia bovis is classically known as a 'small' babesia, usually centrally located in the erythrocyte. The single parasitic form is small and round measuring 1.0 to 1.5 μm. The paired parasitic form of *B. bovis* is typically pear-shaped, usually separated at an obtuse angle, with a rounded distal end. Each parasite measures approximately 1.5×2.5 μm. *Babesia bigemina* single parasites are irregular in shape and much larger, sometimes equalling the diameter of the erythrocyte. The paired parasitic form of *B. bigemina* is joined at an acute angle within the mature erythrocyte, pear-shaped, but with a pointed distal end. Each parasite is approximately 3–5 μm long and 1–1.5 μm wide, and usually has two discrete red-staining dots of nuclear material (*B. bovis* always has only one). However, *B. bovis* and *B. bigemina* both show considerable morphological variation, making it difficult to distinguish one from the other on morphological grounds alone (Callow, 1984; de Vos *et al.*, 2004). Pairs of *B. bovis* that are larger than any *B. bigemina* are occasionally seen.

For *B. divergens*, merozoite size, position inside the erythrocyte and morphological detail are dependent on the host species and have been summarized by Zintl *et al.* (2003)

Diagnosis of babesiosis is made by examination of blood and/or organ smears stained with Giemsa (Böse *et al.*, 1995; Callow, Rogers & de Vos, 1993). For the best results, blood films should be prepared from capillary blood collected, for instance, after pricking the tip of the tail or margin of an ear. The temptation to use blood of the general circulation should be resisted as these specimens may contain up to 20 times fewer *B. bovis* than capillary blood (Callow *et al.*, 1993). In *B. bigemina* infections, parasitized cells are evenly distributed throughout the blood circulation. Thick blood films are 10 times more sensitive and are therefore very useful for the detection of low-level *B. bovis* infections (Böse *et al.*, 1995). These films differ from thin ones in that the blood is not spread over a large area and is not fixed before staining, thus allowing lysis of the red blood cells and concentration of the parasites (Böse *et al.*, 1995).

Serological tests are reviewed by Böse *et al.* (1995) and Bock, Jorgensen & Molloy (2004). These tests are not of value in the clinical stage of the disease but are used for the purposes of research, epidemiological studies and export certification or where vaccine breakdown is suspected.

Böse *et al.* (1995) also reviewed the relative sensitivity of DNA probes, PCR assays, in vitro cultures and subinoculation into susceptible (usually splenectomized) calves to provide a diagnosis. Despite the added sensitivity of these methods and recent advances (Lew & Jorgensen, 2005), stained blood and/or organ smears offer considerable advantages in cost and speed for clinical cases if microscopists with experience in diagnosis of babesiosis are available.

TREATMENT

Reports in the literature refer to a number of effective babesicides (de Vos *et al.*, 2004), but few are now available commercially. Currently, diminazene aceturate and imidocarb dipropionate are the most widely used. Diminazene works rapidly against *B. bovis* and *B. bigemina* at a dose of 3.5 mg/kg intramuscularly. It is well tolerated and will protect cattle from the two diseases for 2 and 4 weeks, respectively (de Vos *et al.*, 2004). Imidocarb is used subcutaneously at a dose of 1.2 mg/kg for treatment, while a dose of 3 mg/kg provides protection from *B. bovis* for 4 weeks and *B. bigemina* for at least 2 months (Taylor & McHardy, 1979). At the high dose, imidocarb also eliminates *B. bovis* and *B. bigemina* from carrier animals. At either dose it can interfere with the development of immunity following live vaccination (de Vos, Dalgliesh & McGregor, 1986). Treatment with long-acting oxytetracycline following vaccination significantly reduces parasitaemia and red blood cell destruction without inhibiting the development of immunity (Jorgensen *et al.*, 1993; Pipano *et al.*, 1987). Oxytetracyclines are not usually sufficient to control virulent field infections.

LIVE VACCINE PRODUCTION

Cattle develop a durable, long-lasting immunity after a single infection with *B. divergens*, *B. bovis* or *B. bigemina*. This feature has been exploited in some countries to immunize cattle against babesiosis (Callow, 1984; de Vos & Jorgensen, 1992; Gray *et al.*, 1989). Methods used to prepare live vaccines against bovine babesiosis have been described or reviewed in some detail (Callow, Dalgliesh & de Vos, 1997; de Vos & Jorgensen, 1992; Pipano, 1995). Most early attempts involved the use of blood from infected carriers (Callow, 1977, 1984; de Vos & Jorgensen, 1992) but, during the past 30 years, more sophisticated techniques were developed to produce standardized live vaccines (Callow *et al.*, 1997). The inherent disadvantages of these vaccines are well known and include the risk of reactions or contamination with pathogenic organisms, sensitization against blood groups and the need for cold chain transportation (Wright & Riddles, 1989). Bock & de Vos (2001) reviewed data available on the efficacy, degree and duration of immunity provided by live vaccines against *B. bovis* and *B. bigemina* infections in Australia. They found that despite the disadvantages, live vaccines provided greater than 95% protection for the life of the animals.

The relative importance of different *Babesia* spp. in different countries dictates the composition of the vaccine. In parts of Africa, *B. bigemina* is the major cause of babesiosis whereas, in Australia, *B. bovis* causes approximately 20 times the economic loss caused by *B. bigemina*. As a result, protection against *B. bovis* has been the main aim of vaccination in Australia for many years although demand for vaccines containing *B. bigemina* has rapidly increased. In 2006 over 85% of vaccine sold contained both species.

Most of the available live vaccines are produced in government-supported production facilities, notably in Australia, Argentina, South Africa, Israel and Uruguay. These vaccines include bovine erythrocytes infected with selected strains. An experimental *B. divergens* vaccine prepared from the blood of infected gerbils *(Meriones unguiculatus)* has also been used in Ireland (Gray & Gannon, 1992), but production ceased in 2002 because of licensing concerns about the safety of a vaccine based on whole blood (J. S. Gray, personal communication, 2003). The risk of contamination of blood-derived vaccine is real (Hugoson, Vennström & Henriksson, 1968; Rogers *et al.*, 1988) and makes stringent quality control essential. Unfortunately, this puts production beyond the means of many countries in endemic regions (de Vos & Jorgensen, 1992). Techniques developed in Australia over many decades have formed the basis for production of live babesia vaccines in most countries where they are used. The following section outlines procedures currently used in Australia.

Origin and purification of strains

Since 1990, three strains of *B. bovis* and one of *B. bigemina* (G strain) have been used to produce vaccines in Australia. Changes in the *B. bovis* vaccine strain were necessary due to periodic increases in the vaccine failure rate above an acceptable background level (Bock *et al.*, 1992, 1995). Candidate low-virulence *B. bovis* isolates were obtained from naturally infected, long-term carrier animals as described by Callow (1977). Contaminating haemoparasites such as *Babesia* spp., *Anaplasma* spp., *Eperythrozoon* spp. and *Theileria buffeli* were eliminated using selective drug treatment combined with rapid passage in calves or culture (Anonymous, 1984; Dalgliesh & Stewart, 1983; Jorgensen & Waldron, 1994; Stewart *et al.*, 1990).

Attenuation of parasites

BABESIA BOVIS

The most reliable way of reducing the virulence of *B. bovis* involves rapid passage of the strain through susceptible splenectomized calves (Callow, Mellors & McGregor, 1979). The mechanism by which attenuation occurs is not fully understood, but may result from selective enrichment of less virulent parasite subpopulations or from downregulation of a virulence gene (Carson *et al.*, 1990; Cowman, Timms & Kemp, 1984; Timms, Stewart & de Vos, 1990). Attenuation is seldom complete and reversion to virulence can occur following passage of parasites through ticks (Timms *et al.*, 1990) or intact cattle (Callow *et al.*, 1979). Attenuation is also not guaranteed, but usually follows after 8 to 20 calf passages (Callow, 1984).

Attenuation of *Babesia* spp. by irradiation has been attempted, but the results were variable (Purnell & Lewis, 1981; Wright *et al.*, 1982). In vitro culture has also been used to attenuate *B. bovis* (Yunker, Kuttler & Johnson, 1987).

BABESIA BIGEMINA

Rapid passage in splenectomized calves is not a reliable means of attenuating *B. bigemina* (Anonymous, 1984), but the virulence of isolates decreases during prolonged residence in latently infected animals. This feature has been used to obtain avirulent strains by splenectomizing latently infected calves and using the ensuing relapse parasites to repeat the procedure (Dalgliesh *et al.*, 1981*a*). A single *B. bigemina* isolate (G strain) has been used in the Australian and South African vaccines since 1972 and the early 1980s, respectively.

Stabilates

The suitability of a strain for use in a vaccine can be determined by challenging vaccinated cattle and susceptible controls with a virulent, heterologous strain. Both safety and efficacy can be judged by monitoring fever, parasitaemia and depression of packed cell volume during the vaccine and challenge reactions (Timms *et al.*, 1983*a*). Any candidate isolate must also be tested for freedom from potential vaccine contaminants.

After testing for virulence, immunogenicity and purity, suitable strains are preserved as master stabilates in liquid nitrogen. Polyvinyl pyrrolidone (MW 40 000 Da) (Vega *et al.*, 1985) is the preferred cryoprotectant as it is not toxic to the parasites at temperatures above 4 °C, allows intravenous inoculation and is safe to use (Standfast & Jorgensen, 1997). Dimethyl sulphoxide is a very effective cryoprotectant but its use was discontinued in Australia in 1991 because of the risk of toxicity to operators, recipient animals and parasites (Dalgliesh, Jorgensen & de Vos, 1990).

Propagation in splenectomized calves

Susceptible splenectomized calves receive inocula from stabilate banks; parasitized blood is collected for production of vaccine when the parasitaemias exceed pre-set limits. Splenectomized calves inoculated with *B. bigemina* stabilate will usually develop a parasitaemia high enough to use for vaccine production, but *B. bovis* requires direct passaging into a second splenectomized calf to produce a sufficiently high parasitaemia for vaccine production. Passaging in splenectomized calves is not recommended for *B. bigemina* as it may result in a reversion to virulence. Vaccine strains of *B. bovis* are not passaged more than 30 times (including the attenuation passages) to safeguard against diminished immunogenicity (Callow & Dalgliesh, 1980).

A sufficient volume of blood can be collected from a 4–6-month-old calf to provide up to 25 000 doses. To do this the calf is sedated, the jugular vein catheterized and blood collected into a closed system using a peristaltic pump. Heparin is a suitable anticoagulant. After collection of the blood, the calf is treated with a babesicide and given supportive therapy. Depending on the volume of blood collected, the calf is also transfused using blood from a suitable donor, or autotransfused using blood collected and stored from the same animal prior to infection. Due to the high cost and limited availability of suitable health-tested donors, calves previously infected with *B. bovis* can subsequently be infected with *B. bigemina* as a source of organisms for vaccine production. With use of such 'recycled' calves, the risk of red cell agglutination, when mixing blood to make multivalent vaccine, can be prevented by washing the *B. bigemina*-infected red cells by serial centrifugation and washing to remove agglutinating antibodies. This does not appreciably affect parasite viability (Standfast *et al.*, 2003). In Australia, quinuronium sulphate (de Vos *et al.*, 2004) is used to treat the primary *B. bovis* infection because it has no residual effect and will not suppress the development of a subsequent *B. bigemina* infection.

Vaccine specifications

FROZEN VACCINE

Frozen vaccine has some very significant advantages over the chilled form, notably a long shelf-life that allows thorough post-production testing of potency and safety before dispatch. Its production also allows for judicious use of suitable contaminant-free donor cattle. Frozen vaccine is the only product available in South Africa and Israel, and demand for it is growing in Australia, reaching 10% of total demand in 2006.

Glycerol is used as a cryoprotectant in Australia in preference to dimethyl sulphoxide because it allows post-thaw storage life of the vaccine of at least 8 hours (Dalgliesh *et al.*, 1990; Jorgensen, de Vos & Dalgliesh, 1989*b*). Parasitized bovine blood is slowly mixed with an equal volume of phosphate buffered saline (PBS) solution containing 3 M glycerol, glucose and antibiotics, equilibrated at 37 °C for 30 minutes, dispensed into 5-ml cryovials and frozen at 10 °C/min. These vials of vaccine concentrate are then stored in liquid nitrogen. Vaccine is prepared by thawing a cryovial in water at 37 °C and then diluting the contents in 50 ml of PBS containing 1.5 M glycerol, glucose and antibiotics to make 25 × 2-ml doses of vaccine with a 10% overfill.

The capability exists in Australia to produce monovalent *B. bovis* and *B. bigemina* vaccines but, since 2001, the only frozen vaccine marketed has been a vaccine concentrate registered as 'Combavac 3 in 1'. It contains erythrocytes from three separate donors infected with *B. bovis*, *B. bigemina* and *Anaplasma centrale*, respectively. The infected erythrocytes are concentrated using a blood concentration method recommended by the Kimron Veterinary Institute in Israel (V. Shkap, personal communication, 1996) and mixed with 3 M glycerol to produce the trivalent concentrate.

Frozen vaccine must be transported in suitably insulated containers with liquid N_2 or solid CO_2 as refrigerant and this limits the ability to supply the vaccine to all destinations. To ensure infectivity, the prepared vaccine must be used within 8 hours of thawing and, once thawed, should not be refrozen. Vaccine prepared with glycerol must not be inoculated intravenously (Dalgliesh, 1972).

A frozen bivalent *B. bovis* and *B. bigemina* vaccine and frozen monovalent *B. bovis* and *B. bigemina* vaccines using dimethyl sulphoxide as the cryoprotectant are produced in South Africa (de Waal & Combrink, 2006) and Israel (Pipano, 1997), respectively.

CHILLED VACCINE

Most of the babesiosis vaccines produced to date have been provided in a chilled form. In Australia alone, 35 million doses were supplied between 1966 and 2005. Reasons for its popularity have been its ease of production even with limited resources, ease of transportation, ease of use and, in Australia at least, low cost. The chilled vaccines currently used in Australia contain 1×10^7 *B. bovis*, 2.5×10^6 *B. bigemina* and 1×10^7 *Anaplasma centrale* organisms per 2-ml dose (Standfast *et al.*, 2003)

To reduce the risk of neonatal haemolytic disease in calves of vaccinated dams, users are advised not to vaccinate cattle repeatedly. Most owners now vaccinate only young stock and seldom more than twice. As a result of this, and since the reduction of the dose rate from 5 ml to 2 ml and introduction of a cell-free diluent (Callow, 1984), no case of acute neonatal haemolytic disease has been confirmed in vaccinated cattle in Australia since 1976.

Chilled vaccine has a very short shelf-life, which is currently 4 days in Australia. Therefore, rapid, reliable means of communication and transport are required to ensure viability of the vaccine at point of use.

Quality assurance

DONOR CALF ORIGIN, QUARANTINE, TESTING AND CERTIFICATION

Australia is free of many of the infectious pathogens and arthropod-borne diseases that affect cattle in other countries (de Vos & Jorgensen, 1992). These potential contaminants therefore do not pose a serious risk to bovine blood-based vaccines produced in Australia. In other countries, testing protocols may be required for such potential contaminants.

The calves to be used in vaccine production in Australia are bred on site using cattle sourced from herds in *Boophilus microplus*-free regions. Breeder cattle are screened for, and maintained free of *B. bovis*, *B. bigemina*, *A. marginale*, bovine leukaemia virus (BLV), bovine immunodeficiency virus, bovine spumavirus (bovine syncytial virus), bovine pestivirus, *Neospora caninum*, *Coxiella burnetii* (Q fever) and *Boophilus microplus*. The cattle are also routinely vaccinated with commercial bacterial and viral vaccines. Calves are produced throughout the year to ensure donors are available at regular intervals. The calves are brought into an insect-free quarantine environment at 7 to 21 days of age and are subject to stringent quarantine and testing over a period of 8 to 12 weeks. During the quarantine period, each calf is also splenectomized. Only when all test results are available are calves cleared for use. Any calves show-ing evidence of infection other than *Theileria buffeli* are rejected. *Theileria buffeli* infections are eliminated using buparvaquone and primaquine phosphate (Stewart *et al.*, 1990).

FROZEN VACCINE

Potency (infectivity) is tested by thawing and diluting five vials of vaccine concentrate, storing the vaccine for 8 hours at 4 °C to 8 °C and then inoculating susceptible groups of cattle. Each batch of vaccine is tested in a group of at least five cattle, and preferably 20 or more. The cattle are monitored for the presence of infection by examination of stained blood smears as well as retrospective serology (Molloy *et al.*, 1998*a*, 1998*b*). Post-production monitoring and testing can be carried out on the product for specific disease agents if required for import/export purposes provided validated tests are avail-able. Serum retention samples and DNA are collected from calves at the time of blood collection and again 2 weeks later and stored for this purpose.

CHILLED VACCINE

As the chilled vaccines have a short shelf-life, undertaking quality assurance on the final product is not possible. There-fore, increased reliance has to be placed on pre-production quality control, especially for obtaining, testing and housing vaccine donors free of harmful infections. Despite precau-tions, one batch of vaccine produced in Australia during 1986 was later shown to be contaminated with BLV (Rogers *et al.*, 1988). The incident pointed to a deficiency in testing proce-dures that relied on serological assays. Major changes were made subsequently to minimize the risk of future contam-ination with BLV and have been progressively enhanced to include other disease agents.

USE OF LIVE VACCINE

Cattle born in vector-infested regions

Because endemic stability is difficult to maintain, most cattle owners in endemic areas of Australia supplement natural exposure by vaccinating calves at weaning age. Vaccination is also recommended if cattle are being moved into or within the endemic area.

Control of outbreaks

Use of vaccine in the face of an outbreak is common prac-tice in Australia. Superimposing vaccination in this way on a natural infection will not exacerbate the condition, but will pre-empt the development of virulent infections in the pro-portion of the herd not yet exposed to field challenge. To prevent further exposure, the group should also be treated with an acaricide capable of preventing tick attachment from the time of diagnosis to 3 weeks after vaccination. Injectable or pour-on formulations of ivermectin and moxidectin (Waldron & Jorgensen, 1999) as well as fluazuron (B. C. Hosking, R. E. Bock, H. R. Schmid and J.-F. Graf, unpub-lished data, 2004), despite being highly effective acaricides, do not prevent transmission of *Babesia* spp.

Clinically affected cattle should be treated as soon as possible with a suitable babesicide. In the case of a severe outbreak, it may be advisable to treat all the cattle with a prophylactic compound (imidocarb or diminazene) and to vaccinate them later when the drug residue will not affect vaccine parasite multiplication.

Susceptible cattle imported into vector-infested country or regions

Large numbers of cattle, predominantly of *Bos taurus* breeds, are being imported into tropical, developing countries to upgrade local livestock industries. In the past, this prac-tice has led to significant losses due to tick-borne diseases including babesiosis (Callow, 1977), if preventative mea-sures were not taken. Vaccination of naive cattle moving from 'tick-free' to endemic areas is usually very effective. This practice has played a crucial role in making the live-stock industries in these parts more sustainable and compet-itive in meeting market demand with regard to breed type. Unfortunately, many countries involved in the importation of cattle do not have access to an effective locally produced vaccine.

K strain *B. bovis* and G strain *B. bigemina* from Australia have been shown experimentally to be protective in South Africa (de Vos, Bessenger & Fourie, 1982; de Vos, Com-brink & Bessenger, 1982); Sri Lanka (D. J. Weilgama, per-sonal communication, 1986), Bolivia (Callow, Quiroga & McCosker, 1976) and Malawi (Lawrence *et al.*, 1993). Vac-cine containing these strains has also been used with benefi-cial results in countries in many parts of the world, including Zimbabwe and Swaziland in Africa, Venezuela and Ecuador in South America, Malaysia and the Philippines in South-east Asia and islands of the Caribbean (authors' unpublished observations). Where local vaccine production is not prac-tical, the feasibility of importing a vaccine strain to protect imported or local cattle should be considered providing that it is effective, of low virulence and free from contaminants.

Hazards of live vaccine use and precautions

SEVERE REACTIONS

The likelihood of vaccine-induced reactions has been reduced with the development of attenuated strains, but there is always a risk of reactions when highly susceptible, adult cattle are immunized. Calves 3 to 9 months of age have a high level of natural resistance and therefore a low risk of reactions to the vaccine. In some countries, such as Argentina, vaccination is only recommended for calves, while in other countries such as Australia and South Africa, adult vaccination can be undertaken provided proper precautions are taken. Cattle with a high *Bos indicus* content rarely show adverse reactions to vaccination. Bock *et al.* (2000) investigated reports of severe reactions to *B. bovis* vaccine and, using PCR assays on the DNA obtained from affected cattle, found that four of the five cases were due to concurrent field infections.

Except in late pregnancy, cows are no more likely to show severe vaccination reactions than any other class of adult stock. However, the consequences of severe reactions are more serious in pregnant cows as the accompanying fever may cause abortion. The risk to the cow and foetus from vaccine reactions is much less than from field infections but close monitoring for reactions is nevertheless essential. Special care should also be taken with large (particularly fat) bulls, as a high fever can cause a temporary (6 to 8 weeks) loss of fertility (Singleton, 1974). In the case of valuable cows and bulls, it is advisable to take rectal temperatures during the reaction times and to treat any showing prolonged elevated temperatures or clinical evidence of disease with a suitable babesicide. There is little field evidence that stress, including nutritional stress, has a significant effect on reaction rates or immunity following *B. bovis* vaccination (Callow & Dalgliesh, 1980).

Low doses of imidocarb or diminazene have been used in some countries to suppress potential vaccine reactions but this practice is not recommended because of the effect it may have on subsequent immunity (Combrink & Troskie, 2004; de Vos *et al.*, 1986). Trials in Israel (Pipano *et al.*, 1987) and Australia (Jorgensen *et al.*, 1993) have shown that oxytetracycline can be used to ameliorate *B. bovis* and *B. bigemina* vaccine reactions without affecting subsequent immunity.

POTENTIAL SPREAD OF *BABESIA* SPP. FOLLOWING VACCINATION

Concern is often expressed that natural spread of infection may occur from vaccinated to unvaccinated cattle. Aus-

tralian evidence suggests it is very unlikely that use of vaccine will introduce infection into a previously uninfected area. A 15-year study of the history of babesiosis outbreaks in Australia found no evidence that vaccination of cattle on a neighbouring property was the cause of a babesiosis outbreak (Callow & Dalgliesh, 1980). On the other hand, inadequate tick control by a neighbour and spread of naturally infected ticks was found to be a contributing cause. Tick numbers appear to be more relevant in transmission of babesia than the presence of animal reservoirs of infection.

The current *B. bovis* strain (Dixie) used in Australia is known to be transmissible by *Boophilus microplus* under laboratory conditions and to increase in virulence following tick transmission (authors' unpublished observations). Despite this, Bock *et al.* (1999a) offered circumstantial evidence that the presence of this strain in vaccinated cattle is unlikely to alter the dynamics of transmission of parasites under field conditions or constitute a significant risk to naive cattle grazing with vaccinated cattle once vaccine-induced parasitaemias have fallen to undetectable levels.

The G strain of *B. bigemina* used in Australia since 1972 (Dalgliesh *et al.*, 1981a) is poorly if at all transmissible by *Boophilus microplus* (Dalgliesh, Stewart & Rodwell, 1981b; Mason, Potgieter & van Rensburg, 1986). Also cattle infected with *B. bigemina* are reported to remain infective for ticks for only 4 to 7 weeks (Johnston *et al.*, 1978; Mahoney, 1969). So if transmission to ticks occurs, it will only be over a short period.

LOSS OF VIABILITY

Stored in liquid nitrogen, frozen vaccine will remain viable for many years, but long-term storage at −80 °C will affect viability (authors' unpublished observations). If 3M glycerol is used as cryoprotectant, thawed vaccine can remain viable for only 8 hours at temperatures ranging from 4 to 30 °C. If dimethyl sulphoxide is used, vaccine should be used within 30 minutes although work in South Africa has indicated that thawed dimethyl sulphoxide-based vaccine remains infective for 8 hours if kept on melting ice (de Waal & Combrink, 2006). Chilled babesia vaccine can remain viable for up to a week if stored at 4 °C. At higher temperatures, viability is lost rapidly.

LACK OF PROTECTION

Since the introduction of a standardized method of production in Australia, live babesiosis vaccines have generally proved to be highly effective (Callow & Dalgliesh, 1980).

In most cases, a single vaccination provided lasting, probably life-long immunity against field infections with antigenically different strains. However, troublesome failures of the Australian *B. bovis* vaccine occurred in 1966, 1976 and again in 1988–90 (Bock *et al.*, 1992), and were thought to be due to loss of immunogenicity brought about by frequent passaging of the vaccine strains in splenectomized calves (Callow & Dalgliesh, 1980) or selection over time of breakthrough field strains (Bock *et al.*, 1992). In each case, the problem was solved by replacing the vaccine strain. A similar loss of immunogenicity in a multi-passaged strain of *B. bovis* was reported in South Africa (de Vos, 1978). To prevent future recurrences of the problem, the number of passages of the vaccine strains of *B. bovis* is limited by frequently reverting to a master stabilate with a low passage number (Callow & Dalgliesh, 1980).

More recently (1992–3), *B. bovis* vaccine failures were again reported in Australia despite restrictions on passage numbers and replacement of the vaccine strain (Bock *et al.*, 1995). The occurrence of these failures did not correlate with time after vaccination. Bock *et al.* (1995) considered eight possible factors, and while the situation is complex and no simple cause is forthcoming, recent research emphasis has been on the immune responsiveness of the host and immunogenicity of vaccine parasite subpopulations (Bock *et al.*, 1995; Dalrymple, 1993; Lew *et al.*, 1997*a*, 1997*b*). The *B. bovis* vaccine strain used in Australia since 1996 has been shown by PCR assay to contain two subpopulations and to be more protective than higher passages of this strain that contained one subpopulation (Bock & de Vos, 2001).

OTHER CONTROL TECHNIQUES

Vector control

Vector control was first used successfully to control and eventually eradicate *Babesia* spp. from the USA (Pegram, Wilson & Hansen, 2000). In Africa, babesiosis forms only part of very important complexes of ticks and tick-borne diseases and intensive, usually government-regulated tick control programmes have been used for many years. The situation on other continents is much less complex than in Africa, but where babesiosis is endemic, disease control rather than eradication is generally the only realistic option. Eradication of the tick vectors is a permanent solution to the problem but is rarely considered practical, environmentally sustainable or economically justifiable on either a national or a local basis.

Natural endemic stability

The concept of natural endemic stability with respect to tick-borne diseases (Mahoney, 1974) is appealing as it offers the potential to minimize the incidence of clinical disease despite reduction of acaricide use (see previous section). Although it is clearly possible to achieve this status naturally for babesiosis under favourable conditions, there is evidence that this state rarely exists in commercial herds in Australia.

In a study of dairy farms in *Boophilus microplus* endemic areas of Australia, Sserugga *et al.* (2003) found 74% of herds belonging to farmers who allowed a 'few' ticks to persist, assuming their cattle would be protected from tick-borne disease, had insufficient exposure to confer herd immunity and a high risk of tick fever outbreaks. Therefore, leaving a 'few' ticks, although it is likely to have some protective effect, cannot be considered a satisfactory approach to controlling the disease. It seems farmers underestimated the numbers of ticks needed or would not allow sufficient ticks on their cattle to achieve endemic stability.

The belief that endemic stability can be managed effectively in the long term is often mistaken as it requires careful manipulation of transmission rates each year for each management group, ideally with routine assessment of seroprevalence. The model for endemic stability was developed in Australia and the Americas where the disease–vector interactions are relatively simple. The African situation is more complex and less predictable, involving four main diseases, several vectors, game reservoirs and a larger range of susceptibility of bovine breeds. Even in Australia, variations in climate, host genotypes and management strategies mean that if natural 'endemic stability' is attainable it is likely to be short-lived.

Endemic stability is often ascribed to a population after only one cross-sectional serological study, and has then been inappropriately extrapolated to devise disease control strategies for districts or even regions. Under current Australian conditions, this concept of endemic stability is not particularly relevant, and use of the term in other regions also requires examination. A more meaningful approach is an economic concept of disease control that incorporates risk management and loss thresholds that will vary with each property, region and/or stock-owner.

Norval, Perry & Young (1992*a*) described endemic stability as a climax relationship between host, agent, vector and environment in which all co-exist with the virtual absence of clinical disease. A serological survey in 1996 of 7067 unvaccinated weaner cattle, 6 to 12 months of age, on 115 properties in officially *Boophilus microplus*-infested areas of northern Australia indicated that the average percentage of animals seropositive for *B. bovis* and *B. bigemina* per herd was only 4% and 11%, respectively (Bock *et al.*, 1997b). Given the generally high *Bos indicus* content and extensive management of these herds, the risk that this represents is difficult to ascertain, but in general appears to be low in line with the Norval *et al.* (1992b) definition of endemic stability. However, where the *Bos taurus* content of a herd was increased, the risk could be very high and significant losses would be expected and have indeed occurred (unpublished observations).

Animal genetics

Certain breeds of cattle, notably *Bos taurus* breeds, are known to be more susceptible to primary *B. bovis* infection (Bock *et al.*, 1997a). Bock *et al.* (1999a) showed purebred *Bos indicus* cattle had a high degree of resistance to babesiosis, but cross-bred cattle were sufficiently susceptible to warrant the use of preventive measures such as vaccination. Genotype also plays a role in the development of protective immunity with *Bos taurus* breeds more likely to show a deficient immunity. An investigation of 62 reports of *B. bovis* vaccine failures in Australia showed that all were in cattle with no greater than 3/8 *Bos indicus* infusion and 85% were in pure *Bos taurus* herds (Bock *et al.*, 1995). As discussed above (see section 'Breed resistance to babesiosis'), the use of cattle with more than 50% *Bos indicus* content greatly reduces the impact of babesiosis.

Integrated control

Livestock industries need to take a pragmatic approach to management of the ticks and diseases associated with them. In the long-term, an effective outcome can be achieved by integrating the strategic use of acaricides and exploitation of natural exposure supplemented with vaccines and the use of tick-resistant breeds of cattle (Norval, Perry & Hargreaves, 1992; Perry *et al.*, 1998).

In vitro culture-derived vaccines

In vitro culture methods reviewed by Pudney (1992) are also used to provide *B. bigemina* and *B. bovis* parasites for vaccines (Jorgensen *et al.*, 1992; Mangold *et al.*, 1996). These techniques are not widely used for production of vaccine, but have proven to be valuable research tools. In countries where it is difficult to obtain sufficient numbers of disease-free, susceptible donor calves, and materials and facilities for

tissue culture are available, in vitro production of vaccine may be a viable option. At the National Institute of Agricultural Technology in Argentina, over 50 000 doses of *B. bovis* and *B. bigemina* vaccine are produced using in vitro culture and sold to cattle producers each year (A. A. Guglielmone, personal communication, 1999). Some studies found neither virulence nor immunogenicity of babesia vaccine strains were appreciably modified by short term (3 months) maintenance in culture (Jorgensen, de Vos & Dalgliesh, 1989a; Timms & Stewart, 1989). However, more recent observations using PCR of polymorphic genetic markers have shown that proportions of *B. bovis* subpopulations changed with long-term continuous cultivation (Lew *et al.*, 1997b). These drifts may not be significant but indications are that protection provided by culture-modified *B. bovis* strains may be inferior to that of parasites not exposed to a culture environment (Bock *et al.*, 1995). Until more information is available, use of long-term cultures in the production of vaccine is not recommended unless facilities are available to monitor changes in parasite populations and to test for immunogenicity.

Non-living vaccines

Non-living vaccines would overcome many of the inherent difficulties in production, transport and use of live vaccines. One of the earliest attempts to induce protective immunity in cattle against *B. bovis* infection using a non-living vaccine was by Mahoney (1967a). The inoculation of cattle with killed parasites mixed with Freund's complete adjuvant partially protected against homologous challenge. A killed *B. divergens* vaccine is prepared in Austria from the blood of infected calves (Edelhofer *et al.*, 1998), but little information is available on the level and duration of the conferred immunity.

Soluble parasite antigen vaccines

Soluble parasite antigens (SPA) of *B. bovis* and *B. bigemina* have been extensively studied and proposed for use as vaccine in developing countries with reports of promising results (Montenegro-James, 1989; Montenegro-James *et al.*, 1992). Dogs vaccinated with SPA from in vitro cultures of *B. canis* and *B. rossi* adjuvanted with saponin were protected against clinical symptoms from heterologous *B. canis* infection from 3 weeks after booster vaccination for at least 6 months (Schetters *et al.*, 2006). Other studies have shown the level and duration of protection conferred by SPA against heterologous *B. bovis* challenge in cattle was considerably less than that of live vaccines (Timms, Stewart & Dalgliesh, 1983*b*). The protective capacity of culture-derived SPA from *B. divergens* was found to be good for homologous and variable for heterologous challenge (Precigout *et al.*, 1991; Valentin *et al.*, 1993). Jorgensen *et al.* (1993) found prevaccination with homologous SPA reduced the parasitaemia and development of anaemia but not the fever following vaccination with live *B. bovis* vaccine.

Recombinant/subunit vaccines

Advances in our understanding of mechanisms of immunity to many protozoa, the availability of *Babesia* spp. genome sequences and the development of molecular tools for generating recombinant vaccines suggest this is the future direction for protozoal vaccine development. However, our incomplete understanding of immune mechanisms to primary and secondary infection, elaborate mechanisms (such as antigenic variation) used by parasites to evade host immunity, and the heterogeneity of the major histocompatibility complex (MHC) class II molecules in cattle, remain obstacles to developing effective vaccines using this technology (Brown *et al.*, 2006*a*; Jenkins, 2001).

Several attempts have been made to develop recombinant or subunit babesia vaccines. The search has focused mainly on variable merozoite surface antigens (VMSA) that are functionally relevant and immunodominant in naturally immune cattle and have functional domains that are conserved among strains. Candidate antigens identified include *B. bovis* VMSA, MSA-1 (Hines *et al.*, 1995) and MSA-2 (Wilkowsky *et al.*, 2003) and rhoptry-associated protein-1 (RAP-1) (Brown *et al.*, 2006*a*; Norimine *et al.*, 2002).

Cattle injected with recombinant MSA-1 produced antibodies that were capable of neutralizing merozoite invasion of erythrocytes, but were not protected against virulent *B. bovis* challenge (Hines *et al.*, 1995). LeRoith *et al.* (2005) found significant genetic variation with a complete lack of antibody cross-reactivity between MSA-1 from vaccine strains and their respective breakthrough isolates. MSA-2c has been identified as highly conserved among *B. bovis* strains and bovine antibodies to recombinant MSA-2c were able to neutralize the invasion of erythrocytes by merozoites (Wilkowsky *et al.*, 2003). However, Berens *et al.* (2005) found proteins encoded by MSA-2 genes were quite diverse both between and within geographical regions and harbour evidence of genetic exchange among other VMSA family members, including MSA-1. This does not augur well for its usefulness as a vaccine candidate.

RAP-1 is a 60-kDa antigen of *Babesia* spp. that is recognized by antibodies and T-cells from naturally immune cattle (Norimine *et al.*, 2002; Rodriguez *et al.*, 1996). Native and recombinant *B. bovis* RAP-1 conferred partial protection against homologous challenge (Wright *et al.*, 1992), but was not protective against heterologous challenge (Brown *et al.*, 2006*a*). Native *B. bigemina* RAP-1 also induces partial protection against challenge infection (McElwain *et al.*, 1991).

Delbecq *et al.* (2006) cited a study (unpublished) in which a recombinant *B. divergens* vaccine based on the merozoite surface protein Bd37 and saponin QuilA induced complete protection against heterologous challenge in a rodent model. 'Complete protection' was not defined nor was the nature of the heterologous challenge. Protective immunity was induced only when hydrophobic moieties were present in the protein. Bd37 recombinant proteins lacking such hydrophobic residues were totally inefficient despite induction of specific antibodies (Delbecq *et al.*, 2006).

Babesia bovis recombinant antigens 12D3 and 11C5 were first investigated as potential vaccine antigens in the mid 1980s, but they did not induce adequate levels of protection against challenge (Wright *et al.*, 1992). A more recent reinvestigation of these antigens again gave equivocal results (Hope *et al.*, 2005).

Results to date suggest that vaccines based on single antigens or even several antigens in combination do not confer the level or duration of cross-protection provided by living vaccines. Unfortunately, the majority of experimental immunization trials use homologous challenge infections that do not reflect the situation in the field where heterologous virulent challenge would occur. Even a multicomponent recombinant vaccine may not provide long-term protection against field strains of *B. bovis* that show considerable genetic variation (Dalrymple, 1993; Lew *et al.*, 1997*a*). The difficulty is to identify antigens that are targeted by the host's protective immune response across all strains, induce an appropriate long-term memory response and deliver them to the animal in a way that is cost-effective. A subunit vaccine that protects against clinical signs but allows for limited parasite replication may be an ideal strategy for protecting susceptible individuals (Jenkins, 2001).

Allred and Al-Khedery (2006) suggested that the antigenic variation of *B. bovis* may offer an 'Achilles heel' whereby small molecular mimics might be developed to block adhesion to the endothelium, inhibit enzymes associated with antigenic variation events or inhibit the export of variable erythrocyte surface antigen (VESA) to the surface of the bovine erythrocyte. If *B. bovis* cytoadherence or antigenic variation was thus blocked the host's adaptive immune response may be able to clear the parasite.

DNA vaccines

Another avenue of research for the development of an alternative to live vaccine is DNA vaccine adjuvants. Recognition of foreign DNA (in particular unmethylated CpG motifs in DNA) by the innate immune system is a relatively recent discovery. CpG motifs in DNA derived from *B. bovis* stimulated B-cells to proliferate and produce IgG (Brown *et al.*, 1998), and activated macrophages to secrete IL-12, TNF-α and nitric oxide. These vaccine adjuvants may therefore provide an important innate defence mechanism to control parasite replication and promote persistent infection to increase the duration of immunity (Shoda *et al.*, 2001).

CONCLUSIONS AND OPPORTUNITIES FOR FUTURE RESEARCH

Consumer concerns about residues associated with the use of diminazene and imidocarb will continue to provide an impetus to the search for alternative methods for treating babesiosis. The complete sequences of *Babesia* spp. genomes should help identify potential drug targets in the future (Vial & Gorenflot, 2006). However, because of the small demand for these drugs, pharmaceutical companies have shown little interest in recent decades as sales are unlikely to justify development costs. This situation is unlikely to change in the foreseeable future.

Live *Babesia* spp. vaccines have an enviable level of efficacy (lifelong immunity in over 95% of animals from one vaccination) but have major shortcomings relating to shelf-life, tick transmissibility, and risk of transmission of adventitious disease agents. As a result, costly quality-assurance programmes are required. Even with a strict quality-assurance programme in place, it is extremely unlikely that many countries would consider importing the vaccines, thus limiting their use to countries where they are already being produced.

Laboratories producing frozen live *Babesia* spp. vaccines still use the techniques reported by Mellors *et al.* (1982) and Dalgliesh *et al.* (1990). Recent scientific and technological achievements offer some promise for the development of improved cryopreserved vaccines. These include new cryoprotectants and combinations of cryoprotectants developed in related organisms (reviewed by Hubalek, 1996), models to quantify cryopreservation efficiency (Pudney,

1992), the development of mouse haemoprotozoan models (Wyatt, Goff & Davis, 1991) and a new generation of programmable freezing machines that minimize the subjectivity of cryopreservation and deliver consistent, reproducible user-defined freezing conditions.

Development of subunit vaccines for the control of babesiosis is the real challenge that will face scientists in the next decade. Initial attempts failed because they were based largely on molecular principles without a proper understanding of the mechanisms involved in protective immunity. Even with recent advances in our understanding of bovine immunology and the elucidation of *Babesia* spp. genomes, progress in vaccine development against babesiosis has been slow and many obstacles remain.

As reviewed by Brown *et al.* (2006*a*, 2006*b*), a protective subunit vaccine will probably require inclusion of multiple antigens or immunogenic epitopes of multiple proteins. Another strategy, opened up by the availability of the babesia genome, is to knock out genes involved in virulence and tick transmissibility.

Progress in vaccine development also requires an improved understanding of the mechanisms of immunity in the bovine host. To this effect, developers of subunit vaccines would benefit from the lessons learnt by the manufacturers of live vaccines, particularly with regard to the loss of immunogenicity of multi-passaged strains. Vaccine developers will also have to factor in the diversity of bovine MHC class II molecules (Brown *et al.*, 2006*a*) and antigenic variation in the parasites (Allred & Al-Khedery, 2004).

In addition, a novel delivery system may be required to minimize the need for repeat vaccination. Numerous options are being investigated and include viral and bacterial vectors, new generation adjuvants, and antigen production in food plants.

The challenges are daunting but considering the number of cattle at risk world-wide, the need is ever present.

ACKNOWLEDGEMENTS

We are grateful to Cordelia Gosman from The Graphics Place for the original artwork used in the life cycle figure (Fig. 13.1).

REFERENCES

Agbede, R. I. S., Kemp, D. H. & Hoyte, H. M. D. (1986). Secretory and digest cells of female *Boophilus microplus*: invasion and development of *Babesia bovis* – light and electron microscope studies. In *Morphology, Physiology and Behavioural Biology of Ticks*, eds. Sauer, J. R. and Hair, A. J., pp. 457–471. New York: John Wiley.

Ahmed, J. S. (2002). The role of cytokines in immunity and immunopathogenesis of piroplasmoses. *Parasitology Research* 88 (Suppl.), S48–S50.

Allred, D. R. (2003). Babesiosis: persistence in the face of adversity. *Trends in Parasitology* 19, 51–55.

Allred, D. R. & Al-Khedery, B. (2004). Antigenic variation and cytoadhesion in *Babesia bovis* and *Plasmodium falciparum*: different logics achieve the same goal. *Molecular and Biochemical Parasitology* 134, 27–35.

Allred, D. R. & Al-Khedery, B. (2006). Antigenic variation as an exploitable weakness of babesial parasites. *Veterinary Parasitology* 138, 50–60.

Allsopp, M. T., Cavalier-Smith, T., De Waal, D. T. & Allsopp, B. A. (1994). Phylogeny and evolution of the piroplasms. *Parasitology* 108, 147–152.

Angus, B. M. (1996). The history of the cattle tick *Boophilus microplus* in Australia and achievements in its control. *International Journal of Parasitology* 26, 1341–1355.

Anonymous (1984). *Ticks and Tick-Borne Disease Control: A Practical Field Manual*, vol 2, *Tick-Borne Disease Control*. Rome: Food and Agriculture Organization of the United Nations.

Auger, M. J. & Ross, J. A. (1992). The biology of the macrophage. In *The Macrophage*, eds. Lewis, C. E. and McGee, J. O. D., pp. 1–57. Oxford, UK: IRL Press.

Babes, V. (1888). Sur l'hémoglobinurie bactérienne du boeuf. *Comptes rendus hebdomadaires des séances de l'Academie des Sciences* 107, 692–694.

Barker, S. C. & Murrell, A. (2002). Phylogeny, evolution and historical zoogeography of ticks: a review of recent progress. *Experimental and Applied Acarology* 28, 55–68.

Barker, S. C. & Murrell, A. (2004). Systematics and evolution of ticks with a list of valid genus and species. *Parasitology* 129, S15–S36.

Berens, S. J., Brayton, K. A., Molloy, J. B., *et al.* (2005). Merozoite surface antigen 2 proteins of *Babesia bovis* vaccine breakthrough isolates contain a unique hypervariable region composed of degenerate repeats. *Infection and Immunity* 73, 7180–7189.

Bock, R. E. & de Vos, A. J. (2001). Immunity following use of Australian tick fever vaccine: a review of the evidence. *Australian Veterinary Journal* 79, 832–839.

Bock, R. E., de Vos, A. J., Kingston, T. G. & McLellan, D. J. (1997a). Effect of breed of cattle on innate resistance to infection with *Babesia bovis*, *B. bigemina* and *Anaplasma marginale*. *Australian Veterinary Journal* 75, 337–340. [Published erratum appears in *Australian Veterinary Journal* (1997) 75, 449.]

Bock, R. E., de Vos, A. J., Kingston, T. G., Shiels, I. A. & Dalgliesh, R. J. (1992). Investigations of breakdowns in protection provided by living *Babesia bovis* vaccine. *Veterinary Parasitology* 43, 45–56.

Bock, R. E., de Vos, A. J., Lew, A. E., Kingston, T. G. & Fraser, I. R. (1995). Studies on failure of T strain live *Babesia bovis* vaccine. *Australian Veterinary Journal* 72, 296–300.

Bock, R. E., de Vos, A. J., Rayner, A. C., *et al.* (1997b). Assessment of the risk of tick fever mortalities in north-western Queensland beef industry. In *Challenging the Boundaries, Proceedings of Annual Conference, Australian Association of Cattle Veterinarians*, pp. 175–182. Brisbane, Australia: Australian Veterinary Association.

Bock, R. E., Jorgensen, W. K. & Molloy, J. B. (2004). Bovine babesiosis. In *Manual of Standards for Diagnostic Tests and Vaccines for Terrestrial Animals*, vol. 1, pp. 507–518. Paris: Office International des Épizooties. Available online at www.oie.int/eng/normes/mmanual/A_00059.htm.

Bock, R. E., Kingston, T. G. & de Vos, A. J. (1999a). Effect of breed of cattle on transmission rate and innate resistance to infection with *Babesia bovis* and *B. bigemina* transmitted by *Boophilus microplus*. *Australian Veterinary Journal* 77, 461–464.

Bock, R. E., Kingston, T. G., Standfast, N. F. & de Vos, A. J. (1999b). Effect of cattle breed on innate resistance to inoculations of *Babesia bigemina*. *Australian Veterinary Journal* 77, 465–466.

Bock, R. E., Lew, A. E., Minchin, C. M., Jeston, P. J. & Jorgensen, W. K. (2000). Application of PCR assays to determine the genotype of *Babesia bovis* parasites isolated from cattle with clinical babesiosis soon after vaccination against tick fever. *Australian Veterinary Journal* 78, 179–181.

Böse, R., Jorgensen, W. K., Dalgliesh, R. J., Friedhoff, K. T. & de Vos, A. J. (1995). Current state and future trends in the diagnosis of babesiosis. *Veterinary Parasitology* 57, 61–74.

Bouattour, A., Darghouth, M. A. & Daoud, A. (1999). Distribution and ecology of ticks (Acari: Ixodidae) infesting livestock in Tunisia: an overview of eight years' field collections. *Parassitologia* 41 (Suppl.), S5–S10.

Brown, W. C. & Logan, K. S. (1992). *Babesia bovis*: bovine helper T cell lines reactive with soluble and membrane antigens of merozoites. *Experimental Parasitology* 74, 188–199.

Brown, W. C. & Palmer, G. H. (1999). Designing blood-stage vaccines against *Babesia bovis* and *B. bigemina*. *Parasitology Today* 15, 275–281.

Brown, W. C., Estes, D. M., Chantler, S. E., Kegerreis, K. A. & Suarez, C. E. (1998). DNA and a CpG oligonucleotide derived from *Babesia bovis* are mitogenic for bovine B cells. *Infection and Immunity* 66, 5423–5432.

Brown, W. C., Logan, K. S., Wagner, G. G. & Tetzlaff, C. L. (1991). Cell-mediated immune responses to *Babesia bovis* merozoite antigens in cattle following infection with tick-derived or cultured parasites. *Infection and Immunity* 59, 2418–2426.

Brown, W. C., Norimine, J., Goff, W. L., Suarez, C. E. & McElwain, T. F. (2006*a*). Prospects for recombinant vaccines against *Babesia bovis* and related parasites. *Parasite Immunology* 28, 315–327.

Brown, W. C., Norimine, J., Knowles, D. P. & Goff, W. L. (2006*b*). Immune control of *Babesia bovis* infection. *Veterinary Parasitology* 138, 75–87.

Brown, W. C., Zhao, S., Woods, V. M., Dobbelaere, D. A. & Rice-Ficht, A. C. (1993). *Babesia bovis*-specific CD4+ T cell clones from immune cattle express either the Th0 or Th1 profile of cytokines. *Revue d'Elevage et de Médecine Vétérinaire des Pays Tropicaux* 46, 65–69.

Caeiro, V. (1999). General review of tick species present in Portugal. *Parassitologia* 41, 11–15.

Callow, L. L. (1977). Vaccination against bovine babesiosis. In *Immunity to Blood Parasites of Man and Animals*, eds. Miller, L. H., Pino, J. A. & McKelvey, J. J. Jr, pp. 121–149. New York: Plenum Press.

Callow, L. L. (1979). Some aspects of the epidemiology and control of bovine babesiosis in Australia. *Journal of the South African Veterinary Association* 50, 353–356.

Callow, L. L. (1984). Piroplasms. In *Animal Health in Australia*, vol. 5, *Protozoal and Rickettsial Diseases*, pp. 121–160. Canberra, Australia: Australian Bureau of Animal Health.

Callow, L. L. & Dalgliesh, R. J. (1980). The development of effective, safe vaccination against babesiosis and anaplasmosis in Australia. In *Ticks and Tick-Borne Diseases, Proceedings of a Symposium held at the 56th Annual Conference of the Australian Veterinary Association*, eds. Johnston, L. A. Y. and Cooper, M. G., pp. 4–8. Townsville, Australia: Australian Veterinary Association.

Callow, L. L. & Hoyte, H. M. D. (1961). Transmission experiments using *Babesia bigemina*, *Theileria mutans*, *Borrelia* sp. and the cattle tick *Boophilus microplus*. *Australian Veterinary Journal* 37, 381–390.

Callow, L. L., Dalgliesh, R. J. & de Vos, A. J. (1997). Development of effective living vaccines against bovine babesiosis: the longest field trial? *International Journal for Parasitology* 27, 747–767.

Callow, L. L., McGregor, W., Parker, R. J. & Dalgliesh, R. J. (1974*a*). The immunity of cattle to *Babesia argentina* after drug sterilization of infections of varying duration. *Australian Veterinary Journal* 50, 6–11.

Callow, L. L., McGregor, W., Parker, R. J. & Dalgliesh, R. J. (1974*b*). Immunity of cattle to *Babesia bigemina* following its elimination from the host, with observations on antibody levels detected by the indirect fluorescent antibody test. *Australian Veterinary Journal* 50, 12–15.

Callow, L. L., Mellors, L. T. & McGregor, W. (1979). Reduction in virulence of *Babesia bovis* due to rapid passage in splenectomized cattle. *International Journal for Parasitology* 9, 333–338.

Callow, L. L., Quiroga, Q. C. & McCosker, P. J. (1976). Serological comparison of Australian and South American strains of *Babesia argentina* and *Anaplasma marginale*. *International Journal for Parasitology* 6, 307–310.

Callow, L. L., Rogers, R. J. & de Vos, A. J. (1993). Tick-borne diseases: cattle pathology and serology. In *Australian Standard Diagnostic Techniques for Animal Diseases*, eds. Corner, L. A. and Bagust, T. J., pp. 1–16. East Melbourne, Australia: CSIRO Information Services.

Carson, C. A., Timms, P., Cowman, A. F. & Stewart, N. P. (1990). *Babesia bovis*: evidence for selection of subpopulations during attenuation. *Experimental Parasitology* 70, 404–410.

Combrink, M. P. & Troskie, P. C. (2004). Effect of diminazene block treatment on live redwater vaccine reactions. *Onderstepoort Journal of Veterinary Research* 71, 113–117.

Court, R. A., Jackson, L. A. & Lee, R. P. (2001). Elevated anti-parasitic activity in peripheral blood monocytes and neutrophils of cattle infected with *Babesia bovis*. *International Journal for Parasitology* 31, 29–37.

Cowman, A. F., Timms, P. & Kemp, D. J. (1984). DNA polymorphisms and subpopulations in *Babesia bovis*. *Molecular and Biochemical Parasitology* 11, 91–103.

Criado-Fornelio, A., Martinez-Marcos, A., Buling-Sarana, A. & Barba-Carretero, J. C. (2003). Molecular studies on *Babesia*, *Theileria* and *Hepatozoon* in southern Europe. II. Phylogenetic analysis and evolutionary history. *Veterinary Parasitology* 114, 173–194.

Dalgliesh, R. J. (1972). Effects of low temperature preservation and route of inoculation on infectivity of *Babesia bigemina* in blood diluted with glycerol. *Research in Veterinary Science* **13**, 540–545.

Dalgliesh, R. J. (1993). Babesiosis. In *Immunology and Molecular Biology of Parasite Infections*, ed. Warren, S. K., pp. 352–383. Oxford, UK: Blackwell Scientific Publications.

Dalgliesh, R. J. & Stewart, N. P. (1982). Some effects of time, temperature and feeding on infection rates with *Babesia bovis* and *Babesia bigemina* in *Boophilus microplus* larvae. *International Journal for Parasitology* **12**, 323–326.

Dalgliesh, R. J. & Stewart, N. P. (1983). The use of tick transmission by *Boophilus microplus* to isolate pure strains of *Babesia bovis*, *Babesia bigemina* and *Anaplasma marginale* from cattle with mixed infections. *Veterinary Parasitology* **13**, 317–323.

Dalgliesh, R. J., Callow, L. L., Mellors, L. T. & McGregor, W. (1981*a*). Development of a highly infective *Babesia bigemina* vaccine of reduced virulence. *Australian Veterinary Journal* **57**, 8–11.

Dalgliesh, R. J., Jorgensen, W. K. & de Vos, A. J. (1990). Australian frozen vaccines for the control of babesiosis and anaplasmosis in cattle: a review. *Tropical Animal Health and Production* **22**, 44–52.

Dalgliesh, R. J., Stewart, N. P. & Callow, L. L. (1978). Transmission of *Babesia bigemina* by transfer of adult male *Boophilus microplus* [letter]. *Australian Veterinary Journal* **54**, 205–206.

Dalgliesh, R. J., Stewart, N. P. & Rodwell, B. J. (1981*b*). Increased numbers of strahlenkörper in *Boophilus microplus* ticks ingesting a blood-passaged strain of *Babesia bigemina*. *Research in Veterinary Science* **31**, 350–352.

Dalrymple, B. P. (1993). Molecular variation and diversity in candidate vaccine antigens from *Babesia*. *Acta Tropica* **53**, 227–238.

de Vos, A. J. (1978). Immunogenicity and pathogenicity of three South African strains of *Babesia bovis* in *Bos indicus* cattle. *Onderstepoort Journal of Veterinary Research* **45**, 119–124.

de Vos, A. J. & Jorgensen, W. K. (1992). Protection of cattle against babesiosis in tropical and subtropical countries with a live, frozen vaccine. In *Tick Vector Biology: Medical and Veterinary Aspects*, eds. Fivaz, B. H., Petney, T. N. & Horak, I. G., pp. 159–174. Heidelberg, Germany: Springer-Verlag.

de Vos, A. J., Bessenger, R. & Fourie, C. G. (1982). Virulence and heterologous strain immunity of South African and Australian *Babesia bovis* strains with reduced pathogenicity. *Onderstepoort Journal of Veterinary Research* **49**, 133–136.

de Vos, A. J., Combrink, M. P. & Bessenger, R. (1982). *Babesia bigemina* vaccine: comparison of the efficacy and safety of Australian and South African strains under experimental conditions in South Africa. *Onderstepoort Journal of Veterinary Research* **49**, 155–158.

de Vos, A. J., Dalgliesh, R. J. & Callow, L. L. (1987). *Babesia*. In *Immune Responses in Parasitic Infections: Immunology, Immunopathology and Immunoprophylaxis*, vol. 3, ed. Soulsby, E. J. L., pp. 183–222. Boca Raton, FL: CRC Press.

de Vos, A. J., Dalgliesh, R. J. & McGregor, W. (1986). Effect of imidocarb dipropionate prophylaxis on the infectivity and immunogenicity of a *Babesia bovis* vaccine in cattle. *Australian Veterinary Journal* **63**, 174–178.

de Vos, A. J., De Waal, D. T. & Jackson, L. A. (2004). Bovine babesiosis. In *Infectious Diseases of Livestock*, vol. 1, eds. Coetzer, J. A. W. & Tustin, R. C., pp. 406–424. Cape Town, South Africa: Oxford University Press.

de Waal, D. T. & Combrink, M. P. (2006). Live vaccines against bovine babesiosis. *Veterinary Parasitology* **138**, 88–96.

Delbecq, S., Hadj-Kaddour, K., Randazzo, S., *et al.* (2006). Hydrophobic moieties in recombinant proteins are crucial to generate efficient saponin-based vaccine against apicomplexan *Babesia divergens*. *Vaccine* **24**, 613–621.

Edelhofer, R., Kanout, A., Schuh, M. & Kutzer, E. (1998). Improved disease resistance after *Babesia divergens* vaccination. *Parasitology Research* **84**, 181–187.

Friedhoff, K. T. (1988). Transmission of *Babesia*. In *Babesiosis of Domestic Animals and Man*, ed. Ristic, M., pp. 23–52. Boca Raton, FL: CRC Press.

Gale, K. R., Waltisbuhl, D. J., Bowden, J. M., *et al.* (1998). Amelioration of virulent *Babesia bovis* infection in calves by administration of the nitric oxide synthase inhibitor aminoguanidine. *Parasite Immunology* **20**, 441–445.

Goff, W. L., Johnson, W. C., Parish, S. M., *et al.* (2002). IL-4 and IL-10 inhibition of IFN-gamma- and TNF-alpha-dependent nitric oxide production from bovine mononuclear phagocytes exposed to *Babesia bovis* merozoites. *Veterinary Immunology and Immunopathology* **84**, 237–251.

Goff, W. L., Johnson, W. C., Parish, S. M., *et al.* (2001). The age-related immunity in cattle to *Babesia bovis* infection involves the rapid induction of interleukin-12, interferon-gamma and inducible nitric oxide synthase mRNA expression in the spleen. *Parasite Immunology* **23**, 463–471.

Goff, W. L., Storset, A. K., Johnson, W. C. & Brown, W. C. (2006). Bovine splenic NK cells synthesize IFN-gamma in response to IL-12-containing supernatants from *Babesia bovis*-exposed monocyte cultures. *Parasite Immunology* **28**, 221–228.

Goff, W. L., Wagner, G. G. & Craig, T. M. (1984). Increased activity of bovine ADCC effector cells during acute *Babesia bovis* infection. *Veterinary Parasitology* **16**, 5–15.

Goff, W. L., Wagner, G. G., Craig, T. M. & Long, R. F. (1982). The bovine immune response to tick-derived *Babesia bovis* infection: serological studies of isolated immunoglobulins. *Veterinary Parasitology* **11**, 109–120.

Goodger, B. V., Wright, I. G. & Mahoney, D. F. (1981). Initial characterization of cryoprecipitates in cattle recovering from acute *Babesia bovis* (*Argentina*) infection. *Australian Journal of Experimental Biology and Medical Science* **59**, 521–529.

Goodger, B. V., Wright, I. G. & Waltisbuhl, D. J. (1985). *Babesia bovis*: the effect of acute inflammation and isoantibody production in the detection of babesial antigens. *Experientia* **41**, 1577–1579.

Gough, J. M., Jorgensen, W. K. & Kemp, D. H. (1998). Development of tick gut forms of *Babesia bigemina in vitro*. *Journal of Eukaryotic Microbiology* **45**, 298–306.

Gray, J. S. & Gannon, P. (1992). Preliminary development of a live drug-controlled vaccine against bovine babesiosis using the Mongolian gerbil, *Meriones unguiculatus*. *Veterinary Parasitology* **42**, 179–188.

Gray, J. S., Langley, R. J., Brophy, P. O. & Gannon, P. (1989). Vaccination against bovine babesiosis with drug-controlled live parasites. *Veterinary Record* **125**, 369–372. [Published erratum appears in Veterinary Record (1989) **125**, 646.]

Hines, S. A., Palmer, G. H., Jasmer, D. P., Goff, W. L. & McElwain, T. F. (1995). Immunization of cattle with recombinant *Babesia bovis* merozoite surface antigen-1. *Infection and Immunity* **63**, 349–352.

Hope, M., Riding, G., Menzies, M., *et al.* (2005). Potential for recombinant *Babesia bovis* antigens to protect against a highly virulent isolate. *Parasite Immunology* **27**, 439–445.

Hoyte, H. M. (1961). Initial development of infectious *Babesia bigemina*. *Australian Veterinary Journal* **8**, 462–466.

Hubalek, Z. (1996). *Cryopreservation of Microorganisms at Ultra-Low Temperatures*. Kvetna, Czech Republic: Academy of Sciences of the Czech Republic.

Hugoson, G., Vennström, R. & Henriksson, K. (1968). The occurrence of bovine leukosis following the introduction of babesiosis vaccination. *Bibliotheca Haematologica* **30**, 157–161.

Jacobson, R. H., Parrodi, F., Wright, I. G., Fitzgerald, C. J. & Dobson, C. (1993). *Babesia bovis*: in vitro phagocytosis promoted by immune serum and by antibodies produced against protective antigens. *Parasitology Research* **79**, 221–226.

Jenkins, M. C. (2001). Advances and prospects for subunit vaccines against protozoa of veterinary importance. *Veterinary Parasitology* **101**, 291–310.

Johnson, W. C., Cluff, C. W., Goff, W. L. & Wyatt, C. R. (1996). Reactive oxygen and nitrogen intermediates and products from polyamine degradation are babesiacidal in vitro. *Annals of the New York Academy of Sciences* **791**, 136–147.

Johnston, L. A., Leatch, G. & Jones, P. N. (1978). The duration of latent infection and functional immunity in Droughtmaster and Hereford cattle following natural infection with *Babesia argentina* and *Babesia bigemina*. *Australian Veterinary Journal* **54**, 14–18.

Jorgensen, W. K. & Waldron, S. J. (1994). Use of *in vitro* culture to isolate *Babesia bovis* from *Theileria buffeli*, *Eperythrozoon wenyoni* and *Anaplasma* spp. *Veterinary Parasitology* **53**, 45–51.

Jorgensen, W. K., Bock, R. E., Kingston, T. G., de Vos, A. J. & Waldron, S. J. (1993). Assessment of tetracycline and *Babesia* culture supernatant as prophylactics for moderating reactions in cattle to live *Babesia* and *Anaplasma* vaccines. *Australian Veterinary Journal* **70**, 35–36.

Jorgensen, W. K., de Vos, A. J. & Dalgliesh, R. J. (1989*a*). Comparison of immunogenicity and virulence between *Babesia bigemina* parasites from continuous culture and from a splenectomized calf. *Australian Veterinary Journal* **66**, 371–372.

Jorgensen, W. K., de Vos, A. J. & Dalgliesh, R. J. (1989*b*). Infectivity of cryopreserved *Babesia bovis*, *Babesia bigemina* and *Anaplasma centrale* for cattle after thawing, dilution and incubation at 30 degrees C. *Veterinary Parasitology* **31**, 243–251.

Jorgensen, W. K., Waldron, S. J., McGrath, J., *et al.* (1992). Growth of *Babesia bigemina* parasites in suspension cultures for vaccine production. *Parasitology Research* **78**, 423–426.

Lawrence, J. A., Malika, J., Whiteland, A. P. & Kafuwa, P. (1993). Efficacy of an Australian *Babesia bovis* vaccine strain in Malawi. *Veterinary Record* **132**, 295–296.

Legg, J. (1935). *The Occurrence of Bovine Babesiellosis in Northern Australia, CSIRO*. Pamphlet No. 56. Melbourne, Australia: CSIRO.

Leroith, T., Brayton, K. A., Molloy, J. B., *et al.* (2005). Sequence variation and immunologic cross-reactivity among *Babesia bovis* merozoite surface antigen 1 proteins from vaccine strains and vaccine breakthrough isolates. *Infection and Immunity* **73**, 5388–5394.

Levine, N. D. (1971). Taxonomy of the piroplasms. *Transactions of the American Microscopical Society* **90**, 2–33.

Levine, N. D. (1985). *Veterinary Protozoology*. Ames, IA: Iowa State University Press.

Lew, A. & Jorgensen, W. (2005). Molecular approaches to detect and study the organisms causing bovine tick borne diseases: babesiosis and anaplasmosis. *African Journal of Biotechnology* **4**, 292–302.

Lew, A. E., Bock, R. E., Croft, J. M., *et al.* (1997a). Genotypic diversity in field isolates of *Babesia bovis* from cattle with babesiosis after vaccination. *Australian Veterinary Journal* **75**, 575–578.

Lew, A. E., Dalrymple, B. P., Jeston, P. J. & Bock, R. E. (1997b). PCR methods for the discrimination of *Babesia bovis* isolates. *Veterinary Parasitology* **71**, 223–237.

Lignières, J. (1903). Bovine babesiosis: new investigations and observations on the multiplicity, the evolution and natural transmission of the parasites involved in the disease and on vaccination. *Archives de Parasitologie* **7**, 398–407.

Lohr, K. F. (1973). Susceptibility of non-splenectomized and splenectomized Sahiwal cattle to experimental *Babesia bigemina* infection. *Zentralblatt für Veterinarmedizin* B **20**, 52–56.

Mackenstedt, U., Gauer, M., Fuchs, P., *et al.* (1995). DNA measurements reveal differences in the life cycles of *Babesia bigemina* and *B. canis*, two typical members of the genus *Babesia*. *Parasitology Research* **81**, 595–604.

Mackenstedt, U., Gauer, M., Mehlhorn, H., Schein, E. & Hauschild, S. (1990). Sexual cycle of *Babesia divergens* confirmed by DNA measurements. *Parasitology Research* **76**, 199–206.

Mahoney, D. F. (1967a). Bovine babesiosis: the immunization of cattle with killed *Babesia argentina*. *Experimental Parasitology* **20**, 125–129.

Mahoney, D. F. (1967b). Bovine babesiosis: the passive immunization of calves against *Babesia argentina* with special reference to the role of complement fixing antibodies. *Experimental Parasitology* **20**, 119–124.

Mahoney, D. F. (1969). Bovine babesiasis: a study of factors concerned in transmission. *Annals of Tropical Medicine and Parasitology* **63**, 1–14.

Mahoney, D. F. (1972). Immune responses to hemoprotozoa. II. *Babesia* spp. In *Immunity to Animal Parasites*, ed. Soulsby, E. J. L., pp. 301–341. New York: Academic Press.

Mahoney, D. F. (1974). The application of epizootiological principals in the control of babesiosis in cattle. *Bulletin of Office International des Epizooties* **81**, 123–138.

Mahoney, D. F. (1986). Studies on the protection of cattle against *Babesia bovis* infection. In *The Ruminant Immune System in Health and Disease*, ed. Morrison, W. I., pp. 539–544. Cambridge, UK: Cambridge University Press.

Mahoney, D. F. & Goodger, B. V. (1969). *Babesia argentina*: serum changes in infected calves. *Experimental Parasitology* **24**, 375–382.

Mahoney, D. F. & Mirre, G. B. (1971). Bovine babesiasis: estimation of infection rates in the tick vector *Boophilus microplus* (Canestrini). *Annals of Tropical Medicine and Parasitology* **65**, 309–317.

Mahoney, D. F. & Mirre, G. B. (1979). A note on the transmission of *Babesia bovis* (syn. *B. argentina*) by the one-host tick, *Boophilus microplus*. *Research in Veterinary Science* **26**, 253–4.

Mahoney, D. F. & Ross, D. R. (1972). Epizootiological factors in the control of bovine babesiosis. *Australian Veterinary Journal* **48**, 292–298.

Mahoney, D. F., Kerr, J. D., Goodger, B. V. & Wright, I. G. (1979a). The immune response of cattle to *Babesia bovis* (syn. *B. argentina*): studies on the nature and specificity of protection. *International Journal for Parasitology* **9**, 297–306.

Mahoney, D. F., Wright, I. G. & Goodger, B. V. (1979b). Immunity in cattle to *Babesia bovis* after single infections with parasites of various origin. *Australian Veterinary Journal* **55**, 10–12.

Mahoney, D. F., Wright, I. G., Goodger, B. V., *et al.* (1981). The transmission of *Babesia bovis* in herds of European and Zebu × European cattle infested with the tick, *Boophilus microplus*. *Australian Veterinary Journal* **57**, 461–469.

Mahoney, D. F., Wright, I. G. & Mirre, G. B. (1973). Bovine babesiasis: the persistence of immunity to *Babesia argentina* and *B. bigemina* in calves (*Bos taurus*) after naturally acquired infection. *Annals of Tropical Medicine and Parasitology* **67**, 197–203.

Mangold, A. J., Vanzini, V. R., Echaide, I. E., *et al.* (1996). Viability after thawing and dilution of simultaneously cryopreserved vaccinal *Babesia bovis* and *Babesia bigemina* strains cultured in vitro. *Veterinary Parasitology* **61**, 345–348.

Mason, T. E., Potgieter, F. T. & van Rensburg, L. (1986). The inability of a South African *Babesia bovis* vaccine strain to infect *Boophilus microplus*. *Onderstepoort Journal of Veterinary Research* **53**, 143–145.

McCosker, P. J. (1981). The global importance of babesiosis. In *Babesiosis*, eds. Ristic, M. & Kreier, J. P., pp. 1–24. New York: Academic Press.

McElwain, T. F., Perryman, L. E., Musoke, A. J. & McGuire, T. C. (1991). Molecular characterization and immunogenicity of neutralization-sensitive *Babesia bigemina* merozoite surface proteins. *Molecular and Biochemical Parasitology* 47, 213–222.

McGuire, T. C., Musoke, A. J. & Kurtti, T. (1979). Functional properties of bovine IgG1 and IgG2: interaction with complement, macrophages, neutrophils and skin. *Immunology* 38, 249–256.

McLeod, R. & Kristjanson, P. (1999). *Final Report of Joint Esys/ILRI/ACIAR TickCost Project: Economic Impact of Ticks and Tick-Borne Diseases to Livestock in Africa, Asia and Australia.* Nairobi, Kenya: International Livestock Research Institute.

Mehlhorn, H. & Shein, E. (1984). The piroplasms: life cycle and sexual stages. *Advances in Parasitology* 23, 37–103.

Mellors, L. T., Dalgliesh, R. J., Timms, P., Rodwell, B. J. & Callow, L. L. (1982). Preparation and laboratory testing of a frozen vaccine containing *Babesia bovis, Babesia bigemina* and *Anaplasma centrale. Research in Veterinary Science* 32, 194–197.

Molloy, J. B., Bowles, P. M., Bock, R. E., *et al.* (1998*a*). Evaluation of an ELISA for detection of antibodies to *Babesia bovis* in cattle in Australia and Zimbabwe. *Preventive Veterinary Medicine* 33, 59–67.

Molloy, J. B., Bowles, P. M., Jeston, P. J., *et al.* (1998*b*). Development of an enzyme-linked immunosorbent assay for detection of antibodies to *Babesia bigemina* in cattle. *Parasitology Research* 84, 651–656.

Montenegro-James, S. (1989). Immunoprophylactic control of bovine babesiosis: role of exoantigens of *Babesia. Transactions of the Royal Society of Tropical Medicine and Hygiene* 83, S85–S94.

Montenegro-James, S., Toro, M., Leon, E. & Guillen, A. T. (1992). Field evaluation of an exoantigen-containing *Babesia* vaccine in Venezuela. *Memorias do Instituto Oswaldo Cruz* 87, S283–S288.

Ndi, C., Bayemi, P. H., Ekue, F. N. & Tarounga, B. (1991). Preliminary observations on ticks and tick-borne diseases in the north west province of Cameroon. I. Babesiosis and anaplasmosis. *Revue d'Elevage et de Médecine Vétérinaire des Pays Tropicaux* 44, 263–265.

Norimine, J., Suarez, C. E., McElwain, T. F., Florin-Christensen, M. & Brown, W. C. (2002). Immunodominant epitopes in *Babesia bovis* rhoptry-associated protein 1 that elicit memory CD4(+)-T–lymphocyte responses in *B. bovis*-immune individuals are located in the amino-terminal domain. *Infection and Immunity* 70, 2039–2048.

Norval, R. A. I., Perry, B. D. & Hargreaves, S. K. (1992*a*). Tick and tick-borne disease control in Zimbabwe: what might the future hold? *Zimbabwe Veterinary Journal* 23, 1–15.

Norval, R. A. I., Perry, B. D. & Young, A. S. (1992*b*). *The Epidemiology of Theileriosis in Africa.* London: Academic Press.

Parker, R. J., Shepherd, R. K., Trueman, K. F., *et al.* (1985). Susceptibility of *Bos indicus* and *Bos taurus* to *Anaplasma marginale* and *Babesia bigemina* infections. *Veterinary Parasitology* 17, 205–213.

Pegram, R. G., Wilson, D. D. & Hansen, J. W. (2000). Past and present national tick control programs: why they succeed or fail. *Annals of the New York Academy of Sciences* 916, 546–554.

Perry, B. D., Chamboko, T., Mahan, S. M., *et al.* (1998). The economics of integrated tick and tick-borne disease control on commercial farms in Zimbabwe. *Zimbabwe Veterinary Journal* 29, 21–29.

Pipano, E. (1995). Live vaccines against hemoparasitic diseases in livestock. *Veterinary Parasitology* 57, 213–231.

Pipano, E. (1997). Vaccines against hemoparasitic diseases in Israel with special reference to quality assurance. *Tropical Animal Health and Production* 29 (Suppl.), S86–S90.

Pipano, E., Markovics, A., Kriegel, Y., Frank, M. & Fish, L. (1987). Use of long-acting oxytetracycline in the immunization of cattle against *Babesia bovis* and *B. bigemina. Research in Veterinary Science* 43, 64–66.

Potgieter, F. T. (1977). The life cycle of *Babesia bovis* and *Babesia bigemina* in ticks and in cattle in South Africa. Unpublished Ph.D. thesis, Rand Afrikaans University, Johannesburg, South Africa.

Potgieter, F. T. & Els, H. J. (1976). Light and electron microscopic observations on the development of small merozoites of *Babesia bovis* in *Boophilus microplus* larvae. *Onderstepoort Journal of Veterinary Research* 43, 123–128.

Potgieter, F. T. & Els, H. J. (1977*a*). The fine structure of intra-erythrocytic stages of *Babesia bigemina. Onderstepoort Journal of Veterinary Research* 44, 157–168.

Potgieter, F. T. & Els, H. J. (1977*b*). Light and electron microscopic observations on the development of *Babesia bigemina* in larvae, nymphae and non-replete females of *Boophilus decoloratus. Onderstepoort Journal of Veterinary Research* 44, 213–231.

Potgieter, F. T. & Els, H. J. (1979). An electron microscopic study of intra-erythrocytic stages of *Babesia bovis* in the brain capillaries of infected splenectomized calves. *Onderstepoort Journal of Veterinary Research* 46, 41–49.

Precigout, E., Gorenflot, A., Valentin, A., *et al.* (1991). Analysis of immune responses of different hosts to *Babesia divergens* isolates from different geographic areas and capacity of culture-derived exoantigens to induce efficient cross-protection. *Infection and Immunity* **59**, 2799–2805. [Published erratum appears in *Infection and Immunity* (1992) **60**, 1728.]

Pudney, M. (1992). Cultivation of *Babesia*. In *Recent Developments in the Control of Anaplasmosis, Babesiosis and Cowdriosis*, ed. Dolan, T. T., pp. 129–140. Nairobi, Kenya: International Laboratory for Research on Animal Diseases.

Purnell, R. E. & Lewis, D. (1981). *Babesia divergens*: combination of dead and live parasites in an irradiated vaccine. *Research in Veterinary Science* **30**, 18–21.

Riek, R. F. (1963). Immunity to babesiosis. In *Immunity to Protozoa*, eds. Garnham, P. C. C., Pierce, A. E. & Roitt, I., pp. 160–179. Oxford, UK: Blackwell Scientific Publications.

Riek, R. F. (1964). The life cycle of *Babesia bigemina* (Smith and Kilborne, 1893) in the tick vector *Boophilus microplus* (Canestrini). *Australian Journal of Agricultural Research* **15**, 802–821.

Riek, R. F. (1966). The life cycle of *Babesia argentina* (Lignières, 1903) (Sporozoa: Piroplasmidea) in the vector *Boophilus microplus* (Canestrini). *Australian Journal of Agricultural Research* **17**, 247–254.

Rodriguez, S. D., Palmer, G. H., McElwain, T. F., *et al.* (1996). CD4+ T-helper lymphocyte responses against *Babesia bigemina* rhoptry-associated protein I. *Infection and Immunity* **64**, 2079–2087.

Rogers, R. J., Dimmock, C. K., de Vos, A. J. & Rodwell, B. J. (1988). Bovine leucosis virus contamination of a vaccine produced *in vivo* against bovine babesiosis and anaplasmosis. *Australian Veterinary Journal* **65**, 285–287.

Rudzinska, M. A., Spielman, A., Lewengrub, S., Trager, W. & Piesman, J. (1983). Sexuality in piroplasms as revealed by electron microscopy in *Babesia microti*. *Proceedings of the National Academy of Sciences of the USA* **80**, 2966–2970.

Sahibi, H., Rhalem, A., Berrag, B. & Goff, W. L. (1998). Bovine babesiosis: seroprevalence and ticks associated with cattle from two different regions of Morocco. *Annals of the New York Academy of Sciences* **849**, 213–218.

Sayin, F., Dincer, S., Karaer, Z., *et al.* (1996). Studies on seroprevalence of *babesia* infection of cattle in Turkey. In *New Dimensions in Parasitology, Proceedings of the 8th International Congress of Parasitology*, ed. Özcel, M. A.,

pp. 505–516. Izmir, Turkey: Turkish Society for Parasitology.

Schetters, T. P. M., Kleuskens, J. A. G. M., Scholtes, N. C., *et al.* (2006). Onset and duration of immunity against *Babesia canis* infection in dogs vaccinated with antigens from culture supernatants. *Veterinary Parasitology* **138**, 140–146.

Shkap, V., Pipano, E., McElwain, T. F., *et al.* (1994). Cross-protective immunity induced by *Babesia bovis* clones with antigenically unrelated variable merozoite surface antigens. *Veterinary Immunology and Immunopathology* **41**, 367–374.

Shoda, L. K., Kegerreis, K. A., Suarez, C. E., *et al.* (2001). DNA from protozoan parasites *Babesia bovis*, *Trypanosoma cruzi*, and *T. brucei* is mitogenic for B lymphocytes and stimulates macrophage expression of interleukin-12, tumor necrosis factor alpha, and nitric oxide. *Infection and Immunity* **69**, 2162–2171.

Shoda, L. K., Palmer, G. H., Florin-Christensen, J., *et al.* (2000). *Babesia bovis*-stimulated macrophages express interleukin-1beta, interleukin-12, tumor necrosis factor alpha, and nitric oxide and inhibit parasite replication in vitro. *Infection and Immunity* **68**, 5139–5145.

Singleton, E. F. (1974). The effect of heat on reproductive function in the bull. Unpublished Ph.D. thesis, University of Queensland, Brisbane, Australia.

Smith, R. D., Molinar, E., Larios, F., *et al.* (1980). Bovine babesiosis: pathogenicity and heterologous species immunity of tick-borne *Babesia bovis* and *B. bigemina* infections. *American Journal of Veterinary Research* **41**, 1957–1965.

Smith, T. & Kilborne, F. L. (1893). Investigations into the nature, causation and prevention of Southern cattle fever. In *9th Annual Report of the Bureau of Animal Industry for the year 1892*, pp. 177–304. Washington, DC: US Government Printing Office.

Sserugga, J. N., Jonsson, N. N., Bock, R. E. & More, S. J. (2003). Serological evidence of exposure to tick fever organisms in young cattle in Queensland dairy farms. *Australian Veterinary Journal* **81**, 147–152.

Standfast, N. F. & Jorgensen, W. K. (1997). Comparison of the infectivity of *Babesia bovis*, *Babesia bigemina* and *Anaplasma centrale* for cattle after cryopreservation in either dimethylsulphoxide (DMSO) or polyvinylpyrrolidone (PVP). *Australian Veterinary Journal* **75**, 62–63.

Standfast, N. F., Bock, R. E., Wiecek, M. M., *et al.* (2003). Overcoming constraints to meeting increased demand for *Babesia bigemina* vaccine in Australia. *Veterinary Parasitology* **115**, 213–222.

Stewart, N. P., de Vos, A. J., McHardy, N. & Standfast, N. F. (1990). Elimination of *Theileria buffeli* infections from cattle by concurrent treatment with buparvaquone and primaquine phosphate. *Tropical Animal Health and Production* 22, 116–122.

Stich, R. W., Shoda, L. K., Dreewes, M., *et al.* (1998). Stimulation of nitric oxide production in macrophages by *Babesia bovis*. *Infection and Immunity* 66, 4130–4136.

Sutherst, R. W. (1987). The dynamics of hybrid zones between tick (Acari) species. *International Journal for Parasitology* 17, 921–926.

Taylor, R. J. & McHardy, N. (1979). Preliminary observations on the combined use of imidocarb and *Babesia* blood vaccine in cattle. *Journal of the South African Veterinary Association* 50, 326–329.

Timms, P. & Stewart, N. P. (1989). Growth of *Babesia bovis* parasites in stationary and suspension cultures and their use in experimental vaccination of cattle. *Research in Veterinary Science* 47, 309–314.

Timms, P., Dalgliesh, R. J., Barry, D. N., Dimmock, C. K. & Rodwell, B. J. (1983*a*). *Babesia bovis*: comparison of culture-derived parasites, non-living antigen and conventional vaccine in the protection of cattle against heterologous challenge. *Australian Veterinary Journal* 60, 75–77.

Timms, P., Stewart, N. P. & Dalgliesh, R. J. (1983*b*). Comparison of tick and blood challenge for assessing immunity to *Babesia bovis*. *Australian Veterinary Journal* 60, 257–259.

Timms, P., Stewart, N. P. & de Vos, A. J. (1990). Study of virulence and vector transmission of *Babesia bovis* by use of cloned parasite lines. *Infection and Immunity* 58, 2171–2176.

Tjornehoj, K. T., Lawrence, J. A., Whiteland, A. P. & Kafuwa, P. T. (1996). Field observations on the duration of immunity in cattle after vaccination against *Anaplasma* and *Babesia* species. *Onderstepoort Journal of Veterinary Research* 63, 1–5.

Trueman, K. F. & Blight, G. W. (1978). The effect of age on resistance of cattle to *Babesia bovis*. *Australian Veterinary Journal* 54, 301–305.

Uilenberg, G. (2006). Babesia: a historical overview. *Veterinary Parasitology* 138, 3–10.

Uilenberg, G., Thiaucourt, F. & Jongejan, F. (2004). On molecular taxonomy: what is in a name? *Experimental and Applied Acarology* 32, 301–312.

Valentin, A., Précigout, E., L'Hostis, M., *et al.* (1993). Cellular and humoral immune responses induced in cattle by vaccination with *Babesia divergens* culture-derived exoantigens correlate with protection. *Infection and Immunity* 61, 734–741.

Vega, C. A., Buening, G. M., Rodriguez, S. D., Carson, C. A. & McLaughlin, K. (1985). Cryopreservation of *Babesia bigemina* for in vitro cultivation. *American Journal of Veterinary Research* 46, 421–423.

Vial, H. J. & Gorenflot, A. (2006). Chemotherapy against babesiosis. *Veterinary Parasitology* 138, 147–160.

Waldron, S. J. & Jorgensen, W. K. (1999). Transmission of *Babesia* spp. by the cattle tick (*Boophilus microplus*) to cattle treated with injectable or pour-on formulations of ivermectin and moxidectin. *Australian Veterinary Journal* 77, 657–659.

Wilkowsky, S. E., Farber, M., Echaide, I., *et al.* (2003). *Babesia bovis* merozoite surface protein-2c (MSA-2c) contains highly immunogenic, conserved B-cell epitopes that elicit neutralization-sensitive antibodies in cattle. *Molecular and Biochemical Parasitology* 127, 133–141.

Wright, I. G. & Goodger, B. V. (1988). Pathogenesis of babesiosis. In *Babesiosis of Domestic Animals and Man*, ed. Ristic, M., pp. 99–118. Boca Raton, FL: CRC Press.

Wright, I. G. & Riddles, P. W. (1989). Biotechnology in tick-borne diseases: present status, future perspectives. In *FAO Expert Consultation of Biotechnology for Livestock Production and Health*, pp. 325–340. Rome: Food and Agriculture Organization of the United Nations.

Wright, I. G., Casu, R., Commins, M. A., *et al.* (1992). The development of a recombinant *Babesia* vaccine. *Veterinary Parasitology* 44, 3–13.

Wright, I. G., Goodger, B. V., Buffington, G. D., *et al.* (1989). Immunopathophysiology of babesial infections. *Transactions of the Royal Society of Tropical Medicine and Hygiene* 83 (Suppl.), S11–S13.

Wright, I. G., Goodger, B. V., Leatch, G., *et al.* (1987). Protection of *Babesia bigemina*-immune animals against subsequent challenge with virulent *Babesia bovis*. *Infection and Immunity* 55, 364–368.

Wright, I. G., Mahoney, D. F., Mirre, G. B., Goodger, B. V. & Kerr, J. D. (1982). The irradiation of *babesia bovis*. II. The immunogenicity of irradiated blood parasites for intact cattle and splenectomized calves. *Veterinary Immunology and Immunopathology* 3, 591–601.

Wyatt, C. R., Goff, W. & Davis, W. C. (1991). A flow cytometric method for assessing viability of intraerythrocytic hemoparasites. *Journal of Immunological Methods* **140**, 23–30.

Yokoyama, N., Okamura, M. & Igarashi, I. (2006). Erythrocyte invasion by *Babesia* parasites: current advances in the elucidation of the molecular interactions between the protozoan ligands and host receptors in the invasion stage. *Veterinary Parasitology* **138**, 22–32.

Yunker, C. E., Kuttler, K. L. & Johnson, L. W. (1987). Attenuation of *Babesia bovis* by in vitro cultivation. *Veterinary Parasitology* **24**, 7–13.

Zintl, A., Mulcahy, G., Skerrett, H. E., Taylor, S. M. & Gray, J. S. (2003). *Babesia divergens*: a bovine blood parasite of veterinary and zoonotic importance. *Clinical Microbiology Reviews* **16**, 622–636.

14 • *Theileria*: life cycle stages associated with the ixodid tick vector

R. BISHOP, A. MUSOKE, R. SKILTON, S. MORZARIA, M. GARDNER
AND V. NENE

INTRODUCTION

The genus *Theileria* comprises tick-transmitted sporozoan protozoa that are the causative agents of a variety of disease syndromes in domestic and wild ruminants, and are collectively responsible for economic losses amounting to hundreds of millions of dollars annually in sub-Saharan Africa and Asia. *Theileria* are unique among protozoa, in that certain species are capable of immortalizing either mammalian lymphocytes, or cells of the monocyte/macrophage lineage that they infect. *Theileria* has been included within a subphylum designated the Apicomplexa, based on the common possession of an apical complex containing secretory organelles involved in invasion, or establishment, in the cells of their mammalian and invertebrate hosts. However the evolutionary and functional equivalence of the apical complex between different genera and hence the taxonomic validity of the Apicomplexa remains unclear. Analysis of 18S ribosomal RNA gene sequences demonstrates that the genus *Theileria* is phylogenetically most closely related to *Babesia*, a genus of tick-borne protozoan infective to the red cells of mammals including domestic livestock, and more distantly to the genus *Plasmodium* which causes malaria in humans and other species of vertebrates (Allsopp *et al.*, 1994). There are similarities, but also significant differences, in features of the life cycle, genome organization and mammalian host immune responses to infection between *Theileria* and *Plasmodium*.

Economically important *Theileria* species that infect cattle and small ruminants are transmitted by ixodid ticks of the genera *Rhipicephalus*, *Amblyomma*, *Hyalomma* and *Haemaphysalis*. *Theileria* species infective to domestic ruminants are summarized in Table 14.1. Globally the most economically important species are *T. parva* and *T. annulata*. *Theileria parva* is transmitted by *Rhipicephalus* ticks, and causes a rapidly fatal lymphoproliferative disease known as East Coast fever (ECF). The disease was estimated to have been

responsible for 170 million dollars of economic loss in 1989 alone (Mukhebi, Perry & Kruska, 1992) and limits introduction of more productive exotic (*Bos taurus*) cattle in much of eastern, central and southern Africa. The primary host of *T. parva* is the African Cape buffalo (*Syncerus caffer*) in which the parasite typically does not cause any disease. *Theileria annulata*, originating from Asian water buffalo (*Bulbulus bubulis*), and transmitted by several *Hyalomma* tick species, is responsible for tropical theileriosis from southern Europe to China, a vast region in which an estimated 250 million cattle are at risk. Both *T. parva* and *T. annulata* induce a transformation-like phenotype in nucleated mammalian host cells, which is a major cause of pathology. Several other non-transforming *Theileria* are also responsible for disease in domestic ruminants, as a result of anaemia induced by the intra-erythrocytic stage. *Theileria sergenti* and *T. buffeli* cause disease and economic loss in cattle in East Asia. *Theileria mutans* may be responsible for disease in cattle in sub-Saharan Africa particularly in mixed infections with other tick-borne pathogens, but the economic impact of this species has yet to be fully assessed. A recent study in Uganda indicated a very high prevalence of *T. mutans* in cattle and provided evidence that on some farms this species was associated within clinical theileriosis in the absence of detectable *T. parva* infections (Oura *et al.*, 2004). *Theileria lestoquardi* is the most pathogenic of several *Theileria* species of small ruminants infective to sheep and goats in northern Africa, and widely across Asia. Most wild bovids in Africa are infected by one or more species of *Theileria* and the epidemiological situation can be complex with East African cattle potentially being infected with up to five *Theileria* species at one time (Norval, Perry & Young, 1992) in addition to other tick-borne pathogens (Oura *et al.*, 2004). Cattle can be infested with several species of tick and more than one species of *Theileria* can also be transmitted by the same tick. For example, the morphologically very similar sporozoites

Ticks: Biology, Disease and Control, ed. Alan S. Bowman and Patricia A. Nuttall. Published by Cambridge University Press.
© Cambridge University Press 2008.

Table 14.1 Theileria *species infective to domestic ruminants, their ixodid tick vectors and geograpical distribution*

Theileria species	Major ixodid vectors	Known distribution

Theileria parva	*Rhipicephalus appendiculatus*
	   R. zambesiensis
	   R. duttoni
		Eastern, central and southern Africa

Theileria annulata	*Hyalomma anatolicum* and other
	   Hyalomma species
		Southern Europe, western, southern and
		   eastern Asia, northern Africa

Theileria mutans	*Amblyomma variegatum* and four
	   other *Amblyomma* species
		Western, eastern, central and southern
		   Africa, Caribbean islands

Theileria velifera	*Amblyomma variegatum* and other
	   Amblyomma species
		Western, eastern, central and southern
		   Africa

Theileria tarurotragi	*Rhipicephalus appendiculatus*
	   R. zambesiensis
	   R. pulchellus
		Eastern, central and southern Africa

Theileria sergenti	*Haemaphysalis* species	Japan and Korea
Theileria buffeli	*Haemaphysalis* species	Europe, Asia, Australia, eastern Africa
Theileria lestoquardi[a]	*Hyalomma* species	Asia and northern Africa
Theileria ovis[a]	*Hyalomma* species	Asia
Theileria separata[a]	*Hyalomma* species	Asia

[a] Indicates *Theileria* species infective to small ruminants; other species are infective to cattle.

of *T. parva* and the normally non-pathogenic *T. taurotragi* can both occur together in *R. appendiculatus* salivary glands, although they can be distinguished by molecular methods based on hybridization to rRNA (Bishop *et al.*, 1994). Additional *Theileria* species that are infective to livestock, and may possibly be pathogenic in specific circumstances, continue to be discovered. The range of *T. buffeli* has been extended to Africa (Ngumi *et al.*, 1994), and a hitherto undescribed *Theileria* that is pathogenic to small ruminants has been discovered in North Western China (Schnittger *et al.*, 2000). A *Theileria* species originally described from schizont-infected lymphocyte culture isolates from Cape buffalo (Conrad *et al.*, 1987; Allsopp *et al.*, 1993) has recently been found in cattle subjected to challenge by ticks that are presumed to have fed on buffalo (R. Bishop & A. J. Musoke, unpublished data). *Theileria* infections have also been associated with bovine fatalities in the United States, and the presence of *Theileria* was confirmed in both *Amblyomma americanum* and *Dermacentor variabilis* by PCR amplification using primers derived from small subunit rRNA (SSU rRNA) gene sequences (Chae *et al.*, 1999).

Most research on *Theileria* to date has been devoted to the life-cycle stages occurring in the mammalian host, particularly the pathogenic, intracellular schizont stages of *T.* *parva* and *T. annulata*, which infect cells of the lymphoid and myeloid lineages and activate host signal transduction pathways, inducing cancer-like syndromes and inhibiting apoptosis in infected cells. Host pathways modulated by *Theileria* infection include those involving Src tyrosine kinases, JNK amino terminal kinase, phosphatidyl inositol 3 kinase and casein kinase II resulting in induction of a variety of host transcription factors. There are several recent comprehensive reviews of this aspect of *Theileria* biology (Dobbelaere & Heussler, 1999; Heussler, 2002; Dobbelaere & Kuenzi, 2004). Little is known about the parasite molecules that are responsible for host cell modification. One group of candidates is the Tashat gene family encoding putative transcription factors that appear to be exported from the parasite and localized to the host nucleus (Swan *et al.*, 1999). The availability of the genome sequences of two *Theileria* species enables more systematic approaches for identifying parasite molecules that manipulate the host cell (reviewed by Shiels *et al.*, 2006). This chapter will concentrate on tick vector-associated stages of *Theileria*. Additionally, salivary gland expressed sequence tag data are now available and provide insights into the sialome of *Rhipicephalus appendiculatus* and *Amblyomma variegatum*. The only tick-expressed stage of *Theileria* for which molecular information is currently available

LIFE CYCLE OF *THEILERIA PARVA*

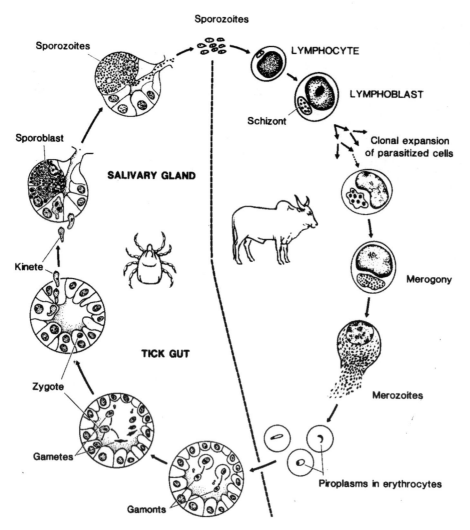

Fig. 14.1. Life cycle of *Theileria parva* in cattle and the ixodid tick vector *Rhipicephalus appendiculatus*.

is the sporozoite, which is secreted from the tick salivary glands into vertebrate host tissues. We include a review of the status of vaccine development based on a sporozoite antigen present in *T. parva*, *T. annulata* and *T. lestoquardi*.

OVERVIEW OF THE *THEILERIA* LIFE CYCLE

Theileria parva life-cycle stages in the tick vector and bovine host have been comprehensively described (Norval, Perry &

Young, 1992) and are summarized in Fig. 14.1. The three-host brown ear tick *Rhipicephalus appendiculatus* is the major vector. This tick is so named because the adults have a preference for feeding on the ears of the mammalian host (Fig. 14.2). *Theileria parva* exhibits a very narrow tick and mammalian host range, and there is no laboratory animal model. *Rhipicephalus appendiculatus* is classified as a three-host tick, because the larvae, nymphs and adults feed on different hosts, which are not necessarily cattle. Transmission is trans-stadial; larval or nymphal instars of the tick acquire

Fig. 14.2. Adult *Rhipicephalus appendiculatus* ticks in situ on the ear of a *Bos indicus* animal.

infections from a blood meal and the pathogen is then transmitted to a new host, after moulting, by nymphs or adults, respectively. Transmission of *Theileria* has not been shown to be transovarian, unlike the related protozoan *Babesia* which is transmitted by the one-host tick *Rhipicephalus* (*Boophilus*). *Hyalomma anatolicum*, the major vector of *T. annulata*, is a two-host tick, with larval and nymphal instars feeding on the same host. Sporogony takes place in the tick salivary glands in response to a combination of increased temperature and components present within the blood meal. Infective sporozoites are released from about day 3 to day 7 of tick feeding. One infected cell may contain 40 000–50 000 sporozoites, but the mechanics of tick feeding result in a phased release of sporozoites from the salivary glands (Shaw & Young, 1994; Shaw, 2002). A tick with a single infected acinar cell can potentially cause severe disease and death. However, the severity of ECF is sporozoite dose dependent and individual animals exhibit different thresholds of susceptibility. Some animals undergo a mild disease reaction and recover due to development of protective cellular immune responses.

Once introduced into the host, *T. parva* sporozoites invade a restricted range of cattle and buffalo cells. In vitro these include purified B-cells, or T-cells which express receptors which are either CD4+, CD8+ or CD4−/CD8− (Baldwin *et al.*, 1988), and also afferent lymph veiled cells and monocytes (Shaw, Tilney & McKeever, 1993), although the latter cell types cannot subsequently be immortalized. The entry process is probably energy dependent and is likely to involve specific receptors located on the host cells (Shaw, 1997, 2002). *Theileria annulata* infects cells of the monocyte/macrophage lineage, although it can also enter B-cells. Other species such as *T. taurotragi*, originally isolated from eland (*Taurotragus oryx*), appear to have a much wider host range, at least in vitro (Stagg *et al.*, 1983). Like viruses transmitted by *R. appendiculatus* (see Chapter 10) components of tick saliva appear to promote *T. parva* pathogen transmission by facilitating sporozoite entry into host leukocytes (Shaw, Tilney & McKeever, 1993). Unlike other sporozoans, *Theileria* sporozoites are not motile. They do not possess all components of the typical apical complex described for other Apicomplexa, entry into the host cells is not orientation specific, and the rhoptries and microspheres are involved in establishment in the host cell subsequent to entry, rather than in the entry process (Shaw, Tilney & Musoke, 1991). Unlike those of other Apicomplexa, *T. parva* sporozoites also appear to lack morphologically distinguishable micronemes (Shaw, Tilney & Musoke, 1991). The newly internalized parasite is surrounded by a host cell membrane, which is rapidly (<15 min) removed by the parasite, thereby evading certain intracellular host cell defence mechanisms. In this respect, *Theileria* is similar to a phylogenetically diverse range of intracellular pathogens, including *Listeria, Shigella* and *Trypanosoma cruzi* (Andrews & Webster, 1991), presumably as a result of independent evolution of this solution to the problem of intracellular survival. Rapid dissolution of the host cell membrane by *T. parva* is closely correlated with the discharge of the contents of the rhoptries and microspheres. The major surface antigen of *T. parva* sporozoites (a 67-kDa protein) and bovine major histocompatibility complex (MHC) class I molecules are implicated in the entry process as monoclonal antibodies (mAbs) to these proteins inhibit sporozoite entry (Dobbelaere *et al.*, 1984; Musoke *et al.*, 1984, 1992; Shaw, Tilney & Musoke, 1991). Because of the restricted host cell specificity it seems unlikely that MHC class I molecules constitute the sole recognition site of sporozoites. Sporozoite organelles that discharge their contents after entry are likely to contain a microtubule nucleating factor as the free intracellular sporozoites are rapidly surrounded by a network of microtubules.

The schizont is a multinucleated, syncytial body. Association of the schizont with the host cell nuclear spindle ensures that daughter host cells remain infected during cytokinesis (Carrington *et al.*, 1995). Although the parasite and host cells divide in synchrony, schizont DNA synthesis occurs as the host cell enters mitosis and is immediately followed by division when the host cell is in metaphase (Irvin *et al.*,

1982). The schizont stage of both *T. parva* and *T. annulata* is the major cause of pathology and disease (reviewed by Irvin & Morrison, 1987). Merogony occurs within infected lymphocytes (Shaw & Tilney, 1992) and merozoites are released by host cell rupture. The merozoites invade red blood cells (RBCs) in a manner similar to that described for sporozoite entry into lymphocytes (Shaw & Tilney, 1995), and differentiate into piroplasms, which, like the intralymphocytic schizont, lie free in the erythrocyte cytoplasm. There is very little multiplication of the piroplasm stage of *T. parva*, although intraerythrocytic multiplication of the piroplasm occurs in *T. annulata* infections, and contributes to the pathology of tropical theileriosis (Irvin & Morrison, 1987). Other species of *Theileria*, especially the non-transforming *T. mutans, T. buffeli* and *T. sergenti*, exhibit high levels of intraerythrocytic merogony resulting in anaemia and pathology due to RBC lysis. Ticks become infected after feeding on blood containing infected RBCs. Morphologically distinct forms of the parasite are found in the tick gut, which is the probable site of gametogenesis. Gametes fuse to form a zygote which invades the cells of the gut epithelium. Kinetes are released into haemolymph and invade cells of the salivary gland. Sporogony is initiated by stimuli received following tick attachment to the mammalian host and appears closely linked with the moulting cycle of the tick. All stages of the life cycle are haploid except for the zygote in the tick gut.

THE *THEILERIA PARVA* AND *THEILERIA ANNULATA* GENOME SEQUENCES

The *T. parva* and *T. annulata* genomes are both approximately 8.3 megabases in size and encode predicted proteomes of 4036 and 3972 proteins, respectively (Gardner *et al.*, 2005; Pain *et al.*, 2005). The genomes generally exhibit a high degree of synteny and homology, with the exception of coding and non-coding domains located at chromosome ends. As in many other microbial eukaryotes there are multicopy open reading frames (ORFs), encoding putatively secreted proteins, associated with the telomeres in both *Theileria* species. Although the organization of these families is similar, there is limited sequence identity between ORFs from this region although in both species the telomere proximal gene families are rich in the amino acid motif SVSP. Excluding the multicopy gene families, *T. parva* and *T. annulata* have 60 and 34 species-specific genes, respectively.

Neither the *T. annulata* or *T. parva* genome contained obvious candidate 'oncogenes', suggesting that the mechanism of 'reversible transformation' of the eukaryotic host cell induced by *Theileria* may differ from those previously described for virus-induced tumours. By comparison with *Plasmodium* spp. the *Theileria* genomes are relatively rich in introns, but non-coding repetitive sequences are infrequent and intergenic sequences are generally short.

The genome sequence reveals that *T. parva* has only two ribosomal gene coding units which are identical in sequence indicating that, unlike *Plasmodium falciparum* (McCutchan *et al.*, 1995), *Theileria* does not express distinct ribosomal RNAs that are differentially expressed between arthropod vector and mammalian host. An interesting feature that is potentially relevant to survival of *Theileria* in the ixodid vector is the presence of the enzymes trehalose-6 phosphate synthetase and trehalose phosphatase in the *T. parva* genome. It is tempting to speculate that this disaccharide plays a role in desiccation resistance of the parasite during the lengthy periods when the tick is waiting to feed on vertebrate hosts.

A striking difference between *T. parva* and *T. annulata* that may be relevant to parasite transmission from the mammalian host to the ixodid vector is an exceptionally rapidly evolving gene family in *T. parva* comprising a tandemly arrayed locus, absent from *T. annulata*, that is apparently transcribed solely in the piroplasm stage. This family has been given the acronym Tpr (*T. parva* repeat) and is unusual in being located in the interior of a chromosome. Additional copies with homologous C-terminal domains (designated Tar in *T. annulata*) are dispersed on all chromosomes in both species. The complete Tpr locus could not be assembled from shotgun sequence data, but non-overlapping contigs of 41 kb and 13 kb and also a 7.0-kb sequence at one end of the array were determined. The organization of 15 Tpr ORFs within the longest contig is shown in Fig. 14.3. Seven of the ORFs appeared to be 'complete' since they contained in-frame ATG codons <50 amino acids from the N-terminus of the protein (labelled in italics in Fig. 14.3) while the other eight lacked in-frame ATG codons within 50 amino acids of the presumptive N-terminus (labelled underneath in bold in Fig. 14.3). This confirmed the organization of the Tpr locus described previously (Baylis *et al.*, 1991). Tpr ORFs have been shown to be transcribed in the intraerythrocytic piroplasm stage of *T. parva* (Baylis *et al.*, 1991; Bishop *et al.*, 1997). Piroplasms do not replicate in the bovine host, and their primary function is assumed to be in transmission of *T. parva* to the tick vector. Ten piroplasm cDNA sequences (1275–1409 bp in length) from an animal infected with the same clone of *T. parva* from which the genome sequence was derived

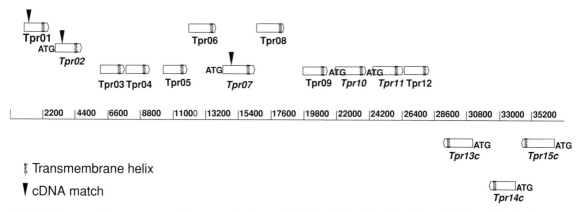

Fig. 14.3. Diagram of the arrangement of 15 open reading frames (ORFs) in a 34-k section of the *T. parva* Tpr locus. 'Complete' ORFS with ATG codons in the first 50 amino acids are labelled in italics, 'incomplete ORFs' are shown in roman. The three ORFs with high similarity (98%–99%) to piroplasm cDNA sequences are highlighted by arrows. Transmembrane helices are indicated by a bar.

(R. Bishop & R. Skilton, unpublished data) differed both from one another and from eight additional sequences (Bishop *et al.*, 1997). Three transcripts exhibited high similarity (98–99%) to ORFs within the array (indicated by arrows in Fig. 14.3). However, none of the 18 transcripts was identical to any genomic Tpr sequence, perhaps indicating post-transcriptional diversification of the RNA. Analysis of the schizont transcriptome using massively parallel signature sequencing (Bishop *et al.*, 2005) revealed signatures originating from two of the dispersed genes containing Tpr-homologous C-terminal domains. Expressed sequence tag data also indicate transcription of the Tar genes in *T. annulata* schizonts (Pain *et al.*, 2005).

The Tpr locus has the features of a system that has evolved for the generation of diversity. The exceptional variability is reminiscent of the Var genes of *P. falciparum*, which encode membrane proteins (PfEMP1) that are expressed on the erythrocyte surface and mediate capillary adherence preventing clearance of infected cells by the spleen of the host (reviewed by Borst *et al.*, 1995). The conserved C-terminal ends of the Tpr ORFs contain predicted membrane-associated helices and two ORFs are predicted to contain signal anchors (Fig. 14.3). This is consistent with insertion of the C-terminal Tpr domain into either the erythrocyte or parasite membrane. Attempts to identify a protein corresponding to the Tpr ORFs in *T. parva* piroplasm-infected erythrocytes have been unsuccessful. However, it is tempting to speculate that the products of this locus,

expressed in the erythrocyte or the tick gut stages, play a role in transmission to the vector, since *T. parva* piroplasms have no other known function.

LIFE CYCLE STAGES AND DEVELOPMENT OF *THEILERIA* IN THE IXODID TICK VECTOR

The life cycle stages of *Theileria* species in their ixodid vectors have been relatively little studied due to the difficulty of culturing them. The available data have been reviewed previously (Mehlhorn & Schein, 1984; Norval, Perry & Young, 1992; Shaw & Young, 1994; Shaw, 2002).

Tick gut stages

The life cycle of *Theileria* in the tick begins with the ingestion of piroplasm-infected erythrocytes within the blood meal. At repletion, millions of infected erythrocytes are ingested by adult ticks, even from animals with low piroplasm parasitaemias. The volumes of blood ingested by larvae and nymphs are smaller and hence the numbers of parasites surviving and developing in the midgut are correspondingly lower. Due to the failure to culture the red blood cell stages of *T. parva*, whether gametocytes (stages that are predetermined to differentiate into gametes in the tick gut) occur in the circulating blood as they do in *Plasmodium* infections is unknown (Sinden *et al.*, 1996). Since there is currently no evidence for sexual differentiation within the mammalian erythrocytic stages, and there is also a marked drop in temperature after the tick completes repletion, it is possible that the processes controlling sexual differentiation in *Theileria* and *Babesia* differ markedly from those of other sporozoan protozoa. It is believed that the vast majority of the ingested piroplasms are rapidly destroyed in the gut lumen, possibly by acid phosphatases secreted by gut epithelial cells (Shaw & Young, 1994). Soon after the completion of feeding, sexual structures designated 'Strahlenkorper' can be observed in smears by light microscopy (Mehlhorn & Schein, 1984). The formation of the zygote in *Theileria* and *Babesia* differs from most other sporozoans, since the gametes appear structurally similar (Shaw, 2002). The spherical zygote invades a gut epithelial cell and differentiates into a motile kinete. While the mechanism of entry of the zygote into gut epithelial cells is uncertain, the developing intracellular parasite is not enclosed in a parasitophorous vacuole, and lies free in the cytoplasm. Therefore, a common feature of all four intracellular stages of *Theileria*, the kinete in the tick gut, the sporozoite in the tick salivary gland and the intralymphocytic and erythrocytic stages in the mammalian host, is that they rapidly escape host membranes and lie free in the cytoplasm. *Theileria* kinetes develop only in the tick gut cells and are typically uninucleate.

Salivary gland stages

The motile kinetes invade salivary glands; this process is very specific in that kinetes have not been observed in other tick organs. Development of kinetes and their appearance in the tick haemocoel is associated with the tick moulting cycle (Shaw, 2002). To reach the salivary gland, the kinete must cross several obstacles, the basal membrane and lamina to exit the gut cells, and additional barriers, allowing entry into the salivary gland acinar cell in which sporogony occurs. Ixodid salivary glands contain four types of acinii: I, II and III are present in both males and females, while type IV is in males only (see Chapter 3). Type I acini are nongranular, and thought to be involved primarily in osmoregulation, whereas, types II, III and IV contain a variety of secretory cell types designated by lower-case letters from a to h. It appears that the motile kinetes have the ability to specifically recognize the type III salivary gland acinar cells within the salivary glands and may also be able to selectively invade e cells within the acinus. It has been suggested that the carbohydrate composition of the type III acinus determines the susceptibility of this cell type to parasite infection (reviewed by Shaw, 2002). *Theileria* develop almost exclusively within the e cells of the type III acinus, although it has been demonstrated that in ticks very highly infected with *T. annulata* the d cells and type II acini can also be infected (Mehlhorn & Schein, 1984).

Sporogony, the only multiplicative process within the tick, occurs within the salivary gland and is typically triggered by feeding, although it has been shown that sporogony can also be completed at high ambient temperatures in a proportion of acini (Young & Leitch, 1982). The number of sporozoites produced during sporogony has been estimated at 30 000–50 000 per infected acinar cell for female *R. appendiculatus* and up to 140 000 for *T. taurotragi* (Norval, Perry & Young, 1992; Shaw & Young, 1994). *Theileria parva* sporozoite numbers are typically lower in male *R. appendiculatus* than in females, and there are also fewer sporozoites in the e cells of nymphal as compared to adult ticks, with potential consequences for the relative importance of different instars in *T. parva* transmission (Ochanda *et al.*, 1996). The ultrastructure of the mature sporozoites has been described (Fawcett, Doxsey & Buscher, 1982). Unlike in other apicomplexan genera there is no clearly defined apical complex. A spherical body enclosed by three or more membranes is located close to the nucleus, and may correspond to the apicoplast described from *Plasmodium* and other genera of Sporozoa (Wilson & Williamson, 1997). As in other life cycle stages of *Theileria* (Shaw & Tilney, 1992; Ebel *et al.*, 1997) structures corresponding to the endoplasmic reticulum and Golgi apparatus of higher eukaryotes are not visible, although proteins associated with the classical secretory pathway are present in *Theileria* (Janoo *et al.*, 1999). There are a number of peripherally located membrane-bounded bodies termed microspheres and also a group of up to six larger, membrane-bounded rhoptries. Both these secretory organelles have been shown to discharge after entry into the host cell, in association with establishment of the parasite (Shaw, Tilney & Musoke, 1991). Specific proteins have been localized to these organelles, including the p104 antigen located in the rhoptries (Iams *et al.*, 1990) and the p150 and PIM antigens in the microspheres (Skilton *et al.*, 1998; C. Wells, A. Musoke & P. Toye, unpublished data).

There is some evidence to suggest that sporozoites are released gradually from an infected cell in a manner similar to the release of secretory granules by apocrine secretion (summarized in Shaw & Young, 1994). This has implications for parasite challenge to the mammalian host and the effectiveness of control strategies using recombinant vaccines for control of *Theileria*.

SEXUAL RECOMBINATION IN *THEILERIA* WITHIN THE IXODID VECTOR

As do *Plasmodium* species, *T. parva* undergoes an obligatory sexual cycle in the tick vector. While the mammalian life cycle stages are haploid, there is a transient diploid phase in the tick gut after fusion of gametes (Gauer *et al.*, 1995). In *T. parva*, according to direct measurements of the DNA content of single cells by fluorescence microscopy, a post-zygotic meiosis occurs in the gut prior to differentiation of the kinete, rapidly restoring the haploid condition. In *T. annulata* by contrast the kinete appears to remain diploid and meiotic division may occur at later stages during sporogony in the salivary glands (Gauer *et al.*, 1995). Laboratory crosses have been performed between *T. parva* Muguga and two other parasite stocks, *T. parva* Uganda and *T. parva* Marikebuni, by feeding ticks on cattle co-infected with either stock (Morzaria *et al.*, 1993). Hybridization of 'stock specific' oligonucleotides derived from *T. parva* Muguga and Uganda (Allsopp *et al.*, 1989; Bishop *et al.*, 1993) revealed that 38% of single infected acini derived from progeny of the Muguga/Uganda cross appeared to be recombinant. Subsequent analysis of a cloned parasite derived from a recombinant acinus revealed mixed profiles using several probes in Southern blot assays. The data suggested that at least four crossover events had occurred on the two smallest chromosomes. In the case of the *T. parva* Muguga/Marikebuni cross, three independent clonal progeny were isolated, and in all a *T. parva* Muguga-type Tpr locus appeared to have been introgressed into a genome in which the majority of other polymorphic loci were *T. parva* Marikebuni type. These may have represented independent occurrences of similar recombination events, since the Tpr genotypes differed from one another and also a previously characterized *T. parva* Marikebuni Tpr (Bishop *et al.*, 2002a).

The extent of recombination in *T. parva* field populations is currently unknown, although as in *Plasmodium* (Paul & Day, 1998; Awadalla *et al.*, 2001) there is genetic heterogeneity between individuals and populations in the field and it seems reasonable to assume that this is attributable at least partially to recombination. The extent of recombination can only be resolved by direct examination of kinetes using PCR analysis of single-copy allelic markers, as has been done for *Plasmodium* oocytes from mosquito populations in the field. In one study, a high percentage of apparently recombinant *P. falciparum* zygotes were found at a Tanzanian study site (Babiker *et al.*, 1994). However, such studies will be technically more difficult for *Theileria* infections in ticks. According to research on *Plasmodium*, the degree of outcrossing may vary according to infection intensity in different epidemiological situations, and even quite high levels of inbreeding may still allow linkage equilibrium between different alleles in *Plasmodium* populations (Paul *et al.*, 1995; reviewed by Paul & Day, 1998). It is also possible that factors such as host and vector movements may be more important in genetic substructuring of *T. parva* populations than the extent of meiotic recombination, and this may differ substantially between cattle and the reservoir host cape buffalo (*Syncerus caffer*). The major determinant of genetic substructuring among nematode parasites of sheep, cattle and white-tailed deer (*Odocoileus virginianus*), was host movement, genetic differentiation of parasite populations being greater in the white-tailed deer, due to more extensive movement of livestock between different localities (Blouin *et al.*, 1995). An issue that has yet to be addressed for *Theileria* or other parasites (Day *et al.*, 1992) is the extent to which additional processes of diversification, including somatic mutation, genome rearrangements and gene conversion during the asexual multiplication phases, in addition to classical recombination during the sexual cycle, generate diversity in parasite populations.

TRANSMISSION DYNAMICS OF *THEILERIA PARVA* AND FACTORS AFFECTING VECTOR COMPETENCE

The transmission dynamics and epidemiology of theileriosis vary in different areas of eastern, central and southern Africa according to a complex interplay of factors, including the level of tick control, cattle genotype and management regime, the proximity of a wildlife reservoir and the interaction of tick and parasite populations with differing genetic composition. Initial transmission dynamic modeling of *T. parva* infections (Medley, Perry & Young, 1993) focused on an epidemiological state described by the term 'endemic stability' (Norval, Perry & Young, 1992) in which African zebu (*Bos indicus*) cattle previously exposed to *T. parva* over several generations, and subjected to little or no tick control, are continuously challenged by infected ticks, and either become immune or die by day 150 of exposure. In endemically stable situations the majority of ticks are believed to exhibit relatively low levels of infection in terms of both prevalence (percentage of ticks infected) and abundance (the mean number of infected acini per tick), both of which are frequently low in these epidemiological circumstances (Young *et al.*, 1986). Laboratory data on experimental tick infections (Buscher & Otim, 1986), and also field data (S. Morzaria, unpublished data) indicate that the abundance of *T. parva* infection can frequently be described by a negative binomial distribution. Therefore the abundance of mature *T. parva* infections in salivary glands is typically overdispersed, with a small proportion of ticks having many infected acini, while a majority may contain only a single infected acinus. The key conclusion from preliminary transmission dynamic modelling work (Medley, Perry & Young, 1993) is that the degree to which animals that have recovered from theileriosis, but remain infected (known as the 'carrier state'), are able to transmit the infection to tick larvae or nymphs is a crucial determinant of the dynamics of infection in a herd. Currently there are no comparative experimental data on the relative tick transmissibility of *T. parva* parasites from carrier animals, relative to those undergoing acute infections. It is known that the level of piroplasm parasitaemia is the most important parameter influencing levels of infection in *R. appendiculatus* ticks experimentally fed on cattle (Young *et al.*, 1996). However, no correlation has been found between the number of zygotes and kinetes present in the tick gut and haemolymph, respectively, and the number of sporoblasts in the salivary glands (Buscher & Otim, 1986). This implies that there is some degree of parasite destruction occurring either within the gut cells, or during kinete migration through the haemolymph. Although the mechanisms are not yet fully understood, there may be an element of *Theileria* specificity, since *R. appendiculatus* haemolymph exhibited no apparent effect on bacteria in vitro (Watt *et al.*, 2001). Arthropod vector immune responses to parasites is an emerging area most extensively researched in *Drosophila* and *Anopheles* mosquitoes (Dimopoulos, 2003). Recent work suggests that certain stocks may not induce a long-term carrier state since even when experimentally infected or vaccinated cattle are persistently infected with *T. parva*, as judged by positivity using a PCR-based assay, the infection may not be experimentally transmissible by ticks (Skilton *et al.*, 2002). Data generated using PCR assays also indicate that animals from endemic areas that are serologically positive for *Theileria* infection (Young *et al.*, 1986) may not necessarily be carriers of *T. parva* parasites based on the inability to detect parasite DNA in these animals (R. Bishop, unpublished data).

Two important parameters in tick-borne disease transmission dynamics are: (1) differences in vector competence for pathogens between different tick stocks; (2) the relative importance of infected nymphs, which infest the mammalian host in higher numbers, as compared to infected adults which

have considerably higher parasite abundance per infected individual but occur in lower numbers on the host, in the transmission of *T. parva* to cattle. It has been shown that different *R. appendiculatus* stocks have varying effectiveness in their ability to transmit two *T. parva* stocks (from Kenya and Zimbabwe) to cattle in the laboratory (Ochanda *et al.*, 1998). These authors also demonstrated interaction between parasite and tick in terms of transmission efficiency, which was dependent on the specific combination of tick and parasite genotype (Ochanda *et al.*, 1998). Furthermore, the offspring of individual ticks within families derived from two different *R. appendiculatus* stocks (Kiambu and Muguga) have been demonstrated to have heritabilities of between 0.4 and 0.6 for susceptibility to infection with *T. parva* (Young *et al.*, 1995), indicating that these differences are partially genetically determined. The basis of such differences has not yet been investigated for ticks, but by analogy with insects may be related to factors such as the levels of expression of nitrous oxide in the gut, or antimicrobial peptides, such as cecropins and serpins, in a variety of tissues. This suggests that selection of tick lines that are refractory or susceptible to *T. parva* infection may be possible. The first attempt to make an objective comparison of transmission of *T. parva* between different instars is that of Ochanda *et al.* (1998). The conclusion of this research was that larval/nymphal transmission is less efficient than nymphal/adult transmission. One probable factor underlying this result is that mean numbers of type III acini were 1736 in adult females, 1346 in adult males but only 87 in nymphae (Ochanda *et al.*, 1996). However, the ratio of larval : nymphal : adult instars is estimated to be 1 : 100 : 1000 (Norval, Perry & Young, 1992). In addition enumeration of the nymphal and larval immature stages is technically much more difficult than for adults. The relative quantitative importance of nymphal versus larval transmission of *T. parva* is therefore not yet resolved (Ochanda *et al.*, 1996). Due to infection by lower numbers of parasites, nymphal transmission may typically result in milder *T. parva* infections and hence be important in the induction of immunity in cattle that subsequently maintain a carrier state.

Vector biology and population dynamics in relation to *Theileria* transmission has been reviewed previously (Norval, Perry & Young, 1992; Shaw & Young, 1994). The ixodid ticks that are most important for transmission of economically important *Theileria* species, namely, *R. appendiculatus*, *H. anatolicum* and *A. variegatum*, have a wide variety of hosts, but all have a preference for feeding on large mammals as adults. However the larvae and nymphs also frequently infest small mammals (mainly carnivores and lagomorphs) in the case of *Rhipicephalus*, but additional vertebrate taxa including reptiles and birds for *Hyalomma* and *Ambylomma* (see Norval, Perry & Young, 1992 for further details). There are often several tick species capable of transmitting the same *Theileria* species. For example in southern Africa, *R. appendiculatus* and *A. variegatum* are replaced by *R. zambeziensis* and *A. hebraeum* as major agents in the transmission of *T. parva* and *T. mutans*, respectively. Geographical variation in the behavior of tick populations can affect transmission. For example in southern Africa, *R. appendiculatus* undergoes diapause during the cold dry season, resulting in a distinct seasonality of transmission, which is different to the situation in eastern Africa (Norval, Perry & Young, 1992; Shaw & Young, 1994). The existence of mildly pathogenic or non-pathogenic *Theileria* that can also be transmitted to domestic livestock, and co-infect the same ticks, complicates the epidemiology. Another issue in relation to vector population dynamics is to what extent infection with *Theileria* may impact on the survival of the vector. Most *T. parva* infections in the field are at low levels and may not significantly affect vector survival. However, *Hyalomma* ticks have been observed with sufficiently high infections of *T. annulata* to reduce their feeding efficiency and survival (Walker *et al.*, 1983).

SPOROZOITE-BASED VACCINES AGAINST *THEILERIA*

Currently management of ECF is primarily effected through control of the tick vector using acaricides, although this is unsustainable in the medium to long term (Tatchell, 1987; Willadsen, 2004). An 'infection and treatment' live vaccine based on the injection of a potentially lethal dose of sporozoites together with a long-acting dose of tetracycline (Radley *et al.*, 1975) has been developed. However the live vaccine has not been widely deployed due to problems related to use of potentially lethal live parasites, the strain-specific immunity induced in a proportion of vaccinated cattle, and most importantly the requirement for a cold chain and skilled veterinarians to ensure effective delivery.

Humoral immune responses directed against surface epitopes on the sporozoite stage, particularly to the major surface antigen, p67, can also be protective, probably by neutralizing infectivity of sporozoites for bovine lymphocytes (Dobbelaere *et al.*, 1984; Musoke *et al.*, 1984, 1992). An immuno-electron micrograph illustrating the distribution of p67 on the sporozoite surface is shown in Fig. 14.4. The

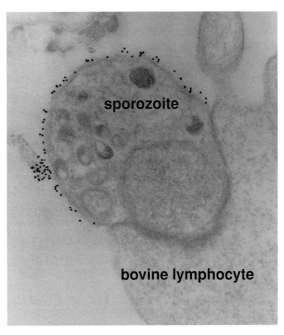

Fig. 14.4. Immuno-electron micrograph showing the distribution of the p67 protein on the surface of the *Theileria parva* sporozoite as it enters a host cell. Bovine leukocytes and a *T. parva* sporozoite preparation were incubated for 5 min and lightly fixed. They were then treated with anti-p67 sera and subsequently a secondary antibody coupled to 10-nm gold particles, prior to embedding and sectioning. Native p67 is visualized as black dots on the sporozoite surface. The p67 protein is uniformly distributed on the 'free' sporozoite surface but is not detectable within the lymphocyte. Scale bar = 0.5 μm. The authors are grateful to Mr Clive Wells for permission to use this image.

gene encoding p67 has been cloned (Nene *et al.*, 1992). A non-soluble full-length bacterially expressed recombinant (Musoke *et al.*, 1992), a soluble bacterially expressed near full-length bacterial form (p635) incorporating a prokaryotic signal sequence (Bishop *et al.*, 2003), and a near full-length form expressed in insect cells (Nene *et al.*, 1995) have consistently been shown to induce approximately 70% protection against lethal needle challenge with sporozoite stabilates, using both *Bos indicus* and *Bos taurus* cattle, in laboratory experiments. Furthermore, initial nucleotide sequence analysis of the p67 gene revealed absolute conservation among cattle-derived parasites (Nene *et al.*, 1996), suggesting that the vaccine should be robust against heterologous parasite challenge in areas where only cattle are present. In order to promote further development and optimization of the p67 recombinant vaccine, neutralizing epitopes within the protein have been identified using Pepscan analysis (Nene *et al.*, 1999). In addition, an 80 amino acid section located near the C-terminus of the protein containing some of the epitopes recognized by the neutralizing mAbs has been demonstrated to provide equivalent levels of protection to the full-length molecule (Bishop *et al.*, 2003). A p67 construct fused C-terminally to green fluorescent protein and expressed in insect cells is the first recombinant form of the protein to be recognized by antibodies raised against the native protein (Kaba *et al.*, 2002) although it has not yet been extensively evaluated in vaccine trials. Several p67 vaccine field trials have been performed in Kenya, using both p635 and the C-terminal 80 amino acid fragment, in order to determine protective efficacy against tick challenge in production systems with different transmission dynamics and levels of tick challenge. These trials resulted in reduction in severe disease of 47–50% (Musoke *et al.*, 2005). Sequence data confirmed the absolute conservation of the p67 gene sequence in 15 schizont isolates from vaccinated and unvaccinated cattle. Given the lack of heterogeneity in the p67 gene among cattle-transmissible parasites, other factors may contribute to the lower levels of protection against *T. parva* sporozoite challenge observed under field conditions. These include (1) differences in the immune responses to vaccination in individual cattle resulting in differences in the threshold of susceptibility between different animals, and (2) stochastic differences in levels of exposure to immunomodulatory molecules secreted by ticks during feeding. The p67 field trial results suggest that the *T. parva* p67 vaccine would be a useful model system to evaluate whether incorporation of tick salivary gland antigen components into a multivalent vaccine might improve the efficacy of a recombinant anti-pathogen vaccine when subjected to a field tick challenge.

Recombinant sporozoite vaccines against other *Theileria* species

The major surface antigen of *T. annulata*, SPAG1, exhibits significant sequence identity (40%) with p67. SPAG1 shares cross-reacting, neutralizing determinants that inhibit invasion of leukocytes in an in vitro assay, and it has been demonstrated that recombinant forms of these proteins exhibit a degree of mutual cross-protection on heterologous challenge (Knight *et al.*, 1996; Boulter & Hall, 1999). By contrast with p67, several *T. annulata* SPAG1 isolates from cattle are polymorphic in deduced amino acid sequence (Katzer, Carrington & Knight, 1994). A homologous gene has also been cloned from the pathogenic small ruminant

parasite *T. lestoquardi* (Skilton *et al.*, 2000). This molecule has been expressed in a recombinant form but not yet evaluated against sporozoite challenge in sheep or goats. The molecule, SLAG1, is closer to SPAG1 (58% identity) than to p67 (40% identity) in amino acid sequence.

EXPRESSED SEQUENCE TAG DATA FROM *R. APPENDICULATUS* AND *A. VARIEGATUM* SALIVARY GLANDS: INSIGHTS INTO PARASITE–VECTOR INTERACTION

In an initial attempt to analyse genes encoding potentially secreted proteins of ixodid tick species that transmit *Theileria*, 2109 non-redundant expressed sequence tags (ESTs) from a total dataset of 3992 sequences were determined from cDNA derived from *A. variegatum* dissected salivary glands (Nene *et al.*, 2002). In addition 9844 ESTs from *T. parva*-infected *R. appendiculatus* salivary glands, dissected 4 days after induction of sporogony, and 9162 ESTs from uninfected salivary glands from ticks fed for the same length of time were determined (Nene *et al.*, 2004). These were clustered and analysed using the TIGR automated annotation pipeline and used to build a gene index (accessible at URL: http://www.tigr). The index contains links to annotation data from other organisms and potential roles in metabolic pathways. Excluding sporozoite sequences identified by BLAST searching against the *T. parva* genome, the *R. appendiculatus* dataset contained a total of 18 422 ESTs and equated to 4971 clusters and 7359 singletons (Nene *et al.*, 2004). Relative to the *A. variegatum* gene index, this indicates diminishing returns in terms of information and suggests that even for a single tick tissue such as the salivary gland, 20 000 ESTs may not represent adequate coverage to gain a comprehensive snapshot of gene expression. This problem could be solved by construction of normalized libraries. A notable feature of the data was that six out of the eight most highly transcribed *A. variegatum* salivary gland genes potentially encoded glycine/proline-rich proteins that may be components of the tick cement cone. Distantly related homologues of a T-cell immunosuppressant protein originally described in *Dermacentor* (Bergman *et al.*, 2000) were also identified, suggesting that certain proteins may be conserved between the sialomes of different tick species.

One aim of the *R. appendiculatus* study was to identify transcripts whose expression is modulated by infection with parasite. Few highly expressed transcripts fell into this category, although cDNAs encoding three putative glycine rich

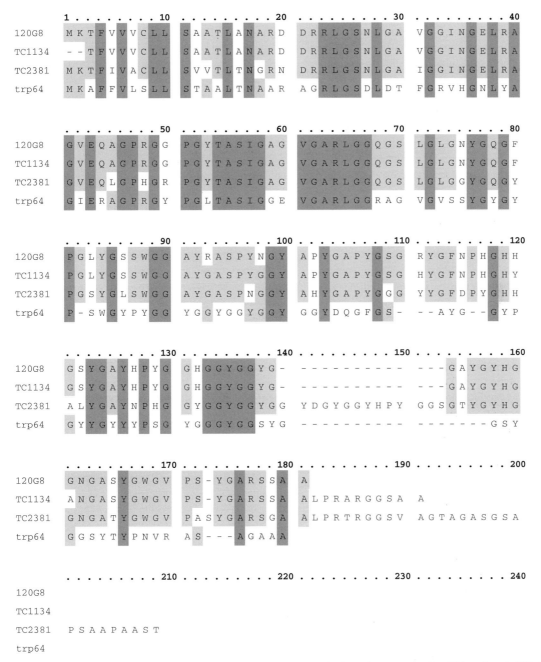

Fig. 14.5. Alignment of the predicted amino acid sequences of the *R. appendiculatus* 64trp cement antigen (Trimnell *et al.*, 2002; GenBank Q8T611), labelled Trp64 (bottom line: line 4) with a homologue isolated using signal sequence trap (Lambson *et al.*, 2005; GenBank CF972301-2316), labelled 120G8 (top line: line 1) with additional homologues within the *R. appendiculatus* gene index generated from the same tick stock as 120G8 (Nene *et al.*, 2004; GenBank CD778192-797237) labelled TC1134 and TC2381 (lines 2 and 3, respectively). The alignment was generated using ClustalX and mismatched bases are indicated by shading.

transcripts were significantly upregulated in the infected salivary glands according to χ^2 analysis ($p < 0.05$). In addition 653 and 558 tentative consensus sequences, comprising one to seven clones, were present in only the infected or uninfected salivary glands, respectively. This may indicate parasite modulation of expression of secreted proteins, but a higher level of transcriptome coverage will be required to confirm this.

Six hundred and twenty ESTs were identified as being derived from the sporozoite, although only approximately 200 were absent from a schizont transcriptome data developed using massively parallel signature sequencing (MPSS) (Bishop *et al.*, 2005). Automated annotation indicated that the majority represented hypothetical or conserved hypothetical proteins. An interesting finding was the identification of homologues of two major *Boophilus microplus* haembinding proteins. These exhibited domains also present in vitellogenins which are major components of yolk. The apovitellin B sequence homologue in *R. appendiculatus* contained a predicted signal peptide, consistent with secretion. 'Signal trap' functional selection was used to isolate additional *R. appendiculatus* salivary gland proteins (Lambson *et al.*, 2005). This indicates that the classical secretory pathway operates in ticks as well as other arthropods, with predicted signal peptides ranging from 15 to 47 amino acids in length. Ten of the functional signal peptides could not be predicted using hidden Markov or neural network algorithms, indicating a requirement for improved algorithms. Both the signal peptide and gene index approaches resulted in identification of homologues (50–60% identity) to the *R. appendiculatus* 64trp cement antigen which is protective against *R. appendiculatus* tick infestation and also blocks pathogen transmission of *Ixodes*-transmitted encephalitis virus in rodent models (Trimnell *et al.*, 2002; Labuda *et al.*, 2006). An alignment of one 64trp homologue identified through the signal trap approach and two additional variants from the *R. appendiculatus* gene index is shown in Fig. 14.5. It is currently not known whether these related 64trp-like proteins also induce protection in rodents.

CONCLUSIONS AND OPPORTUNITIES FOR FUTURE RESEARCH

Although significant information has accumulated regarding the tick-infective stages of *Theileria* over the last 25 years, much of this currently relates to the sporozoite stage. Knowledge of the cell biology of the tick gut and haemolymph stages, including the molecular processes involved in infec-

tion of ticks by the intraerythrocytic stages of *Theileria*, factors triggering differentiation of gametes, and the invasion of the gut epithelia and subsequent development of the motile kinete, is still rudimentary. The availability of two *Theileria* genome sequences, combined with high-resolution analysis of parasite transcriptomes, will enable new approaches to understanding the biology of *Theileria*. Identification of *T. parva* genes that are specifically expressed in sporozoites and piroplasms, combined with mammalian host and vector transcriptome and proteome data should result in novel insights into pathogen–vector–host interaction. This will include mechanisms of action of tick molecules involved in modulation of mammalian immune and inflammatory systems and insights into how tick secreted salivary gland molecules promote *Theileria* transmission. Another research area is genotyping of *Theileria* spp. and their ixodid vectors in relation to differences in vector competence. Molecular markers for genotyping the parasite have recently been developed and applied to *T. parva* population genetics (Oura *et al.*, 2003; Odongo *et al.*, 2006) and development of more comprehensive panels of molecular markers for the vectors will be facilitated by the availability of tick genome sequences. More sophisticated models of transmission dynamics are likely to be developed in the future as the quality of the input data improves. This should in turn increase the accuracy of predicting the impact of control measures.

The emerging discipline of immunoinformatics has the potential to facilitate identification of protective antigens derived from ticks. Antigen identification to date has primarily been empirical. The *R. appendiculatus* 64trp cement antigen induces both protection against tick infestation and pathogen transmission blocking in model hosts (Trimnell *et al.*, 2002; Labuda *et al.*, 2006; reviewed by Nuttall *et al.*, 2006). However, the RIM36 cement antigen that is immunodominant in cattle (Bishop *et al.*, 2002*b*) was ineffective in initial trials as a stand-alone vaccine for controlling tick infestation in either rabbits or cattle (R. Bishop and A. J. Musoke, unpublished data). The possibility of combining the p67 and SPAG sporozoite recombinant proteins with one or more tick antigens within a multivalent vaccine in order to enhance protection against field tick challenge is an attractive future option to explore for the control of theileriosis.

REFERENCES

Allsopp, B., Carrington, M., Bayliss, M. T., *et al.* (1989).
    Improved characterization of *Theileria parva* isolates using

the polymerase chain reaction and oligonucleotide probes. *Molecular and Biochemical Parasitology* **35**, 137–148.

Allsopp, B., Baylis, H. A., Allsopp, M. T. E. P., *et al.* (1993). Discrimination between six species of *Theileria* using oligonucleotide probes which detect small ribosomal RNA sequences. *Parasitology* **107**, 157–165.

Allsopp, M. T. E. P., Cavalier-Smith, T., De Waal, D. T. & Allsopp, B. A. (1994). Phylogeny and evolution of the piroplasms. *Parasitology* **108**, 147–152.

Andrews, N. W. & Webster, P. (1991). Phagolysomal escape by intracellular pathogens. *Parasitology Today* **7**, 335–340.

Awadalla, P., Walliker, D., Babiker, H. & Mackinnon, M. (2001). The question of *Plasmodium falciparum* population structure. *Trends in Parasitology* **17**, 351–353.

Babiker, H., Ranford-Cartwright, L., Currie, D., *et al.* (1994). Random mating in a natural population of the malaria parasite *Plasmodium falciparum*. *Parasitology* **109**, 413–421.

Baldwin, C. L., Black, S. J., Brown, W. C., *et al.* (1988). Bovine T cells, B cells, and null cells are transformed by the protozoan parasite *Theileria parva*. *Infection and Immunity* **56**, 462–467.

Baylis, H. A., Sohal, S. K., Carrington, M., Bishop, R. P. & Allsopp, B. A. (1991). An unusual repetitive gene family in *Theileria parva* which is stage-specifically transcribed. *Molecular and Biochemical Parasitology* **49**, 133–142.

Bergman, D. K., Palmer, M. J., Caimano, M. J., Radolf, J. D. & Wikel, S. K. (2000). Isolation and molecular cloning of a secreted immunosuppressant protein from *Dermacentor andersoni* salivary glands. *Journal of Parasitology* **86**, 516–525.

Bishop, R. P., Sohanpal, B. K., Allsopp, B. A., *et al.* (1993). Detection of polymorphisms among *Theileria parva* stocks using repetitive, telomeric and ribosomal DNA probes and anti-schizont monoclonal antibodies. *Parasitology* **107**, 19–31.

Bishop, R. P., Sohanpal, B. K., Morzaria, S. P., *et al.* (1994). Discrimination between *Theileria parva* and *T. taurotragi* in the salivary glands of *Rhipicephalus appendiculatus* ticks using oligonucleotides homologous to ribosomal RNA sequences. *Parasitology Research* **80**, 259–261.

Bishop, R., Musoke, A., Morzaria, S., Sohanpal, B. & Gobright, E. (1997). Concerted evolution at a multicopy locus in the protozoan parasite *Theileria parva*: extreme divergence of potential protein coding sequences. *Molecular and Cellular Biology* **17**, 1666–1673.

Bishop, R., Geysen, D., Skilton, R., *et al.* (2002a). Genomic polymorphism, sexual recombination and molecular epidemiology of *Theileria parva*. In *World Class Parasites*, vol. 3, *Theileria*, eds. Dobbelaere, D. A. E. & Mckeever, D. J., pp. 23–40. Amsterdam: Kluwer Academic.

Bishop, R., Lambson, B., Wells, C., *et al.* (2002b). A cement protein of the tick *Rhipicephalus appendiculatus*, located in the secretory e cell granules of the type III salivary gland acini, induces strong antibody responses in cattle. *International Journal for Parasitology* **32**, 833–842.

Bishop, R., Nene, V., Staeyert, J., *et al.* (2003). Immunity to East Coast fever induced by a polypeptide fragment of the major surface coat of *Theileria parva* sporozoites. *Vaccine* **21**, 1205–1212.

Bishop, R., Shah, T., Pelle, R., *et al.* (2005). Analysis of the transcriptome of the protozoan *Theileria parva* using MPSS reveals that the majority of genes are transcriptionally active in the schizont stage. *Nucleic Acids Research* **33**, 5503–5511.

Blouin, M. S., Yowell, C. A., Courteney, C. H. & Dame, J. B. (1995). Host movement and the genetic structure of populations of parasitic nematodes. *Genetics* **141**, 1007–1014.

Borst, P., Bitter, W., McCulloch, R., Van Leuwen, F. & Rudenko, G. (1995). Antigenic variation in malaria. *Cell* **82**, 1–4.

Boulter, N. & Hall, R. (1999). Immunity and vaccine development against bovine theilerioses. *Advances in Parasitology* **44**, 41–97.

Buscher, G. & Otim, B. (1986). Quantitative studies on *Theileria parva* in the salivary glands of *Rhipicephalus appendiculatus* adults: quantitation and prediction of infection. *International Journal for Parasitology* **16**, 93–100.

Carrington, M., Allsopp, B., Baylis, H., *et al.* (1995). Lymphoproliferation caused by *Theileria parva* and *Theileria annulata*. In *Molecular Approaches to Parasitology*, eds. Boothroyd, J. C. & Komuniecki, R., pp. 43–56. New York: Wiley–Liss.

Chae, J., Levy, M., Hunt, J. Jr, *et al.* (1999). *Theileria* sp. infections associated with bovine fatalities in the United States confirmed by small-subunit rRNA gene analyses of blood and tick samples. *Journal of Clinical Microbiology* **37**, 3037–3040.

Conrad, P. A., Stagg, D. A., Grootenhuis, J. G., *et al.* (1987). Isolation of *Theileria* parasites from African buffalo (*Syncercus caffer*) and characterization with anti-schizont monoclonal antibodies. *Parasitology* **94**, 413–423.

Day, K. P., Koella, J. C., Nee, S., Gupta, S. & Read, A. F. (1992). Population genetics and dynamics of *Plasmodium falciparum*: an ecological view. *Parasitology* **104**, S35–S52.

Dimopoulos, G. (2003). Insect immunity and its implication in mosquito–malaria interactions. *Cellular Microbiology* **5**, 3–14.

Dobbelaere, D. & Heussler, V. (1999). Transformation of leukocytes by *Theileria parva* and *T. annulata*. *Annual Review of Microbiology* **53**, 1–42.

Dobbelaere, D. & Kaenzi, P. (2004). The strategies of the *Theileria* parasite: a new twist in host–parasite interactions. *Current Opinion in Immunology* **16**, 524–530.

Dobbelaere, D. A. E., Spooner, P. R., Barry, W. C. & Irvin, A. D. (1984). Monoclonal antibodies neutralize the sporozoite stage of different *Theileria parva* stocks. *Parasite Immunology* **6**, 361–370.

Ebel, T., Middleton, J. F. S., Frisch, A. & Lipp, J. (1997). Characterization of a secretory type *Theileria parva* glutaredoxin homologue identified by novel screening procedures. *Journal of Biological Chemistry* **272**, 3042–3048.

Fawcett, D. W., Doxsey, S. & Buscher, G. (1982). Salivary glands of the tick vector of East Coast fever. III. Ultrastructure of sporogony in *Theileria parva*. *Tissue Cell* **14**, 183–206.

Gardner, M. J., Bishop, R., Shah, T., De Villiers, E., Carlton, J. M., Hall, N., Ren, Q., Paulsen, I. T., Pain, A., Berriman, M., Wilson, R. J. M., Sato, Shigeharu., Ralph, S. A., Mann, D. J., Xiong, Zikai., Shallom, S. J., Weidman, J., Jiang, L., Lynn, J., Weaver, B., Shoaibi, A., Domingo, A. R., Wasawo, D., Crabtree, J., Wortman, J. R., Hass, B., Angiuoli, S. V., Creasy, T. H., Lu, C., Suh, B., Silva, J., Utterback, T. R., Feldblyum, T. V., Pertea, M., Allen, J., Nierman, W. C., Taracha, E., Salzberg, S. L., White, O. R., Fitzhugh, H. A., Morzaria, S., Venter, J. C., Fraser, C. M. & Nene, V. (2005). Genome sequences of *Theileria parva*, a bovine pathogen causing a lymphoproliferative disease. *Science* **309**, 134–137.

Gauer, M., Mackenstedt, U., Mehlhorn, H., *et al.* (1995). DNA measurements and ploidy determination of developmental stages in the life cycles of *Theileria annulata* and *T. parva*. *Parasitology Research* **81**, 565–574.

Heussler, V. T. (2002). *Theileria* survival strategies and host cell transformation. In *World Class Parasites*, vol. 3, *Theileria*, eds. Dobbelaere, D. A. E. & McKeever, D. J., pp. 69–84. Amsterdam: Kluwer Academic.

Iams, K. P., Young, J. R., Nene, V., *et al.* (1990). Characterization of the gene encoding a 104-kilodalton microneme-rhoptry protein of *Theileria parva*. *Molecular and Biochemical Parasitology* **39**, 47–60.

Irvin, A. D. & Morrison, W. I. (1987). Immunopathology, immunology and immunoprophylaxis of *Theileria*

Infections. In *Immune Responses in Parasitic Infections*, ed. Soulsby, E. J. L., pp. 223–274. Boca Raton, FL: CRC Press.

Irvin, A. D., Ocama, J. G. & Spooner, P. R. (1982). Cycle of bovine lymphoblastoid cells parasitized by *Theileria parva*. *Research in Veterinary Science* **33**, 298–304.

Janoo, R., Musoke, A., Wells, C. & Bishop, R. P. (1999). A Rab 1 homologue with a novel isoprenylation signal provides insight into the secretory pathway of *Theileria parva*. *Molecular and Biochemical Parasitology* **102**, 131–143.

Kaba, S. A., Nene, V., Musoke, A. J., Vlak, J. M. & van-Oers, M. M. (2002). Fusion of green fluorescent protein improves expression levels of *Theileria parva* sporozoite surface antigen p67 in insect cells. *Parasitology* **125**, 497–505.

Katzer, F., Carrington, M. & Knight, P. (1994). Polymorphism of SPAG-1, a candidate antigen for inclusion in a sub-unit vaccine against *Theileria annulata*. *Molecular and Biochemical Parasitology* **67**, 1–10.

Knight, P. A., Musoke, A. J., Gachanja, J. N., *et al.* (1996). Conservation of neutralizing determinants between the sporozoite surface antigens of *Theileria annulata* and *Theileria parva*. *Experimental Parasitology* **82**, 229–241.

Labuda, M., Trimnell, A. R., Lickova, M., *et al.* (2006). An antivector vaccine protects against a lethal vector-borne pathogen. *PLoS Pathogens* **2**, 001–009.

Lambson, B., Nene, V., Obura, M., *et al.* (2005). Identification of candidate sialome components expressed in tick salivary glands using secretion signal complementation in mammalian cells. *Insect Molecular Biology* **14**, 403–414.

McCutchan, T., Li, J., McConkey, G., Roegers, M. & Waters, A. (1995). The cytoplasmic RNAs of *Plasmodium* spp. *Parasitology Today* **11**, 134–138.

Medley, G. F., Perry, B. D. & Young, A. S. (1993). Preliminary analysis of the transmission dynamics of *Theileria parva* in eastern Africa. *Parasitology* **106**, 251–264.

Mehlhorn, H. & Schein, E. (1984). The piroplasms: life cycle and sexual stages. *Advances in Parasitology* **23**, 37–103.

Morzaria, S. P., Young, J. R., Spooner, P. R., Dolan, T. T. & Bishop, R. P. (1993). *Theileria parva*: a restriction map and genetic recombination. In *Genome Analysis of Protozoan Parasites*, ed. Mozaria, S. P., pp. 67–73. Nairobi: International Laboratory for Research on Animal Diseases.

Mukhebi, A. W., Perry, B. D. & Kruska, R. (1992). Estimated economics of theileriosis control in Africa. *Preventive Veterinary Medicine* **12**, 73–85.

Musoke, A. J., Nantulya, V. M., Rurangira, F. R. & Buscher, G. (1984). Evidence for a common protective antigenic determinant on sporozoites of several *Theileria parva* strains. *Immunology* **52**, 231–238.

Musoke, A. J., Morzaria, S., Nkonge, C., Jones, E. & Nene, V. (1992). A recombinant sporozoite surface antigen of *Theileria parva* induces protection in cattle. *Proceedings of the National Academy of Sciences of the USA* **89**, 514–518.

Musoke, A., Rowlands, J., Nene, V., *et al.* (2005). Subunit vaccines based on the p67 major surface protein of *Theileria parva* sporozoites reduce severity of infection derived from field tick challenge. *Vaccine* **23**, 3084–3095.

Nene, V., Iams, K. P., Gobright, E. & Musoke, A. J. (1992). Characterization of the gene encoding a candidate vaccine antigen of *Theileria parva* sporozoites. *Molecular and Biochemical Parasitology* **51**, 17–28.

Nene, V., Inumaru, S., McKeever, D. J., *et al.* (1995). Characterization of an insect cell-derived *Theileria parva* sporozoite vaccine antigen and immunogenicity in cattle. *Infection and Immunity* **63**, 503–508.

Nene, V., Musoke, A., Gobright, E. & Morzaria, S. (1996). Conservation of the sporozoite p67 vaccine antigen in cattle-derived *Theileria parva* stocks with different cross-immunity profiles. *Infection and Immunity* **64**, 2056–2061.

Nene, V., Gobright, E., Bishop, R., Morzaria, S. & Musoke, A. (1999). Linear peptide specificity of antibody responses to p67 and sequence diversity of neutralizing epitopes: implications for a *Theileria parva* vaccine. *Infection and Immunity* **67**, 1261–1266.

Nene, V., Lee, D., Quackenbush, J., *et al.* (2002). AvGI, an index of genes transcribed in the salivary glands of the ixodid tick *Amblyomma variegatum*. *International Journal for Parasitology* **32**, 1447–1456.

Nene, V., Lee, D., K'anga, S., *et al.* (2004). Genes transcribed in the salivary glands of female *Rhipicephalus appendiculatus* ticks infected with *Theileria parva*. *Insect Biochemistry and Molecular Biology* **34**, 1117–1118.

Ngumi, P. N., Lesan, A. C., Williamson, S. M., *et al.* (1994). Isolation and preliminary characterization of a previously unidentified *Theileria* parasite of cattle in Kenya. *Research in Veterinary Science* **57**, 1–9.

Norval, R. A. I., Perry, B. D. & Young, A. S. (1992). *The Epidemiology of Theileriosis in Africa*. London: Academic Press.

Nuttall, P. A., Trimnell, A. R., Kazimirova, M. & Labuda, M. (2006). Exposed and concealed antigens as vaccine targets for controlling ticks and tick-borne diseases. *Parasite Immunology* **28**, 155–163.

Ochanda, H., Young A. S., Wells, C., Medley, G. F. & Perry, B. D. (1996). Comparison of the transmission of *Theileria* parva between different instars of *Rhipicephalus appendiculatus*. *Parasitology* **113**, 243–253.

Ochanda, H., Young, A. S., Medley, G. F. & Perry, B. D. (1998). Vector competence of seven rhipicephalid tick stocks in transmitting two *Theileria parva* parasite stocks from Kenya and Zimbabwe. *Parasitology* **116**, 539–545.

Odongo, D., Oura, C. A. L., Spooner, P., *et al.* (2006). Linkage disequilibrium and genetic diversity at mini- and micro-satellite loci of *Theileria parva* isolated from cattle in three regions of Kenya. *International Journal for Parasitology* **36**, 937–946.

Oura, C. A. L., Odongo, D., Lubega, G., *et al.* (2003). A panel of microsatellite and minisatellite markers for the characterization of field isolates of *Theileria parva*. *International Journal for Parasitology* **33**, 1641–1653.

Oura, C. A. L., Bishop, R., Lubega, G. W. & Tait, A. (2004). Application of a reverse line blot to the study of haemoparasites in cattle in Uganda. *International Journal for Parasitology* **34**, 603–613.

Pain, A., Renauld, H., Berriman, M., Murphy, L., Yeats, C. A., Weir, W., Kerhornou, A., Aslett, M., Bishop, R., Bouchier, C., Cochet, M., Culson, R. M. R., Cronin, A., De Villiers, E., Fraser, A., Fosker, N., Gardner, M., Goble, A., Griffiths-Jones, S., Harris, D. E., Katzer, F., Larke, N., Lord, A., Maser, P., Mckellar, S., Mooney, P., Morton, F., Nene, V., O'Neil, S., Price, C., Quail, M. A., Rabbinowitsch, E., Rawlings, N. D., Rutter, S., Saunders, D., Seeger, K., Shah, T., Squares, R., Squares, S., Tivey, A., Walker, A. R., Woodward, J., Dobbelaere, D. A. E., Langsley, G., Rajandream, M. A., McKeever, D., Shiels, B., Tait, A., Barrell, B. & Hall, N. (2005). The genome of the host-cell transforming parasite *Theileria annulata* and a comparison with *T. parva*. *Science* **309**, 131–133.

Paul, R. E. L. & Day, K. P. (1998). Mating patterns of *Plasmodium falciparum*. *Parasitology Today* **14**, 197–202.

Paul, R., Packer, M., Walmsley, M., *et al.* (1995). Mating patterns in malaria parasite populations of Papua New Guinea. *Science* **269**, 1709–1711.

Radley, D. E., Brown, C. G. D., Burridge, M. P., *et al.* (1975). East Coast fever. I. Chemoprophylactic immunization of cattle against *Theileria parva* (Muguga) and five theilerial strains. *Veterinary Parasitology* **1**, 35–41.

Schnittger, L., Hong, Y., Jianxun, L., *et al.* (2000). Phylogenetic analysis by rRNA comparison of the highly pathogenic sheep-infecting parasites *Theileria lestoquardi* and a *Theileria* species identified in China. *Annals of New York Academy of Sciences* **916**, 271–275.

Shaw, M. K. (1997). The same but different: the biology of *Theileria* sporozoite entry into bovine cells. *International Journal for Parasitology* 27, 457–474.

Shaw, M. K. (2002). *Theileria* development and host cell invasion. In *World Class Parasites*, vol. 3, *Theileria*, eds. Dobbelaere, D. A. E. & McKeever, D. J., pp. 1–22. Amsterdam: Kluwer Academic.

Shaw, M. K. & Tilney, L. G. (1992). How individual cells develop from a syncytium: merogony in *Theileria parva* (Apicomplexa). *Journal of Cell Science* 101, 109–123.

Shaw, M. K. & Tilney, L. G. (1995). The entry of *Theileria parva* merozoites into bovine erythrocytes occurs by a process similar to sporozoite invasion of lymphocytes. *Parasitology* 111, 455–461.

Shaw, M. K. & Young, A. S. (1994). The biology of *Theileria* species in ixodid ticks in relation to parasite transmission. *Advances in Disease Vector Research* 10, 23–63.

Shaw, M. K., Tilney, L. G. & Musoke, A. J. (1991). The entry of *Theileria parva* sporozoites into bovine lymphocytes: evidence for MHC Class I involvement. *Journal of Cell Biology* 113, 87–101.

Shaw, M. K., Tilney, L. G. & McKeever, D. J. (1993). Tick salivary gland extract and interlukeukin-2 stimulation enhance susceptibility of lymphocytes to infection by *Theileria parva* sporozoites. *Infection and Immunity* 61, 1486–1495.

Shiels, B., Langsley, G., Weir, W., *et al.* (2006). Alteration of host cell phenotype by *Theileria annulata* and *Theileria parva*: mining for manipulators in the parasite genomes. *International Journal for Parasitology* 36, 9–12.

Sinden, R. E., Butcher, G. A., Billker, O. & Fleck, S. L. (1996). Regulation of infectivity of *Plasmodium* to the mosquito vector. *Advances in Parasitology* 38, 53–117.

Skilton, R. A., Bishop, R. P., Wells, C. W., *et al.* (1998). Cloning and characterisation of a 150 kDa microsphere antigen of *Theileria parva* that is immunologically cross-reactive with the polymorphic immunodominant molecule (PIM). *Parasitology* 117, 321–330.

Skilton, R. A., Musoke, A. J., Nene, V., *et al.* (2000). Molecular characterization of a *Theileria lestoquardi* gene encoding a candidate sporozoite vaccine antigen. *Molecular and Biochemical Parasitology* 107, 309–314.

Skilton, R. A., Bishop, R. P., Katende, J. M., Mwaura, S. & Morzaria, S. P. (2002). The persistence of *Theileria parva*

infection in cattle immunized using two stocks which differ in their ability to induce a carrier state: analysis using a novel blood spot PCR assay. *Parasitology* 124, 265–276.

Stagg, D. A., Young, A. S., Leitch, B. L., Grootenhuis, J. G. & Dolan, T. T. (1983) Infection of mammalian cells with *Theileria* species. *Parasitology* 86, 243–254.

Swan, D. G., Phillips, K., Tait, A. & Shiels, B. R. (1999). Evidence for localization of a *Theileria* parasite AT hook DNA-binding protein to the nucleus of immortalized bovine host cells. *Molecular and Biochemical Parasitology* 101, 117–129.

Tatchell, R. J. (1987). Tick control in the context of ECF immunization. *Parasitology Today* 3, 7–10.

Trimnell, A. R., Hails, R. S. & Nuttall, P. A. (2002). Dual action ectoparasite vaccine targeting 'exposed' and 'concealed' antigens. *Vaccine* 20, 3360–3568.

Walker, A. R., Latif, A. A., Morzaria, S. P. & Jongehan, F. (1983). Natural infection rate of *Hyalomma anatolicum* with *Theileria* in Sudan. *Research in Veterinary Science* 35, 87–90.

Watt, D. M., Walker, A. R., Lamza, K. A. & Ambrose, N. C. (2001). Tick–*Theileria* interactions in response to immune activation of the vector. *Experimental Parasitology* 97, 89–94.

Willadsen, P. (2004). Anti-tick vaccines. *Parasitology* 129, S367–S387.

Wilson, R. J. M. & Williamson, D. H. (1997). Extrachromosomal DNA in the Apicomplexa. *Microbiology and Molecular Biology Reviews* 61, 1–16.

Young, A. S. & Leitch, B. L. (1982). Epidemiology of East Coast fever: some effects of temperature on the development of *Theileria parva* in the tick vector, *Rhipicephalus appendiculatus*. *Parasitology* 83, 199–211.

Young, A. S., Leitch, B. L., Newson, R. L. & Cunningham, P. M. (1986). Maintenance of *Theileria parva parva* infections in an endemic area of Kenya. *Parasitology* 93, 9–16.

Young, A. S., Dolan, T. T., Mwakima, F. N., *et al.* (1995). Estimation of heritability of susceptibility to infection with *Theileria parva* in the tick *Rhipicephalus appendiculatus*. *Parasitology* 111, 31–38.

Young, A. S., Dolan, T. T., Morzaria, S. P., *et al.* (1996). Factors influencing infections in *Rhipicephalus appendiculatus* ticks fed on cattle infected with *Theileria parva*. *Parasitology* 113, 255–266.

15 • Characterization of the tick–pathogen–host interface of the tick-borne rickettsia *Anaplasma marginale*

K. M. KOCAN, J. DE LA FUENTE AND E. F. BLOUIN

INTRODUCTION

Anaplasma marginale is a tick-borne pathogen that causes the disease anaplasmosis in cattle (Bram, 1975; Ristic, 1968). This pathogen is classified within the Order Rickettsiales which was recently reorganized into two families, Anaplasmataceae and Rickettsiaceae, based on genetic analyses of 16S rRNA, *groELS* and surface protein genes (Dumler *et al.*, 2001) (Table 15.1). Organisms of the family Anaplasmataceae are obligate intracellular organisms that are found exclusively within membrane-bound vacuoles in the host cell cytoplasm. Phylogenetic analyses consistently supported the formation of four distinct genetic groups of the organisms: (1) *Anaplasma* (96.1% similarity), (2) *Ehrlichia* (97.7%), (3) *Wolbachia* (minimum of 95.6% similarity) and (4) *Neorickettsia* (minimum of 94.9% similarity) (Dumler *et al.*, 2001). The genus *Anaplasma* currently includes the three pathogens of ruminants, *A. marginale*, *A. centrale* and *A. ovis*, together with *A. bovis* (formerly *Ehrlichia bovis*), *A. phagocytophilum* (formerly *E. phagocytophilum*, *E. equi* and the HGE agent), and *A. platys* (formerly *E. platys*). *Aegyptianella*, also included in this genus, was retained as a *genus incertae sedis* due to lack of sequence information.

Anaplasma marginale is distributed worldwide in tropical and subtropical regions of the New World, Europe, Africa, Asia and Australia. Several geographical isolates of *A. marginale* have been identified in North and South America, which differ in morphology, protein sequence, antigenic characteristics and their ability to be transmitted by ticks (Smith *et al.*, 1986; Wickwire *et al.*, 1987; Allred *et al.*, 1990; Rodriguez Camarilla *et al.*, 2000; Palmer, Rurangirwa & McElwain, 2001; de la Fuente *et al.*, 2001*a*, 2001*b*, as reviewed by de la Fuente *et al.*, 2001*c*).

Anaplasma marginale develops persistent infections in mammalian and tick hosts, both of which serve as reservoirs for infection of susceptible hosts. The only known site of replication of *A. marginale* in cattle is bovine erythrocytes (Richey, 1981) (Fig. 15.1). Within these cells membrane-bound inclusion bodies contain from four to eight rickettsiae, and as many as 70% or more of the erythrocytes may become infected during acute infection and disease. Removal of infected cells by the bovine reticuloendothelial system results in mild to severe anaemia and icterus (Richey, 1981). The acute phase of the disease is characterized by weight loss, fever, abortion, lowered milk production and often death.

Mechanical transmission of *A. marginale* occurs when infected blood is transferred to susceptible cattle via blood-contaminated fomites or mouthparts of biting flies (Ewing, 1981). Biological transmission of *A. marginale* is effected by ticks and approximately 20 species of ticks have been incriminated as vectors worldwide (Dikmans, 1950; as reviewed by Ewing, 1981). An updated listing of tick species tested as vectors of *A. marginale* (Table 15.2) lists conflicting reports on transmission of *A. marginale* by some tick species. These inconsistencies may have resulted because some *A. marginale* isolates (i.e. those from Illinois, Florida and California) have subsequently proven not to be transmissible by *Dermacentor* spp. ticks (Smith *et al.*, 1986; Wickwire *et al.*, 1987; de la Fuente *et al.*, 2001*a*, 2002*a*). In general, tick vectors of *A. marginale* include *Rhipicephalus* (*Boophilus*) spp. (hereafter referred to as *Boophilus* spp.), selected *Demacentor* spp., *Ixodes ricinus* and *Rhipicephalus* spp., while *Amblyomma* spp. do not appear to transmit *A. marginale*. Although transovarial transmission has been reported for *D. andersoni* (Howell, Stiles & Moe, 1941), others have not demonstrated this mode of transmission of *A. marginale*. However, transovarial transmission of *A. marginale* by the one-host *Boophilus* ticks has not been thoroughly investigated and may warrant further studies. Recent results on the prevalence of *A. marginale* in wildlife and cattle in Europe suggest that previously unrecognized tick species may be vectors of *A. marginale* in these regions (de la Fuente *et al.*, 2004).

Ticks: Biology, Disease and Control, ed. Alan S. Bowman and Patricia A. Nuttall. Published by Cambridge University Press.

Table 15.1 *Current classification of the Order Rickettsiales*

Order Rickettsiales	
Family Rickettsiaceae:	Obligate intracellular bacteria that grow freely in the cytoplasm of their eukaryotic host cells
Genus *Rickettsia*	
Genus *Orientia*	
Family Anaplasmataceae:	Obligate intracellular bacteria that replicate within membrane-derived vacuoles in the cytoplasm of eukaryotic host cells.

Genus *Anaplasma*
   Anaplasma marginale (type species)
   Anaplasma centrale
   Anaplasma ovis
   Anaplasma bovis (formerly *Ehrlichia bovis*)
   Anaplasma phagocytophilum (formerly *Ehrlichia phagocytophilum*, *E. equi*, HGE agent)
   Anaplasma platys (formerly *Ehrlichia platys*)
   Aegyptianella (genus incertae sedis due to lack of sequence information)
Genus *Ehrlichia*
   Ehrlichia chaffeensis
   Ehrlichia ruminantium (formerly *Cowdria ruminantium*)
   Ehrlichia ewingii
   Ehrlichia ovis
   Ehrlichia canis
   Ehrlichia muris
Genus *Neorickettsia*
   Neorickettsia helminthoeca
   Neorickettsia risticii (formerly *Ehrlichia risticii*)
   Neorickettsia sennetsu (formerly *Ehrlichia sennetsu*)
Genus *Wolbachia*
   Wolbachia pipientis

Recently, Scoles *et al.* (2005*a*) provided evidence that the Florida strain of *A. marginale*, which is not transmissible by ticks, was more efficiently retained in fly mouthparts than the St Maries strain which is tick transmitted. This form of mechanical transmission is considered to be the major route of dissemination for *A. marginale* in areas of Central and South America, and Africa, where tick vectors do not occur (Ewing, 1981; Foil, 1989), and where *Boophilus microplus*, the tropical cattle tick, does not appear to be a biological vector of *A. marginale* (Figueroa *et al.*, 1998; Coronado, 2001). In areas of the United States where geographic isolates of *A. marginale* are not infective for ticks or where ticks have been eradicated by fire ants, mechanical transmission appears to be the major mode of *A. marginale* transmission

(Smith *et al.*, 1986; Wickwire *et al.*, 1987; de la Fuente *et al.*, 2001*a*).

Intrastadial transmission of *A. marginale* is effected by male ticks. Recent studies have demonstrated that male *Dermacentor* ticks may play a role in the biological transmission of *A. marginale* because they become persistently infected with *A. marginale* and can transmit *A. marginale* repeatedly when they transfer among cattle (Kocan *et al.*, 1992*a*, 1992*b*). Male ticks, therefore, also serve as a reservoir of *A. marginale*, along with persistently infected cattle (Kocan *et al.*, 1992*a*, 1992*b*; Ge *et al.*, 1996; Kocan, Blouin & Barbet, 2000).

The developmental cycle of *A. marginale* was described in persistently infected male ticks infected as adults. This

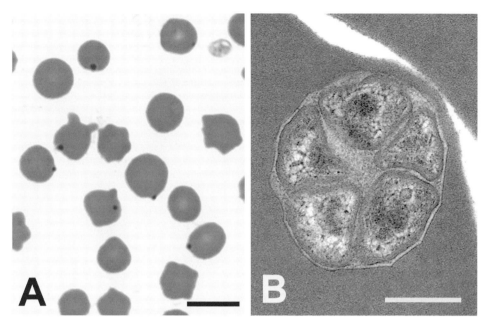

Fig. 15.1. Bovine erythrocytes infected with *Anaplasma marginale*. (A) Inclusion bodies are located at the periphery of the erythrocyte in a stained blood film. Scale bar = 10 μm. (B) Electron micrograph of an *A. marginale* inclusion that contains five organisms. Scale bar = 0.5 μm.

cycle is complex and coordinated with the tick feeding cycle (Kocan, 1986; Kocan *et al.*, 1992*a*, 1992*b*, 2003*a*). Infected erythrocytes taken into ticks with the blood meal provide the source of *A. marginale* infection for tick gut cells (Fig. 15.2A and B). After development of *A. marginale* in tick gut cells, many other tick tissues become infected, including the salivary glands (Fig. 15.2C and D) from where the rickettsiae are transmitted to vertebrates during feeding (Kocan, 1986; Kocan *et al.*, 1992*a*, 1992*b*; Ge *et al.*, 1996). At each site of infection in ticks, *A. marginale* develops within membrane-bound vacuoles or colonies. The first form of *A. marginale* seen within the colony is the reticulated (vegetative) form that divides by binary fission (Fig. 15.3), forming large colonies that may contain hundreds of organisms. The reticulated form then changes into the dense form (Fig. 15.3), which is the infective stage and can survive outside of cells. Cattle become infected with *A. marginale* when the dense form is transmitted during tick feeding via the salivary glands.

Six major surface proteins (MSPs) were identified on *A. marginale* derived from bovine erythrocytes and were found to be conserved on tick- and cell culture-derived organisms (as reviewed by Kocan, Blouin & Barbet, 2000). Three of these MSPs (MSP1a, MSP4 and MSP5) are from single

genes and do not vary antigenically during the multiplication of the bacterium (Barbet *et al.*, 1987; Allred *et al.*, 1990; Visser *et al.*, 1992; Bowie *et al.*, 2002), while the other three (MSP1b, MSP2 and MSP3) are from multigene families and may vary antigenically, most notably in persistently infected cattle (French *et al.*, 1998; French, Brown & Palmer, 1999; Barbet *et al.*, 2000, 2001; Brayton *et al.*, 2001, 2002; Meeus & Barbet, 2001; Bowie *et al.*, 2002). However, recent results demonstrated selection of *msp2* sequence variants in persistently infected ticks (Rurangirwa *et al.*, 1999; Rurangirwa, Stiller & Palmer, 2000; de la Fuente & Kocan, 2001).

MSP1 is a heterodimer composed of two structurally unrelated polypeptides: MSP1a and MSP1b. MSP1a is encoded by *msp1α* (Allred *et al.*, 1990), and is involved in adhesion to bovine erythrocytes and tick cells and transmission of *A. marginale* by *Dermacentor* spp. (McGarey & Allred, 1994; McGarey *et al.*, 1994; de la Fuente *et al.*, 2001*a*, 2001*b*). MSP1b is encoded by at least two genes, *msp1β1* and *msp1β2* (Barbet *et al.*, 1987; Camacho-Nuez *et al.*, 2000; Viseshakul *et al.*, 2000; Bowie *et al.*, 2002) and may be an adhesin for bovine erythrocytes but is not an adhesin for tick cells (McGarey & Allred, 1994; McGarey *et al.*, 1994; de la Fuente *et al.*, 2001*b*). In addition, MSP1a has a neutralization-sensitive epitope (Palmer *et al.*, 1987). The

Table 15.2 *Studies in which ixodid and argasid ticks were tested as vectors of* Anaplasma marginale

Tick species	Transmission[a]	Reference
Ixodid ticks		
Amblyomma americanum	−	Rees, 1934; Sandborn & Moe, 1934; Piercy & Schmidt, 1941
Amblyomma cajennense	−	Rees, 1934; Sanborn & Moe, 1934
Amblyomma maculatum	−	Rees, 1934; Piercy, 1938; Piercy & Schmidt, 1941
Rhipicephalus (*Boophilus*) *annulatus*	+/−	Rees, 1934
	+	Samish, Pipano & Hadonai, 1993
'*Boophilus calcaratus*'	+/−	Sergent *et al.*, 1945
Rhipicephalus (*Boophilus*) *decoloratus*	+	Theiler, 1911, 1912
Rhipicephalus (*Boophilus*) *microplus*	+	Rosenbusch & Gonzalez, 1927; Brumpt, 1931; Bock & de Vos, 1999
Dermacentor albipictus	+	Boynton *et al.*, 1936; Stiller, Leatch & Kuttler, 1981; Stiller & Johnson, 1983; Ewing *et al.*, 1997
	−	Sanborn and Moe, 1934
Dermacentor andersoni	+	Rees, 1933, 1934; Boynton *et al.*, 1936; Sanborn, Stiles & Moe, 1938; Anthony & Roby, 1966; Peterson *et al.*, 1977; Kocan *et al.*, 1981, 1992a, 1992b; Wickwire *et al.*, 1987; Eriks, Stiller & Palmer, 1993; Scoles *et al.*, 2005a, 2005b
	+ (− transovarial)	Anthony & Roby, 1962
	+ transovarial	Howell, Stiles & Moe, 1941
	+/−	Rozeboom, Stiles & Mose, 1940
Dermacentor hunteri	+	Stiller *et al.*, 1999
Dermacentor nitens	−	Sanborn & Moe, 1934; Rees & Avery, 1939
Dermacentor occidentalis	+	Boynton *et al.*, 1936; Anthony & Roby, 1966; Christensen & Howard, 1966; Howarth & Roby, 1972; Stiller & Johnson, 1983
	+ (− transovarial)	Howarth & Hokama, 1973
Dermacentor parumapertus	−	Sanborn & Moe, 1934
Dermacentor variabilis	+	Rees, 1932; Sanders, 1933; Anthony & Roby, 1966; Kocan *et al.*, 1981, 1992a, 1992b
	+ (−transovarial)	Rees, 1934; Anthony & Roby, 1962; Stich *et al.*, 1989
	−	Sanborn & Moe, 1934; Schmidt & Piercy, 1937; Piercy, 1938
	−transovarial	Rees & Avery, 1939
Dermacentor venustus	−	Sanborn & Moe, 1934
Haemaphysalis leporispalustris	−	Sanborn & Moe, 1934
Hyalomma lusitanicum	−	Sergent *et al.*, 1945
Hyalomma mauritanicum	−	Sergent *et al.*, 1945
Hyalomma rufipes	+	Potgieter, 1979
Ixodes pacificus	−	Howarth & Hokama, 1973
Ixodes ricinus	+	Zeller & Helm, 1923; Helm, 1924
	−	Sanborn & Moe, 1934; Piercy, 1938
Ixodes scapularis	+/−	Rees, 1934
	−	Sanborn & Moe, 1934
Ixodes sculptus	−	Rees, 1934
	−	Sandborn & Moe, 1934
Rhipicephalus bursa	+	Brumpt, 1931
	+/−	Sergent *et al.*, 1945
Rhipicephalus evertsi	+	Potgieter, 1979
Rhipicephalus sanguineus	+	Rees, 1930; Parker, 1982
	+/−	Rees, 1934
	− transovarial	Rees & Avery, 1939
	−	Sandborn & Moe, 1934
Rhipicephalus simus	+	Potgieter, 1979; Potgieter *et al.*, 1983
	+ transovarial	Theiler, 1912
Argasid ticks		
Argas persicus	+/−	Howell, Stiles & Moe, 1941
Ornithodoros coriaceus	−	Howarth & Hokama, 1973
'*Ornithodoros megnini*'	−	Sanborn & Moe, 1934; Howell, Stiles & Moe, 1941

[a] +/− indicates that in some trials the authors demonstrated transmission and in other trials they did not.

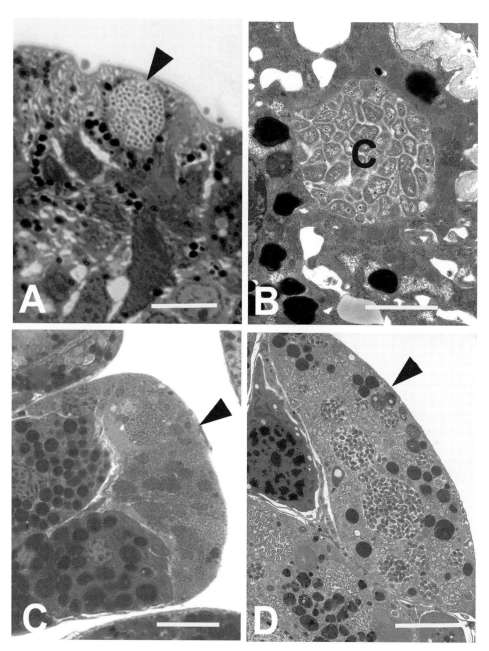

Fig. 15.2. Micrographs of colonies of *Anaplasma marginale* in tick gut and salivary gland cells. (A) Light micrograph of a large colony (arrow) in a tick gut cell. Scale bar = 5 μm. (B) Electron micrograph of a colony (C) in a tick gut cell. Scale bar = 5 μm. (C) Light micrograph of many colonies (arrow) in a salivary gland cell. Scale bar = 10 μm. (D) Electron micrograph of a tick salivary gland cell (arrow) that contains several *A. marginale* colonies. Scale bar = 5 μm.

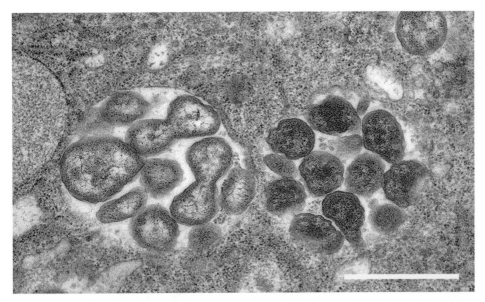

Fig. 15.3. Electron micrographs of the two developmental stages of *Anaplasma marginale* within colonies in tick cells. Reticulated forms within a colony on the left divide by binary fission and dense forms are within a second colony in a cultured tick cell. Scale bar = 1 µm.

molecular weight of MSP1a varies in size among strains of *A. marginale* because of different numbers of tandemly repeated 28–31 amino acid peptides (Allred *et al.*, 1990; de la Fuente *et al.*, 2001*b*, 2002*a*, 2002*e*, 2005, 2007*a*, 2007*b*). *msp1*α is a stable genetic marker for identification of *A. marginale* strains in individual animals during acute and chronic phases of infection and before, during and after tick transmission (Palmer, Rurangirwa & McElwain, 2001; Bowie *et al.*, 2002).

TICK CELL CULTURE: A MODEL FOR THE STUDY OF TICK–*A. MARGINALE* INTERACTIONS

Development of the *A. marginale*/ tick cell culture system

A major impediment in anaplasmosis research was the lack of a laboratory model or a cell culture system. Although cell culture systems were investigated (as reviewed by Kocan *et al.*, 2004), none of these cell culture systems sustained the replication and continuous propagation of *A. marginale*. Our laboratory, in collaboration with Drs U. G. Munderloh and T. J. Kurtti at the University of Minnesota, developed methods for culturing *A. marginale* in an *Ixodes scapularis* tick cell culture which provided research opportunities for studying

the interaction of *A. marginale* and tick cells (Munderloh *et al.*, 1996*a*; Blouin & Kocan, 1998; Blouin *et al.*, 1999).

Bovine erythrocytes infected with *A. marginale* were used to inoculate monolayers of the tick cell line, IDE8, that originally derived from embryos of *I. scapularis*. We documented infection and transformation of *A. marginale* in tick cells from the erythrocyte inclusion body stage to development of large colonies in the cultured tick cells (Munderloh *et al.*, 1996*a*; Blouin & Kocan, 1998; Blouin *et al.*, 1999). The developmental cycle of *A. marginale* in cultured tick cells was similar to that described previously in naturally infected ticks (Blouin & Kocan, 1998) (Fig. 15.4). Cultured *A. marginale* retained its antigenic composition and infectivity for cattle after successive passages (Blouin & Kocan, 1998; Barbet *et al.*, 1999). Infection and development of *A. marginale* in cell culture was found to be predictable, and cell monolayers became completely infected by 10–12 days post exposure (PE), after which cultures developed cytopathic effect as cells began to detach (referred to as 'terminal cultures'). The cell culture-derived organisms proved to be infective for cattle and *Dermacentor* spp. ticks, which fed on those cattle. The six major surface proteins described on erythrocytic *A. marginale* were shown to be conserved on cell culture-derived organisms and the cell culture-derived *A. marginale* isolates retained their antigenic identity as determined by

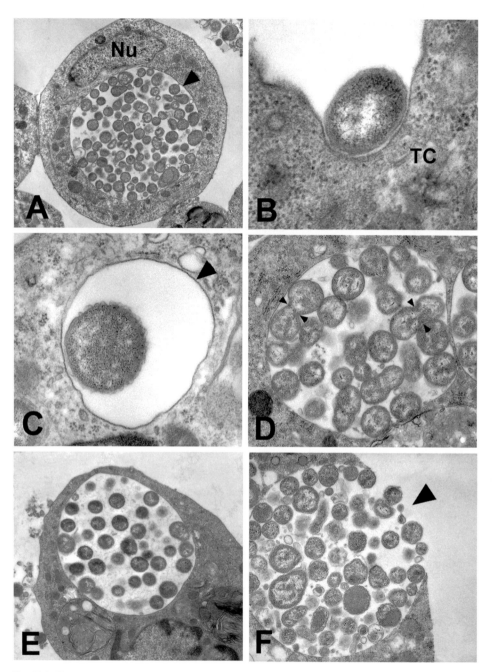

Fig. 15.4. Electron micrographs of *A. marginale* in cultured IDE8 tick cells. (A) Large colony of *A. marginale* adjacent to the nucleus (Nu) within a cultured tick cell. Scale bar = 5 μm. (B) An *A. marginale* adhered to the cell membrane of a cultured tick cell (TC). Scale bar = 0.5 μm. (C) Reticulated form of *A. marginale* within a membrane-bound colony. Scale bar = 0.5 μm. (D) Colony containing reticulated forms of *A. marginale* that are dividing by binary fission (small arrows). Scale bar = 5 μm. (E) Colony that contains dense forms of *A. marginale*. Scale bar = 5 μm. (F) *A. marginale* being released (arrow) from a cultured tick cell. Scale bar = 1 μm.

the molecular weight and sequence of MSP1a (Barbet *et al.*, 1999). Several geographic isolates, including ones originally derived from Virginia, Oklahoma and Oregon, have been propagated in cell culture (Blouin & Kocan, 1998; Blouin *et al.*, 2002*a*). *Anaplasma marginale* derived from cell culture proved to be a potent antigen for use in vaccine preparations (Kocan *et al.*, 2001; de la Fuente *et al.*, 2002*c*) and serological tests (Saliki *et al.*, 1998). Subsequently, this same IDE8 cell line was used in other laboratories to successfully propagate *A. phagocytophilum* (Munderloh *et al.*, 1996*b*, 1999), *E. canis* (Kocan, Munderloh & Ewing, 1998) and *E. (Cowdria) ruminantium* (Bell-Sakyi *et al.*, 2000).

Infection and development of *Anaplasma marginale* in cultured tick cells

The developmental cycle of *A. marginale* in tick cell culture has been described, including the entry and exit of the rickettsiae from the IDE8 cells (Blouin & Kocan, 1998) (Fig. 15.4). Development of *A. marginale* in the cultured tick cells was documented using light and electron microscopy. Infection of IDE8 cells occurred within 15 min PE. Host cell invasion was initiated by the adhesion of the dense form of *A. marginale* (diameter 0.5–0.8 μm) to the host cell membrane. Distinct projections of the host cell plasmalemma adhered along the outer membrane of *A. marginale*. The adhesion between the rickettsiae and tick cell membrane increased along a continuous section forming a depression in the host cell membrane. *Anaplasma marginale* was subsequently enclosed by the host cell membrane and internalized within a vacuole. *A. marginale* transformed into the reticulated (vegetative) form that divides by binary fission. Repeated division resulted in the formation of colonies by 48 h PE that contained hundreds of rickettsiae. The reticulated forms of *A. marginale* subsequently transformed by day 3 PE into the infective or dense forms that were rounded and contained a denser and more uniform distribution of ribosomes and in which DNA fibrils were not readily apparent. On day 4 PE, colony membranes were observed to be fused with the host cell plasmalemma, followed by rupture of the membrane complex. A flap opened in the fused cell membranes that allowed for release of the dense forms from the parasitophorous vacuole without loss of host cell cytoplasm. The released rickettsiae then initiated a new series of infections resulting in host cells containing five or more colonies per cell (Blouin & Kocan, 1998). Tick cell death occurred after most of the cells became infected, resulting in detachment of tick cell monolayers, and cytopathic effect which

was then apparent with light microscopy. The mechanism of *A. marginale* entry into tick cells by endocytosis and exit by a process involving the fusion of the colony and host cell membranes appears to be controlled by both the cell and the parasite (Blouin & Kocan, 1998). Adherence of rickettsiae to the tick cell membrane prior to infection suggests the presence of adhesion molecules on the surface of *A. marginale* that are recognized by tick cell receptors.

Infectivity of *Anaplasma marginale* isolates for tick cells

Many isolates of *A. marginale* from the United States are infective and transmissible by ticks (i.e. Virginia, Oklahoma, Idaho and Oregon) while other isolates (i.e. Florida, Okeechobee, Illinois and California) are not tick transmissible (Smith *et al.*, 1986; Wickwire *et al.*, 1987; de la Fuente *et al.*, 2001*a*, 2002*e*; Kocan *et al.*, 2003*a*). Infectivity of *A. marginale* isolates appears to be directly related to the adhesive properties of the isolate MSP1a expressed in recombinant *E. coli* (de la Fuente *et al.*, 2001*b*, 2002*b*). We demonstrated that three *A. marginale* isolates (Virginia, Oklahoma and Oregon), also shown to be infective for ticks, were infective for the cultured tick cells, while the Florida, Okeechobee and California isolates, not infective for ticks, were also not infective for cultured tick cells (Blouin *et al.*, 2002*a*). Therefore, infectivity for tick cell culture may serve as a predictor for vector competency.

Tick cell culture as a biological assay system

Cultured tick cells have been adapted for use in experimental studies and assays (as reviewed by Blouin *et al.*, 2002*a*). The *A. marginale*/tick cell culture system was adapted for short-term growth in 24-well and 96-well plate formats for use in the development of various assays, including a competitive ELISA and an indirect ELISA for detection and quantification of *A. marginale* in cell cultures (Saliki *et al.*, 1998; Kocan *et al.*, 2001). The cell culture system also provided new research opportunities (as reviewed by Blouin *et al.*, 2002*a*). Biological assays were developed to determine the adhesive properties of selected *A. marginale* MSPs, most notably MSP1a. In addition assays were developed to test inhibition of *A. marginale* infection of tick cells by anti-*A. marginale* antibodies, the efficacy of drugs screened for affecting the multiplication of *A. marginale* in tick cells, and the effect of exogenous compounds, such as tick saliva and bee venom, in infection and transmission of *A. marginale* (as reviewed herein and by de la Fuente *et al.*, 2001*a*, 2001*b*, 2002*a*; Blouin *et al.*, 2002*a*, 2002*b*, 2003; Kocan *et al.*, 2004).

preparations will be needed especially in areas where multiple genotypes have been introduced.

INFECTION EXCLUSION OF *ANAPLASMA MARGINALE* IN CATTLE, TICKS AND CELL CULTURE

Recent studies on *A. marginale*-infected cattle in endemic areas demonstrated that multiple *msp1α* genotypes were present, but that only one genotype was found per individual bovine, suggesting that infection of cattle with other genotypes was excluded (Palmer, Rurangirwa & McElwain, 2001; de la Fuente, Blouin & Kocan, 2003).

Experiments by our group confirmed the phenomenon of infection exclusion of *A. marginale* genotypes in infected bovine erythrocytes, cultured tick cells and naturally infected ticks (de la Fuente *et al.*, 2002*b*; de la Fuente, Blouin & Kocan, 2003; as reviewed by Kocan *et al.*, 2004). Using two tick-transmissible isolates of *A. marginale*, Virginia and Oklahoma, we found that cattle and cell cultures became infected with only one *A. marginale* genotype. When *A. ovis* infections were established in cultures that were subsequently inoculated with the Virginia or Oklahoma isolates, *A. marginale* was excluded. The phenomenon of infection exclusion has also been demonstrated in ticks (de la Fuente, Blouin & Kocan, 2003; as reviewed by Kocan *et al.*, 2004). These results point to the existence of some mechanisms common to tick cells and mammalian host cells during the multiplication of *Anaplasma* spp.

Recently, co-infection of two *A. marginale* strains was demonstrated at low frequency in cattle from a herd with a high prevalence of infection (Palmer *et al.*, 2004). The *A. marginale msp1α* genotypes in cattle infected with these two strains were not closely related, thus apparently allowing for the co-infection. This situation may be similar to that occurring in *A. marginale*–*A. centrale* co-infections (Shkap *et al.*, 2002; Molad *et al.*, 2006).

The phenomenon of infection exclusion of *A. marginale* resulting in one genotype per animal has implications for anaplasmosis epidemiology. If cattle infected with a single *A. marginale* genotype are introduced into an endemic herd, these genotypes may be maintained and most likely also become endemic, if they are transmitted to susceptible cattle. Both persistently infected cattle and ticks could serve as reservoirs of the introduced genotype. Thus restricting cattle movement may prevent establishment of multiple *A. marginale* isolates per geographical area. Novel vaccine

ROLE OF MSP1A AND MSP1B IN INFECTION AND TRANSMISSION OF *ANAPLASMA MARGINALE*

MSP1a and MSP1b form the MSP1 complex. MSP1a is variable in molecular weight among geographic isolates because of different numbers of tandem 28–31 amino acid repeats located in the amino terminal portion of the protein (Allred *et al.*, 1990; de la Fuente *et al.*, 2001*b*, 2001*c*, 2002*a*, 2002*b*). Because of the variation in the repeated portion of the *msp1α*, this gene has been used as a stable genetic marker for identification of *A. marginale* isolates (Barbet *et al.*, 1987, 1999; de la Fuente, van den Bussche & Kocan, 2001). The sequence of *msp1α* is conserved during the multiplication of the rickettsiae in cattle and ticks (Palmer *et al.*, 2001; Bowie *et al.*, 2002). A neutralization-sensitive epitope occurs on the MSP1a tandem repeats (Palmer *et al.*, 1987; Oberle *et al.*, 1988; Palmer, 1989) and is conserved among *A. marginale* isolates (Palmer *et al.*, 1987; de la Fuente, van den Bussche & Kocan, 2001; de la Fuente *et al.*, 2001*c*, 2002*a*). MSP1a was shown to be an adhesin for bovine erythrocytes and both native and cultured tick cells using recombinant *E. coli* expressing MSP1a in microtitre haemagglutination and adhesion recovery assays and by microscopy (McGarey & Allred, 1994; McGarey *et al.*, 1994; de la Fuente *et al.*, 2001*b*). The portion of MSP1a with the tandem repeats was necessary and sufficient to effect adhesion to bovine erythrocytes and tick cells (de la Fuente *et al.* 2002*a*; as reviewed by Kocan *et al.*, 2002). MSP1a mediates infection and transmission of *A. marginale* by *Dermacentor* spp. ticks (de la Fuente *et al.*, 2001*a*) and is involved in immunity to *A. marginale* infection in cattle (Palmer *et al.*, 1987; Brown *et al.*, 2001). Furthermore, the adhesive properties of recombinant MSP1a correlate with the transmissibility of the isolate by *Dermacentor* spp. ticks. Additional studies demonstrated the critical role of the 20th amino acid of the repeat in the interaction of MSP1a with host cell receptors (de la Fuente *et al.*, 2001*a*, 2002*a*). However, analysis of tandemly repeated MSP1a peptides of several geographic isolates of *A. marginale* revealed a complex relationship between the *msp1α* genotype and the tick-transmissible phenotype of the isolate, suggesting that both the sequence and conformation of the repeated peptides influence the adhesive properties of MSP1a (de la Fuente *et al.*, 2002*a*).

MSP1b is polymorphic among geographic isolates of *A. marginale* (Barbet *et al.*, 1987; Camacho-Nuez *et al.*, 2000; Viseshakul *et al.*, 2000; Bowie *et al.*, 2002). Although from a multigene family, only small variations in protein sequences were observed in MSP1b1 and MSP1b2 during the life cycle in cattle and ticks (Bowie *et al.*, 2002). This protein, which forms a complex with MSP1a, is an adhesin for bovine erythrocytes but not for tick cells (McGarey & Allred, 1994; McGarey *et al.*, 1994; de la Fuente *et al.*, 2001*b*).

Both MSP1a and MSP1b are glycosylated (McBride, Xue-Jie Yu & Walker, 2000) which explains the difference between observed (90 kDa for MSP1a and 100 kDa for MSP1b) and predicted molecular weights (63 kDa and 79 kDa, respectively) (Garcia-Garcia *et al.*, 2004*a*). Analysis of the amino acid sequence showed that while MSP1b (Oklahoma isolate) contains only one potential *O*-glycosylation site and seven potential *N*-glycosylation sites, MSP1a (Oklahoma isolate) contains only one potential *N*-glycosylation site and 25 potential *O*-glycosylation sites, particularly in the N-terminal repeated region where Ser/Thr accounts for 32% of the amino acids. Recombinant MSP1a and MSP1b proteins have the same molecular weights as the native proteins, suggesting that post-translational modifications are similar to those in *E. coli*. Carbohydrates were found on both recombinant MSP1a and MSP1b using periodate oxidation and biotin-hydrazide conjugation. Only three sugar residues (glucose, galactose and xylose) were identified by gas chromatography, demonstrating the absence of conserved sugar moieties usually found in *O*-linked and *N*-linked oligosaccharides. Therefore, enzymes that recognize carbohydrate sequences containing amino sugars, such as PNGase F, NANase II, GALase III, HEXase I and *O*-glycosidase DS, were unable to deglycosylate these proteins. Although glycoproteins from both MSP1a and MSP1b contained the same sugars, the molar ratio varied between them. MSP1a had a higher content of glucose and only residual contents of xylose and galactose, while MSP1b had a higher content of xylose and galactose. These differences probably reflect variation in the nature of the glycosylation, since the oligosaccharides should be *O*-linked in MSP1a and *N*-linked in MSP1b. The molecular weight of the N-terminal portion of MSP1a containing the adhesive repeated peptides and the conserved C-terminal region of MSP1a were also larger than predicted. Carbohydrates were detected in both regions, although the degree of modification in the N-terminal region was higher. The significance and biological function of MSP1a and MSP1b glycosylation is unknown but the carbohydrate residues may contribute to the adhesive properties of the MSPs. Glycosylation of membrane proteins may also function in protection of the protein against proteases during development, and could contribute to the generation of antigenic and phenotypic variation of the pathogen.

Expression of MSP1a is differentially regulated in erythrocytic and tick stages of *A. marginale*. Cattle immunized with *A. marginale* from tick cells or bovine erythrocytes produce antibodies against the *A. marginale* MSP5 but a differential antibody response to MSP1a and MSP1b (Kocan *et al.*, 2001; de la Fuente *et al.*, 2002*c*, Garcia-Garcia *et al.*, 2004*b*). Cattle immunized with erythrocyte-derived *A. marginale* elicit an antibody response mainly against MSP1a, while animals immunized with cell culture-derived antigen produce predominantly antibodies to MSP1b. The molecular basis of this differential antibody response was studied by comparing the levels of MSP1a, MSP1b and MSP5 on *A. marginale* harvested from the two host cells. The amount of MSP1b and MSP5 was similar on *A. marginale* from both host cells, but the amount of MSP1a was higher in the erythrocyte-derived *A. marginale*. These differences were also found when RNA transcripts were analysed by RT-PCR, demonstrating that differential expression of MSP1a and MSP1b is regulated at the transcriptional level. Since MSP1a is an *A. marginale* adhesin for tick cells, biological transmission of the pathogen could be enhanced by increased levels of this surface protein. Differences in the expression level of surface exposed molecules may also contribute to phenotypic and antigenic variation of the pathogen.

PHYLOGEOGRAPHY AND EVOLUTION OF *ANAPLASMA MARGINALE*

The phylogenetic relationship and evolution of *A. marginale* isolates is important for understanding the biology and the possibilities for control of anaplasmosis. We chose two *A. marginale* MSPs (MSP1a and MSP4) for phylogenetic analysis. MSPs are involved in interactions with both vertebrate and invertebrate hosts (McGarey and Allred, 1994; McGarey *et al.*, 1994; de la Fuente *et al.*, 2001*b*). They are likely to evolve more rapidly than other genes because they are subjected to selective pressures exerted by host immune systems. For instance, the repeated sequence motifs in *A. marginale* MSP1a are important in host–pathogen interactions by serving as adhesins required for invasion and transmission of *A. marginale*. The role of MSP1a repeats is relevant to tick–pathogen interactions. Therefore, the function of MSP1a may have resulted from distinct evolutionary pressures, specifically from those exerted by ligand–receptor and host–pathogen interactions. Both *msp1α* and *msp4* are stable genetic markers during the multiplication of *A. marginale* (de la Fuente *et al.*, 2001*c*; Bowie *et al.*, 2002).

Phylogenetic analysis of *A. marginale* geographic isolates from the United States was performed using *msp1α* and *msp4* gene and derived protein sequences (de la Fuente, Van Den Bussche & Kocan, 2001). Results of these analyses strongly supported a southeastern clade of *A. marginale* comprising the Virginia and Florida isolates. Furthermore, analysis of 16S rDNA fragment sequences from *D. variabilis*, the tick vector of *A. marginale*, from various areas of the United States supported co-evolution of the vector and pathogen (de la Fuente, van den Bussche & Kocan, 2001).

Phylogenetic studies were also done using New World isolates of *A. marginale* from the United States, Mexico, Brazil and Argentina. Seventeen isolates of *A. marginale* plus two outgroup taxa (*A. centrale* and *A. ovis*) were included for the analysis of MSP4 sequences (de la Fuente *et al.*, 2002*e*). Maximum-parsimony analysis of MSP4 sequences provided phylogenetic information on the evolution of *A. marginale* isolates. Strong bootstrap support was detected for a Latin American clade of *A. marginale* isolates. Moreover, within this Latin American clade, strong bootstrap support was detected for Mexican and South American clades. Isolates of *A. marginale* from the United States also grouped into two clades, a southern clade comprising isolates from Florida, Mississippi and Virginia, and a west–central clade comprising isolates from California, Idaho, Illinois, Oklahoma and Texas. Although little phylogeographic resolution was detected within any of these higher clades, *msp4* sequences appear to be a good genetic marker for inferring phylogeographic patterns of isolates of *A. marginale* on a broad geographic scale. In contrast to the phylogeographic resolution provided by MSP4, DNA and protein sequence variation from MSP1a representing 20 New World isolates of *A. marginale* failed to provide phylogeographic resolution (de la Fuente *et al.*, 2000*d*). Most variation in MSP1a sequences appeared unique to a given isolate. In fact, similar DNA sequence variation in MSP1a was detected within isolates from Idaho and Florida and from Idaho and Argentina. These results suggest that MSP1a sequences may be rapidly evolving and bring into question the use of MSP1a sequences for defining geographic isolates of *A. marginale*. Nevertheless, it was considered necessary to address the possibility that MSP1a may provide phylogeographic information when numerous strains from a given area are included in the analysis.

Eleven *A. marginale* strains from Oklahoma isolated from cattle with anaplasmosis during 2001, plus two previous isolates from Wetumka (Oklahoma isolate: in de la Fuente *et al.*, 2001*c*, 2002*a*, 2002*c*, 2002*d*, 2002*e*; de la Fuente, van den Bussche & Kocan, 2001) and Pawhuska identified in 1997 and 1960s, respectively, were then analysed for MSP1a and MSP4 gene and protein sequences. Maximum parsimony and maximum likelihood (ML) phylogenies of *msp4* sequences of 13 strains from Oklahoma and in comparison with seven Latin American and 12 strains from the United States using as outgroups *A. centrale* and *A. ovis* sequences provided strong bootstrap support for a Latin American clade and, within this clade, support was detected for Mexican and South American clades. Isolates of *A. marginale* from the United States also grouped into two clades, from the southern (isolates from Florida, Mississippi and Virginia) and west–central (isolates from California, Idaho, Illinois, Oregon, Missouri and Texas) states. Both clades contained strains from Oklahoma, suggesting extensive cattle movement in the past. Within Oklahoma, ML analysis of *msp4* sequences provided bootstrap support for east–central and north–central clades, both including strains from Stillwater, OK. Maximum likelihood analysis of codon and amino acid changes over the MSP4 phylogeny evidenced that *msp4* is not under positive selection pressure. In contrast, ML phylogeny of MSP1a DNA and protein sequences of 13 strains from Oklahoma and in comparison with seven Latin American and 13 strains from other parts of the United States demonstrated no geographical clustering and evidenced that the gene is under positive selection pressure.

These results again indicate that MSP1a is not a marker for the characterization of geographic isolates of *A. marginale*, and suggest that the genetic heterogeneity observed among strains of *A. marginale* within Oklahoma, and probably within other endemic regions, could be explained by cattle movement and maintenance of different genotypes by independent transmission events. Therefore, if cattle infected with a new *A. marginale* genotype were imported into a region, the new isolate could become established by mechanical and/or biological transmission to susceptible cattle. In regions with few cattle introductions like Australia, little genotypic variation is found within *A. marginale* isolates (Bock & de Vos, 2001). In regions with extensive cattle movement like Oklahoma, a highly heterogeneous *A. marginale* population would be expected. *msp4* gene sequences appear to be a good genetic marker for evolutionary studies within the genus *Anaplasma* and for inferring phylogeographic patterns of *A. marginale* isolates.

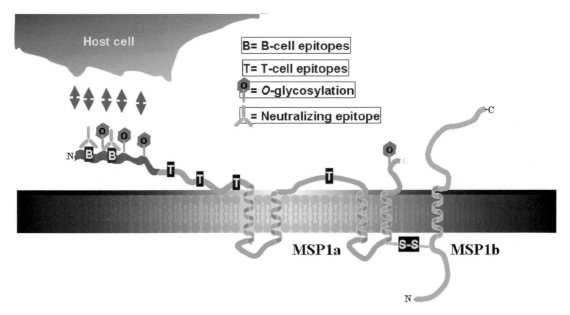

Fig. 15.5. Schematic depicting the interaction of the *A. marginale* MSP1a with the host cell. B-cell epitopes (Garcia-Garcia *et al.*, 2004*c*), neutralization epitopes (Palmer *et al.*, 1987) and *O*-glycosylation sites (Garcia-Garcia *et al.*, 2004*a*) are shown in the extracellular N-terminal region of the protein that contains a variable number of 28–30 repeated peptides (de la Fuente *et al.*, 2001*c*). T-cell epitopes (Brown *et al.*, 2001) are present in the conserved region of the proteins.

An updated analysis of MSP1a repeat sequences of *A. marginale* strains from North and South America, Europe, Asia, Africa and Australia corroborated the genetic heterogeneity of geographic strains of *A. marginale* (de la Fuente *et al.*, 2005, 2007*a*, 2007*b*). The phylogenetic analysis of MSP1a repeat sequences did not result in clusters according to the geographic origin of *A. marginale* strains. The only support (>70%) was provided for clusters containing repeats found exclusively in European (Italian) or American (Mexican and Brazilian) strains (de la Fuente *et al.*, 2007*a*, 2007*b*). Most of the MSP1a repeat sequences were present in strains from a single geographic region, while few sequences were present in strains from two or three of the geographic regions analysed. The phylogenetic analysis of MSP1a repeat sequences suggested tick–pathogen co-evolution and provided evidence of multiple introductions of *A. marginale* strains from various geographic locations worldwide.

The phylogenetic analysis of MSP4 sequences of *A. marginale*, *A. centrale*, *A. ovis* and *A. phagocytophilum* strains derived from infected mammals and ticks in countries from America, Europe and Asia did not provide phylogeographic information but the analysis did differentiate between *Anaplasma* spp. (de la Fuente *et al.*, 2005, 2007*a*,

2007*b*). These results supported those obtained with MSP1a and indicated that while MSP4 is not a good genetic marker for global phylogeographic analysis of *A. marginale* strains, it may still be useful for strain comparison in some regions (de la Fuente *et al.*, 2002*a*).

Tick–pathogen co-evolution probably involves genetic traits of the vector. Recent reports have confirmed the presence of tick receptors for tick-borne pathogens (Pal *et al.*, 2004). The tick receptor for *A. marginale* has not been identified. However, Scoles *et al.* (2005*b*) demonstrated significant variation in *D. andersoni* midgut susceptibility to the pathogen. Futse *et al.* (2003) took a different approach to study tick–*A. marginale* evolutionary adaptations and demonstrated that ticks and *A. marginale* strains retain competence for tick transmission in the absence of vector–pathogen interaction.

PROSPECTS FOR CONTROL OF *ANAPLASMA MARGINALE* BY USE OF PATHOGEN- AND TICK-DERIVED ANTIGENS

The characterization of antigens identified on *A. marginale* and the vaccine preparations developed for the control of infections with *A. marginale* were recently reviewed by Kocan *et al.* (2003) and will not be covered in this review. The recent report of vaccine preparations with antigens derived from *A. marginale* grown in cultured IDE8 cells has opened new possibilities for the development of safe, reproducible vaccine formulations against a broad spectrum of *A. marginale* strains (Kocan *et al.*, 2001; de la Fuente *et al.*, 2002*c*). However, these vaccine formulations have only provided partial protection against infections with *A. marginale*, diminishing the severity and duration of the clinical signs.

These results stress the need for more efficacious vaccines against infections with *A. marginale*. The use of recombinant antigens for the development of subunit vaccines against anaplasmosis has provided promising results (reviewed by Palmer, 1989; Kocan *et al.*, 2000, 2003*b*; de la Fuente *et al.*, 2005). MSP1a, although variable in the number of repeated peptides, induces strong T-cell responses and contains a conserved B-cell epitope in the repeated peptides that is recognized by immunized and protected cattle (Kocan *et al.*, 2003; Garcia-Garcia *et al.*, 2004*c*; de la Fuente *et al.*, 2005 and references therein) (Fig. 15.5).

However, the incorporation of recombinant antigens into vaccines for control of *A. marginale* infections will require a better understanding of the mechanisms governing bacterial infection and transmission and the nature of the protective response developed in persistently infected cattle. Furthermore, the differences in protein expression and structure between rickettsiae grown in tick cells and bovine erythrocytes is crucial for the development of vaccine formulations capable of preventing infection and transmission of the pathogen, the ultimate goal of the vaccine against *A. marginale*. Current results suggest that differences exist between *A. marginale* in tick cells and bovine erythrocytes (Kocan *et al.*, 2003). Further research is needed to define the adaptations of *A. marginale* to cattle and ticks and to define the role of carbohydrates in host–pathogen and tick–pathogen interactions (Garcia-Garcia *et al.*, 2004*a*).

Finally, the inclusion of tick antigens in *A. marginale*-specific vaccines could enhance their protective effect and increase efficacy (Nuttall, 1999; de la Fuente and Kocan, 2003, 2006). This transmission-blocking approach is supported by evidence that host resistance to ticks provides some protection against tick-borne transmission of viruses and *B. burgdorferi* (Wikel *et al.*, 1997). Furthermore, vaccination against *B. microplus* has been demonstrated to contribute to the control of tick-borne diseases (de la Fuente *et al.*, 1998).

Recent experiments have demonstrated that anti-tick vaccines may contribute to the control of tick-borne pathogens by decreasing the exposure of susceptible hosts to ticks and/or by reducing the vector capacity of ticks (de la Fuente and Kocan, 2006; de la Fuente *et al.*, 2006; Labuda *et al.*, 2006). Therefore, the combination of anti-tick antigens with *A. marginale*-derived proteins may provide a means to control *A. marginale* infection and transmission through immunization of cattle.

ACKNOWLEDGEMENTS

The research reviewed herein from the Tick and Tick-Borne Pathogen Vaccine Development Laboratory was supported by project No. 1669 of the Oklahoma Agricultural Experiment Station, the Walter R. Sitlington Endowed Chair for Food Animal Research (K. M. Kocan, College of Veterinary Medicine, Oklahoma State University) and the Oklahoma Center for the Advancement of Science and Technology, Applied Research Grant, AR00(1)-001. Full acknowledgment of collaborators is listed in Kocan *et al.* (2004) *Parasitology* 129, S285–S300.

REFERENCES

Allred, D. R., McGuire, T. C., Palmer, G. H., *et al.* (1990). Molecular basis for surface antigen size polymorphisms and conservation of a neutralization- sensitive epitope in *Anaplasma marginale*. *Proceedings of the National Academy of Sciences of the USA* 87, 3220–3224.

Anthony, D. W. & Roby, T. O. (1962). Anaplasmosis transmission studies with *Dermacentor variabilis* (Say) and *Dermacentor andersoni* (Stiles) (= *D. venustus* Marx) as experimental vectors. In *Proceedings of the 4th National Anaplasmosis Research Conference*, Reno, NV, pp. 78–81.

Anthony, D. W. & Roby, T. O. (1966). The experimental transmission of bovine anaplasmosis by three species of North American ticks. *American Journal of Veterinary Research* 27, 191–198.

Barbet, A. F., Blentlinger, R., Yi, J., *et al.* (1999). Comparison of surface proteins of *Anaplasma marginale* grown in tick cell culture, tick salivary glands, and cattle. *Infection and Immunity* 67, 102–107.

Barbet, A. F., Yi, J., Lundgren, A., *et al.* (2001). Antigenic variation of *Anaplasma marginale*: MSP2 diversity during cyclic transmission between ticks and cattle. *Infection and Immunity* 69, 3057–3066.

Barbet, A. F., Lundgren, A., Yi, J., Rurangirwa, F. R. & Palmer, G. H. (2000). Antigenic variation of *Anaplasma marginale* by expression of MSP2 mosaics. *Infection and Immunity* 68, 6133–6138.

Barbet, A. F., Palmer, G. H., Myler, P. J. & McGuire, T. C. (1987). Characterization of an immunoprotective protein complex of *Anaplasma marginale* by cloning and expression of the gene coding for polypeptide Am105L. *Infection and Immunity* **55**, 2428–2435.

Bell-Sakyi, L. M., Paxton, E. A., Munderloh, U. G. & Sumption, K. J. (2000). Growth of *Cowdria ruminantium*, the causative agent of heartwater, in a tick cell line. *Journal of Clinical Microbiology* **38**, 1238–1240.

Blouin, E. F. & Kocan, K. M. (1998). Morphology and development of *Anaplasma marginale* (Rickettsiales: Anaplasmataceae) in cultured *Ixodes scapularis* (Acari: Ixodidae) cells. *Journal of Medical Entomology* **35**, 788–797.

Blouin, E. F, Barbet, A. F., Yi, J., Kocan, K. M. & Saliki, J. T. (1999). Establishment and characterization of an Oklahoma isolate of *Anaplasma marginale* in cultured *Ixodes scapularis* cells. *Veterinary Parasitology* **87**, 301–313.

Blouin, E. F., de la Fuente, J., Garcia-Garcia, J. C., *et al.* (2002*a*). Use of a cell culture system for studying the interaction of *Anaplasma marginale* with tick cells. *Animal Health Research Reviews* **3**, 57–68.

Blouin, E. F., Kocan, K. M., de la Fuente, J. & Saliki, J. T. (2002*b*). Effect of tetracycline on development of *Anaplasma marginale* in cultured *Ixodes scapularis* cells. *Veterinary Parasitology* **107**, 115–126.

Blouin, E. F., Saliki, J. T., de la Fuente, J., Garcia-Garcia, J. C. & Kocan, K. M. (2003). Antibodies to *Anaplasma marginale* Major Surface Protein 1a and 1b inhibit infectivity for cultured tick cells. *Veterinary Parasitology* **91**, 265–283.

Bock, R. E. & de Vos, A. J. (1999). Effect of cattle on innate resistance to infection with *Anaplasma marginale* transmitted by *Boophilus microplus*. *Australian Veterinary Journal* **77**, 748–751.

Bock, R. E. & de Vos, A. J. (2001). Immunity following use of Australian tick fever vaccine: a review of the evidence. *Australian Veterinary Journal* **79**, 832–839.

Bowie, J. V., de la Fuente, J., Kocan, K. M., Blouin, E. F. & Barbet, A. F. (2002). Conservation of major surface protein 1 genes of the ehrlichial pathogen *Anaplasma marginale* during cyclic transmission between ticks and cattle. *Gene* **282**, 95–102.

Boynton, W. H., Hermes, W. B., Howell, D. E. & Woods, G. M. (1936). Anaplasmosis transmission by three species of ticks in California. *Journal of the American Veterinary Medical Association* **88**, 500–502.

Bram, R. A. (1975). Tick-borne livestock diseases and their vectors. I. The global problem. *World Animal Review* **6**, 1–5.

Brayton, K. A., Knowles, D. P., McGuire, T. C. & Palmer, G. H. (2001). Efficient use of a small genome to generate antigenic diversity in tick-borne ehrlichial pathogens. *Proceedings of the National Academy of Sciences of the USA* **98**, 4130–4135.

Brayton, K. A., Palmer, G. H., Lundgren, A., Yi, J. & Barbet, A. F. (2002). Antigenic variation of *Anaplasma marginale* msp2 occurs by combinatorial gene conversion. *Molecular Microbiology* **43**, 1151–1159.

Brown, W. C., Palmer, G. H., Lewin, H. A. & McGuire, T. C. (2001). CD4(+) T lymphocytes from calves immunized with *Anaplasma marginale* major surface protein 1 (MSP1), a heteromeric complex of MSP1a and MSP1b, preferentially recognize the MSP1a carboxyl terminus that is conserved among strains. *Infection and Immunity* **69**, 6853–6862.

Brumpt, E. (1931). Transmission d'*Anaplasma marginale* par *Rhipicephalus bursa* et par margraopus. *Annuals de Parasitologie* **9**, 4–9.

Camacho-Nuez, J., De Lourdes Muñoz, M., Suarez, C. E., *et al.* (2000). Expression of polymorphic *msp1β* genes during acute *Anaplasma marginale* rickettsemia. *Infection and Immunity* **68**, 1946–1952.

Christensen, J. F. & Howard, J. A. (1966). Anaplasmosis transmission by *Dermacentor occidentalis* taken from cattle in Santa Barbara County, CA. *American Journal of Veterinary Research* **27**, 1473–1475.

Coronado, A. 2001. Is *Boophilus microplus* the main vector of *Anaplasma marginale*? Technical note. *Revista Científica, FCV-LUZ* **11**, 408–411.

de la Fuente, J. & Kocan, K. M. (2001). Expression of *Anaplasma marginale* major surface protein 2 variants in persistently infected ticks. *Infection and Immunity* **69**, 5151–5156.

de la Fuente, J. & Kocan, K. M. (2003). Advances in the identification and characterization of protective antigens for development of recombinant vaccines against tick infestations. *Expert Review of Vaccines* **2**, 583–593.

de la Fuente, J., & Kocan, K. M. (2006). Strategies for development of vaccines for control of ixodid tick species. *Parasite Immunology* **28**, 275–283.

de la Fuente, J., Almazán, C., Blouin, E. F., Naranjo, V. & Kocan, K. M. (2006). Reduction of tick infections with *Anaplasma marginale* and *A. phagocytophilum* by targeting the tick protective antigen subolesin. *Parasitology Research* **100**, 85–91.

de la Fuente, J., Blouin, E. F. & Kocan, K. M. (2003). Infection of ticks with the intracellular rickettsia *Anaplasma marginale* excludes infection with other genotypes. *Clinical and Diagnostic Laboratory Immunology* **10**, 182–184.

de la Fuente, J., Garcia-Garcia, J. C., Blouin, E. F. & Kocan, K. M. (2001*a*). Major surface protein 1a effects tick infection and transmission of the ehrlichial pathogen *Anaplasma marginale*. *International Journal for Parasitology* **31**, 1705–1714.

de la Fuente, J., Garcia-Garcia, J. C., Blouin, E. F. & Kocan, & K. M. (2001*b*). Differential adhesion of major surface proteins 1a and 1b of the ehrlichial cattle pathogen *Anaplasma marginale* to bovine erythrocytes and tick cells. *International Journal for Parasitology* **31**, 145–153.

de la Fuente, J., Garcia-Garcia, J. C., Blouin, E. F., *et al.* (2001*c*). Evolution and function of tandem repeats in the major surface protein 1a of the ehrlichial pathogen *Anaplasma marginale*. *Animal Health Research Reviews* **2**, 163–173.

de la Fuente, J., Garcia-Garcia, J. C., Blouin, E. F. & Kocan, K. M. (2002*a*). Characterization of the functional domain of major surface protein 1a involved in adhesion of *Anaplasma marginale* to host cells. *Veterinary Microbiology* **91**, 265–283.

de la Fuente, J., Garcia-Garcia, J. C., Blouin, E. F., Saliki, J. T. & Kocan, K. M. (2002*b*). Infection of tick cells and bovine erythrocytes with one genotype of the intracellular ehrlichia *Anaplasma marginale* excludes infection with other genotypes. *Diagnostic Laboratory Immunology* **9**, 658–668.

de la Fuente, J. C., Golsteyn Thomas, E. J., van den Bussche, A., *et al.* (2003). Characterization of *Anaplasma marginale* isolated from North American bison. *Applied and Environmental Microbiolology* **69**, 5001–5005.

de la Fuente, J., Kocan, K. M., Garcia-Garcia, J. C., *et al.* (2002*c*). Vaccination of cattle with *Anaplasma marginale* derived from tick cell culture and bovine erythrocytes followed by challenge-exposure by ticks. *Veterinary Microbiology* **89**, 239–251.

de la Fuente, J., Kocan, K. M., Mangold, A. J., *et al.* (2007*a*). Biogeography and molecular evolution of *Anaplasma* species. *Veterinary Parasitology* **119**, 382–390.

de la Fuente, J., Lew, A., Lutz, H., *et al.* (2005). Genetic diversity of *Anaplasma* species major surface proteins and implications for anaplasmosis serodiagnosis and vaccine development. *Animal Health Reviews* **6**, 75–89.

de la Fuente, J., Naranjo, V., Ruiz-Fons, F., *et al.* (2004). Prevalence of tick-borne pathogens in ixodid ticks (Acari: Ixodidae) collected from wild boar (*Sus scrofa*) and Iberian red deer (*Cervus elaphus hispanicus*) in central Spain. *European Journal of Wildlife Research* **50**, 187–196.

de la Fuente, J., Rodriguez, M., Redondo, M., *et al.* (1998). Field studies and cost-effectiveness analysis of vaccination with Gavac against the cattle tick *Boophilus microplus*. *Vaccine* **16**, 366–373.

de la Fuente, J., Ruybal, P., Mtshali, M. S., *et al.* (2007*b*). Analysis of world strains of *Anaplasma marginale* using major surface protein 1a repeat sequences. *Veterinary Microbiology* **119**, 382–390.

de la Fuente, J., van den Bussche, R. A., Garcia-Garcia, J. C., *et al.* (2002*d*). Phylogeography of New World isolates of *Anaplasma marginale* (Rickettsiaceae: Ehrlichieae) based on major surface protein sequences. *Veterinary Microbiology* **88**, 275–285.

de la Fuente, J., van den Bussche, R. A. & Kocan, K. M. (2001). Molecular phylogeny and biogeography of North American isolates of *Anaplasma marginale* (Rickettsiaceae: Ehrlichieae). *Veterinary Parasitology* **97**, 65–76.

de la Fuente, J., van den Bussche, R. A., Prado, T. & Kocan, K. M. (2002*e*). *Anaplasma marginale* major surface protein 1α genotypes evolved under positive selection pressure but are not a marker for geographic isolates. *Journal of Clinical Microbiology* **41**, 1609–1616.

Dikman, G. (1950). The transmission of anaplasmosis. *American Journal of Veterinary Research* **11**, 5–16.

Dumler, J. S., Barbet, A. F., Bekker, C. P. J., *et al.* (2001). Reorganization of the genera in the families Rickettsiaceae and Anaplasmataceae in the order Rickettsiales: unification of some species of *Ehrlichia* with *Anaplasma*, *Cowdria* with *Ehrlichia* and *Ehrlichia* with *Neorickettsia*, descriptions of six new species combinations and designation of *Ehrlichia equi* and "HGE agent" as subjective synonyms of *Ehrlichia phagocytophila*. *International Journal of Systematic Evolutionary Microbiology* **51**, 2145–2165.

Eriks, I. S., Stiller D. & Palmer, G. H. (1993). Impact of persistent *Anaplasma marginale* rickettsemia on tick infection and transmission. *Journal of Clinical Microbiology* **31**, 2091–2096.

Ewing, S. A. (1981). Transmission of *Anaplasma marginale* by arthropods. In *Proceedings of the 7th National Anaplasmosis Conference*, pp. 395–423.

Ewing, S. A., Panciera, R. J., Kocan, K. M., *et al.* (1997). A winter outbreak of anaplasmosis in a non-endemic area of Oklahoma: a possible role for *Dermacentor albipictus*. *Journal of Veterinary Diagnostic Investigation* **9**, 206–208.

Figueroa, J. V., Alvarez, J. A., Ramos, J. A., *et al.* (1998). Bovine babesiosis and anaplasmosis follow-up on cattle relocated in an endemic area for hemoparasitic diseases. *Annals of the New York Academy of Science* **849**, 1–10.

Foil, L. D. (1989). Tabanids as vectors of disease agents. *Parasitology Today* **5**, 88–96.

French, D. M., Brown, W. C. & Palmer, G. H. (1999). Emergence of *Anaplasma marginale* antigenic variants during persistent rickettsemia. *Infection and Immunity* **67**, 5834–5840.

French, D. M., McElwain, T. F., McGuire, T. C. & Palmer, G. H. (1998). Expression of *Anaplasma marginale* major surface protein 2 variants during persistent cyclic rickettsemia. *Infection and Immunity* **66**, 1200–1207. [Published errantum appears in *Infection and Immunity* (1998) **66**, 2400.]

Futse, J. E., Ueti, M. W., Knowles, D. P. Jr & Palmer, G. H. (2003). Transmission of *Anaplasma marginale* by *Boophilus microplus*: retention of vector competence in the absence of vector–pathogen interaction. *Journal of Clinical Microbiology* **41**, 3829–3834.

Garcia-Garcia, J. C., de la Fuente, J., Bell-Eunice, G., Blouin, E. F. & Kocan, K. M. (2004*a*). Glycosylation of major surface protein 1a and its putative role in adhesion of *Anaplasma marginale* to tick cells. *Infection and Immunity* **72**, 3022–3030.

Garcia-Garcia, J. C., de la Fuente, J., Blouin, E. F., *et al.* (2004*b*). Differential expression of the *msp1α* gene of *Anaplasma marginale* occurs in bovine erythrocytes and tick cells. *Veterinary Microbiology* **98**, 261–272.

Garcia-Garcia, J. C., de la Fuente, J., Kocan, K. M., *et al.* (2004*c*). Mapping of B-cell epitopes in the N-terminal repeated peptides of the *Anaplasma marginale* major surface protein 1a and characterization of the humoral immune response of cattle immunized with recombinant and whole organism antigens. *Veterinary Immunology and Immunopathology* **98**, 137–151.

Ge, N. L., Kocan, K. M., Blouin, E. F. & Murphy, G. L. (1996). Developmental studies of *Anaplasma marginale* (Rickettsiales: Anaplasmataceae) in male *Dermacentor andersoni* (Acari: Ixodidae) infected as adults by using non-radioactive in situ hybridization and microscopy. *Journal of Medical Entomology* **33**, 911–920.

Helm, R. (1924). Beitrag zum Anaplasmen-Problem. *Zeitschrift für Infektionskr ankheiten* **25**, 199–226.

Howarth, J. A. & Hokama, Y. (1973). Tick transmission of anaplasmosis under laboratory conditions. In *Proceedings of the 6th National Anaplasmosis Research Conference*, Las Vegas, NV, pp. 117–120.

Howarth, J. A. & Roby, T. O. (1972). Transmission of anaplasmosis by field collections of *Dermacentor occidentalis* Marx (Acarina: Ixodidae). In *76th Meeting of the United States Animal Health Association*, pp. 98–102.

Howell, D. E., Stiles, G. W. & Moe, L. H. (1941). The fowl tick (*Argas persicus*), a new vector of anaplasmosis. *American Journal of Veterinary Research* **4**, 73–75.

Kocan, K. M. (1986). Development of *Anaplasma marginale* in ixodid ticks: coordinated development of a rickettsial organism and its tick host. In *Morphology, Physiology, and Behavioral Ecology of Ticks*, eds. Sauer, J. R. & Hair, J. A., pp. 472–505. Chichester, UK: Ellis Horwood.

Kocan, K. M., Blouin, E. F. & Barbet, A. F. (2000). Anaplasmosis control: past, present and future. *Annals of the New York Academy of Science* **916**, 501–509.

Kocan, K. M., de la Fuente, J., Blouin, E. F. & Garcia-Garcia, J. C. (2002). Adaptation of the tick-borne pathogen, *Anaplasma marginale*, for survival in cattle and tick hosts. *Experimental and Applied Acarology* **28**, 9–25.

Kocan, K. M., de la Fuente, J., Blouin, E. F. & Garcia-Garcia, J. C. (2004). *Anaplasma marginale* (Rickettsiales: Anaplasmataceae): recent advances in defining host-pathogen adaptations of a tick-borne rickettsia. *Parasitology* **12**, S285–S300.

Kocan, K. M., de la Fuente, J., Guglielmone, A. A. & Melendéz, R. D. (2003). Antigens and alternatives for control of *Anaplasma marginale* infection in cattle. *Clinical Microbiology Reviews* **16**, 698–712.

Kocan, K. M., Goff, W. L., Stiller, D., *et al.* (1992*a*). Persistence of *Anaplasma marginale* (Rickettsiales: Anaplasmataceae) in male *Dermacentor andersoni* (Acari: Ixodidae) transferred successively from infected to susceptible calves. *Journal of Medical Entomology* **29**, 657–668.

Kocan, K. M., Hair, J. A., Ewing, S. A. & Stratton, J. G. (1981). Transmission of *Anaplasma marginale* Theiler by *Dermacentor andersoni* Stiles and *Dermacentor variabilis* (Say). *American Journal of Veterinary Research* **42**, 15–18.

Kocan, K. M., Halbur, T., Blouin, E. F., *et al.* (2001). Immunization of cattle with *Anaplasma marginale* derived from tick cell culture. *Veterinary Parasitology* **102**, 151–161.

Kocan, K. M., Munderloh, U. G. & Ewing, S. A. (1998). Development of the Ebony isolate of *Ehrlichia canis* in cultured *Ixodes scapularis* cells. In *79th Conference of Research Workers in Animal Diseases*, Chicago, IL, Abstract 95.

Kocan, K. M., Stiller, D., Goff, W. L., *et al.* (1992*b*). Development of *Anaplasma marginale* in male *Dermacentor andersoni* transferred from parasitemic to susceptible cattle. *American Journal of Veterinary Research* **53**, 499–507.

Labuda, M., Trimnell, A. R., Lickova, M., *et al.* (2006). An antivector vaccine protects against a lethal vector-borne pathogen. *PLoS Pathogens* **2**, e27.

McBride, J. W., Xue-Jie Yu & Walker, D. H. (2000).
Glycosylation of homologous immunodominant proteins of
Ehrlichia chaffeensis and *Ehrlichia canis*. *Infection and
Immunity* **68**, 13–18.

McGarey, D. J. & Allred, D. R. (1994). Characterization of
hemagglutinating components on the *Anaplasma marginale*
initial body surface and identification of possible adhesins.
Infection and Immunity **62**, 4587–4593.

McGarey, D. J., Barbet, A. F., Palmer, G. H., McGuire, T. C.
& Allred, D. R. (1994). Putative adhesins of *Anaplasma
marginale*: major surface polypeptides 1a and 1b. *Infection
and Immunity* **62**, 4594–4601.

Meeus, P. F. & Barbet, A. F. (2001). Ingenious gene generation.
Trends in Microbiology **9**, 353–355.

Molad, T., Mazuz, M. L., Fleiderovitz, L., *et al.* (2006).
Molecular and serological detection of *A. centrale*- and *A.
marginale*-infected cattle grazing within an endemic area.
Veterinary Microbiololgy **113**, 55–62.

Munderloh, U. G., Blouin, E. F., Kocan, K. M., *et al.* (1996*a*).
Establishment of the tick (Acari: Ixodidae)-borne cattle
pathogen *Anaplasma marginale* (Rickettsiales:
Anaplasmataceae) in tick cell culture. *Journal of Medical
Entomology* **33**, 656–664.

Munderloh, U. G., Jauron, S. D., Fingerle, V., *et al.* (1999).
Invasion and intracellular development of the human
granulocytic ehrlichiosis agent in tick cell culture. *Journal of
Clinical Microbiology* **37**, 2518–2524.

Munderloh, U. G., Madigan, J. E., Dumler, J. S., *et al.*
(1996*b*). Isolation of the equine granulocytic ehrlichiosis
agent, *Ehrlichia equi*, in tick cell culture. *Journal of Clinical
Microbiology* **34**, 664–670.

Nuttall, P. A. (1999). Pathogen–tick–host interactions: *Borrelia
burgdorferi* and TBE virus. *Zentralblatt für Bakteriologie* **289**,
492–505.

Oberle, S. M., Palmer, G. H., Barbet, A. F. & McGuire, T. C.
(1988). Molecular size variations in an immunoprotective
protein complex among isolates of *Anaplasma marginale*.
Infection and Immunity **56**, 1567–1573.

Pal, U., Li, X., Wang, T., *et al.* (2004). TROSPA, an *Ixodes
scapularis* receptor for *Borrelia burgdorferi*. *Cell* **119**,
457–468.

Palmer, G. H. (1989). Anaplasmosis vaccines. In *Veterinary
Protozoan and Hemoparasite Vaccines*, ed. Wright, I. G.,
pp. 1–29. Boca Raton, FL: CRC Press.

Palmer, G. H., Knowles, D. P. Jr, Rodriguez, J. L., *et al.*
(2004). Stochastic transmission of multiple genotypically
distinct *Anaplasma marginale* strains in a herd with high
prevalence of *Anaplasma* infection. *Journal of Clinical
Microbiology* **42**, 5381–5384.

Palmer, G. H., Rurangirwa, F. R. & McElwain, T. F. (2001).
Strain composition of the ehrlichia *Anaplasma marginale*
within persistently infected cattle, a mammalian reservoir
for tick transmission. *Journal of Clinical Microbiology* **39**,
631–635.

Palmer, G. H., Waghela, S. D., Barbet, A. F., Davis, W. C. &
McGuire, T. C. (1987). Characterization of a neutralization
sensitive epitope on the AM 105 surface protein of
Anaplasma marginale. *Journal of Parasitology* **17**, 1279–1285.

Parker, R. J. (1982). The Australian brown dog tick
Rhipicephalus sanguineus as an experimental parasite of cattle
and vector of *Anaplasma marginale*. *Australian Veterinary
Journal* **58**, 47–51.

Peterson, K. J., Raleigh, R. J., Stround, R. K. & Goulding,
R. L. (1977). Bovine anaplasmosis transmission studies
conducted under controlled natural exposure in a
Dermacentor andersoni (= *venustus*) indigenous area of
eastern Oregon. *American Journal of Veterinary Research* **38**,
351–354.

Piercy, P. L (1938). *Fifty-First Annual Report*, Texas
Agricultural Experiment Station.

Piercy, P. L. & Schmidt, H. (1941). *Fifty-Fourth Annual
Report*, Texas Agricultural Experiment Station.

Potgieter, F. T. (1979). Epizootiology and control of
anaplasmosis in South Africa. *Journal of the South African
Veterinary Association* **504**, 367–372.

Potgieter, F. T., Kocan, K. M., McNew, R. W. & Ewing, S. A.
(1983). Demonstration of colonies of *Anaplasma marginale*
in the midgut of *Rhipicephalus simus*. *American Journal of
Veterinary Research* **44**, 2256–2261.

Rees, C. W. (1930). Experimental transmission of anaplasmosis
by *Rhipicephalus sanguineus*. *North American Veterinarian* **11**,
17–20.

Rees, C. W. (1932). The experimental transmission of
anaplasmosis by *Dermacentor variabilis*. *Science* **75**,
318–320.

Rees, C. W. (1933). The experimental transmission of
anaplasmosis by *Dermacentor andersoni*. *Parasitology* **21**,
569–573.

Rees, C. W. (1934). *Transmission of Anaplasmosis by Various
Species of Ticks*, US Department of Agriculture Technical
Bulletin No. 418. Washington, DC: US Government
Printing Office.

Rees, C. W. & Avery, J. L. (1939). Experiments on the
hereditary transmission of anaplasmosis by ticks. *North
American Veterinarian* **20**, 35–36.

Richey, E. J. (1981). Bovine anaplasmosis. In *Current
Veterinary Therapy: Food Animal Practice*, ed. Howard, R. J.,
pp. 767–772. Philadelphia, PA: W. B. Saunders.

Ristic, M. (1968). Anaplasmosis. In *Infectious Blood Diseases of Man and Animals*, eds. Weinman, D. & Ristic, M., pp. 478–542. New York: Academic Press.

Rodriguez Camarilla, S. D., Garcia Ortiz, M. A., Hernández Salgado, G., *et al.* (2000). *Anaplasma marginale* inactivated vaccine: dose titration against a homologous challenge. *Comparative Immunology and Microbiology of Infectious Diseases* 23, 239–252.

Rosenbusch, F. & Gonzalez, R. (1927). Die Tristeza Übertragung durch Zecken und dessen Immunitätsprobleme. *Archiv für Protistenkunde* 58, 300–320.

Rozeboom, L. E., Stiles, G. W. & Moe, L. H. (1940). Anaplasmosis transmission by *Dermacentor andersoni* Stiles. *Journal of Parasitology* 26, 95–100.

Rurangirwa, R. T., Stiller, D., French, D. M. & Palmer, G. H. (1999). Restriction of major surface protein 2 (MSP2) variants during tick transmission of the ehrlichia *Anaplasma marginale*. *Proceedings of the National Academy of Sciences of the USA* 96, 3171–3176.

Rurangirwa, F. R., Stiller, D. & Palmer, G. H. (2000). Strain diversity in major surface protein 2 expression during tick transmission of *Anaplasma marginale*. *Infection and Immunity* 68, 3023–3027.

Saliki, J. T., Blouin, E. F., Rodgers, S. J. & Kocan, K. M. (1998). Use of tick cell culture-derived *Anaplasma marginale* antigen in a competitive ELISA for serodiagnosis of anaplasmosis. *Annals of the New York Academy of Science* 849, 273–281.

Samish, M., Pipano, E. & Hadani, A. (1993). Intrastadial and interstadial transmission of *Anaplasma marginale* by *Boophilus annulatus* ticks in cattle. *American Journal of Veterinary Research* 54, 411–414.

Sanborn, C. E. & Moe, L. H. (1934). Anaplasmosis investigations. In *Report of the Oklahoma Agricultural and Mining College and Agricultural Experiment Station, 1932–1934*, pp. 275–279.

Sanborn, C. E., Stiles, G. W. & Moe, L. H. (1938). Anaplasmosis transmission by naturally infected *Dermacentor andersoni* male and female ticks. *North American Veterinarian* 19, 31–32.

Sanders, D. A. (1933). Notes on the experimental transmission of bovine anaplasmosis in Florida. *Journal of the American Veterinary Medical Association* 88, 799–805.

Schmidt, H. & Piercy, P. L. (1937). In *Fiftieth Annual Report*, Texas Agricultural Experiment Station.

Scoles, G. A., Broce, A. B., Lysyk, T. J. & Palmer, G. H. (2005a). Relative efficiency of biological transmission of *Anaplasma marginale* (Rickettsiales: Anaplasmataceae) by *Dermacentor andersoni* (Acari: Ixodidae) compared with mechanical transmission by *Stomoxys calcitrans* (Diptera: Muscidae). *Journal of Medical Entomology* 42, 668–675.

Scoles, G. A., Ueti, M. W. & Palmer, G. H. (2005b). Variation among geographically separated populations of *Dermacentor andersoni* (Acari: Ixodidae) in midgut susceptibility to *Anaplasma marginale* (Rickettsiales: Anaplasmataceae). *Journal of Medical Entomology* 42, 153–162.

Sergent, E., Dontien, A., Parrot, L. & Lestoquard, F. (In Memoriam): (1945). Etudes sur les piroplasmoses bovines. *Institut Pasteur d'Algérie*, p. 816.

Shkap, V., Molad, T, Fish, L. & Palmer, G. H. (2002). Detection of the *Anaplasma centrale* vaccine strain and specific differentiation from *Anaplasma marginale* in vaccinated and infected cattle. *Parasitology Research* 88, 546–552.

Smith, R., Levy, M. G., Kuhlenschmidt, M. S., *et al.* (1986). Isolate of *Anaplasma marginale* not transmitted by ticks. *American Journal of Veterinary Research* 47, 127–129.

Stich, R. W., Kocan, K. M., Palmer, G. H., *et al.* (1989). Transstadial and attempted transovarial transmission of *Anaplasma marginale* Theiler by *Dermacentor variabilis* (Say). *American Journal of Veterinary Research* 50, 1386–1391.

Stiller, D. & Johnson, L. W. (1983). Experimental transmission of *Anaplasma marginale* Theiler by adults of *Dermacentor albipictus* (Packard) and *Dermacentor occidentalis* Marx (Acari: Ixodidae). In *Proceedings of the 87th Annual Meeting of the US Animal Health Association*, pp. 59–65.

Stiller, D., Crosbie, P. R., Boyce, W. M. & Goff, W. E. (1999). *Dermacentor hunteri* (Acari: Ixodidae): experimental vector of *Anaplasma marginale* and *A. ovis* (Rickettsiales: Anaplasmataceae) in calves and sheep. *Journal of Medical Entomology* 36, 321–324.

Stiller, D., Leatch, G. & Kuttler, K. (1981). Experimental transmission of bovine anaplasmosis by the winter tick, *Dermacentor albipictus* (Packard). In *Proceedings of the National Anaplasmosis Conference*, pp. 463–475.

Theiler, A. (1911). Further investigations into anaplasmosis of South African cattle. In *1st Report of the Director of Veterinary Research*, pp. 7–46. Department of Agriculture of the Union of South Africa.

Theiler, A. (1912). Übertragung der Anaplasmosis mittels Zecken. *Zeitschrift für Infektionskrankheiten* 12, 105–116.

Viseshakul, N., Kamper, S., Bowie, M. V. & Barbet, A. F. (2000). Sequence and expression analysis of a surface antigen gene family of the rickettsia *Anaplasma marginale*. *Gene* **253**, 45–53.

Visser, E. S., McGuire, T. C., Palmer, G. H., *et al.* (1992). The *Anaplasma marginale msp5* gene encodes a 19-kilodalton protein conserved in all recognized *Anaplasma* species. *Infection and Immunity* **60**, 5139–5144.

Wickwire, K. B., Kocan, K. M., Barron, S. J., *et al.* (1987). Infectivity of three *Anaplasma marginale* isolates for *Dermacentor andersoni*. *American Journal of Veterinary Research* **48**, 96–99.

Wikel, S. K., Ramachandra, R. N., Bergman, D. K., Burkot, T. R. & Piesman, J. (1997). Infestation with pathogen-free nymphs of the tick *Ixodes scapularis* induces host resistance to transmission of *Borrelia burgdorferi* by ticks. *Infection and Immunity* **65**, 335–338.

Zeller, H. & Helm, R. (1923). Versuche zur Frage der Übertragbarkeit des Texasfiebers auf deutsche Rinder durch die bei uns vorkommenden Zecken *Ixodes ricinus und Haemaphysalis punctata* Cinabarina. *Berliner tierärztlich Wochenschrift* **39**, 1–4.

16 • Emerging and emergent tick-borne infections

S. R. TELFORD III AND H. K. GOETHERT

INTRODUCTION

Human activities continue to change the landscape vastly, altering faunal associations and thereby contact with arthropod vectors, producing circumstances that serve as the basis for the emergence of a vector-borne infection. However, few 'emerging' tick-borne infections are novel. Many (ehrlichiosis, babesiosis) have long been recognized as veterinary health problems. Some rickettsioses may be due to agents that were once thought to be tick endosymbionts. Others, such as the agents of bartonellosis, may form paratenic or dead-end associations with ticks. Some recently identified agents (deer tick virus, *Borrelia lonestari*) are 'in search of an emerging disease'. Emergent epidemiological associations (Masters' disease) are in search of an agent. Finally, apparently well-characterized tick-borne infections, such as Rocky Mountain spotted fever, tularaemia and tick-borne encephalitis, remain neglected by researchers but retain the potential for resurgence. We briefly review the diversity of these infectious agents, identify aetiological enigmas that remain to be solved, and provide a reminder about 'old friends' that should not be forgotten in our pursuit of novelty. We suggest that newly recognized agents or tick–pathogen associations receive careful scrutiny before being declared as potential public health burdens.

REDISCOVERED, BETTER CHARACTERIZED, OR NEW?

Modern approaches to identifying and characterizing infectious agents, using nucleic acid amplification and molecular phylogenetic algorithms, are very powerful (Relman, 2002). However, there are fallacious assumptions that: (1) data accumulated by older ('classical') methods are not as precise and thus not to be trusted; and (2) a DNA or RNA sequence represents something novel if it does not match one that is already present in GenBank or other genetic information databases. The corollary of these fallacies is the idea that we really do not know what those species that have been previously described really represent because genetic information is not available. For example, one might argue that we cannot be sure that Theobald Smith's *Pyrosoma bigeminum* (Smith & Kilborne, 1893) is indeed identical to what we see today as *Babesia bigemina* because his microscope slides no longer exist and we cannot extract DNA from the material for confirmation!

To a certain extent, these fallacies have been formalized by the approach taken by bacteriologists. In 1980, a list of established bacterial taxa was compiled. Names were retained if the reviewing authority thought enough information existed to demonstrate identity. Thus, a bacterial entity that is reported today might be considered new because it was not found in the list of accepted taxa, although it may have been previously described. In contrast, the International Code of Zoological Nomenclature that regulates the taxonomy of parasitic protozoa (and helminths) and their animal hosts, continues to insist that the burden of proof for novelty rests on an exhaustive search for prior descriptions. Viral taxa are evaluated individually as they are identified by appropriate subcommittees within the International Committee on the Taxonomy of Viruses (ICTV) and, more relevant to the field of vector-borne infections, by the American Committee on Arboviruses (ACAV), which maintains a list of registered viral names. Viral taxa that have not been evaluated may be included in the database of registered names but designated as such (in the International Catalogue of Arboviruses). The issue of 'new' agent as opposed to a previously recognized agent that has now been molecularly characterized is more than an historic interest. Much work on the life history or epidemiological significance of a 'new' agent may have already been published. Accordingly, every effort needs to be taken to match a newly

Ticks: Biology, Disease and Control, ed. Alan S. Bowman and Patricia A. Nuttall. Published by Cambridge University Press.
© Cambridge University Press 2008.

recognized entity with one that may have been identified in the past.

It is clear that some changes need to be made to take full advantage of the molecular and genetic methods of identifying pathogenic agents. Polymerase chain reaction sequencing can accomplish in a day what classical bacteriologists might need weeks or months to do. However, PCR use may reduce efforts to achieve a comprehensive understanding of a taxon. Bacteriologists, for example, are now tending to take the 'Candidatus' approach when reporting a potentially new species. The category 'Candidatus' is used to describe prokaryotes for which a formal description, as required by the Bacteriological Code, is not available (Murray & Schleifer, 1994). The name has no nomenclatural standing but is meant to provide a provisional designation to agents that are not yet cultivated or non-cultivatable. Unfortunately, 'Candidatus' is now being used to report DNA sequences that represent putative new taxa in the hope of establishing nomenclatural priority in the absence of attempts to acquire additional information.

The ehrlichioses, rickettsioses and babesiosis represent the full spectrum of 'emerging' tick-borne infections – old, new and rediscovered. With all three, molecular phylogenetic analysis has greatly expanded our understanding of the diversity of the possible aetiological agents. In the following section, we attempt to relate new findings to historical information and identify research needs. We focus on zoonotic infections, although the principles that serve as the basis for this review certainly apply to emerging tick-borne infections of veterinary importance.

Ehrlichiosis

Lyme disease stimulated a renaissance in tick-borne pathogen research in the United States and elsewhere in the world during the 1980s and 1990s. The human ehrlichioses were recognized and emerged as a public health burden during that time in the United States. Three clinically similar, acute-onset febrile illnesses comprising headache, myalgia, rigors and malaise are now known to be caused by tick-transmitted monocytotropic (*Ehrlichia chaffeensis*) and granulocytotropic (*Anaplasma phagocytophilum* and *E. ewingii*) agents closely related to the rickettsiae. Three epidemiological patterns are apparent in human ehrlichiosis in the United States. Human monocytic ehrlichiosis (HME) is due to exposure to Lone Star ticks (*Amblyomma americanum*) and infection by *E. chaffeensis*. Human granulocytic ehrlichiosis (HGE) in the northeastern and northern mid-western United States is associated with exposure to deer ticks (*Ixodes dammini*) and *A. phagocytophilum* or to western blacklegged ticks (*I. pacificus*) and *A. phagocytophilum*. A third epidemiological pattern comprises exposure to either dog ticks (*Dermacentor variabilis*) or Lone Star ticks (*A. americanum*) and infection by *E. ewingii*, previously described as the agent of canine granulocytic ehrlichiosis. A fourth pattern may emerge in large areas of the world with exposure to brown dog ticks (*Rhipicephalus sanguineus*) and infection by *E. canis*. A chronic case of ehrlichiosis due to *E. canis* was described for a Venezuelan patient in 1996; subsequently 20 other cases were identified, a third of which were confirmed by PCR (Perez, Rikihisa & Wen, 1996; Perez *et al.*, 2006).

The spectrum of illness in humans for all three ehrlichioses ranges from completely asymptomatic to fatal, but most individuals experience an acute febrile illness of a week's duration that may spontaneously resolve. Acute HGE infection has been characterized as 'spotless Rocky Mountain Spotted Fever'. Fever ($>37.5\,^{\circ}$C), malaise, rigors, myalgia, sweats and headache are almost universally reported (Bakken *et al.*, 1996b). About a third of cases exhibit nausea, anorexia, arthralgia and cough. Confusion, prostration, diarrhoea, pneumonia and vertigo are less frequently reported; rash is rarely present. Chronic infections or sequelae have not yet been described, although weakness or fatigue appear to persist for as long as 1 month (Bakken *et al.*, 1996b). Tick-borne fever of sheep due to *A. phagocytophilum* is thought to persist for as long as two years if untreated (Foggie, 1951) and is associated with profound immunosuppression (Woldehiwet & Scott, 1993). Human fatalities due to ehrlichiosis appear to be associated with bacterial or fungal secondary infections (Dumler & Bakken, 1995).

Human monocytic ehrlichiosis caused by *E. chaffeensis* was probably first described as Bullis fever (Anigstein & Anigstein, 1975), a well-characterized clinical entity of presumed rickettsial aetiology that had disappeared from the *Merck Manual* and other medical textbooks by the 1950s. Some 750 cases of HME, with eight deaths, were reported by state health departments between 1986 and 1997 (McQuiston *et al.*, 1999). Prospective studies of HME incidence in Army reservists active in Oklahoma and Georgia suggest that it is as common in these states as Rocky Mountain spotted fever, with some five cases per 100 000 (Fishbein *et al.*, 1989; Eng *et al.*, 1990). Deer serve as reservoirs for *E. chaffeensis* (Dawson *et al.*, 1994) and the vector is the Lone Star tick. Incrimination of the vector and reservoir of the agent of HME was based upon laboratory transmission experiments (Ewing *et al.*, 1995) complementing field observations of infection in ticks and deer (Lockhart *et al.*, 1995, 1997*a*, 1997*b*). Deer serve as hosts for all three active stages of the Lone Star tick (larvae, nymphs and adults) although the immatures (larvae, nymphs) may also feed on larger birds, rabbits or raccoons. Prevalence of infection in host-seeking ticks appears to be low, of the order of 1% for either nymphal or adult ticks (Anderson *et al.*, 1993; Murphy *et al.*, 1998). All three stages aggressively attack humans; only nymphs and adults may transmit infection. Larvae – known in many places as 'seed ticks' – may cause numerous (dozens or hundreds of bites from contact with a larval cluster) granulomatous, pruritic lesions.

Human serological reactivity to *E. chaffeensis* has been reported from numerous countries, including Argentina, Mexico, Belgium, Italy, the Netherlands, Portugal, Israel, Burkina Faso, Mozambique, Korea and Thailand (Morais *et al.*, 1991; Pierard *et al.*, 1995; Heppner *et al.*, 1997; Nuti *et al.*, 1998; Gongora-Biachi *et al.*, 1999; Keysary *et al.*, 1999; Ripoll *et al.*, 1999; Groen *et al.*, 2002; Heo *et al.*, 2002). Such reports may reflect exposure to a closely related *Ehrlichia* inasmuch as the only known vector, *A. americanum*, is not present in these countries. It may be that such seroreactivity represents exposure to newly recognized ehrlichiae such as *E. muris* (Kawahara *et al.*, 1999), the 'Anan and HF strains' (Shibata *et al.*, 2000) or the 'Schotti variant' (Schouls *et al.*, 1999; probably the same as 'Candidatus Neoehrlichia shimanensis', Kawahara *et al.*, 2006), all detected initially by PCR from rodent or tick samples; *E. muris*, HF/Anan and 'Candidatus N. shimanensis' have been isolated. 'Candidatus Ehrlichia walkerii', detected by PCR from *I. ricinus* ticks collected in Italy is also within the *Ehrlichia/Cowdria* genogroup (Brouqui *et al.*, 2003) but no information about its biology is available. The actual zoonotic relevance of these diverse agents remains speculative pending their detection in humans. On the other hand, a report of the detection of *E. chaffeensis* DNA in Chinese ticks (Cao *et al.*, 2000*a*) suggests HME is not restricted to the United States, nor to transmission by *A. americanum*. Further work is required to determine the geographical distribution of *E. chaffeensis* and alternative vectors, with careful attention to preventing PCR contamination with reference strains of *E. chaffeensis*.

Infection due to *E. ewingii*, the agent of canine granulocytic ehrlichiosis, was originally detected in dogs in Oklahoma (Ewing & Philip, 1966), but its distribution probably includes much of the southern and central United States. Several cases of human ehrlichiosis caused by *E. ewingii* have been reported from Tennessee, Missouri and Oklahoma, mostly in immunocompromised patients (Buller *et al.*, 1999;

Paddock *et al.*, 2001). The Lone Star tick (*A. americanum*) was incriminated as the vector by experimental transmission between dogs. *Ehrlichia ewingii* DNA has been detected in *Rhipicephalus sanguineus* and *Dermacentor variabilis* from Oklahoma, but their competence as vectors remains unconfirmed (Murphy *et al.*, 1998). Deer support *E. ewingii* infection and are frequently infected, probably because they serve as important hosts for all stages of Lone Star ticks (Yabsley *et al.*, 2002). Because this infection was first recognized in dogs, it seems likely that wild canids (foxes and coyotes) may be involved in the transmission cycle, given their frequent infestation with Lone Star ticks.

The agent of human granulocytic ehrlichiosis (HGE), *A. phagocytophilum*, was first studied in the 1930s as the cause of tick-borne fever of sheep and cattle. A large literature pertains to it as an animal health problem (Woldehiwet & Scott, 1993). For comments on the nomenclature of this agent see Telford & Goethert (2004).

More recently, the rickettsiae have undergone a taxonomic revision based upon 16S rDNA and *groEL* gene sequencing, with ehrlichiae being reclassified within the Anaplasmataceae, distinct from the Rickettsiaceae (Dumler *et al.*, 2001). The agents of tick-borne fever, bovine ehrlichiosis (formerly *E. bovis*) and canine cyclic thrombocytopenia (formerly *E. platys*) are now placed within the genus *Anaplasma*. However, the phylogenetic analysis based on 16S rDNA and *groEL* gene sequences clearly confirms that there are at least four major clades of ehrlichia-like agents (Dumler *et al.*, 2001), a conclusion that is generally accepted based on morphology, life cycle and other criteria. A more conservative taxonomic approach would retain *Anaplasma* and *Cytoecetes* as distinct entities (Fig. 16.1).

An examination of other phylogenetically informative genetic targets (Inokuma *et al.*, 2001), as well as other described entities (see Telford & Goethert, 2004) may stimulate yet another nomenclatural revision in the future as DNA from these agents is collected and analysed. Each additional taxon provides an independent comparison to assess the strengths of the tree topology generated by the existing phylogenetic algorithms (Tenter *et al.*, 2002). Indeed, a recent analysis suggests that there is merit in distinguishing *Cytoecetes* (= *A. phagocytophilum*) from *Anaplasma marginale* and *A. centrale* (Brouqui & Matsumoto, 2007). Pending additional studies with sequences that have been deposited in GenBank since the publication of Dumler *et al.* (2001), HGE (the infection) should continue to be called ehrlichiosis, not human granulocytic anaplasmosis. The name anaplasmosis has been used for more than half a century in the veterinary

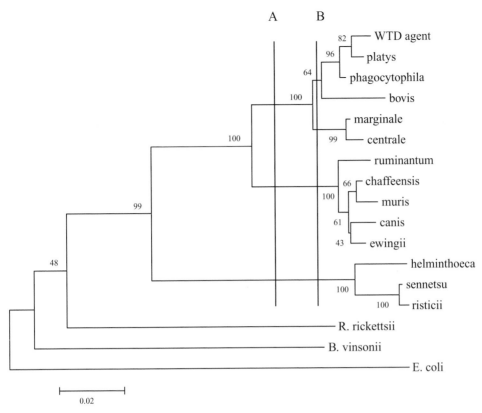

Fig. 16.1. 16S rDNA phylogenetic tree (neighbour-joining algorithm) of ehrlichia-like taxa showing two schemes for making taxonomic decisions. A, as recommended by Dumler *et al.* (2001) subsuming *Cytoecetes* with *Anaplasma*. B, conserving *Cytoecetes*. *Cowdria* could also be retained by moving the cut-off farther to the right, and placing the former *E. bovis* into its own clade; the tropism of *Cowdria* for endothelial cells clearly distinguishes it from the rest of the *Ehrlichia* sensu stricto.

sciences to denote a specific haemolytic disease of ruminants; HGE has no haemolytic component and clinically is difficult to distinguish from the other human ehrlichioses (due to *E. chaffeensis* and *E. ewingii*).

Nearly 500 HGE cases were reported by United States state health departments between 1994 and 1997 with intensive surveillance in Lyme disease endemic sites (McQuiston *et al.*, 1999). Three deaths were reported. The distribution of HGE closely parallels that of Lyme disease. Cases of HGE have been identified in virtually all of the eastern United States. In addition, cases of HGE have been described from northern California, where the western blacklegged tick (*Ixodes pacificus*) serves as the vector (Richter *et al.*, 1996). HGE cases (identified by seroconversion or PCR detection of *A. phagocytophilum*) have been reported from much of Europe, including Belgium, Denmark, Sweden, Slovenia, Spain, and Italy (Petrovec *et al.*, 1997; Lebech *et al.*, 1998;

Nuti *et al.*, 1998; Oteo *et al.*, 2000; Bjoersdorff *et al.*, 2002; Guillaume *et al.*, 2002). Residents of Germany, Norway, United Kingdom, Poland, Israel, and Korea, among others, are sero-reactive (Sumption *et al.*, 1995; Bakken *et al.*, 1996a; Fingerle *et al.*, 1997; Keysary *et al.*, 1999; Heo *et al.*, 2002; Grzeszczuk *et al.*, 2004). Infected ticks or rodents have been reported from virtually all of the countries that have reported human exposure or infection (Cao *et al.*, 2000b; Alekseev *et al.*, 2001; Telford *et al.*, 2002; Kim *et al.*, 2003; Makinen *et al.*, 2003; Skotarzak *et al.*, 2003; Hartelt *et al.*, 2004; Polin *et al.*, 2004; Santos *et al.*, 2004; Ohashi *et al.*, 2005). It seems likely that the potential distribution of HGE matches that of the *I. persulcatus* complex of ticks.

The ecology of HGE is relatively well understood in the eastern United States because the agent tends to share the same white-footed mouse reservoir and deer tick vector as do the aetiological agents of Lyme disease and babesiosis (see Chapter 11). A variety of other reservoir hosts may locally contribute to the force of transmission, including mice, voles, chipmunks, deer or sheep (Tyzzer, 1938; Foggie, 1951; Telford *et al.*, 1996; Belongia *et al.*, 1997; Walls *et al.*, 1997). Under experimental conditions, reservoir infectivity appears to be transient, lasting about 2 weeks (Levin & Fish, 2000), in sharp contrast to *B. burgdorferi* and *B. microti*, the reservoirs of which appear to be infective to the vector for the duration of their life. On the other hand, xenodiagnosis of mice trapped in the autumn revealed that a third of them were infectious to ticks (Telford *et al.*, 1996). The long history of tick-borne fever investigations in the UK neglected the role of reservoirs other than sheep or cattle (Woldehiwet & Scott, 1993), and only recently has *A. phagocytophilum* infection been detected in small rodents there (Ogden *et al.*, 1998; Bown *et al.*, 2003). As with HGE in the United States, the ecology of tick-borne fever may be found to parallel that of the agent of Lyme borreliosis. Indeed, evidence is mounting that, as with *B. burgdorferi* sensu lato, *A. phagocytophilum* may comprise a group of genospecies. Single nucleotide polymorphisms (SNPs) within the 16S rDNA and *groEL* genes of *A. phagocytophilum* seem to associate with strains taken from particular hosts (Massung *et al.*, 2002). European strains appear genetically heterogeneous (Stuen *et al.*, 2002). The infection kinetics of the agent of equine granulocytic ehrlichiosis ('*Ehrlichia equi*') within laboratory mice differs from prototypical HGE, as does its appearance in Giemsa-stained blood smears (author's unpublished observations). The public health significance of members of a putative *A. phagocytophilum* sensu lato may vary greatly.

As with the reports of diverse *Ehrlichia* genogroup agents (see below), we may only speculate as to the public health significance of two members of *Anaplasma* that are frequently detected in ticks. *Anaplasma bovis* (= *E. bovis*) has been reported from diverse metastriate ticks and causes a mild disease of cattle (Donatien & Lestoquard, 1936). Interestingly, it appears to cause fever when inoculated into monkeys (Donatien & Lestoquard, 1940). *Anaplasma bovis* has been detected in cottontail rabbits and their ticks (Goethert & Telford, 2003c) in the northeastern United States, and is increasingly being found in Eurasian tick surveys (e.g. Kim *et al.*, 2003; Parola *et al.*, 2003; Kawahara *et al.*, 2006). Because of its perpetuation in rabbits and cattle across a wide geographical distribution, human exposure might be expected. Similarly, the 'white-tailed deer agent' (Little *et al.*, 1997), which appears to be co-transmitted in the same Lone Star tick–deer system as *E. chaffeensis* (Yabsley *et al.*, 2002),

would be expected to be present in large areas of the eastern United States and infecting an aggressive human-biting tick. However, to date there have been no reports of human infection with either of these ehrlichia-like agents.

Rickettsiosis

Local variants of spotted fever group rickettsioses are increasingly being recognized. Since 1991, 12 'new' infections have been described including Flinders Island spotted fever, Astrakhan fever, African tick-bite fever and Oriental spotted fever (Parola & Raoult, 2000; Parola *et al.*, 2005). Thus, there are 19 established rickettsioses (Table 16.1). Rickettsiae that have previously been characterized as 'harmless' arthropod endosymbionts have now been incriminated as important human pathogens, such as *R. slovaca*, *R. helvetica*, *R. parkeri* and *R. amblyommi* (Raoult *et al.*, 1996, 1997; Parola *et al.*, 2005). Improved methods of rickettsial isolation (particularly the shell-vial centrifugation method: Vestris *et al.*, 2003) and molecular detection have greatly helped in assigning aetiology in suspected rickettsiosis cases. The polymerase chain reaction allows sensitive detection and rapid identification of rickettsiae, perhaps even within a day; previously, cumbersome micro-immunofluorescence assays using a panel of species and genotype-specific monoclonal antibodies were required to identify an isolate, which takes a week or often longer just to propagate. In addition, whereas rickettsial isolation and typing could be performed in only a handful of laboratories with the appropriate biocontainment facilities and experience, PCR may be performed by virtually any laboratory (although there are significant caveats about the reliability of the results depending on experience with the technique, particularly with respect to contamination control). On the other hand, the wide use of PCR sequencing has fostered an epidemic of 'new' rickettsioses. Responding to the potential for a burgeoning list of '*Candidatus*' species, recommendations have been proposed for classfying *Rickettsia* spp. (Fournier *et al.*, 2003a). To designate a *Rickettsia* as new, five genes (16S rRNA and four protein-coding genes) need to be sequenced and compared to those from the existing species. Similar recommendations would be helpful in guiding the taxonomy of other infectious agents.

The incrimination of *R. parkeri* as a zoonotic agent demonstrates the success of these new developments in clarifying the aetiology of rickettsioses. Although what was to later be named *R. parkeri* was recovered from *Amblyomma maculatum* ticks in 1939, it was long thought to be non-pathogenic. Sixty-five years later, a case of 'rickettsialpox' in

Table 16.1 *Old and new rickettsioses: established clinical entities (with more than a couple of described cases) for which rickettsial aetiology is established*

Infection	Distribution	Agent	Main vector[a]	Year recognized[b]	Reference
Rocky Mountain spotted fever	North America	R. rickettsii	Dermacentor andersoni, D. variabilis	1899	Maxey, 1899
Boutonneuse fever (Mediterranean spotted fever)	Mediterranean region, East, Central and southern Africa	R. conorii	Rhipicephalus sanguineus	1910	Conor & Bruch, 1910
African tick-bite fever (South African tick-bite fever)	Southern Africa	R. africae	Amblyomma hebraeum, A. variegatum	1911	Sant' Ana, 1911
Indian tick typhus	Mysore, Kashmir, other states in India?	R. conorii indica	R. sanguineus?	1921	Megaw, 1921
North Asian tick typhus (Siberian tick typhus)	'Asiatic' regions of former USSR, China	R. sibirica	Dermacentor spp., Haemaphysalis concinna	1937	Antonov & Naishtat, 1937
Queensland tick typhus	Eastern Australia	R. australis	Ixodes holocyclus	1946	Andrew, Bonnin & Williams, 1946
Israeli tick typhus	Israel	R. conori israelensis	R. sanguineus?	1974	Goldwasser et al., 1974
Astrakhan fever	Caspian Sea area	R. conori caspia	Rhipicephalus pumilio	1983 (1991)	Tarasevich et al., 1991
Japanese spotted fever	Japan	R. japonica	Dermacentor taiwanensis, Haemaphysalis spp., Ixodes ovatus	1984	Mahara, 1984
Flinders Island spotted fever	Flinders I., Australia	R. honei	Aponomma hydrosauri	1991	Stewart, 1991
Tick-borne lymphadenopathy (TIBOLA)	Southern Europe	R. slovaca	Dermacentor marginatus	1997	Raoult et al., 1997
Far Eastern spotted fever	Asia	R. heilongjiangensis	Dermacentor silvarum	1992	Fan et al., 1999
Australian spotted fever	Australia	'R. marmioni'	Haemaphysalis novaeguineae	2003	Unsworth et al., 2005
No designation	Africa (Morocco, South Africa)	R. aeschlimannii	Hyalomma spp.	2002	Raoult et al., 2002
No designation	Mediterranean	R. massilae	Rhipicephalus spp.	2005	Vitale et al., 2005
No designation	Eastern and southern USA	R. parkeri	Amblyomma maculatum	2004	Paddock et al., 2004
No designation	Eastern USA	R. amblyommi	Amblyomma americanum	1993	Dasch et al., 1993
No designation	Europe, Asia	R. helvetica	Ixodes ricinus, I. ovatus, I. persulcatus	2000	Fournier et al., 2000
No designation	Western USA	R. canadensis	Haemaphysalis leporispalustris?	1970	Bozeman et al., 1970

[a] Question marks refer to probable but not yet demonstrated.
[b] Date when first recognized refers to first published clinical report for the infection; the agent may have been incriminated much later in time.

a southeastern Virgina patient was determined to be due to *R. parkeri*. The agent was successfully isolated in cell culture from a biopsy of the eschar (a characteristic hard plaque covering a dermal ulcer) and identified by PCR sequencing (Paddock *et al.*, 2004). The team that reported this landmark case suggested that the unusual presentation of the case (with multiple eschars, probably reflecting infectious bites by many larval ticks) could explain literature reports of atypical presumed Rocky Mountain spotted fever, suggesting that *R. parkeri* rickettsiosis has occurred and been misidentified. Of note is the report of boutonneuse fever-like infections in Uruguay, where *R. parkeri* has been identified in *Amblyomma* spp. (Venzal *et al.*, 2004), which would indicate that *R. parkeri* rickettsiosis may occur through much of Latin America and the southern United States.

At least two recently recognized rickettsioses appear to be significant public health burdens, with hundreds of cases having been reported (Astrakhan fever) or potentially being transmitted in areas where reporting is poor (African tick-bite fever). Two others (Japanese spotted fever and TIBOLA) may be recognized as more widely distributed given their vector associations. Astrakhan fever is apparently a new syndrome, not having been reported in the comprehensive volume on rickettsial diseases by Russian workers (Zdrodowski & Golinevich, 1960). It was first recognized in 1970 in patients visiting the Caspian Sea area and exposed to *Rhipicephalus pumilio* ticks. A total of 321 cases were reported during active surveillance from 1983 to 1989 (Tarasevich *et al.*, 1991). High fevers, headache, myalgias and non-petechial rash were common findings; only a quarter of the patients had an eschar. The illness appears to run a relatively benign course and no fatalities have been reported. The agent is now considered to be a subspecies of *R. conorii* (Shypnov *et al.*, 2003) although the original analyses suggested that it was distinct from Mediterranean spotted fever (Ereemeva *et al.*, 1994). The enzootic cycle has not yet been described.

Japanese (Oriental) spotted fever also appears to be a new syndrome, although it is possible that it may have been confused with scrub typhus (caused by *Orientia tsutsugamushi*), which was well known by Japanese physicians in the 1920s (Kawamura, 1926). Febrile illnesses with exanthem (a sudden rash) and eschar would have been readily detected. Indeed, Japanese spotted fever was first identified in May–July 1984 in Tokushima prefecture when three patients were diagnosed with presumptive scrub typhus. Scrub typhus cases tend to be found during the autumn and winter in Japan, and the summer presentation served

to suggest a different aetiology. Serum from these patients failed to react as expected in Weil–Felix tests, showing agglutinins against Proteus OX2, which suggested a spotted fever-like infection. The agent was isolated in 1986, and described as *R. japonica* (Uchida, 1993). Patients presented with headache, fever, rigors and a rash that became petechial. Virtually all had eschar and about a third recalled a tick bite (Mahara, 1984). From 1984 to 1995, 144 cases of Japanese spotted fever were reported from southwestern and central Japan, with transmission occuring mainly during the summer (Mahara, 1997). Rickettsiae were detected by immunofluorescence and by *R. japonica*-specific PCR in haemolymph taken from the human-biting ticks *Haemaphysalis flava*, *H. longicornis* and *Ixodes ovatus*, as well as from *H. formosensis*, *H. hystricis* and *Dermacentor taiwanensis*. Vector competence studies have not been performed with these ticks, however, so their relative vectorial capacity remains unknown. The enzootic cycle remains undescribed. *Haemaphysalis hystricis* and *D. taiwanensis* occur in large portions of southern China and Indochina; accordingly, evidence of *R. japonica* infection should be sought wherever these ticks are common.

African tick-bite fever was originally described by Pijper in the 1930s from South African patients. Based on guinea pig cross-protection studies, Pijper (1936) suggested the existence of two tick-transmitted rickettsioses there, one now known as boutonneuse fever caused by *R. conorii* and exposure to *R. sanguineus* ticks, and a milder febrile illness due to exposure to *Amblyomma* ticks. The agent was thought to be a variant of *R. conorii* (*Dermacentroxenus rickettsii* var. *pijperi* (Mason & Alexander, 1939)). Subsequent work (Gear, 1954) refuted the existence of another rickettsiosis in southern Africa, and until the 1990s (Kelly *et al.*, 1991), all cases there were attributed to infection by *R. conorii* and associated with bites of subadult bont ticks (*A. hebraeum*). In 1992, an isolate from a Zimbabwean patient was demonstrated to be identical to one previously isolated from Ethiopian *Amblyomma* spp. (Philip *et al.*, 1966), distinct from *R. conorii*, and was named *R. africae* (Kelly *et al.*, 1996). Much is known about the ecology of the bont tick due to its importance as a vector of heartwater and theileriosis (e.g. Norval, 1977*a*, 1977*b*). Reproductive hosts are large ungulates such as giraffe, rhinoceros, buffalo and cattle; subadults infest a wide range of hosts including all sizes of mammals, birds and reptiles. It seems likely that *R. africae* infection is very common inasmuch as *A. hebraeum* frequently bites humans in southern Africa from Zimbabwe southwards. In 1992, 23% of 169 American soldiers who had participated in a 10-day training exercise in Zimbabwe were infected, as demonstrated by seroconversion to antigens of *R. africae* (Broadhurst *et al.*, 1998). Fever, chills, headache, myalgias, fatigue and lymphadenitis were commonly reported. Nearly all of the infected soldiers had an eschar and a third had more than one eschar, suggesting exposure to multiple bites, probably of subadult ticks. In addition, given the popularity of safaris in southern Africa, African tick-bite fever may be a frequent febrile illness of tourists (Brouqui *et al.*, 1997). *Rickettsia africae* has been detected from *A. variegatum* (Parola *et al.*, 1999), the tropical bont tick, which was introduced from Senegal into the Caribbean island of Guadeloupe in 1828 and spread to much of the West Indies. Thus, 'African' tick-bite fever might be acquired on a Caribbean vacation. In addition, bites by *Hyalomma* spp. may transmit *R. aeschlimanni* (Pretorius & Birtles, 2002) and thus tick-bite fevers in Africa may result from exposure to diverse ticks.

TIBOLA (for tick-borne lymphadenopathy) due to infection by *Rickettsia slovaca* and associated with the bites of adult *Dermacentor marginatus* ticks, seems likely to become increasingly recognized as a public health burden. Rickettsiae first isolated in 1969 from *D. marginatus* (Brezina, Rehacek & Majerska, 1969) were recognized as distinct based on microagglutination and complement-fixation typing experiments, and the name *R. slovaca* was applied (Urvolgyi & Brezina, 1976). Although human exposure was suspected given the biting habits of *D. marginatus*, the role of *R. slovaca* as the cause of illness was only recently described (Raoult *et al.*, 1997). Owing to surveillance for Lyme borreliosis, a series of 86 Hungarian case-patients was described with a distinct syndrome that included a prominent crusting eschar, lymphadenopathy and fever (Lakos & Raoult, 1999). Virtually all cases were bitten by a large tick which attached on the scalp. Two-thirds of the cases were in children younger than 10 years. An eschar, often surrounded by an erythaematous halo, appears to give rise to chronic alopecia at that site. As opposed to Mediterranean spotted fever, a prominent lymphadenopathy (usually occipital or cervical when the bite is on the scalp) accompanies low-grade fever, fatigue, headache and myalgias. Of 13 patients from whom skin or lymph node biopsies were analysed by PCR, 10 were confirmed to have been infected by *R. slovaca*. A similar series of patients was also described from Spain (Oteo & Ibarra, 2002). Interestingly, *R. slovaca* may not be the only rickettsia associated with TIBOLA, as other spotted fever group rickettsia have been found in patients with similar symptoms (Ibarra *et al.*, 2005). *Dermacentor marginatus* is a common tick of lowland steppes and grassland. The

reproductive hosts include a wide variety of ungulates, carnivores and even hares and hedgehogs; subadults feed on insectivores, rodents and mustelids (Pomerantsev, 1959). Its geographical distribution is wide, from central Asia, western Siberia and the Crimea into the Balkans, through southern Europe into Spain and Portugal, and including the Mediterranean islands. Previously this tick was considered a main vector of equine piroplasmosis as well as North Asian tick typhus (due to *R. sibirica*). Given its wide distribution, its generalist feeding habits and its propensity to bite humans, infections transmitted by *D. marginatus* would seem to have a potential for great prevalence.

Flinders Island spotted fever, caused by *R. honei*, poses a zoogeographical enigma. Between 1974 and 1991, the sole physician on this small island off of the southeastern Australian coast identified 26 cases of a spotted fever-like infection (Stewart, 1991). Patients presented with sudden onset of fever, a maculopapular rash with large individual macules, and arthralgia. Serological studies confirmed that Flinders Island spotted fever patients' sera reacted with antigens of the spotted fever group rickettsiae (Graves *et al.*, 1991). Subsequently, isolates were obtained from patients and DNA sequencing of the isolates demonstrated only a distant relationship to *R. australis*, the agent of Queensland tick typhus, which was expected to be closely related. In fact, the most similar sequences were from TT-118, a *Rickettsia* species recovered from a pool of subadult *Ixodes* sp. and *Rhipicephalus* sp. taken from *Rattus rattus* in Thailand (Robertson & Wisseman, 1973). In 1996, the same agent was detected by PCR in adult *Amblyomma cajennense* collected from Texas cattle (Billings *et al.*, 1998*b*). The Flinders Island agent was named *Rickettsia honei* based on the sequencing of four genes as well as serological typing (Stenos *et al.*, 1998). On Flinders Island, the reptile-feeding *Aponomma hydrosauri* is commonly infected by *R. honei* and appears to maintain the agent vertically; its lizard hosts may share burrows with mutton birds, a northern migrant (Graves & Stenos, 2003). Apparently, *A. hydrosauri* will attach to humans and is the vector of Flinders Island spotted fever (Graves & Stenos, 2003). Although one might attribute the presence of *R. honei* in Thailand and southeastern Australia to transport of ticks by migratory birds (such as mutton birds) within Australasian flyways, such a mechanism would be unlikely to explain its presence in Texas. Future analyses should explore whether *R. honei* is a common amblyommine tick 'endosymbiont' associated with febrile infections in humans throughout much of the tropics and subtropics.

Babesiosis

Human babesiosis is increasingly being recognised as a widespread zoonosis due to diverse agents (Telford & Maguire, 2005). The seminal event demonstrating that human babesiosis may comprise diverse organisms occurred in 1992 and stimulated a renewed interest in infection by these sporozoans. An agent designated WA-1, closely related to *B. gibsoni*, a parasite of dogs, infected a Washington State resident (Quick *et al.*, 1993). Two transfusion-related cases (donor and recipient) have since been described (Herwaldt *et al.*, 1997) from Washington, as well as a case in a California infant who had received a transfusion (Kjemtrup *et al.*, 2002). The index WA-1 case was a 41-year-old man who was not immunocompromised. His malaria-like illness was originally thought to be due to *B. microti* but his serum failed to react with antigens of that agent. The parasite was propagated by subinoculating hamsters which died within 10 days. *Babesia microti* rarely kills hamsters, and subsequent sequencing of the 18S rDNA clearly distinguished it from *B. microti* and *B. divergens* (Quick *et al.*, 1993). Comprehensive analyses by ultrastructure, serology, molecular phylogeny, host range and microscopy, along with a literature search for evidence of prior description of a similar organism led to WA-1 being formally described as *B. duncani* (Conrad *et al.*, 2006).

Intensive retrospective and prospective studies on the West Coast of the United States for *B. duncani* cases identified archived blood samples from four babesiosis cases (one fatality) involving splenectomized California residents which were reanalysed by PCR sequencing (Persing *et al.*, 1995). The causative agents' 18S rDNA differed from that from *B. duncani* and *B. microti*. Recent phylogenetic analyses incorporating larger numbers of taxa (Kjemtrup *et al.*, 2000) (Fig. 16.2) suggest that the CA-type babesia appear to be related to newly recognized infections of bighorn sheep (Thomford *et al.*, 1993). Thus, within a decade two distinctive babesial infections were identified from the western United States. The vectors and reservoirs of *B. duncani* and CA-type parasites remain undescribed despite an intensive search. It seems likely that the previously reported Californian babesiosis cases attributed to *B. equi* were actually CA-1 or *B. duncani* infections.

The causative agent of bovine redwater (*B. divergens*) in Europe is currently considered to be solely endemic in Eurasia. Three American residents have been reported to be infected with parasites that are virtually indistinguishable from *B. divergens*, raising the possibility that this agent is not restricted to the Old World. The first case terminated fatally in a Missouri resident in 1995. Parasites identified on the blood smear were indistinguishable from *B. divergens* and a 200-bp region of 18S rDNA was also identical to that of *B. divergens*. Although all evidence pointed to the diagnosis of *B. divergens*, since *B. divergens* was by definition not in North America, the agent was designated new and named MO-1 (Herwaldt *et al.*, 1996). In 2001, another *B. divergens* case occurred in a Kentucky resident. This patient was diagnosed by blood smear with typical *B. divergens* morphology (accole forms, paired pyriforms) and recovered with quinine and clindamycin treatment. Sequencing of the entire 18S rDNA demonstrated its close identity (>99% sequence similarity in the 18S rDNA gene) with the European cattle agent (Beattie, Michelson & Holman, 2002). In 2002, a splenectomized man with acute renal failure and anaemia survived a 40% parasitaemia with a similar agent (Herwaldt *et al.*, 2004) with similar morphological and molecular characteristics.

Recent work on Nantucket Island, Massachusetts has described the maintenance of this same agent (100% identical 18S rDNA sequence to DNA from the Kentucky case) among cottontail rabbits (Goethert & Telford, 2003a). Rabbit sera react with *B. divergens* (Purnell strain) antigen, more so than to the sympatric and closely related *B. odocoilei*. *Babesia divergens* is known to be transmitted transovarially by *Ixodes ricinus* in Europe (Joyner, Davies & Kendall, 1963; Donnelly & Pierce, 1975), and on Nantucket it is similarly maintained by a closely related tick, the rabbit-infesting *Ixodes dentatus*. Five years of field observations reveal that this babesia is endemic on the island (Goethert & Telford, 2003a). Relatively little information is available regarding geographical variation in genes such as 18S rDNA for *Babesia* spp. and therefore the significance of three nucleotide differences out of 1700 bp in its nucleotide sequence (between a reference *B. divergens* strain from Ireland, and the cottontail rabbit parasite) remains unclear. We therefore had conservatively referred to MO-1 and the rabbit parasites as *B. divergens* regardless of the Nearctic location and unusual host. Host range, in particular, is a particularly labile character; the first case of human babesiosis (Škrabalo & Deanovic, 1957) heralded the end of the old view that piroplasms were very host specific. The antigenic similarity of the rabbit agent and the reactivity of sera from the three human cases argued a very close relationship, if not identity, with *B. divergens*. The morphology, with characteristic accole and paired pyriform parasites also argued that the agent was *B. divergens*. However, new evidence, including its morphology and host cell range in vitro (Holman *et al.*, 2005a; Spencer

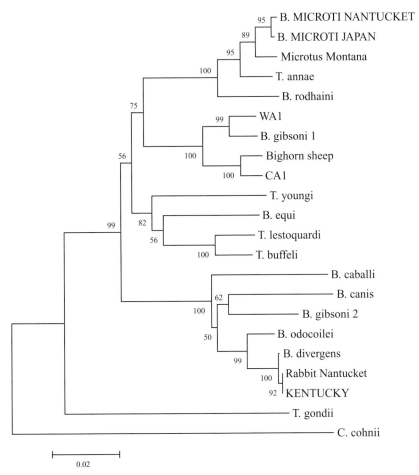

95 ┌ B. MICROTI NANTUCKET
89 └ B. MICROTI JAPAN
— Microtus Montana
— T. annae
— B. rodhaini
99 ┌ WA1
└ B. gibsoni 1
— Bighorn sheep
100 └ CA1
— T. youngi
— B. equi
— T. lestoquardi
— T. buffeli
— B. caballi
— B. canis
— B. gibsoni 2
— B. odocoilei
┌ B. divergens
├ Rabbit Nantucket
└ KENTUCKY
— T. gondii
— C. cohnii

0.02

Fig. 16.2. *Phylogeny of Babesia spp. based on 18S rDNA (neighbour-joining algorithm), indicating relatedness of known human pathogens (in capital letters).*

et al., 2006), and failure to infect cattle (Holman *et al.*, 2005*b*) strongly suggest that the rabbit parasite is distinct from *B. divergens*. Clearly, the field of emerging tick-borne infections benefits from a balance of the old with the new where careful life cycle and morphological studies complement DNA data. In this case, 18S rDNA sequencing could be used to argue for novelty or for conservatively attributing any difference to genetic variation between allopatric populations.

The dogma that babesiosis in splenectomized Europeans is due to *B. divergens* infection has been challenged (Herwaldt *et al.*, 2003). Two cases in splenectomized individuals residing in Italy were attributed to a *Babesia* species that is morphologically similar to *B. divergens* but upon sequencing of the entire 18S rDNA was demonstrated to

be most closely related to *B. odocoilei* which infects American cervids. It is likely that this parasite, dubbed 'EU-1' with implied novelty, is actually *B. capreoli* (Enigk & Friedhoff, 1962) based upon morphological and expected host attributes (both *B. capreoli* and *B. odocoilei* infect deer, and the former is transmitted by *I. ricinus*, the main human-biting tick in northern Italy). Although the suggestion is made that we do not really know which species was the infecting agent for the majority of the 22 reported clinical cases of babesiosis in Europe (Herwaldt *et al.*, 2003), many of these had been confirmed by subinoculation of cattle or jirds or review by experienced microscopists.

Even human babesiosis caused by *B. microti* appears to be more complicated than previously appreciated. Hundreds of cases have been reported from immunocompetent persons from coastal New England and upper midwestern United States sites, with a 5% case fatality rate (Telford & Maguire, 2005). Although *B. microti* is considered to have a Holarctic distribution, few cases outside the United States have been reported (van Peenen *et al.*, 1977; Shih *et al.*, 1997; Tsuji *et al.*, 2001). The lack of cases in Europe is particularly enigmatic because competent reservoir rodents have frequently been found to be infected and are sympatric with a competent vector, *I. ricinus* (Franca, 1910; Krampitz & Baumler, 1978; Healing, 1981; Bajer *et al.*, 2001; Gray *et al.*, 2002). Serological evidence of exposure to *B. microti* has been detected in tick-exposed people (Foppa *et al.*, 2002; Hunfeld *et al.*, 2002) making the absence of reports of clinical illness even more puzzling. Only very recently has a convincing case of autochthonous *B. microti* babesiosis in a German resident been reported (Hildebrandt *et al.*, 2007).

Babesia microti has been considered a homogeneous species throughout its distribution, but historically most studies relied solely on parasite morphology for identification. A genetic analysis of 18S rDNA and β-tubulin genes of *B. microti* (identified by morphology) sampled throughout the United States and Eurasia demonstrates that this organism is a genetically diverse species complex (Fig. 16.2). At least three distinct groups of organisms have been identified: two in rodents and one in medium-sized mammals (Goethert & Telford, 2003*b*). Given the genetic diversity of *B. microti* that is apparent from analysis of a limited number of strains, it may be that particular genotypes are less likely to cause human disease. Indeed, evidence supporting such a hypothesis was recently demonstrated by the first report of *B. microti* babesiosis in Japan: parasites recovered from the transfused patient were typical of those found only in rodents collected from Awaji Island, whereas a different 18S rDNA genotype is commonly found virtually everywhere else in Japan (Tsuji *et al.*, 2001). Interestingly, *B. microti* babesiosis first emerged on Nantucket Island although this protozoan was endemic elsewhere in coastal New England. Perhaps island populations select for intensely transmissible strains.

'Seek and ye shall find' (Matthew 7) was once invoked in an early paper on the epidemiology of human babesiosis (Hoare, 1980) given the wide distribution and diversity of *Babesia* spp. Their role as a confounder for chloroquine-resistant malaria (babesias are not susceptible to chloroquine but are treatable with quinine) in tropical countries was suggested inasmuch as morphological discrimination from the plasmodia within blood smears may be difficult (Young &

Morzaria, 1986). Infection with HIV, now hyperendemic in many tropical areas, may render humans more susceptible to infection by, and disease due to, diverse *Babesia* spp., as do other forms of immune suppression. Piroplasms should routinely be sought as an aetiology for febrile illnesses wherever humans are intensely exposed to ticks.

EPIDEMIOLOGICAL ENTITIES IN SEARCH OF AN AGENT

Perhaps the greatest epidemiological puzzle in tick-borne disease research in the United States is the presence of a Lyme disease mimic in the southern–central states. Another epidemiological entity, which urgently needs intensive study, is the potential for tick-borne rickettsiae to cause cardiac pathology.

Masters' disease

The issue of Lyme disease in the southern and central United States has been greatly controversial. Since the late 1980s, erythema migrans rashes were noted in patients from these areas (many reported by Dr Edwin Masters in Cape Girardeau, MO) (Weder *et al.*, 1989; Masters & Donnell, 1995; Masters *et al.*, 1998; Roberts *et al.*, 1999), and indeed *Borrelia burgdorferi* was detected in *Ixodes* ticks in North Carolina and Alabama as early as 1983 (Pegram *et al.*, 1983). Because of the Centers for Disease Control (CDC) surveillance case definition (at the time) requiring the presence of a known vector, viz., *I. dammini* (deer ticks), great scepticism accompanied such reports because deer ticks only infested sites in the northern United States. *Ixodes dammini* is now considered by most workers to be conspecific with *I. scapularis*, the blacklegged tick (Oliver *et al.*, 1993). However, blacklegged ticks in southern United States sites rarely feed on humans as nymphs, representing a major epidemiological difference that accounts for the low risk even in sites where Lyme disease spirochetes are enzootic (Telford, 1998). In fact, in central and southern United States erythema migrans rashes are associated with *A. americanum*, the Lone Star tick, bites (Masters *et al.*, 1998).

Cases of southern erythema migrans rashes are referred to as masters' disease or STARI (southern tick associated rash illness) to ensure its distinction from bona fide Lyme borreliosis. Other than the rash, the illness is non-specific, relatively mild and is said to resolve spontaneously without sequelae; on the other hand, stoic Missouri farmers who rarely visit physicians will seek treatment for this illness

(Dr Edwin Masters, personal communication). Residents of an endemic site in Maryland reported myalgias, fever and arthralgia (Armstrong *et al.*, 2001). Of 98 residents who reported a rash, 53% recalled that the tick was still attached when the rash developed, in sharp contrast to the erythema migrans of Lyme borreliosis, which appears a week after exposure (Nadelman, Herman & Wormser, 1997). Of 1556 ticks that were saved and submitted by residents who had been bitten in that site, 95% were *A. americanum*, mainly nymphs.

The aetiology of Masters' disease remains undescribed, although a *Borrelia* related to *B. theileri* (the agent of bovine borreliosis, described at the turn of the twentieth century), detected within 1–5% of host-seeking *A. americanum*, and designated '*B. barbouri*' (Rich *et al.*, 2001) or '*B. lonestari*' (Barbour, 1996) may eventually be incriminated as the agent. To date, samples from such patients have generally failed to provide evidence of spirochaetal aetiology: Barbour–Stoenner–Kelly (BSK) cultures of skin biopsies are negative, PCR of such samples in multiple laboratories have been negative (except in one case: James *et al.*, 2001), and sero-conversion to borrelial antigens ambiguous (Wormser *et al.*, 2005). At the very least, some reactivity with spirochaetal antigens (*B. burgdorferi* s.l., *B. hermsi*) would be expected inasmuch as the borreliae are widely cross-reactive. Specific reactivity to *B.* '*lonestari*' has not yet been tested even though the agent has been propagated in vitro using co-cultivation with a tick cell line (Varela *et al.*, 2004). No association with other known tick-borne pathogens (such as *E. chaffeensis*, *R. rickettsii*, *Francisella tularensis*) seems apparent, but other agents within Lone Star ticks that might cause such a rash are spiroplasmas (authors' unpublished data), *Trypanosoma cervi* and *Rickettsia amblyommi*. Assigning aetiology is made even more complex in some sites because both *I. dammini* and *A. americanum* may co-occur (e.g. in New Jersey) and individuals may not accurately identify ticks. *Amblyomma americanum* bites may also provoke a cutaneous reaction that may be misidentified as erythema migrans (Goldman, Rockwell & Richfield, 1952).

Sudden cardiac death and rickettsial infection

Rickettsia helvetica was first described from Swiss *I. ricinus* (Burgdorfer *et al.*, 1979; Beati *et al.*, 1993) as a non-pathogenic member of the spotted fever group rickettsiae. Subsequently this agent was detected in many other *I. ricinus* populations (France, Italy, Bulgaria, Slovenia, Portugal), with prevalence as great as 22% (Nilsson *et al.*, 1999).

In 1999, rickettsia-like organisms were detected by light microscopy in sections of heart tissue taken at autopsy of two young Swedish hockey players. The aetiology of sudden cardiac death was being intensively scrutinized due to an 'epidemic' of sudden cardiac death among orienteers (Wesslen, 2001). Chronic perimyocarditis reminiscent of scrub typhus was detected at autopsy and a search for rickettsiae was initiated. Because *R. helvetica* was known to infect Swedish *I. ricinus* ticks, this agent was the primary candidate as the cause of these putative rickettsial lesions. *Rickettsia helvetica* was detected within myocardial tissue by means of PCR sequencing of the citrate synthase gene (Nilsson, Lindquist & Pahlson, 1999). Rickettsiae were visualized in perimyocardial tissues by immunohistochemistry using anti-Proteus OX (the Weil–Felix reagent, which cross-reacts with rickettsial lipopolysaccharide), used because the investigators lacked specific anti-*R. helvetica* sera. Transmission electron microscopy confirmed the presence of rickettsiae within myocardial endothelium. Sera from these cases were reactive against antigens of *R. helvetica*. In a more recent study, 20% of 84 diseased aortic heart valves tested positive by PCR for *R. helvetica* compared to none of the normal heart valves (Nilsson *et al.*, 2005a, 2005b), suggesting again a disturbing association of this agent with cardiac pathology.

Although two cases do not make an epidemic, more study is urgently required to determine the frequency with which *R. helvetica* (or other tick-borne 'endosymbiotic' rickettsiae) may cause myocarditis and sudden cardiac death given its wide distribution and prevalence in an aggressive human-biting vector. This agent was also associated with cases of sarcoidosis by the same authors (Nilsson *et al.*, 2001), but subsequent independent analyses failed to confirm the suggestion (Planck *et al.*, 2004). In sites where *I. ricinus* is present, *R. helvetica* infection should be part of the differential diagnosis for tick-associated fevers; a recent report indicated that about 9% of sera from an Alsace study population were sero-reactive to *R. helvetica* antigens (Fournier *et al.*, 2000). Infection by *R. helvetica* appears to be relatively mild. Three reports of cases from Sweden that were confirmed by seroconversion and demonstration of the agent in skin biopsies describe a mild febrile illness with myalgia, and eschar but no rash (Nilsson *et al.*, 2005b). Another report of patients from Italy and France describes a mild flu-like illness with no rash that resolves without treatment (Fournier *et al.*, 2004). A prospective study that followed 35 army recruits through their basic field training in Sweden demonstrated that 23% seroconverted, some without noticing any symptoms (Nilsson *et al.*, 2005*b*). *Rickettsia helvetica* has been detected in Japanese *Ixodes persulcatus* (Fournier *et al.*, 2002), and it may be that this agent is the cause of fevers wherever these and related ticks (the main vectors of Lyme borreliosis, granulocytic ehrlichiosis and tick-borne encephalitis) are present.

NEW TICK–PATHOGEN ASSOCIATIONS

Observed 'co-infection' resulting from concurrent babesial, spirochaetal and/or ehrlichial infection in *I. dammini* ticks or mouse reservoirs has emerged as a paradigm for explaining some variations in the clinical spectrum of Lyme borreliosis (Krause *et al.*, 1996). In mouse models, the effect of co-infection on disease has been equivocal. Coinfection with *B. microti* and *B. burgdorferi* has no effect on the course of either infection (Coleman *et al.*, 2005), but co-infection with *B. burgdorferi* and *A. phagocytophilum* increases the spirochaete load in the ears, heart and skin (Holden *et al.*, 2005). However, the frequency of concurrent infection in humans, as opposed to sequential infection, and the attributable effects thereof, remains poorly defined. One prospective study of patients in the United States with erythema migrans estimated that 4% had co-infection with either *A. phagocytophilum* or *B. microti* as determined by positive PCR results or seroconversion (Steere *et al.*, 2003) but these patients were recruited over a large geographical area; it is likely that risk for co-infection is site dependent, due to differences in enzootic transmission parameters.

Co-infection

Early work demonstrating that *I. dammini* nymphs are frequently concurrently infected by *B. burgdorferi* and *B. microti* (Piesman *et al.*, 1986) used the Feulgen reaction for microscopically detecting salivary infections of babesial sporozoites. Nearly 20% of host-seeking nymphal deer ticks were concurrently infected. At that time only *B. microti* was known to have infected *I. dammini* in New England. We now know that *B. odocoilei*, commonly detected in deer throughout the eastern United States, is also present in the northeast (Armstrong *et al.*, 1998) and thus the morphology-based estimates for *B. microti* prevalence may have been inflated. On the other hand, PCR analyses for *B. microti* in unfed ticks tend to underestimate prevalence (S. R. Telford & D. H. Persing, unpublished observations) because of a small number of babesial genomes (sporoblasts or kinetes) within an unfed tick. Dormant sporoblasts reactivate during feeding

to produce thousands of sporozoites, thereby increasing the sensitivity of any detection method when ticks are allowed to 'pre-feed' prior to analysis. Using a combination of pre-feeding, microscopy and PCR with stringent contamination control, we detected only 0–2% of 427 (95% confidence interval 0–0.7%) host-seeking nymphal *I. dammini* that were triply infected (with *B. burgdorferi*, *A. phagocytophilum* and babesia) in four coastal New England sites (authors' unpublished observations), although *B. burgdorferi*/babesia (1.9%, 0.6–3.2%) and *B. burgdorferi*/*A. phagocytophilum* (1.6%, 0.4–2.8%) infection were most common. Concurrent infection in host-seeking ticks, however, seems to be less common than expected. Further observations are required to determine whether the prevalence of concurrent infection reflects an interaction at the level of reservoir immunity (in the vertebrate or in the vector) or that reservoir host species are more diverse and locally determined than the literature suggests.

Bartonellae

Increased scrutiny of ticks, often by the use of eubacterial PCR assays, has suggested that there may be diverse inquilinic ('resident in', without implying a commensal, symbiotic or parasitic association) microbes. In particular, a common commensal of rodents, the bartonellae, have been suggested as another tick-borne infection. *Bartonella* spp. are the aetiological agents of cat scratch fever (*B. henselae*), trench fever (*B. quintana*) and oroya fever (*B. bacilliformis*); their vectors (identified by careful experimental and natural history studies) are fleas, lice and sandflies, respectively. During early investigations of the causes of oroya fever, Noguchi (1926) demonstrated that *B. bacilliformis* could be experimentally transmitted between monkeys by the bites of *D. andersoni* ticks. To our knowledge, this is the only reported demonstration that ticks may serve as vectors of bartonellae.

Because bartonellae are common in the rodents that most frequently serve as hosts for subadult ticks in the *I. persulcatus* species complex (the main vectors of Lyme borreliosis spirochaetes), it should not be surprising to detect evidence of these bacteria within ticks. Trans-stadial survival does not necessarily imply vector competence. Even ungulates appear to be infected by specific bartonellae and indeed *I. ricinus* removed from deer contained *Bartonella* DNA (Schouls *et al.*, 1999). Such evidence may simply represent the presence of these bacteria within the blood meal or non-blood fluids ingested from the dermis as does the recent report of *B. henselae* within ticks removed from humans (Sanogo *et al.*, 2003).

Evidence of frequent bartonella infection has been detected in host-seeking *I. pacificus* (Chang *et al.*, 2001, 2002) but the low annealing temperature used for the PCR assay (42 °C) would appear to invite non-specificity regardless of the design of the primers. Sequencing of the amplicons was not routinely performed and thus the identity of the agents represented by these DNA fragments remains unclear. A more convincing case for an association of bartonellae with ticks was reported for a Lyme borreliosis case in New Jersey from whom *B. henselae* was cultivated. Host-seeking *I. dammini* from the patient's yard yielded *B. henselae* amplicons that were confirmed by sequencing (Eskow, Rao & Mordechai, 2001). Other estimates of prevalence of *Bartonella* in ticks have ranged from 9.8% in *I. ricinus* from France to an incredible 34.5% in *I. dammini* from New Jersey (Adelson *et al.*, 2004; Halos *et al.*, 2005). The vectorial capacity of deer ticks or any other tick for *B. henselae* or any other *Bartonella* spp. remains unproven; such proof would require experimental studies of vector competence and xenodiagnostic studies of field-derived hosts.

Viability, let alone vector competence, should not be inferred from the detection of DNA; indeed, DNA from dead organisms may amplify as well as that from live ones. Although it remains premature to conclude that ticks might transmit bartonellae to humans (or between animals), it is also premature to conclude otherwise. Careful experimental observations such as those demonstrating the role of fleas as the vectors of rodent grahamellae (e.g. Krampitz & Kleinschmidt, 1960) seem crucial to complement molecular epidemiological observations.

Hepatitis C virus

A case of hepatitis C infection was reported to be transmitted by *I. dammini* (Wurzel, Cable & Leiby, 2002). A participant in a Connecticut *B. microti* natural history study developed hepatitis during the study. Hepatitis C virus infection (HCV) was diagnosed based on serology and RT-PCR. The donor denied classical risk factors for acquiring HCV, including occupational exposure to blood (even though the individual was employed as a medical technologist). The authors of the report argued that because *B. microti*-specific immunoglobulin M (IgM) was present in July, this could be interpreted as the patient acquiring an infectious tick bite during June. The demonstration of HCV viraemia between July and August was consistent with acquisition during June. Furthermore, the relatively small risks of acquiring *B. microti* or HCV, as estimated from the blood donor population, would argue that

the 'simultaneous discovery of these two infective events . . . would be highly unusual' (Wurzel *et al.*, 2002). The authors suggested co-transmission of *B. microti* and HCV inasmuch as ticks serve as vectors for other flaviviruses (tick-borne encephalitis). Although the unrelated hepatitis E virus is commonly found in rodents (Favorov *et al.*, 2000) and could theoretically be transmitted by ticks, HCV is known only from hominoid primates (Lemon & Brown, 1995); it would be unusual for *I. dammini* to feed to repletion on a human and survive to moult. Thus, this report is most likely explained by unrecognized occupational exposure to HCV and is unlikely to represent a new tick–microbe association.

Enterovirus

During an investigation of aseptic meningitis cases due to enteroviral infection in Tennessee, a history of tick bite was elicited from several patients. Lone Star ticks and dog ticks (*D. variabilis*) were collected from the counties where the patients lived and tested by RT-PCR for evidence of enteroviral nucleic acids (Freundt *et al.*, 2005). Two of 38 pools of nymphal *A. americanum* yielded amplicons of the correct size and by sequencing were consistent with coxsackievirus or poliovirus (the PCR target of 400 bp in the conserved 5′ untranslated region of these viruses would not be sufficient to allow distinguishing between the various enteroviruses). The authors of this report clearly state that vectorial capacity has not been demonstrated, but noted that a picornavirus (cardiovirus) has been detected in *I. persulcatus* (Lvov *et al.*, 1978) and mosquitoes may experimentally transmit coxsackie A6 virus (Maguire, 1970). Independent confirmation of such findings seems required before any conclusion may be drawn regarding ticks as vectors for enteroviruses.

AGENTS IN SEARCH OF EMERGING DISEASES

The polymerase chain reaction and nucleic acid sequencing provide numerous entities for which matching GenBank accessions are not identifiable. Whether truly newly recognized or simply rediscovered, many such agents are 'in search of an emerging disease'. As responsible public health professionals, we should take all due care to explore aetiological roles following well-established criteria prior to proclaiming to have discovered a 'new' emerging disease. Koch's postulates, the microbiological standard by which an aetiological agent is incriminated, traditionally require (1) that the agent is always associated with the illness, and under circumstances that account for pathology and clinical signs or symptoms of the illness; (2) that it does not occur in healthy individuals; and (3) that the signs and symptoms of the illness may be reproduced by exposure to the pure, in vitro cultivated agent (Evans, 1976). These postulates have been significantly modified through the years to reflect the ever-evolving modes of discovery, culminating in the routine use of PCR (Fredricks & Relman, 1996). Koch's postulates and variants thereof for use at the population (public health) level remain a useful framework for incriminating an aetiological agent – or evaluating the burgeoning literature of emerging diseases.

Many examples of prospecting for 'agents in search of emerging diseases' may be found in the recent literature (e.g. Billings *et al.*, 1998*a*; Shypnov *et al.*, 2001; Simser *et al.*, 2002). Indeed, the use of broad range eubacterial 16S rDNA primers has identified a diverse flora associated with ticks (Martin & Schmidtmann, 1998; Schabereiter-Gurtner, Lubitz & Rolleke, 2003), which may represent soil contaminants of the waxy cuticular surface as well as those that truly infect ticks. We focus on one such report as an example of the questions that are raised when such prospecting is successful. We also call attention to a greatly neglected group of potentially pathogenic microbes, the spiroplasmas, which to date have been routinely ignored when they are detected within ticks.

Rickettsiosis in Southeast Asia

During a search for the biological basis for human serological reactivity to antigens of *E. chaffeensis*, *A. phagocytophilum* and the spotted fever group rickettsiae in Southeast Asia (Parola *et al.*, 2003), two *Ehrlichia* species were detected by PCR sequencing in *Haemaphysalis hystricis* ticks from Vietnam. One of the amplicons was 99.4% identical to *E. chaffeensis* in the large portion of the 16S rDNA that was sequenced. Because *H. hystricis* is known to feed on humans, human exposure to the agents represented by these amplicons might confound epidemiological surveys for evidence of infection by known *Ehrlichia* spp. Serological cross-reactivity is well known among *Ehrlichia* spp. In the same study, two *Rickettsia* species were identified by the same methods from *Dermacentor auratus* and pools of *Dermacentor* larvae from Thailand. Other than their placement in the spotted fever group and *R. bellii* clades, no other information is available regarding these microbes. Isolates were not made. Is '*Candidatus*' the appropriate means of designating the organism represented by the DNA sequence, when no other attempts were made to

acquire other biological information? How different is different when it comes to sequence information? When does one consider a sequence to represent a unique taxon, 'taxon-like' or virtually identical? Are such findings to be considered conservatively as representative of endosymbiotic microbes? Or, because there is human sero-reactivity to known human pathogens whose DNA sequences are very similar to those detected in human biting ticks, is a logical hypothesis to test that they do indeed represent agents in search of an emerging disease? After all, in light of the HIV pandemic and the susceptibility of HIV patients to opportunistic pathogens, one organism's endosymbionts may well become another's pathogen.

Spiroplasmas

The spiroplasmas (Mycoplasmatales) are commonly found within ticks when applying darkfield microscopy for the detection of borreliae, appearing as small (2–4 μm in length) spirochaete- or filament-like organisms with varying degrees of motility. Mycoplasma are the smallest free-living microorganisms, with reproductive units as small as 125 angstroms. They lack a cell wall and are thus highly pleomorphic. Although the vast majority appear to be commensal, mycoplasma are known to cause a number of veterinary syndromes such as contagious bovine pleuropneumonia, chronic respiratory disease of fowl and infectious catarrh of rodents (Hayflick, 1969; Anonymous, 1972). Their role in causing human disease is controversial, other than that of *M. pneumoniae* and *M. genitalium* (*Ureaplasma urealyticum*), the agents of primary atypical pneumonia and non-gonococcal urethritis, respectively. They have been implicated as aetiological agents in erythema multiforme (Ludham, Bridges & Benn, 1964). However, the difficulty with which they are detected within clinical samples and the potential for confusion with common commensal or environmental mycoplasma ensures difficulty in describing their role in the aetiology of an illness.

Spiroplasma are well known as endosymbiotic agents of arthropods and seem to be transmitted by insects to plants, sometimes causing pathology (e.g. citrus canker). *Spiroplasma ixodetis* was described from host-seeking *I. pacificus* ticks collected in Oregon (Tully *et al.*, 1995). *Spiroplasma mirum* was isolated from *Haemaphysalis leporispalustris* (Tully *et al.*, 1983). The relevance of these agents to human health is unknown. Spiroplasmas have been identified by fluorescent antibody in about a third of a sample of German *I. ricinus* (Tenckhoff *et al.*, 1994), and by PCR in a number of pools of unfed *I. ovatus* from Japan (Taroura *et al.*, 2005), but to our knowledge no other published reports exist on their prevalence within ticks. Given the diverse dermatological manifestations associated with tick bites, the role of spiroplasmas should be explored, particularly with lesions that appear similar to erythema multiforme.

AULD LANG SYNE INFECTIONS

In the current research climate that focuses on 'new' emergent infections, there is a risk that 'old' and well-established public health burdens due to ticks have become neglected. In 'olden days' epidemiological entomology was reactive, responding to identify aetiological agents for known outbreaks of disease; now we prospect for agents and attempt to make a case for their importance. Three tick-borne infections for which old acquaintance should never be forgotten are Rocky Mountain spotted fever, tularaemia and tick-borne encephalitis.

Rocky Mountain spotted fever

Rocky Mountain spotted fever (RMSF) was once the most important tick-borne infection in the United States (Harden, 1990) but now appears to be about a tenth as commonly reported as Lyme disease. Before antibiotic treatment was available, the case fatality rate was 20% or greater and even today it is 5% (Walker, 1998). From 1981 to 1995, 9223 RMSF cases were reported in the United States (Dalton *et al.*, 1995); as a comparison, over 100 000 cases of Lyme disease were reported from 1990 to 1999 (Centers for Disease Control, 2006). First described in the Rocky Mountain region (Wood, 1896), the role of ticks as vectors was confirmed by Ricketts during his investigation of the virulent epidemic in the Bitterroot Valley of Montana, where nearly 100 cases had occurred during 1895–1902 with 70% mortality (Ricketts, 1906). The rickettsial aetiology of RMSF was demonstrated by Wolbach (Wolbach, 1919). Interestingly, although hundreds of RMSF cases continue to be reported in the United States, most are from south–central (Oklahoma, Arkansas, Texas and Missouri) and middle Atlantic states (North Carolina, Virginia and Maryland). Relatively few (about 2%) are reported from the western United States (Dalton *et al.*, 1995). Whether the seeming shift in the endemic areas represents human activity, tick density, vagaries of reporting or a true change in the ecology of *R. rickettsii* remains undescribed.

The main vectors of RMSF, *D. andersoni* in the western United States and *D. variabilis* in the eastern and central United States, are widespread and ecologically successful three-host ticks. The reproductive hosts are generally medium-sized mammals such as skunks, raccoons and foxes; subadult ticks tend to feed on rodents such as voles and deer mice. The peri-domestic predilection of the reproductive hosts suggests that RMSF should be as great or greater a public health hazard as is Lyme borreliosis in and around suburban sites. Indeed, a cluster of cases has been reported from around a Manhattan park (Salgo *et al.*, 1988), a distinctly urban site.

Rhipicephalus sanguineus, the brown dog tick, was recently implicated as the vector for an outbreak of 14 confirmed RMSF patients in Arizona, a state where RMSF is rare and *Dermacentor* ticks are uncommon. All the patients had contact with tick-infested dogs and *R. rickettsii* was detected in unfed ticks collected from around a patient's house (Demma *et al.*, 2005). Nearctic populations of *R. sanguineus* have to date been considered to bite humans rarely and thus are irrelevant as vectors; accordingly, little effort has been made to analyse their contribution to ill health. This dogma stood in contrast to the fact that *R. sanguineus* is the main vector for tick typhus, including boutonneuse fever, in the Mediterranean and South Asia, and commonly bites humans. The basis for differential anthropophily and vectorial capacity between Old and New World populations of *R. sanguineus* needs analysis.

The public health burden of the tropical American RMSF variants such as Sâo Paolo typhus and Mexican Fiebre Manchada remain poorly explored. Rocky Mountain spotted fever has been reported from Mexico, Panama, Costa Rica, Colombia, Brazil and Argentina (Ripoll *et al.*, 1999; Galvao *et al.*, 2003). In Brazil, a particularly virulent form of RMSF, with a high case fatality rate, has been described associated with *A. cajennense* (Monteiro, Fonseca & Prado, 1931; deLemos *et al.*, 2001) and the agent has been isolated from or detected by PCR within this important pest species. In Mexico, RMSF-like infection was described in the 1940s (Bustamante, Varela & Ortiz-Mariotte, 1946) and an analysis of a dengue outbreak in the Yucatan suggested that some of the cases may have been due to spotted fever or a similar rickettsiosis (Zavala-Velazquez *et al.*, 1999). Given the extremely wide distribution of *A. cajennense* (from Texas to Argentina) and its well-known capacity as a human biter, it is likely that RMSF is widespread in Latin America, quite underdiagnosed, and its ecology and epidemiology understudied there.

A mechanism for explaining the distribution of *R. rickettsii*-infected ticks (Burgdorfer, Hayes & Mavros, 1981*a*) involving competitive displacement by endosymbiotic rickettsiae ('East side agent', now known as *R. peacocki*: Niebylski *et al.*, 1997*b*) remains one of the most innovative ecological theories ever devised for explaining the regulation of natural microbial populations. Similar hypotheses should be tested for other tick-borne infections in other parts of the world. A better understanding of the transmission dynamics and the nature of such competitive displacement may allow an evaluation of the capacity for RMSF to become resurgent in sites, such as the Rocky Mountain states; or its potential to emerge as a public health burden within the heavily populated and suburbanized northeastern United States.

Tularaemia

Tularaemia has received renewed interest given its placement within Category A – those most likely to be used for bioterrorism – of the United States Select Agent list (Dennis *et al.*, 2001). Studies on *Dermacentor andersoni* in the Bitterroot Valley spotted fever epidemic provided isolates of *Francisella tularensis* (Parker, Spencer & Francis, 1924) and thereby implicated them as vectors. Tularaemia in North America appears to be maintained between ticks and rabbits (Jellison & Parker, 1944), and more than 90% of all American cases appear to be related to rabbit exposure (Gill & Cunha, 1997). In contrast to the main epidemiological features in North America, tularaemia in Eurasia seems more of an environmental infection acquired from agricultural activities such as hay turning, from water contaminated by muskrats or water voles, during the processing of animal products (Pavlovsky, 1966) or, interestingly, by mosquito bites (probably representing contaminative transmission: Hopla, 1974). These differences served as the epidemiological basis for recognizing two distinct types of *F. tularensis*, differing with respect to distribution, reservoirs and virulence (Olsufiev, Emelyanova & Dunayeva, 1959). Type A organisms (also known as *F. tularensis* biovar *tularensis* or *F. tularensis nearctica*) are prevalent in North America but not in Eurasia, maintained in cottontail rabbits, are frequently transmitted by ticks and may cause severe disease (Gill & Cunha, 1997). Type B (*F. tularensis* biovar *palaearctica* or *F. tularensis holarctica*) organisms cause episodic outbreaks (epizootics) in beavers, muskrats and arvicoline rodents in either North America or Eurasia,

may be isolated from water or soil, and cause a milder disease.

Arthropod endosymbionts known as 'Wolbachia' comprise diverse unrelated groups (O'Neill *et al.*, 1992). '*Wolbachia persica*', originally described by Suitor & Weiss (1961) from *Argas* spp., has recently been placed within the genus *Francisella* based on 16S rDNA sequencing (O'Neill *et al.*, 1992; Niebylski *et al.*, 1997*a*; Noda, Munderloh & Kurtti, 1997; Sun *et al.*, 2000; Goethert & Telford, 2005). Interestingly, tissues from ticks infected by these agents appear to be moderately infectious for vertebrates (Suitor & Weiss, 1961). Their roles as potential human pathogens remain to be explored.

The enzootic cycle and vector–pathogen relationship of *F. tularensis*, particularly Type A in North America, remains unexplored. Rabbits (and rodents) are relatively poor reservoir hosts because they quickly succumb to infection, often within days. The rabbit tick (*H. leporispalustris*) is thought to be the main interepizootic reservoir (Jellison, 1974), but alternative enzootic associations such as *D. variabilis* and its hosts have not been thoroughly investigated. Elucidation of the actual mode of transmission requires further work: the few experimental transmission studies that have been done do not definitively identify salivary transmission as opposed to contamination by bacteria-laden faeces deposited during feeding as the route of transmission. The one study suggesting that *F. tularensis* may be introduced by salivation into the dermis during feeding reported on the reduction in tularaemia mortality in tick-bite-sensitized guinea pigs (Bell, Stewart & Wikel, 1979) fed upon by infected ticks. The role of transovarial transmission as a mode of perpetuation (Hopla, 1974) also needs clarification.

Infection of host-seeking *I. ricinus* by *F. tularensis* has been reported from Slovakia and Austria, in addition to the more commonly infected *D. reticulatus* (Vyrostekova *et al.*, 2002). Such findings provide evidence that Palaearctic tularaemia is acquired by tick bite, although most of the infections are associated with exposure to rodent-contaminated hay, muskrats or even mosquito bites (Tarnvik, Sandstrom & Sjostedt, 1996). In North America, *D. variabilis* and *A. americanum* are the only human-biting ticks that have been implicated as zoonotic vectors. As many as 1–5% of *D. variabilis* from an active natural focus on Martha's Vineyard, Massachusetts, contained *F. tularensis tularensis* when analysed by haemolymph test and PCR (Goethert *et al.*, 2004; H. K. Goethert, I. Shani & S. R. Telford, unpublished data).

Tick-borne encephalitis

Tick-borne encephalitis (TBE) was, until recently, the most prevalent tick-borne disease affecting humans. This dubious honour has now been assumed by Lyme borreliosis. However, in many parts of the world (such as vast areas of Russia), TBE is considered the most burdensome vector-borne infection because of its morbidity and mortality. TBE viruses (TBEV) are members of the family Flaviviridae (see Chapter 12). Most mammalian tick-borne flaviviruses (e.g. TBEV, Kyasanur Forest disease virus (KFDV), Omsk haemorrhagic fever virus (OHFV)) are human pathogens. The potential of the seabird tick-borne flaviviruses (e.g. Tyuleniy virus) for human infection is unknown although serosurveys suggest human exposure (Chastel, 1980).

Because infection may occur by inhalation and the case fatality rate can be great for TBEV, OHFV and KFDV infections, Biosafety Level 4 (BSL-4) facilities, practices and procedures are recommended for working with these pathogens within the United States and elsewhere (Richmond & McKinney, 1993). Biosafety Level 3 (BSL-3) practices and procedures are suggested for work with all the other TBE group viruses. These recommendations have tended to limit the number of investigators who may effectively continue to pose questions about their biology. Thus, the placement of TBE in the 'auld lang syne' category may not reflect neglect due to mistaken satisfaction with the state of our knowledge, but rather a perceived logistical inability to initiate new studies.

TBE incidence fluctuates from year to year, with local incidence peaks often seen in 2- to 4-year intervals (Kunz, 1992). These fluctuations are likely to reflect fluctuation in tick and feeding host populations as well as environmental factors, such as temperature and humidity, which may directly affect virus activity. The extent to which global climate change may influence the prevalence of tick-borne infections in general, and TBE in particular, remains to be defined. Predictive risk mapping suggests that TBE foci may disappear from many areas if hot and dry summers become more common as a result of climate change (Randolph & Rogers, 2000). Alteration of vector seasonal activity, however, may be counterbalanced by effects mediated by the extrinsic incubation temperature of the pathogens. More research is needed on the relative importance of these and other factors in enzootic cycles.

TBE incidence has increased within the past decade or so in many European nations. In Latvia, 100 to 300 cases were reported during 1984 to 1992, but a median of 791 cases was reported from 1993 to 1995 (Antykova, 1989; Anonymous, 1995). In southwestern Germany, a total of 78 cases of TBE was reported in the years 1978–84 whereas in 1994 alone, there were 234 (Ackermann, Kruger & Roggendorf, 1986; Kaiser, 1996). Similar trends have been reported elsewhere in Europe, with the exception of Austria, where TBE vaccination is a routine public health measure. Furthermore, new foci of zoonotic TBE transmission are emerging (Treib, 1994; de Marval, 1995) in what were previously non-endemic areas, such as in northern Italy, southern Sweden and Norway, northern Germany and northern Switzerland (Broker & Gniel, 2003). It is not clear whether this apparent emergence corresponds to an increase in the distribution of TBE microfoci, to increased contact between humans and infected ticks, to local weather changes or to a combination of these factors. The fact that those new zoonotic transmission foci tend to appear at the fringes of established transmission areas suggests that this phenomenon may not be caused by increased human exposure alone (Wellmer & Jusatz, 1981; Randolph, 2002). Indeed, *I. ricinus* and TBE appear to have moved to higher elevations in Central Europe, perhaps due to increased climate variability (Daniel *et al.*, 2003), although it should be noted that warmer and dryer weather would probably reduce transmission in currently endemic sites (Randolph & Rogers, 2000).

Due to the overlap of enzootic transmission cycles of different tick-borne pathogens, co-infection with at least one additional agent is relatively common and should always be considered as a possibility. Concurrent Lyme borreliosis, for example, has been shown to be associated with particularly severe TBE manifestations (Oksi *et al.*, 1993). Whether TBEV co-infection with piroplasms or ehrlichiae may modify disease in the human host is not known. Louping ill virus (a member of the TBEV serogroup) infection in sheep is exacerbated by that of tick-borne fever due to *A. phagocytophilum* (Reid, 1986) and thus some effects are anticipated for human coinfection.

The effects of enzootic overlap between closely related pathogens remain poorly explored. Powassan virus (POWV), another member of the TBEV serogroup, is maintained by *I. cookei*, *I. texanus* or other ticks that focus their feeding on woodchucks, skunks and other medium-sized mammals. Such ticks only occasionally bite humans, accounting for the relative scarcity of cases for a virus which is geographically widespread and intensively transmitted: 23–64% of woodchucks were seropositive for POWV in New York and Ontario (Artsob, 1989). In addition, the deer tick, *I. dammini*, the aggressive main vector for Lyme borreliosis in the northeastern United States, is experimentally vector competent for POWV (Costero & Grayson, 1996). This tick naturally maintains a Powassan subtype ('deer tick virus' or DTV: Telford *et al.*, 1997) for which human infection remains to be described. Should deer ticks begin to 'bridge' virulent POWV from the *I. cookei*–woodchuck cycle, it may be that Powassan fever will become more prevalent. On the other hand, DTV may competitively exclude POWV within a site, thereby preventing the more severe infection from becoming a more common public health hazard. It may also be that DTV causes an 'FUO-like' (fever of unknown origin) illness that resolves without sequelae, and that transmission might intensify as has that of Lyme disease over the last decade or two. These issues pose important public health questions as well as some for basic disease ecology.

PROSPECTING FOR EMERGING INFECTIONS

Pathogens, vectors and reservoir hosts exist in predictable ecological assemblages (Pavlovsky, 1966). An extension of this is the idea that microbes exist in guilds, which are the basic units of community structure and represent groups of unrelated taxa that share a common resource (Root, 1967). Fleas, mites, lice and ticks infesting a mouse would comprise a guild of ectoparasites; babesia, trypanosomes, haemogregarines and grahamellae comprise a guild of rodent haematozoa. Such host–pathogen assemblages usually have analogues (ecological equivalents) in each biogeographical region and may be maintained in small enzootic (natural) foci without implying human exposure. Human risk, or the 'zoonotic' condition, denotes overflow or bridging from the enzootic cycle (Spielman & Rossignol, 1984). Emergence of a zoonosis thus requires unusual circumstances of great vector and reservoir density, promoting overflow, intrusion of humans into enzootic foci or the introduction of a man-biting bridge vector. Moreover, because of microbial guilds, rarely does just one agent 'emerge'. An old poster from the Rocky Mountain laboratories stresses the multiple nature of hazards from tick bite (Fig. 16.3), calling the wood tick a 'Pandora's box'.

Tick-borne pathogen guilds

Based upon the guild concept, one could attempt to predict the presence of microbial agents within local tick taxa. For example, for *I. dammini* in North America, we would expect to detect *A. phagocytophilum*, a tick-borne encephalitis-like flavivirus, and an orbivirus in ticks or rodents from sites where Lyme disease spirochaetes, *B. microti*, and the other haemoparasites are present because these agents comprise the guild in European *I. ricinus*. In 1994, the index case-series for human granulocytic ehrlichiosis (HGE) was reported from the vicinity of Duluth, MN (Chen *et al.*, 1994), an area that is thought to be a relict, long-standing site of transmission for deer tick agents (Telford *et al.*, 1993). Rapidly thereafter, using the guild concept, the index case of HGE for New England (Telford *et al.*, 1995) was identified by a specific search for the rickettsial agent in blood samples from *I. dammini*-exposed patients. The success of this approach stimulated a search for a flavivirus, inasmuch as tick-borne encephalitis virus is a prominent member of the microbial guild associated with *I. persulcatus*-like ticks in Eurasia (Telford & Foppa, 2000). Deer tick virus (DTV) was subsequently identified (Telford *et al.*, 1997); molecular and serological characterization now suggests that it is a subtype of Powassan virus (Ebel, Spielman & Telford, 2001). Human infection due to DTV remains undescribed, although prototypic Powassan virus is well-known as a cause of a devastating meningoencephalitis (Artsob, 1989). We have, in fact, argued that DTV may be a less pathogenic subtype inasmuch as prevalence of infection in deer ticks approaches that for HGE or *B. microti*, but severe meningoencephalitis cases have not been frequently reported from our study populations.

Other tick-maintained microbial guilds seem apparent in the northeastern United States (Table 16.2) awaiting attention by researchers. Although diverse ixodid ticks may be found there, we present as an example three that are ecologically successful (widely distributed, abundant) and with some degree of anthropophily. Dog ticks (*D. variabilis*) have been intensively studied due to their main role as vector of RMSF. They are widely distributed in the eastern and central United States, with isolated populations occurring through the Western states. Lone Star ticks (*A. americanum*), a major pest species through the south and central states, appears to be expanding its distribution into New England. Finally, rabbit ticks (*Ixodes dentatus*) may be found wherever cottontail rabbits have been introduced; furthermore, these ticks are readily transported by passerine birds, suggesting the possibility for rapid and extensive dispersal of any agent that they transmit. *Ixodes dentatus* appears to bite humans more frequently than previously thought (Armstrong *et al.*, 2001).

Similar tables can and should be compiled for common human-biting ticks throughout the world, such as *R. sanguineus*, *D. marginatus* and *A. cajennense*. Ecological theory

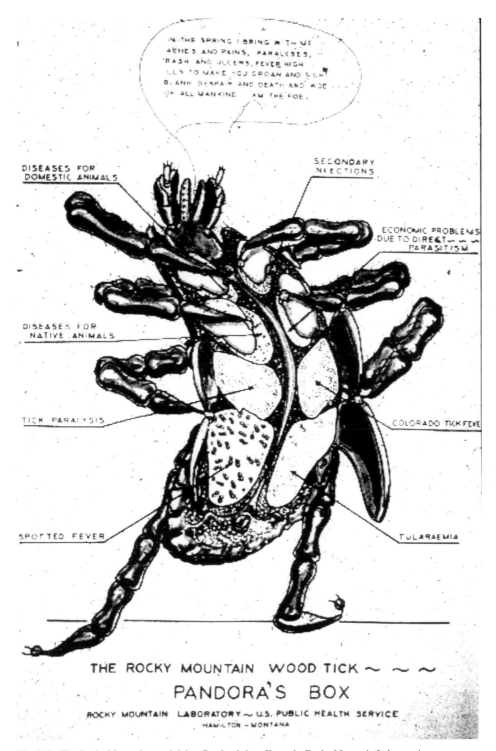

Fig. 16.3. The Rocky Mountain wood tick as Pandora's box. From the Rocky Mountain Laboratories, Hamilton, MT.

Table 16.2 *Microbial guilds presumed associated with ixodid ticks common in New England*

Tick	Microbial guild[a]	Reference
Deer ticks (*Ixodes dammini*)	Spirochaetes: *Borrelia burgdorferi* s.s., *B.* nr. *miyamotoi*	
	Piroplasms: *Babesia microti, B. odocoilei**	
	Ehrlichiae: *Anaplasma phagocytophilum*	
	Rickettsiae: IS agent	Weller *et al.*, 1998
	Arbovirus: Deer tick virus*, Tribec/Eyach?/St Croix River virus* (orbivirus)?	Hubalek, Calisher & Mitter-Mayer, 1987 Chastel *et al.*, 1984 Attoui *et al.*, 2001
Dog ticks (*Dermacentor variabilis*)	Spirochaetes: none described to date	
	Piroplasms: *Babesia lotori?**	Anderson, Magnarelli & Sulzer, 1981
	Ehrlichiae: *Ehrlichia ewingii?*	
	Rickettsiae: *Rickettsia rickettsii*	
	Rickettsiae: *Rickettsia rickettsii*	
	Arbovirus: Colorado Tick fever virus (Coltivirus)	
Lone Star ticks (*Amblyomma americanum*)	Spirochaetes: *Borrelia* 'lonestari'*	
	Piroplasms: *Theileria cervi**	Krinsky & Burgdorfer, 1976
	Ehrlichiae: *Ehrlichia chaffeensis*, WTD agent*	Little *et al.*, 1997
	Rickettsiae: *Rickettsia amblyommi*	Burgdorfer *et al.*, 1981b
	Arbovirus: Lone Star virus (Bunyavirus)*	Kokernot *et al.*, 1969
Rabbit ticks (*Ixodes dentatus*)	Spirochaetes: *Borrelia andersoni**	Marconi, Liveris & Schwartz, 1995
	Piroplasms: *Babesia divergens*	
	Ehrlichiae: *Anaplasma phagocytophilum*	
	Rickettsiae: not described to date	
	Arbovirus: Rabbit syncytial virus (Orbivirus)*	Theil, McCloskey & Scott, 1993
	Connecticut virus (Rhabdovirus)*	Main & Carey, 1980

[a] Question marks indicate that the agent has not been looked for but is likely to be detected; asterisks denote an agent in search of an emerging infection.

provides us with a prospective approach to identifying emergent tick-borne pathogens. Such hypothesis-based, prospective analyses of tick–pathogen–reservoir systems may provide us with an ever-expanding list of aetiological agents 'in search of an emerging illness'. Emergent clinical entities, in turn, may be determined to be due to well-known or poorly characterized tick-borne microbes by focusing our attention on locally abundant ticks and their likely microbial guilds. Rare tick-borne infections may emerge as public health burdens due to changes in tick density and environmental disturbance. In all such situations, the laboratory findings enabled by powerful molecular techniques must be evaluated within an epidemiological and ecological context to determine the actual risk to human health.

ACKNOWLEDGEMENTS

Our work has been funded by grants from the National Institutes of Health (AI 37993, 39002, 53411, 064218) and the Centers for Disease Control and Prevention (U50/CCU 119560).

REFERENCES

Ackermann, R., Kruger, K. & Roggendorf, M. (1986). Spread of early-summer meningoencephalitis in the Federal Republic of Germany. *Deutsche medizinische Wochenschrift* **111**, 927–933.

Adelson, M. E., Rao, R. V. S., Tilton, R. C., *et al.* (2004). Prevalence of *Borrelia burgdorferi*, *Bartonella* spp., *Babesia microti*, and *Anaplasma phagocytophila* in *Ixodes scapularis* ticks collected in northern New Jersey. *Journal of Clinical Microbiology* **42**, 2799–2801.

Alekseev, A., Dubinina, H., Van De Pol, I. & Schouls, L. M. (2001). Identification of *Ehrlichia* spp. and *Borrelia burgdorferi* in *Ixodes* ticks in the Baltic region of Russia. *Journal of Clinical Microbiology* **39**, 2237–2242.

Anderson, B. E., Sims, K. G., Olson, J. G., *et al.* (1993). *Amblyomma americanum*: a potential vector of human ehrlichiosis. *American Journal of Tropical Medicine and Hygiene* **49**, 239–244.

Anderson, J. F., Magnarelli, L. A. & Sulzer, A. J. (1981). Raccoon babesiosis in Connecticut, USA: *Babesia lotori sp. nov. Journal of Parasitology* **67**, 417–425.

Andrew, R., Bonnin, J. M. & Williams, S. (1946). Tick typhus in northern Queensland. *Medical Journal of Australia* **2**, 253–258.

Anigstein, L. & Anigstein, D. (1975). A review of the evidence in retrospect for a rickettsial etiology in Bullis Fever. *Texas Reports in Biology and Medicine* **33**, 201–211.

Anonymous (1972). *Pathogenic Mycoplasmas*. Amsterdam: Elsevier.

Anonymous (1995). Tick-borne encephalitis. *Weekly Epidemiological Record* **70**, 120–122.

Antonov, N. & Naishtat, A. (1937). Far Eastern tick typhus: cases of tropical typhus in the Far East. *Medizina Parazitologiya* **6**, 73–81.

Antykova, L. P. (1989). Epidemiology of tick-borne encephalitis in Leningrad. *Trudy Instituta Imeni Pastera* **65**, 21–25.

Armstrong, P. M., Brunet, L. R., Spielman, A. & Telford, S. R. (2001). Risk of Lyme disease: perceptions of residents of a lone star tick-infested community. *Bulletin of the World Health Organization* **79**, 916–925.

Armstrong, P. M., Katavolos, P., Caporale, D. A., *et al.* (1998). Diversity of *Babesia* infecting deer ticks (*Ixodes dammini*). *American Journal of Tropical Medicine and Hygiene* **58**, 739–742.

Artsob, H. (1989). Powassan encephalitis. In *The Arboviruses: Epidemiology and Ecology*, vol. 4, ed. Monath, T. P., pp. 29–49. Boca Raton, FL: CRC Press.

Attoui, H., Stirling, J. M., Munderloh, U. G., *et al.* (2001). Complete sequence characterization of the genome of the St. Croix River virus, a new orbivirus isolated from the cells of *Ixodes scapularis. Journal of General Virology* **82**, 795–804.

Bajer, A., Pawelczyk, A., Behnke, J. M., Gilert, F. S. & Sinski, E. (2001). Factors affecting the community structure of haemoparasites in bank voles (*Clethrionomys glareolus*) from the Mazury Lake District region of Poland. *Parasitology* **122**, 43–54.

Barbour, A. G. (1996). Does Lyme disease occur in the south? – a survey of emerging tick-borne infections in the region. *American Journal of the Medical Sciences* **311**, 34–40.

Barken, J. S., Krueth, J., Tilden, R. L., Dumler, J. S. & Kristiansen, B. E. (1996a). Serological evidence of human granulocytic ehrlichiosis in Norway. *European Journal of Clinical Microbiology and Infectious Diseases* **15**, 829–832.

Barken, J. S., Krueth, J., Wilson-Nordskog, C., *et al.* (1996b). Clinical and laboratory characteristics of human granulocytic ehrlichiosis. *Journal of the American Medical Association* **275**, 199–205.

Beati, L., Peter, O., Burgdorfer, W., Aeschlimann, A. & Raoult, D. (1993). Confirmation that *Rickettsia helvetica sp. nov.* is a distinct species of the spotted fever group rickettsiae. *International Journal of Systematic Bacteriology* **43**, 521–526.

Beattie, J. F., Michelson, M. L. & Holman, P. J. (2002). Acute babesiosis caused by *Babesia divergens* in a resident of Kentucky. *New England Journal of Medicine* **347**, 697–698.

Bell, J. F., Stewart, S. J. & Wikel, S. K. (1979). Resistance to tick-borne *Francisella tularensis* by tick-sensitized rabbits: allergic klendusity. *American Journal of Tropical Medicine and Hygiene* **28**, 876–880.

Belongia, E. A., Reed, K. D., Mitchell, P. D., *et al.* (1997). Prevalence of granulocytic *Ehrlichia* infection among white-tailed deer in Wisconsin. *Journal of Clinical Microbiology* **35**, 1465–1468.

Billings, A. N., Teltow, G., Weaver, S. C. & Walker, D. H. (1998a). Molecular characterization of a novel *Rickettsia* species from *Ixodes scapularis* in Texas. *Emerging Infectious Diseases* **4**, 305–309.

Billings, A. N., Yu, J., Teel, P. D. & Walker, D. H. (1998*b*). Detection of a spotted fever group rickettsia in *Amblyomma cajennense* (Acari: Ixodidae) in south Texas. *Journal of Medical Entomology* **35**, 474–478.

Bjoersdorff, A., Wittesjo, B., Berglund, J., Massung, R. F. & Eliasson, I. (2002). Human granulocytic ehrlichiosis as a common cause of tick-associated fever in southeast Sweden: report from a prospective clinical study. *Scandinavian Journal of Infectious Diseases* **34**, 187–191.

Bown, K. J., Begon, M., Bennett, M., Woldehiwet, Z. & Ogden, N. H. (2003). Seasonal dynamics of *Anaplasma phagocytophila* in a rodent–tick (*Ixodes trianguliceps*) system, United Kingdom. *Emerging Infectious Diseases* **9**, 63–70.

Bozeman, F. M., Elisberg, B. L., Humphries, J. W., Runcik, K. & Palmer, D. B. (1970). Serologic evidence of rickettsia canada infection of man. *Journal of Infectious Diseases* **121**, 367–371.

Brezina, R., Rehacek, J., Ac, P. & Majerska, M. (1969). Two strains of Rickettsiae of Rocky Mountain Spotted Fever group recovered from *Dermacentor marginatus* ticks in Czechoslovakia. *Acta Virologica* **13**, 142–145.

Broadhurst, L. E., Kelly, D. J., Chan, C. T., *et al.* (1998). Laboratory evaluation of a dot–blot enzyme immunoassay for serologic confirmation of illness due to *Rickettsia conorii*. *American Journal of Tropical Medicine and Hygiene* **58**, 786–789.

Broker, M. & Gniel, D. (2003). New foci of tick-borne encephalitis virus in Europe: consequences for travellers from abroad. *Travel Medicine and Infectious Diseases* **1**, 181–184.

Brouqui, P. & Matsumoto K. (2007). Bacteriology and phylogeny of Anaplasmataceae. In *Rickettsial Diseases*, eds. Raoult, D. & Parola, P., pp. 179–198. New York: Informa Healthcare.

Brouqui, P., Harle, J. R., Delmont, J., *et al.* (1997). African tick-bite fever: an imported spotless rickettsiosis. *Archives of Internal Medicine* **157**, 119–124.

Brouqui, P., Sanogo, Y. O., Caruso, G., Merola, F. & Raoult, D. (2003). *Candidatus* Ehrlichia walkerii: a new *Ehrlichia* detected in *Ixodes ricinus* tick collected from asymptomatic humans in northern Italy. *Annals of the New York Academy of Sciences* **990**, 134–140.

Buller, R. S., Arens, M., Hmiel, S. P., *et al.* (1999). *Ehrlichia ewingii*, a newly recognized agent of human ehrlichiosis. *New England Journal of Medicine* **341**, 148–155.

Burgdorfer, W., Aeschlimann, A., Peter, O., Hayes, S. F. & Philip, R. N. (1979). *Ixodes ricinus*: vector of a hitherto undescribed spotted-fever group agent in Switzerland. *Acta Tropica* **36**, 357–367.

Burgdorfer, W., Hayes, S. F. & Mavros, A. J. (1981*a*). Nonpathogenic rickettsiae in *Dermacentor andersoni*: a limiting factor for the distribution of *Rickettsia rickettsii*. In *Rickettsiae and Rickettsial Diseases*, eds. Burgdorfer, W. & Anacker, R. L., pp. 585–594. New York: Academic Press.

Burgdorfer, W., Hayes, S. F., Thomas, L. A. & Lancaster, J. L. (1981*b*). A new spotted fever group rickettsia from the lone star tick, *Amblyomma americanum*. In *Rickettsiae and Rickettsial Diseases*, eds. Burgdorfer, W. & Anacker, R. L., pp. 595–599. New York: Academic Press.

Bustamante, M. E., Varela, G. & Ortiz-Mariotte, C. (1946). Estudios de fiebre manchada en Mexico. *Revista de Instituto Salubria y Enfermedades Tropicale* **7**, 39–48.

Cao, W.-C., Gao, Y.-M., Zhang, P.-H., *et al.* (2000*a*). Identification *of Ehrlichia chaffeensis* by nested PCR in ticks from southern China. *Journal of Clinical Microbiology* **38**, 2778–2780.

Cao, W.-C., Zhao, Q.-M., Zhang, P.-H., *et al.* (2000*b*). Granulocytic ehrlichiae in *Ixodes persulcatus* ticks from an area in China where Lyme disease is endemic. *Journal of Clinical Microbiology* **38**, 4208–4210.

Centers for Disease Control (2006). *Reported Cases of Lyme Disease By Year, United States, 1991–2005.* Available online at www.cdc.gov/ncidod/dvbid/Lyme/ld_UpClimbLymeDis.htm/

Chang, C. C., Chomel, B. B., Kasten, R. W., Romano, V. & Tietze, N. (2001). Molecular evidence of *Bartonella* spp. in questing adult *Ixodes pacificus* ticks in California. *Journal of Clinical Microbiology* **39**, 1221–1226.

Chang, C. C., Hayashidani, H., Pusterla, N., *et al.* (2002). Investigation of *Bartonella* infection in ixodid ticks from California. *Comparative Immunology Microbiology and Infectious Diseases* **25**, 229–236.

Chastel, C. (1980). Arbovirus transmis par des tiques et associés à des oiseaux de mer: une revue générale. *Médecine Tropicale* **40**, 535–548.

Chastel, C., Main, A. J., Couatarmanach, A., *et al.* (1984). Isolation of Eyach virus from *Ixodes ricinus* and *I. ventalloi* ticks in France. *Archives of Virology* **82**, 161–171.

Chen, S. M., Dumler, J. S., Bakken, J. S. & Walker, D. H. (1994). Identification of a granulocytotropic *Ehrlichia* species as the etiologic agent of human disease. *Journal of Clinical Microbiology* **32**, 589–595.

Coleman, J. L., Levine, D., Thill, C., Kuhlow, C. & Benach, J. L. (2005). *Babesia microti* and *Borrelia burgdorferi* follow independent courses of infection in mice. *Journal of Infectious Diseases* **192**, 1634–1641.

Conor, A. & Bruch, A. (1910). Une fièvre éruptive observée en Tunisie. *Bulletin de la Société de Pathologie Exotique* **3**, 492–496.

Conrad, P. A., Kjemtrup, A. M., Carreno, R. A., *et al.* (2006). Description of *Babesia duncani* n.sp. (Apicomplexa: Babesiidae) from humans and its differentiation from other piroplasms. *International Journal of Parasitology* **36**, 779–789.

Costero, A. & Grayson, M. (1996). Experimental transmission of Powassan virus (Flaviviridae) by *Ixodes scapularis* ticks (Acari: Ixodidae). *American Journal of Tropical Medicine and Hygiene* **55**, 536–546.

Dalton, M. J., Clarke, M. J., Holman, R. C., *et al.* (1995). National surveillance for Rocky Mountain Spotted Fever 1981–1992: epidemiologic summary and evaluation of risk factors for fatal outcome. *American Journal of Tropical Medicine and Hygiene* **52**, 405–413.

Daniel, M., Danielova, V. V., Kriz, B., Jirsa, A. & Nozicka, J. (2003). Shift of the tick *Ixodes ricinus* and tick-borne encephalitis to higher altitudes in Central Europe. *European Journal of Clinical Microbiology and Infectious Diseases* **22**, 327–328.

Dasch, G. A., Kelly, D. J., Richards, A. L. & Sanchez, J. L. (1993). Western Blotting analysis of sera from military personal exhibiting serological reactivity to spotted fever group rickettsiae. *American Journal of Tropical Medicine and Hygiene* **49**, 220 (abstract).

Dawson, J. E., Stallknecht, D. E., Howerth, E. W., *et al.* (1994). Susceptibility of white-tailed deer (*Odocoileus virginianus*) to infection with *Ehrlichia chaffeensis*, the etiologic agent of human ehrlichiosis. *Journal of Clinical Microbiology* **32**, 2725–2728.

De Lemos, E. R. S, Alvarenga, F. B. F., Cintra, M. L., *et al.* (2001). Spotted fever in Brazil: a seroepidemiological study and description of clinical cases in an endemic area in the state of São Paolo. *American Journal of Tropical Medicine and Hygiene* **65**, 329–334.

De Marval, F. (1995). L'encephalite à tiques en Suisse: épidemiologie et prévention. *Médecine et Hygiène* **53**, 224–226.

Demma, L. J., Traeger, M. S., Nicholson, W. L., *et al.* (2005). Rocky Mountain spotted fever from an unexpected tick vector in Arizona. *New England Journal of Medicine* **353**, 587–594.

Dennis, D. T., Inglesby, T. V., Henderson, D. A., *et al.* (2001). Tularaemia as a biological weapon: medical and public health management. *Journal of the American Medical Association* **285**, 2763–2773.

Donatien, A. & Lestoquard, F. (1936). *Rickettsia bovis*, novelle espèce pathogène pour le boeuf. *Bulletin de la Société de Pathologie Exotique* **29**, 1057–1061.

Donatien, A. & Lestoquard, F. (1940). Rickettsiose bovine Algérienne à *R. bovis*. *Bulletin de la Société de Pathologie Exotique* **33**, 245–248.

Donnelly, J. & Pierce, M. A. (1975). Experiments on the transmission of *Babesia divergens* to cattle by the tick *Ixodes ricinus*. *International Journal for Parasitology* **5**, 363–367.

Dumler, J. S. & Barken, J. S. (1995). Ehrlichial diseases of humans: emerging tick-borne infections. *Clinical Infectious Diseases* **20**, 1102–1110.

Dumler, J. S., Barbet, A. F., Bekker, C. P., *et al.* (2001). Reorganization of genera in the families Rickettsiaceae and Anaplasmataceae in the order Rickettsiales: unification of some species of *Ehrlichia* with *Anaplasma*, *Cowdria* with *Ehrlichia* and *Ehrlichia* with *Neorickettsia*, descriptions of six new species combinations and designation of *Ehrlichia equi* and 'HGE agent' as subjective synonyms of *Ehrlichia phagocytophila*. *International Journal of Systematic and Evolutionary Microbiology* **51**, 2145–2165.

Ebel, G. D., Spielman, A. & Telford, S. R. III (2001). Phylogeny of North American Powassan virus. *Journal of General Virology* **82**, 1657–1665.

Eng, T. R., Harkess, J. R., Fishbein, D. B., *et al.* (1990). Epidemiologic, clinical, and laboratory findings of human ehrlichiosis in the United States 1988. *Journal of the American Medical Association* **264**, 2251–2258.

Enigk, K. & Friedhoff, K. (1962). *Babesia capreoli* n. sp., beim Reh (*Capreolus capreolus* L.). *Zentralblatt für Tropenmedizin und Parazitologie* **13**, 8–20.

Ereemeva, M. E., Beati, L., Makarova, V. A., *et al.* (1994). Astrakhan fever rickettsiae: antigenic and genotyping of isolates obtained from human and *Rhipicephalus pumilio* ticks. *American Journal of Tropical Medicine and Hygiene* **51**, 697–706.

Eskow, E., Rao, R. V. S. & Mordechai, E. (2001). Concurrent infection of the central nervous system by *Borrelia burgdorferi* and *Bartonella henselae*: evidence for a novel tick-borne disease complex. *Archives of Neurology* **58**, 1357–1363.

Evans, A. S. (1976). Causation and disease: Henle–Koch postulates revisited. *Yale Journal of Biology and Medicine* **49**, 175–195.

Ewing, S. A. & Philip, C. B. (1966). Ehrlichia-like rickettsiosis in dogs in Oklahoma and its relationship to *Neorickettsia helminthoeca. American Journal of Veterinary Research* **27**, 67–69.

Ewing, S. A., Dawson, J. E., Kocan, A. A., *et al.* (1995). Experimental transmission of *Ehrlichia chaffeensis* (Rickettsiales: Ehrlichieae) among white-tailed deer by *Amblyomma americanum* (Acari: Ixodidae). *Journal of Medical Entomology* **32**, 368–374.

Fan, M. & Y., Zhang, J. Z., Chen, M. & Yu, X. J. (1999). Spotted fever group rickettsioses in China. In *Rickettsiae and Rickettsial Diseases at the Turn of the Third Millennium*, eds. Raoult, D. & Brouqui, P., pp. 247–257. Paris: Elsevier.

Favorov, M. O., Kosoy, M. Y., Tsarev, S. A., Childs, J. E. & Margolis, H. S. (2000). Prevalence of antibody to hepatitis E virus among rodents in the United States. *Journal of Infectious Diseases* **181**, 449–455.

Fingerle, V., Goodman, J. L., Johnson, R. C., *et al.* (1997). Human granulocytic ehrlichiosis in southern Germany: increased seroprevalence in high-risk groups. *Journal of Clinical Microbiology* **35**, 3244–3247.

Fishbein, D. B., Kemp, A., Dawson, J. E., *et al.* (1989). Human ehrlichiosis: prospective active surveillance in febrile hospitalized patients. *Journal of Infectious Diseases* **160**, 803–809.

Foggie, A. (1951). Studies on the infectious agent of tick-borne fever in sheep. *Journal of Pathology and Bacteriology* **63**, 1–15.

Foppa, I. M., Krause, P. J., Spielman, A., *et al.* (2002). Entomologic and serologic evidence of zoonotic transmission *of Babesia microti* in Switzerland. *Emerging Infectious Diseases* **8**, 722–726.

Fournier, P. E., Allombert, C., Supputamongkol, Y., *et al.* (2004). An eruptive fever associated with antibodies to *Rickettsia helvetica* in Europe and Thailand. *Journal of Clinical Microbiology* **42**, 816–818.

Fournier, P. E., Dumler, J. S., Greub, G., *et al.* (2003). Gene sequence-based criteria for identification of new *Rickettsia* isolates and description of *Rickettsia heilongjiangensis* sp. nov. *Journal of Clinical Microbiology* **41**, 5456–5465.

Fournier, P.-E., Fujita, H., Takada, N. & Raoult, D. (2002). Genetic identification of rickettsiae isolated from ticks in Japan. *Journal of Clinical Microbiology* **40**, 2176–2181.

Fournier, P.-E., Grunnenberer, F., Jaulhac, B., Gastinger, G. & Raoult, D. (2000). Evidence of *Rickettsia helvetica* infection in humans, eastern France. *Emerging Infectious Diseases* **6**, 389–392.

Franca, C. (1910). Sur la classification des piroplasmes et description des deux formes de ces parasites. *Archives of the Royal Institute of Bacteriology Camara Pestana Lisbon, Portugal* **3**, 11–18.

Fredricks, D. N. & Relman, D. A. (1996). Sequence-based identification of microbial pathogens: a reconsideration of Koch's postulates. *Clinical and Microbiological Reviews* **9**, 18–33.

Freundt, E. C., Beatty, D. C., Stegall-Faulk, T. & Wright, S. M. (2005). Possible tick-borne human enterovirus resulting in aseptic meningitis. *Journal of Clinical Microbiology* **43**, 3471–3473.

Galvao, M. A. M., Mafra, C. L., Moron, C., Anaya, E. & Walker, D. H. (2003). Rickettsiosis of the genus *Rickettsia* in South America. *Annals of the New York Academy of Sciences* **990**, 57–61.

Gear, J. H. S. (1954). The rickettsial diseases of southern Africa. *South African Journal of Clinical Science* **5**, 158–175.

Gill, V. & Cunha, B. A. (1997). Tularemia pneumonia. *Seminars in Respiratory Infections* **12**, 61–67.

Goethert, H. K. & Telford, S. R. III (2003*a*). Enzootic transmission of *Babesia divergens* in cottontail rabbits on Nantucket Island. *American Journal of Tropical Medicine and Hygiene* **69**, 455–460.

Goethert, H. K. & Telford, S. R. III (2003*b*). What is *Babesia microti? Parasitology* **127**, 301–309.

Goethert, H. K. & Telford, S. R. III (2003*c*). Enzootic transmission of *Anaplasma bovis* in Nantucket cottontail rabbits. *Journal of Clinical Microbiology* **41**, 3744–3747.

Goethert, H. K. & Telford, S. R. III (2005). A new *Francisella* (Beggiatiales: Francisellaceae) inquiline within *Dermacentor variabilis* Say (Acari: Ixodidae). *Journal of Medical Entomology* **42**, 502–505.

Goethert, H. K., Shani, I. & Telford, S. R. III (2004). Genotypic diversity of *Francisella tularensis* infecting *Dermacentor variabilis* ticks from Martha's Vineyard, Massachusetts. *Journal of Clinical Microbiology* **42**, 4968–4973.

Goldman, L., Rockwell, E. & Richfield, D. F. (1952). Histopathological studies on the cutaneous reactions to the bites of various arthropods. *American Journal of Tropical Medicine and Hygiene* **1**, 514–525.

Goldwasser, R. A., Steiman, Y., Klingberg, W., Swartz, T. A. & Klingberg, M. A. (1974). The isolation of strains of rickettsiae of the spotted fever group in Israel and their differentiation from other members of the group by immunofluorescence methods. *Scandinavian Journal of Infectious Diseases* **6**, 53–62.

Gongora-Biachi, R. A., Zavala-Velazquez, J., Castro-Sansores, C. J. & Gonzalez-Martinez, P. (1999). First case of human ehrlichiosis in Mexico. *Emerging Infectious Diseases* **5**, 481 (letter).

Graves, S. & Stenos, J. (2003). *Rickettsia honei*: a spotted fever group rickettsia on three continents. *Annals of the New York Academy of Sciences* **990**, 62–66.

Graves, S., Dwyer, B. W., McColl, B. & McDade, J. E. (1991). Flinders Island Spotted Fever: a newly recognized endemic focus of tick typhus in Bass Strait. II Serological investigations. *Medical Journal of Australia* **154**, 99–104.

Gray, J., Vonstedingk, L. V., Gurtelschmid, M. & Granstrom, M. (2002). Transmission studies *of Babesia microti* in *Ixodes ricinus* ticks and gerbils. *Journal of Clinical Microbiology* **40**, 1259–1263.

Groen, J., Koraka, P., Nur, Y. A., *et al.* (2002). Serologic evidence of ehrlichiosis among humans and wild animals in the Netherlands. *European Journal of Clinical Microbioogy and Infectious Diseases* **21**, 46–49.

Grzeszczuk, A., Stanczak, J., Kubica-Biernat, B., *et al.* (2004). Human anaplasmosis in north-eastern Poland: seroprevalence in humans and prevalence in *Ixodes ricinus* ticks. *Annals of Agricultural and Environmental Medicine* **11**, 99–103.

Guillaume, B., Heyman, P., Lafontaine, S., *et al.* (2002). Seroprevalence of human granulocytic ehrlichiosis infection in Belgium. *European Journal of Clinical Microbiology and Infectious Diseases* **21**, 397–400.

Halos, L., Jamal, T., Maillard, R., *et al.* (2005). Evidence of *Bartonella* sp. in questing adult and nymphal *Ixodes ricinus* ticks from France and co-infection with *Borrelia burgdorferi* sensu lato and *Babesia* sp. *Veterinary Research* **36**, 79–87.

Harden, V. (1990). *Rocky Mountain Spotted Fever: History of a Twentieth-Century Disease*. Baltimore, MD: Johns Hopkins University Press.

Hartelt, K., Oehme, R., Frank, H., *et al.* (2004). Pathogens and symbionts in ticks: prevalence of *Anaplasma phagocytophilum* (*Ehrlichia* sp.), *Wolbachia* sp., *Rickettsia* sp., and *Babesia* sp. in Southern Germany. *International Journal of Medical Microbiology* **293**, 86–92.

Hayflick, L. (1969). *The Mycoplasmatales and the L-phase of Bacteria*. New York: Appleton-Century-Crofts.

Healing, T. D. (1981). Infections with blood parasites in the small British rodents *Apodemus sylvaticus*, *Clethrionomys glareolus* and *Microtus agrestis*. *Parasitology* **83**, 179–189.

Heo, E. J., Park, J. H., Koo, J. R., *et al.* (2002). Serologic and molecular detection of *Ehrlichia chaffeensis* and *Anaplasma phagocytophila* (human granulocytic ehrlichiosis agent) in Korean patients. *Journal of Clinical Microbiology* **40**, 3082–3085.

Heppner, D. G., Wongsrichanalai, C., Walsh, D. S., *et al.* (1997). Human ehrlichiosis in Thailand. *Lancet* **350**, 785–786.

Herwaldt, B. L., Caccio, S., Gherlinzoni, F., *et al.* (2003). Molecular characterization of a non-*Babesia divergens* organism causing babesiosis in Europe. *Emerging Infectious Diseases* **9**, 942–948.

Herwaldt, B. L., De Bruyn, G., Pieniazek, N. J., *et al.* (2004). *Babesia divergens*-like infection, Washington State. *Emerging Infectious Diseases* **10**, 622–629.

Herwaldt, B. L., Kjemtrup, A. M., Conrad, P. A., *et al.* (1997). Transfusion-transmitted babesiosis in Washington State: first reported case caused by a WAI-type parasite. *Journal of Infectious Diseases* **175**, 1259–1262.

Herwaldt, B. L., Persing, D. H., Precigout, E. A., *et al.* (1996). A fatal case of babesiosis in Missouri: identification of another piroplasm that infects humans. *Annals of Internal Medicine* **124**, 643–650.

Hildebrant, A., Hunfeld, K. P., Baier, M., *et al.* (2007). First confirmed autochthonous case of human *Babesia Microti* infection in Europe. *European Journal of Clinical Microbiology and Infectious Diseases* **26**, 595–601.

Hoare, C. A. (1980). Comparative aspects of human babesiosis. *Transactions of the Royal Society of Tropical Medicine and Hygiene* **74**, 143–148.

Holden, K., Hodzic, E., Feng, S., *et al.* (2005). Coinfection with *Anaplasma phagocytophilum* alters *Borrelia burgdorferi* population distribution in C3H/HeN mice. *Infection and Immunity* **73**, 3440–3444.

Holman, P. J., Spencer, A. M., Droleskey, R. E., Goethert, H. K. & Telford, S. R. (2005*a*). *In vitro* cultivation of a zoonotic *Babesia* sp. isolated from eastern cottontail rabbits (*Sylvilagus floridanus*) on Nantucket Island, Massachusetts. *Journal of Clinical Microbiology* **43**, 3995–4001.

Holman, P. J., Spencer, A. M., Telford, S. R. III, *et al.* (2005*b*). Comparative infectivity of *Babesia divergens* and a zoonotic *Babesia divergens*-like parasite in cattle. *American Journal of Tropical Medicine and Hygiene* **73**, 865–870.

Hopla, C. (1974). The ecology of tularemia. *Advances in Veterinary Sciences and Comparative Medicine* **18**, 25–53.

Hubalek, Z., Calisher, C. H. & Mittermayer, T. (1987). A new subtype ('Brezova') of *Tribec orbivirus* (Kemerovo group) isolated from *Ixodes ricinus* males in Czechoslovakia. *Acta Virologica* **31**, 91–92.

Hunfeld, K. P., Lambert, A., Kampen, H., *et al.* (2002). Seroprevalence of *Babesia* infections in humans exposed to ticks in midwestern Germany. *Journal of Clinical Microbiology* **40**, 2431–2436.

Ibarra, V., Portillo, A., Santibanez, S., *et al.* (2005). Debonel/Tibola: is *Rickettsia slovaca* the only etiological agent? *Annals of the New York Academy of Sciences* **1063**, 346–348.

Inokuma, H., Brouqui, P., Drancourt, M. & Raoult, D. (2001). Citrate synthase gene sequence: a new tool for phylogenetic analysis and identification of *Ehrlichia*. *Journal of Clinical Microbiology* **39**, 3031–3039.

James, A. M., Liveris, D., Wormser, G. P., *et al.* (2001). *Borrelia lonestari* infection after a bite by an *Amblyomma americanum* tick. *Journal of Infectious Diseases* **183**, 1810–1814.

Jellison, W. L. (1974). *Tularemia in North America*. Missoula, MT: University of Montana Press.

Jellison, W. L. & Parker, R. R. (1944). Rodents, rabbits and tularemia in North America: some zoological and epidemiological considerations. *American Journal of Tropical Medicine* **25**, 349–362.

Joyner, L. P., Davies, S. F. M. & Kendall, S. B. (1963). The experimental transmission of *Babesia divergens* by *Ixodes ricinus*. *Experimental Parasitology* **14**, 367–373.

Kaiser, R. (1996). Tick-borne encephalitis in southwestern Germany. *Infection* **24**, 398–399.

Kawahara, M., Ito, T., Suto, C., *et al.* (1999). Comparison of *Ehrlichia muris* strains isolated from wild mice and ticks and serologic survey of humans and animals with *E. muris* as antigen. *Journal of Clinical Microbiology* **37**, 1123–1129.

Kawahara, M., Rikihisa, Y., Lin, Q., *et al.* (2006). Genetic variants of *Anaplasma phagocytophilum*, *Anaplasma bovis*, *Anaplasma centrale* and *Ehrlichia* species in wild deer on two major islands in Japan. *Applied Environmental Microbiology* **72**, 1102–1109.

Kawamura, R. (1926). *Studies on Tsutsugamushi Disease (Japanese Flood Fever)*. Cincinnati, OH: Spokesman Printing Co.

Kelly, P. J., Beati, L., Mason, P. R., *et al.* (1996). *Rickettsia africae sp. nov.*, the etiological agent of African tick bite fever. *International Journal of Systematic Bacteriology* **46**, 611–614.

Kelly, P. J., Mason, P. R., Matthewman, L. A. & Raoult, D. (1991). Seroepidemiology of Spotted-Fever Group rickettsial infections in humans in Zimbabwe. *Journal of Tropical Medicine and Hygiene* **94**, 304–309.

Keysary, A., Amram, L., Keren, G., *et al.* (1999). Serologic evidence of human monocytic and granulocytic ehrlichiosis in Israel. *Emerging Infectious Diseases* **5**, 775–778.

Kim, C. M., Kim, M. S., Park, J. H. & Chae, J. S. (2003). Identification of *Ehrlichia chaffeensis*, *Anaplasma phagocytophilum*, and *Anaplasma bovis* in *Haemaphysalis longicornis* and *Ixodes persulcatus* ticks from Korea. *Vector Borne and Zoonotic Diseases* **3**, 17–26.

Kjemtrup, A. M., Lee, B., Fritz, C. L., *et al.* (2002). Investigation of transfusion transmission of a WA1-type babesial parasite to a premature infant in California. *Transfusion* **42**, 1482–1487.

Kjemtrup, A. M., Thomford, J., Robinson, T. & Conrad, P. A. (2000). Phylogenetic relationships of human and wildlife piroplasm isolates in the western United States inferred from the 18S nuclear small subunit RNA gene. *Parasitology* **120**, 487–493.

Kokernot, R. H., Calisher, C. H., Stannard, L. J. & Hayes, J. (1969). Arbovirus studies in the Ohio–Mississippi basin 1964–1967. VII. Lone Star virus, a hitherto unknown agent isolated from the tick *Amblyomma americanum*. *American Journal of Tropical Medicine and Hygiene* **18**, 789–795.

Krampitz, H. E. & Baumler, W. (1978). Occurrence, host range and seasonal prevlance of *Babesia microti* (Franca 1912) in rodents in Southern Germany. *Zeitschrift für Parasitenkunde* **58**, 15–33.

Krampitz, H. E. & Kleinschmidt, A. (1960). Grahamella Brumpt 1911: biologische und morphologische Untersuchungen. *Zeitschrift für Tropenmedizin und Parasitologie* **11**, 336–352.

Krause, P. J., Telford, S. R. III, Spielman, A., *et al.* (1996). Concurrent Lyme disease and babesiosis: evidence for increased severity and duration of illness. *Journal of the American Medical Association* **275**, 1657–1660.

Krinsky, W. L. & Burgdorfer, W. (1976). Trypanosomes in *Amblyomma americanum* from Oklahoma. *Journal of Parasitology* **62**, 824–825.

Kunz, C. (1992). Tick-borne encephalitis in Europe. *Acta Leidensia* **60**, 1–14.

Lakos, A. & Raoult, D. (1999). Tick-borne lymphadenopathy (TIBOLA), a *Rickettsia slovaca* infection? In *Rickettsiae and Rickettsial Diseases at the Turn of the Third Millennium*, eds. Raoult, D. & Brouqui, P., pp. 258–261. Paris: Elsevier.

Lebech, A. M., Hansen, K., Pancholi, P., *et al.* (1998). Immunoserologic evidence of human granulocytic ehrlichiosis in Danish patients with Lyme neuroborreliosis. *Scandinavian Journal of Infectious Diseases* **30**, 173–176.

Lemon, S. M. & Brown, E. A. (1995). Hepatitis C virus. In *Mandell, Douglas and Bennett's Principles and Practice of Infectious Diseases*, 4th edn, eds. Mandell, G. L., Bennett, J. E. & Dolin, R., pp. 1474–1486. New York: Churchill Livingstone.

Levin, M. L. & Fish, D. (2000). Immunity reduces reservoir host competence of *Peromyscus leucopus* for *Ehrlichia phagocytophila*. *Infection and Immunity* **68**, 1514–1518.

Little, S. E., Dawson, J. E., Lockhart, J. M., *et al.* (1997). Development and use of specific polymerase chain reaction for the detection of an organism resembling *Ehrlichia* sp. in white-tailed deer. *Journal of Wildlife Diseases* **33**, 246–253.

Lockhart, J. M., Davidson, W. R., Dawson, J. E. & Stallknecht, D. E. (1995). Temporal association of *Amblyomma americanum* with the presence of *Ehrlichia chaffeensis* reactive antibodies in white-tailed deer. *Journal of Wildlife Diseases* **31**, 119–124.

Lockhart, J. M., Davidson, W. R., Stallknecht, D. E., Dawson, J. E. & Howerth, E. W. (1997*a*). Isolation of *Ehrlichia chaffeensis* from wild white-tailed deer (*Odocoileus virginianus*) confirms their role as natural reservoir hosts. *Journal of Clinical Microbiology* **35**, 1681–1686.

Lockhart, J. M., Davidson, W. R., Stallknecht, D. E., Dawson, J. E. & Little, S. E. (1997*b*). Natural history of *Ehrlichia chaffeensis* (Rickettsiales: Ehrlichieae) in the piedmont physiographic province of Georgia. *Journal of Parasitology* **83**, 887–894.

Ludham, G. B., Bridges, J. B. & Benn, E. C. (1964). Association of Stevens–Johnson syndrome with antibodies for *Mycoplasma pneumoniae*. *Lancet* **i**, 958–959.

Lvov, D. K., Leonova, G. N., Gromashevsky, V. L., *et al.* (1978). Sikhote-Alin virus, a new member of the cardiovirus group (*Picornaviridae*) isolated from *Ixodes persulcatus* ticks in Primorie region. *Acta Virologica* **22**, 458–463.

Maguire, T. (1970). The laboratory transmission of coxsackie A6 virus by mosquitoes. *Journal of Hygiene* **68**, 625–630.

Mahara, F. (1984). Three Weil-Felix (OX2) positive cases with skin eruptions and high fever. *Journal of the American Medical Association* **68**, 4–7.

Mahara, F. (1997). Japanese spotted fever: report of 31 cases and review of the literature. *Emerging Infectious Diseases* **3**, 105–111.

Main, A. J. & Carey, A. B. (1980). Connecticut virus: a new Sawgrass group virus from *Ixodes dentatus* (Acari, Ixodidae). *Journal of Medical Entomology* **17**, 473–476.

Makinen, J., Vuorinen, I., Oksi, J., *et al.* (2003). Prevalence of granulocytic *Ehrlichia* and *Borrelia burgdorferi* sensu lato in *Ixodes ricinus* ticks collected from Southwest Finland and from Vormsi Island in Estonia. *Acta Pathologica, Microbiologica et Immunologica Scandinavica* **111**, 355–362.

Marconi, R. T., Liveris, D. & Schwartz, I. (1995). Identification of novel insertion elements, restriction fragment length polymorphism patterns, and discontinuous 23S rRNA in Lyme disease spirochetes: phylogenetic analyses of rRNA genes and their intergenic spacers in *Borrelia japonica* sp. nov. and genomic group 21038 (*Borrelia andersoni* sp. nov.) isolates. *Journal of Clinical Microbiology* **33**, 2427–2434.

Martin, P. A. & Schmidtmann, E. T. (1998). Isolation of aerobic microbes from *Ixodes scapularis* (Acari: Ixodidae), the vector of Lyme disease in the eastern United States. *Journal of Economic Entomology* **91**, 864–868.

Mason, J. H. & Alexander, R. A. (1939). Studies of the rickettsias of the typhus–Rocky Mountain spotted fever group in South Africa. III. The disease in the experimental animal: cross-immunity tests. *Onderstepoort Journal of Veterinary Animal Industry* **13**, 41–66.

Massung, R. F., Mauel, M., Owens, J. H., *et al.* (2002). Genetic variants of *Ehrlichia phagocytophila*, Rhode Island and Connecticut. *Emerging Infectious Diseases* **8**, 467–472.

Masters, E. J. & Donnell, H. D. (1995). Lyme and/or Lyme-like disease in Missouri. *Missouri Medicine* **92**, 346–353.

Masters, E., Granter, S., Duray, P. & Cordes, P. (1998). Physician-diagnosed erythema migrans and erythema migrans-like rashes following Lone Star tick bites. *Archives of Dermatology* **134**, 955–960.

Maxey, E. E. (1899). Some observations on the so-called spotted fever of Idaho. *Medical Sentinel of Portland* **7**, 433–438.

McQuiston, J. H., Paddock, C. D., Holman, R. C. & Childs, J. E. (1999). The human ehrlichioses in the United States. *Emerging Infectious Diseases* **5**, 635–642.

Megaw, J. W. (1921). A typhus like fever in India, possibly transmitted by ticks. *Indian Medical Gazette* **56**, 361–371.

Monteiro, J. L., Fonseca, F. & Prado, A. (1931). Pesquisas epidemiológicas sobre o typhus exanthematico de São Paolo. *Memorias Instituto Butantan* **6**, 139–173.

Morais, J. D., Dawson, J. E., Greene, C., *et al.* (1991). First European case of ehrlichiosis (letter). *Lancet* **338**, 633–634.

Murphy, G. L., Ewing, S. A., Whitworth, L. C., Fox, J. C. & Kocan, A. A. (1998). A molecular and serologic survey of *Ehrlichia canis*, *E. chaffeensis*, and *E. ewingii* in dogs and ticks from Oklahoma. *Veterinary Parasitology* **79**, 325–339.

Murray, R. G. E. & Schleifer, K. H. (1994). Taxonomic notes: a proposal for recording the properties of putative taxa of procaryotes. *International Journal of Systematic Bacteriology* **44**, 174–176.

Nadelman, R. B., Herman, E. & Wormser, G. P. (1997). Screening for Lyme disease in hospitalized psychiatric patients: prospective serosurvey in an endemic area. *Mount Sinai Journal of Medicine* **64**, 409–412.

Niebylski, M. L., Peacock, M. G., Fischer, E. R., Porcella, S. F. & Schwan, T. G. (1997*a*). Characterization of an endosymbiont infecting wood ticks, *Dermacentor andersoni*, as a member of the genus *Francisella*. *Applied and Environmental Microbiology* 63, 3933–3940.

Niebylski, M. L., Schrumpf, M. E., Burgdorfer, W., *et al.* (1997*b*). *Rickettsia peacockii* sp. nov., a new species infecting wood ticks, *Dermacentor andersoni*, in western Montana. *International Journal of Systematic Bacteriology* 47, 446–452.

Nilsson, K., Lindquist, O., Liu, A. J., *et al.* (1999). *Rickettsia helvetica* in *Ixodes ricinus* ticks in Sweden. *Journal of Clinical Microbiology* 37, 400–403.

Nilsson, K., Lindquist, O. & Pahlson, C. (1999). Association of *Rickettsia helvetica* with chronic perimyocarditis in sudden cardiac death. *Lancet* 354, 1169–1173.

Nilsson, K., Liu, A., Pahlson, C. & Lindquist, O. (2005*a*). Demonstration of intracellular microorganisms (*Rickettsia* spp., *Chlamydia pneumoniae*, *Bartonella* spp.) in pathological human aortic valves by PCR. *Journal of Infection* 50, 46–52.

Nilsson, K., Lukinius, A., Pahlson, C., *et al.* (2005*b*). Evidence of *Rickettsia* spp. infection in Sweden: a clinical, ultrastructural and serological study. *APMIS* 113, 126–134.

Nilsson, K., Pahlson, C., Lukinius, A., *et al.* (2001). Presence of *Rickettsia helvetica* in granulomatous tissue from patients with sarcoidosis. *Journal of Infectious Diseases* 185, 1128–1138.

Noda, H., Munderloh, U. G. & Kurtti, T. J. (1997). Endosymbionts of ticks and their relationship to *Wolbachia* spp. and tick-borne pathogens of humans and animals. *Applied and Environmental Microbiology* 63, 3926–3932.

Noguchi, H. (1926). The experimental transmission of *Bartonella bacilliformis* by ticks (*Dermacentor andersoni*). *Journal of Experimental Medicine* 44, 729–734.

Norval, R. A. I. (1977*a*). Ecology of the tick *Amblyomma hebraeum* Koch in Eastern Cape Province of South Africa. I. Distribution and seasonal activity. *Journal of Parasitology* 63, 734–739.

Norval, R. A. I. (1977*b*). Studies on ecology of the tick *Amblyomma hebraeum* Koch in Eastern Cape-Province of South Africa. II. Survival and development. *Journal of Parasitology* 63, 740–747.

Nuti, M., Serafini, D. A., Bassetti, D., *et al.* (1998). *Ehrlichia* infection in Italy. *Emerging Infectious Diseases* 4, 663–665.

Ogden, N. H., Bown, K., Horrocks, B. K., Woldehiwet, Z. & Bennett, M. (1998). Granulocytic *Ehrlichia* infection in ixodid ticks and mammals in woodlands and uplands of the UK. *Medical and Veterinary Entomology* 12, 423–429.

Ohashi, N., Inayoshi, M., Kitamura, K., *et al.* (2005). *Anaplasma phagocytophilum*-infected ticks, Japan. *Emerging Infectious Diseases* 11, 1780–1783.

Oksi, J., Viljanen, M. K., Kalimo, H., *et al.* (1993). Fatal encephalitis caused by concomitant infection with tick-borne encephalitis virus and *Borrelia burgdorferi*. *Clinical Infectious Diseases* 16, 392–396.

Oliver, J. H. Jr, Owsley, M. R., Hutcheson, H. J., *et al.* (1993). Conspecificity of the ticks *Ixodes scapularis* and *Ixodes dammini* (Acari: Ixodidae). *Journal of Medical Entomology* 30, 54–63.

Olsufiev, N. G., Emelyanova, O. S. & Dunayeva, T. N. (1959). Comparative study of strains of *B. tularense*. II. Evaluation of criteria of virulence of *Bacterium tularense* in the Old and New World and their taxonomy. *Journal of Hygiene, Epidemiology, Microbiology and Immunology* 3, 138–149.

O'Neill, S. L., Giordano, R., Colbert, A. M. E., Karr, T. L. & Robertson, H. M. (1992). 16S ribosomal RNA phylogenetic analysis of the bacterial endosymbionts associatied with cytoplasmic incompatibility in insects. *Proceedings of the National Academy of Sciences of the USA* 89, 2699–2702.

Oteo, J. A. & Ibarra, V. (2002). DEBONEL (*Dermacentor*-borne-necrosis- erythema-lymphadenopathy): a new tick-borne disease? *Enfermedades Infecciosas y Microbiologia Clínica* 20, 51–52.

Oteo, J. A., Blanco, J. R., Martinez de Artola, V. & Ibarra, V. (2000). First report of human granulocytic ehrlichiosis from southern Europe (Spain). *Emerging Infectious Diseases* 6, 430–432.

Paddock, C. D., Folk, S. M., Shore, G. M., *et al.* (2001). Infections with *Ehrlichia chaffeensis* and *Ehrlichia ewingii* in persons coinfected with human immunodeficiency virus. *Clinical Infectious Diseases* 33, 1586–1594.

Paddock, C. D., Sumner J. W., Comer, J. A., *et al.* (2004). *Rickettsia parkeri*: a newly recognized cause of spotted fever rickettsiosis in the United States. *Clinical Infectious Diseases* 38, 805–811.

Parker, R. R., Spencer, R. R. & Francis, E. (1924). Tularemia infection in ticks of the species *Dermacentor andersoni* Stiles in the Bitterroot Valley, Montana. *Public Health Reports* 39, 1057–1073.

Parola, P., Cornet, J.-P., Sanogo, Y. O., *et al.* (2003). Identification of new *Ehrlichia*, *Anaplasma*, and *Rickettsia* genotypes in ticks from the Thai–Myanmar border and Vietnam. *Journal of Clinical Microbiology* 41, 1600–1608.

Parola, P. P. & Raoult, D. R. (2000). New tick-transmitted rickettsial diseases. In *Tick-borne Infectious Diseases: Diagnosis and Management*, ed. Cunha, B. A., pp. 233–250. New York: Marcel Dekker.

Parola, P., Paddock, C. D. & Raoult, D. (2005). Tick-borne rickettsioses around the world: emerging diseases challenging old concepts. *Clinical Microbiology Reviews* 18, 719–756.

Parola, P., Vestris, G., Martinez, D., *et al.* (1999). Tick-borne rickettsiosis in Guadeloupe, the French West Indies: isolation of *Rickettsia africae* from *Amblyomma variegatum* ticks and serosurvey in humans, cattle, and goats. *American Journal of Tropical Medicine and Hygiene* 60, 888–893.

Pavlovsky, E. N. (1966). *Natural Nidality of Transmissible Diseases.* Urbana, IL: University of Illinois Press.

Pegram, P. S. Jr, Sessler, C. N., London, W. L. & Burgdorfer, W. (1983). Lyme disease in North Carolina. *Southern Medical Journal* 76, 740–742.

Perez, M., Bodor M., Zhang, C., Xiong, Q. & Rikihisa, Y. (2006). Human infection with *Ehrlichia canis* accompanied by clinical signs in Venezuela. *Annals of the New York Academy of Sciences* 1078, 110–117.

Perez, M., Rikihisa, Y. & Wen, B. H. (1996). *Ehrlichia* canis-like agent isolated from a man in Venezuela: antigenic and genetic characterization. *Journal of Clinical Microbiology* 34, 2133–2139.

Persing, D. H., Herwaldt, B. L., Glaser, C., *et al.* (1995). Infection with a *Babesia*-like organism in northern California. *New England Journal of Medicine* 332, 298–303.

Petrovec, M., Lotric Furlan, S., Zupanc, T. A., *et al.* (1997). Human disease in Europe caused by a granulocytic *Ehrlichia* species. *Journal of Clinical Microbiology* 35, 1556–1559.

Philip, C. B., Hoogstraal, H., Reiss-Gutfreund, R. & Clifford, C. M. (1966). Evidence of rickettsial disease agents in ticks from Ethiopian cattle. *Bulletin of the World Health Organization* 35, 127–131.

Pierard, D., Levtchenko, E., Dawson, J. E. & Lauwers, S. (1995). Ehrlichiosis in Belgium. *Lancet* 346, 1233–1234.

Piesman, J., Mather, T. N., Telford, S. R. III & Spielman, A. (1986). Concurrent *Borrelia burgdorferi* and *Babesia microti* infection in nymphal *Ixodes dammini*. *Journal of Clinical Microbiology* 24, 446–447.

Pijper, A. (1936). Etude expérimentale comparée de la fièvre boutonneuse et de tick bite fever. *Archives du Institut Pasteur Tunis* 25, 388–401.

Planck, A., Eklund, A., Grunewald, J. & Vene, S. (2004). No serological evidence of *Rickettsia helvetica* infection in Scandinavian sarcoidosis patients. *European Respiratory Journal* 24, 811–813.

Polin, H., Hufnagl, P., Haunschmid, R., Gruber, F. & Ladurner, G. (2004). Molecular evidence of *Anaplasma phagocytophilum* in *Ixodes ricinus* ticks and wild animals in Austria. *Journal of Clinical Microbiology* 42, 2285–2286.

Pomerantsev, B. I. (1959). *Ixodid Ticks (Ixodidae).* Washington, DC: American Institute of Biological Sciences.

Pretorius, A.-M. & Birtles, R. J. (2002). *Rickettsia aeschlimannii*: a new pathogenic spotted fever group rickettsia, South Africa. *Emerging Infectious Diseases* 8, 874 (letter).

Quick, R. E., Herwaldt, B. L., Thomford, J. W., *et al.* (1993). Babesiosis in Washington State: a new species of *Babesia?* *Annals of Internal Medicine* 119, 284–290.

Randolph, S. E. (2002). Predicting the risk of tick-borne diseases. *International Journal of Medical Microbiology* 291 (Suppl. 33), 6–10.

Randolph, S. E. & Rogers, D. J. (2000). Fragile transmission cycles of tick-borne encephalitis virus may be disrupted by predicted climate change. *Proceedings of the Royal Society of London B* 267, 1741–1744.

Raoult, D., Berbis, P., Roux, V., Xu, W. & Maurin, M. (1997). A new tick-transmitted disease due to *Rickettsia slovaca*. *Lancet* 350, 112–113.

Raoult, D., Brouqui, P. & Roux, V. (1996). A new spotted fever group rickettsiosis. *Lancet* 348, 412.

Raoult, D., Fournier, P. E., Abboud, P. & Caron, F. (2002). First documented human *Rickettsia aeschlimannii* infection. *Emerging Infections Diseases* 8, 748–749.

Reid, H. W. (1986). Louping ill. In *The Arboviruses: Epidemiology and Ecology*, vol. 3, ed. Monath, T. P., pp. 117–135. Boca Raton, FL: CRC Press.

Relman, D. A. (2002). Mining the world for new pathogens. *American Journal of Tropical Medicine and Hygiene* 67, 133–134.

Rich, S. M., Armstrong, P. M., Smith, R. D. & Telford, S. R. (2001). Lone star tick-infecting borreliae are most closely related to the agent of bovine borreliosis. *Journal of Clinical Microbiology* 39, 494–497.

Richmond, J. Y. & McKinney, R. W. (1993). *Biosafety in Microbiological and Biomedical Laboratories*, 3rd edn. Washington, DC: US Department of Health and Human Services.

Richter, P. J. Jr, Kimsey, R. B., Madigan, J. E., *et al.* (1996). *Ixodes pacificus* (Acari: Ixodidae) as a vector of *Ehrlichia equi* (Rickettsiales: Ehrlichiae). *Journal of Medical Entomology* 33, 1–5.

Ricketts, H. T. (1906). The transmission of Rocky Mountain spotted fever by the bite of the wood tick (*Dermacentor occidentalis*). *Journal of the American Medical Association* 47, 358 (letter).

Ripoll, C. M., Remondegui, C. E., Ordonez, G., *et al.* (1999). Evidence of rickettsial spotted fever and ehrlichial infections in a subtropical territory of Jujuy, Argentina. *American Journal of Tropical Medicine and Hygiene* **61**, 350–354.

Roberts, D., Wang, X., Kemp, R. G. & Felz, M. W. (1999). Solitary erythema migrans in Georgia and South Carolina. *Archives of Biochemistry and Biophysics* **371**, 326–331.

Robertson, R. G. & Wisseman, C. L. (1973). Tick-borne rickettsiae of the spotted fever group in West Pakistan. II. Serological classification of isolates from West Pakistan and Thailand: evidence for two new species. *American Journal of Epidemiology* **97**, 55–64.

Root, R. B. (1967). The niche exploitation pattern of the blue-gray gnatcatcher. *Ecological Monographs* **37**, 317–350.

Salgo, M. P., Telzak, E. E., Currie, B., *et al.* (1988). A focus of Rocky Mountain spotted fever within New York city. *New England Journal of Medicine* **318**, 1345–1348.

Sanogo, Y. O., Zeaiter, Z., Caruso, G., *et al.* (2003). Detection of *Bartonella henselae* in ticks *Ixodes ricinus* (Acari: Ixodidae) removed from humans in Belluno Province, Italy. *Emerging Infectious Diseases* **9**, 329–332.

Sant'Ana, J. F. (1911). On a disease in man following tick bites and occuring in Lourenço Marques. *Parasitology* **4**, 87–88.

Santos, A. S., Santos-Silva, M. M., Almeida, V. C., Bacellar, F. and Dumler, J. S. (2004). Detection of *Anaplasma phagocytophilum* DNA in *Ixodes* ticks (Acari: Ixodidae) from Madeira Island and Setubal District mainland, Portugal. *Emerging Infectious Diseases* **10**, 1643–1648.

Schabereiter-Gurtner, C., Lubitz, W. & Rolleke, S. (2003). Application of broad range 16S rRNA PCR amplification and DGGE fingerprinting for detection of tick-infecting bacteria. *Journal of Microbiological Methods* **52**, 251–260.

Schouls, L. M., Van De Pol, I., Rijpkema, S. G. & Schot, C. S. (1999). Detection and identification of *Ehrlichia, Borrelia burgdorferi sensu lato*, and *Bartonella* species in Dutch *Ixodes ricinus* ticks. *Journal of Clinical Microbiology* **37**, 2215–2222.

Shibata, S., Kawahara, M., Rikihisa, Y., *et al.* (2000). New *Ehrlichia* species closely related to *Ehrlichia chaffeensis* isolated from *Ixodes ovatus* ticks in Japan. *Journal of Clinical Microbiology* **38**, 1331–1338.

Shih, C. M., Liu, L. P., Chung, W. C., Ong, S. J. & Wang, C. C. (1997). Human babesiosis in Taiwan: asymptomatic infection with a *Babesia microti*–like organism in a Taiwanese woman. *Journal of Clinical Microbiology* **35**, 450–454.

Shypnov, S., Parola, P., Rudakov, N., *et al.* (2001). Detection and identification of spotted fever group rickettsiae in *Dermacentor* ticks from Russia and Central Kazakhstan. *European Journal of Clinical Microbiology and Infectious Diseases* **20**, 903–905.

Shypnov, S. S., Fournier, P.-E., Rudakov, N. & Raoult, D. (2003). 'Candidatus Rickettsia tarasevichae' in *Ixodes persulcatus* ticks collected in Russia. *Annals of the New York Academy of Sciences* **990**, 162–172.

Simser, J. A., Palmer, A. T., Fingerle, V., *et al.* (2002). *Rickettsia monacensis* sp. nov., a spotted fever group rickettsia, from ticks (*Ixodes ricinus*) collected in a European city park. *Applied and Environmental Microbiology* **68**, 4559–4566.

Skotarczak, B., Rymaszewska, A. & Adamska, M. (2003). Polymerase chain reaction in detection of human granulocytic ehrlichiosis (HGE) agent DNA in *Ixodes ricinus* ticks. *Folia Medica Cracoviensia* **44**, 179–186.

Škrabalo, Z. & Deanovic, Ž. (1957). Piroplasmosis in man: report on a case. *Documenta de Medicina Geographica et Tropica* **9**, 11–16.

Smith, T. & Kilborne, F. L. (1893). *Investigation into the Nature, Causation, and Prevention of Texas or Southern Cattle Fever*, Bureau of Animal Industries Bulletin No. 1. Washington, DC: US Department of Agriculture.

Spencer, A. M., Goethert, H. K., Telford, S. R. & Holman, P. J. (2006). In vitro host erythrocyte specificity and differential morphology of *Babesia divergens* and a zoonotic *Babesia* sp. from eastern cottontail rabbits (*Sylvilagus floridanus*). *Journal of Parasitology* **92**, 333–340.

Spielman, A. & Rossignol, P. (1984). Insect vectors and scalars. In *Tropical and Geographic Medicine*, eds. Warren, K. S. & Mahmoud, A. F., pp. 167–183. New York: McGraw-Hill.

Steere, A. C., McHugh, G., Suarez, C., *et al.* (2003). Prospective study of coinfection in patients with erythema migrans. *Clinical Infectious Diseases* **36**, 1078–1081.

Stenos, J., Roux, V., Walker, D. H. & Raoult, D. (1998). *Rickettsia honei* sp. nov., the aetiological agent of Flinders Island spotted fever in Australia. *International Journal of Systematic Bacteriology* **48**, 1399–1404.

Stewart, R. S. (1991). Flinders Island spotted fever: a newly recognized endemic focus of tick typhus in Bass Strait. I. Clinical and epidemiological features. *Medical Journal of Australia* **154**, 94–99.

Stuen, S., Van De Pol, I., Bergstrom, K. & Schouls, L. A. (2002). Identification of *Anaplasma phagocytophilum* (formerly *Ehrlichia phagocytophila*) variants in blood from sheep in Norway. *Journal of Clinical Microbiology* **40**, 3192–3197.

Suitor, E. C. & Weiss, E. (1961). Isolation of a *Rickettsia*-like microorganism (*Wolbachia persica* n. sp.) from *Argas persicus* (Oken). *Journal of Infectious Diseases* **108**, 95–99.

Sumption, K. J., Wright, D. J., Cutler, S. J. & Dale, B. A. (1995). Human ehrlichiosis in the UK. *Lancet* **346**, 1487–1488.

Sun, L. V., Scoles, G. A., Fish, D. & O'Neill, S. L. (2000). *Francisella-like endosymbionts of ticks. Journal of Invertebrate Pathology* **76**, 301–303.

Tarasevich, I. V., Makarova, V. A., Fetisova, N. F., *et al.* (1991). Astrakhan fever: a spotted fever rickettsiosis. *Lancet* **337**, 172–173.

Tarnvik, A., Sandstrom, G. & Sjostedt, A. (1996). Epidemiological analysis of tularaemia in Sweden 1931–1993. *FEMS Immunology and Medical Microbiology* **13**, 201–204.

Taroura, S., Shimada, Y., Sakata, Y., *et al.* (2005). Detection of DNA of '*Candidatus* Mycoplasma haemominutum' and *Spiroplasma* sp. in unfed ticks collected from vegetation in Japan. *Journal of Veterinary Medical Science* **67**, 1277–1279.

Telford, S. R. (1998). The name *Ixodes dammini* epidemiologically justified. *Emerging Infectious Diseases* **4**, 132–134.

Telford, S. R. & Foppa, I. M. (2000). Tickborne encephalitides. In *Tickborne Infectious Diseases, Diagnosis and Management*, ed. Cunha, B. A., pp. 193–214. New York: Marcel Dekker.

Telford, S. R. & Goethert, H. K. (2004). Emerging tick-borne infections: rediscovered and better characterized, or truly new? *Parasitology* **129**, S301–S327.

Telford, S. R. III & Maguire, J. H. (2005). Babesiosis. In *Tropical Infectious Diseases: Prinicples, Pathogens, and Practice, vol. 2*, eds. Guerrant, R., Walker, D. H. & Weller, P. F., pp. 1063–1071. London: Churchill Livingstone.

Telford, S. R., Armstrong, P. M., Katavolos, P., *et al.* (1997). A new tick-borne encephalitis-like virus infecting New England deer ticks, *Ixodes dammini. Emerging Infectious Diseases* **3**, 165–170.

Telford, S. R., Dawson, J. E., Katavolos, P., *et al.* (1996). Perpetuation of the agent of human granulocytic ehrlichiosis in a deer tick-rodent cycle. *Proceedings of the National Academy of Sciences of the USA* **93**, 6209–6214.

Telford, S. R., Gorenflot, A., Brasseur, P. & Spielman, A. (1993). Babesial infections in humans and wildlife. In *Parasitic Protozoa*, vol. 5, ed. Kreier, J. P., pp. 1–47. San Diego, CA: Academic Press.

Telford, S. R. III, Korenberg, E. I., Goethert, H. K., *et al.* (2002). Detection of natural foci of babesiosis and granulocytic ehrlichiosis in Russia. *Journal of Microbiology, Epidemiology and Immunology* **6**, 21–25.

Telford, S. R. III, Lepore, T. J., Snow, P., Warner, C. K. & Dawson, J. E. (1995). Human granulocytic ehrlichiosis in Massachusetts. *Annals of Internal Medicine* **123**, 277–279.

Tenckhoff, B., Kolmel, H. W., Wolf, V. & Lange, R. (1994). Production and characterization of a polyclonal antiserum against *Spiroplasma mirum. Zentralblatt für Bakteriologie* **280**, 409–415.

Tenter, A. M., Barta, J. R., Beveridge, I., *et al.* (2002). The conceptual basis for a new classification of the coccidia. *International Journal for Parasitology* **32**, 595–616.

Theil, K. W., McCloskey, C. M. & Scott, D. P. (1993). Serologic evidence for rabbit syncytial virus in eastern cottontails (*Sylvilagus floridanus*) in Ohio. *Journal of Wildlife Diseases* **29**, 470–474.

Thomford, J. W., Conrad, P. A., Boyce, W. M., Holman, P. J. & Jessup, D. A. (1993). Isolation and *in vitro* cultivation of *Babesia* parasites from free-ranging desert bighorn sheep (*Ovis canadensis nelsoni*) and mule deer (*Odocoileus hemionus*) in California. *Journal of Parasitology* **79**, 77–84.

Treib, J. (1994). First case of tick-borne encephalitis in the Saarland. *Infection* **22**, 368–369.

Tsuji, M., Wei, Q., Zamoto, A., *et al.* (2001). Human babesiosis in Japan: epizootiologic survey of rodent reservoir and isolation of new type of *Babesia microti*-like parasite. *Journal of Clinical Microbiology* **39**, 4316–4322.

Tully, J. F., Rose, D. L., Yunker, C. E., *et al.* (1995). *Spiroplasma ixodetis*. sp. nov., a new species from *Ixodes pacificus* ticks collected in Oregon. *International Journal of Systematic Bacteriology* **45**, 23–28.

Tully, J. F., Whitcomb, R. F., Rose, D. L., Williamson, D. L. & Bove, J. M. (1983). Characterization and taxonomic status of tick spiroplasmas: a review. *Yale Journal of Biology and Medicine* **56**, 599–603.

Tyzzer, E. E. (1938). *Cytoecetes microti*, n.g., n.sp., a parasite developing in granulocytes and infective for small rodents. *Parasitology* **30**, 242–257.

Uchida, T. (1993). *Rickettsia japonica*, the etiologic agent of oriental spotted fever. *Microbiology and Immunology* **37**, 91–102.

Unsworth, N. B., Stenos, J., McGregor, A. R., Dyer, J. R. & Graves, S. R. (2005). Not only 'Flinders Island' spotted fever. *Pathology* **37**, 242–245.

Urvolgyi, J. & Brezina, R. (1976). *Rickettsia slovaca*: new member of spotted fever-group rickettsiae. *Folia Microbiologica* **21**, 503–503.

Van Peenen, P. F., Chang, S. J., Banknieder, A. R. & Santana, F. J. (1977). Piroplasms from Taiwanese rodents. *Journal of Protozoology* **24**, 310–312.

Varela, A. S., Luttrell, M. P., Howerth, E. W., *et al.* (2004). First culture isolation of *Borrelia lonestari*, putative agent of southern tick-associated rash illness. *Journal of Clinical Microbiology* **42**, 1163–1169.

Venzal, J. M., Portillo, A, Estrada-Peña, A., *et al.* (2004). *Rickettsia parkeri* in *Amblyomma triste* from Uruguay. *Emerging Infectious Diseases* **10**, 1493–1495.

Vestris, G., Rolain, J. E., Fournier, P. E., *et al.* (2003). Seven years' experience of isolation of *Rickettsia* spp. from clinical specimens using the shell vial cell culture assay. *Annals of the New York Academy of Sciences* **990**, 371–374.

Vitale, G., Mansueto, S., Rolain, J. M. & Raoult, D. (2005). *Rickettsia massiliae* human isolation. *Emerging Infectious Diseases* **2**, 174–175.

Vyrostekova, V., Khanakah, G., Kocianova, E., Gurycova, D. & Stanek, G. (2002). Prevalence of coinfection with *Francisella tularensis* in reservoir animals of *Borrelia burgdorferi sensu lato*. *Wiener klinische Wochenschrift* **114**, 482–488.

Walker, D. H. (1998). Tick transmitted diseases in the United States. *Annual Review of Public Health* **19**, 237–269.

Walls, J. J., Greig, B., Neitzel, D. F. & Dumler, J. S. (1997). Natural infection of small mammal species in Minnesota with the agent of human granulocytic ehrlichiosis. *Journal of Clinical Microbiology* **35**, 853–855.

Weder, B., Ketz, E., Matter, L. & Schuman, S. H. (1989). Lyme and other tick-borne diseases acquired in South Carolina in 1988: a survey of 1331 physicians. *Journal of Neurology* **236**, 305–306.

Weller, S. J., Baldridge, G. D., Munderloh, U. G., *et al.* (1998). Phylogenetic placement of rickettsiae from the ticks *Amblyomma americanum* and *Ixodes scapularis*. *Journal of Clinical Microbiology* **36**, 1305–1317.

Wellmer, H. & Jusatz, H. J. (1981). Geoecological analyses of the spread of tick-borne encephalitis in Central Europe. *Social Science and Medicine – Medical Geography D* **15**, 159–162.

Wesslen, L. (2001). Sudden cardiac death in Swedish orienteers. Unpublished Ph.D. thesis, Uppsala University, Sweden.

Wolbach, S. B. (1919). Studies on Rocky Mountain spotted fever. *Journal of Medical Research* **41**, 1–197.

Woldehiwet, Z. & Scott, G. R. (1993). Tick-borne (pasture) fever. In *Rickettsial and Chlamydial Diseases of Domestic Animals*, eds. Woldehiwet, Z. & Ristic, M., pp. 233–254. Oxford, UK: Pergamon Press.

Wood, W. W. (1896). *Spotted Fever as Reported from Idaho*. Washington, DC: US Government Printing Office.

Wormser, G. P., Masters, E., Liveris, D., *et al.* (2005). Microbiologic evaluation of patients from Missouri with erythema migrans. *Clinical Infectious Diseases* **40**, 423–428.

Wurzel, L. G., Cable, R. G. & Leiby, D. A. (2002). Can ticks be vectors for hepatitis C virus? *New England Journal of Medicine* **347**, 1724–1725.

Yabsley, M. J., Varela, A. S., Tate, C. M., *et al.* (2002). *Ehrlichia ewingii* infection in white-tailed deer (*Odocoileus virginianus*). *Emerging Infectious Diseases* **8**, 668–671.

Young, A. S. & Morzaria, S. P. (1986). Biology of *Babesia*. *Parasitology Today* **2**, 211–219.

Zavala-Velazquez, J. E., Ruiz Sosa, J., Vado Solis, I., Billings, A. N. & Walker, D. H. (1999). Serologic study of the prevalence of rickettsiosis in Yucatan: evidence for a prevalent spotted fever group rickettsiosis. *American Journal of Tropical Medicine and Hygiene* **61**, 405–408.

Zdrodowski, P. F. & Golinevich, H. M. (1960). *The Rickettsial Diseases*. New York: Pergamon Press.

17 • Analysing and predicting the occurrence of ticks and tick–borne diseases using GIS

M. DANIEL, J. KOLÁŘ AND P. ZEMAN

INTRODUCTION

For many years, scientific research has considered the relationships between the landscape and human health. Increasing rates of environmental changes are dramatically altering patterns of human health at the community, regional and global scales. The emergence of tick-borne diseases (TBD) illustrates the impact that environmental changes can have on human health. Integration of modern geoinformation technologies into landscape epidemiology can contribute significantly to the development and implementation of new disease-surveillance tools. The theory of landscape epidemiology offers the opportunity to use the landscape as a key to the identification of the spatial and temporal distribution of disease risk. Key environmental elements – including elevation, temperature, rainfall and humidity – influence the presence, development, activity and longevity of pathogens, vectors and zoonotic reservoirs of infection, and their interactions with humans (Meade, Florin & Gesler, 1988). The same environmental variables influence distribution of vegetation type as landscape elements and patterns of disease. Remote sensing (RS) from aircraft and satellites can be used to describe landscape elements that influence the patterns and prevalence of disease. In addition, geographical information systems (GIS) provide tools for modelling spatially their occurrence in space and time.

Ticks are ideally suited to GIS and RS applications owing to their close ties with the ecosystem. This relationship is determined by: (1) the type of host–parasite association (most important vector species are three-host ticks); (2) specific requirements of the microclimate; and (3) dependence on clearly defined types of plant associations which both reflect the microclimatic conditions of a habitat occupied by ticks, and also influence them.

Ticks and TBD are highly amenable to predictive mapping. In order to be infected, hosts must enter a habitat in which infected ticks are present. Such areas can be accurately identified topographically (this contrasts with vectors such as mosquitoes that can travel long distances from their breeding places). These traits enable great precision in spatial analyses of the distribution of ticks and TBDs, and in the construction of prognostic maps.

The use of GIS has many implications for landscape epidemiology because it provides the ability to store, integrate, query, display and analyse data from the molecular level to that of satellite resolution through shared spatial components. Field observations on environmental conditions, including vegetation, water and topography, can be combined in a GIS to direct interpretation of RS data and facilitate characterization of the landscape in terms of vector and pathogen prevalence. The associations between disease risk variables (e.g. vector, pathogen and host abundance and distribution) and environmental variables can be quantified using the spatial analysis capabilities of the GIS. Landscape pattern analysis, combined with statistical analysis, allows us to define landscape predictors of disease risk that can be applied to larger regions where field data are unavailable. This makes RS/GIS powerful tools for disease surveillance enabling the prediction of potential disease outbreaks and targeting intervention programmes (CDC, 1994, 1998).

GEOGRAPHICAL INFORMATION SYSTEMS

A GIS is a system of computer software, hardware and data, and personnel to manipulate, analyse and present information that is tied to a spatial location. A GIS further enhances the role of a map. A GIS is not just an automated decision-making system but a tool to query, analyse and map data in support of the decision-making process. Because GIS products can be produced quickly, multiple scenarios can be evaluated efficiently and effectively.

Ticks: Biology, Disease and Control, ed. Alan S. Bowman and Patricia A. Nuttall. Published by Cambridge University Press.
© Cambridge University Press 2008.

Real world description

Our decisions require adequate knowledge about the world. This knowledge, however, is far from complete. Therefore, we have to use incomplete, selected information. The model formed is a set of relationships, conditions and rules describing our understanding of what creates a specific part of our world and how it is functioning. The real world can only be described in terms of a model. There are countless possible approaches to the description and interpretation; consequently, the design of GIS systems varies according to their purpose. Every model is a simplification of the reality because it only includes selected data relevant to the model.

Spatial data

Data used to describe our world must include answers to the key questions: 'What?', 'Where?' and 'When?'. If data are related in such a way that they can answer all three questions, they constitute spatial data.

Depending on the question, two fundamental components characterize spatial data: (1) geometric data or location data, and (2) non-geometric data or descriptive data (attributes). Geographical data are inherently in the form of spatial data. An estimated 80% of all natural sciences data has a spatial component. A GIS uses geographical data in a digital format. Location data usually refer to specified spatial coordination systems of two dimensions (e.g. longitude and latitude) or three dimensions (e.g. the Cartesian orthogonal system used in satellite navigation systems). When a common mapping coordinate system is used, spatial data in GIS are termed geocoded or georeferenced data. Simple geometrical shapes are used to describe location data. A point represents the location of phenomena or objects small enough for such presentation. Sets of connected points create lines indicating the position of linear features like boundaries or communication networks. Closed lines constitute boundaries of an area such as those delineating a country, lake or forest.

Time is the third fundamental component of spatial data. It is a contextual reference. Thus, location and descriptive data relate to a given point in time.

Spatial data in maps and in GIS

Printed maps contain spatial data in a compact and easily accessible format. However, in rapidly changing regions, such maps quickly become outdated. Retrieving information from printed maps is relatively simple for small amounts of information. For larger volumes of data, as required for applications such as environmental assessments, modelling land cover changes, or climate studies, it becomes more difficult and costly. It becomes even more time consuming and cumbersome when data integration from different maps is required.

These drawbacks are overcome by the use of computing power to process large volumes of data for geographical data analysis. The requirements are: (1) all data in digital format and (2) a set of relatively simple instructions (software) for data processing. Generally, there are two approaches to preparing digital spatial data for GIS.

A vector data model stores a table containing coordinates of points together with instructions on which points are alone and which points belong to a common set. Digital data are discrete data and, hence, analogue values are eliminated when a digital data model is applied. Graphical presentation of spatial data is also based on points. To demonstrate that some points belong to one object a line connects the points. The simplest line is a vector in which both start- and end-point are defined. In a vector data model, all lines are represented by chains of vectors, and all areas by polygons. For graphical presentations, discrete values can be transformed into continuous data using mathematical functions to compute the missing values. Other types of function can be used instead of straight lines if they better describe changes of spatial data in space and/or time. Attributes are coded in separate tables using alphanumerical characters as a label for a specific class or category of properties. Association between location data tables and attribute tables is achieved by inserting the same identities in both kinds of tables for identical objects.

A raster data model uses a net of adjacent polygons (termed 'cells') to provide a virtual cover of a given part of a landscape. In its simplest form, cells are identical squares forming an organized grid. Each cell represents one entity of the landscape's surface. Location data for this entity is coded by the cell position in a matrix defined by row and column number. The cells are often called pixels, a term used in display technology for picture elements. Attribute values of the objects that the cells represent are assigned to corresponding pixels. Only one value can be assigned to a single cell, which requires a different raster layer for each thematic topic or period of time. Thematic layers are also used in vector models, but because of differences in attribute manipulation, raster models have more layers than vector models.

An important attribute of GIS is its data-handling capacity; GIS allows the input of data of different origin, content and format, and digital data can be edited and modified relatively easily. Consequently, GIS enables databases to be updated regularly. The system also offers wide flexibility in manipulating the data and displaying it in different formats. Finally, and importantly, GIS allows quantitative methods to be used in spatial data analysis. Analytical methods based on mapped analogue data are primarily qualitative and depend on visual and intuitive assessment of the data. In contrast, GIS permits qualitative methods to be expressed through mathematical relationships computed by software instructions. The real benefit of GIS is the ability to integrate data in models to give a value-added product, customized to meet user needs.

In the long term, the advantages of digital data reward the initial work and financial investment. Rapidly increasing computing power has resulted in a proliferation of GIS applications in environmental analysis during the last 20 years.

Spatial data collection

DATABASE
Geographical or spatial data are stored together in one set of data called a database. Accessibility of the data is dependent upon the way the data are organized in the database. In GIS, the quantity of data is large enough that the form and performance of the database is critical to the overall usefulness of the GIS. Thus, a database management system is installed to facilitate input, storage and retrieval of data. The system determines the response time, storage capacity, flexibility and other parameters affecting the usefulness of a database. Database structures differ primarily according to the facilities or operations they support. The most frequently used database system in GIS, the relational database management system, provides a simple and flexible structure of tables. More advanced structures are object-oriented database systems in which stored units are homogeneous objects.

Databases organized around GIS cores are essential elements of information systems. No current database system completely fulfils the needs of all database applications. Finally, operations required for any type of data manipulation have to be described in software form as a set of relatively simple instructions to be executed by computer. The emphasis in the future will be on improved organization of spatial databases for use in GIS.

DATA SOURCES
The sources of digital geographical data needed for a comprehensive GIS are of a tremendous variety. The variety of data types that can be used in a GIS database increases its applications. However, integration of diverse data for analysis needs to be done in a controllable way to avoid the results of analysis being interpreted incorrectly.

Major data types contributing to GIS databases include: (1) conventional documents; (2) survey measurements of both location and attribute values (satellite global positioning systems can be used for precise location data measurements); (3) conventional photographs, maps, drawings, graphs and other analogue images converted into digital format by either manual digitization of selected geometrical features resulting in a vector data set, or by scanning procedures that produce digital data sets in a raster model; (4) existing digital non-spatial data, e.g. tables or textures; (5) existing spatial databases; (6) remote sensing data acquired by radiometers on satellites or aircraft. Earth observation satellites provide a coherent, objective and regular source of input data for GIS databases. Scanning radiometers or imaging radar can observe large areas of land in a very short time and make digital measurements over the given site repeatedly. Prime databases have now become accessible through the World Wide Web, creating a demand for compatibility of web databases and for universal access.

Spatial data quality

The performance of GIS depends not only on the technical characteristics of hardware and functions offered by software, but primarily on the quality of input data. Digital data are frequently considered to be of higher quality than conventional data, but this is not necessarily the case. Data quality depends on the quality of the data sources: quality of the original input data, the method of data acquisition and on the precision of all stages of data processing. Knowing the quality of data is critical to determining the appropriate use of the data. The most important measures of data quality that affect GIS applications are listed below.

ACCURACY OF LOCATION AND ATTRIBUTE DATA
Positional accuracy and accuracy of continuous attribute variables may deviate from 'true' ground values. The accuracy of discrete attributes is similar to that of classifications used in remote sensing data that are verified by comparing data with randomly selected field measurements or surveys.

TOPOLOGY

Topology requires that logical relationships among data are established. The consistency of the logical relationship is difficult to measure. Two data sets may have inconsistencies arising from slight differences in the position of the same boundary in the two sets, even when each data set is correct according to its level of accuracy. This difference creates a fictional polygon in the database when the data sets are overlaid. Data with unverified topology degrade the quality of the results.

DATA COMPLETENESS

Assessment of completeness is usually limited to reporting the proportion of data available for the area of interest (data coverage). However, qualitative factors of completeness are important to determine the suitability of a data set for an application. Completeness of classification refers to the capability of nomenclature to represent the data. Completeness of verification indicates the number and distribution of measurements used to develop the data.

DATA RESOLUTION

The size of the smallest unit area represented in a data set defines the data resolution. For image or raster data sets, this parameter is termed spatial resolution and corresponds to the size of the picture element (pixel). For maps of vector data sets the resolution is the size of the smallest mapping unit.

DATA TIME

Time is commonly expressed as the date of acquisition of source material. The time component of data quality is critical for dynamic geographical information. It is also important whenever tasks involve data sets collected independently at different times. The relative importance of each quality factor described above depends on the GIS application task. When spatial data analyses are performed manually, assumptions about data quality are made. In the digital environment of GIS, quality standards and the methods to measure them have to be explicitly defined before data entry commences. Unfortunately, the quality of geographical data is often examined only after incorrect conclusions, findings or decisions have been made.

Data analysis

A GIS is used as an interactive analysing tool to support understanding of the environment and successive decision making as well as a database for the storage and retrieval of spatial information. Spatial analysis functions are characteristic features distinguishing a GIS from all other information systems. Ideally, a GIS should have all the functions to allow all conceivable operations with the data. In practice, a range of operations is supported by different GIS products (Table 17.1). These are changing constantly as the development of GIS technology provides growing numbers of increasingly sophisticated functions. The most difficult part of the analysis is defining the task or problem to be addressed and delineating the approaches to solutions. Particular regard must be given to the quality of data and its limitations.

The computing power of modern GIS allows complex analyses to be made, even with large volumes of data. This is particularly important when a model is tested and integration capabilities are needed to obtain quick and reliable analyses of alternative scenarios. Successful spatial analysis using GIS depends on the model selected, type of data relevant to the model and knowledge of how to use GIS technology appropriately. The correct use of GIS certainly requires more than pushing buttons in a GIS menu! Results must always be evaluated for accuracy and content, and alternative approaches used or new data sought when problems arise. Ideally, analytical results should be presented as easily interpreted maps accompanied by an explanatory report.

PRE-GIS ERA: EARLY MAPPING, ANALYSIS AND DISPLAY OF BIOGEOGRAPHICAL AND MEDICAL DATA

In recent years, several fundamental discoveries relating to pathogen transmission by arthropod vectors have revealed the crucial role of ecological conditions in the epidemiology of TBDs. Appreciation of landscape ecology in medical and veterinary medicine led to the development of descriptive, analytical and experimental epidemiology. Previously, the absence of computerized technologies was compensated by vast quantities of empirical data that enabled statistical procedures and the formulation of testable hypotheses. This situation applies to ticks and TBD in the tropics, subtropics and in the temperate zone. The problems of mapping the distribution of ticks are dealt with in the following section. Here we discuss the influence of tick ecology on TBD epidemiology which, in the 'pre-GIS era', resulted in the theory of natural focality of diseases, a concept that still holds true today.

Table 17.1 *Selected software GIS products*

Product	Vendor	Country	Spatial data model
ARC/INFO	ESRI	USA	vector + raster
EPPL7	Minnesota	USA	vector + raster
ERDAS Imagine	ERDAS	USA	raster
GRASS	Open Source	USA	raster + vector
IDRISI Andes	Clarc Univ.	USA	raster + vector
ILWIS	Open Source	Netherlands	raster + vector
Geomatica	PCI Geomatics	Canada	raster + vector
GeoMedia	Intergraph	USA	raster + vector
MacGIS	Univ. of Oregon	USA	raster
Manifold System	Manifold Net Ltd	USA	vector + raster
MapInfo	MapInfo Corp.	USA	vector
Maptitude	Caliper Corp.	USA	vector
SPRING	Open Source	Brasil	raster + vector
TopoL xT	TopoL Software	Czech Republic	vector + raster
TNTmips	Microimages Inc.	USA	raster + vector

Natural focality of diseases

The theory of natural focality of transmissible diseases was formulated by E. N. Pavlovsky. It was based on several years' research in the taiga of the Far East (Khabarovsk region, 1936–38) studying the circulation in nature of the then newly discovered virus of tick-borne encephalitis (TBE), and the transmission pathways to humans. Pavlovsky was also influenced by observations made in expeditions to Central Asia to investigate tick-borne relapsing fever (*Borrelia persica*) transmitted by the argasid tick *Ornithodoros tholozani*. Starting with the assumption that the disease agent and its vector form part of a biocenosis in a given habitat (biotope), Pavlovsky demonstrated that the majority of vector-borne diseases (VBD) are localized in foci located within defined geographical complexes. He called them natural foci of diseases (NFD). The term NFD differs from that of a focus in the general epidemiological sense because it involves a geographical territory that is usually precisely defined and demarcated by certain ecosystems.

Pavlovsky published on this theory of NFD over a 25-year period (Pavlovsky, 1939, 1946, 1948, 1964) and other researchers have contributed (Rosický, 1967; Korenberg, 1983; Kucheruk & Rosický, 1984). Today, NFD is defined as:

> A natural focus of a disease is a geographically demarcated part of the landscape which has formed during the course of natural evolution and contains an association of organisms in which there circulates a disease agent as its integral component, independently of man and domestic animals.

Under certain conditions humans and domestic animals can become involved in the circulation of the infection. Natural focality is therefore a general biological phenomenon wherein the infective agent and its vector have, in the course of evolution, become components of a stabilized ecosystem independently of humans.

FUNDAMENTAL COMPONENTS DETERMINING THE EXISTENCE OF A NATURAL FOCUS OF DISEASE
The fundamental components of a NFD are: (1) the disease agent; (2) susceptible animal hosts; (3) vectors (ticks, or other blood-sucking arthropods); (4) a suitable habitat (plant association); (5) a suitable environment facilitating the agent's circulation. The first three components (together with plants forming a suitable vegetation type) represent an association of micro- and macro-organisms forming a given biocenosis. The location (part of the landscape) in which this association exists is called a biotope (habitat). These components interact resulting in an environment that provides abiotic (e.g. climatic, soil types, etc.) and biotic conditions which support circulation of the infective agent.

Most habitats are clearly delimited in the landscape (e.g. forest, pasture land, clearing, etc.). Contact zones between two or more ecosystems are called ecotones. Often they have an enriched herbal and bushy vegetation providing a greater and more stable food supply, and ample shelter for small mammals (tick hosts). These conditions (particularly microclimatic conditions – see Daniel & Dusbábek, 1994) within ecotones favour the development of vectors and thereby increase opportunities for the circulation of disease agents. A number of ecotones have arisen as a consequence of agricultural activities fragmenting formerly extensive habitats.

Animal hosts of infective agents are often called 'reservoirs'. That term describes hosts in which the agent persists over periods unfavourable for its circulation (winter in temperate zones) and inter-epizootic periods. However, a number of vertebrate hosts do not meet these criteria and may best be classified as major (basic), minor (secondary) and supplementary or chance hosts. For borreliae that cause Lyme disease, Gern et al. (1998) have outlined criteria for the establishment of reservoir status. The discovery of non-viraemic transmission in certain TBDs (Jones et al., 1987) introduced yet another way of classifying hosts and their significance for NFDs.

Detailed research into the circulation of TBD agents has emphasized the importance of vectors as reservoirs (e.g. Ixodes ricinus acting as a reservoir of TBE virus). Factors contributing to the reservoir potential of vectors are their ability to support trans-stadial transmission from one stadium to the next, and transovarial transmission to succeeding filial generations of the vector (Benda, 1958; Danielová et al., 2002b).

An active NFD occurs when all the components of an NFD are operating and circulation of the disease agent is ensured. A human entering an active NFD is at risk of infection from an infected tick. Other transmission pathways may be in play, for example, transmission of TBE virus via the alimentary route by drinking untreated goats' milk or consuming infected milk products (Blaškovič, 1967). A potential focus occurs when any of the components of an NFD are absent or not functioning, the most common being the absence of the infective agent. In such cases there is the risk of rapid transition to an active focus, for example, through the introduction of infected ticks by avian hosts that are vagrants after nesting (Balát & Rosický, 1954), or by the introduction of an infected vertebrate host. An active focus may exist for a long period without manifestation of the human disease. Such an NFD is a latent focus in the epidemiological context (as compared with an active NFD

which is regularly manifested by cases of disease in humans). Changes in latency/activation of a focus explain accounts in the literature of the 'extinction' and 'flare-up' of NFDs. For example, human cases of TBE are absent from certain regions of the Czech Republic although the virus has been demonstrated in a small proportion of I. ricinus ticks in these regions. Conversely, there is currently a renewed incidence of cases of human infection in other regions following several years' absence. The main reason for the apparent latency of these NFDs is the low circulation intensity of the virus among vertebrates (Danielová, Holubová & Daniel, 2002a). Alternatively, the human exposure rate may be low.

The occurrence of disease in humans is just one of several indicators of an NFD in the landscape. Assessment of any given region (mapping, etc.) solely on the basis of absence of human cases of diseases without knowledge of the NFD potential may lead to erroneous conclusions.

Because ticks have limited mobility, the home ranges of their hosts (particularly those of adults) are an important determinant of the extent of NFDs. Major hosts of adult I. ricinus are game animals (e.g. Capreolus capreolus, roe deer) (as in TBE foci in the Czech Republic), and pastured domestic animals. Such foci may be vast in area and cover whole regions (e.g. in the Balkan peninsula and Bulgaria). When deer are the principal host of adult I. ricinus, infections of humans are by tick bite, whereas when cattle and goats are prevalent in NFD, the alimentary route of infection is more common. Both routes of infection are common in the Slovak Republic where adult I. ricinus feed on deer and domestic animals. These differences must be taken into account when assessing the hazard posed to humans in a given area.

THE STRUCTURE OF NATURAL FOCI OF DISEASES
The disease agent, vector and susceptible hosts are evenly distributed within an NFD. Ticks occur in a mosaic-like pattern determined by biotic as well as abiotic factors. Locations that consistently sustain circulation of the disease agent are termed elementary foci (Pavlovsky, 1946, 1948). Such places offer optimal conditions for the vector and its hosts. Elementary foci are often sites where game animals gather and conditions are favourable for small mammals, e.g. sites with mixed vegetation, glades with shrubs, as well as islets of trees in fields. Border zones (ecotones) play an important role. An elementary focus is characteristically structured. Its basis is a nucleus (synonymous with a 'microfocus' or 'hot-spot'), and its centre (axis) may often be a brook (often only periodic) or a hollow with increased humidity and richer vegetation. The nucleus is usually surrounded by a transitional belt forming

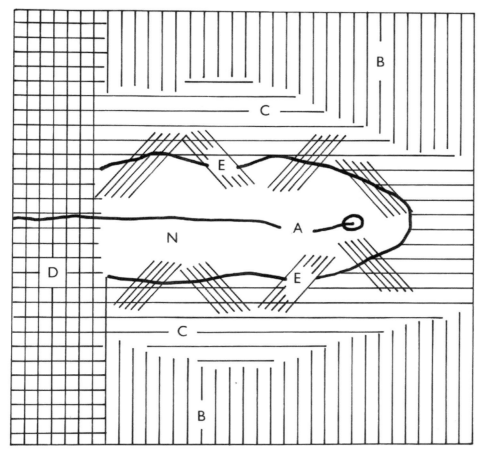

Fig. 17.1. Outline of the structure of a nucleus in a focus of TBE. A, axis of the nucleus; B, spruce monoculture (without any tick occurrence); C, coat of the nucleus; D, grain field (without ticks); E, transitional phytocenoses; N, nucleus. (Reproduced with permission, Daniel & Rosický (1989).)

its boundary (Hejný & Rosický, 1965). The structure and biocenotic status of the nucleus of a long-lived focus of TBE in northeastern Bohemia (Radvan *et al.*, 1960) is depicted in Figs. 17.1 and 17.2.

Depending on the configuration of the terrain and the local distribution of habitats, an elementary focus usually has several nuclei in which the disease agent survives interepizootic periods. Usually the area covered by each nucleus is not great enough to sustain circulation of the agent indefinitely; the activities of alternate nuclei are then crucial. Elementary foci are often the only type of NFD in highly exploited landscapes; identifying them in the terrain is important for epidemiological and epizootological prognoses targeted at timely preventive measures (see Fig. 17.8 below). So-called bioindicators can be used to identify the presence and exact topographical localization of elementary foci in

the field and in cartographical documentation. Bioindicators may be biotic or abiotic. Examples are plant associations, landscape relief, hydrological conditions and slope gradient.

APPLYING THE THEORY OF NATURAL FOCI OF DISEASES IN LANDSCAPE EPIDEMIOLOGY
Detailed analysis of field conditions can inform local healthcare practices but can also extend further research. Korenberg & Kovalevsky (1981) applied their analysis of conditions to the distribution of TBE virus in the whole Euro-Asian continent. Their study is an example of the generalization of empirical data from detailed research in the field conducted on an intercontinental scale, and was a milestone in the 'pre-GIS era'. From approximately 4000 publications, many concerning the Siberian region, they selected

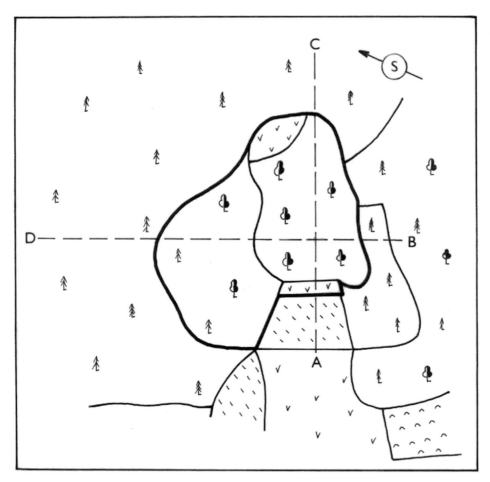

Fig. 17.2. Example of a nucleus in a natural focus of TBE (delineated by thick line) monitored in northeastern Bohemia. Length of tract AC, 235 m; tract BD, 289 m; width of the nucleus proper 167 m, length 141 m (according to Radvan *et al.* (1960); reproduced with permission, Daniel & Rosický (1989)).

studies detailing the major evaluation criteria: geographical position of the focus and corresponding living conditions (delineation of the focus, landscape relief, climate, type of landscape, vegetation cover); vector ecology (for *I. ricinus* and/or *I. persulcatus* – population density, distribution in the terrain, seasonal periodicity of activity, major hosts, infection of ticks with TBE virus); and demographic data (conditions for human contact with ticks, incidence of TBE in humans and its fluctuations on a long-term basis, occurrence of antibodies in the human population). Besides considering conditions that support circulation of the virus in nature, Korenberg & Kovalevsky (1981) highlighted the socio-economic conditions of human populations at risk from infection.

The comparative studies by Korenberg & Kovalevsky (1981) were based on natural foci of TBE. Characteristics of different natural foci were compared and five levels defined: (1) single natural focus; (2) groups of natural foci; (3) classes of natural foci; (4) focal region; (5) groups of focal regions. Across Euro-Asia they identified seven focal regions within the borders of which there exist 20 000–30 000 single NFDs. The greatest proportion is found in Asia (a total of 41 regions in the following five groups: West Siberian, Kazakh–Central Asian, Central Siberia–Transbaikal, Khingan–Amur, and Pacific). In Europe, two groups were recognized: the East European group stretching from the River Visla to the Ural Mountains (with 18 regions that roughly cover the territory of the European part of the former Soviet Union) and the Central European–Mediterranean group of nine regions covering central, western and southern Europe. Each region is characterized by extensive data on vectors (*I. ricinus*, *I. persulcatus* and other tick species supporting TBE virus circulation) derived from 550 bibliographic entries.

Mapping tick distribution

ZOOGEOGRAPHY OF TICKS: SMALL-SCALE MAPS

Contour maps of large territorial areas (countries or whole continents) combined with topographical data provide the basis for mapping tick distribution. Sites of recorded tick occurrence are usually marked with point symbols or alternatively with filled squares covering a network of similar zoogeographical data. For example, the collection of maps of tick incidence (Ixodidae and Argasidae) in France showing 10×10-km grids marked with squares indicating where tick species have been found by flagging in the field or through detection on hosts (Gilot, 1985). The British Isles has been assessed similarly (Martyn, 1988). Location sites are documented in 10-km British and Irish national grids, with texts describing the individual species, the material processed, and museum and other collections.

The distribution of ticks (644 species), based on bibliographical records, has been published in four volumes by G. V. Kolonin (1978, 1981, 1983, 1984). In contour maps with a grid of a density of $10°$ to $20°$, borders of the distribution of individual species are depicted. An important 1 : 8 000 000 scale map gives the distribution of ticks in the temperate zone across the whole Asian continent (Prokhorov *et al.*, 1974). This map uniquely records the results of long-term monitoring of ticks by a team of researchers working in the field. In places not covered in detail, data are interpolated based on knowledge of the biology of each tick species and its relationship with the environment (mainly vegetation type). Ten landscape types are differentiated and mapped along with their corresponding tick associations. Within these landscape types, the distributions of each species are indicated with their approximate densities graded from 1 to 5 on the basis of numbers of host-seeking ticks in the field (determined from flagging by one person over a 1-km tract). The map is supplied with a text describing methodology and the characteristics of each tick species. A supplement of 1 : 32 000 000 scale maps describes the distribution areas of eight epidemiologically important tick species including records of their occurrence outside of their presumed home range. Although the authors stressed that their aim was not to define natural foci of diseases (namely of TBE and rickettsioses), their

map is used for epidemiological assessments. The approach of Prokhorov *et al.* (1974) forms a bridge to the problems of medium- and large-scaled maps.

TICK DISTRIBUTION: MAPS OF MEDIUM AND LARGE SCALE

In the 1970s and 1980s, the problem of drawing maps at medium and large scales for tick distribution studies was taken up by B. Gilot and associates (Gilot, 1985). They extended previously described relationships of ticks with vegetation, and used plant associations as bioindicators. Gilot's conclusions were based on his own vast experience in the field. He set limits for generalizations to ensure that field-collected data did not lose their original value. In many ways he anticipated questions that arose decades later in the application of remote sensing (RS), e.g. spatial resolution, supervised/unsupervised classification, and the adequacy of the scale of prognostic maps. Gilot demonstrated empirically the mosaic-like pattern of tick occurrence for *I. ricinus* that had been previously deduced from observations of environmental conditions.

Gilot (1985) started with a map of 1 : 1 500 000, breaking down the studied territory according to phyto-sociological categories. Such a map enabled predictions of territories that had a high probability of harbouring a certain tick species. However, the map did not enable the determination of population densities or reliably reveal environmental relationships. Therefore Gilot *et al.* (1979) and Gilot, Pautou & Lachet (1981) conducted two vast field experiments with two different map scales in two geographically different regions: a 1 : 35 000 map covering 190 km² of the Jura mountain range, and a 1 : 200 000 map covering 2320 km² of the Rhone valley. The methodology for both maps was unified: an elementary unit of the territory (0.36 km² in the first instance, and 4 km² in the latter) was flagged by one person along several transects for 1 hour. Tick–plant associations were determined by a priori analysis (corresponding to supervised classification in satellite data) and a posteriori analysis (analogous to unsupervised classification), followed by a comparison of the results. Gilot (1985) concluded that, in general, a large magnification scale (1 : 50 000) was best suited to generating a realistic picture of tick distribution. A medium scale (1 : 200 000) is acceptable in regions with a uniform (simple) landscape character; or can be applied for general epidemiological assessment of a larger territory. Large and medium scales were not recommended for territories having complicated topographical conditions (e.g. mountains) where the landscape breaks down into a mosaic of various landscape elements, or when ticks are found within limited areas. In such cases Gilot (1985) recommended the use of maps of small scale (1 : 5000, 1 :2500 or 1 :1000) for delimiting NFDs and for the study of their structure. Gilot pointed out that *I. ricinus* is a suitable species for mapping because of its association with precisely defined environments. For this reason, *I. ricinus* is often used to illustrate points in this chapter.

Other notable applications of medium- and large-scale maps are those of Russian workers who conducted their studies mainly in the temperate zone of Asiatic Russia (namely in the deciduous taiga belt) harbouring NFDs of TBE and rickettsioses. Their work focused on *I. persulcatus*. Reviews of the methods used have been presented by Korenberg (1973), and Vershinina (1985). Another notable review in this area is that of Kitron & Mannelli (1994).

PREDICTIVE MAPPING OF TICKS AND TICK-BORNE DISEASE DISTRIBUTION

Field-collected data on tick occurrence are often lacking. In these situations, tick distribution is predicted based on the spatial distribution of covariates. Predictive mapping can be based on either a priori information about ecophysiological characteristics of the species under study, or on a training database reflecting the relationship between occurrence and covariates ad hoc. The former is based on an experimentally verified dependence of the given tick species or TBD on certain environmental factors, leading to empirical models and expert systems. In the latter approach, application of appropriate mathematical models permits a statistical estimation of the relationship between the occurrence of ticks and/or TBD and the number of covariates in a data set. There are a variety of predictive models available (Guisan & Zimmermann, 2000), none of which can be considered universally the best. An accurate prediction always requires optimization of any given model depending on specific ecological conditions and available data (Robinson, 2000).

Methods most commonly applied in predictions

The methods most commonly applied in tick predictions are summarized in Table 17.2. Various predictive algorithms were used that differ in the type of response variable modelled as well as in the character of predictive data. Species occurrence is primarily determined by macroclimatic factors, hence macroclimate is commonly exploited

in prediction models. An example is CLIMEX, a long-established climate-based model (Sutherst & Maywald, 1985). It belongs among the environmental envelope-type models (Fig. 17.3D). With CLIMEX, the envelope is positively delimited by a 'growth index' which combines favourable temperature and humidity, and negatively delimited by 'stress indices' represented by extremes in temperature and humidity. The output of the model is an 'ecoclimatic index' which is interpreted as an overall measure of climatic suitability for a given species. CLIMEX has found many applications in zoogeography including the prediction of tick occurrence (e.g. Norval *et al.*, 1991; Perry *et al.*, 1991; Sutherst, 2001).

Many deductive methods employ geobotanical indicia to map habitats suitable to ticks. Certain plant associations and their structures reflect macro- and/or mesoclimatic factors, geological and other conditions, but also create microclimatic, nutritional and shelter conditions for the presence of ticks and their hosts. The association of tick population density and a certain type of vegetation is thus a strong predictive criterion of occurrence that can be verified (e.g. Sonenshine, Peters & Stout, 1972; Gilot, Pautou & Moncada, 1975; Estrada-Peña, 2001*a*). Effective application of this principle in landscape epidemiology has been facilitated by the development of RS enabling identification of tick habitats from multispectral images (Hugh-Jones, 1989; Daniel & Kolář, 1990; Hugh-Jones *et al.*, 1992). The vegetation categories discerned in satellite-mediated imagery do not follow the classical phyto-ecological categories obtained by researchers on the ground. Remote sensing classification of tick habitats is more about the structural characteristics of the vegetation but their predictive potential is not diminished compared with ground-based characterization (Duffy *et al.*, 1994; Daniel *et al.*, 1998, 1999; Beck, Lobitz & Wood, 2000).

Mechanistic models base predictions on mathematical formulations of real cause–effect relationships and utilize physiological and/or phenological data of the target species that were assembled in experimental studies. A model of this type is not intended primarily for mapping purposes, but rather for testing theoretical correctness of hypotheses about species response to different environmental conditions. Nevertheless, high correlation between predicted and observed geographical distributions was also reported (Estrada-Peña, 2002*b*).

A strategy different from the empirical approach is the use of covariates expected to be linked with tick occurrence

Table 17.2 *Tick mapping/prediction techniques, data requirements, and selected references*

Algorithm family	Species data (calibration)	Environmental data (predictors)	Reference
Environmental envelope (CLIMEX)	presence	continuous	Sutherst and Maywald, 1985; Perry *et al.*, 1991; Norval *et al.*, 1991
Discriminant analysis	presence/absence	continuous	Randolph, 2000; Cumming, 2000*b*
GLM (linear/logistic regression)	presence/absence or abundance	continuous/categorical	Glass *et al.*, 1994; Cumming, 2000*a*, 2000*b*; De Garine-Wichatitsky, 2000; Guerra *et al.*, 2002; Brownstein *et al.*, 2003
Classification and regression tree	presence/absence	continuous/categorical	Merler *et al.*, 1996
PCA-based (ordination)	presence	continuous	Olwoch *et al.*, 2003
Gower similarity metric	presence	continuous/categorical	Estrada-Peña, 2002*a*, 2003*a*; Estrada-Peña *et al.*, 2005
Deductive (habitat classification, Boolean logic, etc.)	research results/ expert opinion	continuous/categorical	Hugh-Jones *et al.*, 1988; Daniel & Kolář, 1990; Cooper & Houle, 1991; Gilot *et al.*, 1995; Daniel *et al.*, 1998; Estrada-Peña, 2001*a*; Eisen *et al.*, 2005
Mechanistic model	research results (life cycle data)	continuous/categorical	Estrada-Peña, 2002*b*
Spatial interpolation ((co-)kriging)	presence/absence or abundance	continuous	Nicholson & Mather, 1996; Estrada-Peña, 1998, 1999*a*, 1999*b*, 2001*b*

in ad hoc models. Typically these models rely on a relatively small sample of field collected data on tick occurrence ('training' data set) and a set of covariates in the form of digital maps. Besides climatic and vegetational covariates mentioned above, there are geological, soil type, hydrological and host population covariates as well as anthropogenic and other factors co-determining the site or area of occurrence. To envisage the often subtle relationship between tick occurrence or risk of TBD with covariates, certain statistical methods are applied (Fig. 17.3). In principle, statistical prediction can be based either on environmental dependence or spatial dependence. The modelling techniques that are based on species environmental dependence determine relationships between the species occurrence and covariates in environmental space, and then use the spatial pattern of covariates to project it into geographical space. The most common are modifications of generalized linear model (GLM) such as linear and logistic regression, discriminant analysis and classification tree methods. The techniques

based on spatial dependence estimate an area's suitability for a species directly in geographical space. One of such techniques employing autocorrelation is co-kriging (see below).

Bioclimatic zones and prediction scale

Predictive maps can be produced for small territories (e.g. Daniel & Kolář, 1990; de Garine-Wichatitsky, 2000) as well as at pan-continental scales across many bioclimatic zones (e.g. Estrada-Peña, 1999*a*, 1999*b*; Cumming, 2000*a*). The scale and bioclimatic zone markedly influence the choice of predictors. Variables with small spatial variance, namely macroclimatic indices, have nearly no predictive value in small territories, whereas diversified variables such as vegetation type and vegetation structure have limited value in large-scale studies. Many predictive models have been elaborated for the conditions of the tropical and subtropical climate zones of Africa, Australia and South America using mainly macroclimatic variables or their satellite surrogates and

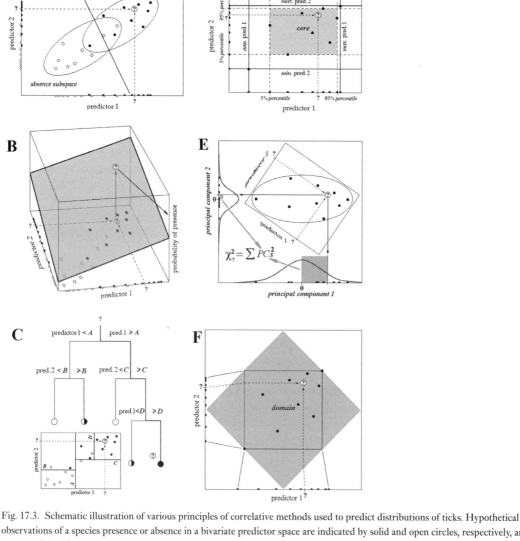

Fig. 17.3. Schematic illustration of various principles of correlative methods used to predict distributions of ticks. Hypothetical observations of a species presence or absence in a bivariate predictor space are indicated by solid and open circles, respectively, and the process of classifying an unsampled site ('?') is traced with arrows. Note that the predictors are not perfectly independent, which violates assumptions of most statistical models, yet is quite common in practice. Further note that the left-hand-side methods (A, B, and C) require both presence and absence training data, while the methods shown at right (D, E and F) make do with the presence data only.
(A) Discriminant analysis bisects the predictor space into two subspaces that best separate the present and absent observations, and classifies an unsampled site according to the subspace in which it falls. (B) GLM fits multidimensional linear regression model to the training dataset, and uses the fitted model to predict the likelihood of presence at an unsampled site. (C) CART generates a dichotomous key-like system of predictors' threshold values that best discriminate between conditions for presence and absence in the training dataset, and uses it to classify an unsampled site. (D) The environmental envelope method of the BIOCLIM type classifies an unsampled site based on whether it falls within or outside the environmental range recorded for the present occurrences disregarding a predefined percentile of the lower and higher values (to eliminate outliers). (E) The PCA-based method rotates axes of the predictor space to neutralize dependencies between predictors, and then, relying on normality of marginal distributions in such transformed training data, uses χ^2 criterion to determine the likelihood of presence at an unsampled site. (F) The Gower metric-based method transforms the predictor space so that the contribution of each predictor is equalized, and classifies an unsampled site according to its proximity in such a space to the most similar present occurrence (alternatively to the mean present occurrence) using point-to-point similarity metric (Gower metric); potential environmental domain of species presence is suggested in grey.

normalized difference vegetation index (NDVI). While climatic variables (monthly temperature and mean precipitation) are generally good predictors of tick occurrence in such conditions, the efficacy of NDVI and vegetation variables generally differs (Cumming, 2002). However, in the temperate zones of Europe and North America, similar studies have demonstrated the predictive potential of vegetation variables for tick and TBD occurrence (e.g. Sonenshine, Peters & Stout, 1972; Gray *et al.*, 1998). Other effective predictors are the type and moisture of soil (e.g. Carey, McLean & Maupin, 1980; Kitron, Bouseman & Jones, 1991; Glass *et al.*, 1995), waterways network (Kitron *et al.*, 1992; Estrada-Peña, 2001*a*), terrain configuration (Merler *et al.*, 1996), game density and even various anthropogenic factors.

Most articles published recently use a similar application of GIS technologies in various regions to analyse TBD caused by different pathogens in human and animal hosts under impact of many environmental factors. Input data differ in the detailed level of information, above all in natural environment conditions. Most attention has been paid primarily to Lyme borreliosis in the USA (e.g. Glavanakov *et al.*, 2001; Frank *et al.*, 2002; Guerra *et al.*, 2002; Baptista *et al.*, 2004; Foley *et al.*, 2005). The distribution of *I. ricinus* in Trentino, Italian Alps and the assessment of areas of potential Lyme borreliosis risk have been similarly investigated (Merler *et al.*, 1996; Rizzoli *et al.*, 2002; Chemini & Rizzoli, 2003).

Species specificity of predictive models

Predictive mapping studies have most commonly involved economically important species in tropical regions (*Rhipicephalus (Boophilus) microplus*, *Amblyomma variegatum* and *Rhipicephalus appendiculatus*) and epidemiologically significant species of the temperate zone (*I. ricinus* in Europe and *I. scapularis* in North America). Predicability of distribution of a particular tick species is influenced by the complexity of its developmental cycle and range of hosts, environmental requirements of both the tick and hosts and other factors (Randolph, 2000; and see Chapter 2).

FROM TICK MAPS TO TICK-BORNE DISEASE PREDICTION

Prediction of TBD risk is dependent on the prediction of occurrence of competent tick species; however, other factors contributing to the existence of natural foci may have a substantial effect. When the influence of such factors is negligible, the potential distribution of TBDs can be estimated on the basis of a simple combination of maps of major vectors with the tools of GIS. In other cases, GIS models should combine additional data sources. Special modelling techniques are required when variations through time of abiotic and biotic factors have an influential effect. Such dynamic factors include tick phenology, cyclic climatic phenomena and population cycles of the hosts. Interactions of these factors can sometimes be complicated (Jones *et al.*, 1998). An elegant technique is the application of Fourier transformation to represent the periodicity of the temporal series of covariates. This approach reduces the amount of data and renders the model relatively easy to present (e.g. Rogers, 2000). Seasonal dynamics of climatic factors plays a key role in one model of TBE distribution (Randolph *et al.*, 1999). The model assumes that virus persistence in a natural focus depends on 'co-feeding' of *I. ricinus* larvae and nymphs which, owing to the phenology of these stages, occurs only where there is a significant annual temperature gradient. This model uses the land surface temperature (LST) index derived from data of the National Oceans and Atmospheric Administration (NOAA) satellites as a surrogate value for near-ground temperature (Green, Rogers & Randolph, 2000; Randolph & Rogers, 2000).

Validation of predictive models

Rapid developments in computer technologies and the increasing supply of RS data have led to wider applications of predictive mapping. Sophisticated GIS software packages offer the user a number of powerful modelling tools. However, their application has the shortcoming of being a 'black box' and can generate maps from input data that do not respect the elementary limitations of the mathematical model applied. Paradoxically, most such outputs look meaningful at first sight. Therefore, it is critical to verify the reliability of such predictions, for which a number of statistical procedures are available (Robinson, 2000).

MAPPING AND ANALYSING TICK-BORNE DISEASE DISTRIBUTION BASED ON EPIDEMIOLOGICAL DATA

Clinical cases are the primary information source of disease occurrence, and cartographical representation of their incidence is often used in exploratory epidemiological studies. Epidemiological observations of morbidity or seroprevalence often consist of Cartesian coordinates of locations of disease cases or counts of cases within arbitrary regions. Procedures for automated plotting of epidemiological data have been widely implemented in standard GIS or specialized software packages such as EPI-MAP and can readily be applied to produce descriptive maps of TBD. These maps vary in form from simple point maps of cases to pictorial representation of counts within tracts (Cliff & Haggett, 1988) and can generally satisfy the demands for decision-making in health resource allocation or in focusing field research. However, for more in-depth analyses of disease distribution, statistical methods are needed. The rationale for switching to the more demanding methodology of spatial statistics are varied. One may wish, for example, to analyse the spatial pattern of TBD cases, or to study the relationship between disease incidence and some explanatory variable, or simply to filter out any random noise from the data. Statistical methods particularly suited to epidemiological data have been developed in the last two decades in response to growing concerns about adverse environmental effects on the health status of populations (Lawson et al., 1999; Lawson, 2001). When adopting these methods for analysing TBD distribution, specific features of these diseases should be taken into account. Unlike other infectious diseases: (1) the factor of contagiousness is absent; (2) cases tend to aggregate based on natural foci; (3) data are scarce as cases are relatively rare; and (4) the population denominator is poorly understood.

Population denominator

Any geographical pattern of disease incidence combines two spatial variables: the intensity of disease risk and the density of the population at risk. To rate the risk from incidence data, some information about the spatial distribution of the population at risk is required. There are two main approaches to this task in spatial epidemiology. The first relies on census tract counts, and the second employs some control events assumed to be uniformly distributed in the population, such as incidence of a common disease. With TBD, however, the risk is associated with a visit to rural locations, so the population at risk does not necessarily reflect residential population density and is not distributed uniformly. Instead, the interplay of behaviour or activities that may lead to exposure to ticks (demographic factors, road networks and landscape features) can shape the population at risk. Moreover, these factors can markedly vary from place to place (Rand et al., 1996). For example, tourism and outdoor recreational activities, which have been recognized as important risk factors in Lyme borreliosis (O'Connell et al., 1998), can exaggerate

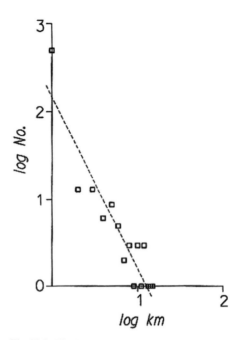

Fig. 17.4. The frequency distribution of distances between residential addresses and places of infection in a set of patients with Lyme borreliosis fitted with power regression illustrates a peri-residential exposure of the population at risk. (Reproduced with permission, Zeman (1997).)

disease incidence in recreationally exploited areas resulting in overestimation of risk. Assessment of the population at risk depends on the scale. Studies of large areas are less likely to be influenced by variations in population denominator, whereas small-area studies require special modelling techniques (e.g. Thomsen, 1991; Diamond, 1992; Whittie, Drane & Aldrich, 1996). For example, paired data can be used on locations of residence and disease acquisition to model patterns of peridomestic exposure (Fig. 17.4).

Analysis of spatial clustering

One important question in spatial epidemiology is whether an observed pattern of cases indicates clusters or whether it might have arisen by chance alone. As stated previously, TBD cases tend to be aggregated so, rather than presence of clusters, it is important to gain an insight into the 'character of dispersion', which may contribute to ecological studies of a particular disease. Several statistical methods for spatial clustering have been adopted to analyse TBD, namely methods for quadrat counts, second-order methods, Moran statistic, and Kulldorff–Nagarwalla statistic. A comprehensive

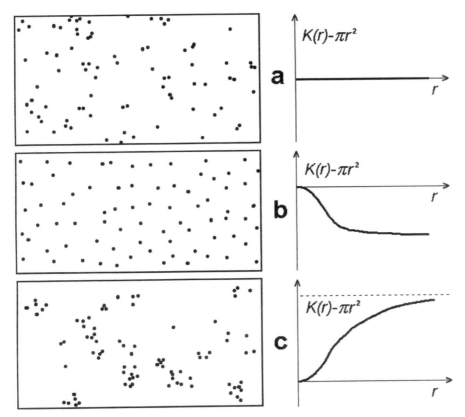

Fig. 17.5. Principal types of spatial distribution: (a) random, (b) regular and (c) aggregated, and corresponding shapes of the $K(r) - \pi r^2$ diagram clearly distinguishing between them. (Reproduced with permission, Zeman, Vitkova & Markvart (1990).)

description of the methods for spatial pattern analysis can be found elsewhere (e.g. Dale *et al.*, 2002).

QUADRAT TECHNIQUE

This method combines the spatial approach with classical statistical methods (Rogers, 1974). Regular grids are placed on an incidence map yielding cell counts of cases which can be analysed by standard methods such as the goodness-of-fit test or regression. Although this is not a true spatial method, as it ignores spatial context of the counts, it is recommended for its simplicity in exploratory studies. When applied to Lyme borreliosis and TBE incidences in an endemic area this technique showed that, although geographically correlated, quadrat counts of Lyme borreliosis cases were nearly random following a Poisson distribution while those of TBE were distinctly aggregated showing a Neyman distribution (Zeman, Vitková & Markvart, 1990). This technique has also been employed to investigate correlations between Lyme borre-

liosis incidence and some explanatory variables (Nicholson & Mather, 1996).

SECOND-ORDER METHODS

These are suited for analysing point-pattern disease maps and are methods based on a mathematical theory of distribution of between-point distances (Diggle, 1983). The crucial statistic is the K function, which characterizes how the number of neighbours varies with distance in an analysed point pattern. A plot of $K(r) - \pi r^2$ versus r (where r is a radius) illustrates the mean excess of cases over a homogeneous Poisson model of the same point density within r about any case. For a 'random' pattern, it coincides with the x-axis (no excess) while positive and negative shifts indicate clustering and regularity, respectively (Fig. 17.5). Another useful statistic is $\sqrt{K(r)/\pi}$, sometimes referred to as the L function. Testing significance in these methods relies on Monte Carlo simulations (Fig. 17.5). An example of applications

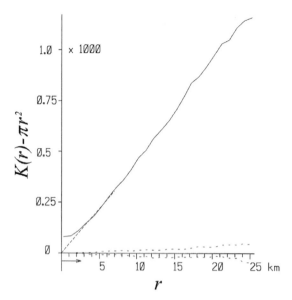

Fig. 17.6. Plot of $K(r) - \pi r^2$ statistic versus radius in km for tick-borne encephalitis cases (solid) along with confidence envelopes of a Monte Carlo model of complete spatial randomness (dotted) documents that spatial clustering is well pronounced in this disease. (Reproduced with permission, Zeman (1997).)

of these methods in the study of TBD includes cattle anaplasmosis (Hungerford, 1991), *I. scapularis*-infested deer (Kitron *et al.*, 1992) and human LB and TBE (Zeman *et al.*, 1990; Zeman, 1997).

MORAN'S *I* STATISTIC

This is a measure of spatial autocorrelation between ordinal or interval values relating to regions or points (Moran, 1950). It is used to test whether values of a variable observed at adjacent geographical locations, such as incidence data in administrative regions of a country, resemble each other more than expected under a randomness model. Classic Moran's *I* does not allow for heterogeneity in the population denominator, which increases the type I error, so several improvements have been proposed. A variant of the population density adjusted estimator was applied to Lyme borreliosis incidence data in Georgia, USA showing spatial autocorrelation between county rates (Oden, 1995). Application of Moran's *I* to Lyme borreliosis morbidity, *I. scapularis* distribution and vegetation cover in Wisconsin showed autocorrelation as well as cross-correlation between these variables (Kitron & Kazimierczak, 1997).

KULLDORFF–NAGARWALLA'S METHOD

This method is a spatial version of the scan statistic with a circular window of variable size systematically sampling the

population in an area of study. The existence of 'clustered circles' is tested such that, for all individuals within the circle, the probability of being a case is significantly higher than for individuals outside the circle. The test statistic is based on the likelihood ratio and achieved via Monte Carlo simulation (Kulldorff & Nagarwalla, 1995). Although general application is possible, this method is particularly suitable for aggregated data such as incidences per census units. A clustering of cases of human granulocytic ehrlichiosis in Connecticut was investigated by this method (Chaput, Meet & Heimer, 2002).

Computing risk maps

Once it is established that disease clustering occurs, it is worth proceeding with a statistical mapping procedure in order to highlight any underlying risk pattern. Most methods are based on a model of continuous space change that can be approximated by a smooth, mathematically defined surface. These techniques have been reviewed elsewhere (e.g. Lam, 1983; Van Beurden & Hilferink, 1993; Lawson, 2001). The literature includes several examples of maps of TBD risk assessed by means of kernel estimator, kriging and splines. Sometimes it is also desirable to compare two spatial patterns of disease cases. By means of the Kelsall–Diggle estimator it is feasible to compute a relative-risk map.

Kernel interpolation at an inserted point is achieved by averaging surrounding values weighted with a distance-decaying function, e.g. Gaussian. This method is applicable for both continuous measures and point pattern data (Bithell, 1990; Lawson & Williams, 1993). It requires judgement concerning a parameter that controls the smoothness of the estimated surface (so-called 'window width'). Even if 'objective' methods for its selection are available (e.g. cross-validation), an empirical 'trial-and-error' approach sometimes leads to maps that are more acceptable from a practical point of view. Kernel smoothing of points of infection was applied to compute the risk maps of Lyme borreliosis and TBE (Zeman, 1997).

Kriging is an interpolation method based on a generalized linear least-squares algorithm, using variograms as weighting functions. Variograms depict the average squared difference between data values as a function of distance and orientation between data locations (Isaaks & Srivastava, 1989). Kriging is a popular and widely implemented method of surface mapping in GIS packages; however, it lacks a unique solution and needs a large amount of interactive input and subjective assessment during the processing. This method was used to create a model that predicts Lyme borreliosis

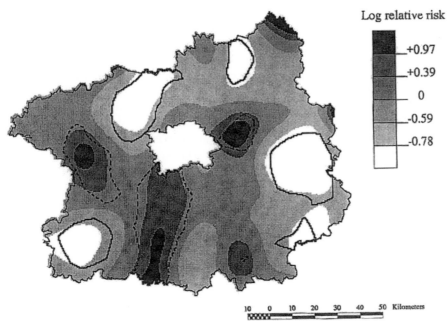

Fig. 17.7. Log relative risk of acquiring Lyme borreliosis through the 'tick' or 'insect' bite according to self-reported data in cases from Central Bohemia computed by the Kelsall–Diggle method; test of non-constant ratio gave *p* value of 0.023, and solid and dashed lines indicate where the 95% tolerance limits were surpassed for the 'insect' and 'tick' reports, respectively. (Reproduced with permission, Zeman (1997).)

transmission risk in Rhode Island, USA based on a survey of nymphal *I. scapularis* population density (Nicholson & Mather, 1996).

Splines are techniques of piecewise polynomial interpolation (Journel & Huibregts, 1978). An interpolated surface is composed of a number of elementary 'tiles' characterized by polynomials of a low degree fitting the enclosed data points and extended to adjacent polynomials so that the whole surface is continuous and smooth. Splines usually tend to intersect the observed values, and require rather subjective tuning up of the control lattice (an analogy with the 'window width' parameter of the kernel estimator). A modification of this method was tested with epidemiological data for TBE (Krejcir, 2000).

The Kelsall–Diggle relative-risk estimation method is a double-kernel smoothing procedure for two point patterns (Kelsall & Diggle, 1995). It is suitable for analysing spatial variation of two types of health events if the question of interest is whether they are distributed equally. If not, an identification of areas of unusually high or low incidence of one of the two types is possible. In a surveillance of Lyme borreliosis, for example, patients were invited to specify likely infection origin as either tick bite or insect bite, or unknown. Spatial

pattern of their residential addresses was then analysed by the Kelsall–Diggle method assuming that all cases originated from tick-bite transmissions even if they were unrecognized (Fig. 17.7). The analysis, however, showed significant differences between the distribution of 'tick' cases and that of 'insect' and 'unknown' cases (Zeman, 1998, 1999).

This is not an exhaustive review of all the work in the field of spatial analysis of TBD data. Some other approaches such as Bayesian and Markoff Chain Monte Carlo (MCMC) methods have been used. The purpose here is to provide an overview of the main analytical approaches currently available which can be employed in TBD research. Despite wide interest in the methodology of GIS and spatial epidemiology, users of the methodologies usually have some statistical knowledge.

SATELLITE REMOTE SENSING

Description of remote sensing data

Remote sensing is an information technology providing information on landscape (i.e. biophysical properties of the real world) through measured values of electromagnetic radiation from the landscape surface.

Electromagnetic radiation is characterized by two parameters: intensity of electromagnetic energy carried by the radiation and wavelength of the radiation. The energy interacts with an object and is reflected or emitted from the object. The energy is modified in a unique way by a given object that distinguishes it from the radiation energy measured at a particular wavelength. It is this energy that is measured and converted into data for subsequent processing. Information about the surface of the Earth can be derived from spatial and spectral distributions of energy emanating from the surface, and also from temporal variations in the distributions.

The optical region of the spectrum (0.3 μm–1000 μm) is commonly used for remote sensing. For practical reasons the optical region is divided into several shorter wavelength intervals of the visible (0.3–0.7 μm), near-infrared (0.7–1.1 μm), middle-infrared (1.1–5 μm), thermal infrared (5–100 μm) and submillimetre infrared (100–1000 μm) region of the spectrum. Due to atmospheric absorption, only wavelengths shorter than about 15 μm are used from the optical spectrum. Other portions of the electromagnetic spectrum are also useful in remote sensing, e.g. waves from the microwave region (0.01 m–1.0 m) penetrate clouds. Remote sensing measurements are particularly discriminating when using several very narrow spectral bands and producing a multispectral data set. In contrast, monospectral data are obtained if only one spectral band is used for measurement (e.g. panchromatic band covering the entire or a substantial part of the visible radiation). The number of channels or bands defines a dimension of the acquired data set. Increased utility of narrow spectral bands has resulted in the progressive narrowing of bands and increased the number of bands used 10- to 100-fold. These hyperspectral remotely sensed data allow more accurate feature identification. Radiation energy is measured using a radiometer. The size of the surface area (pixel) from which a radiometric value is determined defines the spatial resolution, while the number and width of the spectral bands measured define the spectral resolution of the radiometer. Individual pixels captured in continuous lines and columns make up an image. A line of detectors or a mirror enables pixels to be measured in a row, whilst the movement of a platform within the radiometer generates a second dimension of the image. Imaging radiometers are called scanners or multispectral scanners when producing a multispectral data set.

Scanner platforms use various types of carrier, but operational remote sensing systems are based on satellites. At present, Earth-observation satellites circle in specific orbits.

Near-polar, Sun-synchronous orbits carry satellites over both poles. As the Earth rotates below the satellite it brings all sites on the globe within the satellite's field of view. The plane of the orbit is fixed with respect to the Sun. As a result, measurements are taken by the radiometer at approximately the same solar illumination conditions for each daytime overpass. Orbiting around the Earth brings a satellite repeatedly over the same site on the surface. The time interval between two successive measurements of the same site determines the time resolution of the scanner.

Scanners are passive remote sensing systems measuring reflected or emitted radiation originating from a natural source (e.g. Sun, Earth), whereas active remote sensing systems generate radiation from artificial sources within the system (e.g. radar, lidar).

Characteristic properties of remote sensing data

Both optical and microwave systems have their limitations for Earth surface applications. Two major categories of data distortion can be defined: (1) radiometric distortions caused by e.g. atmospheric effects and changes in detector sensitivity; and (2) spatial or geometric distortions due to e.g. oblique view, altitude changes along the orbit, instability of the spacecraft and Earth rotation. The consequence is that repeated measurements taken by different or even the same sensor will have a component of variation owing to the pixel size, variable atmospheric path length, variable atmospheric conditions and the time–space variability of the environmental conditions being measured. These distortions can be identified and corrected by using an array of in situ measurement sites to provide 'ground truth' data.

RADIOMETRIC DATA CORRECTIONS

Measured values of all remote sensing systems are modified by interactions of the radiation with atmospheric particles. Atmospheric water molecules in vapour, liquid and/or solid forms are the primary controlling variable. Optical scanner measurements can even be completely obscured by water particles concentrated in clouds. Other atmospheric particles, such as dust and smoke, offset radiometric measurements of the Earth's surface from true near-surface measured values.

Sensor data output calibration is essential for quantitative use of any radiometric data. The calibration consists of established procedures to relate sensor output data to Earth radiance measurements. Each instrument is tested in the laboratory for stability, linearity and sensitivity to radiance using standard sources of selected electromagnetic waves. Detectors deteriorate in orbit, thus some in-flight calibration may be performed. Calibration data are used to convert the Earth data from digital counts to absolute radiometric values. In-flight calibration is used for infrared measurements because the detectors are sensitive to ageing. Measurements of the visible or near-infrared radiation are not so demanding and pre-flight calibration can be employed. To compare data from different satellites, quantitative calibrated values must be used. Additionally, users of long-term data series must take account of calibration drift.

GEOMETRIC CORRECTION OF IMAGE DATA

Raw data obtained by imaging radiometers differ from maps owing to changes in topography and the orientation of the platform with the scanner. These parameters result in picture element (pixel) size variations over the image and displacement of mutual pixel positions. This makes geometric comparisons with other representations of the same area (e.g. another satellite scene or maps) very difficult. Geometrical transformations must be performed to render raw scanner data or unrectified images comparable with another image or map. During the process of geometrical transformation, the location coordinates of image elements are changed. This operation tags each pixel with its correct latitude and longitude values, removes the geometric distortion, and can even rectify the entire image into a selected map projection. In this geometric form, spatial data from remote sensing are geocoded and ready to be combined with other data in GIS.

GETTING INFORMATION FROM DATA

Satellite sensors are designed to have a direct relationship between the radiometric variable being measured (e.g. radiance or emittance) and the digital number (DN) value. Remote sensing data in absolute radiometric values or as relative numeric digits need to be transformed into the kind of information requested by a specific application. The usual approach is to convert a set of image data into a thematic map. This can be reached through two types of data processing: visual interpretation of the image or digital classification performed by a computer.

Visual analysis discerns image features recognized by the human eye (generally tonal, textural and contextual features). Although visual interpretation is still used widely and can be accurate, it is not without problems. Key concerns are subjectivity in the interpretation, difficulty of handling large multispectral data sets acquired by many remote sensors, and the problem of maintaining a standardized classification.

Computer-based classification techniques can make fuller use of the data set and can be undertaken in a more objective manner. Generally, spectral values are used for assigning a class to measured values, although textural, contextual or polarization features also apply. Digital image classification is commonly used to analyse remotely sensed data because it is readily integrated with other digital spatial data sets within a GIS. Two approaches are used: unsupervised and supervised classification.

UNSUPERVISED CLASSIFICATION

Unsupervised classifiers are effectively clustering algorithms. The classification algorithm groups together pixels with similar properties and assigns them a class label. Since feature values (e.g. spectral responses) are specific for every group of landscape object, clusters may be expected to identify land cover classes. A major problem with unsupervised classification is that there is no guarantee that the clusters will correspond to relevant land cover classes. Therefore, an unsupervised classification is often inappropriate and a supervised classification should be used.

SUPERVISED CLASSIFICATION

Supervised classification uses a priori knowledge of the land cover classes to be identified in the data set. The classification comprises three key stages. Firstly, the training stage in which sites of known land cover class are identified in the image and characterized by given features. The end product of this stage is a set of training statistics that describe the features of the classes to be mapped. Secondly, the training statistics are used in combination with a classification algorithm to allocate each pixel in the image to a land cover class. This typically involves a comparison of the pixel's features with those of the classes derived in the training stage and allocation of the pixel to the class with which it has greatest similarity. In this way, the remotely sensed image is converted into a thematic map depicting the spatial distribution of the land cover classes of interest. The final stage of the classification is the validation stage in which the accuracy of the classification is assessed. This aims to derive a quantitative measure of the accuracy with which land cover has been mapped. The end product of the supervised classification is, therefore, a map of known accuracy that depicts the spatial distribution of land cover classes of interest.

Many factors affect the accuracy of a supervised classification. To date, the majority of digital image classifications have used a conventional statistical classification algorithm. Probabilistic techniques such as the maximum-likelihood classifier and discriminant analysis have been particularly popular. These approaches have firm statistical foundations and allocate each case (e.g. pixel) to the class with which it has the highest probability of membership. Although this is an intuitively appealing approach and can be accurate, the correct application of such classifications requires the satisfaction of several assumptions that are not always tenable and it is sometimes difficult to integrate ancillary data into the analysis. Consequently, considerable attention has been directed to developing alternative classification approaches. This has included the use of a range of non-parametric classifiers, with approaches based on evidential reasoning or neural networks. Comparative studies using a suite of classification methods have shown repeatedly that for most imagery and landscapes, neural networks provide the most accurate classification of land cover.

Future developments will further enhance the novel use of remote sensing techniques in obtaining environmental information about the Earth's habitats. Increased applications for remote sensing techniques result not only from the use of new sensors or improved algorithms, but increasingly from the methodology to merge data from various sources into an end product tailored to a specific need.

Besides sophisticated sensors and effective algorithms for data processing, the success of the final result depends on the collection of spatially or spectral–temporally relevant remote sensing data. In this context, spatially relevant means that the spatial resolution of the sensor data is sufficient to distinguish important habitats; spectral–temporally relevant means the appropriate combination of sensor and acquisition date(s) to identify the habitats in question. Additional aspects include the use of improved sensors with better spatial, spectral and/or temporal characteristics, and the use of better classification algorithms (faster and more accurate) or data fusion techniques for multi-source sensor data. Finally, the use of existing georeferenced information from other sources needs consideration, including knowledge of human activity, soil characteristics, topography and climate data, to generate more accurate classifications.

Satellite remote sensing systems

Data from remote sensing satellites have been used successfully to monitor environmental conditions suited to the emergence and outbreak of infectious diseases. In several projects, the relationships between environmental parameters sensed by satellites (e.g. water, temperature and vegetation cover), the occurrence of disease vectors (e.g. mosquitoes, ticks and flies), disease reservoirs (e.g. deer and rodents) and patterns of human settlements, migration and land use have been established. Based on such relationships, predictive models can be developed to aid public health efforts to control specific diseases.

The earliest family of polar-orbiting meteorological satellites came on-line in the 1960s supplying valuable data for national meteorological services around the world. The best known satellites of this type are the NOAA series of satellites, managed by the US National Oceanic and Atmospheric Administration. Besides meteorology, data from the main imaging sensor, the Advanced Very High Resolution Radiometer (AVHRR), of NOAA satellites have found widespread applications in oceanography, agriculture, hydrology, forestry and other fields. A broader range of satellites, with low spatial resolution for environmental monitoring, has been introduced over the years (Table 17.3).

Typical features of this group are scanners with a large swath width implying a high revisit rate and large pixel size. These properties enable investigations of dynamic phenomena on a global and continental scale. As low-resolution data are the cheapest satellite data available, they are the primary data sources for global change studies and are useful for real-time information where timing is crucial. However, these features have drawbacks for other types of analysis. The geometry of a large swath width results in substantial enlargement of the pixel size at the picture edges compared to pixels in the middle of the line. A pixel size of 1 km for vertical pixels can reach several kilometres for oblique measurements. In addition to the need of geometrical corrections, a pixel of this size often covers too much heterogeneous landscape. When analysing pixels with much inherent variability of classes, it is impossible to assign an average radiometric value for the whole pixel area to just one class. The same applies to other values derived from original measurements (e.g. NDVI, LST). Analysis of remote sensing data requires either defined classes of homogeneous areas in the pixel or a more complex knowledge of radiometric values corresponding to all possible area ratios of class combinations within the pixel. With pixels of 1 km, NOAA-type data cannot be used for local or even regional scale environmental studies.

The opportunity to improve mapping from space of biotic and abiotic components followed the launch of the first Landsat satellite in 1972. The MSS scanner on this US satellite measured 60 m pixels in four spectral bands. For 14 years, Landsat satellites were the only source of remote sensing data with such resolution. French Spot 1 added a new scanner (HRV) providing multispectral data of 20 m pixel and 10 m pixel in panchromatic mode. At present,

Table 17.3 *Satellites providing remote sensing data of low resolution*

Satellite	Sensor	Country	Launch year	Pixel sizes (m)	Swath width (km)	No./types of spectral bands[a]	Temporal resolution (days)
OrbView-2	SeaWiFS	USA	1997	1100–4500	1500–2800	8/V, NIR	1–2
NOAA 17	AVHRR	USA	1998	1100	2800	6/V, NIR, MIR, TIR	0,5
Resurs-O1	MSU-SK	Russia	1998	170–600	600	5/V, NIR, TIR	2–4
Terra	MODIS	USA/Japan	1999	240–1900	370–2300	35/V, NIR, MIR, TIR	2–9
Spot 4,5	Vegetation	France	1998, 2002	1000	2200	4/V, NIR	1
Envisat	MERIS	ESA	2002	300–1200	300–1150	15/V, NIR	3
Aqua	MODIS	USA	2002	250–1000	2300	42/V, NIR, MIR, TIR, MW	2
NOAA 18	AVHRR	USA	2005	1100	2800	6/V, NIR, MIR, TIR	0.5
MetOp-A	AVHRR	ESA	2006	1100	2800	6/V, NIR, MIR, TIR	0.5

[a] NIR, near-infrared; MIR, middle-infrared; TIR, thermal infrared; MW, microwave; V, visible.

there are many potentially useful sensors for mapping vegetation type (Table 17.4). The latest satellite of the Landsat series, Landsat 7, has an advanced ETM+ scanner providing 30 m multispectral data, a 15 m panchromatic band, and an improved 60 m thermal infrared band. The latest Spot 5 satellite has improved spatial resolution in the panchromatic band down to 2.5 m and multispectral data for 10 m pixels. Ikonos, a commercial satellite, has started a new generation of very high-resolution satellites (1 m pixels in the panchromatic band and 4 m in visible and near infrared spectral bands) providing data for detailed studies on large scales.

In addition to the passive systems shown in Tables 17.3 and 17.4, there are also active remote sensing systems in orbit (Table 17.5). Synthetic aperture radars (SAR) are particularly important as they can penetrate cloud and vegetation canopy cover and can detect soil moisture and surface roughness.

APPLICATIONS OF REMOTE SENSING IN THE STUDY OF TICK OCCURRENCE AND IN TICK-BORNE DISEASE EPIDEMIOLOGY

The potential use of aerial photography and other remote sensing techniques in epidemiological studies was first realized in 1970 when the National Aeronautics and Space Administration (NASA) created the Health Applications Office (HAO). NASA coined a new term, 'econoses', to describe diseases dependent on environmental conditions. During the period 1970–76, HAO staff and contractors pub-

lished about 100 scientific reports concerning remote sensing in health-related programmes (Barnes, 1991), including a mosquito habitat verification, but none referred to ticks. In 1977, aerial photographs were used to map habitats of *I. persulcatus* during the building of the Baikal–Amur railway in Asiatic Russia (Kuzikov *et al.*, 1982).

The era of remote sensing in the identification of disease vector habitats commenced with the review of Hugh-Jones (1989). Known sites of *Amblyomma variegatum* infestation were identified in northern St Lucia in the Caribbean using the MSS scanner of Landsat 1. The study was then extended to Guadeloupe where the analysis of Landsat TM imagery identified a series of habitats that appeared to have different tick-carrying capacities (see also Hugh-Jones, 1991). Subsequently, *Dermacentor variabilis* habitats in Orange County, North Carolina were identified using Landsat TM scanner data (Cooper & Houle, 1991), while the distribution and abundance of *Rhipicephalus appendiculatus* in East Africa in the CLIMEX model used data from NOAA satellites (Perry *et al.*, 1991).

The first identification of *I. ricinus* habitats in Europe based on Landsat MSS data was published by Daniel & Kolář (1990). Satellite and aerial imagery were used in the study of *I. scapularis* in the environs of New York (Duffy *et al.*, 1994), and Randolph (1993) used remote sensing data to study the seasonal abundance of *R. appendiculatus* in South Africa. After the mid-1990s, the number of papers on ticks based on remote sensing data increased sharply.

Several landscape features discernible from satellite data have important implications in tick epidemiology, vegetation type being the most important (Table 17.6). Mapping

Table 17.4 *Current optical sensor systems for identifying and mapping vegetation in medium and large scales*

Satellite	Sensor	Country	Launch year	Smallest pixel (m)	Swath width (km)	No./types of spectral bands[a]	Temporal resolution (days)
Landsat 5	TM	USA	1986	30	180	7/V, NIR, MIR, TIR	16
Spot 4	HRV	France	1998	10	60	4/V, NIR	16
Landsat 7	ETM	USA	1998	15	180	8/PAN, V, NIR, MIR, TIR	16
Terra	ASTER	USA	1999	15	60	15/V, NIR, MIR, TIR	16
Ikonos-2	OSA	USA	1999	1	12	4/PAN, V, NIR	2–4
ErosA	PIC	Israel	2000	1.8	11	1/PAN	2–4
QuickBird	BHRC	USA	2001	0.6	12	4/PAN, V, NIR	2–4
Spot 5	HRG	France	2002	2.5	60	5/PAN, V, NIR	7–27
IRS P6	PAN	India	2003	5.6	70	4/V, NIR	5–26
OrbView-3	OHIRIS	USA	2003	1	8	4/PAN, V, NIR	2–3
Kompsat 2	MSC	South Korea	2006	1	15	5/Pan, V, NIR	2–4
Eros B	PIC-2	Israel	2006	0.7	7	1/PAN	2–4
ALOS-1	PRISM	Japan	2006	2.5	35	5/PAV, V, NIR	2–4
WorldView 1	Wv60	USA	2007	0.5	17	1/PAN	2–4

[a] V, visible; NIR, near-infrared; MIR, middle-infrared; TIR, thermal infrared; PAN, panchromatic.

Table 17.5 *Current satellites with synthetic aperture radars*

Satellite	Country	Launch year	Smallest pixel (m)	Swath width (km)	Wavelength/ polarization	Temporal resolution (days)
ERS - 2	ESA	1995	30	100	5, 6 cm/VV	35
Radarsat-1	Canada	1995	8	50	5, 6 cm/HH	2–4
Envisat	ESA	2002	30	100	5, 6 cm/all	35
ALOS-1	Japan	2006	7	70	20 cm	2–5
COSMO-SkyMed 1	Italy	2007	1	1	9, 6 cm/all	5
Radarsat 2	Canada	2007	3	50	5, 6 cm/all	2–4
TerraSAR-X	Germany	2007	1	10	3 cm/all	3

the boundary between vegetation types, or ecotones, is also important as an indicator of habitats for hosts that are critical to the maintenance and transmission of tick-borne diseases. These boundaries may be areas of increased risk for tick–human contact. Additionally, soil moisture is an indicator of suitable tick habitats.

To date, most remote sensing applications for mapping tick-borne diseases risk areas have exploited vegetation types. The distribution of vegetation types results from the combined impact of rainfall, temperature, humidity, topographic effects, soil, water availability and human activities. Vegetation type identification is suited to remote sensing tech-

nology because radiometers on board satellites detect radiation from the thin upper layer of surface objects. Vegetation type is independent of its height above ground. By contrast, the temperature of the top surface of a biomass differs from the enclosed microclimate, and differs in its flux and variance margins (Hugh-Jones, 1991). Temperature values can be derived from radiometric values measured in the thermal infrared part of the spectrum. Even when corrections are applied, the resulting value applies to the top of the vegetation canopy and says nothing of the ground-level microclimate, which is the main factor influencing tick ecology. The extent to which remotely sensed data are used for

Table 17.6 *Potential links between information obtained from remote sensing and tick-borne disease influencing factors*

Remote sensing information	Tick-borne disease influencing factors
Vegetation types (ecotones)	Preferred food sources and habitat for hosts/reservoirs
Forest patches, edges	Habitat requirements of deer and other hosts/reservoirs
Soil moisture	Tick habitat

Source: After Beck, Lobitz & Wood (2000).

studying the spatial and temporal patterns of disease depends on a number of factors. On the whole, inadequate resolution in terms of pixel size and spectral channel or accessibility of satellite data have restricted the use of remote sensing.

A ground resolution threshold of 30 m should be considered as the upper limit for exploring the relationship between vegetation type and disease vectors, reservoirs, and hosts, especially in anthropologically modified landscapes. If vegetation types and forest patches are difficult to discern, the mosaic pattern typical of tick and tick-borne disease distributions is lost (Gilot, 1985).

The number and sophistication of studies using remote sensing and GIS technology have increased dramatically over the last decade. In 1998, NASA's Center for Health Applications of Aerospace Related Technologies (CHAART) evaluated current and planned satellite sensor systems as a first step in enabling human health scientists to determine data relevant for the epidemiological, entomological and ecological aspects of their research, as well as developing remote sensing-based models of transmission risk (Beck *et al.*, 2000). An overview describing the methodology of detecting environmental variables using remote sensing and disease risk forecasting is given by Hay, Randolph & Rogers (2000).

Research involving remote sensing data to study tick-borne disease has focused on investigating the application of satellite data and spatial analysis techniques to identify and map landscape elements that collectively define vector and human population dynamics related to disease transmission risk (Daniel & Kolář, 1990).

A study to develop remote sensing-based models for mapping Lyme disease transmission risk was accomplished during the last decade in the northeastern United States. The first study compared Landsat TM data with canine seroprevalence rate and summarized at the municipality level. The amount of remotely sensed deciduous forest was positively correlated with canine exposure to *Borrelia burgdorferi*. Another study used TM scanner data to map relative tick abundance on residential properties by using TM-derived indices of vegetation greenness and wetness (Dister *et al.*, 1997). The final report of the European Union Concerted Action on Lyme Borreliosis (EUCALB) project (Gray *et al.*, 1998) included recommendations to apply remote sensing and GIS tools in European health services.

Detailed mapping of TBE risk habitats in the Czech Republic marked another step forward in the implementation of remote sensing technology for practical health service (Daniel *et al.*, 1998). This study used long-term records of clinical cases of TBE. Starting from 1951, TBE has been a notifiable disease in the Czech Republic and since 1971 only cases that have been verified in the laboratory have been registered. The records include locations in which tick bites occurred and identified permanent residents of the region under study. Spatially located ground data of TBE incidences documented over the period 1971–2000 in five regions of the Czech Republic (area 52 000 km^2) and Landsat TM data of 30 m resolution were used in the analysis. Detailed classification of the satellite data, supported by observations from field checks, resulted in the definition of nine vegetation types related to various risk levels of *I. ricinus* attack and TBE occurrence. The relationship between risk level and vegetation type, including the importance of ecotones for the circulation of TBE virus, was ascertained. A mathematical model was built that described the epidemiological significance of the structure of the vegetation cover of the landscape to the risk of acquiring TBE. Based on this methodology, risk sites of various levels were determined at the pixel level, and maps printed at 1 : 25 000 scale. An atlas of prognostic risk site maps has been produced (Daniel & Kříž, 2002) containing 20 maps at a 1 : 200 000 scale together with distribution maps of TBE occurrence. An electronic version of these maps enables the scale to be rapidly changed appropriate for the purpose at hand. Several accompanying examples of 1 : 25 000 scale maps are included in the atlas (e.g. Fig. 17.8). The spatial distribution of vector and host species relates not simply to the occurrence of forests or grasslands but also to the pattern of spatial distribution of specific vegetation types and their specific structure. The study demonstrated the epidemiological importance of spatial heterogeneity of the landscape at the level appropriate to the size of tick habitats. Such studies are possible only if ground truth data for disease distribution are available with the required location accuracy.

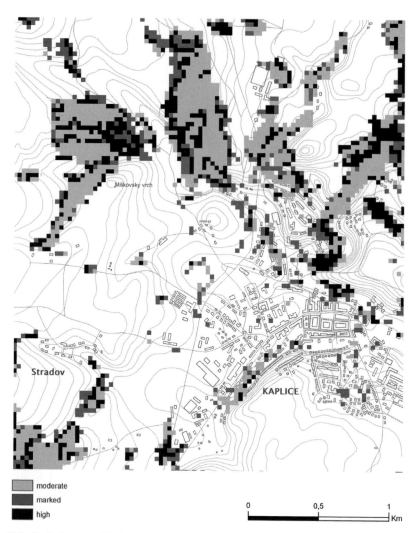

Fig. 17.8. Predictive map of *Ixodes ricinus* high-occurrence habitats and TBE infection risk assessment – a part of TBE natural focus in the surroundings of Kaplice, in the Czech Republic. Risk habitats are dispersed in small 'islands' forming a very typical mosaic pattern of elementary foci. (Satellite data: Landsat TM, 1.9.1991.) The grey scale: (A) Moderate TBE risk – combination of three originally recognized habitat categories: (1) spruce wood cultures of homogeneous structure; (2) coniferous (mostly spruce) wood cultures with heterogeneous structure – both categories essentially without TBE risk; (3) mixed woods mostly with tall stems and prevailing conifers posing TBE risk. (B) Marked TBE risk – combination of three originally recognized habitat categories: (1) young deciduous wood stands of homogeneous structure; (2) tall deciduous woods of homogeneous structure; (3) tall deciduous woods of heterogeneous structure. (C) High TBE risk – combination of three originally recognized habitat categories: (1) deciduous shrubs with fine-grain size of the stand mosaic; (2) mixed wood stands mostly young low and dominated by deciduous trees; (3) young deciduous stands and ecotones with highly heterogeneous structure – peak TBE risk.

Figure 17.8 provides information about the distribution of high-occurrence habitats of *I. ricinus* and TBE infection risk assessment in a part of a highly active TBE natural focus in the surroundings of the town of Kaplice in southern Bohemia, the Czech Republic. This map was prepared for a detailed local orientation (necessary for efficacious preventive measures) and for a spatial structural analysis providing information about environmental conditions of virus circulation in the natural foci. In the original versions (both in printed and electronic copies), nine categories of different risk habitats are recognized and depicted in different colours (Daniel *et al.*, 2004). For black/white reproduction this number has been reduced to three categories, differing in grey tone. The map documents clearly the high dispersion of risk habitats measured in individual pixels surrounding the town of Kaplice closely, sometimes penetrating into the urban residential areas. After generation of this map, positive ixodological and virological verification was performed on the ground (Danielová, Holubová & Daniel, 2002). Moreover, surveys of human activity provided evidence of frequent visits to these 'islands', mainly by young people. An investigation of other southern Bohemian regions with high human TBE incidence demonstrated similar scenarios. Heterogeneity of the landscape resulting in the mosaic distribution of risk habitats (including a high proportion of ecotones) is one of the primary factors causing stable high incidences of TBE in this study area. Estrada-Peña (2003*b*) studied the phenomenon of mosaic distribution of *I. ricinus* by similar methods to GIS in northern Spain. Results of the analysis in south Bohemian TBE foci also demonstrate the importance of detailed spatial resolution of satellite input data. The use of NOAA imagery (1.1 km resolution) exemplifies that such important information about landscape heterogeneity is not detectable at all.

At the country or continental scale, imagery from meteorological satellites has been used to relate the temporal patterns and variations in rainfall and vegetation greenness. Much of this work used data from the AVHRR scanner onboard meteorological NOAA satellites passing over Africa (Rogers & Randolph, 1993) and Europe (Randolph, 2000). In Europe, a cartogram of TBE distribution produced by the pharmaceutical industry (Immuno, 1997) provided the ground truth data set for two areas in central Europe and the Baltic region (Randolph, 2000). Correlation between the ground data and AVHRR data generalized to 8 km pixels has been used as a basis for the TBE prediction map including various possible scenarios in the years 2020–2080 (Randolph & Rogers, 2000). Owing to the characteristics of low-resolution satellites (described above), the quality of these for a landscape with highly diversified habitats may be insufficient for practical use in decision-making.

FUTURE PERSPECTIVES

Utilization of GIS technologies for tick and TBD research has undergone a rapid development under the influence of both the increasing requirements of human and veterinary preventive medicine, and the application of new methods of input data collection. Molecular biology techniques enabling identification of pathogen strains facilitate investigation of the detailed relationships between TBD pathogens, their vectors and animal hosts in natural environments. Recent trends in the elaboration of reliable epidemiological and epizootological TBD data registers, both on national and international scales, and the exchange of information represents another important factor for GIS use in TBD occurrence analyses and risk assessment. One example of this collaboration is the International Tick-Borne Encephalitis Study Group connecting experts from many European countries.

Climate change and its impact on human health bring further application of GIS and RS technologies into public health service. This problem has been studied and evaluated in a book (Menne & Ebi, 2006) summarizing the results of the World Health Organization/European Community project *Climate Change and Adaptation Strategies for Human Health* (cCASHh) including chapters dealing with Lyme borreliosis (Lindgren *et al.*, 2006) and TBE (Daniel *et al.*, 2006). Usage of GIS is emphasized as an important tool for monitoring the changes in the geographical distribution of TBD and the possible explanations for these changes, for risk assessment, and for the determination of preventative measures and adaptation strategies. Vector-borne diseases are highly susceptible to environmental conditions. Climate change influences both the vector biology and, indirectly, also the existence of pathogens within the body of the vector, and the ecology of animal hosts and human behaviour. All of these factors and their interrelationships are detectable by GIS and RS tools.

Technologies now allow distribution of the information in a user-friendly format for the general public, which contributes to prevention and early warnings of impending global health issues. Early warnings are a prerequisite for mitigation measures aimed at reducing the impact of infectious diseases. With the launch of 20 new remote sensing satellites expected by the year 2010, data-collection capabilities will increase considerably. The new generation of research and operational Earth observation satellites will provide users with higher spectral and spatial resolutions. Hyperspectral and high spatial resolution satellite sensor imagery will provide new opportunities for determining environmental conditions favourable to various disease vectors. These improved capabilities, when combined with the increased computing power and spatial modeling capabilities of GIS, should extend the use of remote sensing into operational disease surveillance and control. However, it should not be forgotten that successful use of new remote sensing and GIS technologies for landscape epidemiology is dependent on the user appreciating every part of the geo-information technology chain, i.e. data acquisition, image processing and analysis, database development, field support and spatial statistics and modelling.

The application of GIS and RS allows considerable precision in the spatial analysis of tick and TBD distribution and in the drawing up of maps that are applicable for: (1) analysing the epidemiological situation and planning the management and prevention of TBD within the public health arena; and (2) formulating hypotheses that determine future research. Although RS helps us in formulating hypotheses, the definitive resolution of problems must be sought on the ground. Without verification of RS results in the field, general and definite conclusions cannot be made. As Hugh-Jones (1991) wrote, 'The use of RS and GIS results in more field work, not less, because the truth is on the ground, not in the computer. The advantage is that we know exactly where to go in the field.' Or alternatively put, 'Why should I believe you when you are sitting at the computer terminal with clean boots!?'.

ACKNOWLEDGEMENTS

We are very grateful to Alan S. Bowman and D. Crossley for linguistic and stylistic review of the manuscript.

REFERENCES

Balát, F. & Rosický, B. (1954). Birds in lowland forests, their role and importance for the existence of natural foci of diseases. *Československá Parasitologie* 1, 22–44. (In Czech.)

Baptista, S., Quaresma, A., Aires, T., *et al.* (2004). Lyme borreliosis spirochetes in questing ticks from mainland Portugal. *International Journal of Medical Microbiology* Suppl. 293, 109–116.

Barnes, C. M. (1991). An historical perspective on the applications of remote sensing to public health. *Preventive Veterinary Medicine* 11, 163–166.

Beck, L. R., Lobitz, B. M. & Wood, B. L. (2000). Remote sensing and human health: new sensors and new opportunities. *Emerging Infectious Diseases* 6, 217–227.

Benda, R. (1958). The common tick *Ixodes ricinus* L. as a reservoir and vector of tick-borne encephalitis. I. Survival of the virus (strain B 3) during the development of the tick under laboratory conditions. *Journal of Hygiene, Epidemiology, Microbiology and Immunology* 2, 314–330.

Bithell, J. F. (1990). An application of density estimation to geographical epidemiology. *Statistics in Medicine* 9, 691–701.

Blaškovič, D. (1967). Studies on tick-borne encephalitis. *Bulletin of the World Health Organization* 36 (Suppl. 1), 1–95.

Brownstein, J. S., Holford, T. R. & Fish, D. (2003). A climate-based model predicts the spatial distribution of the Lyme disease vector *Ixodes scapularis* in the United States. *Environmental Health Perspectives* 111, 1152–1157.

Carey, A. B., Mclean, R. G. & Maupin, G. O. (1980). The structure of a Colorado tick fever ecosystem. *Ecological Monographs* 50, 131–151.

CDC (Centers for Disease Control and Prevention) (1994). *Addressing Emerging Infectious Disease Threats: A Strategy for the United States.* Atlanta, GA: US Department of Health and Human Services, CDC.

CDC (Centers for Disease Control and Prevention) (1998). *Preventing Emerging Infectious Diseases: A Strategy for the 21st Century.* Atlanta, GA: US Department of Health and Human Services, CDC.

Chaput, E. K., Meek, J. I. & Heimer, R. (2002). Spatial analysis of human granulocytic ehrlichiosis near Lyme, Connecticut. *Emerging Infectious Diseases* 8, 943–948.

Chemini, C. & Rizzoli, A. (2003). Land use change and biodiversity conservation in the Alps. *Journal of Mountain Ecology* (Suppl.) 7, 1–7.

Cliff, A. D. & Haggett, P. (1988). *Atlas of the Distribution of Diseases: Analytical Approaches to Epidemiological Data.* Oxford, UK: Blackwell Scientific Publications.

Cooper, J. W. & Houle, J. U. (1991). Modelling disease vector habitats using thematic mapper data: identifying *Dermacentor variabilis* habitats in Orange County, North Carolina. *Preventive Veterinary Medicine* 11, 353–354.

Cumming, G. S. (2000a). Using habitat models to map diversity: pan-African species richness of ticks (Acari: Ixodida). *Journal of Biogeography* 27, 425–440.

Cumming, G. S. (2000b). Using between-model comparisons to fine-tune linear models of species ranges. *Journal of Biogeography* 27, 441–455.

Cumming, G. S. (2002). Comparing climate and vegetation as limiting factors for species ranges of African ticks. *Ecology* 83, 255–268.

Dale, M. R. T., Dixon, P., Fortin, M.-J., *et al.* (2002). Conceptual and mathematical relationships among methods for spatial analysis. *Ecography* 25, 558–577.

Daniel, M. & Dusbábek, F. (1994). Micrometeorological and microhabitat factors affecting maintenance and dissemination of tick-borne diseases in the environment. In *Ecological Dynamics of Tick-Borne Zoonoses* eds. Sonenshine,

D. E. & Mather, T. N., pp. 91–138. Oxford, UK: Oxford University Press.

Daniel, M. & Kolář, J. (1990). Using satellite data to forecast the occurence of the common tick *Ixodes ricinus* (L.). *Journal of Hygiene, Epidemiology, Microbiology and Immunology* **34**, 243–252.

Daniel, M. & Kříž, B. (2002). *Tick-Borne Encephalitis in the Czech Republic*, vol. 1, *Predictive Maps of* Ixodes ricinus *Tick High-Occurrence Habitats and a Tick-Borne Encephalitis Risk Assessment in Czech Regions*; vol. 2, *Maps of Tick-Borne Encephalitis Incidence in the Czech Republic in 1971–2000*, Project WHO/EC Climate Change and Adaptation Strategies for Human Health in Europe, EVK-2-2000-0070. Prague: National Institute of Public Health.

Daniel, M. & Rosický, B. (1989). *Medical Entomology and Environment*. Prague: Academia. (In Czech.)

Daniel, M., Danielová, V., Kříž, B. & Beneš, Č. (2006). Tick-borne encephalitis. In *Climate Change and Adaptation Strategies for Human Health*, eds. Menne, B. & Ebi, K. L., pp. 189–205. Darmstadt, Germany: Steinkopff.

Daniel, M., Kolář, J. & Zeman, P. (2004). GIS tools for tick and tick-borne disease occurrence. *Parasitology* **129** (Suppl.), S329–S352.

Daniel, M., Kolář, J., Zeman, P., Pavelka, K. & Sádlo, J. (1998). Predictive map of *Ixodes ricinus* high-incidence habitats and a tick-borne encephalitis risk assessment using satellite data. *Experimental and Applied Acarology* **22**, 417–433.

Daniel, M., Kolář, J., Zeman, P., Pavelka, K. & Sádlo, J. (1999). Tick-borne encephalitis and Lyme borreliosis: comparison of habitat risk assessments using satellite data. *Central European Journal of Public Health* **7**, 35–39.

Danielová, V., Holubová, J. & Daniel, M. (2002a). Tick-borne encephalitis virus prevalence in *Ixodes ricinus* ticks collected in high risk habitats of the South-Bohemian region of the Czech Republic. *Experimental and Applied Acarology* **26**, 145–151.

Danielová, V., Holubová, J., Pejčoch, M. & Daniel, M. (2002b). Potential significance of transovarial transmission in the circulation of tick-borne encephalitis virus. *Folia Parasitologia* **49**, 323–325.

De Garine-Wichatitsky, M. (2000). Assessing infestation risk by vectors: spatial and temporal distribution of African ticks at the scale of a landscape. *Annals of the New York Academy of Sciences* **916**, 223–232.

Diamond, I. (1992). Population counts in small areas. In *Geographical and Environmental Epidemiology: Methods for Small-Area Studies*, eds. Elliot, P., Cuzick, J., English, D. & Stern R., pp. 98–105. Oxford, UK: Oxford University Press.

Diggle, P. J. (1983). *Statistical Analysis of Spatial Point Patterns*. London: Academic Press.

Dister, S. W., Fish, D., Bros, S., Frank, D. H. & Wood, B. L. (1997). Landscape characterization of peridomestic risk for Lyme disease using satellite imagery. *American Journal of Tropical Medicine and Hygiene* **57**, 687–692.

Duffy, D. C., Clark, D. D., Campbell, S. R., *et al.* (1994). Landscape patterns of abundance of *Ixodes scapularis* (Acari: Ixodidae) on Shelter Island, New York. *Journal of Medical Entomology* **31**, 875–879.

Eisen, R. J., Eisen, L. & Lane, R. S. (2005). Remote sensing (Normalized Difference Vegetation Index) classification of risk versus minimal risk habitats for human exposure to *Ixodes pacificus* (Acari: Ixodidae) nymphs in Mendocino County, California. *Journal of Medical Entomology* **42**, 75–81.

Estrada-Peña, A. (1998). Geostatistics and remote sensing as predictive tools of tick distribution: a cokriging system to estimate *Ixodes scapularis* (Acari: Ixodidae) habitat suitability in the United States and Canada from Advanced Very High Radiometer satellite imagery. *Journal of Medical Entomology* **35**, 989–995.

Estrada-Peña, A. (1999a). Geostatistics and remote sensing using NOAA–AVHR satellite imagery as predictive tools in tick distribution and habitat suitability estimations for *Boophilus microplus* (Acari: Ixodidae) in South America. *Veterinary Parasitology* **81**, 73–82.

Estrada-Peña, A. (1999b). Geostatistics as predictive tools to estimate *Ixodes ricinus* (Acari: Ixodidae) habitat suitability in the western Palearctic from AVHRR satellite imagery. *Experimental and Applied Acarology* **23**, 337–349.

Estrada-Peña, A. (2001a). Distribution, abundance, and habitat preferences of *Ixodes ricinus* (Acari: Ixodidae) in northern Spain. *Journal of Medical Entomology* **38**, 361–370.

Estrada-Peña, A. (2001b). Climate warming and changes in habitat suitability for *Boophilus microplus* (Acari: Ixodidae) in Central America. *Journal of Parasitology* **87**, 978–987.

Estrada-Peña, A. (2002a). Increasing habitat suitability in the United States for the tick that transmits Lyme disease: a remote sensing approach. *Environmental Health Perspectives* **110**, 635–640.

Estrada-Peña, A. (2002b). A simulation model for environmental population densities, survival rates and prevalence of *Boophilus decoloratus* (Acari: Ixodidae) using remotely sensed environmental information. *Veterinary Parasitology* **104**, 51–78.

Estrada-Peña, A. (2003a). Climate change decreases habitat suitability for some species (Acari: Ixodidae) in South

Africa. *Onderstepoort Journal of Veterinary Research* **70**, 79–93.

Estrada-Peña, A. (2003*b*). The relationships between habitat topology, critical scales of connectivity and tick abundance *Ixodes ricinus* in a heterogeneous landscape in northern Spain. *Ecography* **26**, 661–671.

Estrada-Peña, A., Sánchez Acedo, C., Quílez, J. & Del Cacho, E. (2005). A retrospective study of climatic suitability for the tick *Rhipicephalus (Boophilus) microplus* in the Americas. *Global Ecology and Biogeography* **14**, 565–573.

Foley, J. E., Queen, E. V., Sacks, B. & Foley, P. (2005). GIS-facilitated spatial epidemiology of tick-borne diseases in coyotes (*Canis latrans*) in northern and coastal California. *Comparative Immunology, Microbiology and Infectious Diseases* **28**, 197–212.

Frank, C., Fix, A. D., Pena, C. A. & Strickland, G. T. (2002). Mapping Lyme disease incidence for diagnostic and preventive decisions, Maryland. *Emerging Infectious Diseases* **8**, 427–429.

Gern, L., Estrada-Peña, A., Frandsen, F., *et al.* (1998). European reservoir hosts of *Borrelia burgdorferi* sensu lato. *Zentralblatt für Bakteriologie* **287**, 196–204.

Gilot, B. (1985). Bases biologiques, écologiques et cartographiques pour l'étude des maladies transmises par les tiques (Ixodidae et Argasidae) dans les Alpes Françaises et leur avant-pays. Unpublished Ph.D. thesis, University of Grenoble, France.

Gilot, B., Pautou, G. & Lachet, B. (1981). La cartographie des populations de tiques exophiles à visée épidémiologique: application à la fièvre boutonneuse méditerranéenne essai à 1/200000 dans la basse valleée du Rhône. *Documents de Cartographie Ecologique (Grenoble)* **34**, 103–111.

Gilot, B., Pautou, G. & Moncada, E. (1975). L'analyse de la végétation appliqué à la détection des populations de tiques exophiles dans le Sud-Est de la France: l'exemple *d'Ixodes ricinus* (Linné, 1758). *Acta Tropica* **32**, 340–347.

Gilot, B., Pautou, G., Moncada, E., Lachet, B. & Christin, J. G. (1979). La cartographie des populations de tiques exophiles par le biaias de la végétation: bases écologiques, intérêt épidémiologique. *Documents de Cartographie Ecologique (Grenoble)* **22**, 65–80.

Glass, G. E., Amerasinghe, F. P., Morgan, J. M. & Scott, T. W. (1994). Predicting *Ixodes scapularis* abundance on white-tailed deer using geographic information systems. *American Journal of Tropical Medicine and Hygiene* **51**, 538–544.

Glass, G. E., Schwartz, B. S., Morgan, J. M., *et al.* (1995). Environmental risk factors for Lyme disease identified with Geographic Information Systems. *American Journal of Public Health* **85**, 944–948.

Glavanakov, S., White, D. J., Caraco, T., *et al.* (2001). Lyme disease in New York State: spatial pattern at a regional scale. *American Journal of Tropical Medicine and Hygiene* **65**, 538–554.

Gray, J. S., Kahl, O., Robertson, J. N., *et al.* (1998). Lyme borreliosis habitat assessment. *Zentralblatt für Bakteriologie* **287**, 211–228.

Green, R. M., Rogers, D. J. & Randolph, S. E. (2000). The use of satellite imagery to predict foci of tick-borne encephalitis. In *Proceedings of the 3rd International Conference Ticks and Tick-Borne Pathogens: Into the 21st Century*, eds. Kazimírová, M., Labuda, M. & Nuttall, P. A., pp. 209–213.

Guerra, M., Walker, E., Jones, C., *et al.* (2002). Predicting the risk of Lyme disease: habitat suitability for *Ixodes scapularis* in the north central United States. *Emerging Infectious Diseases* **8**, 289–297.

Guisan, A. & Zimmermann, N. E. (2000). Predictive habitat distribution models in ecology. *Ecological Modelling* **135**, 147–186.

Hay, S. L, Randolph, S. E. & Rogers, D. J. (eds.) (2000). Remote sensing and Geographical Information Systems in epidemiology. *Advances in Parasitology* **47**, 1–357.

Hejný, S. & Rosický, B. (1965). Beziehungen der Encephalitis zu den natürlichen Pflanzengesellschaften. In *Biosoziologie*, ed. R. Tuexen, pp. 341–347. The Hague, Netherlands: Junk Verlag.

Hugh-Jones, M. (1989). Applications of remote sensing to the identification of the habitats of parasites and disease vectors. *Parasitology Today* **5**, 244–251.

Hugh-Jones, M. (ed.) (1991). Applications of remote sensing to epidemiology and parasitology. *Preventive Veterinary Medicine* **11**, 155–376.

Hugh-Jones, M. E., Barre, N., Nelson, G., *et al.* (1988). Remote recognition of *Amblyomma variegatum* habitats in Guadeloupe using LANDSAT-TM imagery. *Acta Veterinaria Scandinavica* (Suppl.) **84**, 259–261.

Hugh-Jones, M., Barre, N., Nelson, G., *et al.* (1992). Landsat–TM identification of *Amblyomma variegatum* (Acari: Ixodidae) habitats in Guadeloupe. *Remote Sensing of Environment* **40**, 43–55.

Hungerford, L. L. (1991). Use of spatial statistics to identify and test significance in geographic disease patterns. *Preventive Veterinary Medicine* **11**, 273–282.

Immuno (1997). *Tick-Borne Encephalitis (TBE) and its Immunoprophylaxis*. Vienna: Immuno A.G.

Isaaks, E. H. & Srivastava, R. M. (1989). *Applied Geostatistics*. Oxford, UK: Oxford University Press.

Jones, C. G., Ostfeld, R. S., Richard, M. P., Schauber, E. M. & Wolff, J. O. (1998). Chain reactions linking acorns to gypsy moth outbreaks and Lyme disease risk. *Science* 279, 1023–1026.

Jones, L. D., Davies, C. R., Steele, G. M. & Nuttall, P. A. (1987). A novel mode of arbovirus transmission involving a non-viremic host. *Science* 237, 775–777.

Journel, A. G. & Huibregts, C. J. (1978). *Mining Geostatistics*. New York: Academic Press.

Kelsall, J. E. & Diggle, P. J. (1995). Non-parametric estimation of spatial variation in relative risk. *Statistics in Medicine* 14, 2335–2342.

Kitron, U. & Kazmierczak, J. J. (1997). Spatial analysis of the distribution of Lyme disease in Wisconsin. *American Journal of Epidemiology* 145, 558–566.

Kitron, U. & Mannelli, A. (1994). Modeling the ecological dynamics of tick-borne zoonoses. In *Ecological Dynamics of Tick-Borne Zoonoses*, eds. Sonenshine, D. E. & Mather, T. N., pp. 198–239. Oxford, UK: Oxford University Press.

Kitron, U., Bouseman, J. K. & Jones, C. J. (1991). Use of the ARC/INFO GIS to study the distribution of Lyme disease ticks in an Illinois county. *Preventive Veterinary Medicine* 11, 243–248.

Kitron, U., Jones, C. J., Houseman, J. K., Nelson, J. A. & Baumgartner, D. L. (1992). Spatial analysis of the distribution of *Ixodes dammini* (Acari: Ixodidae) on white-tailed deer in Ogle County, Illinois. *Journal of Medical Entomology* 29, 259–266.

Kolonin, G. V. (1978). *World Distribution of Ixodid Ticks* (*Genus* Haemaphysalis). Moscow, USSR: Nauka. (In Russian.)

Kolonin, G. V. (1981). *World Distribution of Ixodid Ticks* (*Genus* Ixodes). Moscow, USSR: Nauka. (In Russian.)

Kolonin, G. V. (1983). *World Distribution of Ixodid Ticks* (*Genera* Hyalomma, Aponomma, Amblyomma). Moscow, USSR: Nauka. (In Russian.)

Kolonin, G. V. (1984). *World Distribution of Ixodid Ticks* (*Genera* Dermacentor, Anocentor, Cosmiomma, Dermacentonomma, Nosomma, Rhipicentor, Rhipicephalus, Boophilus, Margaropus, Anomalohimalaya). Moscow, USSR: Nauka. (In Russian.)

Korenberg, E. I. (1973). Methodological principles of mapping the occurrence of Ixodid ticks. In *Proceedings of 3rd International Congress of Acarology*, eds. Daniel, M. & Rosický, B., pp. 575–577.

Korenberg, E. I. (1983). *What Is a Natural Focus?* Moscow, USSR: Znanie. (In Russian.)

Korenberg, E. I. & Kovalevsky, J. V. (1981). Regional classification of the tick-borne encephalitis area of distribution. *Scientific and Technical Results, Series Medical Geography* 11, 1–148. (In Russian.)

Krejcir, P. (2000). A maximum likelihood estimator of an inhomogeneous Poisson point process intensity using beta splines. *Kybernetika* 36, 455–464.

Kucheruk, V. V. & Rosický, B. (1984). Natural focality of infectious diseases: basic terms and their explanation. *Medical Parasitology and Parasitological Diseases* 2, 7–16. (In Russian.)

Kulldorff, M. & Nagarwalla, N. (1995). Spatial disease clusters: detection and inference. *Statistics in Medicine* 14, 799–810.

Kuzikov, I. V., Korenberg, E. I., Kovalevsky, J. V. & Rodman, L. S. (1982). The principles of average scale mapping of distribution of the ixodid ticks on the basis of aerophoto-materials. *Zoologiocheskii Zhurnal* 61, 1802–1814. (In Russian.)

Lam, N. S. (1983). Spatial interpolation methods: a review. *American Geographer* 10, 129–149.

Lawson, A. B. (2001). *Statistical Methods in Spatial Epidemiology*. Chichester, UK: John Wiley.

Lawson, A. B. & Williams, F. L. R. (1993). Applications of extraction mapping in environmental epidemiology. *Statistics in Medicine* 12, 1249–1258.

Lawson, A. B., Biggeri, A., Böhning, D., *et al.* (1999). *Disease Mapping and Risk Assessment for Public Health*. Chichester, UK: John Wiley.

Lindgren, E. & Jaenson, T. G. T. (2006). Lyme borreliosis in Europe: influences of climate and climate change, epidemiology, ecology and adaptation measures. In *Climate Change and Adaptation Strategies for Human Health*, eds. Menne, B. & Ebi, K. L., pp. 157–188. Darmstadt, Germany: Steinkopff.

Martyn, K. P. (1988). *Provisional Atlas of the Ticks* (*Ixodoidea*) *of the British Isles*. Huntingdon, UK: Institute of Terrestrial Ecology.

Meade, M. S., Florin, J. W. & Gesler, W. M. (1988). *Medical Geography*. New York: Guilford Press.

Menne, B. & Ebi, K. L. (eds.) (2006). *Climate Change and Adaptation Strategies for Human Health*. Darmstadt, Germany: Steinkopff.

Merler, S., Furlanello, C., Chemini, C. & Nicolini, G. (1996). Classification tree methods for analysis of mesoscale distribution of *Ixodes ricinus* (Acari: Ixodidae) in Trentino, Italian Alps. *Journal of Medical Entomology* 33, 888–893.

Moran, P. A. P. (1950). Notes on continuous stochastic phenomena. *Biometrica* 37, 17–23.

Nicholson, M. C. & Mather, T. N. (1996). Methods for evaluating Lyme disease risk using geographic information systems and geospatial analysis. *Journal of Medical Entomology* 33, 711–720.

Norval, R. A. I., Perry, B. D., Kruska, R. & Kundert, K. (1991). The use of climate data interpolation in estimating the distribution of *Amblyomma variegatum* in Africa. *Preventive Veterinary Medicine* 11, 365–366.

O'Connell, S., Granström, M., Gray, J. S. & Stanek, G. (1998). Epidemiology of European Lyme borreliosis. *Zentralblatt für Bakteriologie* 287, 229–240.

Oden, N. (1995). Adjusting Moran's *I* for population density. *Statistics in Medicine* 14, 17–26.

Olwoch, J. M., Rautenbach, C. J. de W., Erasmus, B. F. N., Engelbrecht, F. A. & van Jaarsveld, A. S. (2003). Simulating tick distributions over sub-Saharan Africa: the use of observed and simulated climate surfaces. *Journal of Biogeography* 30, 1221–1232.

Pavlovsky, E. N. (1939). Natural focality of infectious diseases. *Vestnik Akademii Nauk SSSR* 10, 98–108. (In Russian.)

Pavlovsky, E. N. (1946). *Manual of Human Parasitology*, vol. 1. Moscow–Leningrad, USSR: Publishing House of the Academy of Sciences of the USSR. (In Russian.)

Pavlovsky, E. N. (1948). *Manual of Human Parasitology*, vol. 2. Moscow–Leningrad, USSR: Publishing House of the Academy of Sciences of the USSR. (In Russian.)

Pavlovsky, E. N. (1964). *Natural Nidality of Transmissible Diseases in Relation to Landscape Epidemiology of Zooanthroponoses*. Moscow, USSR: Peace Publishers.

Perry, B. D., Kruska, R., Lessard, R., Norval, R. A. I. & Kundert, K. (1991). Estimating the distribution and abundance of *Rhipicephalus appendiculatus* in Africa. *Preventive Veterinary Medicine* 11, 261–268.

Prokhorov, B. B., Baiborodin, V. N., Vershinina, T. A. & Sotchavy, V. V. (1974). *Experience in Ixodid Ticks Mapping on the Territory of Asiatic Russia*. Irkutsk, USSR: Academy of Sciences of the USSR. (In Russian.)

Radvan, R., Hanzák, J., Hejný, S., Rehn, F. & Rosický, B. (1960). Demonstration of elementary foci of tick-borne infections on the basis of microbiological, parasitological and biocenological investigations. *Journal of Hygiene, Epidemiology, Microbiology and Immunology* 4, 81–93.

Rand, P. W., Lacombe, E. H., Smith, R. P., Gensheimer, K. & Dennis, D. T. (1996). Low seroprevalence of human Lyme disease near a focus of high entomologic risk. *American Journal of Tropical Medicine and Hygiene* 55, 160–164.

Randolph, S. E. (1993). Climate, satellite imagery and seasonal abundance of the tick *Rhipicephalus appendiculatus* in Southern Africa: a new perspective. *Medical and Veterinary Entomology* 7, 243–258.

Randolph, S. E. (2000). Ticks and tick-borne disease systems in space and from space. In *Remote Sensing and Geographical Information Systems in Epidemiology*, eds. Baker, J. R., Muller, R. & Rollinson, D., pp. 217–243. London: Academic Press.

Randolph, S. E. & Rogers, D. J. (2000). Fragile transmission cycles of tick-borne encephalitis virus may be disrupted by predicted climate change. *Proceedings of the Royal Society of London B* 267, 1741–1744.

Randolph, S. E., Miklisová, D., Lysý, D., Rogers, D. J. & Labuda, M. (1999). Incidence from coincidence: patterns of tick infestations on rodents facilitate transmission of tick-borne encephalitis virus. *Parasitology* 118, 177–186.

Rizzoli, A., Merler, S., Furlanello, C. & Genchi, C. (2002). Geographical Information Systems and bootstrap aggregation (bagging) of tree-based classifiers for Lyme disease risk prediction in Trentino, Italian Alps. *Journal of Medical Entomology* 39, 485–492.

Robinson, T. P. (2000). Spatial statistics and geographical information systems in epidemiology and public health. In *Remote Sensing and Geographical Information Systems in Epidemiology*, eds. Baker, J. R., Muller, R. & Rollinson, D., pp. 81–127. London: Academic Press.

Rogers, A. (1974). *Statistical Analysis of Spatial Dispersion: The Quadrat Method*. London: Pion.

Rogers, D. J. (2000). Satellites, space, time and the African trypanosomiases. In *Remote Sensing and Geographical Information Systems in Epidemiology*, eds. Baker, J. R., Muller, R. & Rollinson, D., pp. 129–171. London: Academic Press.

Rogers, D. J. & Randolph, S. E. (1993). Distribution of tsetse and ticks in Africa, past, present and future. *Parasitology Today* 9, 266–271.

Rosický, B. (1967). Natural foci of diseases. In *Infectious Diseases: Their Evolution and Eradication*, ed. Cockburn, T. A., pp. 108–126. Springfield, IL: Charles C. Thomas.

Sonenshine, D. E., Peters, A. H. & Stout, J. I. (1972). Rocky Mountain spotted fever in relation to vegetation in the Eastern United States, 1951–1971. *American Journal of Epidemiology* 96, 59–69.

Sutherst, R. W. (2001). The vulnerability of animal and human health to parasites under global change. *International Journal for Parasitology* 31, 933–948.

Sutherst, R. W. & Maywald, G. F. (1985). A computerized system for matching climates in ecology. *Agriculture Ecosystems and Environment* **13**, 281–299.

Thomsen, I. (1991). Population data for small area studies. In *Data Requirements and Methods for Analysing Spatial Patterns of Disease in Small Areas*, pp. 10–14. Copenhagen: World Health Organization. Available online at http://whqlibdoc.who.int/euro/–1993/EUR_ICP_CEH_087_A_1.pdf/

Van Beurden, A. U. C. J. & Hilferink, M. T. A. (1993). *Spatial Analysis with GIS at RIVM: A Background Overview of Spatial Analysis within GIS in research for environment and public health*, Report No. 421503002. Bilthoven, the Netherlands: National Institute of Public Health and Environmental Protection.

Vershinina, T. A. (1985). *Mapping of Ixodid Ticks' Distribution and Seasonal Activity*. Novosibirsk, USSR: Nauka. (In Russian.)

Whittie, P. S., Drane, W. & Aldrich, T. E. (1996). Classification methods for denominators in small areas. *Statistics in Medicine* **15**, 1921–1926.

Zeman, P. (1997). Objective assessment of risk maps of tick-borne encephalitis and Lyme borreliosis based on spatial patterns of located cases. *International Journal of Epidemiology* **26**, 1121–1130.

Zeman, P. (1998). Borrelia-infection rates in tick and insect vectors accompanying human risk of acquiring Lyme borreliosis in a highly endemic region in Central Europe. *Folia Parasitologica* **45**, 319–325.

Zeman, P. (1999). A spatial analysis of uncertain occurence of Lyme borreliosis. *Zentralblatt für Bakteriologie* **289**, 717–719.

Zeman, P., Vitkova, V. & Markvart, K. (1990). Joint occurence of tick-borne encephalitis and Lyme borreliosis in the Central Bohemian region of Czechoslovakia. *Československá Epidemiologie, Mikrobiologie a Imunologie* **39**, 95–105.

18 • Acaricides for controlling ticks on cattle and the problem of acaricide resistance

J. E. GEORGE, J. M. POUND AND R. B. DAVEY

INTRODUCTION

During the nineteenth century, as the number of cattle in the world was increased to feed the human populations of recently industrialized nations, there was a growing awareness of the relationship between infestations of cattle with ticks and disastrous epizootics of disease in herds of cattle. Problems with tick-borne diseases were related to the introduction of improved breeds of cattle into tick-infested areas because of their greater productivity than well-adapted indigenous breeds. Also, cattle infested with ticks and infected with tick-borne disease agents were moved into areas where these tick species had not previously existed (Shaw, 1969).

A severe outbreak of disease in cattle, almost certainly bovine piroplasmosis, occurred in Lancaster County, Pennsylvania, in 1796. Epidemiological evidence indicated a relationship between the disease problem and a recent shipment of cattle into the state from South Carolina, a southern state:

> Experience soon showed that the invariable result following the transportation of southern cattle into the Northern States was the death of all northern cattle along the roads and on the pastures over which the southern cattle had traveled, although the latter animals remained perfectly healthy. In the same way northern cattle taken south almost invariably succumbed to the malady. (Mohler, 1906.)

The disease was called 'Texas fever' or 'cattle fever' and by 1885 resulted in the prohibition of movements of southern cattle into the northern states.

In Australia, cattle that, according to Angus (1996), were 'almost certainly' infested with *Rhipicephalus* (*Boophilus*) *microplus* and infected with 'tick fever' or 'redwater fever' were introduced to the Northern Territory (NT) from Timor, and possibly Bali, some time during the years from 1829 to 1849. There is evidence from archival records that by 1870 tick fever was endemic in the Darwin area. Tick fever and its vector progressively spread eastward and then southward through Queensland. Successive quarantine lines were established by the Queensland government in an effort to contain the problem, but the disease reached southeastern Queensland by 1897, and by 1906 tick fever was in New South Wales in spite of double fencing and strict surveillance.

Much of the complex history of tick-borne diseases of cattle in Africa dates to the colonial settlement of the interior of eastern and southern Africa in the last decades of the nineteenth century. A deadly disease, later determined to be East Coast Fever (ECF), was diagnosed in 1902 in cattle at several locations south of the Zambezi River. The origin of the disease was determined to be importations of cattle from Dar-es-Salaam in Tanzania after the cattle population of much of southern Africa had been destroyed by an epidemic of rinderpest (Lawrence, 1992). Even though there are no records of ECF in eastern Africa before it was identified in 1904, it must have been present for many generations in areas populated by indigenous cattle, particularly in the Lake Victoria Basin and along the coastal strip of eastern Africa (Perry, 1992). Other tick-borne diseases of cattle such as redwater fever, heartwater and anaplasmosis were probably widespread in eastern and southern Africa before the arrival of white settlers and introductions of susceptible cattle in the period from 1885 to 1890 (Lawrence & Norval, 1979; Norval *et al.*, 1984; Perry, 1992).

The economic benefits of resolving questions about the epidemiology and control of tick-borne diseases in the vast cattle-producing areas of eastern and southern Africa, Latin America, Australia and the southern United States motivated research by national and colonial governments in the affected countries plus efforts by international animal health companies to create and market products that provided a means for protecting cattle. The majority of literature on chemical control of ticks documents more than a century of research to test new acaricides for controlling ticks on cattle, strategies for using acaricides, and efforts to mitigate problems of acaricide resistance to all except the most recently developed chemicals. The introduction to this chapter is intended to remind readers of the kinds of problems associated with tick-borne disease agents that are the basis of historical and current needs for technology to control ticks on cattle. The remainder of the chapter represents a selective review of chemical methods for the control of ticks on cattle, the nature of the problem of the evolution of resistance to acaricides, the effects of resistance on the use of acaricides, and the future of chemical methods for the control of ticks on cattle. The excellent review by Taylor (2001) of developments in ectoparasiticides will be of interest to those seeking recent information on chemicals for the control of insects and acarines affecting both livestock (large and small ruminants) and companion animals.

ACARICIDES FOR THE CONTROL OF TICKS ON CATTLE

Even before the seminal discovery by Smith & Kilborne (1893) that proved the role of ticks as vectors of *Babesia*, animal health authorities in the US, Australia and southern Africa were treating cattle with a variety of chemical agents in an effort to control ticks. Early remedies in the US included smearing the legs and sides of cattle with a lard and sulphur mixture, a lard and kerosene combination, cottonseed oil or fish oil. Mixtures of kerosene, cottonseed oil and sulphur; a 10% kerosene emulsion; a mixture of cottonseed oil and crude petroleum; or Beaumont crude oil alone reportedly proved efficacious when applied to cattle two to three times a week with sponges, syringes, brushes, mops or brooms (Francis, 1892; Mohler, 1906). As early as 1895 Australian investigators were immersing cattle in dipping vats containing such things as mineral oil and 'carbolics' (Angus, 1996).

Dipping vats in which cattle were immersed in arsenical solutions revolutionized the control of ticks on cattle, and arsenic quickly replaced other tick control remedies. Angus (1996) attributes the discovery of arsenical solutions in 1896 to an Australian farmer. Shaw (1969) observed that arsenical solutions had been used for over a century to control parasites of sheep before the first reports of their use in 1893 in southern Africa and 1895 in Australia to control ticks on cattle. The Bureau of Animal Industry in the US did not adopt arsenic as its recommended tick control agent until 1910 (Graham

& Hourrigan, 1977). Not only were arsenic dips widely used to control ticks, but they were also key tools in the successful ECF eradication program in South Africa and in the campaign to eradicate *B. annulatus*, *B. microplus* and cattle fever from the US (Graham & Hourrigan, 1977; Lawrence, 1992). The evolution of resistance of ticks to arsenicals, the narrow limits between the effective concentration for tick control and the toxic concentration for cattle, and concerns about toxic residues in animal tissues were major factors for replacing arsenic with synthetic organic insecticides in the decade after World War II ended (Graham & Hourrigan, 1977). Populations of *B. microplus* and *B. decoloratus* developed resistance to arsenic after 1935, and with the lack of an alternative acaricide, *Boophilus* infestations on cattle in parts of the world reached 'enormous' proportions. Relief was not available until the mid-1940s when the first organochlorine products became available (Shaw, 1970).

Organochlorine insecticides were the first synthetic organic insecticides to be marketed and many of them were formulated for the control of ticks on cattle. Dichlorodiphenyltrichloroethane (DDT) and benzenehexachloride (BHC) were the first of this group of chemicals to be used as acaricides (Cobbett, 1947; Maunder, 1949; Whitnall *et al.*, 1951). Dieldrin and aldrin, cyclodiene compounds and toxaphene, a polychloroterpine product, also were widely used for the control of ticks on cattle. In areas such as Australia (Norris & Stone, 1956; Stone & Meyers, 1957) and equatorial and southern Africa (Whitehead, 1958; Baker & Shaw, 1965), cross-resistance of populations of tick species including *B. microplus*, *B. decoloratus* and *Rhipicephalus appendiculatus* to all organochlorines abbreviated the useful life of these chemicals. Organochlorine products for treating livestock are now unavailable or have been withdrawn from the market (Kunz & Kemp, 1994). All of the organochlorine pesticides are persistent in the environment; DDT, BHC and the cyclodienes are especially prone to accumulate in body fat (Ware, 2000).

Unlike the persistent organochlorines, the organophosphate compounds that replaced them were chemically unstable and non-persistent. The organophosphates are generally categorized as the most toxic of all pesticides to vertebrates and are closely related to the nerve gases sarin, soman and tabun (Ware, 2000). The development of organophosphate acaricides was primarily for the control of organochlorine-resistant *Boophilus* ticks which had become common throughout much of the cattle-producing areas of the tropics and subtropics (Shaw, 1970). Ethion, chlorpyrifos, chlorfenvinphos and coumaphos are four of the most widely used organophosphates for the treatment of tick-infested cattle. Carbamate acaricides (e.g. carbaryl and pro-macyl), like the organophosphates, function by inhibiting the target's cholinesterase, but they have very low mammalian and dermal toxicity. Unfortunately, the value of carbamates for the control of ticks was limited because of their cross-resistance with organophosphates (Roulston et al., 1968; Schuntner, Schnitzerling & Roulston, 1971; McDougall & Machin, 1988). Resistance to organophosphates and car-bamates has eliminated or minimized their usefulness in Australia, much of Africa and parts of Latin America (Kunz & Kemp, 1994).

The formamidines, chlordimeform, clenpyrin, chloro-methiuron and amitraz, are members of a small group of chemicals that are effective against ticks. Chlordimeform was introduced in Australia as an additive to organophos-phates in dipping vats to restore their efficacy against an organophosphate-resistant tick strain (Nolan, 1981), but was removed from the market in 1976 because of evidence of car-cinogenicity (Ware, 2000). Results of successful tests of ami-traz for the control of B. microplus on cattle in Australia with an experimental formulation (BTS 27419) were reported in 1971 (Palmer et al., 1971). Subsequent trials with commer-cial amitraz formulations in Australia (Roy-Smith, 1975) and the US (George et al., 1998) proved the efficacy of the aca-ricide against B. microplus. A series of trials executed over a 5-year period in South Africa proved the effectiveness of amitraz for the control of B. decoloratus, R. appendiculatus, R. evertsi and Amblyomma hebraeum (Stanford et al., 1981). Amitraz is unstable in dipping vats, but adding sufficient cal-cium hydroxide or hydrated lime to raise and maintain the pH of the vat solution to 12 insures the stability of the active ingredient (Stanford et al., 1981; George et al., 1998).

Natural pyrethrum, a costly insecticide that is unstable in sunlight, was the predecessor to a series of synthetic pyrethrin-like materials. Compounds in this group of chem-icals were originally called synthetic pyrethroids, but current nomenclature is simply pyrethroids. Pyrethroids have a history of evolution that began in 1949, but the third-generation chemicals, permethrin and fenvalerate, were the first of these materials available for control of ticks on cattle (Davey & Ahrens, 1984; Ware, 2000). Cross-resistance to DDT precluded or abbreviated the use of permethrin and fenvalerate in countries such as Australia and South Africa where DDT resistance had been diagnosed in Boophilus ticks (Nolan, Roulston & Schnitzerling, 1979; Coetzee, Stanford & Davis, 1987). Cypermethrin, deltamethrin and cyhalothrin are examples of fourth-generation cyano-substituted pyrethroids that are effective acaricides (Stubbs, Wilshire & Webber, 1982; Kunz & Kemp, 1994; Aguirre et al., 2000). In Australia, the strategy for registering and using cyano-substituted pyrethroids was influenced by evidence that after selection with permethrin a field strain (Malchi) with a low frequency of resistance to DDT exhibited no enhanced resistance to DDT, but was resistant to perme-thrin. This strain was only slightly resistant to cyperme-thrin and deltamethrin (Nolan et al., 1979). It appeared likely that populations of B. microplus resistant to perme-thrin would evolve rapidly from existing populations having low frequencies of individuals resistant to DDT. Conse-quently, permethrin was not registered, and cypermethrin and deltamethrin were registered for use only at concentra-tions that would be likely to control the most pyrethroid-resistant field strain. A second component of the strategy to delay the evolution of pyrethroid resistance was based on the observation that several organophosphate acaricides would synergize the toxicity to B. microplus of cypermethrin and deltamethrin. The reduction in concentration of a relatively expensive pyrethroid that could be used with a relatively cheap organophosphate synergist provided an efficacious, inexpensive product for the control of organophosphate-resistant tick populations (Schnitzerling, Nolan & Hughes, 1983). Flumethrin, an α-cyano-substituted pyrethroid, was designed for application to cattle as a pour-on, but there is also an emulsifiable concentrate formulation that can be applied as a dip or spray. The active ingredient in the pour-on has a remarkable capacity for spreading rapidly on the skin and hair from points of application along the dorsal line of an animal to all areas of the body. The residual effect of treatment with flumethrin is extended if the pour-on formulation is applied. Flumethrin for the control of both one-host and multi-host tick species on cattle is effective at relatively low concentrations compared to other pyrethroids (Stendel, 1985). The trans-flumethrin isomer is approxi-mately 50 times more toxic to B. microplus than the other most-toxic pyrethroids, cis-cypermethrin and deltamethrin (Schnitzerling, Nolan & Hughes, 1989).

In Australia, the combination products of cypermethrin + chlorfenvinphos and deltamethrin + ethion remain on the market (Jonsson & Matschoss, 1998). Mixtures of different products are also marketed in Latin America. Furlong (1999) listed products consisting of mixtures of cypermethrin + chlorfenvinphos and cypermethrin + dichlorvos among aca-ricides marketed in Brazil. One value of these mixtures may be their possible use for the control of both ticks and the horn fly.

There are two classes of macrocyclic lactones with acaricidal activity. The avermectins are derivatives of the actinomycete *Streptomyces avennitilis* and the milbemycins are derived from fermentation products of *S. hygroscopicus aureolacrimosus* (Lasota & Dybas, 1991). Ivermectin, eprinomectin and doramectin are related to avermectins; moxidectin is the only milbemycin-derived macrocyclic lactone marketed for the control of ticks. Each of the macrocyclic lactones is active systemically in very low doses for the control of ticks. Ivermectin, doramectin and moxidectin treatments, administered as subcutaneous injections, are efficacious for the control of *B. microplus* infestations of cattle (Gonzales *et al.*, 1993; Muniz *et al.*, 1995; Remington *et al.*, 1997; Caproni *et al.*, 1998; George & Davey, 2004). Satisfactory control of *B. microplus* on cattle may also be obtained with pour-on formulations of ivermectin, eprinomectin, doramectin and moxidectin (Davey & George, 2002; George & Davey, 2004). Macrocyclic lactone acaricides are efficacious, but high cost limits their use in cattle production (Kemp *et al.*, 1999).

Fipronil, a phenylpyrazole compound, applied as a pour-on to cattle infested with *B. microplus* and confined in an open-sided barn, had a therapeutic efficacy greater than 99% and a similar degree of persistent protection against larval reinfestation for 8 weeks after the treatment was applied (Davey *et al.*, 1998). Under field conditions with exposure to sunlight and weather, the high degree of persistent efficacy of a single pour-on treatment of fipronil on cattle was reduced by 2 to 3 weeks (Davey *et al.*, 1999). Fipronil is available for the control of ticks in several countries in Latin America, but it has not been registered in the US and some other countries for use on food animals.

Fluazuron, a benzoyl phenyl urea, inhibits chitin formation in *B. microplus*. Most of the benzoyl phenyl ureas including diflubenzuron, lufenuron and flufenoxuron are effective against a wide variety of insects, but fluazuron is an exception and it is efficacious against ticks and some mite species (Taylor, 2001). The adverse consequences for ticks on cattle treated with a pour-on of this acaricide are the reduction of the fecundity and fertility of engorged females to near zero, and mortality of immature ticks because they are unable to moult to the next instar. Efficacy of fluazuron persists for approximately 12 weeks. Because of its characteristic of binding to fat, fluazuron is excreted in milk and it is unnecessary to treat suckling calves. Because of the persistence of residues in fat, it is necessary to withhold treated cattle from human consumption for 6 weeks (Bull *et al.*, 1996).

Spinosad represents a new class of pesticides, the spinosyns. Spinosad is a fermentation metabolite of the actinomycete *Saccharopolyspora spinosa* and has a unique mode of action that involves disruption of the binding of acetylcholine in nicotinic acetylcholine receptors at the postsynaptic cell (Ware, 2000). Spinosad provides about 90% control of *B. microplus* on cattle infested with all three parasitic stages at the time of treatment. Efficacy is greater against nymphal and larval ticks than adults. The product provides excellent persistent efficacy against larval reinfestations of treated cattle for 2 weeks post-treatment (Davey, George & Snyder, 2001). Spinosad's unique mode of action qualifies it as an alternative acaricide to consider for the control of *B. microplus* that are resistant to other chemicals.

Recently published guidelines (Holdsworth *et al.*, 2006) for evaluating the efficacy of acaricides against ticks on ruminants may be a useful reference for laboratories with programmes for evaluation of acaricides.

APPLICATION OF ACARICIDES TO CATTLE

Traditional methods for the delivery of an acaricide treatment to cattle to control ticks required formulation of the acaricide into a form such as an emulsifiable concentrate, wettable powder or flowable product that could be diluted in water and applied to cattle with a hand sprayer, spray race or through immersion of animals in a dipping vat. More recently, treatment possibilities include the use of pour-on products, injectables, an intraruminal bolus, acaricide-impregnated ear tags and pheromone/acaricide-impregnated devices attached in different ways to the host. The effectiveness of an acaricide applied to cattle for the control of ticks depends not only on the degree of toxicity of a chemical, but on the quality, quantity and degree of dispersal of active ingredient deposited on cattle or delivered internally. Whatever the treatment method, adherence to procedures developed by the manufacturer is essential for maximizing the degree of tick control that will occur.

A century of experience with dipping vats has provided solutions to many problems that confound the success of cattle-dipping operations. A variety of factors that include the nature of the formulation, the degree of vat fouling from hair, manure and soil, and the tendency of a product in a dipping vat to strip (i.e. when the concentration of acaricide in the fluid draining from an animal is less than the concentration of active ingredient in the fluid used in treatment) influence the quantity and quality of active ingredient a treatment delivers to the target animal (Schnitzerling & Walker, 1985). To prevent degradation of amitraz in a dipping vat, a pH of approximately 12 must be maintained (Stanford *et al.*, 1981; George *et al.*, 1998). Degradation of coumaphos in a dipping vat to potasan, with its greater oral toxicity to cattle, may occur in fouled dipping vats unless the vats are kept acidified to a pH ≤ 5.5 to prevent blooms of anaerobic bacteria (Davey *et al.*, 1995). Historically, most treatments of cattle with acaricides required application methods that ensured the thorough wetting of the surfaces of an animal with water containing the diluted acaricide formulations. Devices such as hand-held sprayers, spray races or dipping vats were used as means to deliver treatments to cattle. Spray races were generally less effective because of the tendency of ticks to survive on the ears and necks of sprayed animals (Wharton *et al.*, 1970). Regardless of the method of acaricide application, a variety of operational factors such a failure to stir a vat properly after it sits unused for a time, lack of attention to details for replenishing acaricide solutions in a dipping vat, permitting rain to dilute the contents of a vat and failure to apply sufficient spray to completely wet animals are some common problems that minimize the quality of tick control on cattle.

Acaricides such as flumethrin, the macrocyclic lactones, fipronil and fluazuron have physical and chemical attributes that enable their formulation as products that can be delivered to the host as a pour-on (Stendel, 1985; Muniz *et al.*, 1995; Bull *et al.*, 1996; Davey *et al.*, 1998; Davey & George, 2002). A pour-on product is an effective tool for treating small numbers of cattle, but it can also be used to treat large herds. Factors such as cost and resistance of ticks to other acaricides may influence a producer's decision to use a pour-on. Macrocyclic lactone products applied as pour-ons have lower efficacy and are less persistent than flumethrin, fipronil or fluazuron pour-on formulations. Injectable treatments with macrocyclic lactones are more efficacious than treatments with many pour-ons, but the risk of spreading a disease agent within a herd of cattle by contaminated needles must be considered when electing to use this method (Gonzales *et al.*, 1993; Remington *et al.*, 1997; Caproni *et al.*, 1998). The perceived value of persistence in terms of a reduction in the frequency and number of treatments needed to sustain tick control should be weighed against the selection pressure for resistance associated with the declining concentration of residual acaricide.

The costs and inconvenience of mustering cattle regularly for treatments with a parasiticide stimulated research to develop methods for sustaining the delivery of a chemical and extending the duration that control from a single treatment is maintained. The organophosphate systemic insecticide famphur was used in an early unsuccessful effort to develop a practical intraruminal bolus for the control of ticks (Teel, Hair & Stratton, 1979). Four to five boluses releasing 304 mg of active ingredient/bolus per day were required in 180-kg calves to provide the serum levels of 7 mg/kg per day of famphur needed to control *Amblyomma maculatum* and *A. americanum*. In calves, ivermectin is delivered at a rate of 40 µg/kg per day by a prototype of the IVOMEC® SR Bolus, which is designed to function as a mini-osmotic pump. A single one of these boluses in a 168–268-kg calf delivered 12 mg of ivermectin daily for approximately 90 days providing a minimum dose of 40 µg/kg per day. The treatment reduced engorgement success of female *B. decoloratus*, *Hyalomma* spp., *R. appendiculatus* and *R. evertsi evertsi* by >99%, 91%, 95% and 83%, respectively (Soll *et al.*, 1990). In a trial in South Texas against *B. annulatus* on calves weighing approximately 200 kg, the degree of control from treatment with a single IVOMEC® SR Bolus was <30%. Two boluses/calf provided complete control of engorging females for the 20-week trial (Miller, J. A. *et al.*, 2001). However, the cost of a sufficient number of boluses to treat adult cattle could be prohibitive. A bioabsorbable, injectable microsphere formulation containing ivermectin in a poly (lactide-*co*-glycolide) copolymer was used to control ticks on calves kept in a pasture infested with *B. annulatus*. Untreated calves maintained in tick-infested pastures remained heavily infested, but the ivermectin microsphere treatment controlled the ticks on the calves and eradicated the infestation in the treatment pasture within 12 to 15 weeks (Miller, J. A. *et al.*, 1999).

An acaricide-impregnated ear tag placed in each ear of cattle provided a high degree of control of *R. appendiculatus* for up to 160 days after application. Active ingredients in the nine different tags that were tested included the organophosphate propetamphos, several pyrethroids, amitraz and an amitraz + permethrin tag that provided 100% control for the 160 days of one trial. The ear tags had limited efficacy against species such as *B. decoloratus*, *Amblyomma variegatum* and *R. evertsi evertsi* (Young, de Castro & Kiza-Auru, 1985). The cost of tags and the limited protection they offer against ticks other than *R. appendiculatus* minimized the likelihood that ear tags would replace other treatment methods in Africa (Rechav, 1987).

Pheromones released by either male or female ticks attached to a host are the basis of two technologies designed to control different species of tick parasites of livestock (Sonenshine, 2006). One type of device is a combination acaricide and polyvinylchloride resin moulded into spherules coated with pheromone. When these 'tick decoys' are glued to a host, male ticks attracted by the pheromone are killed upon contact with the decoy. Such tick decoys would be potentially useful to control mate seeking males of ticks such as *Amblyomma maculatum* on cattle or *Hyalomma dromedarii* on camels. *Amblyomma variegatum* and *A. hebraeum* are examples of tick species whose unfed nymphs, males and females are attracted to hosts infested by feeding male ticks. Once on a host these ticks aggregate at specific sites where the feeding males are attached. By combining the components of attraction–aggregation–attachment pheromones in a polyvinylchloride matrix with a pyrethroid acaricide and stabilizers a 'tail-tag decoy'was created. Ticks attracted to the device attached to a host's tail would be killed by the acaricide. In spite of the demonstrated efficacy of both 'decoy' devices, Sonenshine (2006) noted that no commercialization of the technologies has occurred.

STRATEGIES FOR THE CONTROL OF TICKS ON CATTLE

The primary interest of cattle-producers in a strategy for the control of ticks on their livestock is likely to be one of profitability although motives for cattle production among small-scale farmers may be different. Assessments of the cost per animal of a control strategy reduced to a comparison of the cost of damage vs. the cost of implementation of a particular control strategy would be expected to indicate the net economic benefit for a producer. How to determine which approach to tick control fits a particular situation and is likely to enhance a producer's income is not a simple problem to resolve. In Australia, through the use of a combination of information from models and data from studies of cattle, an approach for developing control policy guidelines was created (Sutherst et al., 1979). How to use research information and models to create practical solutions for actual problems was the basis for further research that resulted in a set of recommendations for tick-control strategies (Norton, Sutherst & Maywald, 1983). Management options were considered in a context of climate and impact of a component of an integrated tick control strategy on a particular phase of the life cycle of *B. microplus*. Potential elements of integrated tick control approaches consist of the following options: (1) increase the level of host resistance to ticks in herds by stocking pastures with cattle with high levels of heritable

resistance to ticks; (2) employ prophylactic dipping, dipping in response to economic thresholds, or opportunistic dipping when guidelines suggest a particular approach is appropriate; and (3) reduce the host-finding rate of ticks by changing host density or by pasture spelling (i.e. removing animals from a pasture to deny unfed ticks an opportunity for contact with a host).

The excellent research and resulting synthesis of guidelines on tick control by Sutherst et al. (1979) and Norton et al. (1983) is cited in the scientific literature (George, 1990; Nolan, 1990; Sangster, 2001) as a basis for programmes for *B. microplus* control, but it has not been adapted and recast as they intended in a practical form, such as an expert system that producers or advisors could use to prescribe a strategy to fit the needs and production goals of a specific producer. Also, the Norton et al. (1983) recommendations relate primarily to the control of *B. microplus* in Australia, and while the principles have widespread applicability, options are needed to cover situations where more than one tick species, including multi-host species, is the target. Of course, any rational strategy for the control of ticks affecting cattle must feature approaches to prevent rapid selection for resistance to acaricides (Sutherst & Comins, 1979; Nolan, 1990). The need for technology transfer is clear. A programme in Australia to educate dairy farmers and encourage them to adopt sound tick control programmes (Jonsson, 1997; Jonsson & Matschoss, 1998) is an excellent example of the kind of effort needed to help producers with problems of controlling ticks.

RESISTANCE OF TICKS TO ACARICIDES

Since the first report of the development of resistance of *B. microplus* to arsenic in Australia in 1937 (Newton, 1967) and *B. decoloratus* in South Africa in 1939 (Whitehead, 1958), the progressive evolution of resistance of ticks affecting cattle to almost all of the available acaricides has frustrated the efforts of cattle producers to manage ticks and tick-borne diseases affecting their animals. The history of the resistance of ticks to acaricides parallels, with a relatively few years of delay, the introduction of new acaricide products representing several different classes of chemicals. Wharton & Roulston (1970), Solomon (1983) and Kunz & Kemp (1994) provided reviews of the problem and its impact. Selected records of the geographical distribution and the year of documentation of acaricide resistance in populations of tick species important as parasites of cattle are presented in Table 18.1.

Table 18.1 *An overview of occurrences of acaricide resistance in species of ticks that parasitize cattle*

Chemical (approx. date introduced)	Species	Location
Arsenic (1893)	*Boophilus microplus*	Australia, 1936; Argentina, 1936; Brazil, 1948; Colombia, 1948; Uruguay, 1953; Venezuela, 1966
	B. decoloratus	South Africa, 1937; Kenya, 1953; Zimbabwe, 1963; Malawi, 1969
	Amblyomma hebraeum	S. Africa, 1975
	A. variegatum	Zambia, 1975
	Hyalomma rufipes, H. truncatum	S. Africa, 1975
	Rhipicephalus appendiculatus R. evertsi	S. Africa, 1975
DDT (1946)	*B. microplus*	Argentina, 1953; Brazil, 1953; Australia, 1953; Venezuela, 1966; S. Africa, 1979
	B. decoloratus	S. Africa, 1954
Cyclodienes and toxaphene (1947)	*B. microplus*	Australia, 1953; Argentina, 1953; Brazil, 1953; Venezuela, 1966; Colombia, 1966; S. Africa, 1979
	B. decoloratus	S. Africa, 1948; Kenya, 1964; Zimbabwe, 1969; Uganda, 1970
	A. hebraeum	S. Africa, 1975
	A. variegatum	Kenya, 1979
	H. marginatum	Spain, 1967
	H. rufipes, H. truncatum	S. Africa, 1975
	R. appendiculatus	S. Africa, 1964; Zimbabwe, 1966; Kenya, 1968; Tanzania, 1971
	R. evertsi	S. Africa, 1959; Kenya, 1964; Zimbabwe, 1966; Tanzania, 1970
Organophosphorus–carbamate group (1955)	*B. microplus*	Australia, 1963; Brazil, 1963; Argentina, 1964; Colombia, 1967; Venezuela, 1967; S. Africa, 1979; Uruguay, 1983; Mexico, 1986
	A. hebraeum	S. Africa, 1975
	A. variegatum	Tanzania, 1973; Kenya, 1979
	B. decoloratus	S. Africa, 1966; Zambia, 1976
	R. appendiculatus	S. Africa, 1975
	R. evertsi	S. Africa, 1975
Formamidines (1975)	*B. microplus*	Australia, 1981; Brazil, 1995; Colombia, 2000; Mexico, 2002
	Boophilus spp.	S. Africa, 1997
Pyrethroids (1977)	*B. microplus*	Australia, 1978; Brazil, 1989; Mexico, 1994; Venezuela, 1995; Colombia, 1997; Argentina, 2000
	B. decoloratus	S. Africa, 1987
Macrocyclic lactones (1981)	*B. microplus*	Brazil, 2001

Source: Compiled from data in Wharton (1976), Solomon (1983), Aguirre *et al.* (1986, 2000), Ortiz, Santamaria & Fragoso (1994), Coronado (1995), Martins *et al.* (1995), Romero *et al.* (1997), Strydom & Peter (1999), Benavides, Rodríquez & Romero (2000), Martins & Furlong (2001) and Soberanes *et al.* (2002).

Because resistance has progressively eliminated or limited the use of arsenic, chlorinated hydrocarbons, organophosphates, carbamates and pyrethroids, the eventual effect of acaricide resistance on the useful life of the remaining acaricides has been a topic of great concern and much discussion (Nari & Hansen, 1999). Predictably, the spectrum of chemical groups to which ticks have evolved resistance continues to broaden. Resistance to amitraz was first detected in Australia in 1981 when populations of the 'Ulam' amitraz-resistant strain were identified in a few widely spread locations in the country (Nolan, 1981). Identification of the 'Ultimo' strain in Australia in 1992 with its co-resistance to amitraz and all available pyrethroids did not represent a great immediate threat, because its distribution remained limited for several years. By 1999 the spread of the Ultimo strain had accelerated, and its presence at over 50 locations had been diagnosed (Kemp et al., 1999; Kunz & Kemp, 1994). More recently there have been reports of amitraz resistance in populations of *Boophilus* spp. in South Africa (Strydom & Peter, 1999) and *B. microplus* in Brazil (Furlong, 1999), Colombia (Benavides, Rodriguez & Romero, 2000) and Mexico (Soberanes et al., 2002). Macrocyclic lactone resistance of *B. microplus* in Brazil to doramectin with cross-resistance to ivermectin (another avermectin) and moxidectin (a milbemycin) was reported in ticks from one farm. The widespread use of macrocyclic lactone products for parasite control and limited choices of alternative acaricides has caused concern that macrocyclic lactone resistance will become a major problem (Martins & Furlong, 2001). The emergence of resistance in both single- and multi-host ticks in Africa to a variety of acaricides and of *B. microplus* in Latin America and Australia to the organophosphate, pyrethroid, formamidine and macrocyclic lactone acaricides does not, of course, mean that none of the products containing these kinds of active ingredients has any further value. Tick populations susceptible to a variety of acaricides exist and can be controlled, but it is more critical than ever to use existing and improved diagnostic tools to determine where products are still useful and to employ tick control strategies that minimize the rate of selection for resistance.

DIAGNOSIS OF RESISTANCE IN TICKS TO ACARICIDES

A variety of bioassay methods has been developed for assessing the susceptibility of ticks to acaricides, but the ones used most often for tests with organophosphates–carbamates and pyrethroids are the larval packet test (LPT), the lar-

val immersion test (LIT) and the Drummond test (DT) (Kemp et al., 1998). The LPT can also be used for bioassays of fipronil (Miller, R. J. et al., 2001). The LPT was recommended by the Food and Agriculture Organization of the United Nations (FAO) for use as a standard tick bioassay method, but it has not been adopted worldwide. Until recently, satisfactory methods for measuring the susceptibility of ticks to amitraz and macrocyclic lactones were unavailable. A modified LPT that involved the use of formulated amitraz and a nylon fibre substrate instead of filter paper has been used successfully to determine dose–mortality relationships of susceptible and amitraz-resistant strains of *B. microplus* (Miller, Davey & George, 2002). Comparisons of the LPT, LIT and an adult immersion test (AIT) for determining LC_{50} and discriminating doses for macrocyclic lactones against *B. microplus* indicated that the LIT and AIT were likely to provide the most consistent results with all of the macrocyclic lactones (Sabatini et al., 2001). Evaluation of the bioassays for amitraz and the macrocyclic lactones by personnel at a variety of laboratories is needed to confirm their utility. The lapse of time between the identification of a resistance problem and the availability of results from bioassays is a major problem with most existing bioassays. With a one-host tick, such as *B. microplus*, a minimum of about 35 days is required after engorged females are collected and larvae of the appropriate age (7–14 days) are available for testing. If a multi-host tick species is involved, it may require much longer to obtain sufficient numbers of ticks of uniform age to do an analysis. With the AIT test of Sabatini et al. (2001) results can be available in 10 days if a sufficiently large sample of engorged females can be collected from untreated cattle. The difficulty of obtaining the number of engorged females needed for a reliable analysis is likely to limit the value of the AIT for rapid resistance diagnosis except in cases where the frequency of resistant ticks is high enough to minimize the risk of a diagnostic error related to an inadequate sample size.

The need for methods to overcome the limitations of conventional bioassay techniques for resistance diagnosis has stimulated investigations of the potential usefulness of molecular methods. Unique advantages of molecular techniques for diagnosis of resistance are that they are highly specific and sensitive with small quantities of DNA (Sangster et al., 2002). As alternatives to bioassay methods for diagnosis of acaricide resistance in ticks, molecular methods also offer the possibility of obtaining results in one or two days vs. weeks with conventional methods. Sangster et al. (2002) observed that, in spite of the benefits, there are also potential drawbacks with molecular diagnosis, including: (1) tests require detailed knowledge of resistance mechanisms at the molecular level; (2) the identified mechanism must be the predominant one in the field; (3) molecular tests may not be appropriate for all resistance mechanisms; (4) ideally, they need to be offered as a battery of tests so resistance to several available drugs can be measured simultaneously; and (5) PCR technology is relatively complex. Temeyer *et al.* (2006) reviewed literature reporting efforts to determine the molecular basis of organophosphate resistance in *B. microplus* populations in Australia and Mexico that resulted in the isolation of three cDNAs identified putatively as acetylcholine esterase (AChE). In the same paper Temeyer *et al.* reported the first successful expression in the baculovirus system of an AChE cloned from *B. microplus*. Inhibition kinetics of the expressed product revealed sensitivity consistent with that of adult, organophosphate-susceptible neural AChE. A point mutation in an esterase gene was identified in a pyrethroid-resistant *B. microplus* strain from Mexico (Hernandez *et al.*, 2000), but further research (Guerrero, Li & Hernandez, 2002) found that the occurrence of resistance was not associated with the presence of the mutation. In a different pyrethroid-resistant strain of *B. microplus* from Mexico with a target site resistance mechanism, a point mutation was identified in the *para*-type sodium channel gene (He *et al.*, 1999). A PCR diagnostic assay that was created to detect the mutation proved useful for identification of the genotype of ticks with resistance conferred by the mutation of the sodium channel gene (Guerrero *et al.*, 2002). When the PCR assay for the sodium channel mutation was used along with discriminating-dose (DD) bioassays to test pyrethroid-resistant *B. microplus* populations in Mexico, the PCR method identified *R* alleles in some populations in which the frequency of the allele was apparently too low for the DD dose assay to diagnose as resistance. Except in the tick populations with low frequencies of sodium channel mutation, the correlation between the the frequency of mortality from bioassays and the frequency of the *R* allele was significant (Rosario-Cruz *et al.*, 2005). Whether molecular methods for diagnosis of resistance of ticks to acaricides will provide practical alternatives to conventional techniques is a question that requires considerably more research to determine the molecular basis of the multiple forms of resistance that exist, the creation of new diagnostic methods and tests of their sensitivity, specificity, and utility in the field.

Sensitive, reliable diagnostic methods are essential for: (1) recognizing if acaricide resistance is the cause of a tick control failure; (2) determining which acaricide is a suitable alternative to a product when it fails; (3) investigating the epidemiology of resistance; (4) developing control strategies that minimize the rate of selection of resistance genotypes; and (5) developing new acaricides. Often, diagnostic services in a country document the occurrence of resistance of ticks to acaricides and provide some information of the distribution and prevalence of various forms of resistance. Diagnostic laboratories are unlikely to have facilities, finances and staff needed to respond in a timely fashion to the requests of all producers for assistance or to provide more than general information about the geographical distribution of a problem. A major consequence of the lack of adequate information is likely to be poor decisions by producers about which acaricide would be useful to them. After resistance of *B. microplus* to organophosphates was diagnosed in Mexico, the Federal Government allowed pyrethroid products on the market for the first time and their use quickly became common. Many producers switched to a pyrethroid acaricide even though there was no evidence of resistance of the ticks on their cattle to organophosphates (Fragoso *et al.*, 1995). Even if the ticks affecting an individual producer's herd are determined to be resistant to one or more acaricides, it is not possible to relate the degree of resistance indicated by a bioassay to the degree of control expected from proper use of an acaricide. Also, the limited availability and quality of advice on suitable alternative acaricides and strategies for their use hinders implementation of effective responses to problems.

THE MITIGATION OF ACARICIDE RESISTANCE

It is unlikely that it will be possible to prevent the evolution of resistance in tick control programmes that feature the use of acaricides. However, there are options for slowing the rate of selection for resistant individuals and there are a few options that may be used when resistance renders an acaricide ineffective (Nolan, 1990).

The most important fundamental to consider in the design of any control programme is to reduce the number of pesticide treatments to a minimum (Roush, 1993). Sutherst & Comins (1979) made several practical suggestions about resistance management and the use of acaricides: (1) The cost of managing resistant tick populations makes acaricide treatments more expensive and raises the costs of a control programme. Consider financial benefits derived from ignoring light infestations and employing alternative management practices including the use of tick-resistant breeds of cattle or pasture rotation. (2) Decrease acaricide use by increasing the control threshold for initiating control of the spring to autumn generations of ticks. (3) Avoid any unnecessary dipping because the long-term consequences of acaricide treatments are more expensive than the short-term gains. (4) Reduce the adverse effects of disseminating resistant strains of ticks through regional cooperation to impose controls on the movement of cattle. To avoid contact of ticks with cattle having low concentrations of residual acaricide on them, Sutherst & Comins (1979) also encouraged treating at 3-week intervals, especially during the spring or early summer when a large proportion of the tick population is in the parasitic stage. Nolan (1990) emphasized: 'Undue reliance on the illusory economic benefits of residual persistence is implicated as one of the major factors contributing to the early demise of several effective acaricides.' Economic considerations of controlling acaricide-resistant ticks may force producers to be more open to the advantages of using host resistance as a major tool in reducing the number of times they treat with acaricides (Jonsson, 1997).

Maintenance of biosecurity of premises and herds is an ongoing process that should be a routine part of a cattle-producer's efforts to minimize adverse impacts of acaricide resistant ticks on his or her operation. By ensuring that new cattle introduced to a farm or ranch are quarantined, treated with an acaricide and free of ticks before they are turned out to pasture, the risk of introducing a new strain of resistant ticks will be minimized (Sutherst & Comins, 1979; Jonsson & Matschoss, 1998). Unfortunately, a neighbour whose property is adjacent can compromise efforts to maintain good biosecurity. The early restricted distribution of amitraz-resistant *B. microplus* strains in Australia may have been due in part to circumstances that limited the dissemination of resistant ticks on cattle moved for various reasons from the affected farms (Kunz & Kemp, 1994).

The concentration or dose used to control ticks is another operational decision considered important in the selection of individuals with resistance genotypes (Georghiou & Taylor, 1977; Sutherst & Comins, 1979). Theoretically, the use of a high dose when homozygous-resistant individuals are rare would keep their frequency low by removing the more susceptible heterozygous-resistant individuals from the population. While the theoretical basis of the high-dose strategy is sound, the strategy has limitations when applied to the control of pesticide-resistant populations (Roush, 1993). The deficiencies of the high-dose tactic are, according to Roush (1993): economic and environmental limitations on the doses needed; the difficulty of maintaining doses high enough to kill heterozygotes; the deleterious effect of pesticide residues on inward migration of susceptible insects; and the difficulty of maintaining an untreated source for immigrants. Roush (1993) also observed that perhaps one of the greatest flaws of the high-dose strategy is that the dose needed to kill resistant heterozygotes is unknown because the strategy must be applied when the resistance allele frequency is less than 10^{-3}, and it is improbable that this knowledge would be available. A practical limitation on the possible use of a high-dose strategy is the restriction related to the approved (registered) dose for a product. There might be a special situation, such as the official dipping of cattle from Mexico in 3000 ppm coumaphos as a high-dose strategy to protect against organophosphate-resistant *B. microplus* before exportation of the animals to the US, where a high concentration of acaricide might be approved. Even if animal and human safety and environmental issues were not limiting factors for approval of high doses of an acaricide, the higher cost of such a product for general use would probably confer a competitive disadvantage in the market place. In relation to the use of a low-dose strategy, Roush (1993) referred to the 'persistent myth that resistance can be managed by low doses'. Generally, it would not be acceptable or practical to use a strategy that allows a large proportion of the treated individuals to survive to delay the emergence of resistance.

'Rotation' is a term applied to a treatment strategy that alternates the use over time of two or more chemicals with differing modes of action and no potential for cross-resistance (Riddles & Nolan, 1986; Tabashnik, 1990; Roush, 1993). The rotation scheme assumes that the frequency of individuals in the population with resistance to one acaricide will decline during the time the alternative chemical is used. Any decline in the frequency of ticks with resistance genotypes to one of the acaricides depends on a relatively lower degree of fitness of resistant individuals to the alternative acaricide (Tabashnik, 1990). Even though rotation may offer a theoretical advantage in models or laboratory experiments, the tactic must be evaluated in the field, and results can be expected to vary depending on the fitness and mode of inheritance of a particular form of resistance.

The use of mixtures of acaricides to reduce the rate of evolution of acaricide resistance is based on the assumptions that resistance to each acaricide is monogenic; there is no potential for cross-resistance; each acaricide is equally persistent; resistant individuals are rare; and that some of the population remains untreated (Tabashnik, 1990). Also, mixtures and the components of formulations must be compatible and the product must not be toxic to the host (Kemp *et al.*, 1998). Although simulations with the application of mixtures suggested their potential value (Sutherst & Comins, 1979) and several products containing mixtures are on the market, published scientific evidence from field trials of their efficacy and value in the mitigation of resistance of ticks to acaricides is lacking.

Possible countermeasures to exercise once resistance has emerged are limited to attempting to use the acaricide to which resistance has been diagnosed or selecting a new acaricide from a group of chemicals unaffected by the resistance problem (Nolan, 1990). Nolan identified three possibilities to consider if a lack of suitable alternative acaricides favours an effort to continue use of the affected chemical. Increasing the concentration of the acaricide is a tactic that was employed to extend the use of organophosphates in Australia and it could succeed elsewhere. Toxicity to the host and cost are factors that should be considered. The addition of a synergist to the formulation of an acaricide would have potential value if the resistance mechanism in target tick population was known to be detoxification (Nolan & Schnitzerling, 1986). Eradication of the resistant alleles is a possibility if the populations of the resistant ticks are not widespread. It would be necessary to impose strict quarantines and to trace recent movements of cattle from affected premises to ensure that the problem was contained and could be eradicated. If the choice is to select another acaricide, it is obvious that care must be taken to select a product with a mechanism of action that will not be overcome by the resistance mechanism in the ticks against which it will be used. It is important to determine the areas where an acaricide rendered ineffective in some locations by resistance is still effective.

IMPROVED PARADIGMS FOR TICK CONTROL

The evolution of resistance to the majority of the groups of chemicals on the market used to control ticks on cattle and the development of few new products has clouded the future for the chemical control of ticks. Our current problems should force an analysis of the usefulness of available information on tick control and resistance mitigation; how we are applying the technology available for moderating the adverse impacts of tick and tick-borne diseases on beef and dairy operations; and what is needed to help extend the useful life of existing acaricides. First, we need to recognize that basic principles for profitable, well-reasoned tick control/resistance mitigation programmes were developed two decades ago (Sutherst & Comins, 1979; Norton et al., 1983)

and have been cited repeatedly in reviews such as this one. The tick vaccines TickGARD[PLUS] and Gavac are on the market, and knowledge of the value of tick-resistant breeds of cattle has increased (Sangster, 2001), but the principles have not changed. Unfortunately, much of the recent literature documents the problems but provides little evidence of systematic attempts to help producers resolve them. The work by Jonsson (1997) and Jonsson & Matschoss (1998) to determine the attitudes and approaches of dairy farmers in Australia to the problem of tick control and to identify possibilities and obstacles to the use of new methods are models of efforts to help a group of producers improve their approach to tick control. Farmers and ranchers are unlikely to change the way they manage their cattle and parasite problems unless they see convincing evidence that a new approach will confer an economic advantage. They will be most interested in short-term benefits, but they need convincing evidence of the potential for positive long-term outcomes. We have knowledge of the tools available for tick control, but need practical research involving work with producers to understand the most efficacious, cost-effective combinations, how to adapt strategies to specific kinds of cattle operations, and to determine net costs and profits. Such research should lead to the creation of literature and programmes to educate producers and help them make changes in control programmes that will benefit them and help preserve the remaining acaricides. Policy-makers and regulatory authorities, especially in less developed countries, need to be well informed about problems relating not only to acaricide usage and management, but also to questions pertaining to standards for registration, labelling and marketing acaricides and other pesticides.

REFERENCES

Aguirre, J., Sobrino, L., Santamaria, M., et al. (1986). Resistencia de garrapatas en Mexico. In *Seminario Internacional de Parasitologia Animal, Memorias*, eds. Cavazzani, A. H. & Garcia, Z., pp. 282–306. Cuernavaca Morelos, Mexico.

Aguirre, D. H., Viñabal, A. E., Salatin, A. O., et al. (2000). Susceptibility to two pyrethroids in *Boophilus microplus* (Acari: Ixodidae) populations in northwest Argentina: preliminary results. *Veterinary Parasitology* 88, 329–334.

Angus, B. M. (1996). The history of the cattle tick *Boophilus microplus* in Australia and achievements in its control. *International Journal for Parasitology* 26, 1341–1355.

Baker, J. A. F. & Shaw, R. D. (1965). Toxaphene and lindane resistance in *Rhipicephalus appendiculatus*, the brown ear tick of equatorial and southern Africa. *Journal of the South African Veterinary Medical Association* **36**, 321–330.

Benavides, E., Rodríguez, J. L. & Romero, A. (2000). Isolation and partial characterization of the Montecitos strain of *Boophilus microplus* (Canestrini, 1877) multi-resistant to different acaricides. *Annals of the New York Academy of Sciences* **916**, 668–671.

Bull, M. S., Swindale, S., Overend, D. & Hess, E. A. (1996). Suppression of *Boophilus microplus* populations with fluazuron: an acarine growth regulator. *Australian Veterinary Journal* **74**, 468–470.

Caproni, L. Jr, Umehara, O., Moro, E. & Goncalves, L. C. B. (1998). Field efficacy of doramectin and ivermectin against natural infestations of the cattle tick *Boophilus microplus*. *Brazilian Journal of Veterinary Parasitology* **7**, 151–155.

Cobbett, N. G. (1947). Preliminary tests in Mexico with DDT, cube hexachlorocyclohexane (benzene hexachloride), and combinations thereof, for the control of the cattle fever tick, *Boophilus annulatus*. *American Journal of Veterinary Research* **8**, 280–283.

Coetzee, B. B., Stanford, G. D. & Davis, D. A. T. (1987). The resistance spectrum shown by a fenvalerate-resistant strain of blue tick (*Boophilus decoloratus*) to a range of ixodicides. *Onderstepoort Journal of Veterinary Research* **54**, 79–82.

Coronado, A. (1995). Current status of the tropical cattle tick, *Boophilus microplus* in Venezuela. In *Resistencia y Control en Garrapatas y Moscas de Importancia Veterinaria, III Seminario Internacional de Parasitologia Animal*, eds. Rodriquez, S. & Fragoso, H., Acapulco, Guerrero, Mexico, pp. 22–29.

Davey, R. B. & Ahrens, E. H. (1984). Control of *Boophilus* ticks on heifers with two pyrethroids applied as sprays. *American Journal of Veterinary Research* **45**, 1008–1010.

Davey, R. B. & George, J. E. (2002). Efficacy of macrocyclic lactone endectocides against *Boophilus microplus* (Acari: Ixodidae) infested cattle using different pour-on application treatment regimes. *Journal of Medical Entomology* **39**, 763–769.

Davey, R. B., Ahrens, E. H., George, J. E., Hunter, J. E. III & Jeannin, P. (1998). Therapeutic and persistent efficacy of fipronil against *Boophilus microplus* (Acari: Ixodidae) on cattle. *Veterinary Parasitology* **74**, 261–276.

Davey, R. B., Ahrens, E. H., George, J. E., Hunter, J. E. III & Jeannin, P. (1999). Evaluation of a pour-on formulation of fipronil against *Boophilus microplus* (Acari: Ixodidae) under natural South Texas field conditions. *Experimental and Applied Acarology* **23**, 351–364.

Davey, R. B., Ahrens, E. H., George, J. E. & Karns, J. S. (1995). Efficacy of freshly mixed coumaphos suspensions adjusted to various pH levels for treatment of cattle infested with *Boophilus annulatus* (Say) (Acari: Ixodidae). *Preventative Veterinary Medicine* **23**, 1–8.

Davey, R. B., George, J. E. & Snyder, D. E. (2001). Efficacy of a single whole-body spray treatment of spinosad, against *Boophilus microplus* (Acari: Ixodidae) on cattle. *Veterinary Parasitology* **99**, 41–52.

Fragoso, H., Soberanes, N., Ortiz, M., Santamaria, M. & Ortiz, A. (1995). Epidemiologia de la resistencia a ixodicides piretroides en garrapatas *Boophilus microplus* en la Republica Mexicana. In *Resistencia y Control en Garrapatas y Moscas de Importancia Veterinaria, III Seminario Internacional de Parasitologia Animal*, eds. Rodriquez, S. & Fragoso, H., Acapulco, Guerrero, Mexico, pp. 45–57.

Francis, M. (1892). Preventive measures for farm or range use. *Bulletin of the Texas Agriculture Experiment Station* **24**, 253–256.

Furlong, J. (1999). Diagnosis of the susceptibility of the cattle tick, *Boophilus microplus*, to acaricides in Minas, Gerais State, Brazil. In *Control de la Resistencia en Garrapatas y Moscas de Importancia Veterinaria y Enfermedades que Transmiten, IV Seminario Internacional de Parasitologia Animal*, eds. Morales, G., Fragosa, H. & Garcia, Z., Puerto Vallarta, Jalisco, Mexico, pp. 41–46.

George, J. E. (1990). Summing-up of strategies for the control of ticks in regions of the world other than Africa. *Parassitologia* **32**, 203–209.

George, J. E. & Davey, R. B. (2004). Therapeutic and persistent efficacy of a single application of doramectin applied either as a pour-on or injection to cattle infested with *Boophilus microplus* (Acari: Ixodidae). *Journal of Medical Entomology* **41**, 402–407.

George, J. E., Davey, R. B., Ahrens, E. H., Pound, J. M. & Drummond, R. O. (1998). Efficacy of amitraz (Taktic® 12.5% EC) as a dip for the control of *Boophilus microplus* (Canestrini) (Acari: Ixodidae) on cattle. *Preventative Veterinary Medicine* **37**, 55–67.

Georghiou, G. P. & Taylor, C. E. (1977). Operational influences in the evolution of insecticide resistance. *Journal of Economic Entomology* **70**, 653–658.

Gonzales, J. C., Muniz, R. A., Farias, A., Goncalves, L. C. B. & Rew, R. S. (1993). Therapeutic and persistent efficacy of doramectin against *Boophilus microplus* in cattle. *Veterinary Parasitology* **49**, 107–109.

Graham, O. H. & Hourrigan, J. L. (1977). Eradication programs for the arthropod parasites of livestock. *Journal of Medical Entomology* 13, 629–658.

Guerrero, F. D., Li, A. Y. & Hernandez, R. (2002). Molecular diagnosis of pyrethroid resistance in Mexican strains of *Boophilus microplus* (Acari: Ixodidae). *Journal of Medical Entomology* 39, 770–776.

He, H., Chen, A. C., Davey, R. B., Ivie, G. W. & George, J. E. (1999). Identification of a point mutation in the *para*-type sodium channel gene from a pyrethroid-resistant cattle tick. *Biochemical and Biophysical Research Communications* 261, 558–561.

Hernandez, R., He, H., Chen, A. C., *et al.* (2000). Identification of a point mutation in an esterase gene in different populations of the southern cattle tick, *Boophilus microplus*. *Insect Biochemistry and Molecular Biology* 30, 969–977.

Holdsworth, P. A., Kemp, D., Green, P., *et al.* (2006). World Association for the Advancement of Veterinary Parasitology (W.A.A.V.P.) guidelines for evaluating the efficacy of acaricides against ticks (Ixodidae) on ruminants. *Veterinary Parasitology* 136, 29–43.

Jonsson, N. N. (1997). Control of cattle ticks (*Boophilus microplus*) on Queensland dairy farms. *Australian Veterinary Journal* 75, 802–807.

Jonsson, N. N. & Matschoss, A. L. (1998). Attitudes and practices of Queensland dairy farmers to the control of the cattle tick, *Boophilus microplus*. *Australian Veterinary Journal* 76, 746–751.

Kemp, D. H., McKenna, R. V., Thullner, R. & Willadsen, P. (1999). Strategies for tick control in a world of acaricide resistance. In *Control de la Resistencia en Garrapatas y Moscas de Importancia Veterinaria y Enfermedades que Transmiten, IV Seminario Internacional de Parasitologia Animal*, eds. Morales, G., Fragosa, H. & Garcia, Z., Puerto Vallarta, Jalisco, Mexico, pp. 1–10.

Kemp, D. H., Thullner, F., Gale, K. R., Nari, A. & Sabatini, G. A. (1998). *Acaricide Resistance in the Cattle-Ticks* Boophilus microplus *and B. decoloratus: Review of Resistance Data* – Standardization of Resistance Tests and Recommendations for Integrated Parasite Control to Delay Resistance, Report to the Animal Health Services AGAH. Rome: Food and Agriculture Organization.

Kunz, S. E. & Kemp, D. H. (1994). Insecticides and acaricides: resistance and environmental impact. *Revue scientifique et technique de l'Office International des Epizooties* 13, 1249–1286.

Lasota, J. A. & Dybas, R. A. (1991). Avermectins, a novel class of compounds: implications for use in arthropod pest control. *Annual Review of Entomology* 36, 91–117.

Lawrence, J. (1992). History of bovine theileriosis in southern Africa. In *The Epidemiology of Theileriosis in Africa*, eds. Norval, R. A. L, Perry, B. D. & Young, A. S., pp. 1–40. London: Academic Press.

Lawrence, J. A. & Norval, R. A. I. (1979). A history of ticks and tick-borne diseases of cattle in Rhodesia. *Rhodesian Veterinary Journal* 10, 28–40.

Martins, J. R. & Furlong, J. (2001). Avermectin resistance of the cattle tick *Boophilus microplus* in Brazil. *Veterinary Record* 14 July 2001, 64.

Martins, J. R., Correa, B. L., Ceresér, V. H. & Arteche, C. C. P. (1995). A situation report on resistance to acaricides by the cattle tick *Boophilus microplus* in the state of Rio Grande do Sul, southern Brazil. In *Resistencia y Control en Garrapatas y Moscas de Importancia Veterinaria, III Seminario Internacional Parasitologia Animal*, eds. Rodriquez, S. & Fragoso, H., Acapulco, Guerrero, Mexico, pp. 1–6.

Maunder, J. C. J. (1949). Cattle tick control: results achieved in the field with DDT and BHC. *Queensland Agricultural Journal* September, 1–8.

McDougall, K. W. & Machin, M. V. (1988). Stabilization of the carbamate acaricide promacyl in cattle dipping fluid. *Pesticide Science* 22, 307–315.

Miller, J. A., Davey, R. B., Oehler, D. D., Pound, J. M. & George, J. E. (1999). Control of *Boophilus annulatus* (Acari: Ixodidae) on cattle using injectable microspheres containing ivermectin. *Journal of Economic Entomology* 92, 1142–1146.

Miller, J. A., Davey, R. B., Oehler, D. D., Pound, J. M. & George, J. E. (2001). The Ivomec SR bolus for control of *Boophilus annulatus* (Acari: Ixodidae) on cattle in South Texas. *Journal of Economic Entomology* 94, 1622–1627.

Miller, R. J., Davey, R. B. & George, J. E. (2002). Modification of the Food and Agriculture Organization Larval Packet Test to measure amitraz-susceptibility against Ixodidae. *Journal of Medical Entomology* 39, 645–651.

Miller, R. J., George, J. E., Guerrero, F., Carpenter, L. & Welch, J. B. (2001). Characterization of acaricide resistance in *Rhipicephalus sanguineus* (Latrille) (Acari: Ixodidae) collected from the Corozal Army Veterinary Quarantine Center, Panama. *Journal of Medical Entomology* 38, 298–302.

Mohler, J. R. (1906). Texas or tick fever and its prevention. *United States Department of Agriculture Farmer's Bulletin* 258, 1–45.

Muniz, R. A., Hernandez, F., Lombardero, O., *et al.* (1995). Efficacy of injectable doramectin against natural *Boophilus microplus* infestations in cattle. *American Journal of Veterinary Research* **56**, 460–463.

Nari, A. & Hansen, J. W. (1999). *Resistance of Ecto- and Entoparasites: Current and Future Solutions*, Technical Report No. 67 SG/10. Paris: Office International des Epizooties.

Newton, L. G. (1967). Acaricide resistance and cattle tick control. *Australian Veterinary Journal* **43**, 389–394.

Nolan, J. (1981). Current developments in resistance to amidine and pyrethroid tickicides in Australia. In *Tick Biology and Control*, eds. Whitehead, G. B. & Gibson, J. D., pp. 109–114. Grahamstown, South Africa: Tick Research Unit, Rhodes University.

Nolan, J. (1990). Acaricide resistance in single and multi-host ticks and strategies for control. *Parassitologia* **32**, 145–153.

Nolan, J. & Schnitzerling, H. J. (1986). Drug resistance in arthropod parasites. In *Chemotherapy of Parasitic Diseases*, eds. Campbell, W. C. & Rew, R. S., pp. 603–620. New York: Plenum Press.

Nolan, J., Roulston, W. J. & Schnitzerling, H. J. (1979). The potential value of some synthetic pyrethroids for control of the cattle tick (*Boophilus microplus*). *Australian Veterinary Journal* **55**, 463–466.

Norris, K. R. & Stone, B. F. (1956). Toxaphene-resistant cattle ticks (*Boophilus microplus* (Canestrini)) occurring in Queensland. *Australian Journal of Agricultural Research* **7**, 211–226.

Norton, G. A., Sutherst, R. W. & Maywald, G. F. (1983). A framework for integrating control methods against the cattle tick, *Boophilus microplus*, in Australia. *Journal of Applied Ecology* **20**, 489–505.

Norval, R. A. I., Fivaz, B. H., Lawrence, J. A. & Brown, A. F. (1984). Epidemiology of tick-borne diseases of cattle in Zimbabwe. II. Anaplasmosis. *Tropical Animal Health and Production* **16**, 63–70.

Ortiz, E. M., Santamaria, E. M. & Fragoso, S. H. (1994). Resistencia en garrapatas *Boophilus microplus*, a los ixodicidas en Mexico. In *Proceedings of the 14th Pan American Congress on Veterinary Sciences*, eds. Perez Trujillo, J. M. & Gonzales Padilla, E., Acapulco, Guerrero, Mexico, pp. 473–474.

Palmer, B. H., McCarthy, J. F., Kozlik, A. & Harrison, I. R. (1971). A new chemical group of cattle acaricides. *Proceedings of the 3rd International Congress of Acarology*, Prague, pp. 687–691.

Perry, B. D. (1992). History of east coast fever in eastern and central Africa. In *The Epidemiology of Theileriosis in Africa*, eds. Norval, R. A. I., Perry, B. D. & Young, A. S., pp. 41–62. London: Academic Press.

Rechav, Y. (1987). Use of acaricide-impregnated ear tags for controlling the brown ear tick (Acari: Ixodidae) in South Africa. *Journal of Economic Entomology* **80**, 822–825.

Remington, B., Kieran, P., Cobb, R. & Bodero, D. (1997). The application of moxidectin formulations for control of the cattle tick (*Boophilus microplus*) under Queensland field conditions. *Australian Veterinary Journal* **75**, 588–591.

Riddles, P. W. & Nolan, J. (1986). Prospects for the management of arthropod resistance to pesticides. In *Parasitology – Quo Vadit?*, Proceedings of the 6th International Congress of Parasitology, ed. Howell, M. J., pp. 679–687.

Romero, A., Benavides, E., Herrera, C. & Parra, M. H. (1997). Resistencia de la garrapata *Boophilus microplus* a acaricides organofosforados y piretroides sintéticos en el departamento del Huila. *Revista Colombiana de Entomologia* **23**, 9–17.

Rosario-Cruz, R., Guerrero, F. D., Miller, R. J., *et al.* (2005). Roles played by esterase activity and by a sodium channel mutation involved in pyrethroid resistance in populations of *Boophilus microplus* (Acari: Ixodidae) collected from Yucatan, Mexico. *Journal of Medical Entomology* **42**, 1020–1025.

Roulston, W. J., Stone, B. F., Wilson, J. T. & White, L. I. (1968). Chemical control of an organophosphorus- and carbamate-resistant strain of *Boophilus microplus* (Can.) from Queensland. *Bulletin of Entomological Research* **58**, 379–392.

Roush, R. T. (1993). Occurrence, genetics and management of insecticide resistance. *Parasitology Today* **9**, 174–179.

Roy-Smith, F. (1975). Amitraz: Australian field trials against the cattle tick (*Boophilus microplus*). *Proceedings of the 8th British Insecticide and Fungicide Conference*, pp. 565–571.

Sabatini, G. A., Kemp, D. H., Hughes, S., Nari, A. & Hansen, J. (2001). Tests to determine LC_{50} and discriminating doses for macrocyclic lactones against the cattle tick, *Boophilus microplus*. *Veterinary Parasitology* **95**, 53–62.

Sangster, N. C. (2001). Managing parasiticide resistance. *Veterinary Parasitology* **98**, 89–109.

Sangster, N., Batterham, P., Chapman, H. D., *et al.* (2002). Resistance to antiparasitic drugs: the role of molecular diagnosis. *International Journal for Parasitology* **32**, 637–653.

Schnitzerling, H. J. & Walker, T. B. (1985). Factors affecting the performance of acaricides used for control of the cattle tick, *Boophilus microplus*. *Tropical Pest Management* **31**, 199–203.

Schnitzerling, H. J., Nolan, J. & Hughes, S. (1983). Toxicology and metabolism of some synthetic pyrethroids in larvae of susceptible and resistant strains of the cattle tick *Boophilus microplus* (Can.). *Pesticide Science* **14**, 64–72.

Schnitzerling, H. J., Nolan, J. & Hughes, S. (1989). Toxicology and metabolism of isomers of flumethrin in larvae of pyrethroid-suceptible and resistant strains of the cattle tick *Boophilus microplus* (Acari: Ixodidae). *Experimental and Applied Acarology* **6**, 47–54.

Schuntner, C. A., Schnitzerling, H. J. & Roulston, W. J. (1971). Carbaryl metabolism in larvae of organophosphorus and carbamate-susceptible and -resistant strains of cattle tick *Boophilus microplus*. *Pesticide Biochemistry and Physiology* **1**, 424–433.

Shaw, R. D. (1969). Tick control on domestic animals. I. A brief history of the economic significance of tick infestations. *Tropical Science* **11**, 113–119.

Shaw, R. D. (1970). Tick control on domestic animals. II. The effect of modern treatment methods. *Tropical Science* **12**, 29–36.

Smith, T. & Kilborne, F. L. (1893). Investigations into the nature, causation, and prevention of Texas or southern cattle fever. *United States Department of Agriculture, Bureau of Animal Industries Bulletin* **1**, 1–301.

Soberanes, N., Santamaria, M., Fragoso, H. & Garcia, Z. (2002). Primer caso de resistencia al amitraz en garrapata del Ganado *Boophilus microplus* en México. *Técnica Pecuaria en México* **40**, 81–92.

Soll, M. D., Benz, G. W., Carmichael, I. H. & Gross, S. J. (1990). Efficacy of ivermectin delivered from an intraruminal sustained-released bolus against natural infestations of five African tick species on cattle. *Veterinary Parasitology* **37**, 285–295.

Solomon, K. R. (1983). Acaricide resistance in ticks. *Advances in Veterinary Science and Comparative Medicine* **27**, 273–296.

Sonenshine, D. E. (2006). Tick pheromones and their use in tick control. *Annual Review of Entomology* **51**, 557–580.

Stanford, G. D., Baker, J. A. F., Ratley, C. V. & Taylor, R. J. (1981). The development of a stabilized amitraz cattle dip for control of single and multi-host ticks and their resistant strains in South Africa. In *Proceedings of a Conference on Tick Biology and Control*, eds. Whitehead, G. B. & Gibson, J. D., Rhodes University, Grahamstown, South Africa, pp. 143–181.

Stendel, W. (1985). Experimental studies on the tickicidal effect of Bayticol® Pour-on. *Veterinary Medical Report* **2**, 99–111.

Stone, B. F. & Meyers, R. A. J. (1957). Dieldrin-resistant cattle ticks, *Boophilus microplus* (Canestrini), in Queensland. *Australian Journal of Agricultural Research* **8**, 312–317.

Strydom, T. & Peter, R. (1999). Acaricides and *Boophilus* spp. resistance in South Africa. In *Control de la Resistencia en Garrapatas y Moscas de Importancia Veterinaria y Enfermedades que Transmiten, IV Seminario Internacional de Parasitologia Animal*, eds. Morales, G., Fragosa, H. & Garcia, Z., Puerto Vallarta, Jalisco, Mexico, pp. 35–40.

Stubbs, V. K., Wilshire, C. & Webber, G. (1982). Cyhalothrin: a novel acaricidal and insecticidal synthetic pyrethroid for the control of the cattle tick (*Boophilus microplus*) and the buffalo fly (*Haematobia irritans exigua*). *Australian Veterinary Journal* **59**, 152–155.

Sutherst, R. W. & Comins, H. N. (1979). The management of acaricide resistance in the cattle tick, *Boophilus microplus* (Canestrini) (Acari: Ixodidae), in Australia. *Bulletin of Entomological Research* **69**, 519–537.

Sutherst, R. W., Norton, G. A., Barlow, N. D., *et al.* (1979). An analysis of management strategies for cattle tick (*Boophilus microplus*) control in Australia. *Journal of Applied Ecology* **16**, 359–382.

Tabashnik, B. E. (1990). Modeling and evaluation of resistance management tactics. In *Pesticide Resistance in Arthropods*, eds. Roush, R. T. & Tabashnik, B. E., pp. 153–182. New York: Chapman & Hall.

Taylor, M. A. (2001). Recent development in ectoparasiticides. *Veterinary Journal* **161**, 253–268.

Teel, P. D., Hair, J. A. & Stratton, L. G. (1979). Laboratory valuation of a sustained-release famphur bolus against gulf coast and lone star ticks feeding on Hereford heifers. *Journal of Economic Entomology* **72**, 230–233.

Temeyer, K. B., Pruett, J. H., Untalan, P. M. & Chen, A. C. (2006). Baculovirus expression of *Bm*AChE3, a cDNA encoding an acetylcholinesterase of *Boophilus microplus* (Acari: Ixodidae). *Journal of Medical Entomology* **43**, 707–712.

Ware, G. W. (2000). *The Pesticide Book*, 5th edn. Fresno, CA: Thomson Publications.

Wharton, R. H. (1976). Tick-borne livestock diseases and their vectors. V. Acaricide resistance and alternative methods of tick control. *World Animal Review* **20**, 8–15.

Wharton, R. H. & Roulston, W. J. (1970). Resistance of ticks to chemicals. *Annual Review of Entomology* **15**, 381–403.

Wharton, R. H., Roulston, W. J., Utech, K. B. W. & Kerr, J. D. (1970). Assessment of the efficiency of acaricides and their mode of application against the cattle tick *Boophilus microplus*. *Australian Journal of Agricultural Research* **21**, 985–1006.

Whitehead, G. B. (1958). A review of insecticide resistance in the blue tick, *Boophilus decoloratus*, in South Africa. *Indian Journal of Malariology* **12**, 427–432.

Whitnall, A. B. M., Mchardy, W. M., Whitehead, G. B. & Meerholz, F. (1951). Some observations on the control of the bont tick, *Amblyomma hebraeum* Koch. *Bulletin of Entomological Research* **41**, 577–591.

Young, A. S., de Castro, J. J. & Kiza-Auru, P. P. (1985). Control of tick (Acari: Ixodidae) infestation by application of ear tags impregnated with acaricides to cattle in Africa. *Bulletin of Entomological Research* **75**, 609–619.

19 • Anti-tick vaccines

P. WILLADSEN

INTRODUCTION

The case for vaccination as a means of tick control has been made repeatedly and deserves only brief reiteration here. Currently, tick control is heavily dependent on two approaches: the use of chemical pesticides and the use of tick-resistant animals. Chemical pesticides are increasingly problematic for a number of reasons. (1) Resistance to existing pesticides of many chemical classes is widespread and its incidence is increasing (see Chapter 18). The speed with which resistance appears after the release of each new class of chemical is clearly a deterrent to the companies developing such means of parasite control. (2) There is increasing concern about the use of chemicals in all forms of agriculture, both for their potential environmental impact and for their presence in food products. (3) Newer classes of pesticide have tended to be significantly more expensive than their predecessors, an additional deterrent to their application.

Genetically resistant animals, which show a heritable ability to become immunologically resistant to tick infestation, are a vital component of many tick control strategies. They are particularly important in the control of *Rhipicephalus* (*Boophilus*) *microplus* (hereafter referred to as *B. microplus*) on cattle. However, this approach also has difficulties. For the hosts of many tick species, resistance may simply not develop. Even for *B. microplus*, it may be difficult to breed tick resistance while preserving other desirable production characteristics such as high milk yield in dairy cattle.

By contrast, a vaccine can be a non-contaminating, sustainable and cheap technology, potentially applicable to a wide variety of hosts. There are potential limitations as well, first and foremost whether vaccines can be produced which achieve the desired level of efficacy under field conditions. Such issues are discussed in this chapter. From a commercial perspective, a number of other potential advantages of vaccines are less frequently discussed. Briefly, a recombinant vaccine can be produced very cheaply using a multi-purpose manufacturing facility i.e. fermenter; unlike chemical pesticides, dedicated facilities are not required. Secondly, the registration cost of a new vaccine should be substantially less than that of a new pesticide. A significant part of the cost of the development and registration of a new pesticide is in the demonstration of safety towards target and non-target species, in particular, humans. It is highly likely that for a vaccine based on a defined protein antigen, the regulatory hurdles will be significantly less. Thirdly, there are theoretical reasons for expecting that the development of resistance to a vaccine is less likely than for a pesticide. There are examples where a single point mutation in the target molecule is sufficient to render a pesticide ineffective, whereas such point mutations are unlikely to affect vaccine efficacy. This is because, in most cases, a vaccine is likely to target multiple epitopes on each protein antigen. Finally, while the introduction of a tick vaccine for a farming community used to parasite control through pesticides might be initially difficult, in the longer term vaccines should be less subject to some of the serious patterns of misuse seen with pesticides.

Given these advantages, our scientific challenge is to produce truly efficacious vaccines able to provide tick control that is both practical and cost effective.

STAGES IN VACCINE DEVELOPMENT

The development and delivery of a recombinant vaccine can be conveniently broken into a number of consecutive stages: identification and characterization of protective antigens; production of antigens as immunologically effective, recombinant proteins in a commercially viable manner; delivery of antigens in a way that achieves the desired immunological response; validation of the prototype vaccine in a field situation; and finally, delivery of the vaccine to the market. The latter stages of this process are accompanied by much commercial activity in both registration of the product and market identification and development. Each of these stages is addressed in this chapter, emphasizing the scientific phases.

DEMONSTRATION OF THE FEASIBILITY OF VACCINATION

It has been known for about 70 years that partial to strong immunity to tick infestation can be induced by vaccination with a variety of antigenic materials, including whole tick homogenates, salivary glands and salivary gland extracts, tick internal organs, including tick gut material, cement material and so on. This area has been reviewed extensively (e.g. Willadsen, 1980, 1997; Willadsen & Billingsley, 1997; Pruett, 1999; Willadsen & Jongejan, 1999; Nuttall *et al.*, 2006) and will not be reviewed again here. While such experiments do not prove that immunological control will be successful with all ticks, they at least show that immunological control is worthy of further exploration. They are, however, just an initial, small step on the path to vaccine development.

IDENTIFICATION OF PROTECTIVE ANTIGENS

There is no doubt that the current rate-limiting step in the development of anti-tick vaccines, as for most anti-parasite vaccines, lies in the identification of truly efficacious antigens and their expression as effective recombinants. In the past (Willadsen & Kemp, 1988), we have discussed tick antigens as belonging to one of two groups. The first of these includes the antigens involved in naturally acquired resistance to tick infestation, i.e. those immunogenic materials exposed to a host by the normal processes of tick attachment and feeding. The second group of antigens are the 'concealed' antigens, namely those antigens which are not part of the normal host–parasite interaction and which do not under normal circumstances stimulate an immunological response. It is possible to raise a response to these immunogenic molecules by vaccination and for that immunological response to subsequently damage the feeding tick. Typical examples of 'concealed' antigens are the immunogenic proteins located in the gut of the tick where, once an antibody to the protein is raised by vaccination, uptake of specific immunoglobulin during feeding leads to damage to the parasite.

The idea of 'concealed' antigens has two important implications, one positive, the other negative. Many host species fail to develop adequate immunity to a tick infesting them, even after prolonged exposure. The concealed antigen approach then has the positive aspect of offering an alternative approach that may circumvent the factors, whether immunological or physiological, that prevent the natural development of immunity. Even where some immunity does develop, as for *B. microplus* for example, then the concealed antigen vaccines, by operating in a different way and largely on a different stage of the life cycle, substantially enhance the total protective response. The negative aspect of concealed antigens is that natural parasite exposure is unlikely to boost the immune response to the antigens, implying a need for continual boosting through vaccination.

Broadly, three approaches can be used to identify useful vaccine antigens. The first is to use the immunological response of an immune host. Typically, this has meant the study of antigens that elicit an antibody response. The second is to identify tick factors important for the parasite's function or survival and then evaluate these as potential vaccine antigens. The third is the pragmatic one of biochemical fractionation, evaluating progressively simpler protein mixtures by host vaccination and parasite challenge trials. All three approaches have strengths and weaknesses. The immunological approach is often misguided, since it depends on the assumption that the immunological reaction used to select antigens is the one that protects the host against infestation or, at least, that the same antigens are responsible for both 'analytical' and 'protective' responses. This is usually not known. The second, the selection of critical tick factors for evaluation as vaccine candidates, fails simply because we understand so little of what is both really essential to a tick and accessible to the host's immune system. The third, the pragmatic approach, is laborious and time consuming, while the protective effect of crude parasite extracts can easily dissipate into a range of partially protective fractions as purification proceeds. Nevertheless, successes have been delivered with all three approaches. Examples given below include, for the immunological approach, the p29 antigen from *Haemaphysalis longicornis*, for the second approach the serpin HLS1 from *H. longicornis* and for the third, the pragmatic approach, the Bm86 antigen from *B. microplus*.

Rapid developments in molecular technologies are speeding up the identification of antigens and potential antigens. Expressed sequence tag (EST) libraries are now available from a number of tick species and tissues. The DFCI website (http://compbio.dfci.harvard.edu/tgi) is an important resource for this information. For example, over 20 000 ESTs have been described from a normalized *B. microplus* library, derived from multiple tick stages and multiple acaricide susceptible and resistant strains (Guerrero *et al.*,

2005). Salivary gland transcriptomes have been published for *Ixodes*, *Dermacentor*, *Amblyomma* and *Rhipicephalus* spp. (see Chapter 4).

Far more ambitiously, sequencing of the genome of *Ixodes scapularis* is under way (Hill & Wikel, 2005). Sequencing of the *B. microplus* genome has been proposed (Guerrero *et al.*, 2006) and useful resources such as a BAC library and EST collections have been assembled. There are two related challenges to obtaining tick genome sequence. The first is the large size of the genomes, comparable to or even larger than mammalian genomes (Palmer *et al.*, 1994; Ullmann *et al.*, 2005). The genome of *I. scapularis* is estimated to be 2.1×10^9 bp and that of *B. microplus* 7.1×10^9 bp. The second challenge is the preponderance of highly repetitive sequences in those genomes. As noted, 'the most that can be predicted from the three tick genome measurements made to date (*Amblyomma americanum* was examined previously) is that tick genomes will be large, highly variable in size and consist largely of moderately repetitive DNA' (Ullman *et al.*, 2005).

RNA interference (RNAi) in ticks is being used increasingly to study gene function and identify potential vaccine antigens (see Chapter 4, and studies by de la Fuente and co-workers discussed later in this chapter). The application of proteomics to ticks has been slow, which may be due in part to difficulties of interpreting limited sequence information with the limited genomic data available. A start has been made in the proteomic analysis of abundantly expressed proteins from unfed *B. microplus* larvae (Untalan *et al.*, 2005) and of tick salivary secretions (Madden, Sauer & Dillwith, 2002). However, although molecular techniques are powerful, the identification and characterization of effective antigens remains a significant challenge. The difficulty and limited success of all approaches to antigen identification make it important that once a good antigen has been identified in one species of tick it is fully explored for cross-species efficacy of homologous antigens.

While the classification of antigens as either natural or concealed remains broadly valid, particularly in the sense that the two different classes of antigen present different immunological problems in vaccine development, an instructive alternative is to characterize antigens in a more functional sense. The following discussion focuses on protein antigens and potential antigens for which there has been substantial progress in their purification and physicochemical characterization. There is no reason why oligosaccharide or lipid immunogens could not be protective and in fact there is good evidence that some oligosaccharide antigens are very effective, for *B. microplus* at least (Lee & Opdebeeck,

1991; Lee, Jackson & Opdebeeck, 1991). However, neither oligosaccharide nor lipid immunogens have been characterized in any tick, while we have negligible ability to produce them in quantities sufficient for anything more than a small-scale experimental vaccination.

Salivary gland proteins and cement constituents as antigens

Older literature on tick feeding and immunology focused almost exclusively on the tick attachment site and the host–parasite interactions taking place there as well as on the constituents of the cement cone that is to a greater or lesser extent vital for the attachment of the tick to the skin of the host (e.g. Willadsen, 1980). Then, with the focus on the 'concealed' antigen approach to vaccination, this area of research seemed to be neglected for a number of years although it has now been revived (e.g. see Chapters 3 and 4). It is clear, thanks principally to work by Nuttall and colleagues, that the protein complement of the tick salivary gland shows individual variation in addition to the expected species-to-species variation, as well as dynamic changes during the process of tick feeding. These differences are reflected in the antigenic profiles of the ticks (for example Wang & Nuttall, 1994*a*; Lawrie & Nuttall, 2001). Clearly, the salivary gland and saliva are likely to be the vehicles by which factors such as immunomodulators, proteolytic and other hydrolytic enzymes, enzyme inhibitors and modifiers of haemostasis all pass from the tick to the host. These factors are likely to show considerable temporal variability and will be discussed under separate headings. There are other factors that do not fall into such groups, such as the structural components of cement. Mulenga *et al.* (1999) characterized p29, a 29-kDa salivary gland-associated protein from *H. longicornis*, by probing a cDNA library from partially fed adult ticks with sera of rabbits infested with ticks. The amino acid sequence coded for a 277-amino-acid protein including a putative signal peptide. There was significant sequence homology with a number of glycine-rich extracellular matrix proteins or structural proteins and conserved, collagen-like domains. Vaccination with recombinant protein produced in *E. coli* led to a significant reduction in adult female engorgement weight and 40% and 56% mortality of larvae and nymphs post-engorgement, respectively. The group then went on to remove the anti-p29 immunoreactivity from the antiserum and rescreen a library. This procedure identified another two polypeptides, HL34 and HL35, only one of which has so far been expressed as a recombinant protein and tested in vaccination trials in rabbits. This protein, HL34, induced some nymphal and adult mortality (15% and 29%, respectively) and a small reduction in adult engorgement weight. The protein possessed a tyrosine-rich domain, followed by a proline-rich domain and appeared to be induced by feeding. Otherwise, its function remains unknown (Tsuda *et al.*, 2001).

Trimnell, Hails & Nuttall (2002) investigated 64P, a putative cement protein from *Rhipicephalus appendiculatus* of predicted molecular weight 15 kDa and containing a glycine-rich region with similarity to host keratins and collagen. Guinea pigs with acquired immunity to the tick fail to form antibody to it. However, expression of a series of truncated versions of the protein followed by vaccination of tick-naïve hosts showed several of the constructs were able to stimulate significant protection to nymphal and adult infestations. The effects included nymphal mortality as high as 48% and adult mortality up to 70%, with some effects on engorgement weight and egg masses as well. Boosting of antibody titres by tick infestation, coupled with evidence for cross-reactivity with a number of tick tissues, led to the suggestion that this antigen combined the advantages of both 'exposed' and 'concealed' antigens. Curiously, the constructs most effective against nymphal stages had no effect on adults and vice versa though they were portions of the same protein. In a subsequent publication (Trimnell *et al.*, 2005), four recombinant proteins (64TRPs) were used, two full length and two N-terminal fragments. Antigens were applied singly or in combination, with guinea pigs as hosts for *R. sanguineus* and hamsters and rabbits as hosts for *I. ricinus*. Immunological cross-reactivity was examined by Western blotting. Broadly, significant vaccination effects were seen with both tick species and these correlated with cross-reactivity detected on immunoblots. Effects included a reduction in engorgement weight and egg-laying, increased tick mortality, and gut pathology. Under some circumstances, localized skin inflammatory reactions were seen but did not seem to be related to anti-tick effects. Cross-reactivity of antisera with *Amblyomma variegatum* and *B. microplus* was observed but these tick species were not used in challenge experiments. The effect of 64TRPs in partially blocking the transmission of tick-borne encephalitis virus by *I. ricinus* is discussed later.

A third salivary gland and cement cone protein (RIM36) has been characterized from *R. appendiculatus*. It too contains a number of glycine-rich repeat regions and a proline-rich C-terminal region. The protein induces a strong antibody response in tick-exposed cattle, though it is not clear whether the immunological response is protective (Bishop *et al.*, 2002).

The number of salivary gland proteins that have been functionally characterized is small and most of them fall into one or other of the classes to be discussed below. One interesting protein is calreticulin, a calcium-binding protein. The gene from *A. americanum* has been characterized, cloned and expressed; the protein has been shown to be secreted in saliva of both *A. americanum* and *Dermacentor variabilis* (Jaworski *et al.*, 2002). A possible role in host immunosuppression or antihaemostasis has been suggested. Rabbits vaccinated with recombinant tick protein developed necrotic feeding lesions on tick challenge (Jaworski *et al.*, 1995). Calreticulin from *B. microplus* has also been identified, sequenced and expressed. It was found to be poorly immunogenic in cattle (Ferreira *et al.*, 2002).

Screening of an expression library from *I. scapularis* salivary gland with sera from immune guinea pigs identified a number of antigens of interest, including putative anti-complement and histamine-binding (see below) proteins. Most studied was salp25D, an antioxidant protein (Das *et al.*, 2001). The potential of these proteins as vaccine antigens, like the calreticulins, remains to be thoroughly explored. Based on the rationale that salivary gland proteins specifically expressed as a result of feeding could represent antigens of choice for vaccine development, subtractive hybridization of unfed and 4-day-fed *I. scapularis* larvae was used to identify ten differentially expressed genes. These have yet to be assessed in vaccination trials (Xu, Bruno & Luft, 2005).

Hydrolases and their inhibitors

The biochemical literature on hydrolytic enzymes is enormous, reflecting the ease with which they can be investigated as much as their importance. The most studied group, the proteolytic enzymes, also possess a diverse and well-characterized set of specific protein inhibitors. Such enzymes are likely to be key molecules in tick feeding and in the immune response. The role of hydrolases and their inhibitors in tick biology has been the subject of some speculation. A specific group of proteinases and inhibitors, namely those involved in haemostatic mechanisms, is discussed separately.

From a vaccine perspective, serine proteinases for example are attractive because they are intimately involved not only in digestive processes but also in complement activation, blood coagulation and many aspects of the immune system. As such, it is possible that an immune attack on them could be deleterious to a tick (e.g. Mulenga *et al.*, 2001, 2002). The disadvantage of proteinases as antigens is that they occur in very large numbers. For example, the *Drosophilia* genome codes for approximately 400 serine proteinases. In many circumstances, there will probably be a high level of functional redundancy, potentially making vaccination with one or a small subset of them ineffective.

Although there have been attempts to vaccinate against proteinases in other parasites, there has been little experimentation with ticks. An aspartic proteinase precursor from *B. microplus* eggs was shown to confer partial protection (da Silva Vaz *et al.*, 1998). In experiments with expression library immunization using cDNA from *I. scapularis*, one of the two unique cDNAs with putative function that gave some protection was an endopeptidase (Almazán *et al.*, 2003). The most intensively studied proteinase is a membrane-bound carboxydipeptidase from *B. microplus* (Bm91) with sequence similarity to the mammalian angiotensin-converting enzymes and even stronger similarities in biochemical specificity (Riding *et al.*, 1994; Jarmey *et al.*, 1995). The enzyme is located principally in the tick's salivary gland and in vaccination trials was effective as a native protein and, as a recombinant, in further increasing the efficacy of a recombinant Bm86 vaccine (see below) (Willadsen *et al.*, 1996).

Attention has also focused on the inhibitors of proteinases. The double-headed serine proteinase inhibitors from *B. microplus* (Tanaka *et al.*, 1999) appear similar to proteinase inhibitors from the same tick species studied approximately 20 years ago (Willadsen & Riding, 1979, 1980; Willadsen & McKenna, 1983). These offer some immuno-protection against *B. microplus* larvae (Andreotti *et al.*, 1999, 2002). Trypsin inhibitors of *B. microplus* were isolated by affinity chromatography on trypsin-Sepharose and used in Freund's adjuvant to vaccinate *Bos indicus* cattle. Following larval challenge, vaccinated cattle showed a 68% reduction in the number of engorging female ticks and a corresponding reduction in the total egg weight. As the authors point out, the effects of vaccination are different from those seen with the Bm86 vaccine (see below) and may be directed against early larval development. If that is correct, these inhibitors could be an ideal complement to the existing *B. microplus* vaccine. The vaccination 'antigen' contained at least two protein species, as would be expected from older literature and, judging from electrophoresis results, the mixture may have been more complex still. Hopefully the individual inhibitors will be assessed and evaluated as recombinant vaccine candidates.

There has been speculation that a family of high-molecular-weight serine proteinase inhibitors, the serpins, could be target antigens (Mulenga *et al.*, 2001). Recently a conserved serpin amino acid motif was used to clone and express a 378-amino-acid polypeptide from *H. longicornis* that had high sequence similarity to several known serpins. Transcription was induced exclusively in tick midguts by feeding. Vaccination of rabbits with recombinant protein induced 44% and 11% mortality in feeding nymphs and adults, respectively (Sugino *et al.*, 2003). Subsequently, the *H. longicornis* serpin-2 gene (HLS2) was sequenced and shown to code for a 44-kDa protein with the sequence signatures of a serpin. It was expressed in nymphs and adults but not larvae, in partially and fully fed ticks but not unfed ones and in the haemolymph but not midgut and salivary gland. Western blotting was used to provide additional evidence that saliva did not contain native HLS2. Vaccination of rabbits with recombinant protein followed by tick challenge showed a number of significant effects: increases in nymphal and adult mortality, a slight increase in the duration of feeding, and decreases in both engorgement weight and in the percentage of ticks not ovipositing (Imamura *et al.*, 2005).

Similarly, a number of serpin genes have been identified in *R. appendiculatus* (Mulenga *et al.*, 2003). Two serpins from *R. appendiculatus*, RAS-1 and RAS-2, known to be present throughout the life cycle, were expressed in *E. coli* and a mixture used to vaccinate cattle. Following infestation with both nymphs and adult ticks, it was shown that nymphal engorgement was reduced by 61% while adult tick mortality was increased by 28% and 43% for female and male ticks respectively. There was a slight effect on tick engorgement weight. The authors also noted that there were no visible deleterious effects of the vaccination on cattle. It would of course be interesting to know whether the mixture of two serpins was more efficacious than either serpin alone though so far neither antigen has been evaluated as a stand-alone vaccine. Antisera from rabbits repeatedly infested with each of the three instars failed to react with recombinant serpins, which was interpreted to mean that the serpins were not secreted but rather represented concealed antigens (Imamura *et al.*, 2006). The results are some of the most positive so far recorded for vaccination against this tick species using defined, recombinant antigens.

Only two non-proteolytic hydrolases have been investigated in any detail. Del Pino *et al.* (1998) reported that a polyclonal antibody against β-*N*-acetylhexosaminidase from larval extracts, when injected into fully engorged adult female *B. microplus*, inhibited oviposition by 26%. While the result is interesting, the artificiality of the system and the relatively small effect makes the relevance of this observation to vaccine development questionable.

The 5′-nucleotidase from *B. microplus* has been characterized (Liyou *et al.*, 1999) and localized to gut, ovary and, predominantly, on the luminal surface of the Malpighian tubules (Liyou *et al.*, 2000) where a role in purine salvage is likely. It has been tested as a vaccine antigen only as an enzymatically inactive, truncated form expressed in *E. coli* where it appeared to slightly increase the efficacy of a Bm86-based vaccine when co-administered. The small size of both the effect and of the experimental cattle groups ensured that the effect was not statistically significant (Liyou, 1996). In experiments with expression library immunization using cDNA from *I. scapularis* cited above, one of the protective cDNAs was a putative nucleotidase (Almazán *et al.*, 2003). Given this, more thorough evaluation of nucleotidases may be warranted.

Tick-induced host immunomodulation as a source of potential antigens

It is now well established that ticks modulate the immune system of their host in a variety of ways (see, for example, Barriga, 1999; Wikel, 1999; Wikel & Alarcon-Chaidez, 2001; Brossard & Wikel, 2004), while Titus, Bishop & Mejia (2006) catalogue the various immunomodulatory and haemostatic activities identified so far in hard ticks as well as other arthropod salivas. Much of the evidence was obtained using crude salivary gland extracts or, on occasion, through direct tick infestation of the host (see Chapter 9).

If the inhibition or diversion of the host's immune response is critical to tick survival, then it is possible that the tick molecules responsible for such manipulation could themselves be vaccine targets. This has been suggested a number of times, but is still to be validated. The hypothesis here is either that the molecules are so critical to tick survival that their inhibition will lead to tick rejection or death, or that their inhibition by a vaccine-induced immunological response will ablate the parasite's own attempts at immune diversion, allowing the host to mount an effective rather than an ineffective immune response to the parasite. In any case, it is reasonable to assume that immunomodulatory molecules will be mostly secreted and so accessible to the host's immune system. Before these two related hypotheses can be explored, the nature of the parasite's immunomodulatory molecules must first be clarified. Recent work by a number of groups has identified a range of candidates affecting the immune system at various stages between antigen presentation and the effector response (see Chapter 9). These include macrophage migration inhibitory factor (MIF) (Jaworski *et al.*, 2001), potentially a component of an immunomodulatory parasite system that may affect antigen presentation; cytokine-binding proteins that could manipulate leukocytes (Gillespie *et al.*, 2001; Hajnická *et al.*, 2001); Iris, a protein from *I. ricinus* able in recombinant form to inhibit the production of interferon-gamma (IFN-γ) in stimulated peripheral blood mononuclear cells (Leboulle *et al.*, 2002); and a 36-kDa protein from the salivary glands of *Dermacentor andersoni*, Da-p36, the temporal expression of which is consistent with a role in immunosuppression during feeding (Bergman *et al.*, 2000).

Considerable work stemming initially from the laboratory of Nuttall has characterized an interesting group of proteins that specifically bind host immunoglobulin. For a review of early work, see Wang & Nuttall (1994*b*). It has been suggested that these proteins are involved in the specific secretion of ingested host immunoglobulin that has passed from tick gut to haemolymph as a way of avoiding the deleterious effect of anti-tick immunoglobulin that might be taken up with the blood meal. The situation would appear to be more complex than that. At least in the case of *A. americanum*, there is specific uptake of immunoglobulins into tick haemolymph during feeding (Jasinskas, Jaworski & Barbour, 2000), while with both *D. variabilis* and *I. scapularis*, relatively complex patterns of immunoglobulin concentration in tick haemolymph are found (Vaughan, Sonenshine & Azad, 2002). It has been suggested that these immunglobulin-binding proteins could be an appropriate target for vaccine development (Wang & Nuttall, 1999). Vaccination of guinea pigs with a recombinant form of one of these proteins, a male-specific protein from *R. appendiculatus*, had the effect of delaying female engorgement (Wang & Nuttall, 1999). The effects were thus small and indirect. It is interesting that a recent paper (Packila & Guilfoile, 2002) has shown the presence in *I. scapularis* of a related, male-specific protein so they may be a general phenomenon.

The role of complement in protective immunity to ticks is unclear, most of the experimental work in this area having been carried out some time ago (Willadsen, 1980). It is an interesting reflection on the species specificity of the tick–host interaction as well as the probable importance of complement that there is specificity in the ability of salivary gland extracts of different tick species to inhibit host complement, that specificity correlating with host range (Lawrie, Randolph & Nuttall, 1999). Complement inhibition by *B. microplus* proteinase inhibitors has been demonstrated (Willadsen & Riding, 1980) as has the efficacy of these inhibitors in vaccination trials on cattle (Andreotti *et al.*, 1999, 2002; Tanaka *et al.*, 1999; see also above). Valenzuela *et al.* (2000) recently characterized another anti-complement protein from the saliva of *I. scapularis*. It was expressed as a biologically active protein in Cos cells. Interestingly, the sequence is without similarity to known proteins in the databases.

Another group of immunomodulatory proteins of interest are those connected to the effector arm of the immune response. Most interesting among these are the histamine-binding proteins identified in a number of species (Paesen *et al.*, 1999, 2000). These proteins contain two histamine-binding domains, one of low and one of high affinity. Variations are possible. For example, it has also been found that the equivalent molecule from *D. reticulatus* contains a histamine- and a serotonin-binding site (Sangamnatdej *et al.*, 2002). These proteins, structurally closely related to the lipocalins, though showing little similarity in primary amino acid sequence, are likely to be involved in subverting the allergic responses typical of many tick infestations. There is good circumstantial evidence from the older literature that such allergic responses, whether through cutaneous basophil hypersensitivity in guinea pigs as a model host or in more realistic host–parasite systems, can be responsible for tick rejection. This evidence is summarized in Willadsen (1980). Alijamali *et al.* (2003) provided direct evidence for the importance of these proteins. Injection of double-stranded RNA coding for part of a putative *A. americanum* histamine-binding protein reduced the histamine-binding ability of isolated salivary glands as well as leading to aberrant tick feeding. It would be fascinating to see whether vaccination of hosts with the lipocalins would lead to more efficacious, naturally acquired immune rejection.

Another group of proteins with immunomodulatory potential, either through a direct effect on the generation of an immune response, or through inhibition of effector function, are the proteinases and their inhibitors. These have been discussed already.

Other molecules of known function

There is now no doubt that not only salivary gland products and secreted antigens but also the surface of the tick gut are appropriate targets for a protective immune response. It is also increasingly clear that passage of immunoglobulin through the tick, from gut to haemolymph to salivary gland

and host, is a common feature (for example, Wang & Nuttall, 1999). The list of potential protective antigens that might come into contact with host antibody would therefore seem to be enormous. The following lists just a few of the possibilities, where either some vaccination effect has been demonstrated, or molecular characterization of a protein has been coupled to the suggestion that it might be a useful antigen.

Haemostatic mechanisms used by the tick to ensure the success of its blood feeding have been well reviewed (Mans & Neitz, 2004) and suggested as possible targets for immune intervention (Mulenga *et al.*, 2002). Protein anticoagulants, inhibitors of platelet aggregation such as apyrases and inhibitors of fibrinogen receptor function have all been described. Few have been evaluated in vaccination trials. Most work has been done on tick anticoagulants, in part because of their potential biomedical applications. For example, a factor Xa inhibitor from the salivary glands of *Ornithodoros savignyi* has been isolated, characterized and expressed as a recombinant protein (Joubert *et al.*, 1998). Factor Xa inhibitors have also been identified in *R. appendiculatus*, *O. moubata* and *H. truncatum*. Thrombin inhibitors have been described. Iwanga *et al.* (2003) for example have recently described two novel thrombin inhibitors from an *H. longicornis* salivary gland cDNA library. The list can be extended with less well-characterized tick inhibitors, or inhibitors affecting other components of the haemostatic process. These include variabilin, a 47-residue platelet aggregation inhibitor from *D. variabilis* (Wang *et al.*, 1996), disagregin, a fibrinogen receptor antagonist from *O. moubata* (Karczewski, Endris & Connolly, 1994) and savignygrin, an $\alpha_{IIb}\beta_3$ antagonist and platelet disaggregation factor (Mans, Louw & Neitz, 2002). However, what is needed most is an experimental demonstration that such proteins can be effective vaccine antigens.

There has been evidence for a long time that vaccination of a host with tick-derived vitellin could have some effects on ticks. An 80-kDa glycoprotein purified from *B. microplus* larvae, very probably a processed product of vitellogenin, was able to reduce the engorgement percentage, average engorgement weights and egg conversion ratios of adult ticks feeding on vaccinated sheep (Tellam *et al.*, 2002). Similar results were obtained on vaccination with purified vitellin itself. A recombinant form of the protein, incorrectly folded and not glycosylated, had no significant effect as a vaccine antigen, leaving open the question whether the critical protective epitopes were associated with the tertiary structure of the protein or the immunogenic oligosaccharides attached to it in the natural situation. Why antibodies to a storage protein have a deleterious effect on ticks is not obvious, though it may relate to a direct effect on oogenesis. Alternatively, it is known that vitellin binds haem (Tellam et al., 2002) and that this may inhibit the toxicity of haem (Logullo et al., 2002). The toxicity of haem to insects and the importance of detoxification is established (Oliveira et al., 1999).

A completely novel antigen was identified in an elegant series of experiments by Weiss & Kaufman (2004). Two feeding-induced proteins from the gonad of male *Amblyomma hebraeum*, of 16.1 kDa and 11.6 kDa, though separately inactive, together stimulated female engorgement as well as salivatory gland degeneration and partial development of the ovary. This engorgement factor was named 'voraxin'. The proteins were identified from 28 differentially expressed, feeding-induced genes, expressed as recombinants in a baculovirus system and bioassayed in groups through injection of virgin females and subsequent measurement of weight, fluid secretory competence and ovarian weight following tick removal. The process was continued through a succession of trials until the active combination of two was finally identified. When normal mated females were placed on a rabbit which had been immunized with recombinant voraxin, mean weight after 2 weeks was reduced by 72%, and more significantly three-quarters of the ticks did not engorge at all. The result is exciting since it represents a very different approach to vaccination against ticks and a novel target for immune interference in the tick's physiology.

Antigens identified by structured screening

In the absence of a good understanding of the characteristics of a protective antigen, it has often seemed logical to make as few assumptions as possible. Material known to induce protection can be screened for antigens through fractionation to reduce the complexity, eventually identifying a single antigen. Conceptually this is simple, though in practice it can be technically demanding. Such processes have been responsible for several of the antigens already mentioned, and they have contributed to the identification of many. Fractionation and screening as the major approach to antigen identification has been used for both proteins and genes.

First and foremost, the Bm86 antigen from *B. microplus*, the basis of the recombinant vaccines against that tick, was identified through a series of fractionation and vaccination trials without reference to any function, either biochemical or immunological. Given the substantial research devoted to this protein over the last decade, it will be discussed separately. Pragmatic fractionation of *B. microplus* extracts has

also identified other antigens, antigen B (Willadsen, 2001) and a mucin-like protein BMA 7 (McKenna et al., 1998) neither of which has identified functions.

A conceptually similar, gene-based approach was taken by Almazán et al. (2003) using expression library immunization. A cDNA library was prepared from an *I. scapularis* larval cell line using an expression vector containing a strong cytomegalovirus promoter. A total of approximately 4000 cDNA clones were then taken through three cycles of mouse vaccination, using pooled clone collections. In the final trials, cDNA clones were pooled according to putative function (including four pools of unknowns) based on sequence analysis. A number of these showed vaccine efficacies, measured by the inhibition of larval infestation, of up to 57%. Most interesting were two single clones coding for a putative nucleotidase and an endopeptidase. Two pools of 'unknown' cDNAs were also effective. It is interesting that, although attention focused finally on the most efficacious cDNA pools, results of the first two rounds of screening would suggest most pools showed some efficacy. This raises an important question: are there really very large numbers of proteins that, whether expressed as fragments or full-length products, make effective antigens, or are the results somehow reflecting peculiarities of the experimental system? Additionally, several clones induced much higher levels of larval recovery than the controls.

In a second paper, the same group proposed the use of RNAi for the 'high throughput' identification of tick protective antigens (de la Fuente et al., 2005). Double-stranded RNA (dsRNA) was generated both from the same plasmid pools as well as from specific genes or gene fragments of interest and injected directly into *I. scapularis* that were subsequently fed on sheep. Results were consistent with those obtained using expression library immunization (ELI). For example, dsRNA from actin and two previously identified gene fragments reduced tick weight and abolished egg-laying as did some of the RNAi from plasmid pools.

From these observations the group went on to focus on three antigens, 4F8, 4D8 and 4E6, in more detail. The 4F8 sequence has homology with known nucleotidases. 4D8 contains a domain conserved across many species, while 4E6, whose available sequence encodes only 38 amino acids, has only low-level similarity to a *Drosophila* sequence. All three were expressed throughout the tick's life cycle while PCR showed expression of all in at least other *Ixodes* species. 4D8 was present in a more diverse selection of ticks. Each polypeptide was expressed or synthesized and used to vaccinate mice that were subsequently challenged with

I. scapularis larvae. All had effects in inhibiting infestations, while 4D8 and 4E6 also inhibited moulting to nymphs. Overall efficacies were from 62% to 71%. In a further experiment, rabbits were infested simultaneously with *I. scapularis, D. variabilis* and *A. americanum* nymphs and again, effects were seen though the results were very preliminary. The observations were complemented by capillary feeding experiments (Almazán *et al.*, 2005a). The observations were extended by vaccinating groups of sheep with each of the antigens separately as well as with all three simultaneously. Sheep were infested with adult *I. scapularis*. All three antigens reduced the percentage of attached ticks feeding to engorgement, the most effective being 4D8 with a greater than twofold reduction. Similarly all reduced oviposition, the most effective again being 4D8 with a twofold effect. Overall efficacy against adult female ticks ranged from 33% to 71% (Almazán *et al.*, 2005b).

Exploration of the 4D8 antigen has continued rapidly. Sequence conservation in 10 tick species of six genera has been examined and amino acid sequence identities of 60% or greater demonstrated. The effect of silencing the gene using RNAi in adult ticks was observed in five of the species. Striking reductions in tick weight and an increase in mortality were observed in 4D8-silenced ticks when compared with controls, though the most dramatic effect was on oviposition. The effect of RNAi on male and female *D. variabilis* was also examined by microscopy. Salivary glands, guts, spermatagonia and ovaries all showed pathology. The suggestion that the protein has a role in tick development and reproduction is reinforced by studies on 4D8 homologues in *Drosophila melanogaster* and *Caenorhabditis elegans*, reported previously in the literature (de la Fuente *et al.*, 2006a). Another publication reported the effect in *Rhipicephalus sanguineus* of silencing the expression of the homologues of both the 4D8 or subolesin and Bm86 antigens. Injection of unfed female ticks with 4D8 dsRNA subsequently led to decreased tick attachment, survival, feeding and oviposition. The effects of Rs86 dsRNA were less striking, but significant for engorgement weight and oviposition. Silencing both 4D8 and Rs86 simultaneously led to an increase in efficacy measured by all parameters sufficiently strong to suggest synergy. Histological observations were consistent with this, only ticks with both genes silenced showing marked effects, with distended and atrophied guts. The authors suggest that the effects of silencing multiple genes simultaneously may be useful in the identification of combinations of tick antigens for formulation into improved vaccines (de la Fuente *et al.*, 2006b).

These results have led to the proposition that RNAi screening represents a cost-effective and rapid way of screening for antigens. While the value of the technique is considerable, the process identifies only the genes whose downregulation leads to a significant change in phenotype. Whether the actual gene product in the tick is accessible to a host's immunological response would still need to be explored on a case-by-case basis.

Antigens of unknown function

Other antigens tested in reasonable purity include a 39-kDa antigen from *Hyalomma anatolicum* (Sharma *et al.*, 2001) though no sequence information is available. From *H. longicornis*, Mulenga *et al.* (2000) isolated p84, an antigen inducing specific immediate hypersensitivity responses in immune rabbits, and obtained partial sequence information. Its protective efficacy however is unknown. Somewhat more speculatively, *H. longicornis* infecting mice that were producing monoclonal antibodies to a 76-kDa tick gut protein showed reduced hatching of eggs (Nakajima *et al.*, 2003). P27/30, a troponin-I like protein from *H. longicornis* has been expressed as a recombinant protein in *E. coli* and used to vaccinate rabbits and mice. Some effects on the feeding success of *H. longicornis* were seen, though the effects were small (You, 2004, 2005).

Summary: identification of protective antigens

There are clearly a number of antigens worthy of further investigation and a larger number of candidates that would justify at least some evaluation. The biochemical categories into which these antigens fall are notable: principally proteins likely to be structural, simple hydrolytic enzymes and their inhibitors and, as potential though unproven antigens, modulators of the immune system. Antigens with demonstrable effects as native or recombinant proteins are summarized in Tables 19.1 and 19.2.

It is appropriate to ask whether these antigens truly represent the spectrum of potentially efficacious antigens. In principle, there is no obvious reason why drug and vaccine targets should belong to different biochemical classes. Specific targets may, because of their cellular or tissue location, be accessible to a drug but not the immune system. Acetylcholinesterase and acetylcholine receptors are likely to be one such example. In general, a target for one may well be a target for the other. It is notable therefore that two great classes of drug target, namely seven transmembrane segment receptors and ion channels, have not been investigated at all

Table 19.1 *Tick* (B. microplus) *antigens evaluated as native proteins*

Antigen	Reference
Pro-cathespin	Da Silva Vaz *et al.* (1998)
Serine proteinase inhibitors	Andreotti *et al.* (2002)
BMA 7	McKenna *et al.* (1998)
Vitellin	Tellam *et al.* (2002)

as vaccine antigens. This theme was explored in some detail by Sauer, McSwain & Essenberg (1994) who suggested a number of such targets worthy of study. If little has happened in the subsequent decade, this may reflect largely the experimental difficulty of such investigations.

DEVELOPMENT OF A RECOMBINANT VACCINE AGAINST *B. MICROPLUS*

Only once to date has the full sequence of activities necessary for the development of an anti-tick vaccine been carried out, from the initiation of feasibility studies and antigen isolation to the final marketing of a commercial, recombinant vaccine. This has been with the recombinant vaccines against *B. microplus*, based on the Bm86 molecule. The process began in 1981 and the registered vaccine was finally delivered to the market in 1994. It was released in Australia under the trade name TickGARD and subsequently TickGARD Plus, while the same antigen formed the basis of vaccines manufactured in Cuba as Gavac and Gavac Plus. Six years were consumed in the demonstration of the feasibility of the approach and for the isolation of a single and effective native antigen. Subsequent development and registration took about 8 years. This is not unusual by the standards of pesticide or pharmaceutical development. While components of the research could be circumvented or shortened today, this development remains a prototype for such vaccine projects and is discussed further in that light.

All aspects of the development have been described in detail a number of times (Cobon & Willadsen, 1990; Tellam *et al.*, 1992; Willadsen *et al.*, 1995). They will therefore be summarized only briefly.

Following prolonged exposure to *B. microplus*, cattle acquire an immunity that in *Bos taurus* is usually only

Table 19.2 *Recombinant tick antigens*

Antigen	Tick species	Result[a]	Reference
Bm86	*B. microplus*	√	Willadsen *et al.* (1995)
Bm91 (carboxy-dipeptidase)	*B. microplus*	√	Willadsen *et al.* (1996)
5′-Nucleotidase	*B. microplus*	?	Liyou (1996)
Antigen B	*B. microplus*	?	Unpublished
Vitellin	*B. microplus*	×	Tellam *et al.* (2002)
P29	*H. longicornis*	√	Mulenga *et al.* (1999)
HL34	*H. longicornis*	√	Tsuda *et al.* (2001)
HLS1	*H. longicornis*	√	Sugino *et al.* (2003)
HLS2	*H. longicornis*	√	Imamura *et al.* (2005)
P27/30	*H. longicornis*	√	You (2004, 2005)
RAS-1 and -2	*R. appendiculatus*	√	Imamura *et al.* (2006)
64TRP	*R. appendiculatus*	√	Trimnell *et al.* (2002)
			Labuda *et al.* (2006)
Immunoglobulin-binding protein	*R. appendiculatus*	?	Wang & Nuttall (1999)
Calreticulin	*A. americanum*	?	Jaworski *et al.* (1995)
Voraxin	*A. hebraeum*	√	Weiss & Kaufman (2004)
4F8	*I. scapularis*	√	Almazán *et al.* (2005*a*, 2005*b*)
4D8 (subolesin)	*I. scapularis*	√	Almazán *et al.* (2005*a*, 2005*b*)
4F8	*I. scapularis*	√	Almazan *et al.* (2005*a*, 2005*b*)

[a] √ Statistically significant effect demonstrated. ? Effect slight or equivocal. × No demonstrable effect.

partially protective, though with some breeds and in particular *Bos indicus*, the efficacy can be higher. The idea of developing a vaccine to mimic naturally acquired immunity was therefore unattractive; instead, work at CSIRO in the early 1980s investigated the tick's internal organs, in particular the gut, as a source of antigens. The rationale was that vaccination with such antigens, followed by uptake of host immunoglobulin, complement and, to a limited extent, cellular components of the host's immune system might lead to damage to the tick. This followed on a number of earlier reports, particularly that of Allen & Humphreys (1979). Once the efficacy of vaccination with crude extracts of semi-engorged adult female ticks had been demonstrated (Kemp *et al.*, 1986) the identification of the antigens responsible proceeded via a complex series of protein fractionations, efficacy being assessed after each step through vaccination trials in cattle. In 1986, this led to the identification of a key antigen, Bm86 (Willadsen, McKenna & Riding, 1988; Willadsen *et al.*, 1989) and subsequently its gene sequence (Rand *et al.*, 1989). Vaccination with microgram amounts of the native antigen significantly reduced the number of ticks engorging, their weight and the conversion of engorged weight into eggs. Taken together, these effects reduced the yield of eggs resulting from a standard larval infestation by about 90% relative to infestations on control cattle. There were observable effects on the viability of larvae hatching from 'vaccinated eggs' which would further enhance the vaccine's efficacy, but these were rarely measured. Effects of vaccination were predominantly seen on the late stages of the life cycle, that is on adults and post-engorgement survival and egg-laying, a fact that has consequences for the field application of the vaccine.

The Bm86 molecule

The translated sequence coded for a molecule of 650 amino acids, a predicted molecular weight of 71.7 kDa as an unprocessed protein, containing four potential N-linked glycosylation sites, a leader peptide suggestive of transport to the surface of the cell and a single transmembrane segment, located at the C-terminus. In the mature protein, this transmembrane sequence is replaced by a glycosylphosphatidyl inositol anchor (Richardson *et al.*, 1993). The sequence also contained eight epidermal growth factor (EGF)-like domains (Tellam *et al.*, 1992). Unfortunately, there is as yet no similar sequence in the protein databases that might suggest a potential function. Closest is a portion of the Xotch protein, the neurogenic locus Notch-like protein from

Xenopus, where there is a 23% sequence identity and 32% sequence similarity over 620 amino acids. It has been shown that the antigen localizes to the surface of the tick gut digest cells (Gough & Kemp, 1993). The antigen is present throughout the life cycle, from pre-moulting eggs to the engorged adult (P. Willadsen & R. V. McKenna, unpublished data). The fact that vaccine effects are largely confined to the adult ticks may be a consequence of the volume of blood ingested rather than the presence or absence of antigen.

Little has been done to define the protective epitopes in the molecule. Early expression of fragments of the antigen as β-galactosidase fusion proteins showed that several, non-overlapping but large fragments of the molecule could protect (Tellam *et al.*, 1992). The site of at least one of the protective B-cell epitopes was better defined by work of Patarroyo *et al.* (2002). Three peptides, covering amino acids 21–35, 132–145 and 398–411 of the Bm86 molecule, were synthesized as linked single peptides in three variant structures. These immunogens induced antibodies that in two out of three cases reacted strongly with tick gut epithelial cells, and provided protection in excess of 72%. This was claimed to be better than the Cuban recombinant vaccine (see below). The first of these peptides, namely residues 21–35, lies in a region of the molecule expressed as a recombinant protein and found to be ineffective (Tellam *et al.*, 1992) suggesting the protective epitope(s) many be one of the other two peptides. It is possible, even likely, that the molecule possesses additional, uncharacterized B-cell epitopes.

Expression of recombinant antigens

The Bm86 protein in the tick is extensively glycosylated and the glycosylation induces a strong antibody response. Fortunately, there is considerable evidence that this has little or no protective effect (Willadsen & McKenna, 1991), as confirmed by the generation of recombinant proteins with efficacy very close or identical to that of the native antigen (see below). Efficacy of the original, unrefolded *E. coli* β-galactosidase fusion protein (Rand *et al.*, 1989) was improved following a refolding procedure (Tellam *et al.*, 1992). Little further increase in efficacy was observed with protein expressed in insect cells (Tellam *et al.*, 1992). The antigen has also been expressed in three yeasts: *Aspergillus nidulans*, *A. niger* (Turnbull *et al.*, 1990) and *Pichia pastoris* (Montesino *et al.*, 1996; Canales *et al.*, 1997). Expression and evaluation of the recombinant antigen has also been reported from Egypt where, with native cattle, it had very high efficiency (Khalaf, 1999).

Immunology of the *B. microplus* vaccine

An understanding of the immunological basis of a vaccine is important for two reasons. Firstly and most obviously, it is likely that any recombinant vaccine will require some degree of immunological optimization and knowledge of the desirable immunological response should make such optimization more efficient. Less obviously, registration of a vaccine in many countries requires a great deal of (expensive) data on efficacy under various circumstances. If a good immune correlate of protection can be found, it may be a cost-effective surrogate for more difficult or expensive forms of data collection. This was the case for the *B. microplus* vaccine.

Current understanding of the mechanisms underlying vaccine-induced immunity has been summarized previously (Tellam *et al.*, 1992) and no new knowledge has been added over recent years. As might be expected from the proposed mode of action of the vaccine, namely a direct effect of ingested blood on the structural integrity of the tick gut, the reaction is antibody dependent. The efficacy is a linear function of the log anti-Bm86 titre over a wide concentration range (Willadsen *et al.*, 1995; Rodriguez *et al.*, 1995*b*; de la Fuente *et al.*, 1998). The effects of other immune components have been examined using an in vitro tick feeding assay, an imperfect model of the whole host–parasite interaction. Studies with antisera to complex mixtures of antigens show an absence of effect following supplementation with blood leukocytes but an enhancement of gut damage in the presence of complement (Kemp *et al.*, 1989). Antibody to Bm86 alone and in the absence of complement inhibits the endocytotic activity of gut digest cells, though this is unlikely to explain the vaccine's efficacy (Hamilton *et al.*, 1991).

Variability in immunological response to vaccination in an outbred population of domestic animals is one of the critical issues in vaccine development. Sitte *et al.* (2002) studied the link between antibody responses to the earliest version of the Bm86 vaccine and variation in two bovine major histocompatibility complex (MHC) class II alleles. Although the effect of a homozygous deletion in one of the alleles correlated with higher antibody responses, it accounted for only a small proportion of the total animal-to-animal variation in antibody response.

Vaccine efficacy is dependent on achieving and maintaining a strong antibody response. Except for detailed (unpublished) adjuvant trials in Australia, little has been done to sustain high-level antibody responses. These trials evaluated more than 40 different adjuvant formulations in cattle and over 50 in model animals, without identifying anything superior to a variant of a conventional oil formulation (Cobon, 1997). More recently, biodegradable microspheres have been used to deliver a peptide-based Bm86 vaccine, but the efficacy was less than that of a conventional formulation (Sales-Junior *et al.*, 2005). It has been claimed that the Bm86 molecule, expressed in *P. pastoris*, has immunostimulatory activity, for example, increasing antibody responses to infectious rhinotracheitis virus in cattle (Garcia-Garcia *et al.*, 1998). The potential of DNA vaccination with a Bm86-bearing plasmid alone and co-administered with granulocyte-macrophage colony-stimulating factor (GMCSF) and interleukin 1 beta (IL-1β) plasmids has been examined in sheep as a model, without striking results. The possibility of DNA prime and protein boost too was examined, though two vaccinations with recombinant protein remained the most effective schedule (de Rose *et al.*, 1999).

Field application of the Bm86-based vaccines

The mode of action of the vaccine is critical to its on-farm application. From the earliest experiments it was obvious that there was relatively little or no effect on larvae, some effect on nymphs, but a significant effect on adults. There was a reduction in the number of adults engorging to maturity together with high levels of adult mortality post-engorgement, and a significant effect on the egg-laying capacity of the adult ticks which survived. The impact of this mode of action on the field application of the vaccine has been discussed (Willadsen, 2004). In Australia, the vaccine was released with the recommendation that it be combined with the strategic use of acaricides, should cattle be heavily tick-infested at the time of primary or booster vaccination. A simple vaccination schedule, operated in a closed system (that is, without the continual reintroduction of ticks from outside sources) was shown in the first field trials with the prototype of the commercial vaccine to achieve good tick control. This was in the face of heavy initial pasture infestation with larval ticks and in the absence of acaricide usage (Willadsen *et al.*, 1995). In a more realistic field situation, Jonsson *et al.* (2000) subsequently showed that vaccination of Holstein cattle led to a 56% reduction in tick numbers in a field infestation in a single generation, a 72% reduction in reproductive efficiency of the ticks as measured in the laboratory, and a 18.6 kg higher live weight gain in vaccinated cattle over a 6-month period. Curiously, the vaccinated cattle also tended to have lower somatic cell counts in the milk.

Given that under practical field conditions most farmers are likely to use some acaricide jointly with the vaccine, at least in the early years after vaccine adoption, a de facto measure of vaccine efficacy became the reduction in acaricide usage following the introduction of vaccination as a new control measure. The vaccine and acaricide then constitute a simple integrated tick management system. For example, in on-farm trials in 1996–97 on 26 beef cattle properties, each booster vaccination saved on average 2.4 chemical treatments, while a quarter of the cattle properties found that after vaccination they needed no acaricide treatment at all. The combination of vaccine and acaricide, though more complex than traditional practice, was acceptable to 90% of farmers (G. S. Cobon, personal communication). It should have a number of long-term benefits, including a reduced probability of the development of acaricide resistance. Frequency of acaricide resistance correlates positively with frequency of treatment (Jonsson, Mayer & Green, 2000).

There has been a number of reports of the use of Bm86 vaccines in Central and South America. Early results using a Cuban vaccine appeared to be significant statistically, but not very striking in terms of actual impact on tick numbers (Rodriguez et al., 1995b). Somewhat similar results were obtained initially in Brazil (Rodriguez et al., 1995a) where there was approximately a 50% reduction in tick numbers across a number of cattle properties. Subsequent experience has seemed to show an improvement, with figures of 55–100% efficacy in the control of B. microplus infestations in grazing cattle for up to 36 weeks in controlled field trials in Cuba, Brazil, Argentina and Mexico (de la Fuente et al., 1998, 1999). Almost 100% effective control of B. microplus populations resistant to pyrethroids and organophosphates has also been reported using an integrated system of vaccination and amidine acaricide (Redondo et al., 1999). Simulation models for the effect of vaccine and for vaccine–acaricide combinations have been developed (Lodos et al., 1999; Lodos, Boue & de la Fuente, 2000). A later, retrospective analysis of the consequences of the introduction of the Gavac vaccine in Cuba showed an 87% reduction of acaricide treatments, an 82% reduction in the national consumption of acaricides and an overall reduction in the incidence of clinical babesiosis. In part these changes were a consequence of a change from attempted total eradication of ticks prior to the introduction of the vaccine to a system where treatment was carried out only when the number of adult ticks per animal exceeded 10. Despite some complexities in the interpretation, the large number of cattle involved – in excess of half a million – allowed the authors to be confident in interpreting the observations (Valle et al., 2004). These results with the Gavac vaccine are mirrored in more limited reports with the Australian vaccine TickGARD in South America. Hungerford et al. (1995) reported an 89% reduction in pasture contamination with ticks after one season of vaccinations and as a result a reduction in acaricide treatments in a combined programme.

Sequence variation in Bm86

Given the history of pesticide resistance in B. microplus, the question immediately arises whether resistance to the vaccine will occur and, if so, how rapidly. It has certainly been the experience in both Australia and South America that some isolate-to-isolate differences in vaccine susceptibility occur, though there is no evidence for a decrease in efficacy under the selection pressure of vaccination. A related question is the extent of sequence variation in the Bm86 molecule and whether it affects vaccine susceptibility, both within B. microplus and in heterologous protection across species.

As part of the registration process for the Bm86 vaccine in Australia, the efficacy of the vaccine, derived from the Yeerongpilly or Y strain of B. microplus, was assessed against a number of tick isolates. In parallel, a number of Bm86 genes were sequenced by Biotech Australia. Unfortunately, these data are unpublished. However, up to 17 amino acid substitutions in the 660 of the coding sequence have been reported (Cobon, 1997). These formed an apparently random pattern. Bm86 from ticks sourced from Mexico, Venezuela and Argentina has also been sequenced (Cobon et al., 1996; Garcia-Garcia et al., 1999). Using the original sequence as a reference (derived from the Y strain), these showed up to 30 amino acid differences. Only two changes in the Argentinian sequence were not found in the sequences from Mexico and Venezuela, which is relevant to the discussion that follows. Interestingly, the strain from Cuba (presumably the Camcord strain) that formed the basis of the original Gavac vaccine showed only three base changes from the published Y sequence and only one amino acid change (Rodriguez et al., 1994). The most extensive data on sequence variation however was published by Sossai et al. (2005). A 794-bp segment of the gene was sequenced from 30 B. microplus strains obtained from Brazil, Argentina, Uruguay, Venezuela and Colombia. Variations of 3.4–6.1% occurred in the protein sequence relative to the Y strain (source of the original vaccine), and 1.1–4.6% for Bm95 (a variant of Bm86 from an Argentinian strain). Up to 6.1% differences were observed between clones from the same strain.

Table 19.3 *Relationship between vaccine efficacy and Bm86 sequence, relative to Yeerongpilly (Camcord) Bm86*

B. microplus isolate	Sequence difference	Efficacy[a]	Reference
Indooroopilly Y	0% (reference sequence)	89%	Tellam et al. (1992)
Tuxpan (Mexico)	5.7% ($^2/_{35}$)	51%	Garcia-Garcia et al. (1999)
Mora (Mexico)	8.6% ($^3/_{35}$)	58%	Garcia-Garcia et al. (1999)
Mexico	3.3% ($^{22}/_{660}$)	Same as Y	Cobon (1997)
Argentina (field)		55%	Lamberti et al. (1995)
Argentina A	1.6% ($^{10}/_{610}$ or $^{21}/_{610}$)	10%	Garcia-Garcia et al. (2000)
Camcord (Cuba)	0.2% ($^{21}/_{609}$)	84%	Montesino et al. (1996)
Columbia		81%[b]	Patarroyo et al. (2002)

[a] Vaccinations for Indooroopilly, Mexico and Columbia trials used antigens based on the original Yeerongpilly sequence. The others were based on Camcord Bm86.

[b] The immunogen was a 43-residue peptide composed of three peptides from the original Bm86 sequence. The sequence difference in the challenge strain is unknown but likely to be similar to other South American strains.

The critical question is the degree to which sequence variation, which has been recognized for 15 years, correlates with variation in vaccine efficacy which, likewise, has been long recognized. Significant variation in vaccine efficacy (~70% to ~90%) was seen in Australian tick isolates but there was no correlation between the degree of sequence divergence and vaccine efficacy, and no evidence that a single substitution or set of substitutions led to a loss of efficacy. In South America it was found that, although Cuban and Mexican tick isolates showed efficacies ranging from 51% to 88%, a laboratory strain of tick from Argentina, the A strain, showed efficacy close to zero (de la Fuente et al., 1999; Garcia-Garcia et al., 1999). Sequence differences were considered responsible for the variation in vaccine efficacy. Expressed Argentinian Bm86 was found to control the Argentinian laboratory and field strains (Garcia-Garcia et al., 2000). However, it has also been reported that there are no significant differences in vaccine efficacy between Mexican and Y strain Bm86 molecules (Cobon, 1997) assessed against ticks from Australia and Mexico, while the published sequence of Bm95 shows only two amino acid changes which are not found in the Mexican and Venezuelan sequences. Similarly, unpublished trials of 'Y strain' Bm86 against an Argentinian B. microplus showed reasonable efficacy as did earlier trials with Gavac. Interpretation of existing data on vaccine efficacy is made more difficult by the often poor or non-existent information on the tick challenge isolates. The available, fragmentary information on sequence variation and regional vaccine efficacy is listed in Table 19.3

and suggests that a simple link between vaccine efficacy and sequence variability may not exist. This is discussed in more detail in Willadsen (2004). Clearly more experimentation is needed to clarify this important question, though the effort required to do so is substantial. It must be borne in mind always that the efficacy of a vaccine depends on more than the primary amino acid sequence of the antigen.

A final point to note is that three names have now been proposed for protein sequences differing by no more than 6.1%. These are Bm86 (the original designation), Bm95 (Garcia-Garcia et al., 2000), and *Boophilus microplus* Intestinal Protein or BmIP (Sossai et al., 2005). There is no evidence yet for differences in biological function. Such a proliferation of names is inappropriate and likely to be confusing for anyone not intimately familiar with the literature.

Bm86 in other tick species and estimates of cross-protection

The efficacy of B. microplus Bm86-based vaccines has been tested with several tick species. Almost 100% protection against B. annulatus was reported with both Gavac (Fragoso et al., 1998) and TickGARD (Pipano et al., 2003). Against R. appendiculatus and A. variegatum the effects were slight or undetectable (de Vos et al., 2001). With H. anatolicum and H. dromedarii, however, the effects were impressive though only a limited number of cattle were used (de Vos et al., 2001). As for B. microplus, the information on sequence conservation in these species is exceedingly fragmentary (Table 19.4).

Table 19.4 *Other tick species: sequence conservation and efficacy of vaccination with* B. microplus *Bm86*

Tick	% Sequence identity[a]	Efficacy	Reference
B. microplus (Y)	100%	89%	Tellam *et al.* (1992)
B. decoloratus		70%	de Vos *et al.* (2001)
B. annulatus	97% ($^{34}/_{35}$)	100%	Fragoso *et al.* (1998); Pipano *et al.* (2003)
H. longicornis	48% ($^{146}/_{301}$)		Pickering *et al.* (unpublished)
R. sanguineus	67% ($^{488}/_{631}$)		Pickering *et al.* (unpublished)
R. appendiculatus	78% ($^{114}/_{147}$)	~Zero	de Vos *et al.* (2001)
H. anatolicum	63% ($^{402}/_{632}$)	High	de Vos *et al.* (2001)
H. dromedarii		>98%	de Vos *et al.* (2001)
A. variegatum		0%	de Vos *et al.* (2001)

[a] Numbers in brackets give the total number of amino acid identities as a fraction of the total number sequenced.

Even the limited information suggests that cross-protection against other tick species is not simply a function of the degree of sequence conservation. However, it is reasonable to assume that protection with the homologous form of Bm86 in each tick species will be better than that from the heterologous cross-protection measured to date. Cross-protection with Bm86 amply demonstrates that the probability of success is sufficiently high that investigation of cross-species efficacy of any antigen is justified and desirable.

Additivity and synergy: approaches to improved efficacy

Improvements in the efficacy of existing vaccines against *B. microplus* may be achieved through the incorporation of more than one antigen into the vaccine and/or by the joint use of vaccine and pesticide.

Vaccination of cattle with partially purified tick extract is much more efficacious than vaccination with Bm86 alone (Willadsen, McKenna & Riding, 1988). This could be due to differences in adjuvant, to factors like glycosylation which are not replicated in recombinant proteins, or to the presence of multiple antigens. Trials with both recombinant Bm91 and Bm86 in conjunction (Willadsen *et al.*, 1996) or recombinant Bm86 and native BMA 7 (McKenna *et al.*, 1998) have shown that increased efficacy is possible. There is weak evidence that recombinant tick 5'-nucleotidase together with Bm86 may show a similar effect (Liyou, 1996). However, the degree that vaccine efficacy for any tick species is increased through the use of multi-antigen mixtures is an important but little investigated question, and there may even be no effect. For

example, vaccination of sheep with a combination of three *I. scapularis* recombinant or synthetic antigens (4D8, 4F8 and 4E6, described above) did not give results that were superior to the three antigens applied separately (Almazán *et al.*, 2005*b*).

The interaction between vaccine and acaricide, used jointly in a field situation, has been described above. Interestingly, some of the newer macrocyclic lactone acaricides, when applied on vaccinated cattle, seem to show a 10-fold or greater increase in efficacy (Kemp *et al.*, 1999). This must reflect, in part, the additive effect of two independent control measures, but may also be a consequence of increased penetration of pesticide to its target as a result of damage to the tick's gut. Further investigations are needed.

THE BIOLOGY OF VACCINE SUSCEPTIBILITY

Tick–host interactions are both complex and specific. Hence, there is every reason to expect that the efficacy of a vaccine will depend on a number of biological variables that may differ for each antigen and each host. Unfortunately, only fragmentary observations have been reported. For example, sheep are a host for *B. microplus*, though not a good one. Anecdotal evidence suggests that even large larval infestations are likely to yield very few engorged adult ticks. However, transfer of freshly moulted adults onto shaved or clipped sheep skin results in rapid attachment and the successful engorgement and egg-laying of a high proportion of ticks (de Rose *et al.*, 1999). When sheep were immunized with the Bm86 vaccine, the efficacy was 99% or 10-fold better than with cattle (de Rose *et al.*, 1999). This is unlikely to be attributable to antibody titre. Even climatic conditions may influence vaccine performance. Using an in vitro membrane feeding system, gut damage in young adult ticks feeding on a standard anti-Bm86 antiserum was increased by high environmental temperature and low humidity (Hamilton *et al.*, 1991). Whether this has any relevance to field conditons is unknown.

EFFECT OF VACCINATION ON TRANSMISSION OF VECTOR-BORNE PATHOGENS

In principle, tick vaccines could affect transmission of disease agents in at least two ways. By affecting vector numbers, they could directly influence disease incidence. The effects might be positive, or even deleterious if vaccination against ticks prevented the achievement of endemic stability to tick-borne disease. Since it is increasingly clear that pathogen transmission can involve complex interactions between host, vector and disease organism, it is possible that by disturbing the tick, the vaccine also more subtly but more directly affects the disease.

Field data from Cuba show a significant reduction in the incidence of babesiosis due to *Babesia bovis* and anaplasmosis following sustained use of the Bm86 vaccine Gavac. In a retrospective study, prolonged vaccine usage led to a decline in the incidence of tick-borne disease in some but not all areas (de la Fuente *et al.*, 1998) but eventually to a reduction of 97%, a surprising but intriguing result (Valle *et al.*, 2004). More strikingly, there is evidence, though on a smaller scale, that the use of the Australian vaccine in cattle infested with *B. annulatus* prevents the transmission of *Babesia bigemina* and reduces the frequency or severity of disease due to *Babesia bovis* (Pipano *et al.*, 2003). Tentatively this may be attributed to the fact that with this tick species, in contrast to *B. microplus*, the engorgement of both larvae and nymphs, the stages that transmit these pathogens, is severely affected by the vaccine.

Striking results have been reported for 64TRP, a recombinant 15-kDa cement-like protein derived from *R. appendiculatus* (Labuda *et al.*, 2006). As described above, 64TRP affords cross-protection against adult and immature stages of *I. ricinus*. The vaccine's ability to protect mice against tick-borne encephalitis virus was measured using three parameters: transmission, or the number of uninfected nymphs that became infected when fed on immunized mice infested with infected ticks; 'support' or the percentage of

mice challenged with an infected tick that subsequently supported transmission; and survival of mice challenged with a lethal dose of the virus. Immunization with a variety of recombinant 64TRP constructs singly or in combination significantly affected all three parameters, overall performance equalling or exceeding that of a tick-borne encephalitis virus vaccine. As a third treatment, mice were also vaccinated with the TickGARD vaccine. This too reduced transmission and support but did not increase mouse survival. The 64TRP-vaccinated mice showed strong cellular infiltration into the skin on tick infestation, and the effects on viral parameters were tentatively attributed to this reaction. It remains to be determined whether vaccination with the *I. ricinus* homologues of these antigens is more effective than vaccination with the heterologous antigens.

THE FUTURE OF ANTI-TICK VACCINES

The development of tick vaccines has reached a curious stage. Although the field has long passed the proof of concept stage, existing vaccines remain the only examples of commercially available, recombinant vaccines against a parasite. There is adequate if still limited evidence that discoveries in the field are likely to have broad applicability across tick species. Yet progress since the first release of the recombinant vaccines in 1994 has been disappointing. A number of reasons for this limited progress, both scientific and commercial, can be considered.

Scientifically, although the number of candidate vaccine antigens has increased rapidly over recent years, there are still few reports of their assessment in vaccination trials. The portfolio of demonstrably effective antigens therefore remains small. Of greater concern is the fact that, while a number of antigens show significant effects, few are highly efficacious on their own. This suggests multi-antigen vaccines may be required. Reliance on multi-antigen vaccines would involve both important scientific assumptions, as discussed above, as well as commercial challenges.

There is strong imperative to understand what constitutes a protective immunological response by the host. Lack of rapid progress in understanding the biology of tick immunity can be at least partially attributed to the fact that the number of scientists engaged in the field remains very small.

Commercial issues also play a major part in limiting progress. As our knowledge of both protective antigens and protective mechanisms inevitably increases, and as the generic potential of each new antigen become apparent, the cost and time taken in the research phase of vaccine development will decrease. The costs of development from the laboratory to the registered product will remain unchanged. These are sufficiently high to deter a company from developing a vaccine for any but the most lucrative markets, the 'orphan vaccine' phenomenon. This problem is in no way unique to anti-tick vaccines. It applies to the majority of anti-parasite vaccines and to much else besides. The issue has become starkly apparent for tick vaccines, given their success in being the first anti-ectoparasite vaccines to make the full journey from laboratory to the farm. If vaccines for 'minor' but still economically important parasites are ever to reach the farmers of the world, innovative science will need to be matched to creative commercial and regulatory solutions to the orphan issue.

CONCLUSIONS

There is now abundant evidence that vaccination with defined protein antigens is able to induce significant immunity to tick infestation. In a limited number of cases, this immunity has been duplicated by vaccination with recombinant antigens, a critical step on the pathway to commercial vaccine production. The existence of two commercial vaccines has allowed a number of field studies showing that they can make an important contribution to an integrated approach to the control of ticks in the field. Under most circumstances however, the use of a tick vaccine as the single, stand-alone control technology is likely to require more efficacious vaccines than those currently available. Increases in efficacy are most likely to come through the discovery of additional, effective vaccine antigens. The number of antigens with demonstrated effect is increasing, though only slowly, while the number of potential antigens that remain to be evaluated is increasing more quickly. There is limited, though convincing, evidence that some of these antigens will show effective cross-species protection, though in a poorly understood and unpredictable way. The groundwork has been laid; the scientific potential is still to be effectively exploited.

REFERENCES

Allen, J. R. & Humphreys, S. J. (1979). Immunization of guinea pigs and cattle against ticks. *Nature* 280, 491–493.

Aljamali, M. N., Bior, A. D., Sauer, J. R. & Essenberg, R. C. (2003). RNA interference in ticks: a study using histamine binding protein dsRNA in the female tick *Amblyomma americanum*. *Insect Molecular Biology* 12, 299–305.

Almazán, C., Blas-Machado, U., Kocan, K. M., et al. (2005a). Characterization of three *Ixodes scapularis* cDNAs protective against tick infestations. *Vaccine* 23, 4403–4416.

Almazán, C., Kocan, K. M., Bergman, D. K., et al. (2003). Identification of protective antigens for the control of *Ixodes scapularis* infestations using cDNA expression library immunization. *Vaccine* 21, 1492–1501.

Almazán, C., Kocan, K M., Blouin, E. F. & de la Fuente, J. (2005b). Vaccination with recombinant tick antigens for the control of *Ixodes scapularis* adult infestations. *Vaccine* 23, 5294–5298.

Andreotti, R., Gomes, A., Malavazi-Piza, K. C., et al. (2002). BmTI antigens induce a bovine protective immune response against *B. microplus* ticks. *International Immunopharmacology* 2, 557–563.

Andreotti, R., Sampaio, C. A. M., Gomes, A. & Tanaka, A. S. (1999). A serine proteinase inhibitor immunoprotection from *B. microplus* unfed larvae in calves. In *IV Seminario Internacional de Parasitologia Animal*, 20–22 October 1999, Puerto Vallarta, Mexico, pp. 75–86.

Barriga, O. O. (1999). Evidence and mechanisms of immunosuppression in tick infestations. *Genetic Analysis – Biomolecular Engineering* 15, 139–142.

Bergman, D. K., Palmer, M. J., Caimano, M. J., Radolf, J. D. & Wikel, S. K. (2000). Isolation and molecular cloning of a secreted immunosuppressant protein from *D. andersoni* salivary gland. *Journal of Parasitology* 86, 516–525.

Bishop, R., Lambson, B., Wells, C., et al. (2002). A cement protein of the tick *Rhipicephalus appendiculatus*, located in the secretory e cell granules of the type III salivary gland acini, induces strong antibody responses in cattle. *International Journal for Parasitology* 32, 833–842.

Brossard, M. & Wikel, S. K. (2004). Tick immunobiology. *Parasitology* 129, S161–S176.

Canales, M., Enriquez, A., Ramos, E., et al. (1997). Large-scale production in *Pichia pastoris* of the recombinant vaccine Gavac against the cattle tick. *Vaccine* 15, 414–422.

Cobon, G. S. (1997). An anti-arthropod vaccine: TickGARD – a vaccine to prevent cattle tick infestations. In *New Generation Vaccines*, eds. Levine, M. M., Woodrow, G. C., Kaper, J. B. & Cobon, G. S., pp. 1145–1151. New York: Marcel Dekker.

Cobon, G. S. & Willadsen, P. (1990). Vaccines to prevent cattle tick infestations. In *New Generation Vaccines*, eds. Woodrow, G. C. & Levine, M. M., pp. 901–917. New York: Marcel Dekker.

Cobon, G. S., Moore, J. T., Johnston, L. A. Y., et al. (1996). DNA encoding a cell membrane glycoprotein of a tick gut.

US Patent and Trademark Office Appl. No. 325 071 435/240.2.

Das, S., Banerjee, G., Deponte, K., *et al.* (2001). Salp25D, an *Ixodes scapularis* antioxidant, is 1 of 14 immunodominant antigens in engorged tick salivary glands. *Journal of Infectious Diseases* **184**, 1056–1064.

Da Silva Vaz, I. Jr, Logullo, C., Sorgine, M., *et al.* (1998). Immunization of bovines with an aspartic proteinase precursor isolated from *B. microplus* eggs. *Veterinary Immunology and Immunopathology* **66**, 331–341.

de la Fuente, J., Almazán, C., Blas-Machado, U., *et al.* (2006a). The tick protective antigen, 4D8, is a conserved protein involved in modulation of tick blood ingestion and reproduction. *Vaccine* **24**, 4082–4095.

de la Fuente, J., Almazán, C., Blouin, E. F., Naranjo, V. & Kocan, K. M. (2005). RNA interference screening in ticks for identification of protective antigens. *Parasitology Research* **96**, 137–141.

de la Fuente, J., Almazán, C., Naranja, V., Blouin, E. F. & Kocan, K. M. (2006b). Synergistic effect of silencing the expression of tick protective antigens 4D8 and Rs86 in *Rhipicephalus sanguineus* by RNA interference. *Parasitology Research* **99**, 108–113.

de la Fuente, J., Rodriguez, M., Montero, C., *et al.* (1999). Vaccination against ticks (*Boophilus* spp.): the experience with the Bm86-based vaccine Gavac. *Genetic Analysis – Biomolecular Engineering* **15**, 143–148.

de la Fuente, J., Rodriguez, M., Redondo, M., *et al.* (1998). Field studies and cost-effectiveness of vaccination with Gavac against the cattle tick *Boophilus microplus*. *Vaccine* **16**, 366–373.

Del Pino, F. A. B., Brandelli, A., Gonzales, J. C., Henriques, J. A. P. & Dewes, H. (1998). Effect of antibodies against β-*N*-acetylhexosaminidase on reproductive efficiency of the bovine tick *B. microplus*. *Veterinary Parasitology* **79**, 247–255.

De Rose, R., McKenna, R. V., Cobon, G., *et al.* (1999). Bm86 antigen induces a protective immune response against *B. microplus* following DNA and protein vaccination in sheep. *Veterinary Immunology and Immunopathology* **71**, 151–160.

De Vos, S., Zeinstra, L., Taoufik, O., Willadsen, P. & Jongejan, F. (2001). Evidence for the utility of the Bm86 antigen from *B. microplus* in vaccination against other tick species. *Experimental and Applied Acarology* **25**, 245–261.

Ferreira, C. A. S., Da Silva Vaz, I. Jr, Da Silva, S. S., *et al.* (2002). Cloning and partial characterization of a *Boophilus microplus* (Acari: Ixodidae) calreticulin. *Experimental Parasitology* **101**, 25–34.

Fragoso, H., Hoshman-Rad, P., Ortiz, M., *et al.* (1998). Protection against *Boophilus annulatus* infestations in cattle vaccinated with the *B. microplus* Bm86-containing vaccine Gavac. *Vaccine* **16**, 1990–1992.

Garcia-Garcia, J. C., Gonzalez, I. L., Gonzalez, D. M., *et al.* (1999). Sequence variations in the *B. microplus* Bm86 locus and implications for immunoprotection in cattle vaccinated with this antigen. *Experimental and Applied Acarology* **11**, 883–895.

Garcia-Garcia, J. C., Montero, C., Redondo, M., *et al.* (2000). Control of ticks resistant to immunization with Bm86 in cattle vaccinated with the recombinant antigen Bm95 isolated from the cattle tick, *B. microplus*. *Vaccine* **18**, 2275–2287.

Garcia-Garcia, J. C., Soto, A., Nigro, F., *et al.* (1998). Adjuvant and immunostimulating properties of the recombinant Bm86 protein expressed in *Pichia pastoris*. *Vaccine* **16**, 1053–1055.

Gillespie, R. D., Dolan, M. C., Piesman, J. & Titus, R. G. (2001). Identification of an IL-2 binding protein in the saliva of the Lyme Disease vector tick, *Ixodes scapularis*. *Journal of Immunology* **166**, 4319–4326.

Gough, J. M. & Kemp, D. H. (1993). Localization of a low abundance membrane protein (Bm86) on the gut cells of the cattle tick *B. microplus* by immunogold labelling. *Journal of Parasitology* **79**, 900–907.

Guerrero, P. D., Miller, R. J., Rousseau, M. E., *et al.* (2005). BmiGI: a database of cDNAs expressed in *Boophilus microplus*, the tropical/southern cattle tick. *Insect Biochemistry and Molecular Biology* **35**, 585–595.

Guerrero, F. D., Nene, V. M., George, J. E., Barker, S. C. & Willadsen, P. (2006). Sequencing a new target genome: the Southern Cattle Tick, *Boophilus microplus* (Acari: Ixodidae) genome project. *Journal of Medical Entomology* **43**, 9–16.

Hajnická, V., Kocakova, P., Slavikova, M., *et al.* (2001). Anti-interleukin-8 activity of tick salivary gland extracts. *Parasite Immunology* **23**, 483–489.

Hamilton, S. E., Kemp, D. H., McKenna, R. V. & Willadsen, P. (1991). Gut cells of the tick *B. microplus*: the effect of vaccination on digest cells and experiments on blood meal absorption by these cells. In *Modern Acarology*, vol. 1, eds. Dusbabek, F. & Bukva, V., pp. 341–351. The Hague: SPB Academic.

Hill, C. A. & Wikel, S. K. (2005). The *Ixodes scapularis* Genome Project: an opportunity for advancing tick research. *Trends in Parasitology* **21**, 151–153.

Hungerford, J., Pulga, M., Zwtsch, E. & Cobon, G. (1995). Efficacy of TickGARD™ in Brazil. In *Resistencia y Control en Garrapatas y Moscas de Importancia Veterinaria, III Seminario Internacional de Parasitologia Animal*, 11–13 October 1995, Acapulco, Mexico, p. 139.

Imamura, S., Da Silva Vaz, I. Jr, Sugino, M., Ohashi, K. & Onuma, M. (2005). A serine protease inhibitor (serpin) from *Haemaphysalis longicornis* as an anti-tick vaccine. *Vaccine* **23**, 1301–1311

Imamura, S., Namangala, B., Tajima, T., *et al.* (2006). Two serine protease inhibitors (serpins) that induce a bovine protective immune response against *Rhipicephalus appendiculatus* ticks. *Vaccine* **24**, 2230–2237.

Iwanaga, S., Okada, M., Isawa, H., *et al.* (2003). Identification and characterization of novel salivary thrombin inhibitors from the Ixodidae tick, *Haemaphysalis longicornis*. *European Journal of Biochemistry* **270**, 1926–1934.

Jarmey, J. M., Riding, G. A., Pearson, R. D., McKenna, R. V. & Willadsen, P. (1995). Carboxydipeptidase from *B. microplus*: a 'concealed' antigen with similarity to angiotensin-converting enzyme. *Insect Biochemistry and Molecular Biology* **25**, 969–974.

Jasinskas, A., Jaworski, D. C. & Barbour, A. G. (2000). *Amblyomma americanum*: specific uptake of immunoglobulins into tick hemolymph during feeding. *Experimental Parasitology* **96**, 213–221.

Jaworski, D. C., Jasinskas, A., Metz, C. N., Bucala, R. & Barbour, A. G. (2001). Identification and characterization of a homologue of the pro-inflammatory cytokine macrophage migration inhibitory factor in the tick, *Amblyomma americanum*. *Insect Molecular Biology* **10**, 323–331.

Jaworski, D. C., Simmen, F. A., Lamoreaux, W., *et al.* (1995). A secreted calreticulin protein in ixodid tick (*Amblyomma americanum*) saliva. *Journal of Insect Physiology* **41**, 369–375.

Jaworski, D. C., Simmen, F. A., Lamoreaux, W., *et al.* (2002). A secreted calreticulin protein in ixodid tick (*Amblyomma americanum*) saliva. *Journal of Insect Physiology* **41**, 369–375.

Jonsson, N. N., Matschoss, A. L., Pepper, P., *et al.* (2000). Evaluation of TickGARD(PLUS), a novel vaccine against *B. microplus*, in lactating Holstein–Friesian cows. *Veterinary Parasitology* **88**, 275–285.

Jonsson, N. N., Mayer, D. G. & Green, P. E. (2000). Possible risk factors on Queensland dairy farms for acaricide resistance in cattle tick (*B. microplus*). *Veterinary Parasitology* **88**, 79–92.

Joubert, A. M., Louw, A. I., Joubert, F. & Neitz, A. W. H. (1998). Cloning, nucleotide sequence and expression of the gene encoding factor Xa inhibitor from the salivary glands of the tick, *Ornithodoros savignyi*. *Experimental and Applied Acarology* **22**, 603–619.

Karczewski, J., Endris, R. & Connolly, T. M. (1994). Disagregin is a fibrinogen receptor antagonist lacking the asp-gly-arg sequence from the tick, *Ornithodoros moubata*. *Journal of Biological Chemistry* **269**, 6702–6708.

Kemp, D. H., Agbede, R. I. S., Johnston, L. A. Y. & Gough, J. M. (1986). Immunization of cattle against *B. microplus* using extracts derived from adult female ticks: feeding and survival of the parasite on vaccinated cattle. *International Journal for Parasitology* **16**, 115–120.

Kemp, D. H., McKenna, R. V., Thullner, R. & Willadsen, P. (1999). Strategies for tick control in a world of acaricide resistance. In *IV Seminario Internacional de Parasitologia Animal*, 20–22 October 1999, Puerto Vallarta, Mexico, pp. 1–10.

Kemp, D. H., Pearson, R. D., Gough, J. M. & Willadsen, P. (1989). Vaccination against *B. microplus*: localization of antigens on tick gut cells and their interaction with the host immune system. *Experimental and Applied Acarology* **7**, 43–58.

Khalaf, A. S. S. (1999). Control of *B. microplus* ticks in cattle calves by immunization with a recombinant Bm86 glucoprotein antigen preparation. *Deutsche Tierarztliche Wochenschrift* **106**, 248–251.

Labuda, M., Trimnell, A. R., Ličková, M., *et al.* (2006). An antivector vaccine protects against a lethal vector-borne pathogen. *PLoS Pathogens* **2**, 251–259.

Lamberti, J., Signorini, A., Mattos, C., *et al.* (1995). Evaluation of the recombinant vaccine against grazing cattle in Argentina. In *Recombinant Vaccines for the Control of the Cattle Tick*, ed. J. de la Fuente, pp. 205–227. La Habana, Cuba: Elfos Scientiae.

Lawrie, C. H. & Nuttall, P. A. (2001). Antigenic profile of *I. ricinus*: effect of developmental stage, feeding time and the response of different host species. *Parasite Immunology* **23**, 549–556.

Lawrie, C. H., Randolph, S. E. & Nuttall, P. A. (1999). Ixodes ticks: serum species sensitivity of anticomplement activity. *Experimental Parasitology* **93**, 207–214.

Leboulle, G., Crippa, M., Decrem, Y., *et al.* (2002). Characterization of a novel salivary immunosuppressive protein from *I. ricinus* ticks. *Journal of Biological Chemistry* **277**, 10083–10089.

Lee, R. P. & Opdebeeck, J. P. (1991). Isolation of protective antigens from the gut of *B. microplus* using monoclonal antibodies. *Immunology* **72**, 121–126.

Lee, R. P., Jackson, L. A. & Opdebeeck, J. P. (1991). Immune responses of cattle to biochemically modified antigens from the midgut of the cattle tick, *B. microplus*. *Parasite Immunology* **13**, 661–672.

Liyou, N. (1996). An investigation of the 5′-nucleotidase from the cattle tick *B. microplus*. Unpublished Ph.D. thesis. University of Queensland, Brisbane, Australia.

Liyou, N., Hamilton, S., Elvin, C. & Willadsen, P. (1999). Cloning and expression of ecto 5′-nucleotidase from the cattle tick *B. microplus*. *Insect Molecular Biology* **8**, 257–266.

Liyou, N., Hamilton, S., McKenna, R., Elvin, C. & Willadsen, P. (2000). Localization and functional studies on the 5′-nucleotidase of the cattle tick *B. microplus*. *Experimental and Applied Acarology* **24**, 235–246.

Lodos, J., Boue, O. & de la Fuente, J. (2000). A model to simulate the effect of vaccination against *Boophilus* ticks on cattle. *Veterinary Parasitology* **87**, 315–326.

Lodos, J., Ochagavia, M. E., Rodriguez, M. & de la Fuente, J. (1999). A simulation study of the effects of acaricides and vaccination on *Boophilus* cattle-tick populations. *Preventative Veterinary Medicine* **38**, 47–63.

Logullo, C., Moraes, J., Dansa-Petretski, M., *et al.* (2002). Binding and storage of heme by vitellin from the cattle tick, *B. microplus*. *Insect Biochemistry and Molecular Biology* **32**, 1805–1811.

Madden, R. D., Sauer, J. R. & Dillwith, J. W. (2002). A proteomics approach to characterizing tick salivary secretions. *Experimental and Applied Acarology* **28**, 77–87.

Mans, B. J. & Neitz, A. W. H. (2004). Adaptation of ticks to a blood-feeding environment: evolution from a functional perspective. *Insect Biochemistry and Molecular Biology* **34**, 1–17.

Mans, B. J., Louw, A. I. & Neitz, A. W. H. (2002). Disaggregation of aggregated platelets by savignygrin, an $\alpha_{IIb}\beta_3$ antagonist from *Ornithodoros savignyi*. *Experimental and Applied Acarology* **27**, 231–239.

McKenna, R. V., Riding, G. A., Jarmey, J. M., Pearson, R. D. & Willadsen, P. (1998). Vaccination of cattle against the tick *B. microplus* using a mucin-like membrane glycoprotein. *Parasite Immunology* **20**, 325–336.

Montesino, R., Cremata, J., Rodriguez, M., *et al.* (1996). Biochemical characterization of the recombinant *B. microplus* Bm86 antigen expressed by transformed *Pichia pastoris* cells. *Biotechnology and Applied Biochemistry* **23**, 23–28.

Mulenga, A., Sugimoto, C., Sako, Y., *et al.* (1999). Molecular characterization of a *Haemaphysalis longicornis* tick salivary gland-associated 29-kilodalton protein and its effect as a vaccine against tick infestation in rabbits. *Infection and Immunity* **67**, 1652–1658.

Mulenga, A., Sugimoto, C., Ohashi, K. & Onuma, M. (2000). Characterization of an 84 kDa protein inducing an immediate hypersensitivity reaction in rabbits sensitized to *Haemaphysalis longicornis* ticks. *Biochimica et Biophysica Acta* **1501**, 219–226.

Mulenga, A., Sugino, M., Nakajima, M., Sugimoto, C. & Onuma, M. (2001). Tick-encoded serine proteinase inhibitors (Serpins); potential target antigens for tick vaccine development. *Journal of Veterinary Medical Science* **63**, 1063–1069.

Mulenga, A., Tsuda, A., Onuma, M. & Sugimoto, C. (2003). Four serine proteinase inhibitors (serpin) from the brown ear tick, *Rhipicephalus appendiculatus*: cDNA cloning and preliminary characterization. *Insect Biochemistry and Molecular Biology* **33**, 267–276.

Mulenga, A., Tsuda, A., Sugimoto, C. & Onuma, M. (2002). Blood meal acquisition by ticks: molecular advances and implications for vaccine development. *Japanese Journal of Veterinary Research* **49**, 261–272.

Nakajima, M., Yanase, H., Iwanaga, T., *et al.* (2003). Passive immunization with monoclonal antibodies: effects on *Haemaphysalis longicornis* tick infestation of BALB/c mice. *Japanese Journal of Veterinary Research* **50**, 157–163.

Nuttall, P. A., Trimnell, A. R., Kazimirova, M. & Labuda, M. (2006). Exposed and concealed antigens as vaccine targets for controlling ticks and tick-borne diseases. *Parasite Immunology* **28**, 155–163.

Oliveira, M. F., Silva, J. R., Dansa-Petretski, M., *et al.* (1999). Haem detoxification by an insect. *Nature* **400**, 517–518.

Packila, M. & Guilfoile, P. G. (2002). Mating, male *Ixodes scapularis* express several genes including those with sequence similarity to immunoglobulin-binding proteins and metalloproteases. *Experimental and Applied Acarology* **27**, 151–160.

Paesen, G. C., Adams, P. L., Harlos, K., Nuttall, P. A. & Stuart, D. I. (1999). Tick histamine-binding proteins: isolation cloning and three-dimensional structure. *Molecular Cell* **3**, 661–671.

Paesen, G. C., Adams, P. L., Nuttall, P. A. & Stuart, D. I. (2000). Tick histamine-binding proteins: lipocalins with a second binding cavity. *Biochimica et Biophysica Acta* **1482**, 92–101.

Palmer, M. J., Bantle, J. A., Guo, X. & Fargo, W. S. (1994). Genome size and organization in the ixodid tick *Amblyomma americanum*. *Insect Molecular Biology* **3**, 57–62.

Patarroyo, J. H., Portela, R. W., De Castro, R. O., *et al.* (2002). Immunization of cattle with synthetic peptides derived from the *B. microplus* gut protein (Bm86). *Veterinary Immunology and Immunopathology* 88, 163–172.

Pipano, E., Alekceev, E., Galker, F., *et al.* (2003). Immunity against *Boophilus annulatus* induced by the Bm86 (Tick-GARD) vaccine. *Experimental and Applied Acarology* 29, 141–149.

Pruett, J. H. (1999). Immunological control of arthropod ectoparasites: a review. *International Journal for Parasitology* 29, 25–32.

Rand, K. N., Moore, T., Sriskantha, A., *et al.* (1989). Cloning and expression of a protective antigen from the cattle tick *B. microplus*. *Proceedings of the National Academy of Sciences of the USA* 86, 9657–9661.

Redondo, M., Fragoso, H., Ortiz, M., *et al.* (1999). Integrated control of acaricide-resistant *B. microplus* populations on grazing cattle in Mexico using vaccination with Gavac™ and amidine treatments. *Experimental and Applied Acarology* 23, 841–849.

Richardson, M. A., Smith, D. R. J., Kemp, D. H. & Tellam, R. L. (1993). Native and baculovirus-expressed forms of the immunoprotective protein Bm86 from *B. microplus* are anchored to the cell membrane by a glycosylphosphatidyl inositol linkage. *Insect Molecular Biology* 1, 139–147.

Riding, G., Jarmey, J., McKenna, R. V., *et al.* (1994). A protective 'concealed' antigen from *B. microplus*: purification, localization and possible function. *Journal of Immunology* 153, 5158–5166.

Rodriguez, M., Massard, C. L., Henrique da Fonseca, A., *et al.* (1995a). Effect of a vaccination with a recombinant Bm86 antigen preparation on natural infestations of *B. microplus* in grazing dairy and beef pure and cross-bred cattle in Brazil. *Vaccine* 13, 1804–1808.

Rodriguez, M., Penichet, M. L., Mouris, A. E., *et al.* (1995b). Control of *B. microplus* populations in grazing cattle vaccinated with a recombinant Bm86 antigen preparation. *Veterinary Parasitology* 57, 339–349.

Rodriguez, M., Rubiera, R., Penichet, M. L., *et al.* (1994). High level expression of the *B. microplus* Bm86 antigen in the yeast *Pichia pastoris* forming highly immunogenic particles for cattle. *Journal of Biotechnology* 33, 135–146.

Sales-Junior, P. A., Guzman, F., Vargas, M. I., *et al.* (2005). Use of biodegradable PLGA microspheres as a slow release delivery system for the *Boophilus microplus* synthetic vaccine SBm7462. *Veterinary Immunology and Immunopathology* 107, 281–290.

Sangamnatdej, S., Paesen, G. C., Slovak, M. & Nuttall, P. A. (2002). A high affinity serotonin- and histamine-binding lipocalin from tick saliva. *Insect Molecular Biology* 11, 79–86.

Sauer, J. R., McSwain, J. L. & Essenberg, R. C. (1994). Cell membrane receptors and regulation of cell function in ticks and blood-sucking insects. *International Journal for Parasitology* 24, 33–52.

Sharma, J. K., Ghosh, G., Khan, M. H. & Das, G. (2001). Immunoprotective efficacy of a purified 39 kDa nymphal antigen of *Hyalomma anatolicum anatolicum*. *Tropical Animal Health and Production* 33, 103–116.

Sitte, K., Brinkworth, R., East, I. J. & Jazwinska, E. C. (2002). A single amino acid deletion in the antigen binding site of BoLA-DRB3 is predicted to affect peptide binding. *Veterinary Immunology and Immunopathology* 85, 129–135.

Sossai, S., Peconick, A. P., Sales-Junior, P. A., *et al.* (2005). Polymorphism of the *bm*86 gene in South American strains of the cattle tick *Boophilus microplus*. *Experimental and Applied Acarology* 37, 199–214.

Sugino, M., Imamura, S., Mulenga, A., *et al.* (2003). A serine proteinase inhibitor (serpin) from the ixodid tick *Haemaphysalis longicornis*: cloning, and preliminary assessment of its suitability as a candidate for a tick vaccine. *Vaccine* 21, 2844–2851.

Tanaka, A. S., Andreotti, R., Gomes, A., *et al.* (1999). A double headed serine proteinase inhibitor – human plasma kallikrein and elastase inhibitor – from *B. microplus* larvae. *Immunopharmacology* 45, 171–177.

Tellam, R. L., Kemp, D., Riding, G., *et al.* (2002). Reduced oviposition of *B. microplus* feeding on sheep vaccinated with vitellin. *Veterinary Parasitology* 103, 141–156.

Tellam, R. L., Smith, D., Kemp, D. H. & Willadsen, P. (1992). Vaccination against ticks. In *Animal Parasite Control Utilizing Biotechnology*, ed. Yong, W. K., pp. 303–331. Boca Raton, FL: CRC Press.

Titus, R. G., Bishop, J. V. & Mejia, J. S. (2006). The immunomodulatory factors of arthropod saliva and the potential of these factors to serve as vaccine targets to prevent pathogen transmission. *Parasite Immunology* 28, 131–141.

Trimnell, A. R., Davies, G. M., Lissina, O., Hails, R. S. & Nuttall, P. A. (2005). A cross-reactive tick cement antigen is a candidate broad-spectrum tick vaccine. *Vaccine* 23, 4329–4341.

Trimnell, A. R., Hails, R. S. & Nuttall, P. A. (2002). Dual action ectoparasite vaccine targeting 'exposed' and 'concealed' antigens. *Vaccine* 20, 3360–3568.

Tsuda, A., Mulenga, A., Sugimoto, C., *et al.* (2001). cDNA cloning, characterization and vaccine effect analysis of *Haemaphysalis longicornis* tick saliva proteins. *Vaccine* 19, 4287–4296.

Turnbull, I. F., Smith, D. R. J., Sharp, P. J., Cobon, G. S. & Hynes, M. J. (1990). Expression and secretion in *Aspergillus nidulans* and *Aspergillus niger* of a cell surface glycoprotein from the cattle tick *B. microplus*, by using the fungal amdS promoter system. *Applied and Environmental Microbiology* **56**, 2847–2852.

Ullmann, A. J., Lima, C. M. R., Guerrero, F. D., Piesman, J. & Black, W. C. IV (2005). Genome size and organization in the blacklegged tick, *Ixodes scapularis* and the Southern cattle tick, *Boophilus microplus*. *Insect Molecular Biology* **14**, 217–222.

Untalan, P. M., Guerrero, F. D., Haines, L. R. & Pearson, T. W. (2005). Proteome analysis of abundantly expressed proteins from unfed larvae of the cattle tick, *Boophilus microplus*. *Insect Biochemistry and Molecular Biology* **35**, 141–151.

Valenzuela, J. G., Charlab, R., Mather, T. N. & Ribeiro, J. M. C. (2000). Purification, cloning, and expression of a novel salivary anticomplement protein from the tick, *Ixodes scapularis*. *Journal of Biological Chemistry* **275**, 18717–18723.

Valle, M. R., Mèndez, L., Valdez, M., *et al.* (2004). Integrated control of *Boophilus microplus* stocks in Cuba based on vaccination with the anti-tick vaccine Gavac™. *Experimental and Applied Acarology* **34**, 375–382.

Vaughan, J. A., Sonenshine, D. E. & Azad, A. F. (2002). Kinetics of ingested host immunoglobulin G in hemolymph and whole body homogenates during nymphal development of *Dermacentor variabilis* and *Ixodes scapularis* ticks (Acari: Ixodidae). *Experimental and Applied Acarology* **27**, 329–340.

Wang, H. & Nuttall, P. A. (1994*a*). Comparison of the proteins in salivary glands, saliva and haemolymph of *Rhipicephalus appendiculatus* female ticks during feeding. *Parasitology* **109**, 517–523.

Wang, H. & Nuttall, P. A. (1994*b*). Excretion of host immunoglobulin in tick saliva and detection of IgG-binding proteins in tick haemolymph and salivary glands. *Parasitology* **109**, 525–530.

Wang, H. & Nuttall, P. A. (1999). Immunoglobulin-binding proteins in ticks: new target for vaccine development against a blood-feeding parasite. *Cellular and Molecular Life Sciences* **56**, 286–295.

Wang, X., Coons, L. B., Taylor, D. B., Stevens, J. E. Jr & Gartner, T. K. (1996). Variabilin, a novel RGD-containing antagonist of glycoprotein lib-Ilia and platelet aggregation inhibitor from the hard tick *Dermacentor variabilis*. *Journal of Biological Chemistry* **271**, 17785–17790.

Weiss, B. L. & Kaufman, W. R. (2004). Two feeding-induced proteins from the male gonad trigger engorgement of the female tick *Amblyomma hebraeum*. *Proceedings of the National Academy of Sciences of the USA* **101**, 5874–5879.

Wikel, S. (1999). Tick modulation of host immunity: an important factor in pathogen transmission. *International Journal for Parasitology* **29**, 851–859.

Wikel, S. K. & Alarcon-Chaidez, F. J. (2001). Progress toward molecular characterization of ectoparasite modulation of host immunity. *Veterinary Parasitology* **101**, 275–287.

Willadsen, P. (1980). Immunity to ticks. *Advances in Parasitology* **18**, 293–313.

Willadsen, P. (1997). Novel vaccines for ectoparasites. *Veterinary Parasitology* **71**, 209–222.

Willadsen, P. (2001). The molecular revolution in the development of vaccines against ectoparasites. *Veterinary Parasitology* **101**, 353–367.

Willadsen, P. (2004). Anti-tick vaccines. *Parasitology* **129**, S367–S387.

Willadsen, P. & Billingsley, P. F. (1997). Immune intervention against blood-feeding insects. In *The Biology of the Insect Midgut*, eds. Lehane, M. J. & Billingsley, P. F., pp. 323–343. London: Chapman & Hall.

Willadsen, P. & Jongejan, F. (1999). Immunology of the tick–host interaction and the control of ticks and tick-borne diseases. *Parasitology Today* **15**, 258–262.

Willadsen, P. & Kemp, D. H. (1988). Vaccination with 'concealed' antigens for tick control. *Parasitology Today* **4**, 196–198.

Willadsen, P. & McKenna, R. V. (1983). Trypsin chymotrypsin inhibitors from the parasitic tick. *Australian Journal of Experimental Biological and Medical Science* **61**, 231–238.

Willadsen, P. & McKenna, R. V. (1991). Vaccination with 'concealed' antigens: myth or reality? *Parasite Immunology* **13**, 605–616.

Willadsen, P. & Riding, G. A. (1979). Characterization of a proteolytic-enzyme inhibitor with allergenic activity: multiple functions of a parasite-derived protein. *Biochemical Journal* **177**, 41–47.

Willadsen, P. & Riding, G. A. (1980). On the biological role of a proteolytic-enzyme inhibitor from the ectoparasitic tick *B. microplus*. *Biochemical Journal* **189**, 295–303.

Willadsen, P., Bird, P., Cobon, G. S. & Hungerford, J. (1995). Commercialization of a recombinant vaccine against *B. microplus*. *Parasitology* **110**, 43–50.

Willadsen, P., McKenna, R. V. & Riding, G. A. (1988). Isolation from the cattle tick, *B. microplus*, of antigenic material capable of eliciting a protective immunological response in the bovine host. *International Journal for Parasitology* **18**, 183–189.

Willadsen, P., Riding, G. A., McKenna, R. V., *et al.* (1989). Immunologic control of a parasitic arthropod: identification of a protective antigen from *B. microplus*. *Journal of Immunology* **143**, 1346–1351.

Willadsen, P., Smith, D., Cobon, G. & McKenna, R. V. (1996). Comparative vaccination of cattle against *B. microplus* with recombinant antigen Bm86 alone or in combination with recombinant Bm91. *Parasite Immunology* **18**, 241–246.

Xu, Y., Bruno, J. F. & Luft, B. J. (2005). Identification of novel tick salivary gland proteins for vaccine development. *Biochemical and Biophysical Research Communications* **326**, 901–904.

You, M.-J. (2004). Immunization effect of recombinant P27/30 protein expressed in *Escherichia coli* against the hard tick *Haemaphysalis longicornis* (Acari: Ixodidae) in rabbits. *Korean Journal of Parasitology* **42**, 195–200.

You, M.-J. (2005). Immunization of mice with recombinant P27/30 protein confers protection against hard tick *Haemaphysalis longicornis* (Acari: Ixodidae) infestation. *Journal of Veterinary Science* **61**, 47–51.

20 • Anti-tick biological control agents: assessment and future perspectives

M. SAMISH, H. GINSBERG AND I. GLAZER

INTRODUCTION

Since the beginning of the twentieth century investigators have documented numerous potential tick biological control agents, including pathogens, parasitoids and predators of ticks (Jenkins, 1964; Mwangi, 1991; Mwangi et al., 1991; Samish & Rehacek, 1999; Kaaya, 2003; Ostfeld et al., 2006). Several authors have reviewed specific groups of natural enemies of ticks, including pathogens (Lipa, 1971; Hoogstraal, 1977; Chandler et al., 2000), nematodes (Samish, Alekseev & Glazer, 2000a, 2000b; Samish & Glazer, 2001), parasitoids (Cole, 1965; Trjapitzin, 1985; Davis, 1986; Mwangi & Kaaya, 1997; Hu, Hyland & Oliver, 1998; Knipling & Steelman, 2000) and predators (Barre et al., 1991; Mwangi, Newson & Kaaya, 1991; Kok & Petney, 1993; Samish & Alexseev, 2001).

In practice, ticks are controlled at present mostly by chemical acaricides (see Chapter 18). However, biological control is becoming an increasingly attractive approach to tick management because of: (1) increasing concerns about environmental safety and human health (e.g. the gradual increase in use of chemical insecticides in several countries is stimulating the growing market of 'organic' food); (2) the increasing costs of chemical control; and (3) the increasing resistance of ticks to pesticides. To date, biocontrol has been targeted largely at pests of plants, with only a few efforts to introduce biocontrol agents for the control of ticks. Nevertheless, the knowledge and experience accumulated in plant protection will aid in the development of tick biocontrol methods.

Classical biological control includes the recognition, evaluation and importation of a natural enemy from elsewhere, the conservation of local natural enemies and the augmentation of the biocontrol agents. Application methods can include individual inoculations or inundative releases of the natural enemies. Much effort has been applied to control pests by means of biological agents, often as part of integrated pest management (IPM) programmes (De Bach & Rosen, 1991; Van Driesche & Bellows, 1996). During the first half of the twentieth century efforts were made to import parasitoids into the USA for tick control (Larrousse, King & Wolbach, 1928; Cooley & Kohls, 1934; Alfeev, 1946). In addition, oxpeckers have been reintroduced into areas in Africa where these birds had become extinct (Davison, 1963; Grobler, 1976, 1979; Couto, 1994).

Ticks are obligatory blood-sucking arachnids that feed on vertebrates. While argasid ticks (soft ticks) feed for minutes or hours, the ixodid ticks (hard ticks) feed for days to weeks. In most cases, over 90% of the tick's life cycle is spent off their hosts. Accordingly, two general strategies for controlling ticks are in use – on-host and off-host control. On-host control strategies use the vertebrate as bait for ticks, relying upon the relatively high concentration of the pest in a small area, and killing ticks with high potential to propagate. Further, this strategy can benefit from the often more stable environment of the host skin, e.g. temperature or relative humidity (RH) – factors especially important for biocontrol. However, off-host control strategies enable treatments far from the host, under conditions that in some cases are more suitable, e.g. high RH with little ultraviolet (UV) irradiation under trees. Off-host control can minimize exposure of individual vertebrate hosts to pesticides, but large areas of the environment may be exposed to such hazards. In both strategies, environmental hazards associated with tick control can be dramatically reduced if biocontrol techniques are efficiently applied.

In this review, we try to summarize and update published information on biocontrol agents that have the potential to suppress tick populations, emphasizing those not covered in a recent review of the subject (Samish & Rehacek, 1999).

Ticks: Biology, Disease and Control, ed. Alan S. Bowman and Patricia A. Nuttall. Published by Cambridge University Press.

PATHOGENS

The *Bio-Pesticide Manual* (Copping, 2001) lists 96 commercial active ingredients based on microorganisms. Of these, 33 are based on bacteria, 36 on fungi and eight on entomopathogenic nematodes. All three of these groups include potential tick biocontrol agents. Viruses have not yet been used for tick control, but Assenga *et al.* (2006) recently developed a recombinant baculovirus with a tick chitinase gene. Supernatant from an insect cell culture in which the chitinase enzyme was expressed caused mortality in *Haemaphysalis longicornis*, suggesting a potential direction for future biocontrol research. In contrast to the biopesticides used against off-host ticks and most plant pests, development of those targeted at ticks on hosts need to take into account the relatively high temperatures on the skin, the limited ability of the pathogen to penetrate via tick mouthparts while feeding, and, importantly, considerations of safety for people and animals.

Bacteria

Bacteria are commonly found in wild-caught ticks, but most of these bacteria are not considered pathogenic to the ticks (Noda, Munderloh & Kurtti, 1997; Samish, Alekseev & Glazer, 1999a, 1999b). Martin & Schmidtmann (1998) obtained 73 bacterial isolates from field-collected *Ixodes scapularis*, including 11 species of *Bacillus*, mostly in the *B. thuringiensis–B. cereus* species group. Benson *et al.* (2004) also reported diverse bacteria from *I. scapularis*, including a relative of *Wolbachia pipientis*. Members of this group of alphaproteobacteria are known to cause reproductive alterations in insects, so further investigation is warranted to assess the biocontrol potential of this bacterium. Tick haemolymph and cement plugs display bacteriocidal activity (Alekseev *et al.*, 1995; Johns, Sonenshine & Hynes, 1998; Fogaca *et al.*, 1999; Ceraul, Sonenshine & Hynes, 2002) that provides some protection against bacterial attack. Nevertheless, some bacteria show pathogenicity to ticks. For example, *Proteus mirabilis* is pathogenic to *Dermacentor andersoni* (Brown, Reichelderfer & Anderson, 1970). Bacteria also attack *Amblyomma hebraeum*, *Hyalomma marginatum* and *Rhipicephalus evertsi evertsi* (Hendry & Rechav, 1981) apparently causes the blackening disease of *Boophilus decoloratus* and mortality of *Ornithodoros moubata* (Buresova, Fanta & Kopacek, 2006). Brum and colleagues (Brum, Faccini & Do Amaral, 1991; Brum, Teixeira & Da Silva, 1991; Brum & Teixeira, 1992) found the bacterium *Cedecea lapagei*

(Enterobacteriaceae) to be pathogenic to *Rhipicephalus* (*Boophilus*) *microplus*; this bacterium infects ticks via the genital opening and can produce up to 100% mortality under laboratory conditions.

TICK–*BACILLUS THURINGIENSIS* INTERACTIONS

Since ticks ingest primarily host blood, it seems unlikely that *B. thuringiensis*, which attacks the midgut of insects, would successfully enter and cause mortality in ticks. However, Hassanain *et al.* (1997), performing Petri dish tests, found that three commercial varieties of *B. thuringiensis* (*B. t. kurstaki*, *B. t. israelensis* and *B. t. thuringiensis*) produced mortality when sprayed on unfed or engorged adults of *Argas persicus* or *Hyalomma dromedarii*. *Bacillus t. kurstaki* was the most pathogenic (LC$_{50}$ 215–439 µg/ml on day 5 post-infection (PI) for *Argas* and 1200–2344 µg/ml for *Hyalomma*). Spraying *B. thuringiensis* on chickens with adult *A. persicus* ticks or feeding the ticks with the bacteria via capillary caused much higher mortality (up to 72% and 100% respectively) than after dipping in bacterial suspension (up to 32%) (Fedorova, 1988). Zhioua *et al.* (1999a, 1999b) exposed *I. scapularis* to *B. t. kurstaki* by dipping engorged larvae for 30 s in spore suspensions, and reported pathogenicity with an LC$_{50}$ of 1×10^7 spores/ml.

The crystalline δ-endotoxin of *B. thuringiensis* is produced during sporulation and disrupts insect midgut walls (Gill, Cowles & Pietrantonio, 1992). The δ-endotoxin shows specificity to particular insect taxa (Flexner & Belnavis, 2000), so it seems unlikely that this toxin kills ticks by causing toxaemia. Furthermore, the δ-endotoxin of *B. t. kurstaki* has to be activated by an alkaline pH and specific proteases. The extent to which the toxin will be activated in ticks with a midgut pH of 6.8 and intracellular cathepsins (pH optimum 3) (Sonenshine, 1991; Ramamoorthy & Scholl-Meeker, 2001) requires additional investigation.

Ticks ingest host body fluids primarily through their mouthparts, but they also ingest water vapour, through both mouthparts and cuticle, for hydration (Knulle & Rudolph, 1982). Therefore spraying or dipping ticks in highly concentrated bacterial solutions could result in tick mortality if sufficient fluid was ingested to elicit pathogenic effects on the midgut. However, other mechanisms of pathogenicity are also plausible, including: the actions of other toxins such as *B. thuringiensis* exotoxins (Sebesta *et al.*, 1981); negative effects associated with bacterial invasion of the haemocoel (Dubois & Dean, 1995); and the physical blocking of spiracles or other openings by bacterial spores.

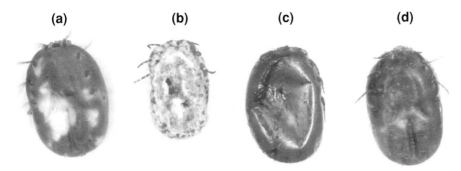

Fig. 20.1. Engorged *Boophilus annulatus* females following infection with *Metarhizium anisopliae* which causes mortality (a and b) and grows out of the cadaver (c), compared with an uninfected tick (d). (Photographs by Dr Michael Samish.)

Fungi

Over 700 species of entomopathogenic fungi have been reported, but only 10 of these have been or are currently being developed for the control of insects (Hajek & St Leger, 1994; Butt, Jackson & Magan, 2001). The most promising fungi belong to the class Deuteromycetes (Fungi Imperfecti). The fungal spores adhere physically to the tick cuticle, attach, germinate (6–24 h PI), form appressoria (24–48 h PI) and then penetrate via the cuticle (2–5 d PI), proliferate in and kill their host (Fig. 20.1), and often emerge (4–9 d PI) and produce new spores (7–9 d PI) (Arruda *et al.*, 2005; Garcia *et al.*, 2005; G. Gindin, D. Ment, I. Glazer & M. Samish, unpublished data). The mortality seems to be caused by massive fungal growth in the haemolymph, where proteases and oxalic acid are released and cause *inter alia* lysis of the internal epidermis, probably mainly of the Malpighian tubes, and rupture of the tick's intestinal wall (Federova, 1988; Suzuki *et al.*, 2003; Garcia, Monteiro & Szabo, 2004; Arruda *et al.*, 2005; Kirkland, Eisa & Keyhani, 2005).

The ability of entomopathogenic fungi to penetrate the cuticle of arthropods, and to kill several stages of the same pest, and also the relatively specific virulence of a single strain to one or a small group of pests, make them good candidates as biocontrol agents. However, fungi also have some disadvantages: they are slow in killing their host, they need high humidity to germinate and to sporulate, they require specific temperature ranges, they are susceptible to UV irradiation, some strains can potentially affect non-target arthropods, and some strains may produce toxins (Ginsberg *et al.*, 2002; Krasnoff *et al.*, 2007). Mass production can be quite costly, and the limited shelf-life of some of the resulting products makes them even more expensive. Many of these disadvantages can be addressed by advanced formulations. Most producers of fungus-based products suggest applica-

tion methods similar to those used for chemical pesticides (Shelton & Roush, 2000). The success of fungi in controlling ticks varies enormously according to fungal strain, formulation, method and timing of application and environmental conditions, and also according to tick species and life stage.

PREVALENCE OF TICK–FUNGUS INTERACTION

In nature, 20 species of fungus have been reported to be associated with ticks. Some 13 species of ticks from seven genera were found to be infected by fungi (Kolomyetz, 1950; Samsinakova, 1957; Steinhaus & Marsh, 1962; Cherepanova, 1964; Krylov, 1972; Samsinakova *et al.*, 1974; Estrada-Peña, Gonzales & Casasolas, 1990; Mwangi, Kaaya & Essumen, 1995; Zhioua *et al.*, 1999a; Guerra *et al.*, 2001). Ticks collected in northeast USA were infected primarily with *Verticillium* spp. and *Beauveria bassiana*; 10 species were isolated in Europe and three in Africa (Cherepanova, 1964; Samsinakova *et al.*, 1974; Mwangi, Kaaya & Essumen, 1995; Kalsbeek, Frandsen & Steenberg, 1995; Zhioua *et al.*, 1999a). Of the engorged female *B. microplus* ticks collected from soil in Brazil, 24.5% were contaminated with *B. bassiana*, 10% with *Metarhizium anisopliae* (Costa *et al.* 2001), and 22% of the *Rhipicephalus sanguineus* nymphs were contaminated with fungi from five genera (Guerra *et al.*, 2001). The percentage of fungus-infected ticks collected from various regions varies widely, mainly according to the stage and species of tick and to the sample site ecology. For instance, 7.5% of the adult *Ixodes ricinus* collected in central Europe in winter were infected by fungi, compared with over 50% of those collected in summer (Samsinakova *et al.*, 1974). In northern Europe, depending on the season 6–32% of engorged *I. ricinus* females were naturally infected by fungi (Kalsbeek, Frandsen & Steenberg, 1995). Only 1.7% of the engorged *Rhipicephalus appendiculatus* females collected in Kenya died from fungal infection (Mwangi, Kaaya & Essumen, 1995), and in the northeastern USA 4.3% of the collected unfed female *I. scapularis* were infected (Zhioua *et al.*, 1999*a*). In nature, a higher percentage of adult ticks seem to be fungus-infected than their pre-imaginal stages, and engorged females seem to be most readily infected (Kalsbeek, Frandsen & Steenberg, 1995; Zhioua *et al.*, 1999*a*).

LABORATORY ASSAYS

When various fungal genera and species, including very many strains' were tested under optimal laboratory conditions, in most cases *M. anisopliae* and *B. bassiana* elicited the highest mortality (Table 20.1) (Guangfu, 1984; Gindin *et al.*, 2001; Samish *et al.*, 2001). *Metarhizium anisopliae* strains were often more virulent than those of *B. bassiana* (Castineiras *et al.*, 1987; Mwangi, Kaaya & Essumen, 1995; Barci, 1997; Kaaya & Hassan, 2000; Gindin *et al.*, 2001; Samish *et al.*, 2001; Sewify & Habib, 2001). However, *M. anisopliae* was rarely isolated from ticks collected in nature (Zhioua *et al.*, 1994; Kaaya & Hassan, 2000; Costa *et al.*, 2001; Fernandes *et al.*, 2004). For ticks, as for other arthropods, fungi originally isolated from a specific host are not necessarily the best candidates for the control of the same arthropod genus or family (Fernandes *et al.*, 2006). Under laboratory conditions, at least 15 ixodid tick species and two argasid ticks were found susceptible to fungi (partially shown in Table 20.1).

Studies comparing the susceptibility of unfed *Boophilus annulatus*, *Hyalomma excavatum* and *R. sanguineus* larvae to 12 fungal strains (from five fungal species) gave the general impression that *Boophilus* and *Hyalomma* larvae showed similar high susceptibility whereas *Rhipicephalus* larvae were more resistant. In contrast, engorged *H. excavatum* females were far more resistant to entomopathogenic fungi than females of the other two tick species (Gindin *et al.*, 2002). Nymphal and adult *Amblyomma maculatum* were much more susceptible to fungi than those of *Amblyomma americanum* whereas *I. scapularis* and *R. sanguineus* were more susceptible than *Dermacentor variabilis* (Kirkland, Cho & Keyhani, 2004; Kirkland, Westwood & Keyhani, 2004). This may be due to fungistatic and/or germination-stimulating compound(s) in some of the tick epicuticles (Kirkland, Cho & Keyhani, 2004; Ment *et al.*, 2006).

Tick eggs, in contrast to most insect eggs, are highly susceptible to fungi (Fig. 20.2), and up to 100% of the eggs exposed to fungi did not hatch (Boichev & Rizvanov, 1960; Gorshkova, 1966; Castineiras *et al.*, 1987; Bittencourt, Massard & Lima, 1994*a*; Monteiro *et al.*, 1998*a*, 1998*b*;

Cameiro *et al.*, 1999; Kaaya, 2000; Gindin *et al.*, 2001, 2002; Paiao, Monteiro & Kronka, 2001; Fernandes *et al.*, 2003). Fungal infection of engorged female ticks often resulted in longer pre-oviposition, oviposition, egg-incubation and average egg-hatching periods, as well as in lowered egg production (Table 20.1). Fungal infection of other tick stages may also cause sub-lethal effects (e.g., reduced meal size, moult percentage) suggesting a relatively prolonged sub-lethal influence of the fungi on the ticks, which may reduce tick populations even more effectively than the direct causes of mortality (Gorshkova, 1966; Bittencourt, Massard & Lima, 1994*b*; Kaaya, Mwangi & Ouna, 1996; Barci, 1997; Correia *et al.*, 1998; Gindin *et al.*, 2001; Samish *et al.*, 2001; Horenbostel *et al.*, 2004). We still lack a detailed description of the delayed sub-lethality process of fungal infections in ticks.

Comparison between the susceptibility of unfed stages of *R. appendiculatus* and *Amblyomma variegatum* or of *H. excavatum* and *R. sanguineus* demonstrated decreasing susceptibility to fungi in progression through the larval, nymph and adult stages (Kaaya, 2000; Samish *et al.*, 2001; Gindin *et al.*, 2002) (see Fig. 20.3). In contrast, unfed *I. scapularis* larvae were less susceptible than unfed adults (Zhioua *et al.*, 1997). Unfed larvae seem to become more resistant to *M. anisopliae* after engorgement and both unfed and engorged larvae became progressively less susceptible during storage (Reis *et al.*, 2001; Gindin *et al.*, 2002; Flor, Kurtti & Munderloh, 2005).

Fungi take several days to kill ticks. For instance, the LT_{50} (time taken to kill half of a tested population) of *M. anisopliae* (at 1×10^7 spores/ml) for eggs, unfed larvae and engorged females of *B. annulatus* generally ranged from <3 to 6 days; in some cases it may have extended to weeks. Engorged larvae of *H. excavatum* and *R. sanguineus* took twice as long to die as unfed larvae of the same species, and it took over 3 weeks for 50% of unfed adults to be killed by these fungi (unpublished observation). Fungal strains differ in their speed of killing ticks and changing spore formulation may strongly decrease mortality time (Polar *et al.*, 2005; G. Gindin, I. Glazer & M. Samish, unpublished data). In addition, the age of spores, the culturing method and medium and storage conditions may also influence the fungal virulence.

FIELD TRIALS: OFF-HOST

An *Abies procera* plantation in Denmark and a forest/shrubby area in the USA were sprayed with *M. anisopliae* spores (4–10×10^{10} spores/m²). Of the unfed *I. ricinus* ticks collected 2 weeks PI, 57% were infected with *M. anisopliae* and 53% of the unfed *I. scapularis* (4 weeks PI) were killed

Table 20.1 *Pathogenicity of fungi to ticks: the effect of entopathogenic fungi on tick species and stages under laboratory conditions, on outdoor ground or while feeding on animal*

Trials in/on	Tick species[a]	Fungus species[b]	Tick stage[c]	Techniques[d]	Results		References
					Mortality up to (%)[f]	Decrease in/of[e]	
Laboratory	A. americanum	B. bassiana	unf F	Not given	100 (F)		Gomathinayagam, Cradock & Needham (2002); Kirkland, Chao & Keyhani (2004)
	A. cajennense	B. bassiana, M. anisopliae	E, unf, eng L, N, A	Im	80 (N, A)	EL, H, WEM	Souza, Reis & Bittencourt (1999a, b); Reis et al. (2001, 2004)
	A. cooperi	B. bassiana, M. anisopliae	N	Im	70 (N)	EN	Reis et al. (2003)
	A. maculatum	B. bassiana, M. anisopliae	N, A	Im	90 (A), 100 (N)		Kirkland, Chao & Keyhani (2004)
	A. variegatum	B. bassiana, M. anisopliae	L, unf, eng N, unf A	Im	100 (L), 80 (N, A)	F, WEM, H	Kaaya (2000); Kaaya & Hassan (2000); Reis et al. (2001); Maranga et al. (2005); Maranga & Kenyatta (2006)
	An. nitens	B. bassiana, M. anisopliae	E, L, eng F	Im	99 (L)	H, Ov, WEM	Monteiro et al. (1998c); Carneiro et al. (1999); Bittencourt et al. (2000); Monteiro, Bittencort & Daemon (2001)
	Ar. persicargas	B. bassiana, M. anisopliae	N, unf, eng A	Im	100 (N, A)		Sewify & Habib (2001); Habbib & Sewify (2002)
	B. annulatus	B. bassiana, M. anisopliae	E, eng F	Im, Con	100 (E, L, A)	WEM	Gindin et al. (1999, 2001, 2002)
	B. microplus	B. bassiana, M. anisopliae, P. fumosoroseus, S. insectorum, V. lecanii	E, L, eng A	Im	92 (E), 100 (L, A)	H, Ov, WEM	Castineiras et al. (1987); Bittencourt et al. (1992, 1994a, 1996, 1997); Correia, Monteiro & Fiorin (1994); Monteiro et al. (1994); Mendes et al. (1995); Rijo (1998); Frazzon et al. (2000); Souza et al. (2000); Nunes et al. (2001); Onofre et al. (2001); Zhioua et al. (2002); Fernandes et al. (2003, 2004); Mythlli et al. (2004); Basso et al. (2005); Paiao, Monteiro & Kronka (2001)
	D. variabilis	B. bassiana, Sc. brevicaulis[f]	A	Im, Top	100 (A)		Gomathinayagam, Cradock & Needham (2002); Yoder, Benoit & Zettler (2003); Kirkland, Westwood & Keyhani (2004)
	H. excavatum	M. anisopliae	unf L	Im, Con	100 (L)		Gindin et al. (2002)
	I. ricinus	B. bassiana	E, L, eng F	Im	100 (L)	H, Ov	Gorshkova (1966)
	I. scapularis	B. bassiana, M. anisopliae	L, N, A	Im, Sp	100 (N, A)	F, EW, WEM	Zhioua et al. (1997); Benjamin, Zhioua & Ostfeld (2002); Kirkland, Westwood & Keyhani (2002); Kirkland, Westwood & Keyhani (2004); Hornbostel et al. (2004, 2005)

(cont.)

Table 20.1 (*cont.*)

Trials in/on	Tick species[a]	Fungus species[b]	Tick stage[c]	Techniques[d]	Results		References
					Mortality up to (%)[e]	Decrease in/of[e]	
	R. appendiculatus	*B. bassiana*, *M. anisopliae*	unf L, N, A, eng A	Im	100 (L), 75 (N, A)	F, WEM	Mwangi, Kaaya & Essumen (1995)
	R. sanguineus	*A. ochraceus*,[g] *B. bassiana*, *M. anisopliae*	E, L, N, eng F	Im, Con, Sp	100 (E, L, N, A)	H, M, WEM	Estrada-Peña, Gonzalez & Casasolas (1990); Barbosa et al. (1997); Monteiro et al. (1998a, 1998b); Samish et al. (2001); Gindin et al. (2002); Kirkland, Westwood & Keyhani (2002); Polar et al. (2005); Prette et al. (2005)
Ground	*A. variegatum*	*B. bassiana*, *M. anisopliae*	L, N, A	Sp	100 (L, N, A)	F, WEM, H	Kaaya, Mwangi & Ouna (1996); Kaaya (2000); Kaaya & Hassan (2000)
	B. annulatus	*M. anisopliae*	A	Sp	100 (A)	Ov, WEM	Samish et al. (2006)
	Ar. persicargus	*B. bassiana*, *M. anisopliae*	N, A	Sp of cloth	100 (N, A)		Sewify & Habib (2001)
	B. microplus	*M. anisopliae*, *M. anisopliae*	unf L, eng F	Sp	94 (L)	H	Bittencourt et al. (2003); Basso et al. (2005)
	I. scapularis	*M. anisopliae*	unf A	Sp	53 (A)		Benjamin, Zhioua & Ostfeld (2002)
	R. appendiculatus	*B. bassiana*, *M. anisopliae*	L, N, A	Sp	100 (L, N, A)	F, WEM, H	Kaaya, Mwangi & Ouna (1996); Kaaya (2000); Kaaya & Hassan (2000)
Animals	*A. variegatum*	*B. bassiana*, *M. anisopliae*	eng A	Sp on ears	Low mortality	F, H	Kaaya, Mwangi & Ouna (1996)
	An. nitens	*B. bassiana*	all parasitic stages	Sp on ears	70 (all parasitic stages)		Bittencourt et al. (2002); Monteiro et al. (2003)
	B. decoloratus	*B. bassiana*, *M. anisopliae*	eng A	Sp	50 (A)	H	Kaaya & Hassan (2000); Kaaya (2000)
	B. microplus	*M. anisopliae*, *V. lecanii*	L, eng F	Sp	99 (all parasitic stages)	R	Castro et al. (1997); Camacho et al. (1998); Correia et al. (1998); Bittencourt et al. (1999)
	R. appendiculatus	*B. bassiana*, *M. anisopliae*	eng A	Sp, Pow Sp on ears	100 (A)	F, H, EW, WEM	Kaaya & Mwangi (1995); Kaaya, Mwangi & Ouna (1996)

[a] Genera: *A*, *Amblyomma*; *An*, *Anocentor*; *Ar*, *Argas*; *B*, *Boophilus*; *D*, *Dermacentor*; *H*, *Hyalomma*; *I*, *Ixodes*; *R*, *Rhipicephalus*.

[b] Genera: *A*, *Aspergillus*; *B*, *Beauveria*; *M*, *Metarhizium*; *P*, *Paecilomyces*; *S*, *Scopulariopsis*; *V*, *Verticillium*.

[c] E, eggs; L, larvae; N, nymphs; A, adults; F, females; unf, unfed stages; eng, engorged stages.

[d] Con, contact of ticks with filter paper impregnated with spore suspension; Im, immersion of ticks in spore suspension; Pow, powdering of ticks by dry spore formulation; Sp, spraying of ticks; Top, topical application of fungi.

[e] EL, ecdysis of larvae; EN, ecdysis of nymphs; EW, engorgement weight; F, fecundity; H, hatchability; M, moulting; Ov, oviposition; R, reproduction; WEM, weigth of egg masses.

[f] Can cause deep fungal infection of humans.

[g] Infection via the anus.

Source: Modified after G. Gindin et al. unpublished data. Table includes only trials resulting in more than 70% mortality in laboratory tests or more than 50% of tick control on ground or on animal tests.

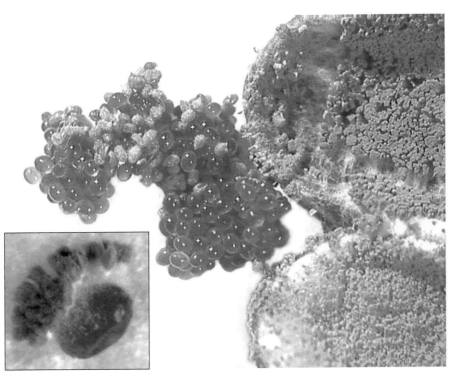

Fig. 20.2. Two engorged *Boophilus annulatus* females after laying several eggs. *Metarhizium anisopliae* growing on the females and (inset, at higher magnification) the eggs. (Photo credit: Dr Michael Samish.)

(Benjamin, Zhioua & Ostfeld, 2002; C. Nielsen, S. Vestergaard, J. Eilenberg & C. Lomer, personal communication). However, spraying *M. anisopliae* against questing nymphs reduced the population by no more than 27% (Hornbostel *et al.*, 2005). When two commercial *B. bassiana* mycoinsecticides were tested against *I. scapularis* nymphs in a residential area, the tick population was reduced by 59–89%, and some 80–90% of the recovered nymphs developed mycoses (Stafford & Kitron, 2002). Ticks inoculated with *B. bassiana* were released in a natural *D. variabilis* microhabitat from which live ticks were collected 1 year later: fungal growth was observed after laboratory incubation of the ticks (T. M. Lucas, L. J. Fielden & J. Hererra, personal communication). When potted grass with unfed adult or nymph *R. appendiculatus* was sprayed with *B. bassiana* spores (1 × 10^9 spores/ml in distilled water) and kept in the field, the mortality reached 96% and 36% for nymphs and adults, respectively. However, spraying with *M. anisopliae* achieved 76% and 64% mortality for nymphs and adults, respectively. Mixture of the two fungal species killed over 99% of the larvae. Similar results were obtained with *A. variegatum* (Kaaya & Mwangi, 1995; Kaaya, Mwangi & Ouna, 1996).

Unfed *A. variegatum* adults dipped in *B. bassiana* or *M. anisopliae* were placed in bags on the grass outdoors in Kenya. Tick mortality was similar for the two fungi, but was about 25–35% higher in the wet season than in the dry season (Maranga *et al.*, 2005).

In Kenya, 5-acre (2-ha) paddocks were seeded with *R. appendiculatus* larvae and sprayed once a month with *M. anisopliae* or *B. bassiana* spores in distilled water (1.2 × 10^9 spores/m^2). During the rainy season the fungi had no effect upon the abundance of the ticks on cattle kept in the paddocks, but during the 3 months after the end of the rainy season *B. bassiana* and *M. anisopliae* reduced the tick population by 80% and 92%, respectively, compared with the control (Kaaya, 2000; Kaaya, Samish & Glazer, 2000).

Buckets with soil and leaves in the open air in Israel were sprayed with *M. anisopliae* spores (1.7 × 10^7 spores/ml) in water with Triton-X. Engorged *B. annulatus* females were added, and the buckets were irrigated. Mortality reached 80–100% within 3 weeks in summer and 40–80% within 4 weeks in winter. Spraying spores before or after adding ticks yielded similar results (Samish *et al.*, 2006). Engorged *B. annulatus* females placed on fungus-treated uncovered, or gravel- or

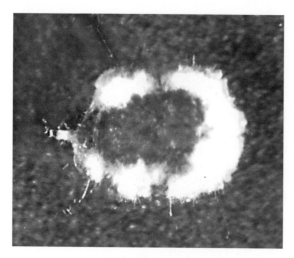

Fig. 20.3. *Beauveria bassiana* growing on an unfed *Rhipicephalus sanguineus* adult. (Photo credit: Dr Michael Samish.)

hay-covered ground caused up to 40% mortality, whereas on leaf- or grass-covered ground the mortality reached 100% (Samish *et al.*, 2006; M. Samish, G. Gindin & I. Glazer, unpublished data). The success of fungi in controlling *B. microplus* on plots covered with *Brachiaria brizantha* plants was greater than on those covered with Tifton 85 (Basso *et al.*, 2005). Grass-covered, 1-m^2 'pasture beds' were each infested with unfed *B. microplus* larvae, and 1 day later were sprayed with up to 6×10^{10} *M. anisopliae* spores/ml. The tick populations were reduced by up to 54% (Bittencourt *et al.*, 2003) or 94% (Basso *et al.*, 2005).

Sackcloth covering poultry houses that were heavily infested with the soft tick *Argas persicus* was sprayed with an *M. anisopliae* spore suspension (5×10^7 spores/ml in water with Tween-80 and 4% oil) and covered with a plastic sheet to maintain high humidity. The tick population was reduced by 53% within 1 week, and after 3 weeks no live ticks could be found (Sewify & Habib, 2001).

Interestingly, use of traps containing *M. anisopliae* spores with pheromone and CO$_2$ in a tick-infested farm gradually reduced the questing tick population during 3 months, and no ticks were recovered in the subsequent 3 months (Maranga & Kenyatta, 2006). The use of tick pheromones and host kairomones to enhance the efficacy of chemical acaricides is discussed in detail elsewhere in this volume (see Chapter 21).

FIELD TRIALS: ON-HOST
Spraying *M. anisopliae* (1×10^8 spores/ml in water with Triton-X–100) on gerbils 1 day after they had been infested

with *R. sanguineus* nymphs reduced the number of engorged nymphs dropping off by 73% (M. Samish, unpublished observations). After 1 year, periodic spraying of spores onto cotton nesting material in rodents' nest boxes on tree trunks modestly affected the number of feeding larvae, and significantly reduced the questing nymph population in the vicinity of some of the treated nests (Hornbostel, Ostfeld & Benjamin, 2005).

Spraying *M. anisopliae* spores (1×10^5 to 1×10^9 spores/ml in water) on cattle infested with *B. microplus* or *B. decoloratus* ticks at various development stages generally caused insignificant or low reductions (up to 50%) of the on-host tick population. However, up to 79% of the females collected from the sprayed cattle died in the laboratory, and their egg mass was reduced by up to nearly 50% (Castro *et al.*, 1997; Correia *et al.*, 1998; Bittencourt *et al.*, 1999; Kaaya & Hassan, 2000). When *B. microplus*-infested cattle were treated one or four times with *Verticillium lecanii* (3.5×10^7/ml, 5 l per head), the tick populations were reduced by 48–79% or 94–99%, respectively (Camacho *et al.*, 1998). In other trials, adult *R. appendiculatus* ticks were allowed to feed in ear bags on rabbits or cattle and were sprayed with *M. anisopliae* spores at dosages of 2.5×10^8 or 1×10^{10} spores per ear, respectively; 30% and 83% of the ticks on rabbits and cattle, respectively, died. In addition, of the eggs laid by *Beauveria*-treated females that previously fed on the rabbits or the cattle, only 0% or 48%, respectively, hatched (Kaaya, Mwangi & Ouna, 1996). Treating *Anocentor nitens*-infested ears of horses with a *B. bassiana* suspension in water resulted in less than 20% tick mortality, whereas their application in a polymerized pulp gel resulted in more than 50% mortality (V. R. E. P. Bittencourt, E. J. Souza, G. L. Costa & A. S. Fagundes, unpublished data). Dipping cattle ears in a spore suspension with 1% Tween-80 a day before infesting them with *A. nitens* larvae did not reduce the yield of engorged females compared with the treatment with Tween-80 only (Monteiro, Bahiense & Bittencourt, 2003).

When *B. bassiana* and *M. anisopliae* spores in water with Triton-X–100 were sprayed on thoroughly washed ears of cattle, live *B. bassiana* and *M. anisopliae* spores were recovered up to 1 and 3 weeks, respectively, after application (Kaaya, Mwangi & Ouna, 1996). However, fungal spores suspended in water and sprayed on cattle were found to be virulent to *Boophilus* ticks for less than a week (Castro *et al.*, 1997; Bittencourt *et al.*, 1999), and most deaths of feeding *R. appendiculatus* occurred 5–10 days after fungal infection (Kaaya, Mwangi & Ouna, 1996).

The optimal temperature for most entomopathogenic fungal strains is 24–27 °C; however, different strains differ in their virulence-retention capability at higher temperatures. The skin temperature of warm-blooded animals is much above the optimal for the fungi, and is influenced by many factors, e.g. animal species, age, area of body, type of food, climate and housing location. A *M. anisopliae* strain that could grow on media at 30–34° C was also more virulent to *B. microplus* feeding on cattle (Polar *et al.*, 2005). The impairment of fungal virulence by elevated temperature may be the reason that strains showing high virulence under laboratory conditions were usually more efficient against ticks on the ground than against feeding ones (Polar *et al.*, 2005). Different tick species and stages feed on skin areas with differing temperatures, and some fungal strains are active at higher temperatures. However, the potential of such strains to become pathogenic to vertebrates must be carefully studied. We have little information on the sub-lethal effect of spores sprayed on feeding ticks under field conditions.

FORMULATION
In most laboratory tests the spores were suspended only in water with a small amount of dispersing agent. Comparison among paraffin oil, palm oil, emulsifiable adjuvant oils, Cropspray and Codacide oil showed that in 10% emulsifiable adjuvant liquid paraffin was the most effective formulation for *M. anisopliae* (Polar *et al.*, 2005). Kaaya & Hassan (2000) and Maranga *et al.* (2005) compared the anti-tick effect of *M. anisopliae* spores suspended in water with 1% Tween-80, with or without 15% peanut oil. Unfed nymph and adult ticks were immersed in the suspension and then kept on vegetation outdoors. The mortality of *A. variegatum* treated with the oil suspension was 30% higher for nymphs and 2.7 times higher for adults than in applications with the water carrier without oil: the corresponding figures for *R. appendiculatus* were 15% for nymphs and 4.3 times higher for adults than in applications with the water without oil. The influence of the oil itself on the ticks has still to be clarified. In another trial, a powder formulation was prepared by mixing *M. anisopliae* spores with various powders (ratio 1:9) and was tested against *R. appendiculatus* adults feeding in ear-bags on cattle. Spore mixtures with powders of millet, maize, sorghum or starch caused 100%, 79%, 64% or 53% mortality respectively (Kaaya & Hassan, 2000). Several of the compounds commonly added to spores may be toxic to ticks: e.g. >0.01% Triton-X, 1% Tween-80, and several oil types (Monteiro, Bahiense & Bittencourt, 2003; Polar *et al.*,

2005). This must be taken into account when suggesting a 'green' control method.

To improve fungal virulence to ticks various mixtures were tested: *M. anisopliae* with *B. bassiana* (Maranga *et al.*, 2005) or a mix of *B. bassiana*; *P. fumoso-roseus* with *B. thuringiensis* (Federova, 1988); or *M. anisopliae* with the chemical pesticide permethrin (Hornbostel *et al.*, 2005). No synergistic effect was found.

Considerable research is still required to develop satisfactory fungus-based acaricides to control feeding ticks. One need is to find fungal strains that are more virulent and specific, especially under on-vertebrate conditions. A second need is to adjust fungal formulation for use against free-living ticks. Substantial knowledge and experience can be obtained from the vast accumulation of information related to the successful use of fungi for control of plant pests. Clearly, the formulation in which the spores are applied is crucial to the level of tick control obtained with fungus-based anti-tick compounds, but very little has been published as yet on the subject.

Entomopathogenic nematodes

Entomopathogenic nematodes (EPNs) of the families Heterorhabditidae and Steinernematidae are known to be parasites of insects. The only free-living stage of the nematode, the third/infective juvenile, actively locates and enters the host via natural openings, and then releases symbiotic bacteria that kill the host insect within 24–72 h. The nematodes then multiply within the host cadaver and 6–18 days PI thousands of infective juveniles are released into the environment. The most common natural habitat of these nematodes is moist ground. The infective juveniles are well adapted to the changing conditions of moisture, temperature, texture and chemical composition associated with different soil types (Glazer, 2001).

The EPNs are known to be pathogenic to over 3000 insect species, although each strain may often be relatively specific to a small group of hosts and thus their effects on most beneficial insects have been found to be negligible (Poinar, 1973; Gaugler, 2002). They are produced commercially on four continents for the control of insect pests of forests and agriculture (Georgis & Manweiler, 1994).

TICK—NEMATODE INTERACTIONS
Entomopathogenic nematodes penetrate engorged female *B. annulatus* ticks almost solely via the anus or genital pore (M. Samish & I. Glazer, unpublished observations).

Heterorhabditid nematodes killed engorged *B. annulatus* females in Petri dishes after less than 2.5 h of exposure, whereas steinernematid nematodes needed more than 4 h to penetrate into ticks (Glazer, Alekseev & Samish, 2001). The injection of a single heterorhabditid nematode into a tick can cause mortality (Glazer *et al.*, 2001). The dosages of EPNs needed to kill 50% or 90% of ticks are comparable to that used commercially in the control of insect pests of plants, but the time required to kill ticks is often relatively long (Samish, Ginsberg & Glazer, 2004).

Tick mortality caused by EPNs seems to be due to the rapid proliferation of the nematode symbiotic bacteria within the ticks, since the nematodes do not go through their natural cycle within ticks, and most infective juveniles die shortly after entry (Mauleon, Barre & Panoma, 1993; Samish, Alekseev & Glazer, 1995; Hill, 1998; Kocan *et al.*, 1998*a*, 1998*b*; Hassanain *et al.*, 1999). Interestingly, when the cuticle of *I. scapularis* was physically slit before nematode infection, the nematodes *Steinernema carpocapsae* and *S. glaseri* reproduced successfully (Zhioua *et al.*, 1995).

LABORATORY ASSAYS

Ticks exposed to nematodes in Petri dishes lined with moist filter paper showed that six genera of Ixodidae and two genera of Argasidae were readily killed by EPNs (Samish *et al.*, 2004). Mauleon, Barre & Panoma (1993) reported that *B. microplus* appeared to be resistant to nematode attack. However Vasconcelos *et al.* (2004) demonstrated high susceptibility (>90% mortality) of this tick species to EPNs. Various tick stages differ substantially in their susceptibility to EPNs, with the fully engorged female ticks being most susceptible and the pre-imaginal stages least sensitive. During feeding, ticks were highly resistant to EPNs and their eggs were totally resistant. Unfed female ticks were killed up to six times as quickly (LT_{50} 1 d for *Rhipicephalus bursa*) than engorged ticks (6 d for *R. bursa*) (Samish, Alekseev & Glazer, 2000*a*, 2000*b*; Samish, Ginsberg & Glazer, 2004). This may be connected to the strong anti-bacterial activity of the tick haemolymph (Ceraul, Sonenshine & Hynes, 2002). Exposure of nematodes to ticks adversely affects the reproduction and egg-laying rate (Vasconcelos *et al.*, 2004; Freitas-Ribeiro *et al.*, 2005).

The many strains of more than 25 nematode species from the families Heterorhabditidae and Steinernematidae display a wide range of characteristics, enabling them to adjust readily to the various different pests in different ecological niches. The 42 nematode strains tested for anti-tick activity showed varying degrees of virulence (Samish *et al.*, 2004).

In laboratory tests, heterorhabditid nematodes were generally more virulent to ticks than steinernematids (Mauleon, Barre & Panoma, 1993; El-Sadawy & Habeeb, 1998; Hill, 1998; Hassanain *et al.*, 1999; Glazer *et al.*, 2001). Nematode strains virulent to one tick stage of one species were found, in most cases, to be highly virulent to other tick species and stages also (El-Sadawy & Habeeb, 1998; Hassanian *et al.*, 1999; Samish *et al.*, 1999*a*, 1999*b*).

FIELD TRIALS

The virulence of nematode strains to ticks as measured in Petri dish tests in the laboratory often differs considerably from that measured in field trials when the nematodes are sprayed on soil. The difference could result from the specific behaviour of each nematode strain in the soil environment, e.g. how deep in the soil they prefer to live, how they react to UV irradiation, low moisture or other factors.

Among nine strains of three nematode species tested against *B. annulatus* ticks on soil, under simulated field conditions, the commercial *S. carpocapsae* Mexican strain was the most effective: a nematode dosage of $50/cm^2$ of this strain killed 100% of the engorged females with an LT_{50} of less than 5 days (Samish *et al.*, 1999*a*).

Environmental conditions strongly influence the pathogenicity of nematodes to ticks. Soils with a high silt concentration or with more than 25% (v/v) manure were found to reduce the anti-tick activity of EPNs relative to that measured on clean sandy soil (Samish, Alekseev & Glazer, 1998; Samish *et al.*, 1999*a*; Alekseev, Glazer & Samish, 2006). The anti-tick activity of EPN strains also varied with temperature (Zhioua *et al.*, 1995): the optimal activity for most strains was 25–28 °C, although some displayed far wider ranges of activity (e.g. 18–34 °C; I. Glazer & M. Samish, unpublished observations). Entomopathogenic nematodes sprayed on humid sandy soil 3 days before ticks were added killed 100% of the ticks, whereas EPNs sprayed on soil containing 25% (v/v) cattle manure resulted in 45% mortality, and on soil containing 40–50% silt only 25% mortality (Alekseev, Glazer& Samish, 2006). Entomopathogenic nematodes sprayed on soil covered with leaf litter or grass were highly efficient in killing ticks compared with those sprayed on uncovered soil. Shading the soil with nets (10–90% shade) and irrigating it twice a day increased tick mortality from nematode attack in comparison with that on unshaded soil or on plots irrigated only once every 2 days (Zangi, 2003).

Nematodes are potentially useful tools for tick control because: (1) engorged ticks are susceptible to some EPNs and also reside in locations that are preferred by many nematode strains; (2) immobile ticks attract mobile nematodes; and (3) spraying or irrigation can be easily used to apply nematodes to a tick habitat. However, the use of nematodes may be limited to defined ecological niches because their pathogenicity is reduced by low humidity or temperature, high concentrations of manure or silt, and by differences in the susceptibility among the various tick stages and species. The wide genetic variation found among the many nematode strains, and presumably in strains yet to be found, means that genetic manipulation of nematodes could increase the range of ecological conditions in which they could be successfully applied against ticks. The development of improved formulations is also important. Finally, in-depth studies are needed to elucidate the interactions between nematodes and ticks under field conditions.

PARASITOIDS

Most parasitoids used in the biological control of insect pests of plants belong to the order Hymenoptera. Over two-thirds of the cases of successful biological control of pest species have been achieved with hymenopteran parasites (De Bach & Rosen, 1991), but only a few species are known to affect ticks. The distinguished entomologist L. O. Howard first described two species of chalcidoid wasps collected from ticks in Texas (Howard, 1907, 1908). These are now both included in the genus *Ixodiphagus* of the family Encyrtidae, which includes seven species, all tick parasites (Trjapitzin, 1985; Davis, 1986; Mwangi *et al.*, 1997).

The most widespread species is *Ixodiphagus hookeri* (synonyms, *Hunterellus hookeri, I. caucurtei*: see Gahan, 1934; Trjapitzin, 1985) (Fig. 20.4) which has been recorded from Asia, Africa, North America and Europe (Hu, Hyland & Mather, 1993). Other species have been reported from these continents and also from Australia (Oliver, 1964; Doube & Heath, 1975; Graf, 1979; Mwangi & Kaaya, 1997). Collection of *I. theileri* from a specimen of the tick *Hyalomma rufipes* that was attached to a migrating bird in Egypt (Kaiser & Hoogstraal, 1958) suggests a possible mode of long-distance transport for these wasps. Most species of *Ixodiphagus* are host-generalists, which have been collected from a variety of hard tick species (Oliver, 1964), and in at least one case (*I. mysorensis*) from soft ticks (*Ornithodoros* sp.) (Mani, 1941). Nevertheless, some degree of host preference has been reported (Bowman, Logan & Hair, 1986). *Ixodiphagus hookeri* uses tick odour cues to find hosts (Takasu *et al.*, 2003).

Fig. 20.4. *Ixodiphagus texanus*, female. (Reproduced from Howard, L. O. (1907). A chalcidid parasite of a tick. *Entomological News* 18(9), 375–378, with permission from Dover Litho Printing Comp., Dover, Delaware, USA.)

Nymphal ticks were parasitized while they were engorging on vertebrates (Smith & Cole, 1943), and parasitoid egg development was found to be associated with ingestion of blood by its host tick (Hu & Hyland, 1998).

The only species that has been released for biological control of ticks is *I. hookeri*. Larrousse *et al.* (1928) released *I. hookeri* (propagated from wasps originally collected in France) on Naushon Island off the coast of Cape Cod, Massachusetts in 1926. The parasites were released as adults, in parasitized *I. scapularis* nymphs, and on mice (with parasitized nymphs attached). Infected nymphs of the target species, *D. variabilis*, were collected the following summer, which suggests that the parasite overwintered successfully. The ticks were less common the year after the parasite was released, but both *D. variabilis* and *I. scapularis* remained common at the site 12 years later, even though the parasite was still present (Cobb, 1942). Smith & Cole (1943) released about 90 000 female *I. hookeri* at two sites on Martha's Vineyard, Massachusetts, but found no evidence of parasite persistence or of tick control. Mwangi *et al.* (1997) released about 150 000 specimens of *I. hookeri* over a 1-year period to control *A. variegatum* on a field with 10 infested cattle in Kenya. They reported a reduction in tick numbers from 44 to two ticks per animal, suggesting that inundative releases can provide tick control at individual sites.

One possible explanation for this pattern can be surmised from recent surveys for *I. hookeri* in southern New England and New York State. This species is present at several sites with extremely dense populations of the tick *I. scapularis*, but it is rare or absent at nearby sites with fewer ticks (Hu *et al.*, 1993; Stafford, Denicola & Magnarelli, 1996; Ginsberg & Zhioua, 1999). Furthermore, sharp and sustained reductions in deer populations that lowered tick numbers at sites in Connecticut resulted in sharply lowered rates of *I. hookeri* parasitism as well (Stafford *et al.*, 2003). Thus, it appears that *I. hookeri* requires high tick densities to persist in the northeastern United States and is not likely to provide tick control under natural conditions. Nevertheless, inundative releases to control ticks in limited areas (e.g. farms, recreation areas) are still potentially feasible (Mwangi *et al.*, 1997). Knipling & Steelman (2000) argued, on the basis of a simple theoretical and economic analysis, that parasitoids have high potential as biocontrol agents for *I. scapularis*. They pointed out that a successful application would require a substantial research effort on parasitoid taxonomy, ecology and behaviour, and an inexpensive production system would be needed (e.g. artificial media for cultivating parasitoids) to produce large numbers of parasites for release. *Ixodiphagus hookeri* can be reared in large numbers in the laboratory at a modest cost in countries with low-priced labour (Mwangi & Kaaya, 1997), but the need to maintain large numbers of ticks and vertebrate hosts could be problematical for commercial production. If additional research confirms the efficacy of parasite release for tick control on individual farms or over broader areas, then improved parasite production methods would have to be developed.

Interestingly, Mather, Piesman & Spielman (1987) found that the *I. scapularis* parasitized by *I. hookeri* on Naushon Island did not carry the pathogen *Borrelia burgdorferi* (causative agent of Lyme borreliosis) and rarely carried *Babesia microti* (agent of human babesiosis), even though these pathogens were common in uninfected ticks at that locale. Thus, wasp infestation might lower pathogen prevalence in ticks, even if it does not control tick numbers. However, the lack of infection in parasitized ticks might have resulted from preferential parasitization of ticks attached to white-tailed deer (*Odocoileus virginianus*) (Stafford, Denicola & Magnarelli, 1996; Hu & Hyland, 1997), a large, conspicuous host that is not a competent reservoir for either pathogen. Furthermore, pathogen-infected ticks are still abundant at many sites where these wasps are present, so wasp parasitism does not, by itself, control the risk of disease. Further study is needed to determine whether wasp infestation influences pathogen infection in ticks.

Ixodiphagus spp. parasitize only ticks, as far as is known. Therefore, non-target effects would presumably be minimal if these parasites were released for tick control. The probable susceptibility of these hymenopteran parasites to many agricultural insecticides would require careful coordination to avoid interference between pesticide applications and parasitoid releases. At present, the lack of efficacy in nature and specific ecological requirements, insufficient information on the efficiency of inundative release, the lack of an efficient method to mass propagate the parasitoids and the unknown cost of a parasitoid control programme render the use of this means for tick control in the near future unlikely.

PREDATORS

Many tick bio-suppressors such as ants, beetles and many bird species are general predators that feed occasionally on ticks, therefore their populations do not depend on the sizes of the tick populations. General predators can sometimes affect the size of a tick population in nature, but manipulating their populations to reduce tick numbers would require large increases in the predator population, which could also cause large changes in populations of non-target species in natural areas (Symondson, Sunderland & Greenstone, 2002). Therefore, general tick bio-suppressors will not be covered in this chapter.

Avians

Some 50 bird species have been reported to eat ticks (Petrischeva & Zhmayeva, 1949; Mwangi *et al.*, 1991; Samuel & Welch, 1991; Petney & Kok, 1993; Verissimo, 1995; Samish & Rehacek, 1999). However, only a few species seem to feed specifically on ticks, and thus only a few would be expected to have a meaningful effect on tick populations.

CHICKENS

Chickens (*Gallus gallus*) confined with cattle in Africa were reported to ingest an average of 338 ticks per bird during 5.5 h. Other experiments found that the birds ate from 9.7 to 81 ticks per bird per hour of foraging. At high tick concentrations, an average of 69% of the ticks were consumed by chickens (Hassan *et al.*, 1991; Hassan, Dipeolu & Munyinyi, 1992; Dreyer, Fourie & Kok, 1997). Chickens are neither tick-specific predators nor obligatory predators, therefore their consumption of ticks depends largely on alternative food availability and the density of the tick population. Thus, chickens are unlikely to reduce tick densities below a certain level. Nevertheless, chickens maintained in any case on small mixed farms can help to reduce tick populations at nearly no cost.

OXPECKERS

Buphagus africanus (the yellow-billed oxpecker) and *B. erythrorhynchus* (the red-billed oxpecker), both native to Africa, are the only birds known to feed specifically on ectoparasites, especially ticks. Oxpecker populations have decreased along with reductions in the numbers of game animals, increased use of bird-poisoning acaricides and possibly also decreased tick populations (Van Someren, 1951; Stutterheim & Stutterheim, 1980; Stutterheim & Brooke, 1981; Robertson & Jarvis, 2000). However, oxpecker reintroduction efforts, an increase in game animals and the introduction of safer acaricides appear to have resulted in increases in these bird populations during the last two decades.

Stomach contents of captured oxpeckers included 16 to 408 ticks per bird (Van Someren, 1951; Bezuidenhout & Stutterheim, 1980). The consumption by young red-billed oxpeckers of larval, nymphal and adult *Boophilus* ticks averaged 1176, 1549 and 1293, respectively, during 6–7 days of exposure to tick-infested cattle. This result was extrapolated to adult red-billed oxpeckers as the equivalent of 12 500 larval or 98 engorged *B. decoloratus* females. Yellow-billed oxpeckers can consume about 10% more ticks than red-billed oxpeckers (Bezuidenhout & Stutterheim, 1980; Stutterheim, Bezuidenhout & Elliott, 1988).

Oxpeckers are visual predators, first plucking the engorged females, then searching large body areas and scissoring and eating the smaller tick stages. Yellow-billed oxpeckers prefer foraging on buffaloes and white rhinos, whereas red-billed oxpeckers prefer other ungulate hosts. Oxpeckers prefer feeding on weak mammals and will feed repeatedly on specific individuals within the same herd, with preference for the hosts with most ticks (Van Someren, 1951; Mooring & Mundy, 1996). *Buphagus* birds may also feed on skin, pieces of meat and blood of the mammalian host; thus they may enlarge wounds or even open partially healed wounds, although most publications suggest that such behaviour is quite rare (Bezuidenhout & Stutterheim, 1980; Stutterheim *et al.*, 1988; Weeks, 1999; McElligott *et al.*, 2004). The Yellow-billed was found to be more aggressive than the red-billed oxpecker.

Several programmes have attempted to reintroduce oxpeckers artificially into areas from which they had previously disappeared (Davison, 1963; Grobler, 1976, 1979; Couto, 1994). Oxpeckers are difficult to propagate in captivity and reintroduction programmes were based on translocation of birds that were captured in the wild. When enough birds were translocated, the oxpeckers established themselves successfully, and calf mortality from tick burdens was reduced (Couto, 1994).

For a successful oxpecker introduction programme, a number of problems must be considered (Mundy & Cook, 1975; Couto, 1994; C. Foggin, personal communication). The tick population should be monitored to assess the importance of the birds and the success of the introduction programme. A suitable location to capture the oxpeckers is needed (e.g. where they feed on tame calves) – mist nets can be used to capture the birds. Oxpeckers are social birds, therefore a minimum of 20 birds is recommended for translocation. It is important to overcome panic among cattle and game animals in the new location when they first encounter the oxpeckers. The new release area should have a suitable climate and should be at least 10 km away from areas treated with bird-poisoning anti-ectoparasitic compounds. Amitraz or pyrethroid-based dips do not seem to be harmful. The new area should include at least 500 head of game animals or of undipped domestic animals, grazing on at least 3000 ha.

Oxpeckers seem well suited to play an important role in IPM programmes for the control of ticks. However, their major value will be as only one part of a broader integrated programme, because substantial tick populations are known to remain in areas populated with oxpeckers (Masson & Norval, 1980; Norval & Lightfoot, 1982).

CONCLUDING REMARKS

The development of anti-tick biological control agents (BCAs) is still in its early research stage. Furthermore, the various steps required for commercialization of these products, including adaptation by companies (production, storage and delivery) and education of consumers (storage, application and evaluation of results), are still in the future. Nevertheless, we believe that the need to develop alternative control methods, and the increasing number of scientists working on anti-tick BCAs will yield useful results. At present, the use of entomopathogenic fungi is believed to be the most promising (Ostfeld *et al.*, 2006) while for unclear reasons too little attention was paid to developing the high control potential existing in the anti-tick parasitoids (Knipling & Steelman, 2000).

The fact that some BCAs and particular strains are far more specific in their selection of target pests than are chemical acaricides (in some cases, specific to individual stages of a given tick species) and that many strains are effective only under specific ecological conditions, gives them considerable advantages over pesticides, because harmful ecological effects are minimized. However, the specificity of some BCAs might require that numerous anti-tick BCAs be developed to deal with tick infestations in different environments.

Partial or total replacement of chemical acaricides with inundative use of tick pathogens and/or parasitoids would require considerable changes in the techniques of producers and suppliers. Products would probably need to be developed for specific small or medium-sized markets, including more pest-specific products with relatively short shelf-lives, and quicker-reacting suppliers. The commercial success of such tailor-made control methods may require on-site instructors or even controllers, with the necessary specific knowledge (which may need to be far more comprehensive than that required to use chemical pesticides). Furthermore, clients will have to accept the slower activity of BCAs compared with chemical controls. Commercial viability might vary, depending on the specific pest, the adaptability of BCA-based approaches and the availability of other effective management techniques.

Biological control of plant pests, by means of parasitoids, predatory mites, viruses, *B. thuringiensis*, bugs, beetles and others, has had several striking successes. These include the use of several enemies/pathogens simultaneously or in a predetermined order. However, only about 5% of all pest problems are treated with biological control methods (Van Dreisch & Bellows, 1996) and many difficulties have to be overcome in order to increase their use (Wysoki, 1998). Relatively few studies have been performed on the existence of promising natural enemies of ticks, or on their use against ticks in most parts of the world. Collaboration between biocontrol experts who have experience in managing plant pests and tick experts could lead to valuable developments in tick biocontrol.

ACKNOWLEDGEMENTS

We wish to thank to Dr Galina Gindin, Dr Amos Navon, Dr Manes Wysoki and Dana Ment for their important critical comments on the manuscript and to Dr Monica Mazuz-Lleshcovich for her work in translating papers published in Portuguese. This work was partially supported by the US Agency for International Development, CDR Grant No. TA-MOU-03-C22-008.

REFERENCES

Alekseev, A. N., Burenkova, L. A., Podboronov, V. M. & Chunikhin, S. P. (1995). Bacteriocidal qualities of ixodid tick (Acarina: Ixodidae) salivary cement plugs and their changes under the influence of a viral tick-borne pathogen. *Journal of Medical Entomology* **32**, 578–582.

Alekseev, E., Glazer, I. & Samish, M. (2006). Effect of soil texture and moisture on the activity of entomopathogenic nematodes against female *Boophilus annulatus* ticks. *BioControl* **51**, 507–518.

Alfeev, N. I. (1946). The utilization of *Hunterellus hookeri* How for the control of the tick, *Ixodes ricinus* L. and *Ixodes persulcatus*. *Review of Applied Entomology* **B34**, 108–109 (abstract).

Arruda, W., Lubeck, I., Schrank, A. & Vainstein, M. H. (2005). Morphological alterations of *Metarhizium anisopliae* during penetration of *Boophilus microplus* ticks. *Experimental and Applied Acarology* **37**, 231–244.

Assenga, S. P., You, M., Shy, C. H., *et al.* (2006). The use of a recombinant baculovirus expressing a chitinase from the hard tick *Haemaphysalis longicornis* and its potential application as a bioacaricide for tick control. *Parasitology Research* **98**, 111–118.

Barbosa, J. V., Daemon, E., Bittencourt, V. R. E. P. & Faccini, J. L. H. (1997). Effect of *Beauveria bassiana* on larvae molting to nympha of *Rhipicephalus sanguineus* (Latreille, 1806) (Acari: Ixodidae). *Brazilian Journal of Veterinary Parasitology* **6**, 53–56.

Barci, L. A. G. (1997). Biological control of the cattle tick *Boophilus microplus* (Acari, Ixodidae) in Brazil. *Arquivos Instituto Biológico, São Paulo* **64**, 95–101.

Barre, N., Mauleon, H., Garris, G. I. & Kermarrec, A. (1991). Predators of the tick *Amblyomma variegatum* (Acari: Ixodidae) in Guadeloupe, French West Indies. *Experimental and Applied Acarology* **12**, 163–170.

Basso, L. M. S., Monteiro, A. C., Belo, M. A. A., *et al.* (2005). Control of *Boophilus microplus* larvae by *Metarhizium anisopliae* in artificially infested pastures. *Pesquisa Agropecuaria Brasileira* **40**, 595–600.

Benjamin, M. A., Zhioua, E. & Ostfeld, R. S. (2002). Laboratory and field evaluation of the entomopathogenic fungus *Metarhizium anisopliae* (Deuteromycetes) for controlling questing adult *Ixodes scapularis* (Acari: Ixodidae). *Journal of Medical Entomology* **39**, 723–728.

Benson, M. J., Gawronski, J. D., Eveleigh, D. E. & Benson, D. R. (2004). Intracellular symbionts and other bacteria associated with deer ticks (*Ixodes scapularis*) from Nantucket and Wellfleet, Cape Cod, Massachusetts. *Applied and Environmental Microbiology* **70**, 616–620.

Bezuidenhout, J. D. & Stutterheim, C. J. (1980). A critical evaluation of the role played by the red-billed oxpecker *Buphagus erythrorhynchus* in the biological control of ticks. *Onderstepoort Journal of Veterinary Research* **47**, 51–75.

Bittencourt, V. R. E. P., Bahiense, T. C., Fernandes, E. K. K. & Souza, E. J. (2003). In-vivo action of *Metarhizium anisopliae* (Metschnikoff, 1879) Sorokin, 1883 sprayed over *Brachiaria decumbens* infested with *Boophilus microplus* larvae (Canestrini, 1887) (Acari: Ixodidae). *Brazilian Journal of Veterinary Parasitology* **12**, 38–42.

Bittencourt, V. R. E. P., Mascarenhas, A. G., Menezes, G. C. R. & Monteiro, S. G. (2000). In vitro action of the fungus *Metarhizium anisopliae* (Metschnikoff, 1879) Sorokin, 1883 and *Beauveria bassiana* (Balsamo) Vuillemin, 1912 on eggs of the tick *Anocentor nitens* (Neummann, 1897) (Acari: Ixodidae). *Revista Brasileiro de Medecina Veterinaria* **22**, 248–251.

Bittencourt, V. R. E. P., Massard, C. L. & Lima, A. F. (1994a). The action *of Metarhizium anisopliae* on eggs and larvae of tick *Boophilus microplus*. *Revisto da Universidade Rural, Serie Ciencias da Vida* **16**, 41–47.

Bittencourt, V. R. E. P., Massard, C. L. & Lima, A. F. (1994b). The action *of Metarhizium anisopliae* on free living stages of *Boophilus microplus*. *Revista da Universidade Rural, Serie Ciencias da Vida* **16**, 49–55.

Bittencourt, V. R. E. P., Massard, C. L., Lima, A. F. & De Lima, A. F. (1992). Use of the fungus *Metarhizium anisopliae* (Metschnikoff, 1879) Sorokin, 1883, in the control of the tick *Boophilus microplus* (Canestrini, 1887). *Arquivos da Universidade Federal Rural do Rio de Janeiro* **15**, 197–202.

Bittencourt, V. R. E. P., Peralva, S. L. F. S., Viegas, E. C. & Alves, S. B. (1996). Effect of *Beauveria bassiana* on eggs haching to larvae of *Boophilus microplus* (Canestrini, 1887) (Acari: Ixodidae). *Brazilian Journal of Veterinary Parasitology* **5**, 81–84.

Bittencourt, V. R. E. P., Souza, E. J., Costa, G. L. & Fagundes, A. S. (2002). Evaluation of a formulation of *Beauveria bassiana* for control of *Anocentor nitens*. *4th International Conference on Ticks and Tick-Borne Pathogens*, Banff, Alberta, Canada, p. 88.

Bittencourt, V. R. E. P., Souza, E. J., Peralva, S. L. F. S., Mascarenhas, A. G. & Alves, S. B. (1997). Effect of

entomopathogenic fungi *Beauveria bassiana* on engorged females of *Boophilus microplus* (Canestrini, 1887) (Acari: Ixodidae). *Brazilian Journal of Veterinary Parasitology* **6**, 49–52.

Bittencourt, V. R. E. P., Souza, E. J., Peralva, S. L. F. S. & Reis, R. C. S. (1999). Efficacy of the fungus *Metarhizium anisopliae* (Metschnikoff, 1887) Sorokin,1883 in field test with bovines naturally infested with the tick *Boophilus microplus* (Canestrini, 1887) (Acari: Ixodidae). *Revista Brasileiro de Medicina Veterinaria* **21**, 78–81.

Boichev, D. & Rizvanov, K. (1960). Relation of *Botrytis cinerea* Pers. to ixodid ticks. *Zoologicheskii Zhurnal Akademija Nauk USSR* **39**, 462–462.

Bowman, J. L., Logan, T. M. & Hair, J. A. (1986). Host suitability of *Ixodiphagus texanus* Howard on five species of hard ticks. *Journal of Agricultural Entomology* **3**, 1–9.

Brown, R. S., Reichelderfer, C. F. & Anderson, W. R. (1970). An endemic disease among laboratory populations of *Dermacentor andersoni* (= *D. venustus*) (Acarina: Ixodidae). *Journal of Invertebrate Pathology* **16**, 142–143.

Brum, J. G. W. & Teixeira, M. O. (1992). Acaricidal activity of *Cedea lapagei* on engorged females of *Boophilus microplus* exposed to the environment. *Arquivos Brasileiros de Medicina Veterinaria i Zootecnologia* **44**, 543–544.

Brum, J. G. W., Faccini, J. L. H. & Do Amaral, M. M. (1991). Infection in engorged females of *Boophilus microplus* (Acari: Ixodidae). II. Histopathology and in vitro trials. *Arquivos Brasileiros de Medicina Veterinaria i Zootecnologia* **43**, 35–37.

Brum, J. G. W., Teixeira, M. O. & Da Silva, E. G. (1991). Infection in engorged females of *Boophilus microplus* (Acari: Ixodidae). I. Etiology and seasonal incidence. *Arquivos Brasileiros de Medicina Veterinaria i Zootecnlogia* **43**, 25–30.

Buresova, V., Franta, Z. & Kopacek, P. (2006). A comparison of *Chryseobacterium indologenes* pathogenicity to the soft tick *Ornithodoros moubata* and hard tick *Ixodes ricinus*. *Journal of Invertebrate Pathology* **93**, 96–104.

Butt, T. M., Jackson, C. W. & Magan, N. (eds.) (2001). *Fungi as Biocontrol Agents: Progress, Problems and Potential*. Wallingford, UK: CAB International.

Camacho, E. R., Navaro, G., Rodriguez, R. M. & Murillo, E. Y. (1998). Effectiveness of *Verticillium lecanii* against the parasitic stage of the tick *Boophilus microplus* (Acari: Metastigmata: Ixodidae). *Revista Colombiana Entomologia* **24**, 67–69.

Carneiro, M. E., Monteiro, S. G. M., Daemon, E. & Bittencourt, V. R. E. P. (1999). Effects of isolate 986 of the fungi *Beauveria bassiana* (Bals.) Vuill., on eggs of the tick *Anocentor nitens* (Neumann, 1897) (Acari: Ixodidae). *Revista Brasileira de Parasitologia* **8**, 59–62.

Castineiras, A., Jimeno, G., Lopez, M. & Sosa, L. M. (1987). Effect of *Beauveria bassiana*, *Metarhizium anisopliae* (Fungi, Imperfecti) and *Pheidole megacephala* (Hymenopthera, Formicidae) on eggs of *Boophilus microplus* (Acarina: Ixodidae). *Revista Salud Animal* **9**, 288–293.

Castro, A. B. A., Bittencourt, V. R. E. P., Daemon, E. & Viegas, E. C. (1997). Efficacy of the fungus *Metarhizium anisopliae* (isolate 959) on the tick *Boophilus microplus* in a stall test. *Revista da Universidade Rural, Serie Ciencias da Vida* **19**, 73–82.

Ceraul, S. M., Sonenshine, D. E. & Hynes, W. L. (2002). Resistance of the tick *Dermacentor variabilis* (Acari: Ixodidae) following challenge with the bacterium *Escherichia coli* (Enterobacteriales: Enterobacteriaceae). *Journal of Medical Entomology* **39**, 376–383.

Chandler, D., Davidson, G., Pell, J. K., *et al.* (2000). Fungal biocontrol of Acari. *Biocontrol Science and Technology* **10**, 357–384.

Cherepanova, N. P. (1964). Fungi which are found on ticks. *Botanicheskii Zhurnal* **49**, 696–699.

Cobb, S. (1942). Tick parasites on Cape Cod. *Science* **95**, 503.

Cole, M. M. (1965). *Biological Control of Ticks by the Use of Hymenopterous Parasites: A Review*. World Health Organization Publication WHO/ebl/43.65, 1–12.

Cooley, R. A. & Kohls, G. M. (1934). A summary on tick parasites. In *Proceedings of the 5th Pacific Science Congress*, pp. 3375–3381.

Copping, L. G. (2001). *The Bio-Pesticide Manual*, 2nd edn. Famham, UK: British Crop Protection Council.

Correia, A. C. B., Fiorin, A. C., Monteiro, A. C. & Verissimo C. J. (1998). Effects of *Metarhizium anisopliae* on the tick *Boophilus microplus* (Acari: Ixodidae) in stabled cattle. *Journal of Invertebrate Pathology* **71**, 189–191.

Correia, A. C. B., Monteiro, A. C. & Fiorin, C. (1994). The effect of *Metarhizium anisopliae* concentrations on *Boophilus microplus* under laboratory conditions. *Simposio de Controle Biologico Anais* **4**, 98–98.

Costa, G. L., Sarquis, M. I. M., De Moraes, A. M. L. & Bittencourt, V. R. E. P. (2001). Isolation of *Beauveria bassiana* and *Metarhizium anisopliae* var *anisopliae* from *Boophilus microplus* tick (Canestrini, 1887), in Rio de Janeiro State, Brazil. *Micropathologia* **154**, 207–209.

Couto, J. T. (1994). Operation oxpecker. *The Farmers*, pp. 10–11.

Da Nunes, T. L. S., Graminha, E. B. N., Maia, A. S., *et al.* (2001). The use of *Sporothrix insectorum* and *Paecilomyces fumosoroseus* against *Boophilus microplus* (Canestrini, 1887): in vitro assay and electronic microscopy. *Semina: Ciencia Agrarias Londrina* **22**, 55–60.

Davis, A. J. (1986). Bibliography of the Ixodiphagini (Hymenoptera, Chalcidoidea, Encyrtidae), parasites of ticks (Acari, Ixodidae), with notes on their biology. *Tijdschrift voor Entomologie* **129**, 181–190.

Davison, E. (1963). Introduction of oxpeckers (*Buphagus africanus* and *B. erythrorhynchus*) into McIlwaine National Park. *Ostrich* **34**, 172–173.

De Bach, P. & Rosen, D. (1991). *Biological Control by Natural Enemies*, 2nd edn. Cambridge, UK: Cambridge University Press.

Doube, B. M. & Heath, A. C. G. (1975). Observations on the biology and seasonal abundance of an encyrtid wasp, a parasite of ticks in Queensland. *Journal of Medical Entomology* **12**, 443–447.

Dreyer, K., Fourie, L. J. & Kok, D. J. (1997). Predation of livestock ticks by chickens as a tick-control method in a resource-poor urban environment. *Ondersterpoort Journal of Veterinary Research* **64**, 273–276.

Dubois, N. R. & Dean, D. H. (1995). Synergism between Cry IA insecticidal crystal proteins and spores of *Bacillus thuringiensis*, other bacterial spores, and vegetative cells against *Lymantria dispar* (Lepidoptera: Lymantridae) larvae. *Environmental Entomology* **24**, 1741–1747.

El-Sadawy, H. A. E. & Habeeb, S. M. (1998). Testing some entomopathogenic nematodes for the biocontrol of *Hyalomma dromedarii* Koch (Acarina: Ixodidae). *Journal of the Union of Arab Biologists, Cairo, A, Zoology* **10**, 1–11.

Estrada-Peña, A., Gonzalez, J. & Casasolas, A. (1990). The activity of *Aspergillus ochraceus* (Fungi) on replete females of *Rhipicephalus sanguineus* (Acari: Ixodidae) in natural and experimental conditions. *Folia Parasitologica* **37**, 331–336.

Federova, S. Z. (1988). *Entomopathogenic Bacteria and Fungi as Regulators of the Numbers of the Tick* Argas persicus *Oken, 1818*. Frunze, Kyrgyzstan: Pensoft Publishers.

Fernandes, V. K. K., Costa, G. L., De Moraes, A. M. L. & Bittencourt, V. R. E. P. (2004). Entomopathogenic potential of *Metarhizium anisopliae* isolated from engorged females and tested in eggs and larvae of *Boophilus microplus* (Acari: Ixodidae). *Journal of Basic Microbiology* **44**, 270–274.

Fernandes, E. K. K., Costa, G. L., De Moraes, M. L., Zahner, V. & Bittencourt, V. R. E. P. (2006). Study on morphology, pathogenicity, and genetic variability of *Beauveria bassiana* isolates obtained from *Boophilus microplus* ticks. *Parasitology Research* **98**, 324–332.

Fernandes, E. K. K., Costa, G. L., Souza, E. J., De Moraes, A. M. L. & Bittencourt, V. R. E. P. (2003). *Beauveria bassiana* isolated from engorged females and tested against eggs and larvae of *Boophilus microplus* (Acari: Ixodidae). *Journal of Basic Microbiology* **43**, 393–398.

Flexner, J. L. & Belnavis, D. L. (2000). Microbial insecticides. In *Biological and Biotechnological Control of Insect Pests*, eds. Rechcigl, J. E. & Rechcigl, N. A., pp. 35–62. Boca Raton, FL: Lewis Publishers.

Flor, L. B., Kurtti, T. J. & Munderloh, U. G. (2005). Characterization of a *Beauveria bassiana* isolate from feral black-legged ticks, *Ixodes scapularis* (Say). In *Proceedings of 38th Annual Meeting of the Society for Invertebrate Pathology*, 7–11 August 2005, Anchorage, Alaska.

Fogaca, A. C., Da Silva, P. I. Jr, Miranda, M. T. M., *et al.* (1999). Antimicrobial activity of a bovine hemoglobin fragment in the tick *Boophilus microplus*. *Journal of Biological Chemistry* **274**, 25330–25334.

Frazzon, A. P. G., Vaz, J. I. S., Masuda, A., Schrank, A. & Vainstein, M. H. (2000). In-vitro assessment of *Metarhizium anisopliae* isolates to control the cattle tick *Boophilus microplus*. *Veterinary Parasitology* **94**, 117–125.

Freitas-Ribeiro, G., Furlong, J., Vasconcelos, V. O., Dolinski, C. & Loures-Ribeiro, A. (2005). Analysis of biological parameters of *Boophilus microplus* Canestrini, 1887 exposed to entomopathogenic nematodes *Steinernema carpocapsae* Santa Rosa and All strains (Steinernema: Rhabditida). *Brazilian Archives of Biology and Technology*. **48**, 911–919.

Gahan, A. B. (1934). On the identities of chalcidoid tick parasites (Hymenoptera). *Washington Entomological Society Proceedings* **36**, 87–88.

Garcia, M. V., Monteiro, A. C. & Szabo, M. P. J. (2004). Colonization and lesions on engorged female *Rhipicephalus sanguineus*, caused by *Metarhizium anisopliae*. *Ciencia Rural* **34**, 1513–1518.

Garcia, M. V., Monteiro, A. C., Szabo, M. J. P., Prette, N. & Bechara, G. H. (2005). Mechanism of infection and colonization of *Rhipicephalus sanguineus* eggs by *Metarhizium anisopliae* as revealed by scanning electron microscopy and histopathology. *Brazilian Journal of Microbiology* **36**, 368–372.

Gaugler, R. (ed.) (2002). *Entomopathogenic Nematology*. Wallingford, UK: CAB International.

Georgis, R. & Manweiler, S. A. (1994). Entomopathogenic nematodes: a developing biological control technology. *Agricultural and Zoological Review* **6**, 63–94.

Gill, S. S., Cowles, E. A. & Pietrantonio, P. V. (1992). The mode of action of *Bacillus thuringiensis* endotoxins. *Annual Review of Entomology* **37**, 615–636.

Gindin, G., Samish, M., Alekseev, E. A. & Glazer, I. (1999). The pathogenicity of entomopathogenic fungi to *Boophilus annulatus* ticks. *Insect Pathogens and Insect Nematodes* **23**, 155–157.

Gindin, G., Samish, M., Alekseev, E. & Glazer, I. (2001). The susceptibility of *Boophilus annulatus* (Ixodidae) ticks to entomopathogenic fungi. *Biocontrol Science and Technology* **11**, 111–118.

Gindin, G., Samish, M., Zangi, G., Mishoutchenko, A. & Glazer, I. (2002). The susceptibility of different species and stages of ticks to entomopathogenic fungi. *Experimental and Applied Acarology* **28**, 283–288.

Ginsberg, H. S. & Zhioua, E. (1999). Influence of deer abundance on the abundance of questing adult *Ixodes scapularis* (Acari: Ixodidae). *Journal of Medical Entomology* **36**, 376–381.

Ginsberg, H. S., Lebrun, R. A., Heyer, K. & Zhioua, E. (2002). Potential nontarget effects of *Metarhizium anisopliae* (Deuteromycetes) used for biological control of ticks (Acari: Ixodidae). *Environmental Entomology* **31**, 1191–1196.

Glazer, I. (2001). Survival biology. *In Entomopathogenic Nematology*, ed. Gaugler, R., pp. 169–187. Wallingford, UK: CAB International.

Glazer, I., Alekseev, E. & Samish, M. (2001). Factors affecting the virulence of entomopathogenic nematodes to engorged female *Boophilus annulatus*. *Journal of Parasitology* **87**, 808–812.

Gomathinayagam, S., Cradock, K. R. & Needham, G. R. (2002). Pathogenicity of the fungus *Beauveria bassiana* (Balsamo) to *Amblyomma americanum* (L.) and *Dermacentor variabilis* (Say) ticks (Acari: Ixodidae). *International Journal of Acarology* **28**, 395–397.

Gorshkova, G. J. (1966). Reduction of fecundity of ixodid tick females by fungal infection. *Vetsnik Leningradskogo University Serie Biologia* **21**, 13–16.

Graf, J. F. (1979). The biology of an encyrtid wasp parasitizing ticks on the Ivory Coast. In *Recent Advances in Acarology*, vol. 1 ed. Rodriguez, J. G., pp. 463–468. New York: Academic Press.

Grobler, J. H. (1976). The introduction of oxpeckers into the Rhodes Matopos National Park. *Honeyguide* **87**, 23–25.

Grobler, J. H. (1979). The re-introduction of oxpeckers *Buphagus africanus* and *B. erythrorhynchus* to Rhodes Matopos National Park, Rhodesia. *Biological Conservation* **15**, 151–158.

Guangfu, T. (1984). Experiment of infection and killing of *Hyalomma detritum* with fungi. *Journal of Veterinary Sciences, China* **7**, 11–13.

Guerra, R. M. S. N. C., Teixeira Filho, W. L., Costa, G. L. & Bittencourt, V. R. E. P. (2001). Fungus isolated from *Rhipicephalus sanguineus* (Acari: Ixodidae), *Cochliomya macellaria* (Diptera: Muscidae) and *Musca domestica* (Diptera: Muscidae), naturally infected on Seropedica, Rio de Janeiro. *Ciencia Animal* **11**, 133–136.

Habib, S. M. & Sewify, G. H. (2002). Biological control of the fowl tick *Argas (Persicargas) persicus* (Laterreille) by the entomopathogenic fungi *Beauveria bassiana* and *Metarhizium anisopliae*. *Egyptian Journal of Biological Pest Control* **12**, 11–13.

Hajek, A. E. & St Leger, R. J. (1994). Interaction between fungal pathogens and insect hosts. *Annual Review of Entomology* **39**, 293–322.

Hassan, S. M., Dipeolu, O. O., Amoo, A. O. & Odhiambo, T. R. (1991). Predation on livestock ticks by chickens. *Veterinary Parasitology* **38**, 199–204.

Hassan, S. M., Dipeolu, O. O. & Munyinyi, D. M. (1992). Influence of exposure period and management methods on the effectiveness of chickens as predators of ticks infesting cattle. *Veterinary Parasitology* **43**, 301–309.

Hassanain, M. A., Derbala, A. A., Abdel-Barry, N. A., El-Sherif, M. A. & El-Sadawy, H. A. E. (1999). Biological control of ticks (Argasidae) by entomopathogenic nematodes. *Egyptian Journal of Biological Pest Control* **7**, 41–46.

Hassanain, M. A., El Garhy, M. F., Abdel-Ghaffar, F. A., El-Sharaby, A. & Megeed, K. N. A. (1997). Biological control studies of soft and hard ticks in Egypt., I. The effect of *Bacillus thuringiensis* varieties on soft and hard ticks (Ixodidae). *Parasitology Research* **83**, 209–213.

Hendry, D. A. & Rechav, Y. (1981). Acaricidal bacteria infecting laboratory colonies of the tick *Boophilus decoloratus* (Acarina: Ixodidae). *Journal of Invertebrate Pathology* **38**, 149–151.

Hill, D. E. (1998). Entomopathogenic nematodes as control agents of developmental stages of the black-legged tick, *Ixodes scapularis*. *Journal of Parasitology* **84**, 1124–1127.

Hoogstraal, H. (1977). Pathogens of Acarina (ticks). In *Pathogens of Medically Important Arthropods*, eds. Roberts, D. W. & Strand, M. A., pp. 337–342. Geneva, Switzerland: World Health Organization.

Hornbostel, V. L., Ostfeld, R. S. & Benjamin, M. A. (2005). Effectiveness of *Metarhizium anisopliae* (Deuteromycetes) against *Ixides scapularis* (Acari: Ixodidae) engorging on *Peromyscus leucopus*. *Journal of Vector Ecology* **30**, 91–101.

Hornbostel, V. L., Ostfeld, R. S., Zhioua, E. & Benjamin, M. A. (2004). Sublethal effects of *Metarhizium anisopliae* (Deuteromycetes) on engorged larval, nymphal, and adult *Ixodes scapularis* (Acari: Ixodidae). *Journal of Medical Entomology* **41**, 922–929.

Hornbostel, V. L., Zhioua, E., Benjamin, M. A., Ginsberg, H. S. & Ostfeld, R. S. (2005). Pathogenicity of *Metarhizium anisopliae* (Deuteromycetes) and permethrin to *Ixodes scapularis* (Acari: Ixodidae) nymphs. *Experimental and Applied Acarology* **35**, 301–316.

Howard, L. O. (1907). A chalcidid parasite of a tick. *Entomological News* **18**, 375–378.

Howard, L. O. (1908). Another chalcidoid parasite of a tick. *Canadian Entomologist* **40**, 239–241.

Hu, R. & Hyland, K. E. (1997). Prevalence and seasonal activity of the wasp parasitoid, *Ixodiphagus hookeri* (Hymenoptera: Encyrtidae) in its tick host, *Ixodes scapularis* (Acari: Ixodidae). *Systematic and Applied Acarology* **2**, 95–100.

Hu, R. & Hyland, K. E. (1998). Effects of the feeding process of *Ixodes scapularis* (Acari: Ixodidae) on embryonic development of its parasitoid, *Ixodiphagus hookeri* (Hymenoptera: Encyrtidae). *Journal of Medical Entomology* **35**, 1050–1053.

Hu, R., Hyland, K. E. & Mather, T. N. (1993). Occurrence and distribution in Rhode Island of *Hunterellus hookeri* (Hymenoptera: Encyrtidae), a wasp parasitoid of *Ixodes dammini*. *Journal of Medical Entomology* **30**, 277–280.

Hu, R., Hyland, K. E. & Oliver, J. H. (1998). A review on the use of *Ixodiphagus* wasps (Hymedoptera: Encyrtidae) as natural enemies for the control of ticks (Acari: Ixodidae). *Systematic and Applied Acarology* **3**, 19–28.

Jenkins, D. W. (1964). *Pathogens, Parasites and Predators of Medically Important Arthropods*, Supplement No. 30. Geneva, Switzerland: World Health Organization.

Johns, R., Sonenshine, D. E. & Hynes, W. L. (1998). Control of bacterial infections in the hard tick *Dermacentor variabilis* (Acari: Ixodidae): evidence for the existence of antimicrobial proteins in tick hemolymph. *Journal of Medical Entomology* **35**, 458–464.

Kaaya, G. P. (2000). Laboratory and field evaluation of entomogenous fungi for tick control. *Annals of the New York Academy of Science* **916**, 559–564.

Kaaya, G. P. (2003). Prospects for innovative methods of tick control in Africa. *Insect Science and Application* **23**, 59–67.

Kaaya, G. P. & Hassan, S. (2000). Entomogenous fungi as promising biopesticides for tick control. *Experimental and Applied Acarology* **24**, 913–926.

Kaaya, G. P. & Mwangi, E. N. (1995). Control of livestock ticks in Africa: possibilities of biological control using the entomogenous fungi *Beauveria bassiana* and *Metarhizium anisopliae*. *Proceedings and Abstracts, Kruger National Park, South Africa* **1995**, 5–16.

Kaaya, G. P., Mwangi, E. N. & Ouna, E. A. (1996). Prospects for biological control of livestock ticks, *Rhipicephalus appendiculatus* and *Amblyomma variegatum*, using the entomogenous fungi *Beauveria bassiana* and *Metarhizium anisopliae*. *Journal of Invertebrate Pathology* **67**, 15–20.

Kaaya, G. P., Samish, M. & Glazer, I. (2000). Laboratory evaluation of pathogenicity of entomopathogenic nematodes to African tick species. *Annals of the New York Academy of Science* **916**, 303–308.

Kaiser, M. N. & Hoogstraal, H. (1958). *Hunterellus theileri* Fiedler (Encyrtidae, Chalcidoidea) parasitizing an African *Hyalomma* tick on a migrant bird in Egypt. *Journal of Parasitology* **44**, 392–392.

Kalsbeek, V., Frandsen, F. & Steenberg, T. (1995). Entomopathogenic fungi associated with *Ixodes ricinus* ticks. *Experimental and Applied Acarology* **19**, 45–51.

Kirkland, B. H., Cho, E. M. & Keyhani, N. O. (2004). Differential susceptibility of *Amblyomma maculatum* and *Amblyomma americanum* (Acari: Ixodidea) to the entomopathogenic fungi *Beauveria bassiana* and *Metarhizium anisopliae*. *Biological Control* **31**, 414–421.

Kirkland, B. H., Eisa, A. & Keyhani, N. O. (2005). Oxalic acid as a fungal acaricidal virulence factor. *Journal of Medical Entomology* **42**, 346–351.

Kirkland, B. H., Westwood, G. S. & Keyhani, N. O. (2004). Pathogenicity of entomopathogenic fungi *Beauveria bassiana* and *Metarhizium anisopliae* to Ixodidae tick species *Dermacentor variabilis, Rhipicephalus sanguineus*, and *Ixodes scapularis. Journal of Medical Entomology* **41**, 705–711.

Knipling, E. F. & Steelman, C. D. (2000). Feasibility of controlling *Ixodes scapularis* ticks (Acari: Ixodidae), the vector of Lyme disease, by parasitoid augmentation. *Journal of Medical Entomology* **37**, 645–652.

Knulle, W. & Rudolph, D. (1982). Humidity relationships and water balance of ticks. In *Physiology of Ticks*, eds. Obenchain, F. D. & Galun, R., pp. 43–70. Oxford, UK: Pergamon Press.

Kocan, K. M., Blouin, E. F., Pidherney, M. S., *et al.* (1998*a*). Entomopathogenic nematodes as a potential biological control method for ticks. *Annals of the New York Academy of Science* **849**, 355–356.

Kocan, K. M., Pidherney, M. S., Blouin, E. F., *et al.* (1998*b*). Interaction of entomopathogenic nematodes (Steinernematidae) with selected species of ixodid ticks (Acari: Ixodidae). *Journal of Medical Entomology* **35**, 514–520.

Kok, O. B. & Petney, T. N. (1993). Small and medium sized mammals as predators of ticks (Ixodoidea) in South Africa. *Experimental and Applied Acarology* **17**, 733–740.

Kolomyetz, U. S. (1950). *Aspergillus fumigatus* as a parasite of ticks. *Priroda* **39**, 64–65.

Krasnoff, S. B., Sommers, C. H., Moon, Y., *et al.* (2007). Production of mutagenic metabolites by *Metarhizium anisopliae. Journal of Agricultural and Food Chemistry* **54**, 7083–7088.

Krylov, B. (1972). Some problems in the study of fungal diseases of the bed bugs *Cimex lectularius* and ticks *Argas persicus*. In *Microbiological Methods for the Control of Poultry Ectoparasites*, pp. 53–55. Frunze, Kyrgyzstan: Ilim.

Larrousse, F., King, A. G. & Wolbach, S. B. (1928). The overwintering in Massachusetts of *Ixodiophagus caucurtei. Science* **67**, 351–353.

Lipa, J. J. (1971). Microbial control of mites and ticks. In *Microbial Control of Insects and Mites*, eds. Burges, H. D. & Hussey, N. W., pp. 357–373. New York: Academic Press.

Mani, M. S. (1941). Studies on Indian parasitic Hymenoptera. I. *Indian Journal of Entomology* **3**, 25–36.

Maranga, R. O. & Kenyatta, J. (2006). Field trials for the control of *Amblyomma variegatum* (Ixodidae) using fungi in pheromone-baited traps. In *Proceedings of 12th International Congress of Acarology*, Amsterdam, p. 116.

Maranga, R. O., Kaaya, G. P., Mueke, J. M. & Hassanali, A. (2005). Effects of combining the fungi *Beauveria bassiana* and *Metarhizium anisopliae* on the mortality of the tick *Amblyomma variegatum* (Ixodidae) in relation to seasonal changes. *Mycopathologia* **159**, 527–532.

Martin, P. A. W. & Schmidtmann, E. T. (1998). Isolation of aerobic microbes from *Ixodes scapularis* (Acari: Ixodidae), the vector of Lyme disease in the eastern United States. *Journal of Economic Entomology* **91**, 864–868.

Masson, C. A. & Norval, R. A. I. (1980). The ticks of Zimbabwe. I. The genus *Boophilus. Zimbabwe Veterinary Journal* **11**, 36–43.

Mather, T. N., Piesman, J. & Spielman, A. (1987). Absence of spirochaetes (*Borrelia burgdorferi*) and piroplasms (*Babesia microti*) in deer ticks (*Ixodes dammini*) parasitized by chalcid wasps (*Hunterellus hookeri*). *Medical and Veterinary Entomology* **1**, 3–8.

Mauleon, H., Barre, N. & Panoma, S. (1993). Pathogenicity of 17 isolates of entomophagous nematodes (Steinernematidae and Heterorhabditidae) for the ticks *Amblyomma variegatum* (Fabricius), *Boophilus microplus* (Canestrini) and *Boophilus annulatus* (Say). *Experimental and Applied Acarology* **17**, 831–838.

McElligott, A. G., Maggini, I., Hunziker, L. & König, B. (2004). Interactions between red-billed oxpeckers and black rhinos in captivity. *Zoo Biology* **23**, 347–354.

Mendes, M. C., Batista-Filho, A., Leite, L. G. & Barci, L. A. G. (1995). The virulence of *Metarhizium anisopliae* to engorged *Boophilus microplus* females under laboratory conditions. *Congresso Brasileiro de Entomologia, Caxambu* **15**, 593.

Ment, D., Gindin, G., Mishoutchenko A., Glazer, I. & Samish, M. (2006). The entomopathogenic fungi–ticks interaction. In *Proceedings of 12th International Congress of Acarology*, Amsterdam, p. 122.

Monteiro, A. C., Correia, A. C. B. & Fiorin, A. C. (1994). Pathogenicity of isolates of *Metarhizium anisopliae* upon *Boophilus microplus* (Acari: Ixodidae) under laboratory conditions. In *4th Simposio de Controle Biologico*, Gramado, Brazil.

Monteiro, S. G., Bahiense, T. C. & Bittencourt, V. R. E. P. (2003). Action of the fungus *Beauveria bassiana* (Balsamo) Vuillemin, 1912 on the parasitic phase of the tick *Anocentor nitens* (Neumann, 1897) Schulze, 1937 (Acari: Ixodidae). *Ciencia Rural* **33**, 559–563.

Monteiro, S. G., Bittencourt, V. R. E. P. & Daemon, E. (2001). Pathogenicity of isolates CG17, EP01 and 986 of the fungi *Beauveria bassiana* on larvae of the tick *Anocentor nitens* (Neumann, 1897) (Acari: Ixodidae) in laboratory. *Revista Brasileira de Ciencia Veterinaria* **8**, 144–146.

Monteiro, S. G. M., Bittencourt, V. R. E. P., Daemon, E. & Faccini, J. L. H. (1998*a*). Effect of the entomopathogenic fungi *Metarhizium anisopliae* and *Beauveria bassiana* on eggs of *Rhipicephalus sanguineus* (Acari: Ixodidae). *Ciencia Rural, Santa Maria* **28**, 461–466.

Monteiro, S. G. M., Bittencourt, V. R. E. P., Daemon, E. & Faccini, J. L. H. (1998*b*). Pathogenicity under laboratory conditions of the fungi *Beauveria bassiana* and *Metarhizium anisopliae* on larvae of the tick *Rhipicephalus sanguineus* (Acari: Ixodidae). *Brazilian Journal of Veterinary Parasitology* **7**, 113–116.

Monteiro, S. G., Carneiro, M. E., Bittencourt, V. R. E. P. & Daemon, E. (1998*c*). Effect of isolate 986 of the fungi

Beauveria bassiana (Bals) Vuill on engorged females of *Anocentor nitens* Neumann, 1897 (Acari: Ixodidae). *Arquivo Brasileiro de Medicina Veterinaria: Zootecnologia*. **50**, 673–676.

Mooring, M. S. & Mundy, P. J. (1996). Factors influencing host selection by yellow-billed oxpeckers at Matoba National Park, Zimbabwe. *African Journal of Ecology* **34**, 177–188.

Mundy, P. J. & Cook, A. W. (1975). Observation of the yellow-billed oxpecker *Buphagus africanus* in northern Nigeria. *Ibis* **117**, 504–506.

Mwangi, E. N. (1991). The role of predators, parasitoids and pathogens in regulating natural populations of the non-parasitic stages of *Rhipicephalus appendiculatus*, Neumann and other livestock ticks, and related aspects of the tick's ecology. Unpublished Ph.D. thesis. Kenyatta University, Nairobi, Kenya.

Mwangi, E. N. & Kaaya, G. P. (1997). Prospects of using tick parasitoids (Insecta) for tick management in Africa. *International Journal of Acarology* **23**, 215–219.

Mwangi, E. N., Dipeolu, O. O., Newson, R. M., Kaaya, G. P. & Hassan, S. M. (1991). Predators, parasitoids and pathogens of ticks: a review. *Biocontrol Science and Technology* **1**, 147–156.

Mwangi, E. N., Hassan, S. M., Kaaya, G. P. & Essuman, S. (1997). The impact of *Ixodiphagus hookeri*, a tick parasitoid, on *Amblyomma variegatum* (Acari: Ixodidae) in a field trial in Kenya. *Experimental and Applied Acarology* **21**, 117–126.

Mwangi, E. N., Kaaya, G. P. & Essumen, S. (1995). Experimental infections of the tick *Rhipicephalus appendiculatus* with entomopathogenic fungi, *Beauveria bassiana* and *Metarhizium anisopliae*, and natural infections of some ticks with bacteria and fungi. *African Journal of Zoology* **109**, 151–160.

Mwangi, E. N., Newson, R. M. & Kaaya, G. P. (1991). Predation of free-living engorged female *Rhipicephalus appendiculatus*. *Experimental and Applied Acarology* **12**, 153–162.

Mythlli, P., Gomathinayagam, S., John, L. & Dhinakar, R. (2004). Biological control of cattle tick, *Boophilus microplus* by entomopathogenic fungus *Beauveria bassiana*. *Pesticide Research Journal* **16**, 11–12.

Noda, H., Munderloh, U. G. & Kurtti, T. J. (1997). Endosymbionts of ticks and their relationship to *Wolbachia* spp. and tick-borne pathogens of humans and animals. *Applied and Environmental Microbiology* **63**, 3926–3932.

Norval, R. A. I. & Lightfoot, C. J. (1982). Tick problems in wildlife in Zimbabwe: factors influencing the occurrence and abundance of *Rhipicephalus appendiculatus*. *Zimbabwe Veterinary Journal* **13**, 11–20.

Oliver, J. H. Jr (1964). A wasp parasite of the possum tick, *Ixodes tasmani*, in Australia. *Pan-Pacific Entomologist* **40**, 227–230.

Onofre, B. O., Miniuk, C. M., De Barros, N. M. & Azevedo, J. L. (2001). Pathogenicity of four strains of entomopathogenic fungi against the bovine tick *Boophilus microplus*. *American Journal of Veterinary Research* **62**, 1478–1480.

Ostfeld, R. S., Amber, P., Hornbostel, V. L., Benjamin, M. A. & Keesing, F. (2006). Controlling ticks and tick-borne zoonoses with biological and chemical agents. *BioScience* **56**, 383–404.

Paiao, J. C. V., Monteiro, A. C. & Kronka, S. N. (2001). Susceptibility of the cattle tick *Boophilus microplus* (Acari: Ixodidae) to isolates of the fungus *Beauveria bassiana*. *World Journal of Microbiology and Biochemistry* **17**, 245–251.

Petney, T. N. & Kok, O. B. (1993). Birds as predators of ticks (Ixodoidea) in South Africa. *Experimental and Applied Acarology* **17**, 393–403.

Petrischeva, P. A. & Zhmayeva, Z. M. (1949). Natural enemies of field ticks. *Zoologichesku Zhurnal* **28**, 479–481.

Poinar, G. O. Jr (1973). *Entomogenous Nematodes: A Manual and Host List of Insect–Nematode Associations*. Leiden, Netherlands: E. J. Brill.

Polar, P., Kairo, M. T. K., Moore, D., Pegram, R. & John, S. A. (2005). Comparison of water, oils and emulsifiable adjuvant oils as formulating agents for *Metarhizium anisopliae* for use in control of *Boophilus microplus*. *Mycopathologia* **160**, 151–157.

Prette, N., Monteiro, A. C., Garcia, M. V. & Soares, V. E. (2005). Pathogenicity of *Beauveria bassiana* isolates towards eggs, larvae and engorged nymphs of *Rhipicephalus sanguineus*. *Ciencia Rural* **35**, 855–861.

Ramamoorthy, R. & Scholl-Meeker, D. (2001). *Borrelia burgdorferi* proteins whose expression is similarly affected by culture temperature and pH. *Infection and Immunity* **69**, 2739–2742.

Reis, R. C. S., Chacon, S. C., Bittencourt, V. R. E. P. & Faccini, J. L. H. (2003). Effect of the fungi *Beauveria bassiana* (Balsamo) and *Metarhizium anisopliae* Sorokin, 1883, on nymphal ecdysis of *Amblyomma cooperi* (Nuttal; Warburton, 1908) (Acari: Ixodidae). *Revista Brasileira de Parasitologia Veterinaria* **12**, 68–70.

Reis, R. C. S., Melo, D. R. & Bittencourt V. R. E. P. (2004). Effects of *Beauveria bassiana* (Bals) Vuill and *Metarhizium*

anisopliae (Metsc) Sorok on engorged females of *Amblyomma cajennense* (Fabricius, 1787) in laboratory conditions. *Arquivos Brasileiros de Medicina Veterinaria i Zootecnologia* **56**, 788–791.

Reis, R. C. S., Melo, D. R., Souza, E. J. & Bittencourt, V. R. E. P. (2001). *In vitro* action of the fungi *Beauveria bassiana* Vuill and *Metarhizium anisopbiae* Sorok on nymphs and adults of *Amblyomma cajenense* (Acari: Ixodidae). *Arquivos Brasileiros de Medicina Veterinaria i Zootecnologia* **53**, 544–547.

Rijo, E. (1998). Biological control of ticks with entomopathogenic fungi. *Revista Pectuaria de Nicaragua* **22**, 17–18.

Robertson, A. & Jarvis, A. M. (2000). Oxpeckers in northeastern Namibia: recent population trends and the possible negative impacts of drought and fire. *Biological Conservation* **92**, 241–247.

Samish, M. & Alekseev, E. A. (2001). Arthropods as predators of ticks (Ixodoidea). *Journal of Medical Entomology* **38**, 1–11.

Samish, M. & Glazer, I. (2001). Entomopathogenic nematodes for the biocontrol of ticks. *Trends in Parasitology* **17**, 368–371.

Samish, M. & Rehacek, J. (1999). Pathogens and predators of ticks and their potential in biological control. *Annual Review of Entomology* **44**, 159–182.

Samish, M., Alekseev, E. A. & Glazer, I. (1995). The development of entomopathogenic nematodes in the tick *B. annulatus*. In *Tick-Borne Pathogens at the Host–Vector Interface: A Global Perspective*, eds. Coons, L. & Rothschild, M., pp. 2–4. Berg-en-Dal, South Africa: Kruger National Park.

Samish, M., Alekseev, E. A. & Glazer, I. (1998). The effect of soil composition on anti-tick activity of entomopathogenic nematodes. *Annals of the New York Academy of Science* **849**, 402–403.

Samish, M., Alekseev, E. A. & Glazer, I. (1999a). Efficacy of entomopathogenic nematode strains against engorged *Boophilus annulatus* females (Acari: Ixodidae) under simulated field conditions. *Journal of Medical Entomology* **36**, 727–732.

Samish, M., Alekseev, E. A. & Glazer, I. (1999b). Interaction between ticks (Acari: Ixodidae) and pathogenic nematodes (Nematoda): susceptibility of tick species at various developmental stages. *Journal of Medical Entomology* **36**, 733–740.

Samish, M., Alekseev, E. A. & Glazer, I. (2000a). Biocontrol of ticks by entomopathogenic nematodes. *Annals of the New York Academy of Science* **916**, 589–594.

Samish, M., Alekseev, E. A. & Glazer, I. (2000b). Mortality rate of adult ticks due to infection by entomopathogenic nematodes. *Journal of Parasitology* 86, 679–684.

Samish, M., Gindin, G., Alekseev, E. & Glazer, I. (2001). Pathogenicity of entomopathogenic fungi to different developmental stages of *Rhipicephalus sanguineus*. *Journal of Parasitology* 87, 1355–1359.

Samish, M., Gindin, G., Gal-Or, S. & Glazer, I. (2006). The potential use of *Metarhizium anisopliae* for the control of ticks under field conditions. In *Proceedings of 12th International Congress of Acarology*, Amsterdam, p. 180.

Samish, M., Ginsberg, H. & Glazer, I. (2004). Biological control of ticks. *Parasitology* 129, S389–S403

Samsinakova, A. (1957). *Beauveria globulifera* (SPEG) Pic. as a parasite of the tick *Ixodes ricinus* L. *Zoologicke Listi* 20, 329–330.

Samsinakova, A., Kalalova, S., Daniel, M., *et al.* (1974). Entomogenous fungi associated with the tick *Ixodes ricinus*. *Folia Parasitologica* 21, 39–48.

Samuel, W. M. & Welch, D. A. (1991). Winter ticks on moose and other ungulates: factors influencing their population size. *Alces* 27, 169–182.

Sebesta, K., Farkas, J., Horska, K. & Vankova, J. (1981). Thuringiensin, the beta-exotoxin of *Bacillus thuringiensis*. In *Microbial Control of Pests and Plant Diseases (1970–1980)*, ed. Surges H. D., pp. 249–281. New York: Academic Press.

Sewify, G. H. & Habib, S. M. (2001). Biological control of the tick fowl *Argas persicargas persicus* by the entomopathogenic fungi *Beauveria bassiana* and *Metarhizium anisopliae*. *Journal of Pest Science* 74, 121–123.

Shelton, A. M. & Roush, R. T. (2000). Resistance to insect pathogens and strategies to manage resistance. In *Field Manual of Techniques in Invertebrate Pathology*, eds. Lacy, L. A. & Kaaya, H. K., pp. 829–845. Dordrecht, Netherlands: Kluwer.

Smith, C. N. & Cole, M. M. (1943). Studies of parasites of the American dog tick. *Journal of Economic Entomology* 36, 469–472.

Sonenshine, D. E. (1991). *Biology of Ticks*, vol. 1. Oxford, UK: Oxford University Press.

Souza, E. J., Reis, R. C. S. & Bittencourt, V. R. E. P. (1999a). Avaliação do efeito in vitro dos fungos *Beauveria bassiana* e *Metarhizium anisopliae* sobre ovos e larvas de *Amblyomma cajennense*. *Revista Brasileira de Ciencia Veterinaria* 8, 127–132.

Souza, E. J., Reis, R. C. S. & Bittencourt, V. R. E. P. (1999b). Effect of contact of the fungi *Beauveria bassiana* and

Metarhizium anisopliae on engorged larvae of *Amblyomma cajennense*. *Revista Brasileira de Ciencia Veterinaria* 6, 84–87.

Souza, E. J., Reis, R. C. S., Melo, D. R., Bahiense, T. C. & Bittencourt, V. R. E. P. (2000). Action of entomopathogenic fungi *Beauveria bassiana* obtained from different sources of isolation, on eggs and larvae of *Boophilus microplus*. *Revista da Universidade Rural, Serie Ciencias Vida* 22, 95–99.

Stafford, K. C. & Kitron, U. (2002). Environmental management for Lyme borreliosis control. In *Lyme Borreliosis Biology, Epidemiology and Control*, eds. Gray, J. S., Kahl, O., Lane, R. S. & Stanek, G., pp. 301–334. Wallingford, UK: CAB International.

Stafford, K. C. III, Denicola, A. J. & Kilpatrick, H. J. (2003). Reduced abundance of *Ixodes scapularis* (Acari: Ixodidae) and the tick parasitoid *Ixodiphagus hookeri* (Hymenoptera: Encyrtidae) with reduction of white-tailed deer. *Journal of Medical Entomology* 40, 642–652.

Stafford, K. C. III, Denicola, A. J. & Magnarelli, L. A. (1996). Presence of *Ixodiphagus hookeri* (Hymenoptera: Encyrtidae) in two Connecticut populations of *Ixodes scapularis* (Acari: Ixodidae). *Journal of Medical Entomology* 33, 183–188.

Steinhaus, E. A. & Marsh, G. A. (1962). Report of diagnoses of diseased insects 1951–1961. *Hilgardia* 33, 349–390.

Stutterheim, C. & Brooke, R. (1981). Past and present ecological distribution of the yellow billed oxpecker in South Africa. *African Journal of Zoology* 16, 44–49.

Stutterheim, C. J. & Stutterheim, I. M. (1980). Evidence of an increase in a red-billed oxpecker population in the Kruger National Park. *South African Journal of Zoology* 15, 284–284.

Stutterheim, I. M., Bezuidenhout, J. D. & Elliott, E. G. R. (1988). Comparative feeding behavior and food preferences of oxpeckers (*Buphagus erythrorhynchus* and *B. africanus*) in captivity. *Onderstepoort Journal of Veterinary Research* 55, 173–179.

Suzuki, E. M., Da Silva, W. W., Costa, G. L. & Bittencourt, V. R. E. P. (2003). Internal organ infection in *Anocentor nitens* (Acari: Ixodidae) engorged females by *Metarhizium anisopliae*. *Brazilian Journal of Veterinary Parasitology* 12, 85–87.

Symondson, W. O. C. Sunderland, K. D. & Greenstone, M. H. (2002). Can generalist predators be effective biocontrol agents? *Annual Review of Entomology* 47, 561–594.

Takasu, K., Takano, S.-I., Sasaki, M., Yagi, S. & Nakamura, S. (2003). Host recognition by the tick parasitoid *Ixodiphagus hookeri* (Hymenoptera: Encyrtidae). *Environmental Entomology* 32, 614–617.

Trjapitzin, V. A. (1985). Natural enemies *of Ixodes persulcatus*. In *Taiga Tick*, Ixodes persulcatus *Schulze (Acarina, Ixodidae): Morphology, Systematics, Ecology, Medical Importance*, ed. Filippova, N. A., pp. 334–347. Leningrad, USSR: Nauka.

Van Driesche, R. G. & Bellows, T. S. J. R. (1996). *Biological Control*. New York: Chapman & Hall.

Van Someren, V. D. (1951). The red billed oxpecker and its relation to stock in Kenya. *East African Agricultural Journal* **17**, 1–11.

Vasconcelos, V. de O., Furlong, J., Freitas, G. M. de, *et al.* (2004). *Steinernema glaseri* Santa Rosa strain (Rhabditida: Steinernematidae) and *Heterorhabditis bacteriophora* CCA strain (Rhabditida: Heterorhabditidae) as biological control agents of *Boophilus microplus* (Acari: Ixodidae). *Parasitology Research* **94**: 201–206

Verissimo, C. J. (1995). Natural enemies of the cattle parasitic tick. *Agropecuaria Catarinense* **8**, 35–37.

Weeks, P. (1999). Interaction between red-billed oxpeckers, *Buphagus erythrorhynchus* and domestic cattle, *Bos taurus*, in Zimbabwe. *Animal Behaviour* **58**, 1253–1259.

Wysoki, M. (1998). Problems and trends of agricultural entomology at the end of the 2nd millennium. *Bollettino del Laboratorio di Entomologia Agria 'Fillipo Silvestri'* **54**, 89–143.

Yoder, J. A., Benoit, J. B. & Zettler, L. W. (2003). Moisture requirements of a soil imperfect fungus, *Scopulariopsis brevicaulis* Bainier, in relation to its tick host. *International Journal of Acarology* **29**, 271–277.

Zangi, G. (2003). Tick control by means of entomopathogenic nematodes and fungi. Unpublished M.Sc. thesis, Hebrew University of Jerusalem, Israel.

Zhioua, E., Browning, M., Johnson, P. W., Ginsberg, H. S. & Lebrun, R. A. (1997). Pathogenicity of the entomopathogenic fungus *Metarhizium anisopliae* (Deuteromycetes) to *Ixodes scapularis* (Acari: Ixodidae). *Journal of Parasitology* **83**, 815–818.

Zhioua, E., Ginsberg, H. S., Humber, R. A. & Lebrun, R. A. (1999a). Preliminary survey for entomopathogenic fungi associated with *Ixodes scapularis* (Acari: Ixodidae) in southern New York and New England, USA. *Journal of Medical Entomology* **36**, 635–637.

Zhioua, E., Heyer, K., Browning, M., Ginsberg, H. S. & Lebrun, R. A. (1999b). Pathogenicity of *Bacillus thuringiensis* variety *kurstaki* to *Ixodes scapularis* (Acari: Ixodidae). *Journal of Medical Entomology* **36**, 900–902.

Zhioua, E., Lebrun, R. A., Ginsberg, H. S. & Aeschlimann, A. (1994). Entomopathogenic nematodes and fungi of *Ixodes scapularis*, the principal vector of the Lyme borreliosis spirochete, *Borrelia burgdorferi*, in north America. In *Proceedings of 6th International Colloquium on Invertebrate Pathology and Microbial Control*, Montpelier, France.

Zhioua, E., Lebrun, R. A., Ginsberg, H. S. & Aeschlimann, A. (1995). Pathogenicity of *Steinernema carpocapsae* and *S. glaseri* (Nematoda: Steinernematidae) to *Ixodes scapularis* (Acari: Ixodidae). *Journal of Medical Entomology* **32**, 900–905.

21 • Pheromones and other semiochemicals of ticks and their use in tick control

D. E. SONENSHINE

INTRODUCTION

In ticks, as in most animals, chemical mediators guide behaviour. These information-bearing compounds, known as semiochemicals, are secreted external to the animal body and, when recognized, direct a specific behavioural response such as food and mate location, escape and other behaviours. Chemical signalling between individuals is clearly one of the earliest types of information exchange to appear in the long history of life on Earth, long before visual or auditory stimuli developed. Indeed, chemical communication via semiochemicals remains the dominant form of communication among many animals. Despite similarities with cell signalling (e.g. cytokines) among the cells of the metazoan animal, or hormones (e.g. ecdysteroids) that stimulate specific physiological responses (e.g. moulting), semiochemicals are fundamentally different in that they are secreted outside of the animal body, are recognized externally and modify the behaviour of the entire individual. With the advances in modern chemistry, biochemistry and molecular biology during the past several decades, a vast literature has accumulated concerning the variety of semiochemicals, their chemical composition, biosynthesis, secretion and perception, and the varying biological roles that these compounds regulate.

Collectively, the repertoire of chemical compounds used within a species or among competing species forms a simple chemical communication system, or chemical language. In many species, this chemical language consists of an ordered hierarchy of specific compounds that are secreted and perceived in a precise, sequential order leading to a desired end result. In others, a single compound (e.g. squalene) may be sufficient to accomplish a specific purpose such as defence against ant predators (Yoder, Pollack & Spielman, 1993a). Occasionally, two or more compounds, often mixed in a specific proportion may induce the maximal behavioural response, e.g. haematin, guanine and xanthine induces clus-

tering in the black-legged tick, *Ixodes scapularis* (Sonenshine et al., 2003).

Semiochemicals are defined by the type of behaviour they mediate, not the specific compound or compounds used to affect that behaviour. Thus, a single compound such as cholesteryl oleate on the cuticular surfaces of female dog ticks, *Dermacentor variabilis*, enables males to recognize those females as suitable mates, whereas a specific mixture of cholesteryl–fatty acid esters is required to achieve the same response in different tick species such as the camel ticks, *Hyalomma dromedarii* (Sonenshine et al., 1991). The same behaviour, male mounting and probing for the genital pore, is accomplished in the two different species but with different compounds. In some cases, the same compound may function in a different role, depending upon the physiological state of the animal and its interaction with other species in its environment. Squalene, an abundant lipid on the surface of mammalian skin, has been reported to be a potent attractant for hungry ticks of some species (Yoder, Stevens & Crouch, 1999) but, as noted above, this same compound may be secreted by fed or feeding ticks when threatened by predatory ants and serve in a defensive role (Yoder et al., 1993b).

Semiochemicals are classified into four major categories, namely: (1) pheromones; (2) allomones; (3) kairomones; and (4) synomones. These categories are based on the benefit that accrues to the animal that secretes them as well as to the animal that perceives them. Pheromones are secreted by individuals of a species that alter the behaviour of other individuals of that same species in a manner that benefits the species. Examples include sex pheromones and assembly pheromones. Often, pheromones are secreted only during certain life stages or periods of development, e.g. females feeding on a vertebrate host. In addition, the recipient animal that perceives the pheromone may only respond when it has reached a specific physiological and developmental stage, e.g. fed or feeding males but not unfed males. This high degree of specificity enhances the value of pheromonal communication, avoiding wasteful metabolic effort and limiting the synthesis, secretion and perception of the pheromone only to those individuals that will achieve the greatest benefit from its use. Allomones are information-bearing compounds or mixtures emitted by individuals of one species that affect the behaviour of individuals of a different species for the benefit of the emitter, e.g. hydrocarbon secretion by ticks to deter ant predators. Kairomones are information-bearing compounds or mixtures released by individuals of one species, detected by individuals of another species that benefit the recipient, e.g. host odours that enable a blood-feeding ectoparasite to locate, recognize and feed on a suitable host. Often, kairomonal compounds are most effective when combined with visual cues, body heat and even sounds made by the vertebrate hosts. Synomones are semiochemicals produced by an individual of one species which, when they contact an individual of a different species, evoke a response favourable to both the emitter and the receiver (Evenden, Judd & Borden, 1999).

Improvements in instrumentation for detecting and identifying semiochemicals and electrophysiological methods for determining the tick's sensory responses to these compounds have provided a greater understanding of how these bioactive substances regulate the behaviour of ticks and other ectoparasites.

This chapter explores the characteristics of the various types of semiochemicals that are fundamental to the biology of ticks. In addition, a review of our knowledge of semiochemical biochemistry, including synthesis and secretion, and how semiochemicals are perceived is also provided. Finally, the practical applications of our accumulating knowledge of tick semiochemicals for development of new products to control ticks, with special emphasis on novel strategies that minimize pesticide use, are addressed.

PHEROMONES

The need to recruit mates for sexual reproduction is critical to the success of any population of sexual animals. In insects and other invertebrates, a series of complex movements in time and space mediated by pheromones guides the process and minimizes wasteful non-specific encounters. An impressive variety of pheromones has been described in insects, covering the diverse behaviours found in this vast assemblage of species. Examples include varieties of pheromones, food

finding (foraging or trail-marking) pheromones, arrestment pheromones, alarm pheromones, nest-building pheromones and many others (Wheeler, 1976; Ayasse, Paxton & Tengo, 1995; Fauvergue, Hopper & Antolin, 1995; Roelofs, 1995; Stowe et al., 1995). An exhaustive list is beyond the scope of this review. In the case of ticks, a substantial body of new knowledge has accumulated since Berger's (1972) discovery of 2,6-dichlorophenol as a sex pheromone of the lone star tick *Amblyomma americanum*, the first pheromone discovered in these acarines. Since then, four different types of pheromones have been identified in ticks. These include: (1) arrestment (= assembly) pheromones; (2) attraction–aggregation–attachment pheromones; (3) sex pheromones; and (4) primer pheromones. Synomones, although well known in Lepidoptera and perhaps other arthropods (Evenden et al., 1999), have not been reported from ticks. Consequently, they will not be considered further in this review.

Arrestment (= assembly) is defined as the cessation of kinetic activity, a response that reduces the distance between individuals that perceive the stimulus in their environment (Cardé & Baker, 1984) and leads to clusters of individuals in their natural environment. These pheromones are widespread in ticks, having been found in both hard (Ixodidae) and soft ticks (Argasidae). The attraction–aggregation–attachment pheromones attract appetitive ticks (i.e. ticks attempting to feed) to a tick-infested host and induce them to cluster together at a single location before probing and attaching to the host skin. Sex pheromones comprise compounds or mixtures of compounds that mediate the various phases of mate recruitment, mate selection and, ultimately, insemination and fusion of gametes between the mating partners. Occasionally, sex pheromones and arrestment or aggregation–attachment pheromones have been confused. By definition, a sex pheromone is a semiochemical emitted by individuals of one sex that mediates the sexual behaviour of the opposite sex. Primer pheromones mediate physiological functions, e.g. reducing tick fecundity in response to overcrowding (Khalil, 1984; see also Chapter 8). Little else is known about the existence of primer pheromones in ticks. Consequently, this review will be concerned only with the first three types of pheromones.

Arrestment (assembly) pheromones

Although arrestment pheromones are widespread in both argasid and ixodid ticks, they are best known in argasid ticks, where they lead to the formation of tick clusters in caves, under ledges, cracks and crevices where the ticks hide in, or near, the nests of their hosts. These clusters are thought to enhance mating and host-finding success (Sonenshine, 1985). First described from *Argas* ticks by Leahy, Vandehay & Galun (1973), arrestment behaviour has since been reported from at least 14 species of soft ticks as well as several hard ticks including *Ixodes ricinus* (Graf, 1978), *I. holocyclus* and *Aponomma concolor* (Treverrow, Stone & Cowie, 1977), *Hyalomma dromedarii* (Leahy, Hajkova & Bourchalova, 1981), *Rhipicephalus appendiculatus* and *Amblyomma cohaerans* (Otieno *et al.*, 1985), *R. evertsi* (Gothe & Neitz, 1985) and *I. scapularis* (Allan & Sonenshine, 2002). As described by Leahy *et al.* (1975), clustering of soft ticks, *Ornithodoros moubata* (*sensu lato*) was initiated when individuals crawled over a substrate contaminated with excreta and body exudates from other ticks. Gradually, the individual ticks became akinetic, and remained in close contact with one another in tight clusters. Arrestment behaviour was so intense that an investigator could lift a large cluster with a pair of forceps without the individuals dispersing! A noteworthy feature of the arrestment response was its interspecific nature. *Ornithodoros moubata* males formed clusters in response to ticks of the same genus, as well as to other species such as *O. tholozani* and even between genera, e.g. *O. moubata* with *Argas persicus* and vice versa. Despite these interspecific interactions, these responses are still regarded as pheromone behaviour by most experts in the field (e.g. Dusbábek *et al.*, 1991) although they could also be classified as synomones. Some differences were noted between the sexes; *O. moubata* females clustered more rapidly in response to *O. moubata* male extracts than female extracts. In *I. scapularis*, nymphs and adults also clustered in response to cast larval skins (Allan & Sonenshine, 2002). Clustering behaviour is believed to have survival value in that the individuals forming the mass accumulate in sites favourable for avoiding stressful environmental conditions, and where they are more likely to encounter hosts. In caves or crevices, this often leads to clusters forming in cool, sheltered niches near the entrances. In prostriate ticks, such as *I. ricinus*, clusters of host-seeking ticks are often found on vegetation and are believed to favour contact with passing hosts (Graf, 1974); up to 70% of females found in the vegetation are mated. Similar clusters have been found for the black-legged tick, *I. scapularis*, in the USA.

Purines are the major component of the arrestment pheromone in most of the ticks examined. Purines are abundant in tick faeces. In *I. ricinus*, ticks respond to faeces from freshly moulted ticks of either sex and to faeces-contaminated filter papers (Grenacher *et al.*, 2001). The most abundant component of the excreta in this tick is guanine. Other purines (e.g. xanthine and hypoxanthine) have been reported from the excreta from several other tick species. In *A. persicus*, guanine comprised 89.8–98.6% of the purines in the tick's excreta, while the remainder comprised hypoxanthine (1–5%) and xanthine (up to 9.0%). Traces of uric acid and guanosine also were found in the faeces of some of the argasid ticks, but these compounds did not induce an arrestment response in the bioassays. Mixtures of xanthine and guanine (1 : 25) or adenine : xanthine : guanine (1 : 1 : 25) most closely approximate the natural response (Dusbábek *et al.*, 1991). Guanine was shown to be the primary stimulant for *A. persicus* and several other tick species and was found to be active at concentrations as low as 8×10^{-12} moles/cm^2 (Otieno *et al.*, 1985). In *I. ricinus*, each of the faecal components guanine, xanthine, uric acid and 8-azaguanine (a bacterial breakdown product of guanine) presented individually caused arrestment responses by individual male ticks. However, the mixture of these compounds was 100-fold more effective in stimulating the arrestment response than any individual compound presented separately (an interesting finding since the authors failed to detect either xanthine or uric acid as natural constituents in the faeces of this tick) (Grenacher *et al.*, 2001). In *I. scapularis*, ticks arrested in response to guanine, hypoxanthine, xanthine, inosine and haematin. The strongest response, however, was to a mixture of guanine, xanthine and adenine (in a ratio of 25 : 1 : 1) and was similar to the response to cast skins (Allan & Sonenshine, 2002). Further study of the chemical components of *I. scapularis* faeces and faecal contaminants showed the presence of haematin, guanine and xanthine, but no evidence of other purines. The ratio of guanine to xanthine was 10.6 : 1 in extracts of nymphal skins, but only 0.95 : 1 in extracts of larval skins. Haematin proved especially important as an arrestment stimulus for adults of this species. Video tracking showed that adult *I. scapularis* were significantly more likely to assemble in response to tick exudates containing these compounds in as little as 3 h as compared to the controls (Sonenshine *et al.*, 2003). Clustering of *I. ricinus* and *I. scapularis* in vegetation facilitates contact between the sexes. As noted previously in the argasid ticks, arrestment behaviour is also believed to enhance host-finding opportunities and possibly even protect the ticks against desiccation (Yoder & Knapp, 1999; Kiszewski, Matuschka & Spielman, 2001). The use of this new knowledge for developing novel types of pheromone-assisted tick control products will be discussed below.

Volatiles also appear to contribute to clustering behaviour in the natural environment. Volatiles from female *Argas walkerae* stimulated engorged conspecific males to cluster on surfaces impregnated with volatile extracts from these females (Neitz & Gothe, 1984). Similarly, male and female *R. evertsi* were attracted to, and induced to assemble in response to, water-soluble volatiles collected from unfed males (Gothe & Neitz, 1985). Grenacher *et al.* (2001) reported that *I. ricinus* responded to faeces-contaminated filter papers enclosed in a bronze mesh, but much slower (24 h) than when they were allowed contact with these surfaces. They found evidence of ammonia emanating from the tick faeces. Ammonia is a common by-product of the degradation of nitrogenous wastes and a potent attractant for ticks and many other haematophagous arthropods (Haggart & Davis, 1979, 1981*a*; Steullet & Guerin, 1994). Other volatiles, yet to be discovered, may also contribute to the tick arrestment responses.

Although widespread, arrestment pheromones may not be present in all ticks. Taylor *et al.* (1987) found no evidence of such behaviour in the American ticks *Dermacentor variabilis* or *D. andersoni*.

Tick clustering behaviour appears to involve a two-step process: (1) attraction to a volatile source and (2) arrestment in response to various purines. Purines have very low vapour pressure and, consequently, are not attractants. Ammonia and perhaps other volatiles emanating from tick faeces gradually attract free-living, unfed adults and even nymphs to the point source. Tactile contact with the purines, especially guanine and xanthine, triggers the arrestment response and causes the ticks to cease activity, forming a cluster. Further studies are needed to determine the concentration and range over which the volatile attractant (e.g. ammonia) is effective and whether any other tick-originated volatiles (e.g. CO_2) are also attractive.

Attraction–aggregation–attachment pheromones

So named because they attract unfed males and females from grassy meadows, duff and sandy shelters of their natural environment, these pheromones also stimulate the attracted ticks to aggregate following contact with the vertebrate host and feed close together (Fig. 21.1). The attraction–aggregation–attachment (AAA) pheromone is a mixture of organic volatiles secreted solely by feeding males but attractive to both male and female ticks. Thus, although male-originated, it is not a sex pheromone because it attracts both sexes. In some tick species, males will attack cattle, buffalo and other large ungulates irrespective of the pres-

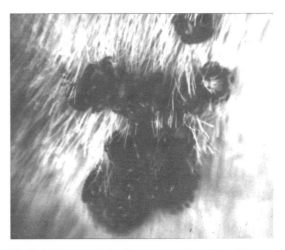

Fig. 21.1. Photograph showing an aggregation of bont ticks, *Amblyomma hebraeum*, feeding on a cow in Zimbabwe. (Photo credit: Dr Suman Mahan, Harare, Zimbabwe; Heartwater Research Project, Department of Veterinary Pathology, University of Florida, Gainesville, FL, USA.)

ence of females, but female ticks appear to require the AAA pheromone, without which they will not attach and feed. This pheromone occurs only in certain species of the genus *Amblyomma* in which the adults feed on large ungulates. In Africa, these include *Amblyomma hebraeum*, *A. variegatum*, *A. lepidum*, *A. gemma*, and *A. marmoreum*. *Amblyomma variegatum* also infests cattle and other livestock in islands in the Caribbean. In North America, the only species that has been reported to exhibit this type of pheromone-dependent behaviour is the Gulf Coast tick, *A. maculatum* (Obenchain, 1984; Sonenshine, 1993).

Compared to the low emission rates of the volatile sex pheromone (2,6-dichlorophenol), the kinetics of AAA pheromone secretion are remarkable. The pheromonal components are secreted in large quantities, e.g. several micrograms of *o*-nitrophenol or methyl salicylate per tick per hour (Diehl *et al.*, 1991; Pavis & Barré, 1993; Price *et al.*, 1994) and are attractive for up to 3 m from a tick-infested host. However, CO_2 emanating from the host is essential for maximum activity. Carbon dioxide is believed to act as a non-specific excitant, arousing ticks sheltering in the natural environment. Once excited, they respond to the pheromone which signals the presence of attached, feeding ticks and provides directional information that enables the host-seeking ticks to discriminate male-infested from uninfested hosts (Norval, Andrew & Yunker, 1989). In *A. hebraeum* and *A. variegatum*, males, females and even nymphal ticks are attracted.

Females rarely attach to uninfested hosts (Norval, Andrew & Yunker, 1989) unless they encounter the male-originated pheromone. In some species, attached males move their hind legs vigorously when females, attracted by the pheromone, approach nearby and grasp the females if they make contact, thereby facilitating aggregation, attachment and subsequent conspecific mating (Obenchain, 1984). Although aggregations may form anywhere on the host body, *A. hebraeum* and *A. variegatum* usually persist mostly in the perianal region, groin, around the udders of cows or anteriorly on the head and neck or in the front axillary area where they unlikely to be removed by grooming.

The chemical composition of the AAA pheromone was first elucidated for the tropical bont tick, *A. variegatum*, by Schöni *et al.* (1984) who showed that it consists of a mixture of three organic volatiles, *o*-nitrophenol (=2-nitropenol), methyl salicylate and nonanoic acid. They reported that the pheromone consists of a blend of these three compounds in a distinct ratio of 2:1:8, respectively. According to these authors, *o*-nitrophenol stimulates searching and aggregating behaviour, while methyl salicylate and *n*-nonanoic acid were believed to stimulate attachment. These same three compounds also are produced by the bont tick, *A. hebraeum*, a close relative in Zimbabwe and South Africa. However, except for parts of Zimbabwe where they are sympatric, their ranges do not overlap. Following attachment and feeding, *o*-nitrophenol and methyl salicylate were secreted at relatively high rates, from approximately 0.6–1.8 µg/h for *o*-nitrophenol and 0.02–0.6 µg/h for methyl salicylate. Although the ratios varied, the mean ratio for these two volatiles after 12 days of feeding was 4:1 *o*-nitrophenol:methyl salicylate. Emission rates for *A. hebraeum* were somewhat lower, from 0.23 to 0.28 µg/h for *o*-nitrophenol. Methyl salicylate was not detected in the emissions (Diehl *et al.*, 1991). Subsequent studies revealed additional components, especially 2,6-dichlorophenol and benzaldehyde (Lusby *et al.*, 1991; Price *et al.*, 1994). Certain of these compounds, especially *o*-nitrophenol and methyl salicylate, are effective as long-range attractants, attracting ticks from as far as 4 m from the release site (Norval *et al.*, 1991a). Methyl salicylate, *o*-nitrophenol and 2,6-dichlorophenol were strong stimulants for inducing the aggregation response by unfed female *A. variegatum*. Methyl salicylate and *o*-nitrophenol presented individually were as effective for inducing attraction as the crude extract. In contrast, none of the naturally occurring phenols or volatile acids was as effective as the natural extract in stimulating aggregations by unfed female *A. hebraeum* (Norval *et al.*, 1992a).

In addition, *o*-nitrophenol and methyl salicylate stimulated attachment by *A. variegatum* while *o*-nitrophenol and 2,6-dichlorophenol stimulated attachment by *A. hebraeum*, suggesting differences in the selective response by these two species. These findings indicate that the two tick species are capable of distinguishing differences in the composition of the natural pheromone, differences which may enable them to avoid interspecific mating. Substantial differences in the relative abundance and secretion rates of these compounds, as well as another volatile, benzaldehyde, are known to occur (Price *et al.*, 1994) and these differences may facilitate the formation of species-specific aggregations in areas where ranges of the two species overlap. In both species, *o*-nitrophenol is secreted in large quantities and serves as a primary long-range attractant, attracting adults of both species more or less equally (Norval *et al.*, 1991a). It also stimulates aggregation in *A. variegatum*, and aggregation and attachment in *A. hebraeum*. However, differences in the amounts of methyl salicylate, abundant in *A. variegatum*, but virtually absent in *A. hebraeum*, and benzaldehyde, abundant in *A. hebraeum* but virtually absent in *A. variegatum*, suggest differences in the AAA pheromone which may explain the formation of species-specific aggregations that have been observed in nature, even though many of the individual components are attractive to both species (Price *et al.*, 1994). In addition, recent evidence shows that 1-octen-3-ol, isolated from *A. variegatum* and *A. hebraeum*, is attractive to *A. variegatum* and may also contribute to the aggregation response (McMahon *et al.*, 2001). Other differences also occur. Apps, Viljoen & Pretorius (1988) reported the occurrence of heptanoic, octanoic and 2-methyl propanoic acids in effluents collected from feeding *A. hebraeum* males but their role as pheromones, if any, is unknown. Nothing is known about the AAA pheromone of the other African *Amblyomma* species or the North American Gulf Coast tick, *A. maculatum*.

Nymphal *A. hebraeum* and *A. variegatum* also attach to hosts in response to the AAA pheromone or to its individual components. Among the naturally occurring components presented individually, *o*-nitrophenol, methyl salicylate and 2-methyl propanoic acid induced attachment responses for *A. hebraeum* that were as strong or stronger than the naturally occurring mixture. For *A. variegatum*, only *o*-nitrophenol and methyl salicylate induced attachment responses similar to that induced by the natural mixture (Norval *et al.*, 1991b). In both species, nymphs responded to a narrower range of volatile compounds than the adults (Norval *et al.*, 1992b). Again, as with the adults, there is clear evidence that the two species discriminate among the naturally occurring volatiles.

A male-produced attachment pheromone also was found in *Amblyomma cajennense*, an important pest of bovines and other livestock in the New World. However, ticks attaching in response to this pheromone did not form clusters around the pre-attached, feeding males. The composition of the pheromone was not determined (Rechav, Goldberg & Fielden, 1997).

The AAA pheromone is produced and secreted by the large dermal glands (Type 2) located on the ventral surfaces of the males of these ticks. Both *o*-nitrophenol and methyl salicylate were identified by HPLC in extracts of these glands dissected from fed male ticks (Diehl *et al.*, 1991) (Fig. 21.2). Similar large dermal glands also occur on the scutum of metastriate ticks where they secrete a hydrocarbon-rich fluid containing squalene and perhaps other compounds implicated in defensive reactions against predators (Yoder, Pollack & Spielman, 1993*a*; Yoder *et al.*, 1993*b*). According to Rechav *et al.* (1977), *A. hebraeum* males commence pheromone secretion when attached at least 5 days and continue to maximum levels by 8–9 days of continuous feeding. In studies with *A. variegatum*, pheromone secretion did not begin until males were attached for at least 5 days. Thereafter, attractiveness of males for unfed females was greatest between 14 and 23 days of male engorgement. In bioassays performed on goats, females continued to be attracted to males attached as long as 50 days. A great deal of variation was observed in the day-to-day production of the individual pheromone components as well as differences between individual ticks (Pavis & Barré, 1993).

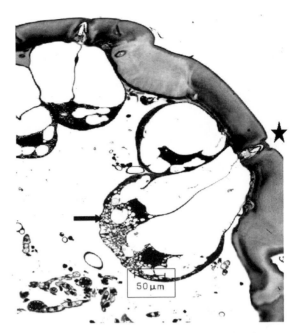

Fig. 21.2. Light micrograph illustrating a histological section through the ventral region of a male *Amblyomma variegatum* showing the large dermal glands (Type II dermal glands). These glands were reported to be site of synthesis and secretion of the attraction–aggregation–attachment pheromone secreted during feeding (Diehl *et al.*, 1991). The arrow indicates the cells of a large dermal gland; the asterisk indicates the pore of the dermal gland. (Photo credit: Dr P. A. Diehl, University of Neuchatel, Neuchatel, Switzerland.)

Sex pheromones of ixodid ticks

By definition, sex pheromones are compounds or mixtures of compounds secreted by individuals of one sex that are attractive to individuals of the opposite sex. Sex pheromones are important components of a complex process of courtship behaviour that leads to insemination of receptive females and, consequently, an essential element in the preservation of species integrity. Although kinetic activity, visual signals and other non-chemical cues may also contribute to the courtship process in many animals, most ticks depend upon a hierarchy of chemical messages to regulate their species-specific mating behaviour. Rather than a simple event comprising only one or two steps, mating is a complex sequence of discrete behavioural events arranged in a temporal sequence, each of which is mediated by different pheromones. This hierarchy of sexual selection procedures is the basis for the high degree of species-specific selection that maximizes conspecific mating.

Most of our knowledge of tick mating behaviour comes from studies of metastriate ticks of the family Ixodidae (all Ixodidae except the genus *Ixodes*). In the metastriates, sexual activity occurs solely during blood-feeding on hosts while in *Ixodes*, mating can occur off the host by unfed adults. Metastriate ticks are sexually immature when they emerge from the nymphal moult, and remain so until they attach to a host and begin feeding. Once feeding begins, spermatogenesis and oogenesis are initiated (see Chapter 8), and sex pheromone secretion soon follows. Three types of sex pheromones have been described: (1) an attractant sex pheromone (ASP); (2) a mounting sex pheromone (MSP); and (3) a genital sex pheromone (GSP), each of which mediates different aspects of the courtship process. Details of the process of pheromone-guided mating behaviour are summarized in the accompanying diagram (Fig. 21.3).

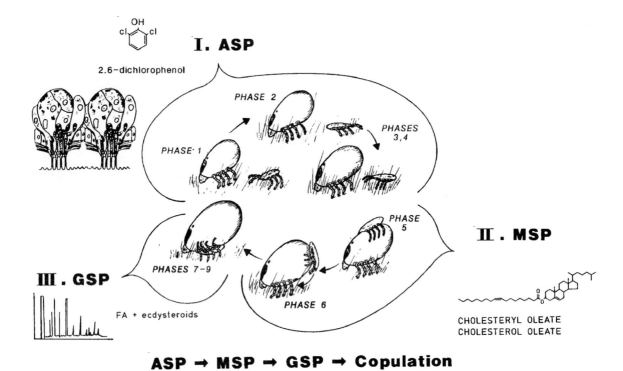

ASP → MSP → GSP → Copulation

Fig. 21.3. Hypothetical model illustrating the sequence of behavioural events that occur during sexual courtship in a representative metastriate ixodid tick, *Dermacentor variabilis*. Phase 1, feeding female secretes volatile attractant sex pheromone, 2,6-dichlorophenol, exciting males feeding nearby. Phase 2, sexually excited male detaches and commences searching behaviour in response to sex attractant odour. Phases 3 and 4, sexually excited male orients to and walks toward the pheromone-emitting source and locates the pheromone-secreting female. These first four phases are mediated by 2,6-dichlorophenol. Phase 5, the male contacts the candidate female and detects the presence of cholesteryl oleate, the mounting sex pheromone. Phase 6, the male recognizes the female as a suitable mate and climbs onto the female's dorsal surface, turns posteriorly and crawls over the posterior end of the female's opisthosoma (tip-over behaviour) and onto the ventral surface. This behaviour constitutes the male mounting response. Phases 7–9, the male locates the female gonopore, positions itself under the female with legs intertwined, and inserts its chelicerae into the female's genital pore. Identification of the gonopore and genital probing with the chelicerae is mediated by the genital sex pheromone, comprising a mixture of long-chain fatty acids and ecdysteroids. Following successful identification of the female as a conspecific mate, the spermatophore is formed and copulation ensues. ASP, attractant sex pheromone; FA, fatty acids; GSP, genital sex pheromone; MSP, mounting sex pheromone. (Reproduced with permission from *Annual Review of Entomology* **30** (1985) and *Modern Acarology*.)

ATTRACTANT SEX PHEROMONE (ASP)

2,6-Dichlorophenol is the only proven ASP in metastriate ticks, although the possible role of other unknown volatiles is not excluded. This compound has been reported from six genera of ticks, including 15 different species. Fed or feeding males that recognize this compound become excited, detach from the host skin and crawl over the host in search of the pheromone-emitting females. Unfed males can detect the pheromone but do not respond to it (Haggart & Davis, 1981*b*). In contrast to the male-originated AAA pheromone discussed above, all of the known tick sex-attractant pheromones are produced by the females. In some species, e.g. *D. variabilis*, 2,6-dichlorophenol may also be produced by males and stored in their pheromone glands, but it is not secreted. In some other species, males produce and secrete this compound for other uses, e.g. as an attractant and attachment stimulus in *A. variegatum* and *A. hebraeum* (Norval *et al.*, 1992*a*; Price *et al.*, 1994) and even as an attractant stimulus for immature ticks (Yoder & Stevens, 2000). These findings should not alter our recognition of its primary role in sexual communication among adult ticks.

2,6-Dichlorophenol is the pheromone that initiates the mate-finding process in most ixodid tick species. Females commence biosynthesis of the pheromone soon after emerging from the nymphal moult, whereupon it remains stored in oily droplets within the lobes of the foveal glands. These paired glands are located under the dorsal body cuticle where they are attached by ducts to the dorsal foveae. The dorsal foveae, located in the middle of the alloscutum just posterior to the scutum, contain clusters of slit-like pores (Fig. 21.4). Foveal glands occur in both males and females, although the glands in the females are larger and contain larger quantities of 2,6-dichlorophenol than those of the males. When female ticks attach and commence feeding, pheromone secretion is initiated. The vesicles within the cells of the foveal glands break down, and the pheromone-containing oils migrate into the ducts. The pheromone-containing oils pass out of the glands via these openings onto the cuticular surface, releasing the volatile 2,6-dichlorophenol into the atmosphere. This provides a simple type of controlled release of the pheromone from the tick's body surface.

The regulation of attractant sex pheromone secretion is discussed below (see section 'Pheromone glands'). For a more detailed description of the ultrastructure and physiology of the foveal glands, see Sonenshine (1985).

A curious anomaly was the finding of 2,6-dichlorophenol in the cattle tick, *Rhipicephalus* (*Boophilus*) *microplus* (hereafter referred to as *B. microplus*) in all of its different active life stages, but no evidence that it functions as a sex pheromone. Male *B. microplus* do not show the characteristic orientation and searching responses seen in males of other metastriate ticks during their courtship activities. Just how the males identify the females in the cattle ticks is unknown (de Bruyne & Guerin, 1998). This observation of 2,6-dichlorophenol in the various stages of this tick is consistent with Yoder *et al.*'s (2002) hypothesis linking chlorophenol production to water conservation.

MOUNTING SEX PHEROMONES (MSP)

These are contact sex pheromones produced by feeding females that enable males to identify the females as suitable mating partners. Even though attracted by ASP from a feeding female, the excited male will not mate with a female unless it recognizes this pheromone upon contact with the female body surface. Recognition of MSP mediates dorsal mounting, tip-over (ventral turning) and genital searching behaviour (Fig. 21.3, phases 5 and 6). The critical importance of this pheromone was demonstrated in studies on *D. variabilis* by Hamilton & Sonenshine (1988), who showed

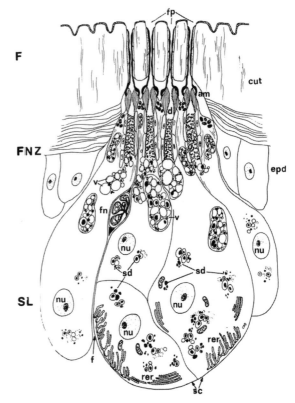

Fig. 21.4. Diagrammatic reconstruction of the sex pheromone gland showing the secretory lobes, ductular zone and the foveal pores. In unfed ticks, the cells of the secretory lobes are filled with numerous tiny vesicles containing 2,6-dichlorophenol stored in lipid droplets. In feeding females, the individual vesicles rupture, releasing the oily droplets. The droplets migrate towards the ducts where they coalesce into large masses and pass through the ducts and out onto the external surface of the tick body. am, ampulla; cut, cuticle; d, duct; epd, epidermal cell; F, fovea; fn, foveal nerve; FNZ, foveal neck zone; fp, foveal pore; nu, nucleus; rer, rough endoplasmic reticulum; sc, secretory cells; sd, secretory droplets; SL, secretory lobe; v, vesicle. (Reproduced from Sonenshine (1991) with permission from Oxford University Press, New York.)

that sexually excited males would not attempt mating with females that had been washed with a lipid solvent (hexane), even when 2,6-dichlorophenol was added. However, when lipid extracts made from fed females were applied, the mounting response was restored. Sexually excited males even attempted to mate with beads and other 'dummy females' coated with a mixture of the MSP extract and 2,6-dichlorophenol. Chemical studies showed that MSP of *D. variabilis* was cholesteryl oleate, a fatty acid ester of cholesterol. Comparison with other tick species showed a more complex picture, with a mixture of cholesteryl esters serving as the MSP instead of a single compound. Greatest differences were found between tick genera (Sonenshine et al., 1991). Male ticks showed only a limited ability to distinguish between the MSP extracts or artificial mixtures, i.e. the MSP enables the male to recognize the female as a possible mating partner but does not guarantee species-specific identification. Subsequent studies showed that the same class of compounds functions as the mounting sex pheromone in the brown ear tick *Rhipicephalus appendiculatus* (Hamilton et al., 1994) as well as in the brown dog tick *R. sanguineus* and the camel tick *Hyalomma dromedarii* (Sobby et al., 1994). In the camel tick, chemical studies showed the presence of four relatively abundant cholesteryl esters, cholesteryl acetate, laurate, linoleate and oleate, as well as trace amounts of cholesteryl palmitate and cholesteryl stearate. Bioassays with sexually active males showed strong responses to hexane-washed, delipidized females treated, in addition to 2,6-dichlorophenol, with each of these four cholesteryl esters. Cholesteryl acetate and cholesteryl oleate gave the strongest responses, not significantly different from the natural controls. However, when these compounds were incubated with esterases, the males no longer responded to the individual fatty acids or cholesterol released by the digestive enzymes. These findings provide support for the hypothesis that the males were responding to the cholesteryl esters, not to the fatty acids or cholesterol alone.

GENITAL SEX PHEROMONE (GSP)

This little-known pheromone occurs in at least a few species of closely related ticks, e.g. *D. variabilis* and *D. andersoni*. During the courtship process, identification of this pheromone by the sexually excited males stimulates synthesis and extrusion of the spermatophore and subsequent insemination. Thus, this pheromone mediates the final stages of the courtship process, stages 7–9 in Fig. 21.3. Identification of this pheromone minimizes the occurrence of interspecific matings when individuals of both species infest the same host. Chemical studies showed that the pheromone occurs in the vestibular portion of the vagina. In *D. variabilis*, GSP consists of a mixture of long–chain (C_{14-20}), saturated fatty acids, and the steroid 20-hydroxyecdysone (Allan et al., 1988; Taylor, Sonenshine & Phillips, 1991). Courting males crawling over the female's ventral body are guided to the genital pore when they detect these compounds with their chelicerae. Identification of the pheromone stimulates the males to insert their chelicerae into the vestibular vagina. In this location, the males soon receive further positive rein-

forcement as they detect even greater concentrations of the pheromone components. This in turn stimulates the males to synthesize and transfer the sperm-filled spermatophore into the female's vulva, using their chelicerae to complete the process. Examination of the fine structure of the cheliceral digits has revealed the presence of tiny pores innervated by sensory neurons. If the digits were ablated, the copulatory response was lost. The potency of the GSP is such that sexually excited males will even copulate with neutered females, i.e. females from which the vestibular vagina was excised, if GSP was placed in the residual gonopore (Allan et al., 1988). In *D. andersoni*, free fatty acids and ecdysteroids were much more abundant in the vestibular vagina and on the external genital surface than in *D. variabilis*. An interesting corollary was that *D. andersoni* males exhibited a much higher response threshold for the differing concentrations of the same compounds than did *D. variabilis* males. These differences in fatty acid and ecdysteroid concentrations and the corresponding differences in response thresholds are believed to explain the ability of the males of the two different species to discriminate their conspecific mates (Allan, Phillips & Sonenshine, 1989). Comparison with ticks of the genus *Amblyomma* showed evidence for a low level of GSP in *A. americanum* but not *A. maculatum* (Allan, Phillips & Sonenshine, 1991).

The sex pheromone(s) of prostriate ticks (genus *Ixodes*) has not been identified. Ticks of this genus lack foveal glands and no evidence of 2,6-dichlorophenol has been found in extracts. In a recent study of the lipophilic compounds extracted from the body surface of adult *Ixodes persulcatus*, no evidence of substituted phenols was found, although cholesterol, cholesterol derivatives and other lipids were identified. However, whether any of these compounds act as sex pheromones for regulating *I. persulcatus* mating behaviour was not determined (Tkachev et al., 2000).

Sex pheromones of argasid ticks

Sex pheromone mediation of mating behaviour also occurs in argasid ticks. In ticks of the genus *Ornithodoros*, the sex pheromone is secreted in the coxal fluid of adult females several days after feeding. The females must also be moving in order to attract sexually active males. Males seeking mates recognize the ambulatory females, make contact and detect the pheromone. The pheromone also initiated courtship behaviour by males of different *Ornithodoros* species (Schlein & Gunders, 1981; Mohamed et al., 1990). Males did not respond to other males or nymphal ticks. In *O. savignyi*, unfed females were unattractive to males, although unfed males responded strongly to attractive females. In females, sex pheromone activity increased gradually to a maximum 6 days after feeding. Washing the female's body surface or sealing the coxal orifices eliminated sex attractant activity. Thus far, the chemical composition of the pheromone has not been identified.

ALLOMONES

When ticks such as *D. variabilis* are disturbed by grasping, especially the legs, large pores on the dorsolateral surfaces (large dermal glands = dermal glands Type II) secrete a hydrocarbon-based fluid that is rich in squalene (Yoder *et al.*, 1993*b*; Yoder & Domingus, 2003). Previously termed sagittiform sensilla because of their presumed sensory functions (Dinnik & Zumpt, 1949), they are widespread on the bodies of metastriate ticks (Fig. 21.5) but absent in prostriate ticks (genus *Ixodes*) and argasid ticks. The amount of secretion is substantial, averaging approximately 2% of tick body weight. When attacked by fire ants (*Solenopsis invicta*), unfed adult ticks (*D. variabilis* or *A. americanum*) were observed to secrete waxy material that protected ticks from predation. Depletion of defensive waxes (repeated leg grasping) or immobility (e.g. fully engorged ticks) made ticks vulnerable to ant attack (Yoder, Pollack & Spielman 1993*a*), and this is supported by field observations of ticks in ant-infested areas (Wilkinson, 1970; Harris & Burns, 1977; Barré, Garris & Lorvelec, 1997). Prostriate ticks, *I. scapularis*, or argasid ticks, *O. moubata*, which lack large dermal glands, also were attacked immediately by foraging ants which dismembered them and ate their appendages. Beetles (*Tenebrio molitor*), however, attacked and ate any tick that they encountered. Ants also ate beetles. Therefore, when ants were presented with both ticks and beetles, ant predation on beetles protected or rescued many ticks that would otherwise have been eaten by beetles. This was the first report of an allomone in ticks. Ticks apparently lack the ability to synthesize squalene from precursor compounds and instead acquire it from their hosts during blood-feeding. Subsequently, it appears that the ticks sequester large quantities of this compound in the large dermal glands, where it constitutes up to 25% of glandular secretions. Other hydrocarbons identified in the tick defence secretion (namely C_{20}, C_{24} and a methyl branched C_{25} alkane) are as effective as squalene in protecting ticks against attack by ants (J. Yoder, personal communication), implying that this allomone functions because it is hydrocarbon-based, not necessarily because it is squalene. Other chemicals may also play a role as defensive chemicals in ticks. Yoder,

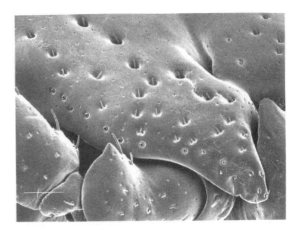

Fig. 21.5. Apertures of large wax glands on the margin of the scutum of an adult female *Dermacentor variabilis*. The gland associated with these pores secretes a squalene-rich exudate when the tick legs are pressure-stimulated. Bar = 100 μm. (Photo credit: Dr Jay Yoder. Reproduced from Yoder *et al.* (1993*b*) *Journal of Insect Physiology* **39**, 291–296 with permission of the editor.)

Wittenberg & Blomquist (1997) noted that larval ticks, which lack squalene, avoid predation by ants, suggesting squalene is not the only component of the tick allomone. However, to date no other defensive compound has been identified in ticks. The tick defence secretion does not have an obnoxious repellent odour; rather its role in predator avoidance is subtle and serves to temporarily neutralize predator aggressiveness. Interestingly, features of the tick allomone are comparable to the majority of allomones in insects (Blum, 1985).

KAIROMONES

In order to obtain blood for their nutritional needs, ticks must be able to detect the presence of hosts and recognize locations on the host body suitable for feeding. Kairomones are compounds emanating from prospective hosts, e.g. CO_2 in animal breath, compounds on host skin such as NH_3, lactic acid, or compounds deposited on vegetation by passing animals, that stimulate tick appetence behaviour.

Perhaps the best-known kairomones that attract host-seeking ticks are CO_2 and NH_3, compounds typically found in animal breath. Tropical bont ticks (*A. variegatum*), for example, are aroused from their resting state by increases in the CO_2 concentration above ambient levels (Steullet & Guerin, 1992*a*). For these ticks and many other species, CO_2 is a non-specific, general excitant. Once stimulated, the ticks respond to differences in the concentration gradient that leads them to the CO_2 source. Ammonia is also known to attract ticks, e.g. *R. sanguineus* (Haggart & Davis, 1981*a*).

Upon contact with the host body, several compounds on the host skin are attractive and apparently stimulate appetent behavior. In addition to CO_2 and NH_3 noted previously, the lipid squalene is a major component of the compounds found on human skin (Yoder, Atwood & Stevens, 1998) as well as urea, butyric acid and lactic acid. Human sweat is rich in aliphatic carboxylic acids, in addition to some of the other compounds noted above. One class of carboxylic acids, the C_4–C_6, 2-oxocarboxylic acids are highly attractive to mosquitoes (*Anopheles gambiae*) (Healy *et al.*, 2002). Whether such sweat compounds also attract ticks is unknown.

A blend of different host-derived odorants was found to be highly attractive to both *I. ricinus* and *B. microplus*. However, a phenolic fraction extracted from bovine odours stimulated *B. microplus* but not *I. ricinus*. No single odorant was as attractive to the ticks as a blend simulating the natural host odours. Two compounds in particular, 1-octen-3-ol and *o*-nitrophenol, compounds characteristically found in bovine odours, evoked a significantly stronger response from *B. microplus* than *I. ricinus*. Such differences in tick sensitivity to different host odour profiles may contribute to an understanding of how ticks recognize their preferred hosts and proceed to feed on them (Osterkamp *et al.*, 1999). Prostriate ticks also recognize and respond to substances deposited by their hosts on vegetation. Black-legged ticks, *I. scapularis*, both males and females, exhibited an arrestment response when they contacted objects contaminated with substances rubbed from the pelage of the tarsal and interdigital glands of white-tailed deer (*Odocoileus virginianus*). Substances collected from the preorbital glands of deer elicited a strong arrestment response among *I. scapularis* females. Extracts from other deer body regions had little or no effect (Carroll, Mills & Schmidtmann, 1996). Similar responses were observed with *Ixodes neitzi* that encounter vegetation contaminated by substances from klipspringer antelope, a favoured host for these ticks in southern Africa. The ticks cluster on vegetation marked with emanations of the animal's preorbital glands, which enhances the opportunities for future encounters with these hosts (Rechav *et al.*, 1978). In both cases, these deer substances are believed to act as kairomones that, together with CO_2, account for the accumulation of host-seeking ticks along animal trails.

PHEROMONE GLANDS

Glands associated with pheromone production have been identified in metastriate ixodid ticks but not in prostriate ixodids. The coxal glands are believed to serve as the source

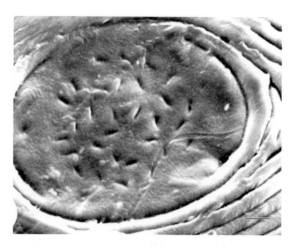

Fig. 21.6. Scanning electron micrograph of one of the paired foveae dorsales from an adult female *Dermacentor variabilis*. This image illustrates the numerous slit-like pores in each fovea. Oily droplets containing the sex-attractant pheromone, 2,6-dichlorophenol, ooze from these pores, releasing the volatile pheromone to the atmosphere. (Reproduced from Sonenshine (1991) with permission from Oxford University Press, New York.)

of the sex pheromone in argasid ticks (Schlein & Gunders, 1981). 2,6-Dichlorophenol is produced in the foveal glands, small paired glands located under the foveae dorsales on the dorsal alloscutum (i.e. region immediately posterior to the scutum) (Figs. 21.4 and 21.6). Evidence for the role of these glands in the production, storage and secretion of 2,6-dichlorophenol was obtained by experimental methods and electron microscopy, as summarized by Sonenshine (1991). In unfed females, the dominant feature of the cells that make up the lobes of the gland is the large number of vesicles scattered throughout the cytoplasm, each containing an oil droplet. The phenol is sequestered in these oil droplets. When examined by X-ray microanalysis, the oil droplets but not other parts of these cells were found to contain large amounts of chlorine bound in organic form (Sonenshine *et al.*, 1983). Feeding initiates the disruption of storage vesicles, allowing the 2,6-dichlorophenol-containing oil droplets to migrate to the ducts near the apices of the glandular cells. From here, masses of oil droplets accumulate in the ducts of foveal glands and disperse out of the tiny slit-like pores of each fovea onto the external body surface. The foveal glands are innervated by small nerves. The finding of abundant neurosecretory granules in the axons of the foveal nerves as well as the inhibition of 2,6-dichlorophenol secretion following treatment with catecholamine antagonists suggest that pheromone secretory activity is mediated, at least in part, by the neurosecretory pathway. Other studies summarized by Sonenshine (1991) also indicated a stimulatory role for the steroid 20-hydroxyecdysone.

In contrast to the foveal glands, virtually nothing is known about the biosynthesis or regulation of secretory activity by the ventrally located Type II dermal glands in *Amblyomma* species, the glands responsible for the secretion of the AAA pheromone. The source of the mounting sex pheromone is unknown, but presumed to be the ubiquitous Type I dermal glands scattered over the body surface. The source of the genital sex pheromone also is unknown. Two likely candidates are (1) the lobular accessory gland, which surrounds the cuticle-lined vestibular region of the vagina or (2) the tubular accessory glands at the junction of the vestibular and cervical regions of the vagina.

PERCEPTION OF SEMIOCHEMICALS BY TICKS

Ticks, like other animals, bear numerous sensory organs which enable them to sense chemicals in their environment as well as neural networks for interpreting and responding to the information received. In addition to the numerous setiform sensilla (= hair-like setae) on the legs and most of the tick's body that function as mechanosensilla, ticks have clusters of specialized chemosensory sensilla that detect chemical compounds in their environment. Two major types of chemosensilla occur: olfactory sensilla and gustatory sensilla.

Ticks extend their forelegs like antennae when actively exploring their environment or responding to stimuli from a stationary perch. This behaviour enables olfactory sensilla located on the foreleg tarsi to detect odours. Olfactory sensilla are located in the Haller's organ, a prominent sensory apparatus on the dorsal surface of the tarsus, while others are located anterior and posterior to this structure (Fig. 21.7). The single-wall (SW) olfactory sensilla are easily recognized by the innumerable tiny pores all over the surface when examined by high-magnification scanning electron microscopy (SEM) (Fig. 21.8). The pores open directly into the lymph cavity within the sensillum, allowing compounds absorbed from the atmosphere on the surface to migrate into the interior and stimulate the dendrites to produce nerve impulses. Most metastriate ticks that have been studied have six or seven setiform sensilla in the anterior capsule, one or two of which are the large, SW multiporose olfactory sensilla. In *D. variabilis*, this is represented by the large setiform sensillum,

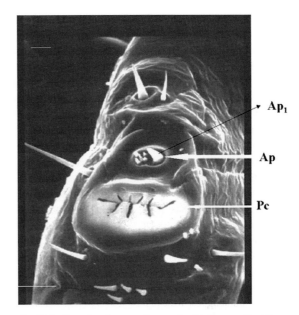

Fig. 21.7. Scanning electron micrograph of the tarsus of leg I in *Dermacentor variabilis* showing the Haller's organ. Arrows show the location of olfactory sensilla. Ap, anterior pit; Ap_1, multiporose olfactory sensilla; P_c, posterior capsule. Scale bar = 50 μm.

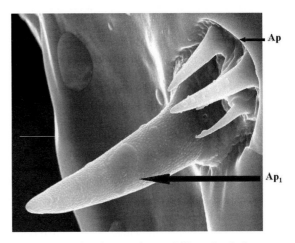

Fig. 21.8. Scanning electron micrograph illustrating the large olfactory sensillum (Ap_1) in the anterior pit of Haller's organ surrounded by other sensilla. Ap, anterior pit; Ap_1, olfactory sensillum. Scale bar = 5 μm.

Ap_1, shown in Fig. 21.7 and Fig. 21.8. Anterior to the Haller's organ are other prominent setiform sensilla, at least one of which is an olfactory sensillum. These multiporose olfactory sensilla detect the sex pheromone 2,6-dichlorophenol and other substituted phenols. Moreover, different neurons among the large number of sensory neurons that innervate any particular sensilla are responsive to different compounds, e.g. NH_3 as well as substituted phenols (Haggart & Davis, 1981a). Several multiporose sensilla also occur in the Haller's organ capsule. In argasid ticks, the number of anterior pit sensilla among the different species is more variable (from five to 11) but again there are one or two SW multiporose olfactory sensilla.

In addition to its impressive morphological complexity, the Haller's organ presents an amazing complex of functionally diverse neurons that innervate the few visible setiform sensilla. For example, NH_3 is detected by two sensilla on the tarsus of the brown dog tick, *R. sanguineus*, one located in the anterior pit, the other in the cluster of four sensilla distal to the Haller's organ (Holscher, Gearhart & Barker, 1980). In the lone star tick, *A. americanum*, 2,6-dichlorophenol is detected by the large multiporose olfactory sensillum (AP_1). Other neurons detect CO_2 (see below). In *A. variegatum*, Haller's organ sensilla also detect H_2S (Steullet & Guerin, 1992b). In view of the multiple sensory neuron innervation of the AP_1 sensillum, it is likely that it detects other odorants as well (Haggart & Davis, 1981a). The response characteristics of the typical olfactory sensilla such as the pheromone detecting AP_1 are phasic–tonic, i.e. the neuronal spike activity accelerates rapidly when exposed to the odorant at concentrations slightly above threshold, and then decays gradually. Such neurons are sensitive to extremely low concentrations of products, e.g. parts per billion. Once maximum spike activity has occurred, exposure to higher concentrations of the odorant has no effect. This type of sensory response is exceptionally well suited for detection of distant odours. In the case of CO_2, differences in odour concentration can be determined by two types of sensilla. *Amblyomma variegatum* has two CO_2-sensitive receptors, one excited by and the other inhibited by CO_2. Both exhibit phasic–tonic responses. The CO_2-excitable receptor responds to a relatively wide range of concentrations above 0.1% (atmospheric concentrations are approximately 0.04%). In contrast, the CO_2-inhibitable receptor is extremely sensitive to a much narrower range of concentrations slightly above ambient, and small increases in the range of 10–20 ppm above ambient are sufficient to inhibit this receptor. This combination of receptor responses is thought to enable the tick to identify small changes in CO_2 concentrations above ambient and recognize the increasingly higher concentrations it encounters as it approaches a vertebrate host (Steullet & Guerin, 1992a; Perritt, Couger & Barker, 1993).

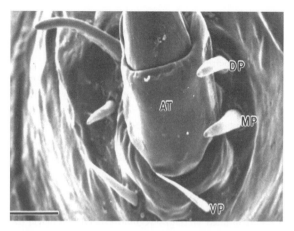

Fig. 21.9. Scanning electron micrograph illustrating the three pairs of gustatory sensilla at the base of the claw apotele on the leg I tarsus of a *Dermacentor variabilis* male tick. Ventral view. AT, apotele; DP, dorsal pair of claw sensilla; MP, middle (lateral) pair of claw sensilla; VP, ventral pair of claw sensilla. Scale bar = 25 μm. (Reproduced from Phillips & Sonenshine (1993) with permission from Kluwer Academic Publishers, Van Godewijckstraat 30, P.O. Box 17, 3300 AA Dordrecht, The Netherlands.)

Gustatory sensilla also occur in the Haller's organ on tarsus I as well as elsewhere on the leg I tarsus and on the terminal segments of the palps. Gustatory sensilla are recognized by presence of a single pore at the tip end of the shaft and smooth walls. Studies by Phillips & Sonenshine (1993) showed (by ablation or surface coating) that the tip pore sensilla at the end of the leg I tarsus (claw sensilla), located at the base of the claw apotele, are used by the male ticks to detect cholesteryl esters (mounting sex pheromone) during mating behaviour (Fig. 21.9). Three pairs of claw sensilla occur, arranged in a semicircle around the claw apotele. The dorsal and middle pairs of sensilla are mechanogustatory, while the ventral pair is solely mechanosensory. Small sensilla located on the terminal segments of the palps form a sensory field, serving as gustatory and/or mechanogustatory sensilla to detect compounds on the skin.

The chelicerae also bear sensilla on the cutting edges of the digits. In addition to a papilla (*B. microplus*) or conical depression (*D. variabilis*) that is a mechanoreceptor, pores occur that are innervated by sensory neurons. The cheliceral digits respond to physiological concentrations of NaCl, ATP and glutathione, compounds normally present in host blood. In addition to detecting these phago-stimulants, the cheliceral digits of sexually active *D. variabilis* males also respond to extracts of the conspecific female vulva and to physiological concentrations of ecdysone and 20-hydroxyecdysone. In contrast, sexually active *D. andersoni* males responded to 20-hydroxyecdysone but not ecdysone (Taylor *et al.*, 1991). When the cheliceral digits were excised, sexually active males attempted to mate with normal, unaltered females. They inserted their mouthparts in the female's genital pore, but were unable to form spermatophores and failed to copulate (Sonenshine *et al.*, 1984). These findings suggest that the cheliceral digits play an important role in mating, as well as feeding behaviour.

APPLICATIONS TO TICK CONTROL

Control of ticks on livestock, domestic fowl and pets has depended almost entirely on the large-scale administration of toxic substances onto the external body surfaces of these hosts or by the administration of systemics (see Chapter 18). Used as acaricides, these toxicants also have been dispersed in large quantities in the natural environment to kill ticks clinging to vegetation along roadsides, trails or surrounding homes and gardens where they might attack humans and their companion animals. The discovery of insecticidal compounds with low mammalian toxicity such as the pyrethroids or avermectins and improvements in the methods of delivery have enhanced their efficacy for tick control but at greatly increased cost. Moreover, the rapid development of insecticidal resistance to each new compound, the increased cost of new product registration with regulatory agencies and opposition from environmental activists have limited their usefulness for tick control. As a consequence, efforts to develop tick control strategies using tick semiochemicals in combination with acaricides have attracted considerable interest.

The earliest recorded attempt to use tick pheromones to assist in tick control was by Gladney *et al.* (1974). These workers combined an extract of the aggregation–attachment pheromone from male Gulf Coast ticks, *A. maculatum*, with an acaricide (isobenzan) and deposited it onto a bovine. Female ticks were lured to the treated site, attached nearby and were killed. Similar results were obtained in South Africa when an extract of fed male bont ticks, *A. hebraeum*, was combined with an acaricide (toxaphene) and applied to cattle. The resulting combination lured nymphal and adult ticks to the treated areas where they attached and died (Rechav & Whitehead, 1978). In both cases, however, efficacy was short-lived and it soon became apparent that, to be effective, the pheromone must be delivered continuously by means of a slow-release device. In addition, the pheromone–acaricide delivery system must be optimized to exploit those characteristics of the target species, e.g. host location behaviour or mating behaviour where its use would be most effective. This review will examine three specific types of pheromone-assisted tick control devices that have been developed in recent years that exploit different kinds of tick behaviour, namely, arrestments, confusants (= disrupt mating behaviour) and 'attract and kill' devices.

Arrestment pheromone-impregnated device for *Ixodes* ticks

A patented device (Allan, Sonenshine & Burridge, 2002) incorporating purines from the faecal wastes of the prostriate tick *I. scapularis* into oily droplets released from a pump sprayer was prepared for delivery to vegetation. The oily droplets (Last Call™, IPM Technologies, Portland, OR) adhere to vegetation where *I. scapularis* quest for hosts. The incorporation of the arrestment pheromone components guanine and xanthine along with an acaricide (Permethrin) causes ticks that encounter the droplets to cling to the contaminated surfaces where they acquire a lethal dose of the acaricide. Laboratory studies showed that the incorporation of the arrestment pheromone increased the mortality from 70% for the device with acaricide alone to 95% for pheromone–acaricide mixtures (S. A. Allan, personal communication).

Confusants (mating disruption)

One of the reasons that ticks are difficult to kill is that they remain fixed (attached) in specific locations on the animal body. Large amounts of highly toxic acaricides must be spread over the host's body to reach the stationary ticks. Males of some tick species, however, are more susceptible when they become excited by the female sex attractant, 2,6-dichlorophenol, detach and crawl over the surface searching for females. A confusant exploits this mate-searching behaviour by creating a ubiquitous background of 2,6-dichlorophenol, exciting the males to search but minimizing their ability to locate females as the emitting source. In one experiment, a water emulsion of gelatin-microencapsulated 2,6-dichlorophenol was combined with an acaricide (Propoxur) and applied to tick-infested dogs. The gelatin microcapsules adhered to the hairs, providing a slow-release mechanism that persisted for many days. Significantly more males than females were killed by the treatment, and most of the surviving females died without laying eggs. Oviposition was reduced to less than 10% of the amount observed with the controls (Sonenshine, Taylor & Corrigan, 1985).

'Attract and kill' devices

TICK DECOY

This device was developed by Hamilton and Sonenshine (Hamilton & Sonenshine, 1989) to attract *D. variabilis* mate-seeking males to plastic spherules containing small quantities of toxicant and kill them. The plastic spherules made of common plastic, polyvinyl chloride (PVC), served as female mimics, i.e. decoys. These devices were impregnated with the sex attractant pheromone 2,6-dichlorophenol and the organophosphorus acaricide Propoxur. Following manufacture, the plastic devices were coated with the *D. variabilis* mounting sex pheromone cholesteryl oleate. When completed, the PVC decoys were attached with cement to the hair coat of a tick-infested rabbit at a rate of 10 decoys per naturally attached female tick. Sexually excited males were released onto the rabbit and allowed to search for mates, either decoys or live females. Within 30 minutes, 89% of the males were found in the mating posture on the decoys; the remainder were attached to the host skin adjacent to these devices. All of the males were dead, presumably as a result of the lengthy exposure to the acaricide-emitting decoys. In contrast, when the experiment was repeated with decoys with 2,6-dichlorophenol but no cholesteryl oleate, only 20% of the males attempted to mate with the decoys and only 36% were dead after 30 minutes. Clearly, the presence of the mounting sex pheromone (cholesteryl oleate) was essential to increase the contact time with the acaricide emitting source if the ticks were to acquire a lethal dose (Table 21.1). The advantage of this tick control strategy is expressed in its ability to disrupt tick reproduction. Although targeted against mate-seeking males, surviving females failed to engorge to repletion. Most females died and the few that dropped off failed to lay eggs. Thus, no F_1 generation capable of infesting new hosts resulted from this treatment. When used in a confined environment, e.g. kennel, barn, etc., this strategy may be expected to lead to total eradication of the tick infestation with only a few consecutive applications (assuming no new introductions). When the same procedure was applied to tick-infested cows in a barn, the results were similar. Most of the sexually active males were lured to the decoys instead of the naturally occurring females; 92.3% died within 1 hour and all were dead within 1.5 hours (Sonenshine, Hamilton & Lusby, 1992). In Egypt, this treatment method was used to control infestations of *H. dromedarii*, the vector of tropical theileriosis, on camels. Application of decoys impregnated with pheromone + cyfluthrin showed an efficacy of 85.3% vs. the controls. Sexually active males migrated rapidly to the decoys and some even mounted these devices for attempted copulation (Adbel-Rahman, Fahmy & Aggour, 1998).

BONT TICK DECOY

A modified version of the decoy strategy was adapted for use on cattle and other livestock attacked by the African bont ticks, *A. variegatum* and *A. hebraeum*, the major vectors of the heartwater-causing *Anaplasma* (= *Cowdria*) *ruminantium*. In this application, a mixture of the known AAA pheromonal components *o*-nitrophenol, methyl salicylate and 2,6-dichlorophenol as well as the proven artificial attractant phenylacetaldehyde (Norval *et al.*, 1992*a*) were impregnated into plastic tags attached to the tails of cattle (Fig. 21.10). Also included was a pyrethroid, either Cyfluthrin or Flumethrin (Bayer). The pheromone components emerging from the plastic tags attracted bont ticks from the surrounding vegetation. Attracted to the treated animals, the ticks formed aggregations on the animal body adjacent to, or near, the decoys, acquired lethal doses of the acaricides and died. In a large-scale trial in Zimbabwe with hundreds of animals, including a control group, tick control for cattle treated with cyfluthrin-impregnated tags was 94.9% for the first 3-month trial and 99.3% for a second 3-month trial. Efficacy using flumethrin-impregnated tags was slightly lower, 87.5% and 95.1%, respectively. Tags impregnated with α-cypermethrin (FMC) were much less effective (Norval *et al.*, 1996) (Table 21.2).

The efficacy of the bont tick decoy was evaluated against tropical bont ticks, *A. variegatum*, on cattle in the island of Guadeloupe in the Caribbean. Pheromone- or pheromone–acaricide-impregnated tags were applied to both the neck and tails of cattle in this study. The efficacy, expressed as tick mortality on treated vs. untreated cattle, was similar for the pheromone plus acaricide-impregnated tags and the tags with acaricides alone. However, cattle with pheromone-impregnated tags had significantly greater proportions of ticks on the hindquarters and front regions as compared to untreated cattle, indicating that ticks aggregated in response to the attractants. Chemical analysis showed detectable levels of acaricides on all body regions of the animals throughout the trial. Emission rates for the pheromone components from the tags were most rapid during the first 4 weeks after the tags were installed. On average, the concentrations of the pheromone components in the tags were 78% lower at week 4 than at the time of manufacture. Thereafter, emission rates were slower, averaging 51.5% for all pheromone components during the remaining 9 weeks. Small quantities of

Table 21.1 *Pheromone–pesticide decoys kill male ticks (D. variabilis) and prevent mating when administered at a ratio of 10 : 1 decoys : live female ticks on rabbits*

	Type of treatment[a]							
	(1) Both pheromones + acaricide		(2) Both pheromones, no acaricide		(3) 2,6-Dichlorophenol + acaricide		(4) 2,6-Dichlorophenol, no acaricide	
Location of males on host	Percent of sexually active ♂♂ > release ± s.d. (hours > ♂♂ release)							
	0–0.5	48	0–0.5	48	0–0.5	48	0–0.5	48
Mating with decoys[b]	89.0* ± 3.3	–	73.0 ± 4.8	14.6 ± 1.2	20.0 ± 4.0	0.0	23.0 ± 5.8	10.0 ± 0.3
Attached beside females	11.0 ± 3.3	–	17.0 ± 1.1	70.8 ± 1.4	24.0 ± 3.8	2.0* ± 2.0	14.0 ± 2.1	25.0 ± 0.1
Mating with live females[c]	0.0	–	0.0	8.3 ± 0.6	6.0 ± 1.6	0.0	19.0 ± 3.3	7.5 ± 1.0
Attached elsewhere	0.0	–	10.0	6.3 ± 0.6	50.0 ± 0.4	2.0 ± 5.5	44.0 ± 1.3	57.0 ± 1.0
Unattached	0.0	–	0.0	0.0	0.0	96.0 ± 4.8	0.0	0.0
Dead	100	–	0.0	0.0	36.0 ± 4.3	98.0 ± 1.3	0.0	0.0

* Males dead; two males deposited spermatophores onto decoys before they died.

[a] Two treatments were done with both pheromones, i.e. decoys impregnated with 2,6-dichlorophenol and coated with cholesteryl oleate. In group 1, decoys were impregnated with an acaricide (Propoxur); in group 2, no acaricide was included. Two other treatments were done with only one pheromone, i.e. decoys impregnated with 2,6-dichlorophenol alone. In group 3, decoys were impregnated with an acaricide (Propoxur); in group 4, no acaricide was included.

[b] Mating with decoys = males physically on or in direct contact with the plastic decoys.

[c] Attached beside decoys = males attached to host skin adjacent to decoys but not in direct contact with these devices.

Fig. 21.10. Photograph showing a bont tick decoy attached to the tail of a cow in Zimbabwe. These pheromone–acaricide-impregnated tags proved highly effective in controlling infestations of bont tick, *Amblyomma hebraeum*, for as long as 3 months. (Photo credit: Dr Suman Mahan, Harare, Zimbabwe; Heartwater Research Project, Department of Veterinary Pathology, University of Florida, Gainesville, FL, USA.)

pheromone components, from 1.4% to 2.6%, remained in the plastic tags when the trials were terminated at 13 weeks (Allan *et al.*, 1998).

FUTURE INVESTIGATIONS

Interest in the use of tick semiochemicals to aid in tick control has increased as a result of the several successful demonstrations, described above. In addition, continued opposition of environmental activists to the large-scale use of acaricides and insecticides in general has stimulated interest in alternative approaches to tick control. Although none of the patented technologies have been commercialized to date, negotiations are in progress with different companies to license and market the tick decoy for use on livestock and the arrestment pheromone-impregnated droplets (Last Call™, IPM Technologies, Inc., Portland, OR) for treatment of vegetation to control ticks transmitting *Borrelia burgdorferi* sensu stricto, the agent of Lyme disease. New research is being directed to enhancing the effectiveness of the latter device by the incorporation of NH_3- and CO_2-emitting substances to broaden its attraction for *I. scapularis*.

In addition to the need to continue research on the efficacy of the existing pheromone-based technologies described above, other avenues also exist that can be exploited for possible use in tick control. One concept that has already received some attention is to combine extracts of the tarsal glands of deer with an acaricide in a suitable matrix (e.g. Last Call™) for delivery to tick-infested vegetation. This concept exploits the use of kairomones that attract black-legged

Table 21.2 *Control*[a] *of* A. hebraeum *ticks on cattle using tail tags impregnated with AAA pheromone and cyfluthrin, flumethrin or* α-*cypermethrin during field trials in Zimbabwe*

Stage/Treatment	Week of trial					
	2	4	6	8	10	12
Males						
Cyfluthrin	99.6	97.4	99.5	98.0	99.1	99.6
Flumethrin	95.3	87.9	97.2	92.3	91.7	91.8
α-Cypermethrin	67.2	64.8	32.5	64.4	68.8	74.3
Females						
Cyfluthrin	100	94.3	100	98.5	98.8	100
Flumethrin	100	94.3	100	95.1	97.1	93.5
α-Cypermethrin	6.5	56.4	19.3	71.3	71.3	74.8

[a] Percent control determined by the formula $C = 100 - (T/U \times 100)$ where U is the mean number of ticks present on control cattle and T is the mean number of ticks present on treated cattle.

Source: From Norval *et al.* (1996) with permission from Kluwer Academic Publishers, Van Godewijckstraat 30, P.O. Box 17, 3300 AA Dordrecht, the Netherlands.

ticks, *I. scapularis*. Preliminary investigations currently in progress suggest that this strategy may be useful for controlling *I. scapularis* in their natural habitat. Another avenue involves the development of vaccines to target the odorant binding proteins that bind and transfer pheromone compounds to the dendrites in the olfactory and gustatory receptors that detect the pheromones. Finally, molecular methods may prove useful for determining the enzymatic steps in the biosynthetic pathway responsible for pheromone biosynthesis. Knowledge of pheromone biosynthesis offers alternative opportunities for disrupting pheromone secretion.

ACKNOWLEDGEMENTS

I express my gratitude to Dr Jay Yoder, Wittenberg University, Springfield, OH for his assistance in reviewing and editing this chapter.

REFERENCES

Abdel-Rahman, M. S., Fahmy, M. M. & Aggour, M. G. (1998). Trials for control of ixodid ticks using pheromone–acaricide tick decoys. *Journal of the Egyptian Society of Parasitology* **28**, 551–557.

Allan, S. A. & Sonenshine, D. E. (2002). Evidence of an assembly pheromone in the black-legged deer tick, *Ixodes scapularis*. *Journal of Chemical Ecology* **28**, 15–27.

Allan, S. A., Barré, N., Sonenshine, D. E. & Burridge, M. J. (1998). Efficacy of tags impregnated with pheromone and acaricides for control of *Amblyomma variegatum*. *Medical and Veterinary Acarology* **12**, 141–150.

Allan, S. A., Phillips, J. S. & Sonenshine, D. E. (1989). Species recognition elicited by differences in composition of the genital sex pheromone in *Dermacentor variablis* and *Dermacentor andersoni* (Acari: Ixodidae). *Journal of Medical Entomology* **26**, 539–546.

Allan, S. A., Phillips, J. S. & Sonenshine, D. E. (1991). *Amblyomma americanum* and *Amblyomma maculatum* (Acari: Ixodidae): role of genital sex pheromones. *Experimental and Applied Acarology* **11**, 9–21.

Allan, S. A., Phillips, J. S., Taylor, D. E. & Sonenshine, D. E. (1988). Genital sex pheromones of ixodid ticks: evidence for the role of fatty acids from the anterior reproductive tract in mating of *Dermacentor variablis* and *Dermacentor andersoni*. *Journal of Insect Physiology* **34**, 315–323.

Allan, S. A., Sonenshine, D. E. & Burridge, M. J. (2002). Tick pheromones and uses thereof. United States Patent Office, Patent No. 6,331,297.

Apps, P. J., Viljoen, H. W. & Pretorius, V. (1988). Aggregation pheromones of the bont tick *Amblyomma hebraeum*: identification of candidates for bioassay. *Onderstepoort Journal of Veterinary Research* **55**, 135–137.

Ayasse, M., Paxton, R. J. & Tengo, J. (1995). Mating behavior and chemical communication in the order Hymenoptera. *Annual Review of Entomology* **46**, 31–78.

Barré, N., Garris, G. I. & Lorvelec, O. (1997). Field sampling of the tick *Amblyomma variegatum* (Acari: Ixodidae) on pastures in Guadeloupe: attraction of CO_2 and/or tick pheromones and conditions of use. *Experimental and Applied Acarology* **21**, 95–108.

Berger, R. S. (1972). 2,6-Dichlorophenol, sex pheromone of the lone star tick. *Science* **177**, 704–705.

Blum, M. S. (1985). *Fundamentals of Insect Physiology*. New York: John Wiley.

Cardé, R. T. & Baker, T. C. (1984). Sexual communication with pheromones. In *Chemical Ecology of Insects*, eds. Bell, W. C. & Cardé, R. T., pp. 355–383. Sunderland, MA: Sinauer Associates.

Carroll, J. F., Mills, G. D. & Schmidtmann, E. T. (1996). Field and laboratory responses of adult *Ixodes scapularis* (Acari: Ixodidae) to kairomones produced by white-tailed deer. *Journal of Medical Entomology* **33**, 640–644.

De Bruyne, M. & Guerin, P. M. (1998). Contact chemostimuli in the mating behaviour of the cattle tick, *Boophilus microplus*. *Archives of Insect Biochemistry and Physiology* **39**, 65–80.

Diehl, P. A., Guerin, P. M., Vlimant, M. M. & Steullet, P. (1991). Biosynthesis, production site and emission rates of the aggregation–attachment pheromone in males of two *Amblyomma* ticks. *Journal of Chemical Ecology* **17**, 833–847.

Dinnik, J. & Zumpt, F. (1949). The integumentary sense organs of the larvae of the Rhipicephalinae (Acarina). *Psyche* **56**, 1–17.

Dusbábek, F., Simek, P., Jegorov, A. & Troska, J. (1991). Identification of xanthine and hypoxanthine as components of assembly pheromone in excreta of argasid ticks. *Experimental and Applied Acarology* **11**, 307–316.

Evenden, M. L., Judd, J. R. & Borden, J. H. (1999). A synomone imparting distinct sex pheromone communication channels for *Choristoneura rosaceana* (Harris) and *Pandemis limitata* (Robinson) (Lepidoptera: Tortricidae). *Chemoecology* **9**, 73–80.

Fauvergue, X., Hopper, K. R. & Antolin, M. F. (1995). Mate finding via a trial sex pheromone by a parasitoid wasp. *Proceedings of the National Academy of Sciences of the USA* **92**, 900–904.

Gladney, W. J., Gabbe, R. R., Ernst, S. E. & Oehler, D. D. (1974). The Gulf Coast tick: evidence of a pheromone produced by males. *Journal of Medical Entomology* **11**, 303–306.

Gothe, R. & Neitz, A. W. H. (1985). Investigation into the participation of male pheromones of *Rhipicephalus evertsi evertsi* during infestation. *Onderstepoort Journal of Veterinary Research* **52**, 6–70.

Graf, J. F. (1974). Ecologie et éthologie d'*Ixodes ricinus* L. en Suisse (Ixodoidea: Ixodidae). III. Copulation, nutrition et ponte. *Acarologia* **16**, 636–642.

Graf, J. F. (1978). Ecologie et éthologie d'*Ixodes ricinus* L. en Suisse (Ixodoidea: Ixodidae). III. *Bulletin de la Société Entomologique Suisse* **51**, 241–253.

Grenacher, S., Kröber, T., Guerin, P. M. & Vlimant, M. (2001). Behavioral and chemoreceptor cell responses of the tick, *Ixodes ricinus* to its own faeces and faecal constituents. *Experimental and Applied Acarology* **25**, 641–660.

Haggart, D. A. & Davis, E. E. (1979). Electrophysiological responses of two types of ammonia-sensitive receptors on the first tarsi of ticks. In *Recent Advances in Acarology*, ed. Rodriguez, J. G., pp. 421–425. New York: Academic Press.

Haggart, D. A. & Davis, E. E. (1981*a*). Ammonia-sensitive neurons on the first tarsi of the tick, *Rhipicephalus sanguineus*. *Journal of Insect Physiology* **26**, 51–523.

Haggart, D. A. & Davis, E. E. (1981*b*). Neurons sensitive to 2,6-dichlorophenol on the tarsi of the tick, *Amblyomma americanum* (Acari: Ixodidae). *Journal of Medical Entomology* **18**, 187–193.

Hamilton, J. G. C., Papadopoulos, E., Harrison, S. J., Lloyd, C. M. & Walker, A. R. (1994). Evidence for a mounting sex pheromone in the brown ear tick *Rhipicephalus appendiculatus*, Neuman 1901 (Acari: Ixodidae). *Experimental and Applied Acarology* **18**, 331–338.

Hamilton, J. G. C. & Sonenshine, D. E. (1988). Evidence for the occurrence of mounting sex pheromone on the body surface of female *Dermacentor variabilis* (Say) (Acari: Ixodidae). *Journal of Chemical Ecology* **14**, 401–410.

Hamilton, J. G. C. & Sonenshine, D. E. (1989). Methods and apparatus for controlling arthropod populations. United States Patent Office, Patent No. 4,884,361.

Harris, W. G. & Burns, E. C. (1977). Predation of the lone star tick by the imported fire ant. *Environmental Entomology* **1**, 362–365.

Healy, T. P., Copland, M. J., Cork, A., Przyborowska, A. & Halket, J. (2002). Landing responses *of Anopheles gambiae* elicited by oxocarboxylic acids. *Medical and Veterinary Entomology* **16**, 126–132.

Holscher, K. H., Gearhart, H. L. & Barker, R. W. (1980). Electrophysiological responses of three tick species to carbon dioxide in the laboratory and field. *Annals of the Entomological Society of America* **73**, 288–292.

Khalil, G. M. (1984). Fecundity-reducing pheromone in *Argas* (*Persicargas*) *arboreus* (Ixodoidea: Argasidae). *Parasitology* **88**, 395–402.

Kiszewski, A. E., Matuschka, F. R. & Spielman, A. (2001). Mating strategies and spermiogenesis in ixodid ticks. *Annual Review of Entomology* **46**, 167–182.

Leahy, M. G., Hajkova, Z. & Bourchalova, J. (1981). Two female pheromones in the metastriate ticks, *Hyalomma dromedarii* (Acarina: Ixodidae). *Acta Entomologica Bohemoslavia* **78**, 224–230.

Leahy, M. G., Karuhize, G., Mango, C. C. & Galun, R. (1975). An assembly pheromone and its perception in the tick, *Ornithodoros moubata* (Murray) (Acari: Argasidae). *Journal of Medical Entomology* **12**, 284–287.

Leahy, M. G., Vandehay, R. & Galun, R. (1973). Assembly pheromones in the soft tick, *Argas persicus* (Olsen). *Nature* **246**, 515–517.

Linthicum, K. J. & Bailey, C. L. (1994). Ecology of Crimean–Congo hemorrhagic fever. In *Ecological Dynamics of Tick-Borne Zoonoses*, eds. Sonenshine, D. E. & Mather, T. N., pp. 392–437. New York: Oxford University Press.

Lusby, W. R., Sonenshine, D. E., Yunker, C. E., Norval, R. A. I. & Burridge, M. J. (1991). Comparison of known and suspected pheromonal constituents in males of the African ticks *Amblyomma hebraeum* Koch and *Amblyomma variegatum* (Fabricius). *Experimental and Applied Acarology* **13**, 143–152.

McMahon, C., Guerin, P. M. & Syed, Z. 2001. 1-Octen-3-ol isolated from bont ticks attracts *Amblyomma variegatum*. *Journal of Chemical Ecology* **27**, 471–486.

Mohamed, F. S. A., Khalil, G. M., Marzouk, A. S. & Roshdy, M. A. (1990). Sex pheromone recognition of mating behavior in the tick *Ornithodoros* (*Ornithodoros*) *savignyi* (Audouin) (Acari: Argasidae). *Journal of Medical Entomology* **27**, 288–294.

Neitz, A. W. H. & Gothe, R. (1984). Investigations into the volatility of female pheromones and the aggregation-inducing property of guanine in *Argas* (*Persicargas*) *walkerae*. *Onderstepoort Journal of Veterinary Research* **54**, 197–201.

Norval, R. A. I., Andrew, H. R. & Yunker, C. E. (1989). Pheromone mediation of host selection in bont ticks (*Amblyomma hebraeum* Koch). *Science* **243**, 364–365.

Norval, R. A. I., Peter, T., Meltzer, M. I., Sonenshine, D. E. & Burridge, M. J. (1992*b*). Response of the ticks *Amblyomma hebraeum* and *A. variegatum* to known or potential components of the aggregation–attachment pheromone. IV. Attachment stimulation of nymphs. *Experimental and Applied Acarology* **16**, 247–253.

Norval, R. A. I., Peter, T., Yunker, C. E., Sonenshine, D. E. & Burridge, M. J. (1991*a*). Response of the ticks *Amblyomma hebraeum* and *A. variegatum* to known or potential components of the aggregation–attachment pheromone. I. Long-range attraction. *Experimental and Applied Acarology* **13**, 11–18.

Norval, R. A. I., Peter, T., Yunker, C. E., Sonenshine, D. E. & Burridge, M. J. (1991*b*). Response of the ticks *Amblyomma hebraeum* and *A. variegatum* to known or potential components of the aggregation–attachment pheromone. II. Attachment stimulation. *Experimental and Applied Acarology* **13**, 19–26.

Norval, R. A. I., Peter, T., Yunker, C. E., Sonenshine, D. E. & Burridge, M. J. (1992*a*). Response of the ticks *Amblyomma hebraeum* and *A. variegatum* to known or potential components of the aggregation–attachment pheromone. III. Aggregation. *Experimental and Applied Acarology* **16**, 237–245.

Norval, R. A. I., Sonenshine, D. E., Allan, S. A. & Burridge, M. J. (1996). Efficacy of pheromone–acaricide impregnated tail-tag decoys for control of bont ticks, *Amblyomma hebraeum*, on cattle in Zimbabwe. *Experimental and Applied Acarology* **20**, 31–46.

Obenchain, F. D. (1984). Behavioral interactions between the sexes and aspects of species specificity pheromone mediated aggregation and attachment in *Amblyomma*. In *Acarology*, vol. 1, eds. Griffith, D. A. & Bowman, C. E., pp. 387–392. Chichester, UK: Ellis Horwood.

Osterkamp, J., Wahl, U., Schmalfuss, G. & Haas, W. (1999). Host-odour recognition in two tick species is coded in a blend of vertebrate volatiles. *Journal of Comparative Physiology A* **185**, 59–67.

Otieno, D. A., Hassanali, A., Obenchain, F. D., Steinberg, A. & Galun, R. (1985). Identification of guanine as an assembly pheromone of ticks. *Insect Science Applications* **6**, 667–670.

Pavis, C. & Barré, N. (1993). Kinetics of male pheromone production by *Amblyomma variegatum* (Acari: Ixodidae). *Journal of Medical Entomology* **30**, 961–965.

Perritt, D. W., Couger, G. & Barker, R. W. (1993). Computer-controlled olfactometer system for studying behavioral responses of ticks to carbon dioxide. *Journal of Medical Entomology* **30**, 571–578.

Phillips, J. S. & Sonenshine, D. E. (1993). Role of the male claw sensilla in perception of female mounting sex pheromone in *Dermacentor variabilis*, *Dermacentor andersoni* and *Amblyomma americanum*. *Experimental and Applied Acarology* **17**, 631–653.

Price, T. L. Jr, Sonenshine, D. E., Norval, R. A. I., Yunker, C. E. & Burridge, M. J. (1994). Pheromonal composition of two species of African *Amblyomma* ticks: similarities, differences and possible species specific components. *Experimental and Applied Acarology* **18**, 37–50.

Rechav, Y. & Whitehead, G. B. (1978). Field trials with pheromone–acaricide mixtures for control of *Amblyomma hebraeum*. *Journal of Economic Entomology* **71**, 149–151.

Rechav, Y., Goldberg, M. & Fielden, L. J. (1997). Evidence for attachment pheromone in the Cayenne tick (Acari: Ixodidae). *Journal of Medical Entomology* **34**, 234–237.

Rechav, Y., Norval, R. A. I., Tannock, J. & Colborne, J. (1978). Attraction of the tick *Ixodes neitzi* to twigs marked by the klipspringer antelope. *Nature* **275**, 310–311.

Rechav, Y., Parolis, H., Whitehead, G. B. & Knight, M. M. (1977). Evidence of an assembly pheromone(s) produced by males of the bont tick *Amblyomma hebraeum* (Acarina: Ixodidae). *Journal of Medical Entomology* **14**, 71–78.

Roelofs, W. L. (1995). Chemistry of sex attraction. *Proceedings of the National Academy of Sciences of the USA* **92**, 44–49.

Schlein, Y. & Gunders, A. E. (1981). Pheromone of *Ornithodoros* spp. (Argasidae) in the coxal fluid of female ticks. *Parasitology* **83**, 467–471.

Schöni, R., Hess, E., Blum, W. & Ramstein, K. (1984). The aggregation–attachment pheromone of the tropical bont tick *Amblyomma variegatum* Fabricius (Acari: Ixodidae): isolation, identification and action of its components. *Journal of Insect Physiology* **30**, 613–618.

Sobby, H., Aggour, M. G., Sonenshine, D. E. & Burridge, M. J. (1994). Cholesteryl esters on the body surface of the camel tick *Hyalomma dromedarii* (Koch, 1844) and the brown dog tick, *Rhipicephalus sanguineus* (Latreille, 1806). *Experimental and Applied Acarology* **18**, 265–280.

Sonenshine, D. E. (1985). Pheromones and other semiochemicals of the Acari. *Annual Review of Entomology* **30**, 1–28.

Sonenshine, D. E. (1991). *Biology of Ticks*, vol. 1. Oxford, UK: Oxford University Press.

Sonenshine, D. E. (1993). *Biology of Ticks*, vol. 2. Oxford, UK: Oxford University Press.

Sonenshine, D. E., Adams, T., Sallan, S. A., McLaughlin, J. R. & Webster, F. X. (2003). Chemical composition of some components of the arrestment pheromone of the black-legged tick, *Ixodes scapularis* (Acari: Ixodidae) and their use in tick control. *Journal of Medical Entomology* **40**, 849–859.

Sonenshine, D. E., Hamilton, J. G. C. & Lusby, W. R. (1992). Use of cholesteryl esters as mounting sex pheromone in combination with 2,6-dichlorophenol and pesticides to control populations of hard ticks. United States Patent Office, Patent No. 5,149,526.

Sonenshine, D. E., Hamilton, J. G. C., Phillips, J. S. & Lusby, W. R. (1991). Mounting sex pheromone: its role in regulation of mate recognition in the Ixodidae. In *Modern Acarology*, eds. Dusbabek, F. & Bukva, V., vol. 1, pp. 69–78. The Hague: SPB Academic Publishing.

Sonenshine, D. E., Homsher, P. J., Dees, W. H., Carson, K. A. & Wang, V. B. (1984). Evidence of the role of the cheliceral digits in the perception of genital sex pheromones during mating in the American dog tick, *Dermacentor variabilis* (Say). *Journal of Medical Entomology* **21**, 296–306.

Sonenshine, D. E., Khalil, G. M., Homsher, P. J., *et al.* (1983). Development, ultrastructure and activity of the foveal glands and foveae dorsales of the camel tick, *Hyalomma dromedarii* (Acari: Ixodidae). I. *Journal of Medical Entomology* **20**, 424–439.

Sonenshine, D. E., Taylor, D. & Corrigan, G. (1985). Studies to evaluate the effectiveness of sex pheromone impregnated formulations for control of populations of the American dog tick *Dermacentor variabilis* (Say) (Acari: Ixodidae). *Experimental and Applied Acarology* **1**, 23–34.

Steullet, P. & Guerin, P. M. (1992*a*). Perception of breath components by the tropical bont tick *Amblyomma variegatum* Fabricius (Ixodidae). I. CO_2-excited and CO_2-inhibited receptors. *Journal of Comparative Physiology A* **170**, 665–676.

Steullet, P. & Guerin, P. M. (1992*b*). Perception of breath components by the tropical bont tick *Amblyomma variegatum* Fabricius (Ixodidae). II. Sulfide receptors. *Journal of Comparative Physiology A* **170**, 677–685.

Steullet, P. & Guerin, P. M. (1994). Identification of vertebrate volatiles stimulating olfactory receptors on tarsus I of the tick *Amblyomma variegatum* Fabricius (Ixodidae). II. Receptors outside the Haller's organ capsule. *Journal of Comparative Physiology A* **174**, 39–47.

Stowe, M. K., Turlings, T. C., Loughrin, J. H., Lewis, W. J. & Tumlinson, J. H. (1995). The chemistry of eavesdropping, alarm and deceit. *Proceedings of the National Academy of Sciences of the USA* **92**, 23–28.

Taylor, D., Phillips, J. S., Allan, S. A. & Sonenshine, D. E. (1987). Absence of assembly pheromones in the hard ticks, *Dermacentor variabilis* and *Dermacentor andersoni* (Acari: Ixodidae). *Journal of Medical Entomology* **24**, 628–632.

Taylor, D., Sonenshine, D. E. & Phillips, J. S. (1991). Ecdysteroids as a component of the genital sex pheromone in two species of hard ticks, *Dermacentor variabilis* (Say) and *Dermacentor andersoni* Stiles (Acari: Ixodidae). *Experimental and Applied Acarology* **12**, 275–296.

Tkachev, A. V., Dobrotvorsky, A. K., Vjalkov, A. I. & Morozov, S. V. (2000). Chemical composition of lipophylic compounds from the body surface of adult *Ixodes persulcatus* ticks (Acari: Ixodidae). *Experimental and Applied Acarology* **24**, 145–158.

Treverrow, N. L., Stone, B. F. & Cowie, M. (1977). Aggregation pheromone in two Australian hard ticks, *Ixodes holocyclus* and *Aponomma concolor*. *Experientia* **33**, 680–682.

Wheeler, J. W. (1976). Insect and mammalian pheromones. *Lloydia* **39**, 53–59.

Wilkinson, P. R. (1970). Factors affecting the distribution and abundance of the cattle tick in Australia: observations and hypotheses. *Acarologia* **3**, 492–508.

Yoder, J. A. & Domingus, J. L. (2003). Identification of hydrocarbons that protect ticks (Acari: Ixodidae) against fire ants (Hymenoptera: Formicidae) but not lizards (Squamata: Polychrotidae), in allomonal defense secretion. *International Journal of Acarology* **29**, 87–91.

Yoder, J. A. & Knapp, D. C. (1999). Cluster-promoted water conservation by larvae of the American dog tick, *Dermacentor variabilis* (Acari: Ixodidae). *International Journal of Acarology* **25**, 5–57.

Yoder, J. A. & Stevens, B. W. (2000). Attraction of immature stages of the American dog tick (*Dermacentor variabilis*) to 2,6-dichlorophenol. *Experimental and Applied Acarology* **24**, 159–164.

Yoder, J. A., Atwood, A. D. & Stevens, B. W. (1998). Attraction to squalene by ticks (Acari: Ixodidae): first demonstration of a host-derived attractant. *International Journal of Acarology* **24**, 143–147.

Yoder, J. A., Hanson, P. E., Pizzuli, J. L., Sanders, C. I. & Domingus, J. L. (2002). Sex pheromone production and its relationship to water conservation: studies on a trichlorophenol in the American dog tick, *Dermacentor variabilis* (Acari: Ixodidae). *International Journal of Acarology* (submitted).

Yoder, J. A., Pollack, R. J. & Spielman, A. (1993a). An ant-diversionary secretion of ticks: first demonstration of an acarine allomone. *Journal of Insect Physiology* **39**, 42–435.

Yoder, J. A., Pollack, R. J., Spielman, A., Sonenshine, D. E. & Johnston, D. E. (1993b). Secretion of squalene by ticks. *Journal of Insect Physiology* **39**, 291–296.

Yoder, J. A., Stevens, B. W. & Crouch, K. C. (1999). Squalene: a naturally abundant mammalian skin secretion and long distance tick-attractant. *Journal of Medical Entomology* **36**, 526–529.

Yoder, J. A., Wittenberg, T. L. & Blomquist, G. J. (1997). Dietary contribution to the defense secretion of ixodid ticks. In *Acarology IX, Proceedings*, eds. Mitchell, R., Horn, D. J., Needham, G. R. & Welbourn, W. C., pp. 713–714. Columbus, OH: Ohio Biological Survey.

Index

Page entries for headings with subheadings refer only to general aspects of that topic; page entries in **bold** refer to figures/tables